# Lecture Notes in Computer Science    5028

Commenced Publication in 1973
Founding and Former Series Editors:
Gerhard Goos, Juris Hartmanis, and Jan van Leeuwen

Arnold Beckmann   Costas Dimitracopoulos
Benedikt Löwe (Eds.)

# Logic and Theory
# of Algorithms

4th Conference on Computability in Europe, CiE 2008
Athens, Greece, June 15-20, 2008
Proceedings

 Springer

Volume Editors

Arnold Beckmann
Swansea University
Singleton Park, Swansea, SA2 8PP, United Kingdom
E-mail: a.beckmann@swansea.ac.uk

Costas Dimitracopoulos
University of Athens
University Campus, Ano Ilisia, 15771 Athens, Greece
E-mail: cdimitr@phs.uoa.gr

Benedikt Löwe
Universiteit van Amsterdam
Plantage Muidergracht 24, 1018 TV Amsterdam, The Netherlands
E-mail: bloewe@science.uva.nl

Library of Congress Control Number: 2008928722

CR Subject Classification (1998): F.1, F.2.1-2, F.4.1, G.1.0, I.2.6, J.3

LNCS Sublibrary: SL 1 – Theoretical Computer Science and General Issues

ISSN       0302-9743
ISBN-10    3-540-69405-6 Springer Berlin Heidelberg New York
ISBN-13    978-3-540-69405-2 Springer Berlin Heidelberg New York



Springer is a part of Springer Science+Business Media

springer.com

© Springer-Verlag Berlin Heidelberg 2008
Printed in Germany

Typesetting: Camera-ready by author, data conversion by Scientific Publishing Services, Chennai, India
Printed on acid-free paper       SPIN: 12278411       06/3180       5 4 3 2 1 0

# Preface

CiE 2008: Logic and Theory of Algorithms
Athens, Greece, June 15–20, 2008

*Computability in Europe* (CiE) is an informal network of European scientists working on computability theory, including its foundations, technical development, and applications. Among the aims of the network is to advance our theoretical understanding of what can and cannot be computed, by *any* means of computation. Its scientific vision is broad: computations may be performed with discrete or continuous data by all kinds of algorithms, programs, and machines. Computations may be made by experimenting with any sort of physical system obeying the laws of a physical theory such as Newtonian mechanics, quantum theory, or relativity. Computations may be very general, depending on the foundations of set theory; or very specific, using the combinatorics of finite structures. CiE also works on subjects intimately related to computation, especially theories of data and information, and methods for formal reasoning about computations. The sources of new ideas and methods include practical developments in areas such as neural networks, quantum computation, natural computation, molecular computation, computational learning. Applications are everywhere, especially, in algebra, analysis and geometry, or data types and programming. Within CiE there is general recognition of the underlying relevance of computability to physics and a broad range of other sciences, providing as it does a basic analysis of the causal structure of dynamical systems.

This volume, *Logic and Theory of Algorithms*, is the proceedings of the fourth in a series of conferences of CiE that was held at the University of Athens, June 15–20, 2008.

The first three meetings of CiE were at the University of Amsterdam in 2005, at the University of Wales Swansea in 2006, and at the University of Siena in 2007. Their proceedings, edited in 2005 by S. Barry Cooper, Benedikt Löwe, and Leen Torenvliet, in 2006 by Arnold Beckmann, Ulrich Berger, Benedikt Löwe, and John V. Tucker, and in 2007 by S. Barry Cooper, Benedikt Löwe, and Andrea Sorbi, were published as *Springer Lecture Notes in Computer Science*, Volumes 3526, 3988, and 4497, respectively.

CiE and its conferences have changed our perceptions of computability and its interface with other areas of knowledge. The large number of mathematicians and computer scientists attending these conference had their view of computability theory enlarged and transformed: they discovered that its foundations were deeper and more mysterious, its technical development more vigorous, its applications wider and more challenging than they had known. The Athens meeting promised to extend and enrich that process.

The annual CiE conference, based on the Computability in Europe network, has become a major event, and is the largest international meeting focused on computability theoretic issues. The series is coordinated by the CiE Conference Series Steering Committee:

Arnold Beckmann (Swansea)
Paola Bonizzoni (Milan)
S. Barry Cooper (Leeds)
Benedikt Löwe (Amsterdam, Chair)
Elvira Mayordomo (Zaragoza)
Dag Normann (Oslo)
Peter van Emde Boas (Amsterdam)

We will reconvene 2009 in Heidelberg and 2010 in Ponta Delgada (Açores).

## Structure and Programme of the Conference

The conference was based on invited tutorials and lectures, and a set of special sessions on a range of subjects; there were also many contributed papers and informal presentations. This volume contains 25 of the invited lectures and 33% of the submitted contributed papers, all of which have been refereed. There will be a number of post-proceedings publications, including special issues of *Theory of Computing Systems*, *Archive for Mathematical Logic*, and *Journal of Algorithms*.

### Tutorials

John V. Tucker (Swansea), *Applied Computability: A European Perspective*
Moshe Y. Vardi (Houston TX), *Logic, Automata, Games, and Algorithms*

### Opening Lecture

Keith Devlin (Stanford CA), *The Computing Species*

### Invited Plenary Talks

Rosalie Iemhoff (Utrecht), *Kripke Models for Constructive Set Theory*
Antonina Kolokolova (St. John's NL), *Many Facets of Complexity in Logic*
Johann Makowsky (Haifa), *Uniform Algebraic Reducibilities Between
    Parameterized Numeric Graph Invariants*
Dag Normann (Oslo), *50 Years of Continuous Functionals*
Prakash Panangaden (Montréal QC), *Domain Theory and the Causal Structure
    of Space-Time*

Christos Papadimitriou (Berkeley CA), *On Nash, Brouwer, and Other Non-Constructive Proofs*

Jiří Wiedermann (Prague) and Jan van Leeuwen (Utrecht), *How We Think of Computing Today*

**Special Sessions**

**Algorithms in the History of Mathematics,** organized by Jens Høyrup and Karine Chemla

Andréa Bréard (Lille), *A Summation Algorithm from 11th Century China: Possible Relations Between Structure and Argument*

Marouane Ben Miled (Tunis): *Preuves et Algorithmes Dans Un Contexte Algébrique, Entre le 9e et le 12e Siècle,*

Harold Edwards (New York), *Kronecker's Algorithmic Mathematics*

Jens Høyrup (Roskilde), *The Algorithm Concept – Tool for Historiographic Interpretation or Red Herring?*

**Formalising Mathematics and Extracting Algorithms from Proofs,** organized by Henk Barendregt and Monika Seisenberger

John Harrison (Hillsboro), *New (Un)Decidability Results from the Formalization of Mathematics*

Pierre Letouzey (Paris), *Extraction in Coq: An Overview*

Lawrence C. Paulson (Cambridge), *The Relative Consistency of the Axiom of Choice Mechanized Using Isabelle/ZF*

Christophe Raffalli (Le Bourget du Lac), *Krivine's Realizability: From Storage Operators to the Intentional Axiom of Choice*

**Higher-Type Recursion and Applications,** organized by Ulrich Berger and Dag Normann

Lars Kristiansen (Oslo), *Recursion in Higher Types and Resource Bounded Turing machines*

John Longley (Edinburgh), *Interpreting Localized Computational Effects Using Higher Type Operators*

Ralph Matthes (Toulouse), *Recursion on Nested Datatypes in Dependent Type Theory*

Colin Riba (Sophia Antipolis), *Rewriting with Unions of Reducibility Families*

**Algorithmic Game Theory,** organized by Elias Koutsoupias and Bernhard von Stengel

Constantinos Daskalakis (Berkeley CA), *Computing Equilibria in Large Games We Play*

Hugo Gimbert (Bordeaux), *Solving Simple Stochastic Games*

Rahul Savani (Warwick), *A Simple P-Matrix Linear Complementarity Problem for Discounted Games*

Troels Bjerre Sørensen (Aarhus), *Deterministic Graphical Games Revisited*

**Quantum Algorithms and Complexity,** organized by Viv Kendon and Bob Coecke

Dan Browne (London), *A Classical Analogue of Measurement-Based Quantum Computation?*
Jiannis K. Pachos (Leeds), *Why Should Anyone Care About Computing with Anyons?*
Peter Richter (Orsay), *The Quantum Complexity of Markov Chain Monte Carlo*
Matthias Christandl (Cambridge), *A Quantum Information-Theoretic Proof of the Relation Between Horn's Problem and the Littlewood-Richardson Coefficients*

**Biology and Computation,** organized by Natasha Jonoska and Giancarlo Mauri

Alessandra Carbone (Paris), *Genome Synthesis and Genomic Functional Cores*
Matteo Cavaliere (Trento), *Computing by Observing*
Erzsébet Csuhaj-Varjú (Budapest), *P Automata: Membrane Systems as Acceptors*
Mark Daley (London ON), *On the Processing Power of Protozoa*

## Organization and Acknowledgements

The conference CiE 2008 was organized by: Dionysis Anapolitanos (Athens), Arnold Beckmann (Swansea), Costas Dimitracopoulos (Athens, Chair), Michael Mytilinaios (Athens) †, Thanases Pheidas (Heraklion), Stathis Zachos (Athens and New York NY).

The Programme Committee was chaired by Arnold Beckmann and Costas Dimitracopoulos:

| | |
|---|---|
| Luigia Aiello (Rome) | Elvira Mayordomo (Zaragoza) |
| Thorsten Altenkirch (Nottingham) | Franco Montagna (Siena) |
| Klaus Ambos-Spies (Heidelberg) | Michael Mytilinaios (Athens) † |
| Giorgio Ausiello (Rome) | Mogens Nielsen (Aarhus) |
| Arnold Beckmann (Swansea) | Isabel Oitavem (Lisbon) |
| Lev Beklemishev (Moscow) | Catuscia Palamidessi (Palaiseau) |
| Paola Bonizzoni (Milan) | Thanases Pheidas (Heraklion) |
| Stephen A. Cook (Toronto ON) | Ramanujam (Chennai) |
| Barry Cooper (Leeds) | Andrea Schalk (Manchester) |
| Costas Dimitracopoulos (Athens) | Uwe Schöning (Ulm) |
| Rod Downey (Wellington) | Helmut Schwichtenberg (Munch) |
| Elias Koutsoupias (Athens) | Alan Selman (Buffalo NY) |
| Orna Kupferman (Jerusalem) | Andrea Sorbi (Siena) |
| Sophie Laplante (Orsay) | Ivan Soskov (Sofia) |
| Hannes Leitgeb (Bristol) | Christopher Timpson (Oxford) |
| Benedikt Löwe (Amsterdam) | Stathis Zachos (Athens and New York NY) |

We are delighted to acknowledge and thank the following for their essential financial support: Bank of Greece, Graduate Program in Logic and Algorithms (MPLA), Hellenic Ministry of Education, John S. Latsis Foundation, National and Kapodistrian University of Athens, Rizareio Foundation, The Elsevier Foundation.

The high scientific quality of the conference was possible through the conscientious work of the Programme Committee, the special session organizers, and the referees. We are grateful to all members of the Programme Committee for their efficient evaluations and extensive debates, which established the final programme. We also thank the following referees:

Jesus Aranda
Andreas Abel
Peter Aczel
Klaus Aehlig
Pilar Albert
Spyridon
   Antonakopoulos
Luis Antunes
Toshiyasu Arai
Argimiro Arratia
Albert Atserias
David Aubin
Olivier Bournez
John Baez
Evangelos Bampas
Freiric Barral
Ulrich Berger
Riccardo Biagioli
Guillaume Bonfante
Andrey Bovykin
Vasco Brattka
Mark Braverman
Thomas Brihaye
Gerth Brodal
Dan Browne
Manuel Campagnolo
Douglas Cenzer
Arkadevn
   Chattopadhyay
Panagiotis Cheilaris
Luca Chiarabini
Chi Tat Chong
Petr Cintula
Félix Costa
Paola D'Aquino
Bembé Daniel
Olivier Danvy

Bhaskar DasGupta
Constantinos
   Daskalakis
Adam Day
Alberto Dennunzio
Ilias Diakonikolas
Arnaud Durand
Martin Dyer
Mirna Dzamonja
Martin Escardo
Michael Fellows
Fernando Ferreira
Jean-Christophen
   Filliatre
Joe Fitzsimons
Gaëlle Fontaine
Enrico Formenti
Dimitris Fotakis
Pierluigi Frisco
Hristo Ganchev
Parmenides
   Garcia Cornejo
William Gasarch
Yiannis
   Giannakopoulos
Sergey Goncharov
Joel David Hamkins
Aram Harrow
Chrysafis Hartonas
Bastiaan Heeren
Christoph Heinatsch
Denis Hirschfeldt
John Hitchcock
Peter Hoyer
Simon Huber
Franz Huber
Martin Hyland

Hans Hyttel
Eyke Hüllermeier
Rosalie Iemhoff
Natasha Jonoska
Reinhard Kahle
Eirini Kaldeli
Christos Kapoutsis
Basil Karadais
Jarkko Kari
Elham Kashefi
Viv Kendon
Thomas Kent
Iordanis Kerenidis
Bakhadyr Khoussainov
Peter Koepke
George Koletsos
Konstantinos Kollias
Costas Koutras
Simon Kramer
Natalio Krasnogor
Werner Kuich
Piyush P. Kurur
Geoffrey LaForte
Michael Lampis
Troy Lee
Alberto Leporati
David Lester
Paolo Liberatore
Kamal Lodaya
Maria Lopez-Valdes
Salvador Lucas
Jack H. Lutz
Guillaume Malod
Evangelos Markakis
Euripides Markou
Giancarlo Mauri
Alexander Meduna

Klaus Meer
Emanuela Merelli
Wolfgang Merkle
Dale Miller
Peter Bro Miltersen
Eugenio Moggi
Antonio Montalban
Malika More
Philippe Moser
Luca Motto Ros
Anders Möller
Margherita Napoli
Vincent Nesme
Phuong Nguyen
Long Nguyen
Andre Nies
Karl-Heinz Niggl
Tobias Nipkow
Christos Nomikos
Martin Ochoa
Carlos Olarte
Martin Otto
Aris Pagourtzis
Jens Palsberg
Prakash Panangaden
Katia Papakonstan
   -tinopoulou
Nikos Papaspyrou
Gheorghe Paun
A. Pavan
Sylvain Perifel

Ion Petre
Mauro Piccolo
George Pierrakos
Natacha Portier
Katerina Potika
Florian Ranzi
Diana Ratiu
Kenneth Regan
Jan Reimann
Gerhard Reinelt
Martin Roetteler
Jeremie Roland
David Rydeheard
Markus Sauermann
Ruediger Schack
Stefan Schimanski
Peter Schuster
Monika Seisenberger
Domenico Senato
Sunil Easaw Simon
Reed Solomon
Ludwig Staiger
Frank Stephan
Lutz Strassburger
Aaron Stump
Kristian Støvring
Kohei Suenaga
Nik Sultana
S.P. Suresh
Deian Tabakov
Aris Tentes

Sebastiaan Terwijn
Neil Thapen
P.S. Thiagarajan
Jacobo Toran
Edmondo Trentin
Trifon Trifonov
Tarmo Uustalu
Pierre Valarcher
Frank Valencia
Peter van Emde Boas
Angelina Vidali
Nicolas Vieille
Anastasios Viglas
Giuseppe Vizzari
Paul Voda
Vladimir Vyugin
Andreas Weiermann
Pascal Weil
Philip Welch
Thomas Wilke
Joost Winter
Ronald de Wolf
Guohua Wu
Yue Yang
Alberto Zanardo
Liyu Zhang
Martin Ziegler
Albert Ziegler
Vittorio Amos Ziparo

We thank Andrej Voronkov for his EasyChair system which facilitated the work of the Programme Committee and the editors considerably.

April 2008

Arnold Beckmann
Costas Dimitracopoulos
Benedikt Löwe

# Michael Mytilinaios (1948–2007)

We are saddened to report that Michael Mytilinaios, one of the members of the Programme Committee of the conference *Computability in Europe 2008*, died on March 12, 2007. In this obituary, we present aspects of his research and teaching in memory of Professor Mytilinaios. It has been prepared with the valuable help of our colleagues John C. Cavouras (Athens), Panos Katerinis (Athens), and Theodore A. Slaman (Berkeley CA).

## 1 Research

Michael Mytilinaios received his doctorate in 1985 from the University of Chicago, where he worked with Theodore Slaman and William Tait. Mytilinaios had a deep and abiding respect for mathematics, more as an intellectual discipline than as a technical medium. His metamathematical inclinations were well reflected in his research topics.

Though he enjoyed problem solving, see [8] for the solution of a problem of Downey and [7] for the solution of a problem of Brown and Simpson, Mytilinaios wrote primarily on foundational aspects of recursion theory. Following earlier unpublished work of Stephen Simpson, Mytilinaios in his Ph.D. thesis [5] and later with collaborators Marcia Groszek and Slaman [3, 4, 6] studied the correlation between the combinatorial levels of the priority method with the hierarchy of subsystems of Peano Arithmetic as developed by Paris and Kirby [9]. Fixing the base theory $P^- + I\Sigma_0$, the Kirby and Paris schemes $B\Sigma_n$, bounding for $\Sigma_n$ formulas, and $I\Sigma_n$, induction for $\Sigma_n$ formulas, stratify first order Peano Arithmetic with proper implications $I\Sigma_{n+1} \implies B\Sigma_{n+1} \implies I\Sigma_n$.

Considerable expertise was available for the analysis of priority methods in weak systems. In $\alpha$-recursion theory, priority methods had been implemented in $L_\alpha$, a $\Sigma_1$-admissible initial segment of $L$. The contexts were similar to the extent that being $\Sigma_1$-admissible is analogous to satisfying $\Sigma_1$-induction and the analogy led to early results. Later results depended more on properties of models of arithmetic, such as definable cuts and standard systems.

By a variation on the Sacks-Simpson method used to solve Post's problem in $L_\alpha$, Simpson had shown that $I\Sigma_1$ is sufficient to prove that the original Friedberg-Muchnik construction does produce a pair of recursively enumerable sets of incomparable Turing degree. Mytilinaios's thesis analyzes the next level of the priority method, finite injury constructions with no a priori bound on the actions of individual strategies, as exemplified by the Sacks Splitting Theorem. As originally devised, such arguments make explicit use of $I\Sigma_2$. Mytilinaios adapted the Shore blocking method from $\alpha$-recursion theory to prove the surprising theorem that the Splitting Theorem is provable in $I\Sigma_1$.

Looking further to infinite injury, the paradigm example is the construction of a high recursively enumerable set. Mytilinaios and Slaman showed that the existence of a recursively enumerable set whose Turing degree is neither low nor complete cannot be proven from $B\Sigma_2$ [6], and showed that $I\Sigma_2$ is sufficient to prove the existence of a high recursively enumerable set [3]. Subsequently, Chong and Yang [2] showed that the existence of a high recursively enumerable set is equivalent to $I\Sigma_2$ over the base theory of $B\Sigma_2$. Looking even higher, Mytilinaios and Slaman showed that for each $n$, the existence of an incomplete recursively enumerable set that is neither $low_n$ nor $high_{n-1}$, while true, cannot be established in $P^- + B\Sigma_{n+1}$. Consequently, no bounded fragment of first order arithmetic establishes the facts that the $high_n$ and $low_n$ jump hierarchies are proper on the recursively enumerable degrees.

The cross-referencing of priority methods with fragments of arithmetic attracted substantial attention among recursion theorists. Among the other notable contributors are Chong Chi Tat, Joseph Mourad, Richard Shore, Yang Yue, and Hugh Woodin. See [1] for further information.

The foundational study initiated by Simpson and Mytilinaios is one of the central chapters in the story of the recursively enumerable sets and their Turing degrees.

## 2  Teaching and Organization

Michael Mytilinaios joined the Department of Informatics of the Athens University of Economics and Business in 1991, serving it until his death in 2007. During this period he taught the following undergraduate courses in the Department: *logic, computability and complexity*; *discrete mathematics, special topics in discrete mathematics, linear algebra*. Michael also taught the following postgraduate courses: *applied cryptography*, in his department; *computability*, in co-operation with Yiannis N. Moschovakis, in the Graduate Program in Logic and Algorithms of the University of Athens (MPLA); and *epistemology and teaching of mathematics*, in the Graduate Program in Teaching and Methodology of Mathematics of the University of Athens. Michael was an active participant in the weekly MPLA Seminar until his death. He wrote lecture notes on *Logic, Computability and Complexity*, and *Discrete Mathematics*.

Michael was a member of the organizing committee of the second PLS (Pan-hellenic Logic Symposium) and LC 2005 (Logic Colloquium 2005) and a member of the program committee of the second to fifth PLS.

Throughout the period he served at the Department of Informatics, his colleagues considered him as the best teacher in the department because of his patience, his devotion to teaching, his high moral values, his willingness to help students, and the clarity of his lectures.

His early death was a terrible loss not only to his students and his colleagues, but also to the entire university community.

# References


1. Chong, C.T., Yang, Y.: Recursion theory on weak fragments of Peano arithmetic: a study of definable cuts. In: Proceedings of the Sixth Asian Logic Conference (Beijing, 1996), pp. 47–65. World Sci. Publ., River Edge (1998)
2. Chong, C.T., Yang, Y.: $\Sigma_2$ induction and infinite injury priority argument. I. Maximal sets and the jump operator. J. Symbolic Logic 63(3), 797–814 (1998)
3. Groszek, M., Mytilinaios, M.: $\Sigma_2$-induction and the construction of a high degree. In: Recursion theory week (Oberwolfach, 1989). Lecture Notes in Math., vol. 1432, pp. 205–221. Springer, Berlin (1990)
4. Groszek, M.J., Mytilinaios, M.E., Slaman, T.A.: The Sacks density theorem and $\Sigma_2$-bounding. J. Symbolic Logic 61(2), 450–467 (1996)
5. Mytilinaios, M.E.: Priority Arguments and Models of Arithmetic. PhD thesis, The University of Chicago (1985)
6. Mytilinaios, M.E., Slaman, T.A.: $\Sigma_2$-collection and the infinite injury priority method. J. Symbolic Logic 53(1), 212–221 (1988)
7. Mytilinaios, M.E., Slaman, T.A.: On a question of Brown and Simpson. In: Computability, enumerability, unsolvability. London Math. Soc. Lecture Note Ser., vol. 224, pp. 205–218. Cambridge Univ. Press, Cambridge (1996)
8. Mytilinaios, M.E., Slaman, T.A.: Differences between resource bounded degree structures. Notre Dame J. Formal Logic 44(1), 1–12 (2004)
9. Paris, J.B., Kirby, L.A.S.: $\Sigma_n$-collection schemas in arithmetic. In: Logic Colloquium 1977 (Proc. Conf., Wrocław, 1977). Stud. Logic Foundations Math., vol. 96, pp. 199–209. North-Holland, Amsterdam (1978)

# Table of Contents

# Deterministic Graphical Games Revisited*

Daniel Andersson, Kristoffer Arnsfelt Hansen,
Peter Bro Miltersen, and Troels Bjerre Sørensen

University of Aarhus, Datalogisk Institut, DK-8200 Aarhus N, Denmark
{koda,arnsfelt,bromille,trold}@daimi.au.dk

**Abstract.** We revisit the *deterministic graphical games* of Washburn. A deterministic graphical game can be described as a simple stochastic game (a notion due to Anne Condon), except that we allow arbitrary real payoffs but disallow moves of chance. We study the complexity of solving deterministic graphical games and obtain an almost-linear time comparison-based algorithm for computing an equilibrium of such a game. The existence of a linear time comparison-based algorithm remains an open problem.

## 1 Introduction

Understanding rational behavior in *infinite duration games* has been an important theme in pure as well as computational game theory for several decades. A number of central problems remain unsolved. In pure game theory, the existence of near-equilibria in general-sum two-player stochastic games were established in a celebrated result by Vieille [14,15], but the existence of near-equilibria for the three-player case remains an important and elusive open problem [2]. In computational game theory, Condon [4] delineated the efficient computation of positional equilibria in *simple stochastic games* as an important task. While Condon showed this task to be doable in **NP ∩ coNP**, to this day, the best deterministic algorithms are not known to be of subexponential complexity. To the computer science community, the problem of computing positional equilibria in simple stochastic games is motivated by its hardness for finding equilibria in many other natural classes of games [17], which again implies hardness for tasks such as model checking the $\mu$-calculus [5], which is relevant for the *formal verification* of computerized systems.

### 1.1 Simple Stochastic Games

A simple stochastic game [4] is given by a graph $G = (V, E)$. The vertices in $V$ are the *positions* of the game. Each vertex belongs either to player *Max*, to player *Min*, or to *Chance*. There is a distinguished *starting position* $v_0$. Furthermore, there are a number of distinguished *terminal positions* or just *terminals*, each

---

* Research supported by *Center for Algorithmic Game Theory*, funded by The Carlsberg Foundation.

A. Beckmann, C. Dimitracopoulos, and B. Löwe (Eds.): CiE 2008, LNCS 5028, pp. 1–10, 2008.

labeled with a *payoff* from Min to Max.[1] All positions except the terminal ones have outgoing arcs. The game is played by initially placing a token on $v_0$, letting the token move along a uniformly randomly chosen outgoing arc when it is in a position belonging to Chance and letting each of the players decide along which outgoing arc to move the token when it is in a position belonging to him. If a terminal is reached, then Min pays Max its payoff and the game ends. Infinite play yields payoff 0. A *positional strategy* for a player is a selection of one outgoing arc for each of his positions. He plays according to the strategy if he moves along these arcs whenever he is to move. It is known (see [4]) that each position $p$ in a simple stochastic game can be assigned a *value* Val($p$) so that:

1. Max has a positional strategy that, regardless of what strategy Min adopts, ensures an expected payoff of at least Val($p$) if the game starts in $p$.
2. Min has a positional strategy that, regardless of what strategy Max adopts, ensures that the expected payoff is at most Val($p$) if the game starts in $p$.

The value of the game itself is the value of $v_0$. Condon considered the complexity of computing this value. It is still open if this can be done in polynomial time. In the present paper, we shall look at some easier problems. For those, we want to make some distinctions which are inconsequential when considering whether the problems are polynomial time solvable or not, but important for the more precise (almost linear) time bounds that we will be interested in in this paper.

- A *weak solution* is Val($v_0$) *and* a positional strategy for each player satisfying the conditions in items 1 and 2 above for $p = v_0$.
- A *strong solution* is the list of values of *all* positions in the game *and* a positional strategy for each player that for *all* positions $p$, ensures an expected payoff of at least/most Val($p$) if the game starts in $p$.

In game theory terminology, weak solutions to a game are *Nash equilibria* of the game while strong solutions are *subgame perfect equilibria*. Figure 1 illustrates that the distinction is not inconsequential. In the weak solution to the left, Max

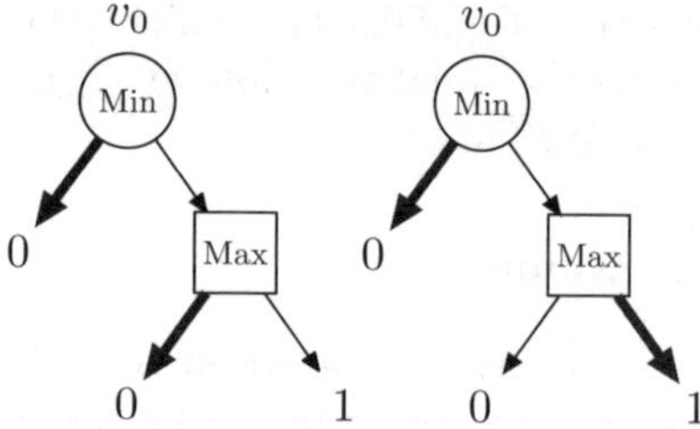


**Fig. 1.** The left solution is weak, the right is strong

---

[1] In Condon's original paper, there were only two terminals, with the payoffs 0 and 1. The relaxation to arbitrary payoffs that we adopt here is fairly standard.

(if he gets to move) is content to achieve payoff 0, the value of the game, even though he could achieve payoff 1. Note that the game in Figure 1 is *acyclic*. In contrast to the general case, it is of course well known that a strong solution to an acyclic game can be found in linear time by straightforward dynamic programming (known as *backwards induction* in the game theory community). We shall say that we weakly (resp. strongly) solve a given game when we provide a weak (resp. strong) solution to the game. Note that when talking about strong solutions, the starting position is irrelevant and does not have to be specified.

## 1.2 Deterministic Graphical Games

Condon observed that for the case of a simple stochastic game with *no Chance positions* and only 0/1 payoffs, the game can be strongly solved in *linear time*. Interestingly, Condon's algorithm has been discovered and described independently by the artificial intelligence community where it is known under the name of *retrograde analysis* [13]. It is used routinely in practice for finding optimal strategies for combinatorial games that are small enough for the game graph to be represented in (internal or external) memory and where dealing with the possibility of cycling is a non-trivial aspect of the optimal strategies. The best known example is the construction of tables for *chess endgames* [8]. Condon's algorithm (and retrograde analysis) being linear time depends crucially on the fact that the games considered are win/lose games (or, as is usually the case in the AI literature, win/lose/draw games), i.e., that terminal payoffs are either 0 or 1 (or possibly also $-1$, or in some AI examples even a small range of integers, e.g., [12]). In this paper we consider the algorithmic problem arising when *arbitrary real payoffs* are allowed. That is, we consider a class of games similar to but incomparable to Condon's simple stochastic games: We disallow chance vertices, but allow arbitrary real payoffs. The resulting class were named *deterministic graphical*[2] *games* by Washburn [16].

Some simple examples of deterministic graphical games are given in Figure 2. In (a), the unique strong solution is for Min to choose right and for Max to choose left. Thus, the outcome is infinite play. In (b), the unique strong solution is for Min to choose right and for Max to choose right. The values of both vertices are 1, but we observe that it is *not* a sufficient criterion for correct play to choose a vertex with at least as good a value as your current vertex. In particular,

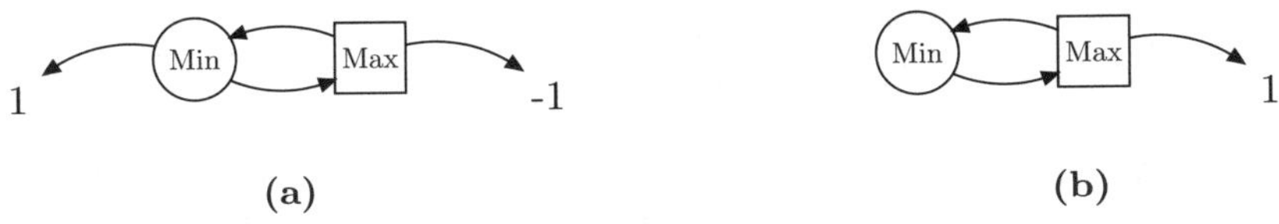

(a)               (b)

**Fig. 2. (a)** Infinite play equilibrium. **(b)** All values are 1, but one choice is suboptimal.

---

[2] There is no relation to the more recent concept of "graphical games" — a succinct representation for multi-player games [9].

according to this criterion, Max could choose left, but this would lead to infinite play and a payoff of 0, which is a suboptimal outcome for Max.

Washburn [16] gives an algorithm for computing a strong solution to a deterministic graphical game, but its running time is cubic in the size of the game. We observe below that if a *sorted* list of the payoffs (with pointers to the corresponding terminals of the game) is given in advance, optimal strategies can again be found in linear time without further processing of the payoffs. From this it follows that a deterministic graphical game with $n$ payoffs and $m$ arcs in the graph can be strongly solved in time $O(n \log n + m)$ by a *comparison-based* algorithm. The main question we attempt to approach in this paper is the following:

**Main Question.** *Can a deterministic graphical game be (weakly or strongly) solved in linear time by a comparison-based algorithm?*

We believe this to be an interesting question, both in the context of game solving (deterministic graphical games being a very simple yet non-trivial natural variant of the general problem) and in the context of the study of comparison-based algorithms and comparison complexity. This paper provides neither a positive nor a negative answer to the question, but we obtain a number of partial results, described in the next subsection.

## 1.3   Our Results

Throughout this section we consider deterministic graphical games with $n$ denoting the number of terminals (i.e., number of payoffs) and $m$ denoting the total size (i.e., number of arcs) of the graph defining the game. We can assume $m \geq n$, as terminals without incoming arcs are irrelevant.

**Strategy Recovery in Linear Time.** The example of Figure 2 (b) shows that it is not completely trivial to obtain a strong solution from a list of values of the vertices. We show that this task can be done in linear time, i.e. time $O(m)$. Thus, when constructing algorithms for obtaining a strong solution, one can concentrate on the task of computing the values Val$(p)$ for all $p$. Similarly, we show that given the value of just the starting position, a weak solution to the game can be computed in linear time.

**The Number of Comparisons.** When considering comparison-based algorithms, it is natural to study the number of comparisons used separately from the running time of the algorithm (assuming a standard random access machine). By an easy reduction from sorting, we show that there is no comparison-based algorithm that *strongly* solves a given game using only $O(n)$ comparisons. In fact, $\Omega(n \log n)$ comparisons are necessary. In contrast, Mike Paterson (personal communication) has observed that a deterministic graphical game can be *weakly* solved using $O(n)$ comparisons and $O(m \log n)$ time. With his kind permission, his algorithm is included in this paper. This also means that for the case of weak solutions, our main open problem cannot be solved in the negative using current lower-bound techniques, as it is not the number of comparisons that is the bottleneck. Our lower bound uses a game with $m = \Theta(n \log n)$ arcs. Thus, the following interesting open question concerning only the comparison complexity remains: *Can a deterministic graphical game be* strongly *solved using $O(m)$ comparisons?* If resolved in the negative, it will resolve our main open problem for the case of strong solutions.

**Almost-Linear Time Algorithm for Weak Solutions.** As stated above, Mike Paterson has observed that a deterministic graphical game can be weakly solved using $O(n)$ comparisons and $O(m \log n)$ time. We refine his algorithm and obtain an algorithm that weakly solves a game using $O(n)$ comparisons and only $O(m \log \log n)$ time. Also, we obtain an algorithm that weakly solves a game in time $O(m + m(\log^* m - \log^* \frac{m}{n}))$ but uses a superlinear number of comparisons. For the case of *strongly* solving a game, we have no better bounds than those derived from the simple algorithm described in Section 1.2, i.e., $O(m + n \log n)$ time and $O(n \log n)$ comparisons. Note that the bound $O(m + m(\log^* m - \log^* \frac{m}{n}))$ is linear in $m$ whenever $m \geq n \log \log \ldots \log n$ for a constant number of 'log's. Hence it is at least as good a bound as $O(m + n \log n)$, for any setting of the parameters $m, n$.

## 2   Preliminaries

**Definition 1.** *A* deterministic graphical game (DGG) *is a digraph with vertices partitioned into sets of non-terminals $V_{\mathrm{Min}}$ and $V_{\mathrm{Max}}$, which are game positions where player* Min *and* Max*, respectively, chooses the next move (arc), and terminals $T$, where the game ends and* Min *pays* Max *the amount specified by $p : T \to \mathbf{R}$.*  □

For simplicity, we will assume that terminals have distinct payoffs, i.e., that $p$ is injective. We can easily simulate this by artificially distinguishing terminals with equal payoffs in some arbitrary (but consistent) fashion. We will also assume that $m \geq n$, since terminals without incoming arcs are irrelevant.

**Definition 2.** *We denote by $\mathrm{Val}_G(v)$ the value of the game $G$ when the vertex $v$ is used as the initial position and infinite play is interpreted as a zero payoff. This will also be called "the value of $v$ (in $G$)".*  □

*Remark 1.* That such a value indeed exists will follow from Proposition 1. We shall later see how to construct optimal *strategies* from vertex values.

**Definition 3.** *To* merge *a non-terminal $v$ with a terminal $t$ is to remove all outgoing arcs of $v$, reconnect all its incoming arcs to $t$, and then remove $v$.*  □

The definitions of a *strong* and a *weak* solution are as stated in the introduction. The following algorithm is a generalization of Condon's linear time algorithm [4] for solving deterministic graphical games with payoffs in $\{0, 1\}$. That algorithm is known as *retrograde analysis* in the AI community [13], and we shall adopt this name also for this more general version.

**Proposition 1.** *Given a DGG and a permutation that orders its terminals, we can find a strong solution to the game in linear time and using no further comparisons of payoffs.*

*Proof.* If all payoffs are 0, then all values are 0 and every strategy is optimal.

Suppose that the minimum payoff $p(t)$ is negative. Any incoming arc to $t$ from a Max-vertex that is not the only outgoing arc from that vertex is clearly suboptimal and can be discarded. Each other incoming arc is an *optimal choice* for its source vertex, which can therefore be merged with $t$. Symmetric reasoning applies when the maximum payoff is positive.                                       □

This immediately yields the *sorting method* for strongly solving DGGs: First sort the payoffs, and then apply Proposition 1.

**Corollary 1.** *A DGG with $m$ arcs and $n$ terminals can be strongly solved in $O(m + n \log n)$ time.*                                       □

**Definition 4.** *To* merge *a terminal $s$ with another terminal $t$ is to reconnect all incoming arcs of $s$ to $t$ and then remove $s$. Two terminals are* adjacent *if their payoffs have the same sign and no other terminal has a payoff in between.*       □

The following lemma states the intuitive fact that when we merge two adjacent terminals, the only non-terminals affected are those with the corresponding values, and they acquire the same (merged) value.

**Lemma 1.** *If $G'$ is obtained from the DGG $G$ by merging a terminal $s$ with an adjacent terminal $t$, then for each non-terminal $v$, we have*

$$\mathrm{Val}_{G'}(v) = \begin{cases} \mathrm{Val}_G(v) & \textit{if } \mathrm{Val}_G(v) \neq \mathrm{Val}_G(s), \\ \mathrm{Val}_{G'}(t) & \textit{if } \mathrm{Val}_G(v) = \mathrm{Val}_G(s). \end{cases} \tag{1}$$

*Proof.* Consider all DGGs with a fixed structure (i.e., underlying graph) but with varying payoffs. Since a strong solution can be computed by a comparison-based algorithm (Proposition 1), the value of any particular position $v$ can be described by a min/max formula over the payoffs. The claim of the lemma can be seen to be true by a simple induction in the size of the relevant formula.   □

By repeatedly merging adjacent terminals, we "coarsen" the game. Figure 3 shows an example of this. The partitioning method we shall use to construct coarse games in this paper also yields sorted lists of their payoffs. Hence, we shall be able to apply retrograde analysis to solve them in linear time.

**Corollary 2.** *The signs of the values of all vertices in a given DGG can be determined in linear time.*

*Proof.* Merge all terminals with negative payoffs into one, do likewise for those with positive payoffs, and then solve the resulting coarse "win/lose/draw" game by retrograde analysis.                                       □

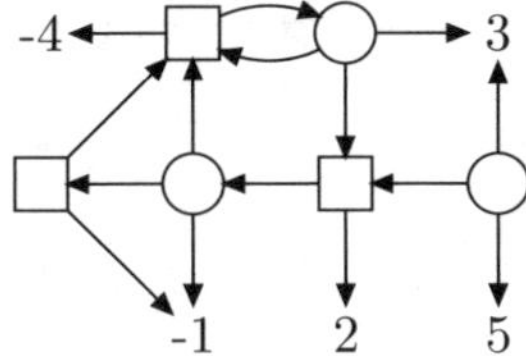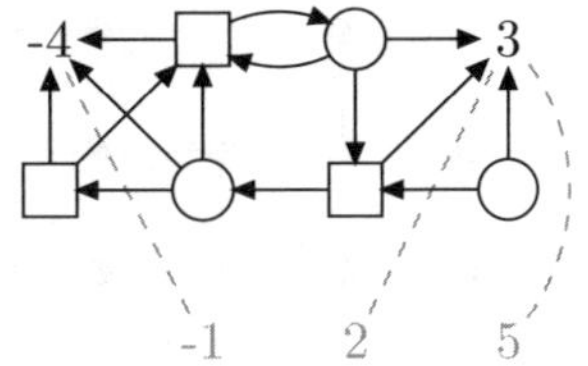

**Fig. 3.** Coarsening by merging $\{-4, -1\}$ and $\{2, 3, 5\}$

Clearly, arcs between vertices with different values cannot be part of a strong solution to a game. From this, the following lemma is immediate.

**Lemma 2.** *In a DGG, removing an arc between two vertices with different values does not affect the value of any vertex.* □

*Remark 2.* Corollary 2, Lemma 2, and symmetry together allow us to restrict our attention to games where all vertices have positive values, as will be done in subsequent sections.

**Proposition 2.** *Given the value of the initial position of a DGG, a weak solution can be found in linear time. If the values of* all *positions are known, a strong solution can be found in linear time.*

*Proof.* In the first case, let $y$ be the value of initial position $v_0$. We partition payoffs in at most five intervals: $(-\infty, \min(y, 0))$, $\{\min(y, 0)\}$, $(\min(y, 0), \max(y, 0))$, $\{\max(y, 0)\}$ and $(\max(y, 0), \infty)$. We merge all terminals in each of the intervals, obtaining a game with at most five terminals. A strong solution for the resulting coarse game is found in linear time by retrograde analysis. The pair of strategies obtained is then a weak solution to the original game, by Lemma 1.

In the second case, by Lemma 2, we can first discard all arcs between vertices of different values. This disintegrates the game into smaller games where all vertices have the same value. We find a strong solution to each of these games in linear time using retrograde analysis. Combining these solutions in the obvious way yields a strong solution to the original game, by Lemma 2. □

## 3 Solving Deterministic Graphical Games

### 3.1 Strongly

For solving DGGs in the strong sense, we currently know no asymptotically faster method than completely sorting the payoffs. Also, the number of *comparisons* this method performs is, when we consider bounds *only depending on the number of terminals $n$*, optimal. Any sorting network [10] can be implemented by an

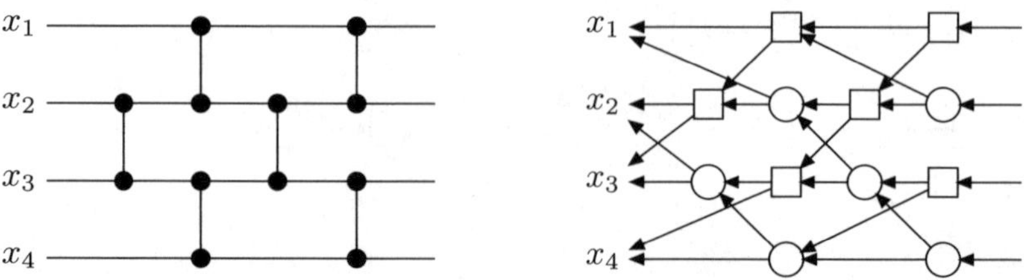

**Fig. 4.** Implementing a sorting network by a deterministic graphical game

acyclic DGG, by simulating each comparator by a Max-vertex and a Min-vertex. Figure 4 shows an example of this. Thus, we have the following tight bound.

**Proposition 3.** *Strongly solving a DGG with $n$ terminals requires $\Theta(n \log n)$ comparisons in the worst case.*    □

Implementing the asymptotically optimal AKS-network [1] results in a game with $\Theta(n \log n)$ vertices and arcs. Thus, it is still consistent with our current knowledge that a game can be strongly solved using $O(m)$ comparisons.

## 3.2   Weakly

The algorithms we propose for weakly solving DGGs all combine coarsening of the set of payoffs with retrograde analysis. By splitting the work between these two operations in different ways, we get different time/comparison trade-offs. At one extreme is the sorting method. At the other, we partition the payoffs around their median (which can be done in linear time by Blum *et al.* [3]), use retrograde analysis to solve the coarse game obtained by merging the terminals in each half, and then discard the irrelevant half of the terminals (the one *not* containing the value of the starting vertex) and all vertices with the corresponding values. This method, which is due to Mike Paterson, uses the optimal $O(n)$ comparisons, but requires $\Theta(\log n)$ iterations, each with a worst case running time of $\Theta(m)$.

$O(n)$ **Comparisons and** $O(m \log \log n)$ **Time.** To improve the running time of Paterson's algorithm, we stop and sort the remaining terminals as soon as this can be done in $O(n)$ time. The number of comparisons is still $O(n)$. As noted in Section 2, we may assume that all vertices have positive values.

*Algorithm.* Given a DGG $G$ with $m$ arcs, $n$ terminals, and starting position $v_0$, do the following for $i = 0, 1, 2, \ldots$

1. Partition the current set of $n_i$ terminals around their median payoff.
2. Solve the coarse game obtained by merging the terminals in each half.
3. Remove all vertices that do not have values in the half containing $\mathrm{Val}_G(v_0)$.
4. Undo step 1 for the half of $v_0$.

When $n_i \log n_i \leq n$, stop and solve the remaining game by the sorting method.

*Analysis.* Steps 1–4 can be performed in $O(m)$ time and $O(n_i)$ comparisons. The number of iterations is $O(\log n - \log f(n))$, where $f(n)$ is the inverse of $n \mapsto n \log n$, and since this equals $O(\log \log n)$ we have the following.

**Theorem 1.** *A DGG with $m$ arcs and $n$ terminals can be weakly solved in $O(m \log \log n)$ time and $O(n)$ comparisons.* $\qquad\qquad\square$

**Almost-Linear Time.** We can balance the partitioning and retrograde analysis to achieve an almost linear running time, by a technique similar to the one used in [7] and later generalized in [11].[3] Again, we assume that all vertices have positive values.

*Algorithm.* Given a DGG $G$ with $m$ arcs, $n$ terminals, and starting position $v_0$, do the following for $i = 0, 1, 2, \ldots$

1. Partition the current set of $n_i$ terminals into groups of size at most $n_i/2^{m/n_i}$.
2. Solve the coarse game obtained by merging the terminals in each group.
3. Remove all vertices having values outside the group of $\mathrm{Val}_G(v_0)$.
4. Undo step 1 for the group of $v_0$.

When $n_i/2^{m/n_i} < 1$, stop and solve the remaining game by the sorting method.

*Analysis.* All steps can be performed in $O(m)$ time. For the first step we can do a "partial perfect quicksort", where we always partition around the median and stop at level $\lceil m/n_i \rceil + 1$.

To bound the number of iterations, we note that $n_i$ satisfies the recurrence

$$n_{i+1} \leq n_i/2^{m/n_i}, \tag{2}$$

which by induction gives

$$n_i \leq \frac{n}{\underbrace{b^{b^{b^{\cdot^{\cdot^{\cdot^{b}}}}}}}_{i}} \tag{3}$$

where $b = 2^{m/n}$. Thus, the number of iterations is $O(\log_b^* n)$, where $\log_b^*$ denotes the number of times we need to apply the base $b$ logarithm function to get below 1. This is easily seen to be the same as $O(1 + \log^* m - \log^* \frac{m}{n})$. We have now established the following.

**Theorem 2.** *A DGG with $m$ arcs and $n$ terminals can be weakly solved in $O(m + m(\log^* m - \log^* \frac{m}{n}))$ time.* $\qquad\qquad\square$

*Remark 3.* When $m = \Omega(n \log^{(k)} n)$ for some constant $k$, this bound is $O(m)$.

**Acknowledgements.** We are indebted to Mike Paterson for his contributions to this work. We would also like to thank Uri Zwick and Gerth Brodal for helpful discussions.

---

[3] Note, however, that while the technique is similar, the problem of solving deterministic graphical games does not seem to fit into the framework of [11].

# References


1. Ajtai, M., Komlós, J., Szemerédi, E.: An $O(n \log n)$ sorting network. In: Proceedings of the 15th Annual ACM Symposium on the Theory of Computing, pp. 1–9 (1983)
2. Aumann, R.J.: Presidential address at the First International Congress of the Game Theory Society. Games and Economic Behavior 45, 2–14 (2003)
3. Blum, M., Floyd, R.W., Pratt, V., Rivest, R.L., Tarjan, R.E.: Linear time bounds for median computations. In: Proceedings of the 4th Annual ACM Symposium on the Theory of Computing, pp. 119–124 (1972)
4. Condon, A.: The complexity of stochastic games. Information and Computation 96, 203–224 (1992)
5. Emerson, E.A., Jutla, C.S., Sistla, A.P.: On model checking for the $\mu$-calculus and its fragments. Theor. Comput. Sci. 258(1-2), 491–522 (2001)
6. Everett, H.: Recursive games. In: Kuhn, H.W., Tucker, A.W. (eds.) Contributions to the Theory of Games. Annals of Mathematical Studies, vol. III(39), Princeton University Press, Princeton (1957)
7. Gabow, H.N., Tarjan, R.E.: Algorithms for two bottleneck optimization problems. J. Algorithms 9, 411–417 (1988)
8. Heinz, E.A.: Scalable Search in Computer Chess: Algorithmic Enhancements and Experiments at High Search Depths. Morgan Kaufmann Publishers Inc., San Francisco (1999)
9. Kearns, M., Littman, M.L., Singh, S.: Graphical models for game theory. In: Proceedings of the Conference on Uncertainty in Artificial Intelligence, pp. 253–260 (2001)
10. Knuth, D.E.: The Art of Computer Programming, 3rd edn. Sorting and Searching, vol. 3. Addison-Wesley, Reading (1997)
11. Punnen, A.P.: A fast algorithm for a class of bottleneck problems. Computing 56, 397–401 (1996)
12. Romein, J., Bal, H.: Solving the game of awari using parallel retrograde analysis. IEEE Computer 36(10), 26–33 (2003)
13. Thompson, K.: Retrograde analysis of certain endgames. Journal of the International Computer Chess Association 9(3), 131–139 (1986)
14. Vieille, N.: Two-player stochastic games I: A reduction. Israel Journal of Mathematics 119, 55–91 (2000)
15. Vieille, N.: Two-player stochastic games II: The case of recursive games. Israel Journal of Mathematics 119, 93–126 (2000)
16. Washburn, A.R.: Deterministic graphical games. Journal of Mathematical Analysis and Applications 153, 84–96 (1990)
17. Zwick, U., Paterson, M.S.: The complexity of mean payoff games on graphs. Theoretical Computer Science 158(1-2), 343–359 (1996)

# Program Schemes with Deep Pushdown Storage

Argimiro Arratia[1,*] and Iain A. Stewart[2]

[1] Dpto. de Matemática Aplicada, Facultad de Ciencias, Universidad de Valladolid,
Valladolid 47005, Spain
`arratia@mac.uva.es`
[2] Department of Computer Science, Durham University, Science Labs, South Road,
Durham DH1 3LE, U.K.
`i.a.stewart@durham.ac.uk`

**Abstract.** Inspired by recent work of Meduna on deep pushdown automata, we consider the computational power of a class of basic program schemes, $NPSDS_s$, based around assignments, while-loops and non-deterministic guessing but with access to a deep pushdown stack which, apart from having the usual push and pop instructions, also has deep-push instructions which allow elements to be pushed to stack locations deep within the stack. We syntactically define sub-classes of $NPSDS_s$ by restricting the occurrences of pops, pushes and deep-pushes and capture the complexity classes **NP** and **PSPACE**. Furthermore, we show that all problems accepted by program schemes of $NPSDS_s$ are in **EXPTIME**.

## 1 Introduction

In automata theory, there is a variety of machine models, both restricting and enhancing pushdown automata. However, rarely is there a model based on modifications of pushdown automata capturing a class of languages lying between the classes of the context-free and the context-sensitive languages. In an attempt to remedy this situation, Meduna [10] recently introduced deep pushdown automata and showed that these models coincide with Kasai's state grammars [9], and thus give rise to an infinite hierarchy of languages lying between the classes of the context-free and the context-sensitive languages. Meduna's deep pushdown automata pop and push items from and to their stack in the standard way; however, they also have the facility to insert (but not read) strings at some position within the stack.

Inspired by Meduna's machine model, we consider in this paper the formulation of program schemes with access to a deep pushdown stack. Program schemes work on arbitrary finite structures (rather than just strings) and are more computational in flavour than are formulae of logics studied in finite model theory and descriptive complexity, yet they remain amenable to logical manipulation. The concept of a program scheme originates from the 1970's with work of, for example, Constable and Gries, Friedman, and Hewitt and Paterson [3,6,11], and

---

* Supported by grants Ramón y Cajal (MEC+FEDER-FSE); MOISES (TIN2005-08832-C03-02) and SINGACOM (MTM2004-00958), MEC–Spain.

A. Beckmann, C. Dimitracopoulos, and B. Löwe (Eds.): CiE 2008, LNCS 5028, pp. 11–21, 2008.

complexity-theoretic considerations of such program schemes were subsequently studied by, for example, Harel and Peleg, Jones and Muchnik, and Tiuryn and Urzyczyn [7,8,17].

As mentioned above, in this paper it is our intention to consider the computational power of basic program schemes when augmented with a deep pushdown stack (program schemes with standard stacks have been considered in [2,15]). A deep pushdown stack is such that the usual pops and pushes are available as well as an additional instruction which allows elements to be written to locations deep in the stack but only so that the depth at which such a location lies (from the top of the stack) is bounded by some polynomial in the size of the input structure; that is, deep-pushes cannot be to locations too deep within the stack. Note that we only have deep-pushes, not deep-pops; for if we had both then we would have, essentially, access to a stack the top (polynomial) portion of which could be used as an array (program schemes with arrays have been considered in [14,16]). Our goal is to classify the computational power of basic program schemes with deep pushdown stacks in comparison with the standard complexity classes of computational complexity.

The results of [2] show that any problem in **P** can be accepted by some program scheme from our class of program schemes with access to a deep pushdown stack, which we call $\mathrm{NPSDS}_s$. It turns out that we can (syntactically) define sub-classes of the class of program schemes $\mathrm{NPSDS}_s$, obtained by restricting how the occurrences of pops, pushes and deep-pushes are structured, capturing the complexity classes **NP** and **PSPACE**. Furthermore, we show that all problems accepted by program schemes of $\mathrm{NPSDS}_s$ are in **EXPTIME**.

Let us end this introduction by remarking that there do exist programming languages with facilities for the manipulation of elements deep within a stack. One such is the programming language Push [12], specifically defined for use in genetic and evolutionary computational systems, where the instructions YANK and SHOVE allow deep access to stack elements, by means of integer indices.

Due to space limitations we omit the proofs of our results and other technical details. The complete paper is available in [1].

## 2   Basic Definitions

Throughout, a *signature* $\tau$ is a finite tuple of relation and constant symbols. The binary relation symbol *succ* (which never appears in any signature) is only ever interpreted as a *successor relation*, that is, as a relation of the form: $\{(u_0, u_1), (u_1, u_2), \ldots, (u_{n-2}, u_{n-1}) : u_i \neq u_j, \text{ for } i \neq j\}$, on some domain of size $n$, and 0 (resp. *max*) is only ever interpreted as the minimal (resp. maximal) element $u_0$ (resp. $u_{n-1}$) of the successor relation *succ*.

In [2], a class of program schemes based around the usage of a stack was defined. A *program scheme* $\rho \in \mathrm{NPSS}_s$ involves a finite set $\{x_1, x_2, \ldots, x_k\}$ of *variables*, for some $k \geq 1$, and is over a signature $\tau$. It consists of a finite sequence of *instructions* where each instruction is one of the following:

- an *assignment instruction* of the form $x_i$ := $y$, where $i \in \{1, 2, \ldots, k\}$ and where $y$ is a variable from $\{x_1, x_2, \ldots, x_k\}$, a constant symbol of $\tau$ or one of the special constant symbols 0 and *max* (which do not appear in any signature);
- a *guess instruction* of the form **guess** $x_i$, where $i \in \{1, 2, \ldots, k\}$;
- a *while instruction* of the form **while** $\varphi$ **do** $\alpha_1$; $\alpha_2$; $\ldots$; $\alpha_q$; **od**, where $\varphi$ is a quantifier-free formula over $\tau \cup \langle succ, 0, max \rangle$ whose free variables are from $\{x_1, x_2, \ldots, x_k\}$ and where each of $\alpha_1, \alpha_2, \ldots, \alpha_q$ is another instruction of one of the forms given here (note that there may be nested while instructions);
- a *push instruction* of the form **push** $x_i$, where $i \in \{1, 2, \ldots, k\}$;
- a *pop instruction* of the form $x_i$ := **pop**, where $i \in \{1, 2, \ldots, k\}$;
- an *accept* (resp. a *reject*) instruction **accept** (resp. **reject**).

A program scheme $\rho \in \mathrm{NPSS}_s$ over $\tau$ takes a $\tau$-structure $\mathcal{A}$ as input. The program scheme $\rho$ computes on $\mathcal{A}$ in the obvious way except that:

- the binary relation *succ* is taken to be any successor relation on $|\mathcal{A}|$, and 0 (resp. *max*) is the minimal (resp. maximal) element of this successor relation;
- initially, every variable takes the value 0;
- the pop and push instructions provide access to a stack which is initially empty;
- execution of the instruction **guess** $x_i$ non-deterministically assigns an element of $|\mathcal{A}|$ to the variable $x_i$.

An input $\tau$-structure $\mathcal{A}$, together with a chosen successor relation *succ*, is *accepted* by $\rho$ if there is at least one computation of $\rho$ on input $\mathcal{A}$ (and with regard to the chosen successor relation) leading to an accept instruction (if ever a pop of an empty stack is attempted in some computation then that computation is deemed rejecting). However, we only consider program schemes $\rho$ of $\mathrm{NPSS}_s$ that are successor-invariant, where *successor-invariant* means that every input structure is such that it is either accepted no matter which successor relation is chosen or rejected no matter which successor relation is chosen. Such successor-invariant program schemes accept classes of structures closed under isomorphisms, *i.e.*, problems. Henceforth, we assume that all our program schemes are successor-invariant, and we equate (a class of) program schemes with the (class of) problems they accept.

We extend our class of program schemes $\mathrm{NPSS}_s$ to the class of program schemes $\mathrm{NPSDS}_s$ by allowing deep-push instructions:

- a *deep-push instruction* is of the form **dpush**$(\mathbf{y}, z)$, where $\mathbf{y}$, the *index*, is a tuple of variables from $\{x_1, x_2, \ldots, x_k\}$ and constant symbols from $\tau \cup \langle 0, max \rangle$ and where $z$ is a variable from $\{x_1, x_2, \ldots, x_k\}$ or a constant symbol from $\tau \cup \langle 0, max \rangle$.

Suppose that $\rho$ is some program scheme with deep-push instructions and that $\mathcal{A}$ is input to $\rho$, with associated successor relation *succ* (and constants 0 and $max$). Suppose that $\mathtt{dpush(y,}z)$ is a deep-push instruction appearing in $\rho$, where $\mathbf{y}$ is an $m$-tuple. The $m$-tuples of $|\mathcal{A}|^m$ are ordered lexicographically as

$$(0,0,\ldots,0),(0,0,\ldots,u_1),(0,0,\ldots,u_2),\ldots,(max,max,\ldots,max),$$

where $0,u_1,u_2,\ldots,max$ is the linear order encoded within the successor relation $succ$. Associate these $m$-tuples with the integers $0,1,\ldots,n^m-1$, respectively. Denote the integer associated with the tuple $\mathbf{u}$ as $\sharp\mathbf{u}$, and denote the tuple of values associated with the integer $i \in \{0,1,\ldots,n^m-1\}$ as $i\sharp$. When the instruction $\mathtt{dpush(y,}z)$ is executed, the variables of $\mathbf{y}$ each hold a value from $|\mathcal{A}|$; so, $\sharp\mathbf{y} \in \{0,1,\ldots,n^m-1\}$. The instruction $\mathtt{dpush(y,}z)$ pushes the value of $z$ to the stack location $\sharp\mathbf{y}$ (the *offset*) from the top of the stack, with the top of the stack being the location 0 from the top of the stack, the next location down being the location 1 from the top of the stack and so on; in particular, the deep-push overwrites the value in the stack location $\sharp\mathbf{y}$ with the value of $z$. Thus, the instruction has access to the top $n^m$ locations of the stack for pushing *but not for popping*. Only pops of the top element of the stack are allowed (in the usual way). If ever a deep-push attempts to push a value to some location below the bottom location of the stack then that particular computation is deemed rejecting. As usual, we only ever work with successor-invariant program schemes of $\mathrm{NPSDS}_s$.

Let us introduce some syntactic sugar. First, we employ the usual if-then-else instructions directly, as such instructions can easily be constructed using while instructions. Second, in describing our program schemes we use a convenient shorthand in relation to the offsets and indices in deep-push instructions. According to our syntax, the offset of a deep-push instruction is given via a tuple of variables, the index, with the offset value calculated according to the successor relation accompanying an input structure. Rather than include routine 'housekeeping' code, we allow ourselves to describe offset values in an integer-style. For example, suppose that we wished to use an offset of $2n-2$ in a deep-push instruction (where $n$ is the size of the input structure). Strictly speaking, we should use a pair of variables, $(x_1,x_2)$, and a portion of code to implement this. However, we abbreviate this code with the instruction $\mathtt{dpush((}2n\ \mathtt{-}\ 2\mathtt{)}\sharp\mathtt{,}y\mathtt{)}$. In general, if $i$ is any positive integer then we write $i\sharp$ to denote a tuple of variables (of the required length) encoding the offset of value $i$. All of the integer offsets appearing in our program schemes are such that they can easily be computed with an appropriate portion of code. Thus, for example, the instruction $\mathtt{dpush(((}n\ \mathtt{-}\ \sharp count\mathtt{)}\ \mathtt{+}\ \sharp y\mathtt{)}\sharp\mathtt{,}y\mathtt{)}$ is shorthand for a portion of code which builds values for a tuple of 2 variables which encodes an integer equal to $n$ minus the integer currently encoded by the current value of the variable $count$ plus the integer encoded by the current value of the variable $y$. We also use the above shorthand in assignments, *e.g.*, $(count,y) := (\sharp(count,y)\ \mathtt{+}\ 1)\sharp$. Furthermore, we abuse our notation by writing, for example, $count := 2n\sharp$ or $\mathtt{while}\ \sharp x < 2n-1\ \mathtt{do}$, where what we should really write is $(count_1,count_2) := 2n\sharp$ and $\mathtt{while}\ \sharp(x_1,x_2) < 2n-1$. (As can be seen, we often use names for variables different to $x_1,x_2,\ldots$)

*Example 1.* Consider the following program scheme $\rho$ of $\text{NPSDS}_s$ over the signature $\tau = \langle E, C, D \rangle$, where $E$ is a binary relation symbol and $C$ and $D$ are constant symbols (so, an input structure can be considered as a digraph with two specified vertices):

```
1      guess w;                     ** push some 0's onto the stack **
2      while w = 0 do
3         push 0;
4         guess w;
5      od;
6      count := 0;                  ** count counts the number of guessed **
7      dpush(count,C);              ** vertices in a path making sure that **
8      dpush((n + ♯C)♯,max);        ** n are guessed and registered        **
9      while count ≠ max do
10        count := (♯count + 1)♯;
11        guess x;
12        dpush(count,x);
13        dpush((n + ♯x)♯,max);
14     od;
15     count := 0;                  ** count counts the number of vertices **
16     x := pop;                    ** popped when checking the validity   **
17     while count ≠ max do         ** of the guessed path                 **
18        count := (♯count + 1)♯;
19        y := pop;
20        if ¬E(x,y) then reject; fi;
21        x := y;
22     od;
23     if x ≠ D then reject; fi;
24     count := 0;                  ** count counts the number of vertex **
25     w := pop;                    ** registrations checked so far       **
26     if w ≠ max then reject; fi;
27     while count ≠ max do
28        count := (♯count + 1)♯;
29        w := pop;
30        if w ≠ max then reject; fi;
31     od;
32     accept;
```

Essentially, on an input digraph of size $n$, $\rho$ begins by building a stack of 0s and then guesses a path of $n$ vertices, starting with vertex $C$, and stores this path in the top $n$ locations in the stack. When a vertex $u$ is guessed, this is registered by deep-pushing $max$ to the stack location with offset $n + \sharp u$. After this guessing phase, the path is popped from the stack and checked as to whether it is a valid path (ending in $D$). If so then the registrations are then checked to confirm that all vertices appear once on the path. Hence, an input structure is accepted by $\rho$ iff $C \neq D$ and there is a Hamiltonian path from vertex $C$ to vertex $D$.          □

# 3  Our Results

## 3.1  Restricted Program Schemes

It turns out that restricting the values pushed by push and deep-push instructions to 0 and $max$ does not limit the problems accepted by our program schemes; we call such program schemes *boolean* program schemes. Let us denote the class of program schemes where push instructions must be of the form $\mathtt{push}\ 0$ or $\mathtt{push}\ max$ and where deep-push instructions must be of the form $\mathtt{dpush(y,0)}$ or $\mathtt{dpush(y},max)$ by $\mathrm{NPSDS}_s^b$.

**Proposition 1.** *Any problem accepted by a program scheme of $NPSDS_s$ can be accepted by a program scheme of $NPSDS_s^b$.*

We also restrict the syntax of program schemes of $\mathrm{NPSDS}_s$ (and $\mathrm{NPSDS}_s^b$) by limiting the usage of pops, pushes and deep-pushes in phases of a program scheme. A *batch* of instructions is a well-formed sequence of instructions so that if one of these instructions is a while-do instruction (resp. if instruction) then the corresponding while-od instruction (resp. fi instruction) must also appear in the batch. Let $\rho$ be a program scheme of $\mathrm{NPSDS}_s$. If $\rho$ can be written (as a concatenation of instructions) as a batch of instructions $\rho_0$, followed by a batch of instructions $\rho_1$, and so on, finally ending with a batch of instructions $\rho_k$, then we write $\rho = (\rho_0, \rho_1, \ldots, \rho_k)$. The allowed usage of pops, pushes and deep-pushes in any batch of instructions is signalled as follows:

- if pops are allowed (resp. disallowed) then we signal this with $o$ (resp. $\bar{o}$);
- if pushes are allowed (resp. disallowed) then we signal this with $u$ (resp. $\bar{u}$);
- if deep-pushes are allowed (resp. disallowed) then we signal this with $d$ (resp. $\bar{d}$).

Thus, if pops and pushes are allowed in some batch of instructions $\rho_i$, but not deep-pushes, then we write $\rho_i \in \mathrm{NPSDS}_s(ou\bar{d})$. We adapt our notation to situations where a program scheme is the concatenation of a sequence of batches of instructions by detailing the allowed usage of pops, pushes and deep-pushes in each batch by a sequence of parameters. So, for example, if $\rho = (\rho_0, \rho_1, \rho_2)$ where $\rho_0 \in \mathrm{NPSDS}_s(ou\bar{d})$, $\rho_1 \in \mathrm{NPSDS}_s(\bar{o}u\bar{d})$ and $\rho_2 \in \mathrm{NPSDS}_s(o\bar{u}d)$, then we write $\rho \in \mathrm{NPSDS}_s(ou\bar{d}, \bar{o}u\bar{d}, o\bar{u}d)$. The above applies equally to program schemes of $\mathrm{NPSDS}_s^b$.

The proof of Proposition 1 is such that it does not alter the interleaving of pops, pushes and deep-pushes in the simulating program scheme. So, for example, the problem accepted by some program scheme of $\mathrm{NPSDS}_s(ou\bar{d}, \bar{o}u\bar{d}, o\bar{u}d)$ can be accepted by some program scheme of $\mathrm{NPSDS}_s^b(ou\bar{d}, \bar{o}u\bar{d}, o\bar{u}d)$.

## 3.2  Capturing NP

We can syntactically capture the complexity class **NP** with a sub-class of program schemes of $\mathrm{NPSDS}_s$.

**Proposition 2.** *Every program scheme $\rho = (\rho_0, \rho_1, \rho_2)$ in $NPSDS_s(ou\overline{d}, \overline{ou}d,$ $ou\overline{d})$ accepts a problem in* **NP**.

Denote by $\text{NPSDS}_s^{+b}$ the sub-class of program schemes of $\text{NPSDS}_s^b$ where any instruction involving a deep-push must be of the form $\mathtt{dpush(y,}max\mathtt{)}$; that is, only the value $max$ can be deep-pushed to a stack location. We can actually use our program scheme for the problem HP in Example 1, in tandem with results due to Dahlhaus [5] and Stewart [13], to show that the class of program schemes $\text{NPSDS}_s^{+b}(ou\overline{d}, \overline{ou}d, ou\overline{d})$ actually contains **NP**.

**Proposition 3.** *For every problem $\Omega$ in* **NP**, *there exists a program scheme of $NPSDS_s^{+b}(ou\overline{d}, \overline{ou}d, ou\overline{d})$ accepting $\Omega$.*

Propositions 2 and 3 immediately yield the following corollary.

**Corollary 1.** $NPSDS_s^{+b}(ou\overline{d}, \overline{ou}d, ou\overline{d}) = NPSDS_s(ou\overline{d}, \overline{ou}d, ou\overline{d}) = $ **NP**.

### 3.3   Beyond NP

Let us demonstrate the power of program schemes of $\text{NPSDS}_s$ and outline a program scheme accepting the complement of the problem in Example 1.

*Example 2.* Let $\tau = \langle E, C, D \rangle$, with $E$ a relation symbol of arity 2 and $C$ and $D$ constant symbols. The problem co-HP is defined as

$$\{\mathcal{A} : \mathcal{A} \text{ is a } \tau\text{-structure and there is not a Hamiltonian path in the digraph}$$
$$\text{with edges given by the relation } E^{\mathcal{A}} \text{ from vertex } C^{\mathcal{A}} \text{ to vertex } D^{\mathcal{A}}\}.$$

There exists a program scheme $\rho$ of $\text{NPSDS}_s$ accepting co-HP.

We outline what the program scheme $\rho$ that accepts co-HP does. The program scheme $\rho$ begins by simply non-deterministically building a 'sufficiently tall' stack of 0's. The top $n$ elements of the stack (the 'top batch') are regarded as the current potential Hamiltonian path of vertices in the input digraph; initially, of course, this path is $0, 0, \ldots, 0$. The next $n$ elements of the stack (the 'middle batch') are regarded as work-space that will enable us to verify that there are no repeated vertices in the current path. The next $n$ elements of the stack (the 'lower batch') will be such that they will contain the lexicographically-next potential Hamiltonian path.

The top element (of the top batch) is popped from the stack and, using a deep-push, the name of this vertex, $u$, say, is registered in location $\sharp u$ of the middle batch by writing $max$ (note that the top batch of elements now consists of $n - 1$ elements). A check is also made to verify that $u$ is, in fact, the constant $C$. We iteratively pop elements from the stack and verify that the current path is indeed a path in our input digraph, registering these elements, as above, as we proceed. Also, alongside our checking of the current potential Hamiltonian path, we build the lexicographically-next path in the lower batch of stack elements. Essentially, as we pop the vertices of the current path from the top batch of the stack, we look for the location *place* that holds an element that is not equal to *max* but where every location in the top batch lower than this location *place* holds an element equal to *max*; the element in this location *place* is to be incremented by 1 and all elements in the top-batch at lower locations are to be set to 0 to get the lexicographically-next potential Hamiltonian path. This is done, using deep-pushes, so that the lexicographically-next path appears in the lower batch. Having popped all the vertices of the current path from the stack, we verify that the last element popped was $D$ and we use our registrations to verify that the path is indeed Hamiltonian. If not then we proceed as above except that what was the lower batch is now the top batch and we are working with the lexicographically-next potential Hamiltonian path, as our current path. We continue until all potential Hamiltonian paths have been checked or until we find a Hamiltonian path. (For the detailed program see [1].)                $\square$

## 3.4   Capturing PSPACE

Having demonstrated the power of program schemes of $NPSDS_s$, we now turn to capturing the complexity class **PSPACE**. The technique used in Example 2 is used in the proofs of subsequent results.

**Proposition 4.** *Any polynomial-space Turing machine can be simulated by a program scheme of $NPSDS_s(\overline{o}u\overline{d}, o\overline{u}d)$ so that the stack of the program scheme encodes the evolution of the work-tape of the Turing machine. Thus,* **PSPACE** $\subseteq NPSDS_s(\overline{o}u\overline{d}, o\overline{u}d)$.

In fact, more can be said. The proof of Proposition 4 is such that the program scheme $\rho$ constructed is a program scheme of $NPSDS_s^{+b}(\overline{o}u\overline{d}, o\overline{u}d)$.

**Corollary 2.** *Any polynomial-space Turing machine can be simulated by a program scheme of $NPSDS_s^{+b}(\overline{o}u\overline{d}, o\overline{u}d)$ so that the stack of the program scheme encodes the evolution of the work-tape of the Turing machine. Thus,* **PSPACE** $\subseteq NPSDS_s^{+b}(\overline{o}u\overline{d}, o\overline{u}d)$.

**Proposition 5.** *Every program scheme $\rho$ in $NPSDS_s^{+b}(u\overline{o}\overline{d}, o\overline{u}d)$ accepts a problem in* **PSPACE**.

**Corollary 3.** $NPSDS_s^{+b}(\overline{o}u\overline{d}, o\overline{u}d) = NPSDS_s(u\overline{o}\overline{d}, o\overline{u}d) = $ **PSPACE**.

## 3.5   Beyond PSPACE

Hitherto, we have been focussing on specific restrictions of $NPSDS_s$. We now examine the computational complexity of problems accepted by arbitrary program schemes of $NPSDS_s$. Our basic technique in the proof of Proposition 6 is inspired by that used in [2] to show that basic program schemes with an ordinary stack only accept problems in **P**, which in turn was inspired by Cook's construction involving pushdown automata in [4]. We require some definitions before we present our results.

**Definition 1.** *Let $\rho$ be a program scheme of $NPSDS_s$, where all deep-pushes occur with an index of at most $k$, and let $\mathcal{A}$ be an input structure of size $n$. An* extended ID *$(\alpha, \mathbf{s})$ of $\rho$ on input $\mathcal{A}$ is an ID $\alpha$ together with a tuple $\mathbf{s}$ of between $0$ and $n^k$ elements of $|\mathcal{A}|$. A pair of extended IDs, $((\alpha, \mathbf{s}), (\beta, \mathbf{t}))$, is called* realizable *if there is a computation of $\rho$ on $\mathcal{A}$ starting in configuration $[\alpha, \mathbf{s}]$ and ending in configuration $[\beta, \mathbf{t}]$, where $|\mathbf{s}| = |\mathbf{t}|$ and the stack associated with any interim configuration has height at least $|\mathbf{s}|$.*

**Definition 2.** *Let $((\alpha, \mathbf{s}), (\beta, \mathbf{t}))$ and $((\alpha', \mathbf{s}'), (\beta', \mathbf{t}'))$ be two pairs of extended IDs. These pairs yield the pair of extended IDs $((\alpha, \mathbf{s}), (\beta'', \mathbf{t}''))$ if one of the following two rules can be applied.*

*(a)* *The instruction associated with $\beta$ is a push, and the execution of this instruction transforms the configuration $[\beta, \mathbf{t}]$ into the configuration:*
   - *$[\alpha', \mathbf{s}']$, if $|\mathbf{t}| < n^k$;*
   - *$[\alpha', (u_0, \mathbf{s}')]$, if $|\mathbf{t}| = n^k$ (in which case $\mathbf{t}$ is a prefix of $(u_0, \mathbf{s}')$).*

   *The instruction associated with $\beta'$ is a pop, and the execution of this instruction transforms the configuration $[\beta', \mathbf{t}']$ or $[\beta', (u_0, \mathbf{t}')]$ (depending upon whether $|\mathbf{t}| < n^k$ or $|\mathbf{t}| = n^k$, respectively, above) to the configuration $[\beta'', \mathbf{t}'']$.*
*(b)* *$(\beta, \mathbf{t}) = (\alpha', \mathbf{s}')$ and either $(\beta'', \mathbf{t}'') = (\beta', \mathbf{t}')$ or the instruction associated with $\beta'$ is neither a pop nor a push and execution of this instruction transforms the configuration $[\beta', \mathbf{t}']$ to the configuration $[\beta'', \mathbf{t}'']$.*

*We refer to rules $(a)$ and $(b)$ as the* yield rules.

**Proposition 6.** *Let $\rho$ be some program scheme of $NPSDS_s$ and let $\mathcal{A}$ be some input structure. Any pair of realizable extended IDs of $\rho$ on input $\mathcal{A}$ can be obtained from the set $Y = \{((\alpha, \mathbf{s}), (\alpha, \mathbf{s})) : (\alpha, \mathbf{s})$ is an extended ID of $\rho$ on input $\mathcal{A}\}$ by iteratively applying the yield rules.*

The complexity class **EXPTIME** consists of those problems solvable by a deterministic algorithm running in time $O(2^{p(n)})$, for some polynomial $p(n)$. Using Proposition 6 in tandem with the well-known result that the path-system problem can be solved in polynomial-time, we obtain the following.

**Corollary 4.** *Any problem accepted by a program scheme of $NPSDS_s$ is in* **EXPTIME**.

## 4    Conclusions

In this paper, we have initiated the study of a high-level model of computation in which there is access to a deep pushdown stack, and we have managed to capture the complexity classes **NP** and **PSPACE** by restricting our model whilst showing that an arbitrary program scheme accepts a problem in **EXPTIME**. Whilst we have shown that there is a considerable increase in computational power obtained by replacing stacks with deep pushdown stacks in a basic class of program schemes (assuming $\mathbf{P} \neq \mathbf{PSPACE}$!), we have as yet been unable to ascertain exactly the computational complexity of the problems accepted by program schemes of $\text{NPSDS}_s$. At the moment, all we know is that $\mathbf{PSPACE} \subseteq \text{NPSDS}_s \subseteq \mathbf{EXPTIME}$.

Looking at Example 2 where we show that the complement of the Hamiltonian path problem is accepted by a program scheme of $\text{NPSDS}_s$, we can see no obvious restriction of the general class of program schemes so that we capture the complexity class $\mathbf{co\text{-}NP}$ (the complement of $\mathbf{NP}$). Note that our program scheme in Example 2 is clearly in $\text{NPSDS}_s(\overline{ou}\overline{d}, o\overline{u}d)$ yet by Corollary 3, $\text{NPSDS}_s(\overline{ou}\overline{d}, o\overline{u}d)$ captures $\mathbf{PSPACE}$. It would be interesting to capture $\mathbf{co\text{-}NP}$, and more generally the classes of the Polynomial Hierarchy, by restricting the program schemes of $\text{NPSDS}_s$.

# References


1. Arratia-Quesada, A., Stewart, I.A.: On the power of deep pushdown stacks, `http://www.dur.ac.uk/i.a.stewart/Abstracts/DeepPushdownStacks.htm`
2. Arratia Quesada, A.A., Chauhan, S.R., Stewart, I.A.: Hierarchies in classes of program schemes. Journal of Logic and Computation 9, 915–957 (1999)
3. Constable, R., Gries, D.: On classes of program schemata. SIAM Journal of Computing 1, 66–118 (1972)
4. Cook, S.A.: Characterizations of pushdown machines in terms of time-bounded computers. Journal of the Association for Computing Machinery 18, 4–18 (1971)
5. Dahlhaus, E.: Reduction to NP-complete problems by interpretations. In: Börger, E., Hasenjaeger, G., Rödding, D. (eds.) Proc. of Logic and Machines: Decision Problems and Complexity. LNCS, vol. 171, pp. 357–365. Springer, Heidelberg (1984)
6. Friedman, H.: Algorithmic procedures, generalized Turing algorithms and elementary recursion theory. In: Gandy, R.O., Yates, C.M.E. (eds.) Logic Colloquium, pp. 361–390. North-Holland, Amsterdam (1971)
7. Harel, D., Peleg, D.: On static logics, dynamic logics, and complexity classes. Information and Control 60, 86–102 (1984)
8. Jones, N.D., Muchnik, S.S.: Even simple programs are hard to analyze. Journal of Association for Computing Machinery 24, 338–350 (1977)
9. Kasai, T.: An hierarchy between context-free and context-sensitive languages. Journal of Computer and System Sciences 4, 492–508 (1970)
10. Meduna, A.: Deep pushdown automata. Acta Informatica 42, 541–552 (2006)
11. Paterson, M., Hewitt, N.: Comparative schematology. In: Record of Project MAC Conf. on Concurrent Syst. and Parallel Comput., pp. 119–128. ACM Press, New York (1970)
12. Spector, L., Klein, J., Keijzer, M.: The Push3 execution stack and the evolution of control. In: Beyer, H.-G., O'Reilly, U.-M. (eds.) Proc. of Genetic and Evolutionary Computation Conference, pp. 1689–1696. ACM Press, New York (2005)
13. Stewart, I.A.: Using the Hamiltonian path operator to capture NP. Journal of Computer and System Sciences 45, 127–151 (1992)
14. Stewart, I.A.: Program schemes, arrays, Lindström quantifiers and zero-one laws. Theoretical Computer Science 275, 283–310 (2002)

15. Stewart, I.A.: Using program schemes to logically capture polynomial-time on certain classes of structures. London Mathematical Society Journal of Computation and Mathematics 6, 40–67 (2003)
16. Stewart, I.A.: Logical and complexity-theoretic aspects of models of computation with restricted access to arrays. In: Cooper, S.B., Kent, T.F., Löwe, B., Sorbi, A. (eds.) Proc. of Computation and Logic in the Real World, Third Conference on Computability in Europe (CiE 2007), pp. 324–331 (2007)
17. Tiuryn, J., Urzyczyn, P.: Some relationships between logics of programs and complexity theory. Theoretical Computer Science 60, 83–108 (1988)

# Herbrand Theorems and Skolemization
## for Prenex Fuzzy Logics*

Matthias Baaz[1] and George Metcalfe[2,**]

[1] Institute of Discrete Mathematics and Geometry, Technical University Vienna,
Wiedner Hauptstrasse 8-10, A-1040 Wien, Austria
baaz@logic.at
[2] Department of Mathematics, Vanderbilt University
1326 Stevenson Center, Nashville TN 37240, USA
george.metcalfe@vanderbilt.edu

**Abstract.** Approximate Herbrand theorems are established for first-order fuzzy logics based on continuous $t$-norms, and used to provide proof-theoretic proofs of Skolemization for their Prenex fragments. Decidability and complexity results for particular fragments are obtained as consequences.

**Keywords:** Herbrand Theorem, Skolemization, Fuzzy Logics.

## 1   Introduction

Herbrand's theorem and Skolemization are fundamental tools in the proof theory of first-order Classical Logic, forming the theoretical basis for automated reasoning methods such as Resolution. Extending these tools to non-classical logics is an important but challenging task, often requiring considerable ingenuity. For example, in Intuitionistic Logic, the most general forms of Herbrand's theorem and Skolemization fail but may be rescued either by restricting to fragments or extending the language (see e.g. [3]).

In this paper, we investigate these issues for fuzzy logics, the broader purpose being to provide a basis for applications such as fuzzy logic programming [15] and fuzzy description logics [14,8]. Fuzzy logics, featured prominently in [6], are defined on the real unit interval $[0, 1]$ with conjunction and implication connectives interpreted by left-continuous $t$-norms and their residua, and quantifiers $\forall$ and $\exists$, by suprema and infima. Among logics based on continuous $t$-norms, Łukasiewicz Logic, Gödel Logic, and Product Logic are considered fundamental since any continuous $t$-norm is an ordinal sum of the corresponding three $t$-norms [11]. Also important in this context is Basic Logic [6], characterized by validity for all continuous $t$-norms.

Herbrand's theorem and Skolemization were established for the Prenex fragment of first-order Gödel Logic in [5,1]. The proof-theoretic approach of [5] was also extended to obtain similar results for the logic of left-continuous $t$-norms (Monoidal $t$-Norm Logic) in [2]. For other logics based on continuous $t$-norms, the situation is more complicated, however. This is related to the fact that first-order Gödel Logic is the only

---

* Research Supported by FWF Project P17503-N12.
** Corresponding author.

A. Beckmann, C. Dimitracopoulos, and B. Löwe (Eds.): CiE 2008, LNCS 5028, pp. 22–31, 2008.

recursively axiomatizable member of the family; first-order Łukasiewicz Logic and Product Logic being $\prod_2$-complete and non-arithmetical respectively, and the complexity of the other members lying somewhere between (we refer to [13,10,7] for details).

In the case of Łukasiewicz Logic, although Herbrand's theorem fails even for existential formulas, an "approximate Herbrand theorem" may be obtained instead [12,4]. Essentially, for a valid existential formula, there exist Herbrand disjunctions for successive approximations to validity: for any $r < 1$ a disjunction exists that always takes a value greater than $r$. This result is used in [4] to define Gentzen-style proof systems for the logic and to give a proof-theoretic proof of Skolemization. The proof-theoretic treatment allows any system of functions to be considered as Skolem functions obeying constraints (such as commutativity), giving greater flexibility for replacing functions in formulas with quantified terms.

Here we extend the techniques for Gödel Logic [1] and Łukasiewicz Logic [4] to the Prenex fragments of other fuzzy logics based on continuous $t$-norms. More concretely, we show first that Gödel Logic is the only such logic admitting a Herbrand theorem for its Prenex fragment. We then show that a wide class of logics admit an approximate Herbrand theorem for their Prenex fragments, and use these results – as for Łukasiewicz Logic – to obtain proof-theoretic proofs of Skolemization. As consequences, we obtain $\prod_2$-membership for the Prenex fragments of these logics, and decidability for the $\forall\exists$-fragments. Since Łukasiewicz Logic (alone among fuzzy logics) admits a full quota of quantifier shifts and hence an equivalent Prenex form for all formulas, these last two corollaries extend also to the whole logic [4].


## 2   Fuzzy Logics

Let us begin by recalling definitions and pertinent features of the logics that we are interested in, referring the reader to [6] for further details and motivation. A *continuous t-norm* is a commutative, associative, increasing function $* : [0,1]^2 \to [0,1]$ satisfying $1 * x = x$ for all $x \in [0,1]$ that is continuous in the usual sense. Its *residuum* is the unique function $\to_* : [0,1]^2 \to [0,1]$ satisfying $x \leq y \to_* z$ iff $x * y \leq z$. The functions $\min$ and $\max$ can be expressed in terms of $*$ and $\to_*$; i.e. $\min(x,y) = x * (x \to_* y)$ and $\max(x,y) = \min((x \to_* y) \to_* y, (y \to_* x) \to_* x)$.

The most important examples of continuous $t$-norms and their residua are displayed in Table 1. These $t$-norms play a special role: *any* continuous $t$-norm is locally isomorphic to one of these three. More precisely, let $([a_i, b_i])_{i \in I}$ be a countable family of non-overlapping proper subintervals of $[0,1]$ (i.e. $a_i < b_i$ and $(a_i, b_i) \cap (a_j, b_j) = \emptyset$ for all $i, j \in I$ where $i \neq j$); let $(*_i)_{i \in I}$ be a family of $t$-norms; and let $(f_i)_{i \in I}$ be a family of order-isomorphisms from $[a_i, b_i]$ onto $[0,1]$. Then the *generalized ordinal sum* $\sum_{i \in I}([a_i, b_i], *_i, f_i)$ is the unique function $* : [0,1]^2 \to [0,1]$ defined by:

$$u * v = \begin{cases} f_k^{-1}(f_k(u) *_k f_k(v)) & \text{if } u, v \in [a_k, b_k] \\ \min(u, v) & \text{otherwise} \end{cases}$$

The Łukasiewicz and Product $t$-norms are generalized ordinal sums of just one $t$-norm, while the Gödel $t$-norm is the ordinal sum of the empty family. More generally:

**Table 1.** Fundamental $t$-norms and their residua

| Name | $t$-Norm | Residuum |
|---|---|---|
| Łukasiewicz | $x *_\text{Ł} y = \max(0, x + y - 1)$ | $x \to_\text{Ł} y = \min(1, 1 - x + y)$ |
| Gödel | $x *_\text{G} y = \min(x, y)$ | $x \to_\text{G} y = \begin{cases} 1 \text{ if } x \leq y \\ y \text{ otherwise} \end{cases}$ |
| Product | $x *_\text{P} y = x \cdot y$ | $x \to_\text{P} y = \begin{cases} 1 & \text{if } x \leq y \\ y/x & \text{otherwise} \end{cases}$ |

**Theorem 1 ([11]).** *Each continuous t-norm is the generalized ordinal sum of (isomorphic) copies of the Łukasiewicz and Product t-norms.*

For a continuous $t$-norm $*$ with residuum $\to_*$, the logic $\mathsf{L}(*)$ is based on a usual first-order language with countable sets of predicate symbols, (object) constants, function symbols (with positive arity), and variables; quantifiers $\forall$ and $\exists$; binary connectives $\odot$ and $\to$; and a (logical) constant $\bot$, where:

$$\neg A =_{def} A \to \bot \qquad A \wedge B =_{def} A \odot (A \to B)$$
$$\top =_{def} \neg\bot \qquad A \vee B =_{def} ((A \to B) \to B) \wedge ((B \to A) \to A)$$

Terms $t, s$ and formulas $A, B$ are built inductively from the elements of this language in the usual manner, adopting standard notions of subformulas, scope, and free and bound variables. A sequence of terms $t_1, \ldots, t_n$ is often written $\bar{t}$, denoting a formula $A$ with variables among $\bar{x}$ by $A(\bar{x})$, and $A$ with each $x_i$ replaced by $t_i$ for $i = 1 \ldots n$ by $A(\bar{t})$. Function-free, quantifier-free, one-variable, and propositional formulas are those built using no function symbols, no quantifiers, just one variable, and no quantifiers or variables, respectively. We write $(Q\bar{x})A(\bar{x})$ for a formula $(Q_1 x_1) \ldots (Q_n x_n)A(x_1, \ldots, x_n)$ where $Q_i \in \{\forall, \exists\}$ for $i = 1 \ldots n$, and recall that a Prenex formula is of the form $(Q\bar{x})P(\bar{x})$ where $P$ is quantifier-free.

$\mathsf{L}(*)$-*interpretations* $\mathcal{I} = (\mathcal{D}, v_\mathcal{I})$ consist of a non-empty set $\mathcal{D}$ and a valuation function $v_\mathcal{I}$ that maps constants and object variables to elements of $\mathcal{D}$; $n$-ary function symbols to functions from $\mathcal{D}^n$ into $\mathcal{D}$; and $n$-ary predicate symbols to functions from $\mathcal{D}^n$ into $[0, 1]$. As usual, $v_\mathcal{I}$ is extended to all terms inductively by the condition $v_\mathcal{I}(f(t_1, \ldots, t_n)) = v_\mathcal{I}(f)(v_\mathcal{I}(t_1), \ldots, v_\mathcal{I}(t_n))$ for any $n$-ary function symbol $f$ and terms $t_1, \ldots, t_n$. For an $n$-ary predicate symbol $p$ and terms $t_1, \ldots, t_n$:

$$v_\mathcal{I}(p(t_1, \ldots, t_n)) = v_\mathcal{I}(p)(v_\mathcal{I}(t_1), \ldots, v_\mathcal{I}(t_n))$$

For a variable $x$ and element $d \in \mathcal{D}$, let $v_\mathcal{I}[x \leftarrow d]$ be the valuation obtained from $v_\mathcal{I}$ by changing $v_\mathcal{I}(x)$ to $d$. Then $v_\mathcal{I}$ is extended to all formulas by $v_\mathcal{I}(\bot) = 0$ and:

$$v_\mathcal{I}(A \odot B) = v_\mathcal{I}(A) * v_\mathcal{I}(B)$$
$$v_\mathcal{I}(A \to B) = v_\mathcal{I}(A) \to_* v_\mathcal{I}(B)$$
$$v_\mathcal{I}((\forall x)A(x)) = \inf\{v_\mathcal{I}[x \leftarrow d](A(x)) : d \in \mathcal{D}\}$$
$$v_\mathcal{I}((\exists x)A(x)) = \sup\{v_\mathcal{I}[x \leftarrow d](A(x)) : d \in \mathcal{D}\}$$

where for the defined connectives $\wedge$ and $\vee$, we get:

$$v_{\mathcal{I}}(A \wedge B) = \min(v_{\mathcal{I}}(A), v_{\mathcal{I}}(B)) \qquad v_{\mathcal{I}}(A \vee B) = \max(v_{\mathcal{I}}(A), v_{\mathcal{I}}(B))$$

$A$ is $\mathsf{L}(*)$-*valid*, written $\models_{\mathsf{L}(*)} A$, if $v_{\mathcal{I}}(A) = 1$ for all $\mathsf{L}(*)$-interpretations $\mathcal{I}$.

It will also be crucial in this paper to consider the following notions of "approximate validity" for $\rhd \in \{>, \geq\}$ and $r \in [0, 1]$:

$$\models_{\mathsf{L}(*)}^{\rhd r} A \quad \text{if} \quad v_{\mathcal{I}}(A) \rhd r \text{ for all } \mathsf{L}(*)\text{-interpretations } \mathcal{I}.$$

For convenience, we will adopt the following terminology:

- we call every $\mathsf{L}(*)$ based on a continuous $t$-norm $*$, a *fuzzy logic*.
- we will speak of the *Łukasiewicz* or *Product components* of $[0, 1]$ for a fuzzy logic $\mathsf{L}(*)$ meaning the intervals where $*$ behaves like these $t$-norms.
- we call any component $[a, 1]$, the *top component* for $*$, and if $*$ has a top Łukasiewicz or Product component, we call $\mathsf{L}(*)$ an *Ł-fuzzy logic* or *P-fuzzy logic*, respectively.
- if $*$ acts as min on some $[a, 1]$ for $a < 1$ then $\mathsf{L}(*)$ is called a *G-fuzzy logic*.

The particular logics based on the fundamental $t$-norms are known as *Łukasiewicz Logic* Ł, *Gödel Logic* G, and *Product Logic* P. Finally, we note that *Basic Logic* BL may be characterized as the logic based on all continuous $t$-norms; i.e. $\models_{\mathsf{BL}} A$ iff $\models_{\mathsf{L}(*)} A$ for each continuous $t$-norm $*$.[1]

## 3 Herbrand Theorems

We first recall some basic notions. For a formula $A$, let $\mathcal{C}(A)$ and $\mathcal{F}(A)$ be the constants and function symbols occurring in $A$, respectively, adding a constant if the former is empty. For $c \in \mathcal{C}(A)$ and unary function symbol $f \in \mathcal{F}(A)$, we let $f^0(c) = c$ and $f^{k+1}(c) = f(f^k(c))$ for $k \in \mathbb{N}$. The *Herbrand universe* $\mathcal{U}(A)$ of $A$ is the set of ground terms built using $\mathcal{C}(A)$ and $\mathcal{F}(A)$; i.e. $\mathcal{U}(A) = \bigcup_{n=0}^{\infty} \mathcal{U}_n(A)$ where:

$$\mathcal{U}_0(A) = \mathcal{C}(A)$$
$$\mathcal{U}_{n+1}(A) = \mathcal{U}_n(A) \cup \{f(t_1, \ldots, t_k) : t_1, \ldots, t_k \in \mathcal{U}_n(A) \text{ and } f \in \mathcal{F}(A) \text{ with arity } k\}$$

Letting $\mathcal{P}(A)$ be the predicate symbols of $A$, the *Herbrand base* $\mathcal{B}(A)$ is defined as:

$$\mathcal{B}(A) = \{p(t_1, \ldots, t_k) : t_1, \ldots, t_k \in \mathcal{U}(A) \text{ and } p \in \mathcal{P}(A) \text{ with arity } k\}$$

Note that as in Classical Logic, formulas $(\exists \bar{x})P(\bar{x})$ where $P$ is quantifier-free are valid for a fuzzy logic iff they are valid over the Herbrand universe $\mathcal{U}(P)$. Hence in such cases, valuations $v_{\mathcal{I}}$ for an interpretation $\mathcal{I}$ can (and will, throughout this paper) be identified with a member of $[0, 1]^{\mathcal{B}(P)}$.

For Gödel Logic, the following existential Herbrand theorem has been established:

---

[1] Note that in the literature (e.g. [6]), fuzzy logics are often introduced via axiomatic systems, while here we rather take the semantic perspective as primary.

**Theorem 2 ([5,1]).** $\models_\mathsf{G} (\exists \bar{x}) P(\bar{x})$ *where $P$ is quantifier-free iff:*

$$\models_\mathsf{G} \bigvee_{i=1}^{n} P(\bar{t}_i) \quad \text{for some } \bar{t}_1, \ldots, \bar{t}_n \in \mathcal{U}(P)$$

The idea for the harder left-to-right direction in [1] is to build an infinite tree inductively by considering at level $l$ all possible linear orderings of the first $l$ members of an enumeration of $\mathcal{B}(P)$ (noting that in G only the order of the truth values matters). Suppose that at some level $l$, for every linear ordering, all G-valuations $v_\mathcal{I}$ satisfying that ordering give a value 1 to $P(\bar{t})$ for some $\bar{t} \in \mathcal{U}(P)$. Then the disjunction of these $P(\bar{t})$ (one for each ordering) must be G-valid. On the other hand, if there is no such level, then using König's lemma there exists an infinite branch: this can be used to find an interpretation contradicting the validity of $(\exists \bar{x}) P(\bar{x})$.

However, this proof does not work in general for fuzzy logics; indeed:

**Proposition 1.** *The only fuzzy logic having the existential Herbrand Theorem is* G.

*Proof.* Consider the following formula:

$$F = (\exists x)[((p(x) \to (p(f(x)) \vee q)) \to q) \to q]$$

We first claim that $\models_\mathsf{L} F$ for any fuzzy logic L. Consider $v_\mathcal{I} \in [0,1]^{\mathcal{B}(F)}$ and $c \in \mathcal{U}(F)$. If $v_\mathcal{I}(p(f^k(c))) \leq v_\mathcal{I}(p(f^{k+1}(c)))$ or $v_\mathcal{I}(p(f^k(c))) \leq v_\mathcal{I}(q)$ for any $k \in \mathbb{N}$, then easily $v_\mathcal{I}(F) = 1$. Hence we can assume that:

$$v_\mathcal{I}(q) < \ldots < v_\mathcal{I}(p(f^k(c))) < \ldots < v_\mathcal{I}(p(f(c))) < v_\mathcal{I}(p(c))$$

Suppose that $v_\mathcal{I}(p(f^{k+1}(c)))$ is not in the same component as $v_\mathcal{I}(q)$ for some $k \in \mathbb{N}$. Then $v_\mathcal{I}(p(f^k(c)) \to (p(f^{k+1}(c)) \vee q))$ is not in the same component as $v_\mathcal{I}(q)$ so $v_\mathcal{I}((((p(f^k(c)) \to (p(f^{k+1}(c)) \vee q)) \to q) \to q) = v_\mathcal{I}(q \to q) = 1$. Assume then that $v_\mathcal{I}(q)$ and $v_\mathcal{I}(p(f^{k+1}(c)))$ for all $k \in \mathbb{N}$ are in the same component. Using the interpretation of $\to$ in Product and Łukasiewicz components to shift quantifiers:

$$\begin{aligned}
v_\mathcal{I}(F) &\geq v_\mathcal{I}((\exists x)(p(x) \to p(f(x)))) \\
&\geq v_\mathcal{I}((\exists x)(\forall y)(p(x) \to p(y))) \\
&= v_\mathcal{I}((\forall x)(p(x)) \to (\forall y)(p(y))) \\
&= 1
\end{aligned}$$

But now if the existential Herbrand theorem holds for L, then for some $n$:

$$\models_\mathsf{L} \bigvee_{i=1}^{n} [((p(f^{i-1}(c)) \to (p(f^i(c)) \vee q)) \to q) \to q]$$

If L has at least one component $[a, b]$, i.e. it is not Gödel Logic, then we can let $v(q) = a'$ for some $a' > a$ and $v_\mathcal{I}(p(f^i(c))) = b - i(b - a')/n$ for $i = 1 \ldots n$, giving:

$$v_\mathcal{I}(p(f^{i-1}(c))) > v_\mathcal{I}(p(f^i(c))) \quad \text{for } i = 1 \ldots n$$

It follows that $v_\mathcal{I}(\bigvee_{i=1}^{n}[((p(f^{i-1}(c)) \to (p(f^i(c)) \vee q)) \to q) \to q]) < 1$. But this is a contradiction, so the existential Herbrand theorem fails. $\qquad \square$

# 4  Approximate Herbrand Theorems

Although the Herbrand theorem fails in all but Gödel Logic of our fuzzy logics, in many cases we can obtain an approximate version. Rather than find a valid Herbrand disjunction for a valid existential formula, we find Herbrand disjunctions that are approximately valid to a degree arbitrarily close to 1. To be more precise, consider the following result for Łukasiewicz Logic:

**Theorem 3 ([12,4]).** $\models_{\text{Ł}} (\exists \bar{x}) P(\bar{x})$ where $P$ is quantifier-free iff for all $r < 1$:

$$\models_{\text{Ł}}^{>r} \bigvee_{i=1}^{n} P(\bar{t}_i) \text{ for some } \bar{t}_1, \ldots, \bar{t}_n \in \mathcal{U}(P)$$

Let us just sketch the topological proof for the trickier left-to-right direction, referring to [16] for terminology. First we observe that the set of valuations $[0,1]^{\mathcal{B}(P)}$ is compact with respect to the product topology, using the Tychonoff theorem in the case that $\mathcal{B}(P)$ is countably infinite. Fix $r < 1$. Then we note that since the connectives of Łukasiewcz Logic are all continuous, for each $\bar{t} \in \mathcal{U}(P)$, valuations $v_{\mathcal{I}}$ such that $v_{\mathcal{I}}(P(\bar{t})) \leq r$ form a closed subset of $[0,1]^{\mathcal{B}(P)}$. But if $\models_{\text{Ł}} (\exists \bar{x}) P(\bar{x})$, then the intersection of all these sets must be empty. So by the finite intersection property for compact spaces, there is a finite intersection that is empty; i.e. $\bar{t}_1, \ldots, \bar{t}_n \in \mathcal{U}(P)$ such that $\models_{\text{Ł}}^{>r} \bigvee_{i=1}^{n} P(\bar{t}_i)$.

This approach generalizes to other logics with connectives interpreted by continuous functions on a compact space. However, Łukasiewicz Logic is the only fuzzy logic with this property: the residuum of any $t$-norm that is not order-isomorphic to the Łukasiewicz $t$-norm is not continuous. Instead we combine this approach with the strategy outlined above for Gödel Logic. Let us say that a quantifier-free formula $P(\bar{x})$ for a fuzzy logic L is L-continuous on a compact set $V \subseteq [0,1]^{\mathcal{B}(P)}$ if for all $\bar{t} \in \mathcal{U}(P)$ the function $H : V \to [0,1]$ defined by $H(v_{\mathcal{I}}) = v_{\mathcal{I}}(P(\bar{t}))$ is continuous on $V$ with respect to the product topology.

**Lemma 1.** Let L be a fuzzy logic and let $P(\bar{x})$ be a quantifier-free formula L-continuous on a compact subset $V$ of $[0,1]^{\mathcal{B}(P)}$ where $\models_{\text{L}} (\exists \bar{x}) P(\bar{x})$. Then for all $r \in [0,1]$, there exist $\bar{t}_1, \ldots, \bar{t}_n \in \mathcal{U}(P)$ such that $v_{\mathcal{I}}(\bigvee_{i=1}^{n} P(\bar{t}_i)) > r$ for all $v_{\mathcal{I}} \in V$.

*Proof.* For each $\bar{t} \in \mathcal{U}(P)$, let:

$$S(\bar{t}) = \{v_{\mathcal{I}} \in V : v_{\mathcal{I}}(P(\bar{t})) \leq r\} \quad \text{and} \quad S = \{S(\bar{t}) : \bar{t} \in \mathcal{U}(P)\}$$

Each $S(\bar{t})$ is a closed subset of $V$ by the L-continuity of $P(\bar{x})$. If $\models_{\text{L}} (\exists \bar{x}) P(\bar{x})$, then there is no $v_{\mathcal{I}} \in V$ such that $v_{\mathcal{I}}(P(\bar{t})) \leq r$ for all $\bar{t} \in \mathcal{U}(P)$, so the intersection of $S$ must be empty. Hence by the finite intersection property for compact spaces, there exist $S(\bar{t}_1), \ldots, S(\bar{t}_n)$ with an empty intersection. I.e. for every $v_{\mathcal{I}} \in V$, $v_{\mathcal{I}}(P(\bar{t}_i)) > r$ for some $i \in \{1, \ldots, n\}$. So $v_{\mathcal{I}}(\bigvee_{i=1}^{n} P(\bar{t}_i)) > r$ for all $v_{\mathcal{I}} \in V$. $\qquad\square$

**Theorem 4.** Let L be an Ł-fuzzy logic and let $P(\bar{x})$ be a quantifier-free formula. Then $\models_{\text{L}} (\exists \bar{x}) P(\bar{x})$ iff for all $r < 1$:

$$\models_{\text{L}}^{>r} \bigvee_{i=1}^{n} P(\bar{t}_i) \text{ for some } \bar{t}_1, \ldots, \bar{t}_n \in \mathcal{U}(P)$$

*Proof.* If $\models_{\mathsf{L}}^{>r} \bigvee_{i=1}^{n} P(\bar{t}_i)$ for all $r < 1$, then $\models_{\mathsf{L}}^{>r} (\exists \bar{x})P(\bar{x})$ for all $r < 1$, so clearly $\models_{\mathsf{L}} (\exists \bar{x})P(\bar{x})$. For the other direction, let $A_1, A_2, \ldots$ be an enumeration of the subformulas of $P(\bar{t})$ for all $\bar{t} \in \mathcal{U}(P)$. We also enumerate the components of $\mathsf{L}$ as $[a, 1]$, $[a_1, b_1], [a_2, b_2], \ldots$ where $a < 1$, each $a_i < b_i$, and (since $\mathsf{L}$ is Ł-fuzzy) $[a, 1]$ is a Łukasiewicz component. Fix $a < r < 1$.

We construct a tree $T$ labelled with constraints on $v_{\mathcal{I}} \in [0, 1]^{\mathcal{B}(P)}$. Each node of $T$ at level $l$ is labelled with a constraint $K$ on the formulas $A_1, \ldots, A_l$ that:

(i)  places each $A_i$ between, before, after, or into one of the first $l$ components;
(ii) linearly orders or identifies as equal the $A_i$s not placed in $[a, 1]$ by (i).

We say that $v_{\mathcal{I}}$ satisfies $K$ at level $l$ if $v_{\mathcal{I}}(A_i)$ obeys the restrictions of $K$ (i.e. being in the right order and appropriate components, etc.) for $i = 1 \ldots l$. A constraint $K'$ extends $K$ if every $v_{\mathcal{I}}$ satisfying $K'$ also satisfies $K$.

The tree $T$ is constructed inductively as follows:

(1)  The root of $T$ is at level 0 and labelled with the empty constraint.
(2)  Let $w$ be a node at level $l$ labelled with a constraint $K$. Suppose that there exist $\bar{t}_1, \ldots, \bar{t}_n \in \mathcal{U}_l(P)$ such that $v_{\mathcal{I}}(\bigvee_{i=1}^{n} P(\bar{t}_i)) > r$ for all $v_{\mathcal{I}}$ satisfying $K$. Then $w$ is a leaf node of $T$. Otherwise, for each level $l + 1$ constraint $K'$ extending $K$, a successor node $w'$ labelled with $K'$ is appended to $w$ at level $l + 1$.

Notice that the nodes at any given level $l$ of $T$ partition $[0, 1]^{\mathcal{B}(P)}$ into distinct chunks. Moreover, for every $v_{\mathcal{I}} \in [0, 1]^{\mathcal{B}(P)}$, there is a branch of $T$ such that $v_{\mathcal{I}}$ satisfies the constraints at every node of the branch. Two cases arise:

(a)  $T$ is *finite*. Let $w_1, \ldots, w_n$ be the leaf nodes of $T$, where $v_{\mathcal{I}}(\bigvee_{i=1}^{m_j} P(\bar{t}_{ij})) > r$ for all $v_{\mathcal{I}}$ satisfying the constraint at $w_j$ for $j = 1 \ldots n$. Then since the union of $v_{\mathcal{I}} \in [0, 1]^{\mathcal{B}(P)}$ satisfying one of these constraints is the whole of $[0, 1]^{\mathcal{B}(P)}$, $\models_{\mathsf{L}}^{>r} \bigvee_{j=1\ldots n}^{i=1\ldots m_j} P(\bar{t}_{ij})$ as required.

(b)  $T$ is *infinite*. Then by König's lemma, $T$ has an infinite branch. We identify a constraint $K$ as the union of all the constraints on the branch. Let $V$ be the members of $[0, 1]^{\mathcal{B}(P)}$ satisfying this extended constraint and take some $v_0 \in V$. We consider the set $V_0 = \{v \in V : \text{for all } A, \text{ either } v(A) = v_0(A) \text{ or } v_0(A) \in [a, 1]\}$. That is, we fix the values of the formulas constrained to be in components other than $[a, 1]$, noting that the particular values of these formulas (with respect to the constraint $K$) does not affect the values of formulas in $[a, 1]$. Clearly $V_0$ is a compact subset of $[0, 1]^{\mathcal{B}(P)}$. Moreover, the interpretations of the connectives are continuous functions when restricted to $V_0$, since $[a, 1]$ is a Łukasiewicz component. Hence $P(\bar{x})$ is L-continuous on $V_0$. But then by Lemma 1, there exist $\bar{t}_1, \ldots, \bar{t}_n \in \mathcal{U}(P)$ such that $v_{\mathcal{I}}(\bigvee_{i=1}^{n} P(\bar{t}_i)) > r$ for all $v_{\mathcal{I}} \in V_0$, and hence also (by an easy induction) for all $v_{\mathcal{I}} \in V$. But this contradicts the construction of the tree.     $\square$

This approximate Herbrand theorem has a nice corollary. Let $F = (\forall \bar{x})(\exists \bar{y})P(\bar{x}, \bar{y})$ where $P(\bar{x}, \bar{y})$ is both quantifier-free and function-free. Then $\models_{\mathsf{L}} F$ iff $\models_{\mathsf{L}} (\exists \bar{y})P(\bar{c}, \bar{y})$

for some new constants $\bar{c}$. Let $\mathcal{C}$ be the (finite) set of constants occurring in $(\exists\bar{y})P(\bar{c},\bar{y})$, adding one if the set is empty. Using the previous theorem:

$$\models_{\mathsf{L}} F \text{ iff for all } r<1, \models_{\mathsf{L}}^{>r} \bigvee_{i=1}^{n} P(\bar{c},\bar{t}_i) \text{ for some } \bar{t}_1\dots\bar{t}_n \in \mathcal{C} \text{ iff } \models_{\mathsf{L}} \bigvee_{\bar{a}\in\mathcal{C}} P(\bar{c},\bar{a})$$

But checking validity in a propositional fuzzy logic is decidable, so we have established the following:

**Corollary 1.** *The function-free $\forall\exists$-fragment of any Ł-fuzzy logic is decidable.*

A natural question to ask at this point is whether the approximate Herbrand theorem holds for all fuzzy logics. The answer is no. For fuzzy logics based on a $t$-norm $*$ with a top Product component or an interval $[a,1]$ where $a < 1$ and $*$ acts like min (P-fuzzy logic or G-fuzzy logic), assignments to values close to 1 can be "shifted downwards".

**Proposition 2.** *The approximate Herbrand theorem fails for any* P-*fuzzy logic or* G-*fuzzy logic except* G.

*Proof.* Let L be any P-fuzzy logic or G-fuzzy logic except G, with top component $[a,1]$ in the former case and the $t$-norm acting like min on $[a,1]$ in the latter case. Fix some $r \in (a,1)$. By Proposition 1, the usual Herbrand theorem fails. So we have $\models_{\mathsf{L}} (\exists\bar{x})P(\bar{x})$ and for any $F = \bigvee_{i=1}^{n} P(\bar{t}_i)$ with $\bar{t}_1,\dots,\bar{t}_n \in \mathcal{U}(P)$, $v_{\mathcal{I}}(F) < 1$ for some L-interpretation $\mathcal{I}$. But now we can define a new L-interpretation $\mathcal{I}'$ by scaling downwards values of atoms in $(a,1]$ such that for any propositional formula $G$ involving only atoms from $F$, either $v_{\mathcal{I}'}(G) < r$ or $v_{\mathcal{I}'}(G) = 1$, and $v_{\mathcal{I}'}(G) = 1$ iff $v_{\mathcal{I}}(G) = 1$. So in particular, $v_{\mathcal{I}'}(F) \leq r$. $\qquad\square$

## 5  Skolemization

We now use the approximate Herbrand theorem in a rather neat way: to provide a proof-theoretic proof of Skolemization for the Prenex fragments of a wide range of fuzzy logics. Note that while for Classical Logic, the usual process involves removing existential quantifiers and preserving satisfiability, here we follow common terminology for fuzzy logics (see e.g. [1]) and remove universal quantifiers, hoping rather to preserve validity.

Let $A$ be a Prenex formula and assume harmlessly that the $i$th occurrence of $\forall$ is labelled $\forall^i$ and that no function symbol $f_i$ of any arity occurs in $A$. Then the *Skolem form* $A^S$ of $A$ is defined by induction as follows:

(1)  If $A$ is of the form $(\exists\bar{x})P(\bar{x})$ where $P$ is quantifier-free, then $A^S$ is $(\exists\bar{x})P(\bar{x})$.
(2)  If $A$ is of the form $(\exists\bar{x})(\forall^i y)B(\bar{x},y)$, then $A^S$ is $((\exists\bar{x})B(\bar{x},f_i(\bar{x})))^S$.

Our aim is to prove that $\models_{\mathsf{L}} A$ iff $\models_{\mathsf{L}} A^S$. The key step in establishing the more difficult right-to-left direction of this equivalence, is to show that if a Herbrand disjunction for $A^S$ is approximately valid, then $A$ is valid to the same degree. We first recall the following lemma from [4], which gives an algorithmic procedure for obtaining $A$ from any Herbrand disjunction of $A^S$ using simple rules acting on multisets of formulas. We denote an arbitrary multiset of formulas by $\Gamma$, the multiset sum of $\Gamma_1$ and $\Gamma_2$ by $\Gamma_1 \uplus \Gamma_2$, and use square bracket notation $[\,:\,]$ for multiset construction:

**Lemma 2 ([4]).** *Let $(\exists \bar{x})P^F(\bar{x})$ be the Skolem form of $(Q\bar{y})P(\bar{y})$. Then $[(Q\bar{y})P(\bar{y})]$ is derivable from any finite non-empty sub-multiset of $[P^F(\bar{t}) : \bar{t} \in \mathcal{U}(P^F)]$ using:*

$$\frac{\Gamma \uplus [(\forall x)A(x)]}{\Gamma \uplus [A(t)]} \qquad \frac{\Gamma \uplus [(\exists x)A(x)]}{\Gamma \uplus [A(s)]} \qquad \frac{\Gamma \uplus [A]}{\Gamma \uplus [A, A]} \qquad \frac{\Gamma \uplus [A]}{\Gamma}$$

*where in the leftmost rule, $t$ is any ground term not occurring in $\Gamma$ or $A$.*

But now notice that for any fuzzy logic, if we interpret a multiset as a disjunction of its members, then each of these rules is sound with respect to non-strict approximate validity. That is, for a rule with premise $[A_1, \ldots, A_n]$ and conclusion $[B_1, \ldots, B_m]$, $\models_\mathsf{L}^{\geq r} A_1 \vee \ldots \vee A_n$ implies $\models_\mathsf{L}^{\geq r} B_1 \vee \ldots \vee B_m$. Hence by a simple induction on the height of a derivation using these rules:

**Proposition 3.** *Let $(\exists \bar{x})P^F(\bar{x})$ be the Skolem form of $(Q\bar{y})P(\bar{y})$. Then for any fuzzy logic $\mathsf{L}$ and $r \in [0, 1]$:*

$$\text{if } \models_\mathsf{L}^{\geq r} \bigvee_{i=1}^{n} P^F(\bar{t}_i) \text{ for some } \bar{t}_1, \ldots, \bar{t}_n \in \mathcal{U}(P^F), \text{ then } \models_\mathsf{L}^{\geq r} (Q\bar{y})P(\bar{y})$$

We can now establish Skolemization for the Prenex formulas of logics by combining this last result with the approximate Herbrand theorem.

**Theorem 5.** *Let $(Q\bar{y})P(\bar{y})$ be a Prenex formula with Skolem form $(\exists \bar{x})P^F(\bar{x})$. Then for any Ł-fuzzy logic $\mathsf{L}$:*

$$\models_\mathsf{L} (Q\bar{y})P(\bar{y}) \quad \textit{iff} \quad \models_\mathsf{L} (\exists \bar{x})P^F(\bar{x})$$

*Proof.* The left-to-right direction follows easily using standard quantifier properties of $\mathsf{L}$. For the right-to-left-direction, suppose that $\models_\mathsf{L} (\exists \bar{x})P^F(\bar{x})$. If $\mathsf{L}$ is an Ł-fuzzy logic, then by Theorem 4, for all $r < 1$, $\models_\mathsf{L}^{\geq r} \bigvee_{i=1}^{n} P^F(\bar{t}_i)$ for some $\bar{t}_1, \ldots, \bar{t}_n \in \mathcal{U}(P^F)$. By Proposition 3, for all $r < 1$, $\models_\mathsf{L}^{\geq r} (Q\bar{y})P(\bar{y})$. Hence $\models_\mathsf{L} (Q\bar{y})P(\bar{y})$ as required. $\quad\square$

Skolemization also allows us to extend the approximate Herbrand theorem to the whole Prenex fragment of Ł-fuzzy logics: we just use Theorem 5 to find an appropriate existential formula, and apply Theorem 4.

**Corollary 2.** *Let $\mathsf{L}$ be an Ł-fuzzy logic and let $(Q\bar{y})P(\bar{y})$ be a Prenex formula with Skolem form $(\exists \bar{x})P^F(\bar{x})$. Then $\models_\mathsf{L} (Q\bar{y})P(\bar{y})$ iff $(\exists \bar{x})P^F(\bar{x})$ iff for all $r < 1$:*

$$\models_\mathsf{L}^{\geq r} \bigvee_{i=1}^{n} P^F(\bar{t}_i) \text{ for some } \bar{t}_1, \ldots, \bar{t}_n \in \mathcal{U}(P^F)$$

Moreover, since the problem of determining the approximate validity of a disjunction of propositional formulas is decidable (see e.g. [9]), we obtain the following result.

**Corollary 3.** *The Prenex fragment of any Ł-fuzzy logic is in $\prod_2$.*

# References

1. Baaz, M., Ciabattoni, A., Fermüller, C.G.: Herbrand's theorem for prenex Gödel logic and its consequences for theorem proving. In: Nieuwenhuis, R., Voronkov, A. (eds.) LPAR 2001. LNCS (LNAI), vol. 2250, pp. 201–216. Springer, Heidelberg (2001)
2. Baaz, M., Ciabattoni, A., Montagna, F.: Analytic calculi for monoidal t-norm based logic. Fundamenta Informaticae 59(4), 315–332 (2004)
3. Baaz, M., Iemhoff, R.: The skolemization of existential quantifiers in intuitionistic logic. Annals of Pure and Applied Logic 142(1-3), 269–295 (2006)
4. Baaz, M., Metcalfe, G.: Herbrand's theorem, skolemization, and proof systems for Łukasiewicz logic (submitted),
   `http://www.math.vanderbilt.edu/people/metcalfe/publications`
5. Baaz, M., Zach, R.: Hypersequents and the proof theory of intuitionistic fuzzy logic. In: Clote, P.G., Schwichtenberg, H. (eds.) CSL 2000. LNCS, vol. 1862, pp. 187–201. Springer, Heidelberg (2000)
6. Hájek, P.: Metamathematics of Fuzzy Logic. Kluwer, Dordrecht (1998)
7. Hájek, P.: Arithmetical complexity of fuzzy logic – a survey. Soft Computing 9, 935–941 (2005)
8. Hájek, P.: Making fuzzy description logic more general. Fuzzy Sets and Systems 154(1), 1–15 (2005)
9. Hanikova, Z.: A note on the complexity of tautologies of individual t-algebras. Neural Network World 12(5), 453–460 (2002)
10. Montagna, F.: Three complexity problems in quantified fuzzy logic. Studia Logica 68(1), 143–152 (2001)
11. Mostert, P.S., Shields, A.L.: On the structure of semigroups on a compact manifold with boundary. Annals of Mathematics 65, 117–143 (1957)
12. Novák, V.: On the Hilbert-Ackermann theorem in fuzzy logic. Acta Mathematica et Informatica Universitatis Ostraviensis 4, 57–74 (1996)
13. Ragaz, M.E.: Arithmetische Klassifikation von Formelmengen der unendlichwertigen Logik. PhD thesis, ETH Zürich (1981)
14. Straccia, U.: Reasoning within fuzzy description logics. Journal of Artificial Intelligence Research 14, 137–166 (2001)
15. Vojtás, P.: Fuzzy logic programming. Fuzzy Sets and Systems 124, 361–370 (2001)
16. Willard, S.: General Topology. Dover (2004)

# Decidability of Hybrid Logic
# with Local Common Knowledge
# Based on Linear Temporal Logic LTL


Sergey Babenyshev and Vladimir Rybakov

Manchester Metropolitan University, Manchester, U.K.
Siberian Federal University, Krasnoyarsk, Russia
sergei.babyonyshev@gmail.com, V.Rybakov@mmu.ac.uk



**Abstract.** Our paper[1] considers a hybrid $\text{LTL}_{\mathcal{ACK}}$ between the multi-agent logic with the local common knowledge operation and an extended version of the linear temporal logic $\mathcal{LTL}$. The logic is based on the semantics of Kripe/Hintikka models with potentially infinite runs and the time points represented by clusters of states with agents' accessibility relations. We study the satisfiability problem for $\text{LTL}_{\mathcal{ACK}}$ and related decidability problem. The key result is an algorithm which recognizes theorems of $\text{LTL}_{\mathcal{ACK}}$ (so we show that $\text{LTL}_{\mathcal{ACK}}$ is decidable), which, as a consequence, also solves the satisfiability problem. Technique is based on verification of validity for special normal reduced forms of rules in models of double exponential in the size of rules.

**Keywords:** linear temporal logic, multi-agent logics,hybrid logics, relational Kripke-Hintikka models, decidability algorithms, satisfiability.


## 1 Introduction

Hybrid logics nowadays is an active area of research in non-classical Mathematical Logic, Computer Science (CS) and Artificial Intelligence (AI) [1]. They are usually introduced by combination (fusion) of several non-classical logics. We choose as a background logic for our research the linear temporal logic LTL. Temporal logics and, in particular, LTL, is currently the most widely used specification formalism for reactive systems. They were first suggested to be used for specifying properties of programs in late 1970's (cf. Pnueli [2]). The most used temporal framework is the linear-time propositional temporal logic LTL, which has been studied from various viewpoints (cf. e.g. Manna and Pnueli [3,4], Clark E. et al. [5]). Temporal logics have numerous applications to safety, liveness and fairness, to various problems arising in computing. Model checking for LTL formed a noticeable trend in applications of logic in Computer Science, which uses, in particular, applications of automata theory (cf. Vardi [6,7]).


---

[1] This research is supported by Engineering and Physical Sciences Research Council (EPSRC), U.K., grant EP/F014406/1.



A. Beckmann, C. Dimitracopoulos, and B. Löwe (Eds.): CiE 2008, LNCS 5028, pp. 32–41, 2008.

Temporal logics themselves can be considered as special cases of hybrid logics, e.g. as a bimodal logic with some additional laws imposed on interaction of modalities to emulate the flow of time. The mathematical theory dedicated to study of various aspects of interaction of temporal operations (e.g. axiomatizations of temporal logics) and to construction of effective semantic theory based on Kripke/Hintikka-like models and temporal Boolean algebras, formed a highly technical branch in non-classical logics (cf. e.g. van Benthem [8,9], Goldblatt [10], Gabbay and Hodkinson [11], Hodkinson [12]).

Another component of our hybrid logic is a multi-agent knowledge logic. The multi-agent logics is another active area in AI and Knowledge Representation (cf. Bordini et al. [13], Dix et al. [14], Hoek et al. [15], Fisher [16], Hendler [17], Kacprzak [18], Wooldridge [19]). An agent is an autonomous entity that acts in a world, by interacting with its environment and with other agents. Multi-Agent Systems — systems that may include many such entities — are becoming more and more popular in Computer Science and Artificial Intelligence. This is because, firstly, they offer a convenient metaphor for modeling the reality around us as a world inhabited by such autonomous, active, possibly intelligent elements, secondly, they can be used as a methodology that enables design and implementation of large systems in a really modular way. In a sense, multi-agent logics came from a particular application of multi-modal logics to reasoning about knowledge (cf. Fagin et al. [20, 21], Halpern and Shore [22]), where modal-like operations $K_i$ (responsible for knowledge of individual agents) were embedded in the language of propositional logic. These operations are intended to model effects and properties of agents' knowledge in changing environment, If we consider a hybrid with multi-agent logic, an immediate question is which logic should be used as the background logic. There were many approaches, but, if the logic used as the basis is too expressive — the undecidability phenomenon can occur (cf. Kacprzak [18] with reduction of the decidability to the domino problem). If the basic logic is just the boolean logic, and the agents are autonomous, decidability for the standard systems usually can be obtained by standard techniques, e.g. filtration, (cf. [20, 21]).

In this paper we attempt to implement a multi-agent knowledge logic with local common-knowledge operator based on the linear temporal logic LTL. The result is a multi-agent logic $\text{LTL}_{\mathcal{ACK}}$. We address the problems of satisfiability and decidability of $\text{LTL}_{\mathcal{ACK}}$. Our paper proposes an algorithm which recognizes theorems of $\text{LTL}_{\mathcal{ACK}}$ (so we show that $\text{LTL}_{\mathcal{ACK}}$ is decidable). This algorithm works in three steps. First, it reduces a formula in the language of $\text{LTL}_{\mathcal{ACK}}$ to an inference rule, then the rule to a special normal reduced form, and then it checks validity of the reduced form in special finite models of size efficiently bounded on the size of the original formula. Also we discuss some possible variations of $\text{LTL}_{\mathcal{ACK}}$ where the proposed technique might work.

## 2   Preliminaries, Definitions, Notation

Our hybrid logic is based on an extension of the linear temporal logic LTL. The language of LTL augments the one of Boolean logic by specifically temporal (modal-like) operations Until and Next. Our logic also employs the agents' knowledge operations and local common-knowledge operation. To justify the choice of the language we start with introducing of semantic objects upon which our logic will be based. These are the following Kripe/Hintikka-like models with linear discrete time.

The frame $\mathcal{N}_C := \langle \bigcup_{i \in N} C(i), R, R_1, \ldots R_m, Next \rangle$ is a tuple, where $N$ is the set of natural numbers, $C(i), i \in N$ are some pairwise disjoint nonempty sets, $R, R_1, \ldots, R_m$ are binary relations, emulating agents' accessibility. For all elements $a$ and $b$ from $\bigcup_{i \in N} C(i)$

$$aRb \iff [a \in C(i) \text{ and } b \in C(j) \text{ and } i < j] \text{ or } [a, b \in C(i) \text{ for some } i];$$

any $R_j$ is a reflexive, transitive and symmetric relation (i.e an equivalence relation), and

$$\forall a, b \in \bigcup_{i \in N} C(i),\ aR_j b \implies [a, b \in C(i) \text{ for some } i];$$

$$a\ Next\ b \iff \exists i[a \in C(i)\ \&\ b \in C(i+1)].$$

As usual for LTL, these frames are intended to model the reasoning (or computation) in discrete time, so each $i \in N$ (any natural number $i$) is the time index for a cluster of states arising at a step in current computation.

Any $C(i)$ is a set of all possible states in the time point $i$, and the relation $R$ models discrete current of time. Relations $R_j$ are intended to model knowledge-accessibility relations of agents in any cluster of states $C(i)$ at the time point $i$. Thus any $R_j$ is supposed to be $S5$-like relation, i.e an equivalence relation. The reasoning (computation) is supposed to be concurrent for all agents — after a step, a new cluster of possible states appears, and the agents will be given new access rules to the information in the new time cluster of states. The $Next$ relation is the standard one — it describes all states available in the next time point cluster.

To model logical laws of reasoning (computation) represented by structures $\mathcal{N}_C$ we propose the following language extending the language of the standard LTL. It uses the language of Boolean logic, the operations $\mathbf{N}$ (next), $\mathbf{U}$ (until)) and the new operations $\mathbf{U}_w$ (weak until) and $\mathbf{U}_s$ (strong until), the unary agents' knowledge operations $\mathbf{K}_j$, $1 \leq j \leq m$ and an additional unary operation $\mathbf{CK}$ for local common knowledge. Formation rules for formulas are as usual. The intended meaning of the operations is as follows.

$\mathbf{K}_j \varphi$ means: the agent $j$ *knows* $\varphi$ at the current state of a time cluster;

$\mathbf{CK}\varphi$ means that $\varphi$ is of local *common knowledge* at the current state of a time cluster, or that all agents *should know* $\varphi$;

$\mathbf{N}\varphi$ has meaning: $\varphi$ holds in the *all states of next time cluster* ;

$\varphi \mathbf{U} \psi$ can be read: $\varphi$ holds until $\psi$ becomes true;

$\varphi \mathbf{U}_w \psi$ has meaning: $\varphi$ weakly holds until $\psi$ becomes true;

$\varphi \mathbf{U}_s \psi$ has meaning: $\varphi$ strongly holds until $\psi$ becomes true;

For any collection of propositional letters Prop and any frame $\mathcal{N}_C$, a valuation in $\mathcal{N}_C$ is a mapping which assigns truth values to elements of Prop in $\mathcal{N}_C$. Thus, for any $p \in$ Prop, $V(p) \subseteq \bigcup_{i \in N} C(i)$. We will call $\langle \mathcal{N}_C, V \rangle$ a *model* (a Kripke/Hinikka model). For any such model $\mathcal{M}$, a valuation can be extended from propositions of Prop to arbitrary formulas as follows (for $a \in \mathcal{N}_C$, we write $(\mathcal{N}_C, a) \Vdash_V \varphi$ to say that the formula $\varphi$ is true at $a$ in $\mathcal{N}_C$ w.r.t. $V$)

$$\forall p \in \text{Prop}, \quad (\mathcal{M}, a) \Vdash_V p \iff a \in V(p);$$

$$(\mathcal{M}, a) \Vdash_V \varphi \wedge \psi \iff (\mathcal{M}, a) \Vdash_V \varphi \wedge (\mathcal{M}, a) \Vdash_V \psi;$$

$$(\mathcal{M}, a) \Vdash_V \neg \varphi \iff not[(\mathcal{M}, a) \Vdash_V \varphi];$$

$$(\mathcal{M}, a) \Vdash_V \mathbf{K}_j \varphi \iff \forall b[(a \ R_j \ b) \Rightarrow (\mathcal{M}, b) \Vdash_V \varphi]$$

(so $\mathbf{K}_j \varphi$ says that $\varphi$ holds at all states accessible for the agent $j$);

$$(\mathcal{M}, a) \Vdash_V \mathbf{CK} \varphi \iff \exists i[a \in C(i) \ \& \ \forall b \in C(i) : (\mathcal{M}, b) \Vdash_V \varphi].$$

Thus $\varphi$ is of local *common knowledge* for agents in the time cluster $C(i)$ iff $\varphi$ holds in all states of $C(i)$. It is easy to accept that this is a reasonable understanding of *local common knowledge*: the local common knowledge is a fact, which is valid at all possible states of the current time point. Thus, in our formalism, $\mathbf{CK}$ plays role of the *local universal modality*.

$$(\mathcal{M}, a) \Vdash_V \mathbf{N} \varphi \Leftrightarrow \forall b[(a \ \text{Next} \ b) \Rightarrow (\mathcal{M}, b) \Vdash_V \varphi];$$

$$(\mathcal{M}, a) \Vdash_V \varphi \mathbf{U} \psi \Leftrightarrow \exists b[(aRb) \wedge ((\mathcal{M}, b) \Vdash_V \psi) \wedge$$

$$\forall c[(aRcRb) \& \neg(bRc) \Rightarrow (\mathcal{M}, c) \Vdash_V \varphi]];$$

$$(\mathcal{M}, a) \Vdash_V \varphi \mathbf{U}_w \psi \Leftrightarrow \exists b[(aRb) \wedge ((\mathcal{M}, b) \Vdash_V \psi) \wedge \forall c[(aRcRb) \& \neg(bRc) \&$$

$$\& (c \in C(i)) \implies \exists d \in C(i)(\mathcal{M}, d) \Vdash_V \varphi]];$$

$$(\mathcal{M}, a) \Vdash_V \varphi \mathbf{U}_s \psi \Leftrightarrow \exists b[(aRb) \wedge b \in C(i) \wedge \forall c \in C(i)((\mathcal{M}, c) \Vdash_V \psi) \wedge$$

$$\forall c[(aRcRb) \& \neg(bRc) \Rightarrow (\mathcal{M}, c) \Vdash_V \varphi]];$$

**Definition 1.** *For a Kripke structure $\mathcal{M} := \langle \mathcal{N}_C, V \rangle$ and a formula $\varphi$, we say that (i) $\varphi$ is satisfiable in $\mathcal{M}$ (denotation $\mathcal{M} \Vdash_{Sat} \varphi$) if there is a state $b$ of $\mathcal{M}$ ($b \in \mathcal{N}_C$) where $\varphi$ is true: $(\mathcal{M}, b) \Vdash_V \varphi$. (ii) $\varphi$ is valid in $\mathcal{M}$ (denotation $\mathcal{M} \Vdash \varphi$) if, for any $b$ of $\mathcal{M}$ ($b \in \mathcal{N}_C$), the formula $\varphi$ is true at $b$ $(\mathcal{M}, b) \Vdash_V \varphi$.*

**Definition 2.** *For a Kripke frame $\mathcal{N}_C$ and a formula $\varphi$, we say that $\varphi$ is satisfiable in $\mathcal{N}_C$ (denotation $\mathcal{N}_C \Vdash_{Sat} \varphi$) if there is a valuation $V$ in the frame $\mathcal{N}_C$ such that $\langle \mathcal{N}_C, V \rangle \Vdash_{Sat} \varphi$. $\varphi$ is valid in $\mathcal{N}_C$ (i.e. $\mathcal{N}_C \Vdash \varphi$) if $not(\mathcal{N}_C \Vdash_{Sat} \neg \varphi)$.*

**Definition 3.** *The logic* $\mathrm{LTL}_{\mathcal{ACK}}$ *is the set of all formulas which are valid in all frames* $\mathcal{N}_C$.

Thus, the logic $\mathrm{LTL}_{\mathcal{ACK}}$ is defined semantically, and in this paper we do not consider axiomatic systems for $\mathrm{LTL}_{\mathcal{ACK}}$. A formula $\varphi$ is said to be a theorem of $\mathrm{LTL}_{\mathcal{ACK}}$ if $\varphi \in \mathrm{LTL}_{\mathcal{ACK}}$. A formula $\varphi$ in the language of $\mathrm{LTL}_{\mathcal{ACK}}$ is *satisfiable* iff there is a valuation $V$ in the Kripke frame $\mathcal{N}_C$ which makes $\varphi$ satisfiable: $\langle \mathcal{N}_C, V \rangle \Vdash_{Sat} \varphi$. Obviously that a formula $\varphi$ is satisfiable iff $\neg\varphi$ is not a theorem of $\mathrm{LTL}_{\mathcal{ACK}}$: $\neg\varphi \notin \mathrm{LTL}_{\mathcal{ACK}}$, and vise versa, $\varphi$ is a theorem of $\mathrm{LTL}_{\mathcal{ACK}}$ ($\varphi \in \mathrm{LTL}_{\mathcal{ACK}}$) if $\neg\varphi$ is not satisfiable.

The computation of truth values of the operation $\mathbf{U}$ works in frames $\mathcal{N}_C$ in the standard way, but the operation $\mathbf{U}_w$ drastically differs from the standard $\mathbf{U}$. It is sufficient for $\varphi$ to be true only at a certain state of all time clusters before $\psi$ will become true at a state. And the strong until $-\varphi\mathbf{U}_s\psi-$ means that there is a time point $i$, where the formula $\psi$ is true at all states in the time cluster $C(i)$, and $\varphi$ holds in all states of all time points $j$ preceding $i$. Using operations $\mathbf{U}$ and $\mathbf{N}$ we can define all standard temporal and modal operations. For instance, modal operations may be trivially defined: $(i)$ $\Diamond\varphi \equiv \mathit{true}\mathbf{U}\varphi \in \mathrm{LTL}_{\mathcal{ACK}}$; $(ii)$ $\Box\varphi \equiv \neg(\mathit{true}\mathbf{U}\neg\varphi) \in \mathrm{LTL}_{\mathcal{ACK}}$;. The standard temporal operation $\mathbf{F}\varphi$ ($\varphi$ *holds eventually*, which, in terms of modal logic, means $\varphi$ is possible (denotation $\Diamond\varphi$)), can be described as $\mathit{true}\mathbf{U}\varphi$. The temporal operation $\mathbf{G}$, where $\mathbf{G}\varphi$ means $\varphi$ *holds henceforth*, can be defined as $\neg\mathbf{F}\neg\varphi$.

The suggested temporal operations together with knowledge operations and common knowledge add more expressive power to the language. For instance, the formula $\Box\neg K_1\neg\varphi$ says that, for any future time cluster and for any state $a$ of this cluster the knowledge $\varphi$ is *discoverable* for agent 1, i.e., the agent has access to a state $b$ where $\varphi$ holds.

The new temporal operations $\mathbf{U}_s$ and $\mathbf{U}_w$ bring new unique features to the language. For instance the formula $\Box_w\varphi := \neg(\top\mathbf{U}_s\neg\varphi)$ codes the *weak necessity*: it says that in any time cluster $C(i)$ there is a state where $\varphi$ is true. The formula $\neg(\varphi\mathbf{U}_w\Box\varphi) \wedge \Diamond\Box\varphi$ says that, since a time point $i$, $\varphi$ holds in all states, but before $i$ $\varphi$ is false in a state of any time cluster. The formula

$$\varphi\mathbf{U}_w\mathbf{CK}\psi,$$

says that $\varphi$ is true in at least one state of any future time cluster from the current one, till up the formula $\psi$ will be true at *all* states of some future time cluster.

Such properties are hard or impossible to express in terms of the standard modal or temporal operations. Therefore the logic seems to be interesting and we devote the rest of the paper to finding an efficient algorithm to check satisfiability in $\mathrm{LTL}_{\mathcal{ACK}}$ and to the proof of decidability of $\mathrm{LTL}_{\mathcal{ACK}}$.

## 3 Decidability Algorithm for $\mathrm{LTL}_{\mathcal{ACK}}$

The basic instrument we will use to show decidability is implicit modeling of the universal modality (just not-nested first-order universal quantifier at

Kripke/Hintikka frames) by means of converting formulas to inference rules. Our approach is based on our techniques to handle inference rules (cf. [23, 24, 25, 26, 27, 28, 29, 30, 31, 32, 33]). Here we extend our previous research for Logic of Discovery in Uncertain Situations presented in [33] and research at a hybrid of LTL and knowledge logic without interaction for agents (autonomous ones) reported recently at the workshop on Hybrid Logics (2007, Dublin). Recall, a (sequential) (inference) rule is an expression

$$\mathbf{r} := \frac{\varphi_1(x_1, \ldots, x_n), \ldots, \varphi_m(x_1, \ldots, x_n)}{\psi(x_1, \ldots, x_n)},$$

where $\varphi_1(x_1, \ldots, x_n), \ldots, \varphi_m(x_1, \ldots, x_n)$ and $\psi(x_1, \ldots, x_n)$ are some formulas constructed out of letters $x_1, \ldots, x_n$. Letters $x_1, \ldots, x_n$ are variables of $\mathbf{r}$, we use notation $x_i \in Var(\mathbf{r})$ to say $x_i$ is a variable of $\mathbf{r}$.

**Definition 4.** *A rule $\mathbf{r}$ is said to be* valid *in a Kripke model $\langle \mathcal{N}_C, V \rangle$ with the valuation $V$ (we will use notation $\mathcal{N}_C \Vdash_V \mathbf{r}$) if $[\forall a\ ((\mathcal{N}_C, a) \Vdash_V \bigwedge_{1 \leq i \leq m} \varphi_i)] \Rightarrow \forall a\ ((\mathcal{N}_C, a) \Vdash_V \psi)$. Otherwise we say $\mathbf{r}$ is* refuted *in $\mathcal{N}_C$, or refuted in $\mathcal{N}_C$ by $V$, and write $\mathcal{N}_C \not\Vdash_V \mathbf{r}$. A rule $\mathbf{r}$ is* valid *in a frame $\mathcal{N}_C$ (notation $\mathcal{N}_C \Vdash \mathbf{r}$) if, for any valuation $V$ of $Var(\mathbf{r})$, $\mathcal{N}_C \Vdash_V \mathbf{r}$*

For any formula $\varphi$ we can transform it in the rule $x \to x/\varphi$ and employ the technique of reduced normal forms for inference rules as follows. First, it is trivial that

**Lemma 1.** *A formula $\varphi$ is a theorem of $\mathrm{LTL}_{\mathcal{ACK}}$ iff the rule $(x \to x/\varphi)$ is valid in any frame $\mathcal{N}_C$.*

A rule $\mathbf{r}$ is said to have the *reduced normal form* if $\mathbf{r} = \varepsilon_c/x_1$ where

$$\varepsilon_c := \bigvee_{1 \leq j \leq m} (\bigwedge_{1 \leq i \leq n} [x_i^{t(j,i,0)} \wedge (\mathbf{N}x_i)^{t(j,i,1)} \wedge \bigwedge_{1 \leq k \leq m, k \neq i} (x_i \mathbf{U} x_k)^{t(j,i,k,0)} \wedge$$

$$\bigwedge_{1 \leq k \leq m, k \neq i} (x_i \mathbf{U}_w x_k)^{t(j,i,k,1)} \wedge \bigwedge_{1 \leq k \leq m, k \neq i} (x_i \mathbf{U}_s x_k)^{t(j,i,k,2)} \wedge$$

$$\bigwedge_{1 \leq s \leq m} (\neg \mathbf{K}_s \neg x_i)^{t(j,i,s,3)} \wedge \mathbf{CK} x_i^{t(j,i,2)}]),$$

and all $x_s$ are certain letters (variables), $t(j, i, z), t(j, i, k, z) \in \{0, 1\}$ and, for any formula $\alpha$ above, $\alpha^0 := \alpha$, $\alpha^1 := \neg \alpha$.

**Definition 5.** *Given a rule $\mathbf{r}_{nf}$ in the reduced normal form, $\mathbf{r}_{nf}$ is said to be a normal reduced form for a rule $r$ iff, for any frame $\mathcal{N}_C$, $\mathcal{N}_C \Vdash \mathbf{r} \Leftrightarrow \mathcal{N}_C \Vdash \mathbf{r}_{nf}$.*

Similar to Lemma 3.1.3 and Theorem 3.1.11 from [26], it follows

**Theorem 1.** *There exists an algorithm running in (single) exponential time, which, for any given rule $\mathbf{r}$, constructs its normal reduced form $\mathbf{r}_{nf}$.*

38    S. Babenyshev and V. Rybakov

Decidability of LTL$_{ACK}$ will follow (Lemma 1) if we will find an algorithm recognizing rules in the reduced normal form which are valid in all frames $\mathcal{N}_C$. The first necessary observation is

**Lemma 2.** *A rule* $\mathbf{r}_{nf}$ *in the reduced normal form is refuted in a frame* $\mathcal{N}_C$ *if and only if* $\mathbf{r}_{nf}$ *can be refuted in a frame of kind* $\mathcal{N}_C$ *with time clusters of size linear from* $\mathbf{r}_{nf}$.

Key instrument for description of our algorithm is the following special *bent* finite Kripke models. For any frame $\mathcal{N}_C$ and any natural numbers $k, m$, where $m > k > 1$, the frame $\mathcal{N}_C(k, m)$ has the structure:

$$\mathcal{N}_C(k,m) := \langle \bigcup_{1 \leq i \leq m} C(i), \ R, R_1, \ldots R_n, Next \rangle,$$

where $R$ is the accessibility relation from $\mathcal{N}_C$ extended by pairs $(x, y)$, where $x \in C(i), y \in C(j))$ and $i, j \in [n, m]$, so $xRy$ holds for all such pairs. Any relation $R_j$ is simply transferred from $\mathcal{N}_C$, and $Next$ is the relation from $\mathcal{N}_C$ extended by $\forall a \in C(m) \forall b \in C(k)(a \ Next \ b = true)$.

If we arc given with a valuation $V$ of letters from a formula $\varphi$ in $\mathcal{N}_C(k, m)$, the truth values of $\varphi$ can be defined at elements of $\mathcal{N}_C(k, m)$ by the rules similar to the ones for frames $\mathcal{N}_C$ above (actually just in accordance with standard computation of truth values for time operations and knowledge modalities). We display these modified rules below:

$$(\mathcal{N}_C(k, m), a) \Vdash_V \varphi \mathbf{U} \psi \Leftrightarrow \exists b[(aRb) \wedge (\mathcal{N}_C(k, m), b) \Vdash_V \psi) \wedge$$

$$\forall c[(aRcRb) \& \neg(bRc) \Rightarrow (\mathcal{N}_C(k, m), c) \Vdash_V \varphi]],$$

$$(\mathcal{N}_C(k, m), a) \Vdash_V \varphi \mathbf{U}_w \psi \Leftrightarrow \exists b[(aRb) \wedge ((\mathcal{N}_C(k, m), b) \Vdash_V \psi) \wedge$$

$$\forall c[(aRcRb) \& \neg(bRc) \ \& \ (c \in C(i)) \Rightarrow \ \exists d \in C(i)(\mathcal{N}_C(k, m), d) \Vdash_V \varphi]],$$

$$(\mathcal{N}_C(k, m), a) \Vdash_V \varphi \mathbf{U}_s \psi \Leftrightarrow \exists b[(aRb) \wedge b \in C(i) \wedge \forall c \in C(i)$$

$$((\mathcal{N}_C(k, m), c) \Vdash_V \psi) \wedge \forall c[(aRcRb) \& \neg(bRc) \Longrightarrow (\mathcal{N}_C(k, m), c) \Vdash_V \varphi]].$$

Based on Lemma 2 as the basis, we obtain the key lemma:

**Lemma 3.** *A rule* $\mathbf{r}_{nf}$ *in the reduced normal form is refuted in a frame* $\mathcal{N}_C$ *if and only if* $\mathbf{r}_{nf}$ *can be refuted in a frame* $\mathcal{N}_C(k, m)$, *where the size of the frame* $\mathcal{N}_C(n, m)$ *is double exponential in* $\mathbf{r}_{nf}$.

From Theorem 1, Lemma 1 and Lemma 3 we immediately derive

**Theorem 2.** *The logic* LTL$_{ACK}$ *is decidable. The algorithm for testing a formula to be a theorem in* LTL$_{ACK}$ *consists in verification of validity of rules in reduced normal form in frames* $\mathcal{N}_C(n, m)$ *of size double-exponential in the size of reduced normal forms.*

Using developed technique we may approach other hybrid logics based on LTL. One of those is the variant of $\text{LTL}_{\mathcal{ACK}}$, where we consider the new operation $\mathbf{N}_w$ – weak next with interpretation

$$(\mathcal{M}, a) \Vdash_V \mathbf{N}_w \varphi \iff \exists b[(a \, \text{Next} \, b) \implies (\mathcal{M}, b) \Vdash_V \varphi],$$

and the logic with this new operation again will be decidable. Also, for example, we can consider a new operation $Next_w$ on frames $\mathcal{N}_C$ as a restriction of Next, formally

$$\forall a, b \in \bigcup_{i \in N} C(i), a \, \text{Next}_w \, b \implies [\exists i \in N : a \in C(i) \text{ and } b \in C(i+1)];$$

$$\forall a \in \bigcup_{i \in N} C(i)[a \in C(i) \implies \exists b \in C(i+1) : (a \, \text{Next}_w \, b)$$

$$\& \, \forall c \in C(i) \forall d \in C(i+1)((c \, \text{Next}_w \, d) \iff (a \, \text{Next}_w \, d))].$$

Our method of proving satisfiability and decidability for $\text{LTL}_{\mathcal{ACK}}$ works for this case again. In addition some restrictions on agents' accessibility relations $R_i$ may be considered, say, by introduction a hierarchy between $R_i$'s. This hierarchy may be an arbitrary desirable one (of kind $R_i \subseteq R_j$ for supervision). Our method deals with these logics as well and allows to prove their decidability.

## 4 Conclusion, Future Work

We suggest a method to prove decidability of the hybrid logic $\text{LTL}_{\mathcal{ACK}}$ and some similar logics. These instruments are flexible and may work for other various non-classical logics originated in the area of pure Mathematical Logic, AI or CS. There are many open venues for the future research on logic $\text{LTL}_{\mathcal{ACK}}$ and its variants. The issues of axiomatization and complexity are some of them. Logics obtained from $\text{LTL}_{\mathcal{ACK}}$ by introducing operations *Since* and *Previous* based on $C(i)$ with $i \in Z$ or $i \in N$ may be also interesting.

## References


1. Blackburn, P., Marx, M.: Constructive interpolation in hybrid logic. Journal of Symbolic Logic 68(2), 463–480 (2003)
2. Pnueli, A.: The temporal logic of programs. In: Proc. of the 18th Annual Symp. on Foundations of Computer Science, pp. 46–57. IEEE, Los Alamitos (1977)
3. Manna, Z., Pnueli, A.: Temporal Verification of Reactive Systems: Safety. Springer, Heidelberg (1995)
4. Manna, Z., Pnueli, A.: The Temporal Logic of Reactive and Concurrent Systems: Specification. Springer, Heidelberg (1992)
5. Clarke, E., Grumberg, O., Hamaguchi, K.P.: Another look at LTL model checking. In: Dill, D.L. (ed.) CAV 1994. LNCS, vol. 818. Springer, Heidelberg (1994)

6. Daniele, M., Giunchiglia, F., Vardi, M.: Improved automata generation for linear temporal logic. In: Halbwachs, N., Peled, D.A. (eds.) CAV 1999. LNCS, vol. 1633. Springer, Heidelberg (1999)
7. Vardi, M.: An automata-theoretic approach to linear temporal logic. In: Proceedings of the Banff Workshop on Knowledge Acquisition, Banff 1994 (1994)
8. van Benthem, J.: The Logic of Time. Kluwer, Dordrecht (1991)
9. van Benthem, J., Bergstra, J.: Logic of transition systems. Journal of Logic, Language and Information 3(4), 247–283 (1994)
10. Goldblatt, R.: Logics of Time and Computation. CSLI Lecture Notes 7 (1992)
11. Gabbay, D., Hodkinson, I.: An axiomatisation of the temporal logic with until and since over the real numbers. Journal of Logic and Computation 1(2), 229–260 (1990)
12. Hodkinson, I.: Temporal Logic. In: Temporal Logic and Automata, ch. II, vol. 2, pp. 30–72. Clarendon Press, Oxford (2000)
13. Bordini, R.H., Fisher, M., Visser, W., Wooldridge, M.: Model checking rational agents. IEEE Intelligent Systems 19, 46–52 (2004)
14. Dix, J., Fisher, M., Levesque, H., Sterling, L.: Editorial. Annals of Mathematics and Artificial Intelligence 41(2–4), 131–133 (2004)
15. van der Hoek, W., Wooldridge, M.: Towards a logic of rational agency. Logic Journal of the IGPL 11(2), 133–157 (2003)
16. Fisher, M.: Temporal development methods for agent-based systems. Journal of Autonomous Agents and Multi-Agent Systems 10(1), 41–66 (2005)
17. Hendler, J.: Agents and the semantic web. IEEE Intelligent Systems 16(2), 30–37 (2001)
18. Kacprzak, M.: Undecidability of a multi-agent logic. Fundamenta Informaticae 45(2–3), 213–220 (2003)
19. Wooldridge, M., Weiss, G., Ciancarini, P. (eds.): AOSE 2001. LNCS, vol. 2222. Springer, Heidelberg (2002)
20. Fagin, R., Halpern, J., Moses, Y., Vardi, M.: Reasoning About Knowledge. MIT Press, Cambridge (1995)
21. Fagin, R., Geanakoplos, J., Halpern, J., Vardi, M.: The hierarchical approach to modeling knowledge and common knowledge. International Journal of Game Theory 28(3), 331–365 (1999)
22. Halpern, J., Shore, R.: Reasoning about common knowledge with infinitely many agents. Information and Computation 191(1), 1–40 (2004)
23. Rybakov, V.: A criterion for admissibility of rules in the modal system S4 and the intuitionistic logic. Algebra and Logic 23(5), 369–384 (1984)
24. Rybakov, V.: Rules of inference with parameters for intuitionistic logic. Journal of Symbolic Logic 57(3), 912–923 (1992)
25. Rybakov, V.: Hereditarily structurally complete modal logics. Journal of Symbolic Logic 60(1), 266–288 (1995)
26. Rybakov, V.: Admissible Logical Inference Rules. Studies in Logic and the Foundations of Mathematics, vol. 136. Elsevier Sci. Publ., North-Holland (1997)
27. Rybakov, V., Kiyatkin, V., Oner, T.: On finite model property for admissible rules. Mathematical Logic Quarterly 45(4), 505–520 (1999)
28. Rybakov, V.: Construction of an explicit basis for rules admissible in modal system s4. Mathematical Logic Quarterly 47(4), 441–451 (2001)
29. Rybakov, V.: Logical consecutions in intransitive temporal linear logic of finite intervals. Journal of Logic and Computation 15(5), 633–657 (2005)

30. Rybakov, V.: Logical consecutions in discrete linear temporal logic. Journal of Symbolic Logic 70(4), 1137–1149 (2005)
31. Rybakov, V.: Linear Temporal Logic with Until and Before on Integer Numbers, Deciding Algorithms. In: Grigoriev, D., Harrison, J., Hirsch, E.A. (eds.) CSR 2006. LNCS, vol. 3967, pp. 322–334. Springer, Heidelberg (2006)
32. Rybakov, V.: Until-since temporal logic based on parallel time with common past. In: Artemov, S.N., Nerode, A. (eds.) LFCS 2007. LNCS, vol. 4514, pp. 486–497. Springer, Heidelberg (2007)
33. Rybakov, V.: Logic of discovery in uncertain situations — deciding algorithms. In: Apolloni, B., Howlett, R.J., Jain, L. (eds.) KES 2007, Part II. LNCS (LNAI), vol. 4693, pp. 950–958. Springer, Heidelberg (2007)

# Pure Iteration and Periodicity
## A Note on Some Small Sub-recursive Classes


Mathias Barra

Dept. of Mathematics, University of Oslo, P.B. 1053, Blindern, 0316 Oslo, Norway
georgba@math.uio.no
http://folk.uio.no/georgba



**Abstract.** We define a hierarchy $\mathcal{IT} = \bigcup_n \mathcal{IT}^n$ of small sub-recursive classes, based on the schema of *pure iteration*. $\mathcal{IT}$ is compared with a similar hierarchy, based on primitive recursion, for which a collapse is equivalent to a collapse of the small Grzegorczyk-classes. Our hierarchy does collapse, and the induced relational class is shown to have a highly periodic structure; indeed a unary predicate is decidable in $\mathcal{IT}$ iff it is definable in Presburger Arithmetic. The concluding discussion contrasts our findings to those of Kutylowski [12].


## 1   Introduction and Notation

**Introduction:** Over the last decade, several researchers have investigated the consequences of banning successor-like functions from various computational frameworks. Examples include Jones [5,6]; Kristiansen and Voda [9] (functionals of higher types) and [10] (imperative programming languages); Kristiansen and Barra [8] (idc.'s and $\lambda$-calculus) and Kristiansen [7] and Barra [1] (idc.'s).

This approach—banning all growth—has proved successful in the past, and has repeatedly yielded surprising and enlightening results. This paper emerge from a broad and general study of very small *inductively defined classes of functions*.

In 1953, A. Grzegorczyk published the seminal paper *Some classes of recursive functions* [4]; an important source of inspiration to a great number of researchers. Several questions were raised, some of which have been answered, some of which remain open to date. Most notorious is perhaps the problem of the statuses of the inclusions $\mathcal{E}^0_\star \overset{?}{\subseteq} \mathcal{E}^1_\star \overset{?}{\subseteq} \mathcal{E}^2_\star$ (see definitions below).

In 1987, Kutylowski presented various partial results on the above chain in *Small Grzegorczyk classes* [12], a paper which deals with function classes based on variations over the schema of *bounded iteration*.

The paper at hand is the result of a first attempt to combine the work of Kutylowski,with the successor-free approach. The results presented here does not represent a final product. However, in addition to being interesting in their own right—by shedding light on the intrinsic nature of *pure iteration* contrasted to *primitive recursion*—perhaps they may provide a point of origin for a viable route to the answer to the above mentioned problem.


A. Beckmann, C. Dimitracopoulos, and B. Löwe (Eds.): CiE 2008, LNCS 5028, pp. 42–51, 2008.

**Notation:** Unless otherwise specified, a *function* means a function $f : \mathbb{N}^k \to \mathbb{N}$ The *arity* of $f$, denoted $\mathrm{ar}(f)$, is then $k$. The notation $f^{(n)}(x)$ has the usual meaning: $f^{(0)}(x) \overset{\text{def}}{=} x$ and $f^{(n+1)}(x) \overset{\text{def}}{=} f(f^{(n)}(x))$. $\mathsf{P}$ is the *predecessor function*: $\mathsf{P}(x) \overset{\text{def}}{=} \max(0, x-1)$; $\mathsf{S}$ is the *successor function*: $\mathsf{S}(x) \overset{\text{def}}{=} x+1$; and $\mathsf{C}$ is the *case function*: $\mathsf{C}(x, y, z) \overset{\text{def}}{=} x$ if $z = 0$ and $y$ else. $\mathcal{I}$ is the set of *projections*: $\mathsf{I}_i^k(\vec{x}) \overset{\text{def}}{=} x_i$; $\mathcal{N}$ is the set of all *constant functions* $\mathsf{c}(x) \overset{\text{def}}{=} c$.

An *inductively defined class of functions* (idc.), is generated from a set $\mathcal{X}$ whose elements are called the *initial*, *primitive* or *basic* functions, as the least class containing $\mathcal{X}$ and closed under the *schemata, functionals* or *operations* of some set OP of functionals. We write $[\mathcal{X}; \mathrm{OP}]$ for this class[1]. Familiarity with the schema *composition*, denoted COMP, is assumed. We write $h \circ \vec{g}$ for the composition of functions $\vec{g}$ into the function $h$.

A $k$-ary *predicate* is a subset $R \subseteq \mathbb{N}^k$. Predicates are interchangeably called *relations*. A set of predicates which is closed under finite intersections and complements, is called *boolean* or *an algebra*.

A *partition of* $\mathbb{N}^k$ is a *finite* collection of predicates $\mathbb{P} = \{P_i\}_{i \leq n}$, satisfying **(i)** $\bigcup_i P_i = \mathbb{N}^k$, and **(ii)** $i \neq j \iff P_i \cap P_j = \emptyset$ is called The *algebra generated by* $\mathbb{P}$ is the smallest algebra containing $\mathbb{P}$. It can easily be shown that the algebra generated by $\mathbb{P}$ is the set of all unions of sets from $\mathbb{P}$, and that the algebra is finite.

When $R = f^{-1}(0)$, the function $f$ is referred to as *a characteristic function of* $R$, and is denoted $\chi_R$. Note that $\chi_R$ is not unique.

Let $\mathcal{F}$ be a set of functions. $\mathcal{F}_\star$ denotes the set of *relations of* $\mathcal{F}$, viz. those subsets $R \subseteq \mathbb{N}^k$ with $\chi_R \in \mathcal{F}$ for some $\chi_R$. Formally

$$\mathcal{F}_\star^k = \left\{ f^{-1}(0) \subseteq \mathbb{N}^k \,\middle|\, f \in \mathcal{F}^k \right\} \quad \text{and} \quad \mathcal{F}_\star = \bigcup_{k \in \mathbb{N}} \mathcal{F}_\star^k \, ,$$

where $\mathcal{F}^k$ denotes the $k$-ary functions of $\mathcal{F}$.

Whenever a symbol occurs under an arrow, e.g. $\vec{x}$, we usually do not point out the length of the list—by convention it shall be $k$ for variables and $\ell$ for functions unless otherwise specified.


## 2   Background and Motivation

The standard schema of $\ell$-*fold simultaneous primitive recursion*, denoted $\mathrm{PR}^\ell$ defines $\ell$ functions $\vec{f}$ from functions $\vec{h}$ and $\vec{g}$ when

$$f_i(\vec{x}, y) = \begin{cases} g_i(\vec{x}) & , \text{if } y = 0 \\ h_i(\vec{x}, y-1, \vec{f}(\vec{x}, y-1)) & , \text{if } y > 0 \end{cases} ,$$

for $1 \leq i \leq \ell$. A less studied schema is that of $\ell$-*fold simultaneous (pure) iteration*, denoted $\mathrm{IT}^\ell$, where functions $\vec{f}$ are defined from functions $\vec{h}$ and $\vec{g}$ by

$$f_i(\vec{x}, y) = \begin{cases} g_i(\vec{x}) & , \text{if } y = 0 \\ h_i(\vec{x}, \vec{f}(\vec{x}, y-1)) & , \text{if } y > 0 \end{cases} .$$

---

[1] This notation is adopted from Clote [2], where an idc. is called a *function algebra*.

for $1 \leq i \leq \ell$. We omit the superscript when the iteration or recursion is 1-fold. We refer to the $\ell$ in $\mathrm{IT}^\ell$ as the *iteration width*.

It is well-known that in most contexts $\mathrm{IT}^1$ is equivalent to $\mathrm{PR}^\ell$ for arbitrary $\ell \in \mathbb{N}$. More precisely $\left[\mathcal{I} \cup \{0, \mathsf{S}, \mathsf{P}\}; \mathrm{COMP}, \mathrm{IT}^1\right] = \left[\mathcal{I} \cup \mathcal{N} \cup \{\mathsf{S}\}; \mathrm{COMP}, \mathrm{PR}^\ell\right]$. Quite a lot of coding is required, and proofs are technical. For this, and more on these schemata consult e.g. Rose [13].

For the schemata of *bounded $\ell$-fold simultaneous primitive recursion* ($\mathrm{BPR}^\ell$) and *bounded simultaneous $\ell$-fold pure iteration* ($\mathrm{BIT}^\ell$) one requires an additional function $b(\vec{x}, y)$, such that $f_i(\vec{x}, y) \leq b(\vec{x}, y)$ for all $\vec{x}, y$ and $i$. To the authors best knowledge, $\mathrm{BPR}$ was introduced under the name of *limited* primitive recursion by Grzegorczyk in [4], where the schema was used to define his famous hierarchy $\mathcal{E} \stackrel{\text{def}}{=} \bigcup_n \mathcal{E}^n$. A modern rendering of the original definitions, would be e.g. $\mathcal{E}^n \stackrel{\text{def}}{=} [\mathcal{I} \cup \mathcal{N} \cup \{\mathsf{S}, A_n\}; \mathrm{COMP}, \mathrm{BPR}]$, where the function $A_n$ is the $n^{\text{th}}$ Ackermann branch. If we let $\mathcal{PR}$ denote the set of all primitive recursive functions, one of the main results of [4] is that $\mathcal{E} = \mathcal{PR}$. While it is known that the hierarchy of functions is strict at all levels, and that the hierarchy of relational classes $\mathcal{E}^n_\star$ is strict from $n = 2$ and upwards, the problem of whether any of the inclusions $\mathcal{E}^0_\star \subseteq \mathcal{E}^1_\star \subseteq \mathcal{E}^2_\star$ are proper or not remains open to date.

In 1987 Kutyłowski experimented with variations over bounded iteration classes, and published his results in *Small Grzegorczyk classes* [12]. More precisely, Kutyłowski defined classes $I^n \stackrel{\text{def}}{=} [\mathcal{I} \cup \mathcal{N} \cup \{\mathsf{P}, A_n, \mathsf{max}\}; \mathrm{COMP}, \mathrm{BIT}]$, viz. $I^n$ is essentially[2] $\mathcal{E}^n$ with $\mathrm{BIT}$ substituted for $\mathrm{BPR}$. That $I^n = \mathcal{E}^n$ for $n \geq 2$ is fairly straightforward. However, for $n = 0, 1$, a solution of $I^n \stackrel{?}{\subseteq} \mathcal{E}^n$ or $I^1_\star \stackrel{?}{\subseteq} I^2_\star$ are as hard as for the problem of $\mathcal{E}^0_\star \stackrel{?}{\subseteq} \mathcal{E}^1_\star \stackrel{?}{\subseteq} \mathcal{E}^2_\star$, and so the equivalence of $\mathrm{BIT}$ and $\mathrm{BPR}$ is an open problem in this context.

Some interdependencies are nicely summarised in [12]. Of particular interest to us, are Theorems **D** and **F** from [12], which establishes that $I^0_\star = I^1_\star$; and that $I^0_\star = \mathcal{E}^0_\star \Leftrightarrow \mathcal{E}^0_\star = \mathcal{E}^2_\star$ respectively.

Consider the classes $\mathcal{L}^n \stackrel{\text{def}}{=} \left[\mathcal{I} \cup \mathcal{N}; \mathrm{COMP}, \mathrm{PR}^{n+1}\right]$. In Kristiansen and Barra [8] the hierarchy $\mathcal{L} \stackrel{\text{def}}{=} \bigcup_n \mathcal{L}^n$ is defined. It follows from earlier results in Kristiansen [7] that $\mathcal{L}_\star = \mathcal{E}^2_\star$, and more recently Kristiansen and Voda [11] established $\mathcal{L}^0_\star = \mathcal{E}^0_\star$. Furthermore, the hierarchy $\mathcal{L}$ collapses iff $\mathcal{E}^0_\star = \mathcal{E}^2_\star$ (see [8]). Hence the following diagram: $\mathcal{L}^0_\star = \mathcal{E}^0_\star \subseteq \mathcal{E}^1_\star \subseteq \mathcal{E}^2_\star = \mathcal{L}_\star$. Note that $\mathcal{L}^n$ contains no increasing functions, and that the recursive schema is unbounded.

The paper at hand emerges from a desire to investigate the *intersection* of the two approaches: *both omitting the increasing functions, and replacing* $\mathrm{PR}^n$ *with* $\mathrm{IT}^n$.

## 3   The Hierarchy $\mathcal{IT}$

**Definition 1.** *For* $n \in \mathbb{N}$, $\mathcal{IT}^n \stackrel{\text{def}}{=} \left[\mathcal{I} \cup \mathcal{N}; \mathrm{COMP}, \mathrm{IT}^{n+1}\right]$; $\mathcal{IT} \stackrel{\text{def}}{=} \bigcup_{n \in \mathbb{N}} \mathcal{IT}^n$.

---

[2] The appearance of '$\mathsf{max}$' is inessential, and can be added also to $\mathcal{E}^n$ without altering the induced relational class.

The next lemma is rather obvious, and we skip the proof.

**Lemma 1.** *A function is* nonincreasing *if, for some $c \in \mathbb{N}$ we have $f(\vec{x}) \leq \max(\vec{x}, c)$ for all $\vec{x} \in \mathbb{N}^k$. We have that all $f \in \mathcal{IT}^n$ are nonincreasing.* $\square$

We next define a family of very simple and regular partitions, which will be useful for describing the relational classes induced by the $\mathcal{IT}^n$.

Henceforth, let $0 < \mu \in \mathbb{N}$. For $0 \leq m < \mu$ let $m + \mu\mathbb{N}$ have the usual meaning, i.e. '*the equivalence class of $m$ under congruence modulo $\mu$*'. The standard notation $x \equiv_\mu y \overset{\text{def}}{\Leftrightarrow} |x - y| \in \mu\mathbb{N}$ will also be used. It is well-known that viewed as a relation on $\mathbb{N}$, '$\equiv_\mu$' defines an *equivalence relation*, and that equivalences induce partitions. Hence the following definition.

**Definition 2.** *Let $\mathbb{M}(\mu, 0) \overset{\text{def}}{=} \{m + \mu\mathbb{N} \mid 0 \leq m < \mu\}$ be the partition of $\mathbb{N}$ induced by '$\equiv_\mu$'. For $a \in \mathbb{N}$, let $\mathbb{M}(\mu, a)$ denote the partition obtained from $\mathbb{M}(\mu, 0)$ by placing each of the number $0$ through $a - 1$ in their own equivalence class as singletons[3], and let $x \equiv_{\mu,a} x'$ have the obvious meaning.*

*Let '$\equiv_{\mu,a}^k$' be the coordinate-wise extension of '$\equiv_{\mu,a}$' to an equivalence on $\mathbb{N}^k$, i.e. $\vec{x} \equiv_{\mu,a}^k \vec{x}' \overset{\text{def}}{\Leftrightarrow} \forall i\, (x_i \equiv_{\mu,a} x_i')$, and let $\mathbb{M}^k(\mu, a)$ denote the corresponding partition[4].*

*Set $\mathfrak{M}^k \overset{\text{def}}{=} \{\mathbb{M}^k(\mu, a) \mid \mu, a \in \mathbb{N}\}$ and finally*

$$\mathfrak{M}_\star^k = \bigcup_{\mu, a \in \mathbb{N}} \mathbb{A}^k(\mu, a) \quad \text{and} \quad \mathfrak{M}_\star = \bigcup_{k \in \mathbb{N}} \mathfrak{M}_\star^k \,,$$

*where $\mathbb{A}^k(\mu, a)$ is the smallest algebra containing $\mathbb{M}^k(\mu, a)$.*

So a predicate $P \in \mathfrak{M}_\star^k$ is a union of $\mathbb{M}^k(\mu, a)$-classes for some $\mu, a$. We suppress $\mu$ and $a$ when convenient.

If we set[5] $\mathbb{M}(\mu, a) \preceq \mathbb{M}(\mu', a') \overset{\text{def}}{\Leftrightarrow} \mu | \mu' \wedge a \leq a'$, it is easy to see that '$\preceq$' defines a *partial order* on $\mathfrak{M}$. When $\mathbb{M} \preceq \mathbb{M}'$, we say that $\mathbb{M}'$ is *a refinement of* $\mathbb{M}$. Hence, for any pair of partitions there is a *least common refinement*, given by $\mathbb{M}(\mu_1, a_1), \mathbb{M}(\mu_2, a_2) \preceq \mathbb{M}(\mathsf{lcm}(\mu_1, \mu_2), \max(a_1, a_2))$, and which itself is an element of $\mathfrak{M}$.

**Definition 3.** *Let $\mathbb{M}^k = \{M_1, \ldots, M_\ell\}$. A function $f$ is an $\mathbb{M}^k$-case, if there exist $f_1, \ldots, f_\ell \in \mathcal{I} \cup \mathcal{N}$ such that*

$$f(\vec{x}) = \begin{cases} f_1(\vec{x}) \,, \vec{x} \in \mathbb{M}_1 \\ \quad \vdots \qquad \vdots \\ f_\ell(\vec{x}) \,, \vec{x} \in \mathbb{M}_\ell \end{cases}$$

---

[3] Thus e.g. $\mathbb{M}(3, 2) \overset{\text{def}}{=} \{\{0\}, \{1\}, 3\mathbb{N} \setminus \{0\}, (1 + 3\mathbb{N}) \setminus \{1\}, (2 + 3\mathbb{N})\}$. Note that $\mathbb{M}(1, a)$ is the partition consisting of the singletons $\{n\}$, for $n < a$, and $\{a, a + 1, a + 2, \ldots\}$, and $\mathbb{M}(1, 0) = \{\mathbb{N}\}$.

[4] Hence $\mathbb{M}^k(\mu, a)$ consists of all 'boxes' $M_1 \times \cdots \times M_k$ where each $M_i \in \mathbb{M}(\mu, a)$.

[5] '$n | m$' means '$n$ *divides* $m$', and '$\mathsf{lcm}$' denotes '*the least common multiple*'.

*An $\mathfrak{M}$-case, is an $M^k$-case for some $M^k \in \mathfrak{M}^k$. Let $\mathcal{M}$ denote the set of all $\mathfrak{M}$-cases, and let $\mathcal{M}^k$ denote the k-ary functions in $\mathcal{M}$.*

Informally, an $M^k(\mu, a)$-case is arbitrary on $\{0, \ldots, a-1\}^k$, and then 'class-wise', or 'case-wise', a projection or a constant on each of the finitely many equivalence classes. We write $[\vec{x}]_{\mu,a}$, for the equivalence class of $\vec{x}$ under $\equiv^k_{\mu,a}$ (omitting the subscript when convenient). Henceforth, when $f$ is an $\mathfrak{M}$-case, a subscripted $f$ is tacitly assumed to be an element of $\mathcal{I} \cup \mathcal{N}$, and we also write $f(\vec{x}) = f_{[\vec{x}]}(\vec{x}) = f_M(\vec{x})$ when $[\vec{x}] = M \in M^k$. Yet another formulation is that[6] $f$ is an $M$-case iff $f \restriction_M \in \mathcal{I} \cup \mathcal{N}$ for each $M \in M$.

The proposition below is easy to prove.

**Proposition 1. (i)** *If $f$ is an M-case, and $M \preceq M'$, then $f$ is an $M'$-case;* **(ii)** *If $\{M_i\}_{i \leq \ell} \in \mathfrak{M}^k$, and $f(\vec{x}, \vec{y}) = f_i(\vec{x}, \vec{y}) \Leftrightarrow \vec{x} \in M_i$, then $f$ is an $\mathfrak{M}$-case;* **(iii)** *For all predicates $P \in \mathfrak{M}_\star$, there is some $\mathfrak{M}$-case $f$ such that $P = f^{-1}(0)$. Conversely, if $f$ is an $\mathfrak{M}$-case, then $f^{-1}(0) \in \mathfrak{M}_\star$; Hence* **(iv)** *$\mathcal{M}_\star = \mathfrak{M}_\star$.*      $\square$

Our main results are all derived from the theorem below.

**Theorem 1.** $\mathcal{IT}^0 = \mathcal{IT} = \mathcal{M}$ .

That is, the hierarchy $\mathcal{IT}^n$ collapses, and its functions are exactly the $\mathfrak{M}$-cases.

We begin by establishing some basic facts about $\mathcal{IT}^0$. It is straightforward to show that characteristic functions for the intersection $P \cap P'$ and the complement of $P$ are easily defined by composition from $\mathsf{C}$, $\mathcal{I} \cup \mathcal{N}$, $\chi_P$ and $\chi_{P'}$. Observing that $\mathsf{C} \in \mathcal{IT}^0$—since it is definable by iterating $h = \mathsf{I}^3_2$ on $g = \mathsf{I}^2_1$—we conclude that $\mathcal{IT}^0_\star$ is an algebra. Furthermore, given any finite partition $\mathbb{P} = \{P_i\}_{i \leq n}$ of $\mathbb{N}^k$, once we have access to $\mathsf{C}$ and $\{\chi_{P_i}\}_{i \leq n}$, we can easily construct a $\mathbb{P}$-case, by nested compositions. That is, when $\{\chi_{P_i}\}_{i \leq n} \subseteq \mathcal{IT}^0_\star$, then any $\mathbb{P}$-case belongs to $\mathcal{IT}^0$.

Thus, a proof the inclusion $\mathcal{M} \subseteq \mathcal{IT}^0$ is reduced to proving that all $M(\mu, a)$ belong to $\mathcal{IT}^0_\star$.

**Lemma 2.** $2\mathbb{N}$, $1 + 2\mathbb{N} \in \mathcal{IT}^0_\star$ .

*Proof.* Since $\mathcal{IT}^0_\star$ is an algebra, it is sufficient to show that $2\mathbb{N} \in \mathcal{IT}^0_\star$. Consider the function $f(x) = \begin{cases} 0 & , x = 0 \\ \mathsf{C}(1, 0, f(x-1)) & , x > 0 \end{cases}$ .

By induction on $n$, we show that $f(2n) = 0$ and $f(2n+1) = 1$. For the induction start, by definition $f(0) = 0$, and thus $f(1) = \mathsf{C}(1, 0, f(0)) = 1$.

**Case (n+1):** Since $2(n+1) - 1 = 2n + 1$, we obtain

$$f(2(n+1)) \stackrel{\text{def}}{=} \mathsf{C}(1, 0, f(2n+1)) \stackrel{\text{IH}}{=} \mathsf{C}(1, 0, 1) \stackrel{\text{def}}{=} 0 \ , \tag{$\dagger$}$$

Clearly $f(2(n+1)+1) = \mathsf{C}(1, 0, f(2(n+1)) \stackrel{\dagger}{=} \mathsf{C}(1, 0, 0) = 1$, so $f = \chi_{2\mathbb{N}}$.      $\square$

---

[6] $f \restriction_M$ denotes *the restriction of $f$ to $M$.*

Viewed as a binary predicate, '$\equiv_\mu$' is the set $\{(x,y)\,|\,|x-y| \in \mu\mathbb{N}\}$. (So '$\equiv_1$' is all of $\mathbb{N}^2$, and '$\equiv_0$' is 'equality'. Recall the restriction $\mu > 0$.) A reformulation of LEMMA 2 is thus that the unary predicates '$x \equiv_2 0$' and '$x \equiv_2 1$' both belong to $\mathcal{IT}_\star^0$. Below the idea of the last proof is iterated for the general result.

**Lemma 3.** *For all $\mu$, the binary predicate '$x \equiv_\mu y$' is in $\mathcal{IT}_\star^0$.*

*Proof.* The proof is by induction on $\mu \geq 2$; the case $\mu = 1$ is trivial. LEMMA 2 effects induction start, since $x \equiv_2 y \Leftrightarrow \mathsf{C}(\chi_{2\mathbb{N}}(y), \chi_{1+2\mathbb{N}}(y), \chi_{2\mathbb{N}}(x)) = 0$.

**Case** $\mu + 1$: By the i.h. the partition $\mathbb{M}(\mu, 0)$ is in $\mathcal{IT}_\star^0$: for $m < \mu$ we have $\chi_{(m+\mu\mathbb{N})}(x) = \chi_{\equiv_\mu}(x, m)$. Thus $\mathbb{M}(\mu, 1)$ also belongs to $\mathcal{IT}_\star^0$, since we have $x \in \mu\mathbb{N} \setminus \{0\} \Leftrightarrow \mathsf{C}(1, \chi_{\mu\mathbb{N}}(x), x) = 0$ and $\chi_{\{0\}}(x) = \mathsf{C}(0, 1, x)$. Next, consider the $\mathbb{M}(\mu, 1)$-case

$$f(x) = \begin{cases} \mu & , x \in \{0\} \\ m - 1 & , x \in m + \mu\mathbb{N} \setminus \{0\} \text{ for } 1 \leq m \leq \mu \end{cases} .$$

Note that **(i)** $f \upharpoonright_{\{0,\dots,\mu\}}$ is the permutation $(\mu\ 0\ 1 \cdots (\mu - 1))$. Moreover, for all $n \in \mathbb{N}$ and $m \in \{0, \dots, \mu\}$, we have that **(ii)** $f^{(n(\mu+1))}(m) = m$; and **(iii)** $f^{(m)}(m) = 0$. Hence, if $y \equiv_{\mu+1} m'$ and $0 \leq m, m' < \mu + 1$, then

$$f^{(n(\mu+1)+m')}(m) = f^{(n(\mu+1))}\big(f^{(m')}(m)\big) \overset{\text{(ii)}}{=} f^{(m')}(m) = 0 \overset{\text{(i)\&(iii)}}{\Leftrightarrow} m = m' .$$

Thus, the function $f^{(y)}(m)$ is a characteristic function for $m + (\mu + 1)\mathbb{N}$. This function is clearly definable in $\mathcal{IT}^0$ from $f$, constant functions and $\text{IT}^1$. Since $\mathcal{IT}_\star^0$ is boolean, and since $x \equiv_{\mu+1} y \Leftrightarrow \bigvee_{m=0}^{\mu} (x \equiv_{\mu+1} m \wedge y \equiv_{\mu+1} m)$, we are done. $\qquad\square$

**Corollary 1.** *For all $n \in \mathbb{N}$, the predicate '$x = n$' is in $\in \mathcal{IT}_\star^0$.*

*Proof.* For $n = 0$ the proof is contained in the proof of LEMMA 3. Secondly, for $n > 0$, we have that the $\mathbb{M}(n + 1, 0)$-case $h(x) = \begin{cases} n + 1 & , \text{if } x \equiv_{n+1} 0 \\ m + 1 & , \text{if } x \equiv_{n+1} m \end{cases}$ is in $\mathcal{IT}^0$. Consider the function $f(x) \overset{\text{def}}{=} h^{(x)}(1)$. Since $0 < n$ implies $1 \not\equiv_{n+1} 0$, we obtain $0 \leq x < n \Rightarrow h^{(x)}(1) = x + 1 > 0$ and $h^{(n)}(1) = 0$. By definition $h(0) = h(n + 1) = n + 1 > 0$. But then $f(x) = 0$ iff $x = n$, viz. $f = \chi_{\{n\}}$. $\qquad\square$

**Proposition 2.** $\mathcal{M} \subseteq \mathcal{IT}^0$ .

*Proof.* Combining LEMMA 3, COROLLARY 1, and an appeal to the boolean structure of $\mathcal{IT}_\star^0$ yields $\forall \mu, a \in \mathbb{N}\,\big(\mathbb{M}(\mu, a) \subseteq \mathcal{IT}_\star^0\big)$ . Hence, any $\mathfrak{M}$-case is definable in $\mathcal{IT}^0$. The proposition now follows from PROPOSITION 1 **(iii)**. $\qquad\square$

It is trivial that $\mathcal{IT}^0 \subseteq \mathcal{IT}$; thus the inclusion $\mathcal{IT} \subseteq \mathcal{M}$ would complete a full proof of THEOREM 1. Since the basic functions $\mathcal{I} \cup \mathcal{N}$ of $\mathcal{IT}^n$ obviously belong to $\mathcal{M}$, we have reduced our task to proving closure of $\mathcal{M}$ under COMP and $\text{IT}^n$.

**Lemma 4.** $\mathcal{M}$ *is closed under composition.*

*Proof.* Consider $h \in \mathcal{M}^\ell$, and $g^1, \ldots, g^\ell \in \mathcal{M}^k$. Because $f \in \mathcal{M}^k$ means that for some $\mathbb{M}$, $f$ is a case on the *coordinate-wise extension* of $\mathbb{M}$ to $\mathbb{M}^k$, we can find $\mathbb{M}(\mu_h, a_h)$ and $\mathbb{M}(\mu_j, a_j)$ for $1 \le j \le \ell$ corresponding to $h$ and the $g^j$'s respectively, and for which we may find a common refinement $\mathbb{M}$. Hence we may consider the $g^j$'s $\mathbb{M}^k$-cases, and $h$ an $\mathbb{M}^\ell$-case. It is easy to show that $\vec{x} \equiv^k \vec{x}' \Rightarrow \vec{g}(\vec{x}) \equiv^\ell \vec{g}(\vec{x}')$, when $\vec{g}$ are $\mathbb{M}^k$-cases. Formally $\forall M \in \mathbb{M}^k \exists M' \in \mathbb{M}^\ell \, (g(M) \subseteq M')$. But then, where $h \restriction_{M'} = h_{M'} \in \mathcal{I} \cup \mathcal{N}$, and $g^j \restriction_M = g_M^j \in \mathcal{I} \cup \mathcal{N}$, we have $f \restriction_M = h_{M'} \circ \vec{g_M} \in \mathcal{I} \cup \mathcal{N}$ since $\mathcal{I} \cup \mathcal{N}$ is closed under composition. Since $f$ is an $\mathbb{M}$-case, precisely when its restriction to each $\mathbb{M}$-class belongs to $\mathcal{I} \cup \mathcal{N}$, the conclusion of the lemma now follows. $\qquad\square$

*Remark 1.* Note the property of $\mathcal{M}$-functions which enters the proof: when $\vec{g}$ are $\mathbb{M}^k$-cases, they *preserve equivalence*: $\vec{x} \equiv^k \vec{x}' \Rightarrow \vec{g}(\vec{x}) \equiv^\ell \vec{g}(\vec{x}')$.

**Lemma 5.** $\mathcal{M}$ *is closed under simultaneous pure iteration.*

*Proof.* Let $\ell$ be arbitrary, and let $h^1, \ldots, h^\ell \in \mathcal{M}^{k+\ell}$, $g^1, \ldots, g^\ell \in \mathcal{M}^k$. As before—considering a least common refinement if necessary—assume w.l.o.g. that the $h^j$'s are $\mathbb{M}^{k+\ell}(\mu, a)$-cases, and the $g^j$'s are $\mathbb{M}^k(\mu, a)$-cases. Moreover, let $C$ equal the value of the largest constant function corresponding to some $g_i^j$ or $h_i^j$, and note that because separating off more singletons does not necessitate the introduction of any new constant functions, there is no loss of generality in assuming $a > C$.

It is sufficient to show that for arbitrary $M \in \mathbb{M}^k(\mu, a)$, we have that each $f^j \restriction_{M \times \mathbb{N}}$ is a case on the $(k+1)^{\text{st}}$ coordinate, viz. that there is $\mathbb{M}(\mu_M, a_M) \in \mathfrak{M}$ such that $f \restriction_{M \times \mathbb{N}} (\vec{x}, y) = f_{[y]_{\mu_M, a_M}}(\vec{x}, y)$.

If this can be shown, each $f^j$ will be an $\mathbb{P}^{k+1}$-case for any common refinement $\mathbb{P} \in \mathfrak{M}$ of $\mathbb{M}(\mu, a)$ and the finitely many $\mathbb{M}(\mu_M, a_M)$'s.

Thus let $M \in \mathbb{M}(\mu, a)$ be arbitrary. Write $f_y^j(\vec{x})$ for $f^j(\vec{x}, y)$, and $\vec{f_y}$ for $f_y^1, \ldots f_y^\ell$. It is easy to show that $f^j(\vec{x}, y) \le \max(\vec{x}, C)$. Hence, for fixed $\vec{z} \in M$, the sequence of $\ell$-tuples $\left\{\vec{f_y}\right\}_{y \in \mathbb{N}}$ is contained in $\{0, \ldots, \max(\vec{z}, C)\}^\ell$. By the *pigeon hole principle* there are indices $a_M$ and $b_M$ such that $C < a_M < b_M$ and such that $\vec{f_{a_M}}(\vec{z}) = \vec{f_{b_M}}(\vec{z})$. If we let these indices be minimal and set $\mu_M = b_M - a_M$, this observation translates to:

$$\vec{f_y}(\vec{z}) = \begin{cases} \vec{f_y}(\vec{z}) & \text{, if } y < a_M \\ \vec{f_{a_M + m}}(\vec{z}) & \text{, if } a_M \le y \in m + \mu_M \mathbb{N} \text{ and } 0 \le m < \mu_M \end{cases} .$$

So far we have that $f$ restricted to $\{\vec{z}\} \times \mathbb{N}$ is an $\mathbb{M}(\mu_M, a_M)$-case on the $(k+1)^{\text{st}}$ variable.

Secondly, for arbitrary $\vec{x} \in \mathbb{N}^k$, and by exploiting the fact that $\mathbb{M}$-cases preserve $\mathbb{M}$ in the sense of REMARK 1, it is straightforward to show by induction on $y$ that the sequence of $\mathbb{M}^\ell(\mu, a)$-classes $[\vec{f_0}(\vec{x})]_{\mathbb{M}^\ell}, [\vec{f_1}(\vec{x})]_{\mathbb{M}^\ell}, [\vec{f_2}(\vec{x})]_{\mathbb{M}^\ell}, \ldots$ is uniquely determined by the $\mathbb{M}^k(\mu, a)$-class of $\vec{x}$.

Let $\vec{z}, \vec{z}' \in M$. Define a map $\phi_{\vec{z}'}^{\vec{z}} : \mathbb{N} \to \mathbb{N}$, by $\phi(x) = \begin{cases} x & , x \notin \{\vec{z}\} \\ z_i' & , x = z_i \end{cases}$ . Below, we simply write $\phi$ for $\phi_{\vec{z}'}^{\vec{z}}$, and $\phi(\vec{x})$ for $\phi(x_1), \ldots, \phi(x_\ell)$. Obviously $\phi(\vec{z}) = \vec{z}'$. We next show by induction on $y$ that $\left\{ \phi\left(\vec{f}_y(\vec{z})\right) \right\}_{y \in \mathbb{N}} = \left\{ \vec{f}_y(\vec{z}') \right\}_{y \in \mathbb{N}}$ .

**Induction start:** We have

$$f_0^j(\vec{z}) = g_M^j(\vec{z}) = \begin{cases} z_i & , \text{if } g_M^j = \mathsf{I}_i^k \\ c & , \text{if } g_M^j = \mathsf{c} \end{cases} \Rightarrow \phi(f_0^j(\vec{z})) = \begin{cases} \phi(z_i) = z_i' & , \text{if } g_M^j = \mathsf{I}_i^k \\ \phi(c) \overset{\dagger}{=} c & , \text{if } g_M^j = \mathsf{c} \end{cases} .$$

The equality marked (†) is justified thus: Since $a > C \geq c$, we infer that the $\mathbb{M}(\mu, a)$-class of such $c$ is in fact $\{c\}$. Hence, if $c \in \{\vec{z}\}$, say $c = z_i$, then $z_i' = z_i$, and so $\phi(c) = c$. If $c \notin \{\vec{z}\}$, then $\phi(c) = c$ by definition.

**Induction step:** Recall that the $\mathbb{M}^\ell(\mu, a)$-class of $\vec{f}_y(\vec{z})$ and $\vec{f}_y(\vec{z}')$ coincide, and set $M_y = M \times [\vec{f}_y(\vec{z})]_{\mathbb{M}^\ell(\mu,a)}$. Then

$$f_{y+1}^j(\vec{z}) = h^j(\vec{z}, \vec{f}_y(\vec{z})) = h_{M_y}^j(\vec{z}, \vec{f}_y(\vec{z})) = \begin{cases} z_i & , \text{if } h_{M_y}^j = \mathsf{I}_i^{k+\ell} \text{ and } 1 \leq i \leq k \\ f_y^i(\vec{z}) & , \text{if } h_{M_y}^j = \mathsf{I}_{k+i}^{k+\ell} \text{ and } 1 \leq i \leq \ell \\ c & , \text{if } h_{M_y}^j = \mathsf{c} \end{cases}$$

For $f_{y+1}^j(\vec{z}')$, simply add primes to the $z$'s above. By invoking the i.h. for the case of $h_{M_y}^j = \mathsf{I}_{k+i}^{k+\ell}$, the conclusion follows by the same argument employed in the induction start.

Since $f(\vec{z}, y)$ is an $\mathbb{M}(\mu_M, a_M)$-case restricted to $\{\vec{z}\} \times \mathbb{N}$, and since we have indeed shown that for $\vec{z}' \in \mathbb{M}$, if $f(\vec{z}, y) = z_i$, then $f(\vec{z}', y) = z_i'$, and similarly, if $f(\vec{z}, y) = c$ then $f(\vec{z}', y) = c$, we are done.  $\square$

**Proposition 3.** $\mathcal{IT} \subseteq \mathcal{M}$ .  $\square$

**Theorem 2.** $\mathcal{IT}_\star^0 = \mathcal{IT}_\star = \mathcal{M}_\star = \mathfrak{M}_\star$ .  $\square$

THEOREM 2 is a direct corollary to THEOREM 1, the proof of which is completed by PROPOSITION 3.

We now have a complete characterisation of the induced relational class in terms of very simple and regular finite partitions of $\mathbb{N}$. Before the concluding discussion, we include a result which relates $\mathcal{IT}_\star$ to Presburger Arithmetic.

### 3.1   A Note on Presburger Arithmetic and $\mathcal{IT}_\star$

Let $\mathfrak{Pr}\mathfrak{A}$ be the $1^{\text{st}}$-order language $\{0, \mathsf{S}, +, <, =\}$ with the intended structure $\mathfrak{N}_A \overset{\text{def}}{=} (\mathbb{N}, 0, \mathsf{S}, +, <)$—the natural numbers with the usual order, successor and addition. Many readers will recognize $\mathfrak{Pr}\mathfrak{A}$ as the language of *Presburger Arithmetic*, see e.g. Enderton [3, pp. 188–193]. Let $\mathfrak{Pr}\mathfrak{A}_\star$ denote the predicates definable by a $\mathfrak{Pr}\mathfrak{A}$-formula.

Consider the following [3, p. 192, Theorem 32F]: *A set of natural numbers D belongs to $\mathfrak{Pr}\mathfrak{A}_\star$ iff it is eventually periodic, where eventually periodic means* that for some $\mu, a \in \mathbb{N}$ we have $n > a \;\Rightarrow\; (n \in D \Leftrightarrow n + \mu \in D)$. Since this is exactly what it means for $D$ to be in $\mathfrak{M}$, if we let $\mathfrak{Pr2l}^{\mathrm{u}}_{\star}$ and $\mathcal{IT}^{\mathrm{u}}_{\star}$ denote the *unary* predicates of the respective classes, we immediately see that $\mathfrak{Pr2l}^{\mathrm{u}}_{\star}$ and $\mathcal{IT}^{\mathrm{u}}_{\star}$ coincide.

However, this result does not hold for higher arities, since we have the following corollary to THEOREM 1:

**Corollary 2.** $\chi_<, \chi_= \notin \mathcal{IT}$

*Proof.* If $\chi_< \in \mathcal{IT} = \mathcal{M}$, then it is an $\mathbb{M}(\mu, a)$-case for some $\mu, a \in \mathbb{N}$. Clearly $a \equiv_\mu a + \mu$. Hence, we have $1 = \chi_<(a, a) = \chi_<(a, a + \mu) = 0$; a contradiction. Similarly, $0 = \chi_=(a, a) = \chi_=(a, a + \mu) = 1$. $\qquad\square$

As both '*equals*' and '*less than*' are primitive to $\mathfrak{Pr2l}_\star$, we obtain:

**Theorem 3.** (i) $\mathcal{IT}^{\mathrm{u}}_\star = \mathfrak{Pr2l}^{\mathrm{u}}_\star$; (ii) $\mathcal{IT}_\star \subsetneq \mathfrak{Pr2l}_\star$. $\qquad\square$

## 4  Discussion and Directions for Further Research

Consider the assertion: 'iteration is inherently *weak* and *periodic*'. What we have shown beyond doubt is that iteration is weak and periodic when working alone. Secondly we have shown with THEOREM 1 that iteration width, or simultaneity—in the very weak context of this paper—does not add any computational strength. Contrasted to the recursion-based hierarchy $\mathcal{L}$, which does not collapse unless $\mathcal{E}^0_\star = \mathcal{E}^2_\star$, we see that in other weak contexts simultaneity may in fact be stronger.

Recall that $I^0$ is essentially $\mathcal{IT}^0$ with predecessor and successor. In what follows we omit the explicit mention of $\mathcal{I} \cup \mathcal{N}$ and COMP. Let $\mathcal{PIT} \stackrel{\mathrm{def}}{=} [\{\mathsf{P}\}\,;\mathrm{IT}]$, and so we have

$$\mathcal{IT}_\star \subsetneq \mathcal{PIT}_\star \stackrel{?}{\subseteq} I^0_\star \stackrel{\mathrm{Kut}}{=} I^1_\star \stackrel{?}{\subseteq} \mathcal{E}^0_\star\ .$$

Note that iteration width is restricted to 1, and that if two-fold iteration over $\mathsf{P}$ is allowed, one obtains at least one-fold primitive recursion. The first, proper containment follows from e.g. $\chi_= \in \mathcal{PIT}$.

The schema of *bounded minimalisation*, denoted BMIN, defines a function $f$ from functions $g_1, g_2$ by $f(\vec{x}, y) = \mu_{z \le y}\,[g_1(\vec{x}, z) = g_2(\vec{x}, z)]$. Clearly, if an idc. $\mathcal{F}$ includes BMIN, the resulting $\mathcal{F}_\star$ will be closed under bounded quantification. Even though BMIN and BIT are hard to compare directly, the following fact is quite enticing. It is shown in Barra [1] that $[\{\mathsf{P}\}\,;\mathrm{BMIN}]_\star = [\{\mathsf{P},\mathsf{S}\}\,;\mathrm{BMIN}]_\star$, and that $[\{\dot-\}\,;\mathrm{BMIN}]_\star = [\{+\}\,;\mathrm{BMIN}]_\star = \mathfrak{Pr2l}_\star$. That is, there is no loss of predicates by simply removing $\mathsf{S}$ in the above context. A natural open question is thus how close $\mathcal{PIT}$ and $I^0$ are?

We feel that some evidence has been mounted in support of the opening statement of this section. In a context such as Kutyłowski's $I^2$, bounded iteration *is* equivalent to that of bounded primitive recursion. But what about $I^0$? By Kutyłowski [12], we also know that equality between $\mathcal{E}^0_\star$ and $\mathcal{E}^2_\star$ relies on $I^0_\star$, and hence $I^1_\star$, being equal to $\mathcal{E}^0_\star$.

In light of the paper at hand, is there a chance that the equality $I_\star^0 = I_\star^1$ is due—not to the ability of iteration to 'raise $I_\star^0$ up to the level of $I_\star^1$'—but rather a case of iteration being so weak as to 'lower $I_\star^1$ to the level of $I_\star^0$'?

Since the characteristic function of e.g. the primes is obviously non-periodic, and belongs to $\mathcal{E}^0$, at some stage between $\mathcal{IT}^0$ and $I^0$, one must be able to escape the periodicity inherent in iteration. Can one exploit the periodic behavior of functions defined by pure iteration in order to prove $I_\star^0 \neq \mathcal{E}_\star^0$? Or—perhaps via a successful attempt at escaping periodicity—can one obtain the strength of recursion? whence $I_\star^0 = \mathcal{E}_\star^0$ would follow. Since both possibilities would provide a solution to many long standing conundrums of sub-recursion theory, further investigations should be well worth the effort.

# References


1. Barra, M.: A characterisation of the relations definable in Presburger Arithmetic. In: Proceedings of TAMC 2008. LNCS, vol. 4978, pp. 258–269. Springer, Heidelberg (2008)
2. Clote, P.: Computation Models and Function Algebra. In: Handbook of Computability Theory. Elsevier, Amsterdam (1996)
3. Enderton, H.B.: A mathematical introduction to logic. Academic Press, Inc., San Diego (1972)
4. Grzegorczyk, A.: Some classes of recursive functions, in Rozprawy Matematyczne, No. IV, Warszawa (1953)
5. Jones, N.D.: LOGSPACE and PTIME characterized by programming languages. In: Theoretical Computer Science, vol. 228, pp. 151–174 (1999)
6. Jones, N.D.: The expressive power of higher-order types or, life without CONS. J. Functional Programming 11, 55–94 (2001)
7. Kristiansen, L.: Neat function algebraic characterizations of LOGSPACE and LINSPACE. Computational Complexity 14(1), 72–88 (2005)
8. Kristiansen, L., Barra, M.: The small Grzegorczyk classes and the typed $\lambda$-calculus. In: Cooper, S.B., Löwe, B., Torenvliet, L. (eds.) CiE 2005. LNCS, vol. 3526, pp. 252–262. Springer, Heidelberg (2005)
9. Kristiansen, L., Voda, P.J.: The surprising power of restricted programs and Gödel's functionals. In: Baaz, M., Makowsky, J.A. (eds.) CSL 2003. LNCS, vol. 2803, pp. 345–358. Springer, Heidelberg (2003)
10. Kristiansen, L., Voda, P.J.: Complexity classes and fragments of C. Information Processing Letters 88, 213–218 (2003)
11. Kristiansen, L., Voda, P.J.: The Structure of Detour Degrees. In: Proceedings of TAMC 2008. LNCS, vol. 4978, pp. 148–159. Springer, Heidelberg (2008)
12. Kutyłowski, M.: Small Grzegorczyk classes. J. London Math. Soc (2) 36, 193–210 (1987)
13. Rose, H.E.: Subrecursion. Functions and hierarchies. Clarendon Press, Oxford (1984)

# Programming Experimental Procedures for Newtonian Kinematic Machines

E.J. Beggs and J.V. Tucker

School of Physical Sciences,
Swansea University,
Singleton Park,
Swansea, SA2 8PP,
United Kingdom

**Abstract.** By *experimental computation* we mean the idea of computing a function by experimenting with some physical equipment. To analyse the functions computable by experiment, we are developing a methodology that chooses a precise specification of a physical theory $T$ and derives precise descriptions of the procedures and equipment the theory allows. As a case study, we choose a fragment $T$ of Newtonian kinematics and describe a language $EP(T)$, and some of its extensions, for expressing *experimental procedures* allowed by $T$. The languages for experimental procedures are similar to imperative programming languages that express algorithmic procedures. We show that $EP(T)$ can define all functions on the rational numbers that are definable by algorithms.

## 1 Introduction

Imagine using a physical system to perform calculations. To calculate a partial function $f : X \rightarrow Y$ we assemble some equipment and follow an experimental procedure in which

(i) input data $x \in X$ are used to determine initial conditions of the physical system;

(ii) the system operates for a finite or infinite time; and

(iii) the system's behaviour is observed, measurements are taken, and output data $y \in Y$ are obtained at certain times.

We say that the partial function $y = f(x)$ is calculated by *experimental computation*. We can expect to compute functions on continuous data, such as the set $\mathbb{R}$ of real numbers, as well as functions on discrete data, such as on the set $\mathbb{N}$ of natural numbers. The *practice* of experimental computation is general and old.

To design and understand experimental computation we need physical theories to specify equipment and model its behaviour. To perform a calculation, we choose a physical theory $T$, which contains laws about physical properties of physical objects, and rules and operations for configuring and manipulating them. We assemble a piece of equipment and formulate a procedure using the

A. Beckmann, C. Dimitracopoulos, and B. Löwe (Eds.): CiE 2008, LNCS 5028, pp. 52–66, 2008.

rules and basic operations of $T$. In general, the basic actions allowed by $T$ can be scheduled in infinitely many ways to form procedures to operate any equipment governed by $T$. Thus, the physical theory $T$ can be used to *specify* an infinite set of equipment and an infinite set of experimental procedures.

*What can be computed by experimental procedures applied to given equipment?*

An experimental procedure is similar to an algorithmic procedure, especially those inspired by idealised machines. Algorithmic procedures are analysed and used by coding them as programs, using different types of programming languages.

In this paper, we introduce and explore the question

*How can we use a programming language to formalise experimental procedures?*

We fix a small fragment $T$ of Newtonian Kinematics and partially formalise a programming language $EP(T)$ to specify the class of all experimental procedures based on $T$. The language $EP(T)$ is based upon simple kinematic principles but, as we will show, can express significant and intriguing experimental procedures. The purpose of the language $EP(T)$ is to formalise the "programming" of equipment satisfying the specification $T$. In particular, the functions computed by a given piece of equipment can be studied using all possible experimental procedures in $EP(T)$. In describing the language $EP(T)$ we will, of necessity, focus on syntax and take liberties with semantics. The idea of deriving a programming language for a physical theory is general. The languages are imperative in style.

We will suggest ways of extending $EP(T)$ with new constructs, defined both physically and mathematically. Finally, we consider the data types used for measurement and we show that

**Theorem 1.** *A function computable by an algorithm on the rational numbers $\mathbb{Q}$ is computable by an experimental procedure in $EP(T)$ for the free motion of particles in 1-dimension, using integer measurements only.*

That computers can be implemented using Newtonian kinematics has been observed, e.g., by using a billiard ball model by Fredkin and Toffoli [10].

The structure of the paper is this. In Section 2 we introduce the language $EP(T)$, describing its syntax and semantics and giving examples of its procedures. In Section 3 we look at some extensions of language. In Section 4 we show how **while** programs on the data type of rational numbers can be translated into procedures of $EP(T)$. Finally, in Section 5 we discuss some new directions.

This paper belongs to the series Beggs and Tucker [2], [3], [4], [5], which contain a survey of problems, explanations of experimental computation and the methodology, and many related results. Technically, only elementary Newtonian mechanics is needed to follow our arguments here. However, some knowlededge of the following is necessary: classical computability theory, especially its conceptual foundations (e.g., Odifreddi [12], Griffor [9], Stoltenberg-Hansen and Tucker [14], [15]); its extension to continuous data (e.g., Pour-El and Richards [13], Stoltenberg-Hansen and Tucker [16], Tucker and Zucker [18], [19], Weihrauch [20]), and the different approaches to understanding the physical basis of computation.

We wish to thank José Félix Costa, Peter Mosses and our anonymous referees for comments on this paper.

## 2   Experimental Procedures and Languages

### 2.1   Experimental Procedures and Pseudocode

Consider a simple example. Imagine a kinematic machine $M$ which decides the membership of a set $A \subseteq \mathbb{N}$. Suppose it does this by the following experimental procedure:

"Given some number $n \in \mathbb{N}$ the operator of $M$ chooses a particle $P(n)$ and position $X(n)$ and projects the particle with a velocity $V(n)$. Then the operator waits for a time $T(n)$ and if the particle returns before this time then declares that $n \in A$ and otherwise that $n \notin A$."

Procedures of this kind are a "design pattern" that might be called *project and wait*. They can be applied to a number of kinematic systems, since the equipment is not specified; see the bagatelle machines, marble runs, etc. in Beggs and Tucker [2], [3], [4], [5]. We can express the experimental procedure in the following experimental pseudocode:

**exp-pseudocode** Project and wait;
place particle mass $M(n)$ radius $R(n)$ at point $X(n)$;
start clock $t$;
project particle at $X(n)$ with velocity $V(n)$;
**wait** $T(n)$ units and do nothing;
**if** particle in tray **then** return "$n \in A$" **else** return "$n \notin A$"
**end** Marble run.

To analyse such a procedure, a first step is to express the experimental procedure precisely. To turn the pseudocode into code, we must clarify the

(a) individual actions, including parameters defining the mass and size of particles and velocities;
(b) accuracy of measurements;
(c) sequencing of actions that define the procedure.

Specifically, let $T$ be a fragment of Newtonian kinematics based on particles moving in a space, possibly containing objects and a force field, and, in particular, allowing the following *primitive experimental actions*:

(i) place a particle in space;
(ii) project a particle in space;
(iii) observe a particle in space;
(iv) measure time on a clock; and

We will model each type of action by instructions and sequence them using simple control constructs, such as iteration: in this way we determine languages for programming Newtonian kinematic machines. We will define a language by outlining its syntax and exploring its semantics with the help of examples.

## 2.2   A Simple Language $EP(T)$ for Experimental Procedures

We will define a simple language $EP(T)$ for expressing experimental procedures for simple Newtonian kinematic systems.

**Instructions.** The capabilities of the language are defined by the following 10 *general hypotheses* that define a fragment $T$ of physical theory.

**Data.** Measurements of the physical notions of position, size, mass, velocity etc. are made using a *data type of measurements*. The data type is based upon a subset $M$ of numbers. The data type has very limited operations and tests: we will assume that *we can record measurements and compare them*. We will work with two choices of ordered sets of numbers, namely: $M = \mathbb{Z}$, the set of integers, and $M = \mathbb{Q}$, the set of rational numbers. The data type of measurements are the structures

$$(\mathbb{Z}|0, \geq) \text{ and } (\mathbb{Q}|0, \geq).$$

We consider approximate measurements, to within certain accuracies $\epsilon \in M$, and exact measurements, the special case $\epsilon = 0$. We also need the Booleans for comparisons.

**Variables.** We have variables for the numbers in $M$ and for the Booleans. These are the two basic types of the language. We may allow the nature of the variables to be documented by comments *space, mass, ... .*

**Introducing and Naming Particles.** There is an unlimited supply of particles of any mass and radius. There is an unlimited supply of names or identifiers from a set $N$ to label the particles. Particles are chosen for possible use in a system and named by an instruction of the form

$$p := particle(m, r)$$

which chooses a particle of mass $m \in M$ and radius $r \in M$ and names it $p \in N$, if no other particle has the name $p$, and places it in a store ready for use. If $r = 0$ then $p$ is a point particle. This instruction is similar to a declaration of identifiers.

**Removing Particles.** Particles are removed from use in a system by an instruction of the form

$$remove(p)$$

which removes the particle named $p$ in the system and places it in the store ready for reuse.

**Space.** The particles move in a space $S$ that has a coordinate system $C$ to name points and regions in space. $C$ has a metric $d\colon C \times C \to \mathbb{R}$ to measure a notion of distance between points. We will take space $S$ to be represented by a coordinate system $C$ derived from $M^n$ for $n = 1, 2, 3$.

**Projection.** There are devices for projecting a named particle from any point in space with any velocity to arbitrary accuracy. These implement the instruction

$$project(p, x, \epsilon_x, v, \epsilon_v)$$

which projects a particle named $p$ from point $x \in M^n$, to within an accuracy of $\epsilon_x \in M$, and with velocity $v \in M^n$, to within an accuracy of $\epsilon_v \in M$. This is subject to the two conditions that such a particle $p$ exists and the point $x$ is available. Projecting a particle without error margins can be rewritten

$$project(p, x, v)$$

for convenience.

**Placement.** Placing a stationary particle from the store at any point is a special case of projection with $v = 0$ and $\epsilon_v = 0$. We rewrite the instruction $project(p, x, \epsilon_x, 0, 0)$ as

$$place(p, x, \epsilon_x)$$

which places a stationary particle named $p$ at point $x$ within an accuracy of $\epsilon_x$, subject to the two conditions above.

**Observation.** There are devices for finding the position of a particle in a system with an error margin. These implement the operation

$$x := pos(p, \epsilon)$$

which assigns to $n$ variables $x$ the $n$ $C$-coordinates from $M^n$ of the unique particle named $p$ with an error margin of $\epsilon$; specifically, the coordinates of the particle named $p$ are in the ball

$$B(pos(p, \epsilon), \epsilon) = \{x \in M^n : d(x, pos(p, \epsilon)) < \epsilon\}$$

where $d$ is the metric on the coordinate system $M^n$.

**Time.** There is a clock to measure time. We allow the instruction

$$\textbf{wait time } t,$$

which does nothing for $t \in M$ units of time. We can assume time is discrete or continuous.

**Measurements.** Each measurement is a number from $M$. Physical quantities can be measured in any units.

**Control.** In an experimental procedure instructions above are scheduled by control constructs. We assume the scheduling of experimental procedures by the three constructs of sequencing, conditional branching and iteration:

$$E_1; E_2,$$
$$\textbf{if } b \textbf{ then } E_1 \textbf{ else } E_2,$$
$$\textbf{while } b \textbf{ do } E.$$

involving experimental procedures $E, E_1, E_2$ and test $b$. The tests are made form the comparison relations $\geq$ on $M$ and $=$ on Booleans.

To these we add some language conventions concerning declarations of data types and variables. These declarations we will illustrate by example rather than discuss in general.

## 2.3  Behaviour of Procedures

There are much more difficult questions about what happens when the instructions are carried out. We make an important assumption on performing procedures:

**Finite Execution.** In any finite time interval only finitely many instructions of an experimental procedure can be completed.

In a specific language based on these hypotheses, normally it will be sufficient to assume the following:

**Duration of Instructions.** There is a lower bound for the duration of all the instructions.

For ease and elegance of calculations, we will further assume that

**Fast Instructions.** All the instructions are fast relative to the evolution of the system and, hence, their finite duration can be neglected.

Alternately, if we chose not to neglect durations we could postulate specific execution times for each instruction and keep precise track of these times in our calculations. This temporal book-keeping would lead to changes in the timing of various procedures.

**Accuracy.** Note that in this elementary language, we already introduce explicit tolerances and error notions. For example, we use a function $position(p, \epsilon)$ that returns the position of a particle only to within a given accuracy. To see that this is physically sound, consider the measurement of position. A standard method is to use a camera, with a suitable lens and film resolution for the pre-specified accuracy, and simply take a picture at the given time. Now, suppose to the contrary that we postulated an *exact test* that asked "Is the particle in a given region?". This is *not* physically sound. Using a camera we would take a photograph, and could find that, up to the resolution of the camera, the particle was on the boundary of the region. We would then have to take another photo at a finer resolution to determine if the particle was in or out of the region. We might have to take new photos many given times at successively increasing resolutions to determine the answer, and this whole process could involve a large or even unbounded time, which is not practically acceptable.

**Time.** In Newtonian mechanics it is assumed that there is an *absolute universal time.* The **wait** instruction is defined directly in terms of this time.

**Iteration.** With the **while** statement in the language $EP(T)$, we must rule out infinitely many operations being carried out in finite time, and that the Finite Execution Hypothesis is satisfied. A simple way to ensure that the **while** statement always takes a finite time is by assuming that *after performing the test the* **while** *instruction waits for constant delay time $\delta > 0$.*

Measurements do not affect the system. This means that if we insert an extra measurement of the position of a particle this does not affect subsequent measurements.

## 2.4   Examples

To give an impression of the language, here are some examples.

**Example 1: Marble Run.**   Here is an example of a project and wait: the marble run procedure from [4]. Let the data type $M$ of measurement be $\mathbb{Q}$. The coordinate system for the space $\mathbb{R}^2$ is that of polar coordinates $(l, \theta)$. The procedure does not involve approximations.

```
exp-procedure Marble run;
data
input n : nat;
output b : bool;
system x : space; m : mass; r : radius; p : particle; v : velocity; c : clock;
begin
initial calculations
m := M(n);
r := R(n);
x := X(n);
v := 1;
experiment
p := particle(m, r)
project(p, x, v);
wait time 3;
(l, θ) := pos(p, 1/10);
if 3/2 ≥ l then b := true else b := false
end Marble run.
```

The variable declarations above are merely comments not types.

**Example 2: Observations of Solar System.** For many centuries people studied the sky to find the secret of planetary motion. The data were the angular positions of the planets at given times (in historically appropriate form). This gave rise to theories to try to fit the data of Ptolemy and Brahe by Copernicus and Kepler, which Newton abstracted into his law of gravity. This was all done with no ability to affect the system being measured; one recently we really can create particles with a given position and velocity (e.g., space probes). The observation of a solar system can be expressed by a procedure with a body of the following form:

```
experiment
sun := particle(m_s, r_s);
earth := particle(m_e, r_e);
moon := particle(m_m, r_m);
project(sun, x_s, v_s);
project(earth, x_e, v_e);
project(moon, x_m, v_m);
wait time 1 year;
(l_s, θ_s) := pos(sun, 1/10);
(l_e, θ_e) := pos(earth, 1/10);
(l_m, θ_m) := pos(moon, 1/10);
```

To generate a series of observations of positions over time we need to repeat these instructions.

## 2.5  Routines

Here are routines for testing if a particle is in a region and measuring its average speed, both with an error margin. In these experimental procedures we call new functions on the data type of measurement $M$ which will be justified in the next sections.

**Proposition 1.** *The location of a particle $p$ in the open ball $B(x, r)$ can be partially determined with error margin $\epsilon$ using $EP(T)$.*

*Proof.* The answer can be given by a procedure with the following body:

```
experiment location
p := particle(m, r)
y := pos(p, ε/2)
if d(x, y) < r − ε then b := true;
if d(x, y) > r + ε then b := false;
if r − ε ≤ d(x, y) ≤ r + ε then don't know;
end location
```

This test procedure, which returns 3 values, can be denoted: $isparticle?(p, x, r, \epsilon)$.

**Proposition 2.** *In any system, the average speed of a particle over a period $\delta$ can be measured with an error margin of $\epsilon$ using $EP(T)$.*

*Proof.* The average speed of a particle over a period $\delta$ in a system can be found using a procedure with the following body:

```
experiment speed
p := particle(m, r);
x := pos(p, εδ/2);
wait time δ;
y := pos(p, εδ/2);
v := d(x, y)/δ;
end average speed
```

## 3   Restrictions and Extensions of the Language for Experimental Procedures

We can restrict and extend the language $EP(T)$ with new constructs.

### 3.1   Physical Extensions

**Exactness.** Despite the argument about soundness in 2.3, for mathematical interest, we can require that instructions are exact, wherein each instruction uses error $\epsilon = 0$. Let this exact programming language be

$$EP_{\epsilon=0}(T) \subset EP(T).$$

The use of exact instructions was seen in [4], [5].

We can also strengthen the hypotheses of $EP(T)$. There are also quite a variety of changes possible to the theory $T$.

**Particles.** There are other shapes for objects to project such as polyhedra. The use of triangular projectiles was discussed in [5].

**Time.** There are other natural instructions for time. Suppose there are clocks to measure time and implement the instructions

$$start(c),$$
$$t := read(c),$$
$$stop(c),$$

which declares and initialises a clock $c$, reads the time on clock $c$ and stores in location $t$, and stops the clock $c$, respectively. With these new instructions we can iterate experimental procedures for a fixed period or indefinitely in time.

The bounded **wait** instruction is easily implemented using the clocks and **while** instruction, e.g.,

$start(c)$;
**while** $read(c) < t$ **do** skip;
$stop(c)$

This waits for time "$t$ up to an error of $\delta$" which means some time in the interval $[t, t + \delta]$. Also, unbounded waiting is possible, e.g.,

$start(c)$;
**while** $true$ **do** $skip$;
$stop(c)$.

Furthermore, time might be continuous rather than discrete.

## 3.2   Mathematical Extensions

**Data.** In place of $\mathbb{Z}$ and $\mathbb{Q}$ there are other choices for $M$, including: finite extensions of the integers and rationals, such as $\mathbb{Z}(\pi)$ and $\mathbb{Q}(\pi)$, or the set $\mathbb{R}$ of reals. The number systems for the physical quantities could all be distinct (e.g., continuous state and discrete time).

**Calculators.** The data type $M$ of measurements can be enriched with operations and tests, for which there are many possible choices. For example, in case $M = \mathbb{Z}$, we can extend the data type $M$ to the ordered ring with equality

$$M = (\mathbb{Z}|0, 1, +, -, \cdot, =, \geq);$$

and, in the case $M = \mathbb{Q}$, we can extend the data type $M$ to the ordered field with equality

$$M = (\mathbb{Q}|0, 1, +, -, \cdot, ^{-1}, =, \geq).$$

Composing these operations make algebraic formulae called *terms* or *expressions*. Suppose there are calculators for evaluating these algebraic formulae on the data type $M$ of measurements. Then we may add to the language $EP(T)$ the instructions

$$y := f(x)$$

which calculate the value $f(x)$ of the formula $f$ on the argument $x$ and stores it in location $y$.

The impact of this mathematical extension depends upon what operations are assumed on $M$. The control constructs (especially the **while** statement) in $EP(T)$

turns many calculators into universal computers, which can compute *all* functions definable by algorithms[1].

**Computers.** If, in place of calculators, we assume there are computers capable of computing all algorithms over the data in $M$ then we interfere with the comparison of algorithmic and experimental procedures. However, such extensions have a use in making a language for integrating analogue and digital computation.

## 4    Experimental Computation with Rational Numbers

The data type of measurement $M$ seems a relatively simple component in the design of $EP(T)$, when introduced in Section 2. Its purpose is to record and compare measurements. Later, in Section 3, we saw that adding operations to $M$, turning $M$ into a calculator, had computational consequences and needed care. Here we will examine in more detail the influence of $M$ on the computational powers of $EP(T)$. First, we will show how to implement basic integer and rational arithmetic using experimental procedures.

**Theorem 2.** *There is a 1-dimensional Newtonian machine, based upon the motion of free particles on the line, which can be programmed by a procedure of $EP(T)$, with the set $M = \mathbb{Z}$ of integers as data type of measurement, to perform the addition of integers.*

*Proof.* We can add two integers $n$ and $m$ by the following exact experimental procedure:

**exp-procedure** Addition
$project(p, n, 0, m, 0)$
**wait time** 1
$pos(p, 0)$
**end**

It is enough to place particles at integer positions and fire them at integer velocities and to wait a positive integer time (all without error). Addition does not need even a test on $M$! For the subtraction and multiplication of integers we need to add a test $\geq 0$ and a *changesign* operation to the set of integers.

**Theorem 3.** *Using the physical machine in 2, we can implement ring operations $+, -, \times$ and the tests $=$ for integers by experimental procedures in $EP(T)$, equipped with data type of measurement $M = (\mathbb{Z}|changesign, \geq)$.*

---

[1] In general, for an *arbitrary* abstract data type, if we add arrays to the **while** language we have a programming language that is universal and capable of programming all deterministic algorithms based on the operations of the data type, i.e. a general purpose computer for that data type (Tucker and Zucker [18]). However, since our data type $M$ is not arbitrary but based on the integer $\mathbb{Z}$, rational $\mathbb{Q}$ or real numbers $\mathbb{R}$ then the **while** construct is sufficient without arrays (Tucker and Zucker [18]).

*Proof.* Addition follows from Theorem 2. Subtraction is implemented by combining the *changesign* on $M$ and the experimental procedure for addition in the proof of Theorem 2, since $n - m = n + changesign(m)$.

To multiply $n$ and $m$ we use the following exact experimental procedure:

**exp-procedure** Multiplication
$project(p, 0, 0, n, 0)$
**if** $m \geq 0$ **then** (**wait time** $m$) **else** (**wait time** $changesign(m)$)
**if** $m \geq 0$ **then** $(pos(p, 0))$ **else** $(changesign(pos(p, 0)))$
**end**

With this simple data type of measurement $M = (\mathbb{Z}|changesign, \geq)$ we can build the ordered field of rational numbers:

**Theorem 4.** *Using the physical machine in 2, we can implement all of $+, -,$ $\times, ^{-1}$ and the test $=$ for rational numbers by experimental procedures in $EP(T)$ with data type of measurement $M = (\mathbb{Z}|changesign, \geq)$.*

*Proof.* We represent the rational $\frac{n}{m}$ as a pair $(n, m)$ of integers with $m > 0$.

We can derive experimental procedures to implement addition, subtraction and multiplication of rational numbers from the usual formulae for their definition in this representation, i.e. $(n, m) \pm (a, b) = (n\,b \pm a\,m, m\,b)$ and $(n, m).(a, b) = (n\,a, m\,b)$.

To implement the inverse on $(n, m)$, we need to test for $n = 0$. This is done using the given test $n \geq 0$ and *changesign*. Now if $n \neq 0$ then we swap $n$ and $m$, changing the signs of both if necessary. If $n = 0$ the inverse procedure can give an exception. Testing equality on rationals is given by $(n, m) = (a, b)$ if, and only, if $n\,b = a\,m$ and testing $(n, m) \geq 0$ is just $n \geq 0$.

We have used $EP_{\epsilon=0}(T)$ with its exact measurements. Consider the experimental calculations with errors. In the case of addition, if we assume that the *pos* command takes the real position and rounds it to the nearest integer, then we could use non-zero errors in this procedure. In the case of multiplication, it is more difficult to introduce the possibility of errors (at least errors in the velocity), as we would have to use a formula to estimate the error; this is somewhat self-defeating.

**Theorem 5.** *Let $f : \mathbb{Q}^n \to \mathbb{Q}^m$ be computable by an algorithmic procedure. Then $f$ is computable by an experimental procedure of $EP(T)$ with data type of measurement $M = (\mathbb{Z}|changesign, \geq)$.*

*Proof.* The computable functions on $\mathbb{Q}$ can be defined by the **while** program language $While(\Sigma)$ applied to the algebra $A = (\mathbb{Q}|0, 1, =, -, ., ^{-1}, changesign, \geq , =)$ with signature $\Sigma$. The theorem follows from the construction of a translator or compiler $c$ from $While(\Sigma)$ to $EP(T)$. The definition of this mapping $c : While(\Sigma) \to EP(T)$ is by structural induction on **while** programs. The basis case consists of assignments of the form $x := e$, where $e$ is an expression or formula over $\Sigma$, and follows from the following lemma.

**Lemma 3.** *There is a compiler that translates any expression over $\Sigma$ into an experimental procedure over of $EP(T)$ over $M = (\mathbb{Z}|changesign, \geq)$.*

*Proof.* From Theorem 4, we know there exist experimental procedures for every basic operation named in $\Sigma$. These can be combined by sequencing to create experimental procedures for arbitrary expressions. The compiler is defined by structural induction on terms using these observations.

The induction step for Theorem 5 is straight-forward since we included the three basic program forming operations of sequencing, conditional, and iteration in $EP(T)$.

## 5   Concluding Remarks

In contrast to algorithmic computation, the idea of experimental computation is not understood. It is a challenge to create a theory of experimental computation. At present, physical aspects of computation are being mapped through the search for particular examples of experimental computation and unconventional technologies for computing. Some attempts at formulating general principles arc being made in order to argue about general processing capabilities (e.g., the analysis in Akl [1] is designed to rule out universal machines on broad physical grounds). It is too early to expect anything other than a variety of agendas, approaches and debates, possibly based upon misunderstanding. We do not know yet how to theorise about experimental computation; all issues are subtle, both mathematically, physically, and philosophically; and we all express ourselves with difficulty.

We are developing a methodology for theorising about experimental computation that uses the concepts of *experimental procedure* and *equipment*: see [4], [5], [6]. Theoretical intuitions about someone making experiments turn out to be strikingly similar to intuitions about algorithms and computers. However, the primitive actions are very different and they are dependent on physical theory for their form and meaning. Our belief that languages such as $EP(T)$ are useful tools for a systematic theoretical investigation of experimental computation using logical methods must be tested by further research. In this paper we have proposed a syntax for the language $EP(T)$. Syntax has the power to isolate and organise important notions and processes, as we have seen again and again in algebra, logic, linguistics, and programming. Syntax shapes the exploration of diverse semantic models and the search for verisimilitude. However, our discussion is not complete without tackling other aspects such as semantics. Semantics will depend upon interpretations of physical theory, few of which are free of philosophical problems.

The analysis of experimental computation using languages for expressing experimental procedures seems to be new. Given a language syntax, like $EP(T)$, for experimental procedures, one can explore formally

(i) operational ideas about experiments in a physical theory;

(ii) logical properties of a physical theory;

(iii) comparisons of experimental procedures by translations;

(iv) properties of data and their physical representation;

(v) adding more complex physical constructs: for example, in the case of $EP(T)$, not allowing the particles in a system to have unique names; involving non-deterministic choice in control; taking time and other measurements to be continuous data;

(vi) adding more complex physical constructs, such as introducing typing systems for physical concepts, and even physical units (see [7]);

The theory of programming languages has a wealth of questions and technical insights that are relevant to developing language for experimental computation. The use of simple programming languages such as $EP(T)$ will help shape a general theory. Applications with digital-analogue computation and complexity theory is a possibility (after [6]).

But our earlier examples show that the notion of equipment in Newtonian mechanics also needs to be analysed. We have proposed

$$Experimental\ computation = Experimental\ procedure + Equipment.$$

A formal theory is needed that constrains the architecture and construction of mechanical systems and formally defines notions of constructible equipment. We have outlined in [4] the problems of designing languages for the specification of constructible equipment and of combining them with languages like $EP(T)$ for experimental procedures to make complete languages for experimental computation. To compare experimental computation with digital computation we need the equation:

$$Digital\ computation\ language = Algorithmic\ language + Hardware\ description\ language.$$

This is unfamiliar in programming language theory because its primary aim is to formulate algorithms that are independent of machines. We expect that research on languages for experimental computation will lead to new ideas and techniques in programming language theory.

# References


1. Akl, S.G.: Conventional or unconventional: Is any computer universal? In: Adamatzky, A., Teuscher, C. (eds.) From Utopian to Genuine Unconventional Computers, pp. 101–136. Luniver Press, Frome (2006)
2. Beggs, E. J., Tucker, J.V.: Computations via experiments with kinematic systems, Research Report 4.04, Department of Mathematics, University of Wales Swansea, March 2004 or Technical Report 5-2004, Department of Computer Science, University of Wales Swansea (March 2004)
3. Beggs, E.J., Tucker, J. V.: Embedding infinitely parallel computation in Newtonian kinematics. Applied Mathematics and Computation 178, 25–43 (2006)

4. Beggs, E. J., Tucker, J.V.: Can Newtonian systems, bounded in space, time, mass and energy compute all functions? Theoretical Computer Science 371, 4–19 (2007)
5. Beggs, E.J., Tucker, J. V.: Experimental computation of real numbers by Newtonian machines. Proceedings Royal Society Series A 463, 1541–1561 (2007)
6. Beggs, E.J., Costa, J.F., Loff, B., Tucker, J.V.: Computational complexity with experiments as oracles (in preparation, 2008)
7. Chen, F., Rosu, G., Venkatesan, R.P.: Rule-based analysis of dimensional safety. In: Nieuwenhuis, R. (ed.) RTA 2003. LNCS, vol. 2706. Springer, Heidelberg (2003)
8. Geroch, R., Hartle, J.B.: Computability and physical theories. Foundations of Physics 16, 533–550 (1986)
9. Griffor, E. (ed.): Handbook of Computability Theory. Elsevier, Amsterdam (1999)
10. Fredkin, E., Toffoli, T.: Conservative logic. International Journal of Theoretical Physics 21, 219–253 (1982)
11. Kreisel, G.: A notion of mechanistic theory. Synthese 29, 9–24 (1974)
12. Odifreddi, P.: Classical Recursion Theory. Studies in Logic and the Foundations of mathematics, vol. 129. North-Holland, Amsterdam (1989)
13. Pour-El, M.B., Richards, J.I.: Computability in Analysis and Physics, Perspectives in Mathematical Logic. Springer, Berlin (1989)
14. Stoltenberg-Hansen, V., Tucker, J.V.: Effective algebras. In: Abramsky, S., Gabbay, D., Maibaum, T. (eds.) Handbook of Logic in Computer Science. Semantic Modelling, vol. IV, pp. 357–526. Oxford University Press, Oxford (1995)
15. Stoltenberg-Hansen, V., Tucker, J.V.: Computable rings and fields. In: Griffor, E.R. (ed.) Handbook of Computability Theory, pp. 363–447. Elsevier, Amsterdam (1999)
16. Stoltenberg-Hansen, V., Tucker, J. V.: Concrete models of computation for topological algebras. Theoretical Computer Science 219, 347–378 (1999)
17. Stoltenberg-Hansen, V., Tucker, J. V.: Computable and continuous partial homomorphisms on metric partial algebras. Bulletin for Symbolic Logic 9, 299–334 (2003)
18. Tucker, J. V., Zucker, J.I.: Computable functions and semicomputable sets on many sorted algebras. In: Abramsky, S., Gabbay, D., Maibaum, T. (eds.) Handbook of Logic for Computer Science, vol. V, pp. 317–523. Oxford University Press, Oxford (2000)
19. Tucker, J. V., Zucker, J.I.: Abstract versus concrete computation on metric partial algebras. ACM Transactions on Computational Logic 5, 611–668 (2004)
20. Weihrauch, K.: Computable Analysis, An introduction. Springer, Heidelberg (2000)
21. Yao, A.: Classical physics and the Church Turing thesis. Journal ACM 50, 100–105 (2003)

# Linear, Polynomial or Exponential? Complexity Inference in Polynomial Time

Amir M. Ben-Amram[1,*], Neil D. Jones[2], and Lars Kristiansen[3]

[1] School of Computer Science, Tel-Aviv Academic College, Israel
[2] DIKU, the University of Copenhagen, Denmark
[3] Department of Mathematics, University of Oslo, Norway
amirben@mta.ac.il, neil@diku.dk, larskri@iu.hio.no

**Abstract.** We present a new method for inferring complexity properties for imperative programs with bounded loops. The properties handled are: polynomial (or linear) boundedness of computed values, as a function of the input; and similarly for the running time.

It is well known that complexity properties are undecidable for a Turing-complete programming language. Much work in program analysis overcomes this obstacle by relaxing the correctness notion: one does not ask for an algorithm that correctly decides whether the property of interest holds or not, but only for "yes" answers to be sound. In contrast, we reshaped the problem by defining a "core" programming language that is Turing-incomplete, but strong enough to model real programs of interest. For this language, our method is the first to give a certain answer; in other words, our inference is both sound and complete.

The essence of the method is that every command is assigned a "complexity certificate", which is a concise specification of dependencies of output values on input. These certificates are produced by inference rules that are compositional and efficiently computable. The approach is inspired by previous work by Niggl and Wunderlich and by Jones and Kristiansen, but use a novel, more expressive kind of certificates.

**Keywords:** implicit computational complexity, polynomial time complexity, linear time complexity, static program analysis.

## 1   Introduction

Central to the field of Implicit Computational Complexity (ICC) is the following fundamental observation: it is possible to restrict a programming language syntactically so that the admitted programs will possess a certain complexity, say polynomial time. Such results lead to a sweet dream: the "complexity-certifying compiler," that will warn us whenever we compile a non-polynomial algorithm.

Since (as is well known) deciding such a property precisely for any program in a Turing-complete language is impossible, the goal in this line of research is

---

* Research performed while visiting DIKU, the University of Copenhagen, Denmark.

A. Beckmann, C. Dimitracopoulos, and B. Löwe (Eds.): CiE 2008, LNCS 5028, pp. 67–76, 2008.

to extend the capabilities of the certifying compiler as much as possible, while taking into account the price: in other words, explore the tradeoff between the completeness of the method and the method's complexity. At any rate, we insist on the certification being sound—no bad programs shall pass.

The method presented in this paper is a step forward in this research program. It can be used to certify programs as having a running time that is polynomial in certain input values; we also present a variant for certifying linear time. Further, we determine which variables have values that are polynomially (or linearly) bounded. The latter is, in fact, the essential problem: since we will only consider bounded loops, deriving bounds on the running time amounts to deriving a bound on the counters that govern the loops.

Our work complements previous research by Kristiansen and Niggl [KN04], Niggl and Wunderlich [NW06], and Jones and Kristiansen [JK08]. All methods apply to a structured imperative language. The last two, in particular, take the form of a compositional calculus of certificates—to any command a "certificate" is assigned which encodes the certified properties of the command, and a certificate for a composite command is derived from those of its parts. We keep this elegant structure, but use a new kind of certificates designed to accurately discern phenomena such as *accumulator variables* in loops (think of a command X := X+Y within a loop) as well as *self-multiplying variables* (X := X+X).

Normally, in theoretical research such as this, one does not bother with algorithms for analysing a full-featured practical programming language, but considers a certain *core language* that embodies the features of algorithmic interest. It is well known that one can strip a lot of "syntactic sugar" out of any practical programming language to obtain such a core which is still Turing-complete, and once a problem is solved for the core, it is as good as solved for the full language (up to implementing the appropriate translations). We go a step further by proposing that if certain features are *beyond the scope* or our analysis, getting rid of them is best done in the passage to the core language. Here is the prototypical example: Very often, program analyses take a conservative approach to modelling conditionals: both branches are treated as possible. Since our analysis also does so, we include in our core language only the following form for the conditional command: if ? then C1 else C2.

Here C1, C2 represent commands, while the question mark represents that the conditional expression is hidden. In the core language, this command has a *non-deterministic* semantics. Thus, the passage to the core language is an *abstracting* translation: it abstracts away features that we overtly leave out of the scope of our analysis. Our point of view is that it is beneficial to separate the concern of parsing and abstracting practical programs (the *front end*) from the concern of analysing the core language (the *back end*). A simple-minded abstraction (e.g., really just hiding the conditionals) is clearly doable, so there should be no doubt that a front end for a realistic language *can* be built. Current *static analysis* technology allows the construction of sophisticated front ends. Our theoretic effort will concentrate on the "back end"—analysing the core language. The reader

$$X \in \text{Variable} \quad ::= \quad X_1 \mid X_2 \mid X_3 \mid \ldots \mid X_n$$
$$e \in \text{Expression} \quad ::= \quad \texttt{X} \mid \texttt{(e + e)} \mid \texttt{(e * e)}$$
$$C \in \text{Command} \quad ::= \quad \texttt{skip} \mid \texttt{X:=e} \mid C_1\texttt{;}C_2 \mid \texttt{loop X \{C\}}$$
$$\mid \quad \texttt{if ? then C else C}$$

**Fig. 1.** Syntax of the core language. Variables hold nonnegative integers.

may want to peek at Figure 1, showing the syntax of the language. Its semantics
is almost self-explanatory, and is made precise in the next section.

The main result in this paper is a proof that the problems of polynomial and
linear boundedness are decidable for the core language. Our certification method
solves this problem completely: for example, for the problem of polynomial running time, we will certify a core-language program *if and only if* its time is
polynomially bounded. Furthermore, the analysis itself takes *polynomial time.*

*A brief comparison with previous work.* Both previous work we mentioned do not
use a core language to set a clear abstraction boundary. However, [JK08] treats
(implicitly) the same core language. Its inferences are sound, but incomplete, and
its complexity appears to be non-polynomial (this paper is, essentially, a journal
version of [JK05]—where completeness was wrongly claimed). The (implicit) core
language treated by [NW06] can be viewed as an extension of our core language.
When applied to our language, their method too is sound but incomplete (its
complexity is polynomial-time, like ours).

## 2  Problem Definition

The *syntax* of our core language is described in Figure 1. In a command $\texttt{loop X \{C\}}$,
variable X is not allowed to appear on the left-hand side of an assignment in the
loop body C.

*Data.* It is most convenient to assume that the only type of data is nonnegative
integers. More generality is possible but will not be treated here.

*Command semantics.* As already explained, the core language is nondeterministic. The if command represents a nondeterministic choice. The *loop command*
$\texttt{loop } X_\ell \texttt{ \{C\}}$ repeats C a number of times bounded by the value of $X_\ell$. Thus, it
is also nondeterministic, and may be used to model different kinds of loops (for-
loops, while-loops) as long as a bounding variable can be statically determined.

While the use of bounded loops restricts the computable functions to the
primitive recursive class, this is still rich enough to make the problem challenging
(and it can still be pushed to undecidability, if we give up the abstraction and
include conventional, deterministic conditionals, such as an equality test[1]).

---

[1] Undecidability for such a language can be proved by a reduction from Hilbert's 10th
problem.

The formal semantics associates with every command C over variables $X_1, \ldots, X_n$ a relation $[\![C]\!] \subseteq \mathbb{N}^n \times \mathbb{N}^n$. In the expression $\vec{x}[\![C]\!]\vec{y}$, vector $\vec{x}$ (respectively $\vec{y}$) is the store before (after) the execution of C.

The semantics of skip is the identity. The semantics of an assignment leaves some room for variation: either the precise value of the expression is assigned, or a nonnegative integer bounded by that value. The latter definition is useful for abstracting non-arithmetic expressions that may appear in a real-life program. Because our analysis only derives monotone increasing value bounds, this choice does not affect the results. Finally, composite commands are described by the straight-forward equations:

$$[\![C_1; C_2]\!] = [\![C_2]\!] \circ [\![C_1]\!]$$
$$[\![\texttt{if ? then } C_1 \texttt{ else } C_2]\!] = [\![C_1]\!] \cup [\![C_2]\!]$$
$$[\![\texttt{loop } X_\ell \texttt{ \{C\}}]\!] = \{(\vec{x}, \vec{y}) \mid \exists i \leq x_\ell : \vec{x}[\![C]\!]^i \vec{y}\}$$

where $[\![C]\!]^i$ represents $[\![C]\!] \circ \cdots \circ [\![C]\!]$ ($i$ occurrences of $[\![C]\!]$).

For every command we also define its *step count* (informally referred to as running time). For simplicity, the step count of an atomic command is defined as 1. The step count of a loop command is the sum of the step counts of the iterations taken. Because of the nondeterminism in if and loop commands, the step count is also a relation. We also refer to the *iteration count*, which only grows by one each time a loop body is entered. The iteration count is linearly related to the step count, but is easier to analyse.

*Goals of the analysis.* Our *polynomial-bound calculus* (Section 3) reveals, for any given command, which variables are bounded throughout any computation by a polynomial in the input variables[2]. The *linear-bound calculus* (Section 4) identifies linearly-bounded variables instead. Finally, Section 5 extends these methods to characterize commands where the maximum step count is bounded polynomially (respectively, linearly).

As a by-product, the analysis reveals which inputs influence any specific output variable.

*An example.* In the following program, all variables are polynomially bounded (we invite the reader to check); this is not recognized by the previous methods. In the next section we explain how the difficulty illustrated by this example was overcome.

```
loop X_5 {
  if ? then { X_3 := X_1+X_2; X_4 := X_2 }
       else { X_3 := X_2;    X_4 := X_1+X_2 };
  X_1 := X_3 + X_4;
}
```

---

[2] Thus, as pointed out by one of the reviewers, the title of this paper is imprecise: we distinguish polynomial growth from *super-polynomial* one, be it exponential or worse.

# 3   A Calculus to Certify Polynomial Bounds

Our calculus can be seen as a set of rules for *abstract interpretation* of the core language, in the sense of [Cou96][3]. The maximal output value resulting of a given command C is some function $f$ of the input values $x_1, \ldots, x_n$. There are infinitely many possible functions; we map each one into an *abstract value* of which there are finitely many. Each abstract value $V$ is associated with a *concretisation* $\gamma(V)$ which is a (possibly infinite) set of functions such that $f$ is bounded by one of them.

Let $\mathbb{D} = \{0, 1, 1^+, 2, 3\}$ with order $0 < 1 < 1^+ < 2 < 3$. Informally, $\mathbb{D}$ is a set of *dependency types*, describing how a result depends on an input, as follows:

| value | 3 | 2 | $1^+$ | 1 | 0 |
|---|---|---|---|---|---|
| dependency type | at least exponential | polynomial | additive | copy | none |

The notation $[x = y]$ below denotes the value 1 if $x = y$ and 0 otherwise.

**Definition 1.** *Let $V \in \mathbb{D}^n$. The concretisation $\gamma(V)$ includes all functions defined by the following rules, and none others. (1) If there is an $i$ such that $V_i = 1$, then $\gamma(V)$ includes $f(\vec{x}) = x_i$. (2) $\gamma(V)$ includes all polynomials of form $(\sum_i a_i x_i) + P(\vec{x})$, where $a_i \leq [V_i = 1^+]$, and $P$ is a polynomial of non-negative coefficients depending only on variables $x_i$ such that $V_i = 2$. (3) If there is an $i$ such that $V_i = 3$, then $\gamma(V)$ is the set of all $n$-ary functions over $\mathbb{N}$.*

A core-language expression e obviously describes a polynomial and it is straight-forward to obtain a minimal vector $\alpha(\mathsf{e})$ such that $\gamma(\alpha(\mathsf{e}))$ includes that polynomial.

A basic idea (going back to [NW06]) is to approximate the relation $\vec{x}[\![\mathsf{C}]\!]\vec{y}$ by a set of vectors $V_1, \ldots, V_n$ that describe the dependence of $y_1, \ldots, y_n$ respectively on $\vec{x}$. We combine the vectors into a matrix $M \in \mathbb{D}^{n \times n}$ where column $j$ is $V_j$. Thus, $M_{ij}$ is the dependency type of $y_j$ on $x_i$. A complementary and useful view is that $M_{ij}$ describes a *data-flow* from $x_i$ to $x_j$. In fact, $M$ can be viewed as a bipartite, labeled digraph where the left-hand (source) side represents the input variables and the right-hand (target) side represents the output. The set of arcs $\mathbf{A}(M)$ is the set $\{i \to j \mid M_{ij} \neq 0\}$. A list of arcs may also be more readable than a matrix. For example, consider the command $\mathsf{loop\ X_3\ \{X_1 := X_1 + X_2\}}$ or the command $\mathsf{X_1 := X_1 + X_3 * X_2}$. Both are described (in the most precise way) by the following collection of arcs: $\mathsf{X_2} \xrightarrow{1} \mathsf{X_2}, \ \mathsf{X_3} \xrightarrow{1} \mathsf{X_3}, \ \mathsf{X_1} \xrightarrow{1^+} \mathsf{X_1}, \ \mathsf{X_2} \xrightarrow{2} \mathsf{X_1}, \ \mathsf{X_3} \xrightarrow{2} \mathsf{X_1}$

or, as a matrix, $\begin{bmatrix} 1^+ & 0 & 0 \\ 2 & 1 & 0 \\ 2 & 0 & 1 \end{bmatrix}$.

Such matrices/graphs make an elegant abstract domain (or "certificates") for commands because of the ease in which the certificate for a composite command

---

[3] Abstract interpretation is a well-developed theoretical framework, which can shed light on our algorithm. However in this paper we avoid relying on prior knowledge of abstract interpretation.

can be derived. Let $\sqcup$ denote the LUB operation on $\mathbb{D}$, and extend it to matrices (elementwise). If $M_1, M_2$ describe commands $\mathtt{C_1}, \mathtt{C_2}$, it is not hard to see that $M_1 \sqcup M_2$ describes `if ? then C`$_1$` else C`$_2$. For transferring the sequential composition of commands to matrices, we define the following operation: $a \cdot b$ is 0 if either $a$ or $b$ is 0, and otherwise is the largest of $a$ and $b$. Intuitively, composition should be represented by matrix product using $\cdot$ as "multiplication." The problem, though, is what result an "addition" $\oplus$ of $1^+$s should be. Consider the commands

```
C1  =  X₂ := X₁; X₃ := X₁
C2  =  X₁ := X₂ + X₃
```

The graph representing `C1` has arcs $\mathtt{X_1} \overset{1}{\to} \mathtt{X_2}$, $\mathtt{X_1} \overset{1}{\to} \mathtt{X_3}$ and the graph for `C2` has $\mathtt{X_2} \overset{1^+}{\to} \mathtt{X_1}$, $\mathtt{X_3} \overset{1^+}{\to} \mathtt{X_1}$. Hence entry $M_{11}$ of the matrix for `C1; C2` has value $1 \cdot 1^+ \oplus 1 \cdot 1^+ = 1^+ + 1^+$. And for this example, the right answer is 2, because the command doubles $\mathtt{X_1}$ (and in a loop, $\mathtt{X_1}$ will grow exponentially). However, the graph for `C1'` $=$ `if ? then X`$_2$` := X`$_1$` else X`$_3$` := X`$_1$ also includes the arcs $\mathtt{X_1} \overset{1}{\to} \mathtt{X_2}$, $\mathtt{X_1} \overset{1}{\to} \mathtt{X_3}$, but when `C1'` is combined with `C2`, no doubling occurs.

Our conclusion is that matrices are just not enough, and in order to allow for compositional computation, our certificates retain additional information. Basically, we add to the matrices another piece of data which distinguishes between a pair of 1's that arises during a single computation path (as in `C1`) and a pair that arises as alternatives (as in `C1'`). We next move to the formal definitions.

### 3.1  Data Flow Relations

*A few notations*: We use $\mathbf{A^1}(M)$ to denote the set of arcs labeled by $\{1, 1^+\}$. For any set $S$, $C_2(S)$ is the set of 2-sets (unordered pairs) over $S$. For $M \in \mathbb{D}^{n \times n}$, we define $r(M)$ to be $C_2(\mathbf{A^1}(M))$. The identity matrix $I$ has 1 on the diagonal and 0 elsewhere.

A *dataflow relation*, or DFR, is a pair $(M, R)$ where $M \in \mathbb{D}^{n \times n}$ (and has the meaning described above) and $R \subseteq C_2(\mathbf{A^1}(M))$. Thus, $R$ consists of pairs of arcs.

For compactness, instead of writing $\{i \to j, i' \to j'\} \in R$ we may write $R(i, j, i', j')$.

*Definitions: Operations on matrices and DFRs.*

1. $A \otimes B$ is $(\sqcup, \cdot)$ matrix product over $\mathbb{D}$.

2. $(M_1, R_1) \sqcup (M_2, R_2) \overset{def}{=} (M_1 \sqcup M_2, (R_1 \cup R_2) \cap C_2(\mathbf{A^1}(M_1 \sqcup M_2)))$.

3. $(M, R) \cdot (M', R') \overset{def}{=} (M'', R'')$, where:

$$M'' = (M \otimes M') \sqcup \{i \overset{2}{\to} j \mid \exists s \neq t. R(i, s, i, t) \wedge R'(s, j, t, j)\}$$
$$R'' = \{\{i \to j, i' \to j'\} \in C_2(\mathbf{A^1}(M'')) \mid \exists s, t. R(i, s, i', t) \wedge R'(s, j, t, j')\}$$
$$\cup \{\{i \to j, i \to j'\} \in C_2(\mathbf{A^1}(M'')) \mid \exists s. (i, s) \in \mathbf{A^1}(M) \wedge R'(s, j, s, j')\}$$
$$\cup \{\{i \to j, i' \to j\} \in C_2(\mathbf{A^1}(M'')) \mid \exists s. R(i, s, i', s) \wedge (s, j) \in \mathbf{A^1}(M')\}.$$

Observe how the rule defining $M''$ uses the information in the $R$-parts to identify computations that double an input value by adding two copies of it, a situation described by *the diamond*:

$$\begin{array}{ccc} & s & \\ i \nearrow & & \searrow j \\ & t \nearrow & \end{array}$$

The reader may have guessed that if this situation occurs in analysing a program, the two meeting arcs will necessarily be labeled with $1^+$.

**Proposition 1.** *The product is associative and distributes over* $\sqcup$, *i.e.,*
$$(M, R) \cdot ((M_1, R_1) \sqcup (M_2, R_2)) = ((M, R) \cdot (M_1, R_1)) \sqcup ((M, R) \cdot (M_2, R_2)).$$

4. Powers: defined by $(M, R)^0 = (I, r(I))$ and $(M, R)^{i+1} = (M, R)^i \cdot (M, R)$.

5. Loop Correction: for a DFR $(M, R)$, define $LC_\ell(M, R) = (M', R')$ where $M'$ is identical to $M$ except that:
   (a) For all $j$ such that $M_{jj} \geq 2$, $M'_{\ell j} = 3$;
   (b) For all $j$ such that $M_{jj} = 1^+$, $M'_{\ell j} = M_{\ell j} \sqcup 2$;
   and $R' = R \cap C_2(\mathbf{A}^1(M'))$.
   Remarks: Rule (a) reflects the exponential growth that results from multiplying a value inside a loop. If $\mathsf{X}_j$ is doubled, it will end up multiplied by $2^{x_\ell}$. Rule (b) reflects the behaviour of *accumulator variables*. Intuitively, $M_{jj} = 1^+$ reveals that some quantity $y$ is added to $\mathsf{X}_j$ in the loop. Therefore, the effect of the loop will be to add $x_\ell \cdot y$, hence the correction to $M_{\ell j}$.

6. Loop Closure: For a given $\ell$, the loop closure with respect to $\mathsf{X}_\ell$ is the limit $(M^*, R^*)$ of the series $(M_i, R_i)$ where:

$$(\star) \qquad \begin{aligned} (M_0, R_0) &= (I, r(I)) \\ (M_1, R_1) &= (M_0, R_0) \sqcup (M, R) \\ (M_{i+1}, R_{i+1}) &= (LC_\ell(M_i, R_i))^2 \end{aligned}$$

This closure can be computed in a finite (polynomial) number of steps because the series is an increasing chain in a semilattice of polynomial height. A more natural specification of the closure may be the following: $(M^*, R^*)$ is the smallest DFR that is at least $(I, r(I))$ and is fixed under both multiplication by $(M, R)$ and under Loop Correction (the meaning of "smallest" and "at least" has yet to be made precise). However, the definition above (as a limit) is the useful one from the algorithmic point of view.

### Calculation of DFRs

The following inference rules associate a DFR with every core-language command. The association of $(M, R)$ with command $\mathsf{C}$ is expressed by the judgment $\vdash \mathsf{C} : M, R$.

(Skip)
$$\frac{}{\vdash \mathtt{skip} : I, r(I)}$$

(Assignment)
$$\frac{\alpha(\mathsf{e}) = V}{\vdash \mathsf{X}_i := \mathsf{e} : M, r(M)}$$

where $M$ is obtained from $I$ by replacing the $i$th column with $V$.

(Choice)
$$\frac{\vdash \mathsf{C}_1 : M_1, R_1 \qquad \mathsf{C}_2 : M_2, R_2}{\vdash \mathtt{if?then}\,\mathsf{C}_1\,\mathtt{else}\,\mathsf{C}_2 : (M_1, R_1) \sqcup (M_2, R_2)}$$

(Sequence)
$$\frac{\vdash \mathsf{C}_1 : M_1, R_1 \qquad \mathsf{C}_2 : M_2, R_2}{\vdash \mathsf{C}_1; \mathsf{C}_2 : (M_1, R_1) \cdot (M_2, R_2)}$$

(Loop)
$$\frac{\vdash \mathsf{C} : M, R}{\vdash \mathtt{loop}\,\mathsf{X}_\ell\{\mathsf{C}\} : (M^*, R^*)}$$

where $(M^*, R^*)$ is the loop closure of $(M, R)$ with respect to $\mathsf{X}_\ell$.

The DFR for a command can always be computed in time polynomial in the size of the command (i.e., the size of its abstract syntax tree). This is done bottom up, so (since the calculus is deterministic) every node is treated once. The work per node is the application of one of the above rules, each of which is polynomial-time.

## 4   Certifying Linear Bounds

We present an adaption of the method to certify linear bounds on variables. Essentially the only change is that when something is deduced to be non-linear, it is labeled by a 3. Thus, 2's only describe linear bounds. This is summarized here:

| value | 3 | 2 | $1^+$ | 1 | 0 |
|---|---|---|---|---|---|
| dependency type | nonlinear | linear | additive | copy | none |

For example: the command $\mathsf{X}_1 := \mathsf{X}_1 + 2*\mathsf{X}_2$ is described by $\mathsf{X}_2 \xrightarrow{1} \mathsf{X}_2$, $\mathsf{X}_2 \xrightarrow{2} \mathsf{X}_1$, $\mathsf{X}_1 \xrightarrow{1^+} \mathsf{X}_1$.

The calculus $\vdash_{lin}$ for linear bounds deviates from the polymiality calculus (judgments $\vdash$) only as follows.

1. In abstracting expressions into vectors $V \in \{0, 1, 1^+, 2\}^n$, we treat linear expressions as before, while every occurrence of multiplication $\mathsf{e}_1 * \mathsf{e}_2$ creates a Type-3 dependence on all variables in $\mathsf{e}_1$ and $\mathsf{e}_2$. Formally, the abstraction and concretisation functions $\alpha, \gamma$ are replaced with appropriate $\alpha^{lin}$ and $\gamma^{lin}$.
2. For a DFR $(M, R)$, define $LC^{lin}_\ell(M, R) = (M', R')$ where $M'$ is identical to $M$ except that: for all $j$ such that $M_{jj} \in \{1^+, 2\}$, $M'_{\ell j} = 3$; and $R' = R \cap C_2(\mathbf{A}^1(M'))$. $LC^{lin}$ replaces $LC$ in the calculation of the loop closure.

## 5   Analysing Running Time

Recall that the *iteration count* grows by one each time a loop body is entered. To certify that it is polynomially (or linearly) bounded, it suffices to include an extra variable in the program that sums up the values of loop counters, and certify that this variable is so bounded. We can extend our calculi to implicitly include this variable, so that there is no need to actually modify the program. This is achieved as follows.

1. Matrices become of order $(n+1)\times(n+1)$. Thus the identity matrix $I$ includes the entry $\mathsf{X}_{n+1}\xrightarrow{1}\mathsf{X}_{n+1}$ which reflects preservation of the extra variable.
2. The inference rule for the loop command is modified to reflect an implicit increase of the extra variable. The rule thus becomes:

$$(\text{Loop}) \qquad \frac{\vdash \mathsf{C} : M, R}{\vdash \texttt{loop}\,\mathsf{X}_\ell\{\mathsf{C}\} : M', R'}$$

where $M', R'$ are obtained from the loop closure $(M^*, R^*)$ by:

$$M' = M^* \sqcup \{\ell \xrightarrow{1^+} (n+1),\ (n+1) \xrightarrow{1^+} (n+1)\}$$
$$R' = R^* \cup \{\{i \to j, \ell \to n+1\} \mid i \to j \in \mathbf{A}^1(M)\}$$

We assume that the reader can see that the implicit $\mathsf{X}_{n+1}$ is treated by these rules just as a variable that actually accumulates the loop counters.

## 6   Concluding Remarks

We have presented a new method for inferring complexity bounds for imperative programs by a compositional program analysis. The analysis applies to a limited, but non-trivial *core language* and proves that the properties of interest are decidable. We believe that this core-language framework is important for giving a robust yardstick for a project: we have a completeness result, so we can say that we achieved our goal (which, in fact, evolved from the understanding that previous methods did not achieve completeness for such a language).

This work is related, on one hand, to the very rich field of program analysis and abstract interpretation. These are typically targeted at realistic, Turing complete languages, and integrating our ideas with methods from that field is an interesting direction for further research, whether one considers expanding our approach to a richer core language, or creating clever front-ends for realistic languages.

Another connection is with Implicit Computational Complexity. In this field, a common theme is to capture complexity classes. Our core language cannot be really used for computation, but one can define a more complete language that has a simple, complexity-preserving translation to the core language. For example, let us provide the language with an input medium in the form of a binary string, and with operations to read it, as well as reading its length; and then with a richer arithmetic vocabulary, including appropriate conditionals. We obtain a "concrete" programming language $L_{concrete}$ that has an evident abstracting translation $\mathcal{T}$ into the core language, and it is not hard to obtain results such as: $f : \mathbb{N} \to \mathbb{N}$ is PTIME-computable if and only if it can be computed by an $L$ program $p$ such that $\mathcal{T}p$ is polynomial-time (and hence recognized so by our method). We leave the details to the full paper. Note that no procedure for inferring complexity will be complete for $L_{concrete}$ itself, precisely because it is powerful enough to simulate Turing machines.

Finally, let us point out some directions for further research:

- Extending the core language. For instance, our language does not include constants as do the languages considered in [KN04, NW06]. Neither does it include data types of practical significance such as strings (these too are present in the latter works).
- Investigating the design of front ends for realistic languages.
- Extending the set of properties that can be decided beyond the current selection of linear and polynomial growth rates and running times.

*Acknowledgement.* We are grateful to the CiE reviewers for some very detailed and helpful reviews.

# References

[Cou96]  Cousot, P.: Abstract interpretation. ACM Computing Surveys 28(2), 324–328 (1996)

[JK05]   Jones, N.D., Kristiansen, L.: The flow of data and the complexity of algorithms. In: Cooper, S.B., Löwe, B., Torenvliet, L. (eds.) CiE 2005. LNCS, vol. 3526, pp. 263–274. Springer, Heidelberg (2005)

[JK08]   Jones, N.D., Kristiansen, L.: A flow calculus of mwp-bounds for complexity analysis. ACM Trans. Computational Logic (to appear)

[KN04]   Kristiansen, L., Niggl, K.-H.: On the computational complexity of imperative programming languages. Theor. Comput. Sci. 318(1-2), 139–161 (2004)

[NW06]   Niggl, K.-H., Wunderlich, H.: Certifying polynomial time and linear/polynomial space for imperative programs. SIAM J. Comput 35(5), 1122–1147 (2006)

# A Summation Algorithm from 11<sup>th</sup> Century China

## Possible Relations Between Structure and Argument


Andrea Bréard

Université des Sciences et Technologies de Lille, U.F.R de Mathématiques,
Laboratoire Paul Painlevé
`andrea.breard@math.univ-lille1.fr`
and
REHSEIS - UMR7596 Recherches Epistémologiques et Historiques sur les Sciences
Exactes et les Institutions Scientifiques



**Abstract.** Mathematical writings in China relied entirely on the algorithmic mode to express sequences of operations, to justify the correctness of these, and to bring mathematical objects in relation one to another. In this paper, I shall use one example to show how the structural elements in an algorithm convey a mathematical meaning and can be interpreted in the light of the ancient Chinese geometrical tradition. The example stems from an 11<sup>th</sup> century text by Shen Gua 沈括 and calculates the number of kegs of wine piled up in the form of a truncated pyramid with a rectangular base.

**Keywords:** Shen Gua, Song dynasty, Sum of squares.


## 1 Shen Gua's Text

Jotting 301 in Shen Gua's *Brush Talks from the Dream Pool* (*Meng xi bi tan* 夢溪筆談, 11<sup>th</sup> century) calculates the total number of wine kegs when piled up in layers in the shape of the frustum of a rectangular pyramid:[3]

> Among mathematical procedures there are methods to calculate volumes [in units of] *chi*. For categories like the *chumeng* 芻甍[4], the *chutong* 芻童[5], the square pond (*fangchi* 方池), an obscure gorge (*ming gu* 冥谷), the *qiandu* 塹堵[6], the *bienao* 鱉臑[7], the cone with a circular base

---

[3] Translated by the author following Shen and Hu [4], vol. 2 pp. 574-5. For a detailed discussion of this jotting and its historical significance see Bréard [1] and Bréard [2], chapter 3.1.

[4] The *chumeng* represents a kind of geometric solid belonging to the class of triangular prisms.

[5] The frustum of a rectangular pyramid.

[6] A rectangular half parallelepiped. In the edition Shen and Hu [4], vol. 2 p. 574 we find the character *qian* 塹 whereas in the *Nine Chapters on Mathematical Procedures* the character *qian* with the water radical at its left side is used. See [3], problems 5.5-5.7 and 5.14.

[7] A tetrahedron.


A. Beckmann, C. Dimitracopoulos, and B. Löwe (Eds.): CiE 2008, LNCS 5028, pp. 77–83, 2008.

(*yuanzhui* 圓錐) and the *yangma* 陽馬[8] the [external] form of the object is perfectly complete. Only a single procedure, one for volumes with interstices (*xi ji* 隙積), does not exist yet. Among the objects whose parallelepiped volumes are calculated globally by the ancient methods are the following:

We have the cube (*li fang* 立方). It is explained as having six surfaces that are all squares. The corresponding procedure is to multiply twice [the side of a square] by itself, thus you obtain it [the volume].

We have the *qiandu*. It is explained as being similar to an earthen wall. Two sides are battered and the two front sides are on an equal level. The corresponding procedure is to add the upper and lower breadth, to take away half of this and to take [the result] as its breadth. Multiply this with the straight height.

Another [procedure] is to take the straight height as *gu* 股[9], to reduce the lower breadth with the upper breadth and to take half of what remains as *gou* 句[10]. Calculate the hypotenuse [with the procedure] *gougu*[11] and take it as the oblique height.

We have the *chutong*. It is explained as being similar to a *dou*  that has been turned upside-down. All of its four sides are battered. The corresponding procedure is to double the upper length, to incorporate it additionally to the lower length and to multiply this with the upper breadth. Double the lower length, incorporate it additionally to the upper length and multiply this with the lower breadth. Add the two positions and multiply this with the height. Divide by six.

As for 'volumes with interstices', these are explained as: when accumulating these, interstices appear. For categories like the piling up of blocks to make the layers of an altar (*tan* 壇) and the accumulation of kegs of wine in a wine shop, although they resemble 'a *dou*  that has been turned upside-down', with 'all of its four sides battered', but because they have deficient and empty spaces, if we calculate using the method for the *chutong*, we will constantly underestimate their [real] number. I have thought about this and obtained the following: One uses the method for the *chutong*, the result gives the upper line [of the counting board]. Furthermore on the lower line, one lays down the lower breadth, and subtracts the upper breadth from it. What remains shall be multiplied by the height, divided by 6 and added to the upper line. [Numerical commentary:] Let us suppose one has accumulated kegs of wine. In the uppermost layer there are 2 kegs of wine each in breadth and length. In the lowermost layer there are 12 kegs of wine each. Layer by layer are in progression one to another. If one begins by the progression in which the

---

[8] A pyramid with a rectangular base.

[9] *Gu*, the longer side of a right-angled triangle, the height.

[10] *Gou*, the shorter side of a right-angled triangle, the base.

[11] The procedure *gougu* expresses a similar relation in a right-angled triangle between the three sides *gou*, *gu* and *xian* 弦 (the hypotenuse) as the theorem of Pythagoras.

two uppermost layers are one to another, and counts until one gets to 12, then one should obtain 11 layers. One solves this with the method for the *chutong*. Doubling the length of the upper layer gives 4. This added to the lower length gives 16. This multiplied with the upper breadth makes 32. Furthermore, if one doubles the two lower lengths, one obtains 16. Added to the upper length gives 46. Multiplied with the upper breadth, one obtains 312. By adding the two positions [on the counting board], one obtains 344. This multiplied with the height gives 3784. Once more one lays out 12, the lower breadth. Reduced by the upper breadth, one has a remainder of 10. his multiplied by the height gives 110. Added to the upper line gives 3894. Division by 6 gives 649. This represents the number of wine kegs. If one solved [the problem] with the [method for a] *chutong*, one would obtain the [volume] product of a full parallelepiped (*shi fang zhi ji* 實方之積). The procedure for the volume with interstices reveals the volume of what is not exhausted [by the procedure for the continuous *chutong*] at the junction of the corners and adds what exceeds outwards (*xian hejiao bu jin yi chu xian ji* 見合角不盡益出羨積)."

There are two numerical errors in the Song dynasty (AD 960 – 1279) edition of Shen Gua's text. Zhao Yushi 趙與時 (1175-1231) mentions the problem in his *Records written after the guest has left* (*Bintui lu* 賓退錄):

"In scroll 18 of the Guangling edition of the *Brush Talks from the Dream Pool* the commentary to the procedure for accumulations of wine kegs, it should read '24' instead of 'Furthermore, if one doubles the two lower lengths, one obtains 16'. Instead of 'Added to the upper length gives 46' it should read '26'. Scholars and officials who understand mathematics are rare. This is the reason why nobody has ever mentioned these errors."[12]

## 2  Shen Gua's Algorithm

The structure of the above algorithm suggests that there are two separate blocks of operations, which contribute to the correct result when added up. The first block is a series of operations equivalent to the procedure for the continuous *chutong*, the second block, performed on a different line of the counting board, corresponds to an additional term, necessary for adapting the formula for the continuous case to the discrete case.

$$A = \underbrace{[(2d + c) \cdot b + (2c + d) \cdot a] \cdot h \div 6}_{chutong} + \underbrace{(a - b) \cdot h \div 6}_{\text{corrective term}} \tag{1}$$

The question is how Shen Gua did derive this algorithm. Assuming that the text is not corrupted substantially, I would argue that we can make certain hypothesis about his "discovery" of the procedure for a *chutong* with interstices,

---

[12] Translated according to Shen and Hu [4], vol. 1 preface p. 25.

80     A. Bréard

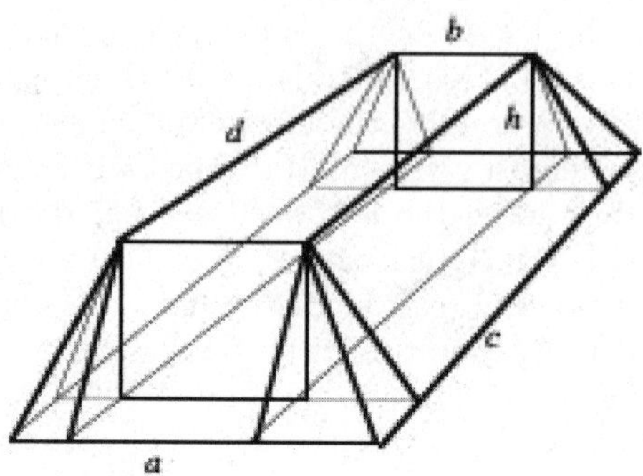

**Fig. 1.** *Chutong*

when we look at the structure of the algorithm given in the above jotting. My argument that the structure of the algorithm here is theoretically significant is supported by the fact, that the algorithm given in its general form – transcribed here as in equation ( 1 ) – does not correspond to the actual calculations for the numerical example performed by Shen Gua. This brings the algorithmic procedure ( 1 ) into a different, more theoretically oriented, light. In his calculations as resumed in ( 2 ) he aims rather at efficiency by dividing only once:

$$A = \Big( [(2 \cdot 2 + 12) \cdot 2 + (2 \cdot 12 + 2) \cdot 12] \cdot 11 + [12 - 2] \cdot 11 \Big) \div 6 \qquad (2)$$

The sequence of examples in the first section of Shen Gua's text suggests that he might have thought of a decomposition of the pile of wine kegs into several blocks of different geometric shapes. Shen lists three procedures: one for the volume of the cube and two for a transversal cut of the *qiandu*, the former giving its surface, the latter calculating its oblique side. But the cube and the *qiandu* are precisely those geometric shapes for which the algorithmic procedure to calculate the continuous volume still holds in the case where each unit of volume is represented by a discrete unit element. The volume becomes then equal to the total number of unit elements piled up in the shape of a cube or a *qiandu*.

A decomposition of the continuous *chutong* has been made explicit earlier in a 263 AC commentary by Liu Hui to the Chinese canonical *Nine Chapters of Mathematical Procedures* (*Jiu zhang suan shu* 九章算術)[13]: a central parallelepiped, four blocks of half cuboids on the sides (*qiandu* 塹堵), and four blocks of pyramids with a square base in the corners; named *yangma* 陽馬 in traditional Chinese mathematics.

As becomes clear from the numerical example given at the end of Shen Gua's jotting, he has imagined an accumulation of discrete objects where breadth and length diminishes by one element from one layer to the next above: there are 11

---

[13] Translated and critically edited in Chemla and Guo [3].

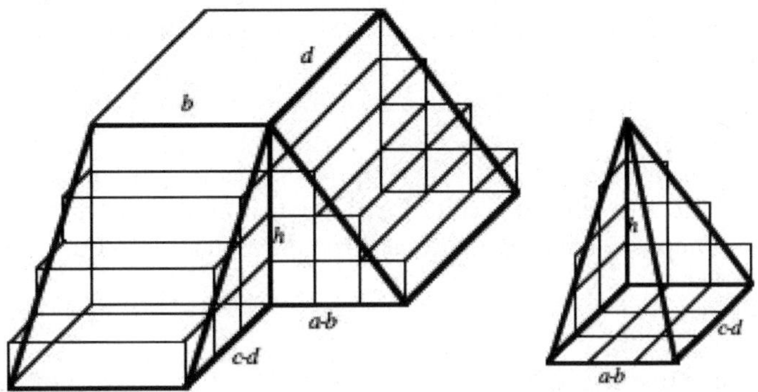



**Fig. 2.** Decomposition of a *chutong* in the discrete case

layers, with 2 kegs of wine in the upper layer and 12 at the bottom. Figure 2 shows how one can conceive of an accumulation equivalent to that of a *chutong* by pushing the elements into one corner.[14] In that case, only for the blocks in the corners, the *yangma*, a corrective term is necessary if one wants to use the procedures of the continuous volumes for the discrete case. As mentioned above, for the cube and the *qiandu* the procedures for the continuous case remain valid.

It is unclear how Shen Gua derived the corrective term for the discretized *yangma* in the corners, but judging from slightly later sources the general pro-

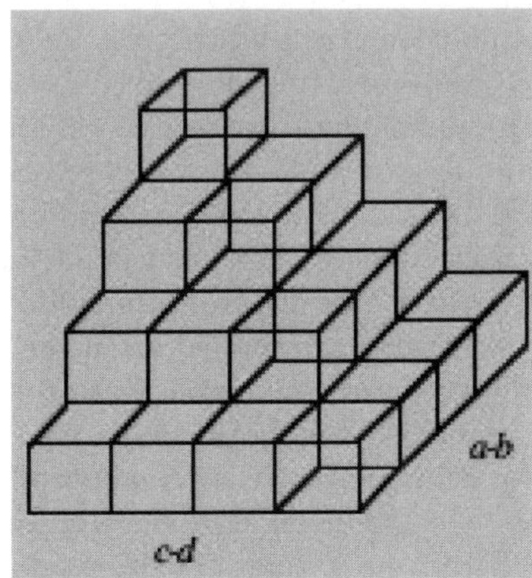



**Fig. 3.** Sum of squares represented with discrete elements

---

[14] I have shown elsewhere that structural elements of such considerations started to play a role in the development of algorithms for the calculation of finite arithmetic series as early as in the *Nine Chapters*. See Bréard [2], pp. 85-6.

cedure for the sum of squares might well already have been known by then.
It is algorithmically described in Yang Hui's *Detailed Explanations to the Nine
Chapters on Mathematical Methods* (*Xiang jie jiu zhang suan fa* 詳解九章算法,
1261)[15] as well as in Zhu Shijie's *Jade Mirror of Four Unknowns* (*Si yuan yu
jian* 四元玉鑑, 1303).

The corrective term in ( 1 ) then appears when one compares the sum of
squares of integers to the procedure for the continuous *yangma* given as early as
in the *Nine Chapters of Mathematical Procedures* from Han dynasty (206 BC –
220 AD):

$$\text{Number of wine kegs in a corner} = 1 + 2^2 + 3^2 + \ldots + \underbrace{(a-b)}_{=(h-1)} \cdot \underbrace{(c-d)}_{=(h-1)}$$

$$\sum_{k=1}^{h-1} k^2 = (h-1) \cdot h \cdot (h-1+\frac{1}{2}) \div 3$$

$$\underbrace{(h-1)}_{=(a-b)} \cdot h \cdot (\underbrace{h-1}_{=(c-d)} + \frac{1}{2}) \div 3 = \underbrace{(a-b) \cdot (c-d) \cdot h \div 3}_{yangma} + \underbrace{(a-b) \cdot h \div 6}_{\text{corrective term}}$$

## 3   Conclusion

The algorithmic description of the procedure of a finite accumulation of discrete
elements in the shape of a known geometric solid analyzed here shows well how
procedures in traditional mathematical writings in China reveal the underlying
rationale. In Shen Gua's case, operating on two different lines on the counting
board corresponds to two operational blocks: on one line the procedure for the
continuous volume is laid out, on another line a corrective term for the discrete
case is calculated. This partition reflects the historical lineage of the treatment of
finite arithmetical series back to algorithms for calculating continuous geometric
solids. As early as in the *Nine Chapters on Mathematical Procedures*, one finds
mathematical problems concerning grain piled up in the form of a cone either on
the ground, against a wall or leaning in a corner. Their volume then is calculated
without correction of the algorithm for the circular cone, but gradually a separate
field of inquiry concerning related algorithms for discrete decompositions and -
mathematically speaking - the summation of the sum of finite arithmetic series
emerges. Shen Gua's algorithm in jotting 301 stands at the watershed between
the pure geometrical algorithms for solid volumes and the algebraic treatment
of sums of finite arithmetic series in Zhu Shijie's *Jade Mirror of Four Unknowns*
(*Si yuan yu jian* 四元玉鑒) in 1303.

---

[15] See Bréard [2], chapter 3.2.

# References


[1] Bréard, A.: Shen Gua's Cuts. Taiwanese Journal for Philosophy and History of Science 10, 141–162 (1998)
[2] Bréard, A.: Re-Kreation eines mathematischen Konzeptes im chinesischen Diskurs: Reihen vom 1. bis zum 19. Jahrhundert. Steiner, Stuttgart (1999)
[3] Chemla, K., Guo, S.: Les neuf chapitres sur les procédures mathématiques. Dunod, Paris (2004)
[4] Shen, G., Hu, D. (eds.): Mengxi bitan jiaozheng (in Chinese) (Critical Edition of Brush Talks from the Dream Pool), 2 vols. Shanghai guji chubanshe, Shanghai (1987)

# Sequential Automatic Algebras

Michael Brough, Bakhadyr Khoussainov, and Peter Nelson

Department of Computer Science, University of Auckland

**Abstract.** A *sequential automatic algebra* is a structure of the type $(A; f_1, \ldots, f_n)$, where $A$ is recognised by a finite automaton, and functions $f_1, \ldots, f_n$ are total operations on $A$ that are computed by input-output automata. Our input-output automata are variations of Mealy automata. We study some of the fundamental properties of these algebras and provide many examples. We give classification results for certain classes of groups, Boolean algebras, and linear orders. We also introduce different classes of sequential automatic algebras and give separating examples. We investigate linear orders considered as sequential automatic algebras. Finally, we outline some of the basic properties of sequential automatic unary algebras.

## Introduction

The main contribution of this paper is the introduction of the various notions of sequential automatic algebras as natural sub-classes of the class of automatic structures. As such these algebras enjoy all the decidability and model-theoretic properties possessed by automatic structures. The goal is twofold. One is to provide examples and investigate fundamental properties of sequential automatic algebras. The other is to show that sequential automatic algebras have certain algebraic and algorithmic advantages as opposed to general automatic structures.

**Algebras** are structures of the form $(A; f_1, \ldots, f_n)$ where each $f_i$ is a total operation on $A$. Usually operations are replaced by their graphs (in finite model theory for example). This transforms the algebra into a purely relational structure. Automata can then be used to recognize the graphs of the operations; this loses the input-output behavior of the functions because an automaton recognising the graph of an operation does not necessarily compute the output in a sequential manner. Here we propose to use input-output automata to represent operations of algebras in order to capture the input-output behavior of operations. Our input-output automata will be variations of Mealy automata.

Let $\Sigma$ be an alphabet, and let $\Sigma_\Box = \Sigma \cup \{\Box\}$, where $\Box \notin \Sigma$. An $n$-**variable sequential Mealy automaton** $\mathcal{M}$ is a tuple $(Q, q_0, \Delta, O)$, where $Q$ is the finite set of states, $q_0 \in Q$ the initial state, $\Delta : Q \times \Sigma_\Box^n \to Q \times \Sigma_\Box$ is the transition function, and $O : Q \to \Sigma^\star$ the final output function. Note that the automaton is *deterministic*. Such an automaton processes inputs of the form $(w_1, \ldots, w_n)$, where each $w_i \in \Sigma^\star$, and outputs a string from $\Sigma^\star$ as follows. Think of the automaton as having $n$ input tapes, with $w_i$ on the $i$th tape, and one output tape on which it writes symbols from $\Sigma_\Box$. The automaton moves each of its $n$ heads

A. Beckmann, C. Dimitracopoulos, and B. Löwe (Eds.): CiE 2008, LNCS 5028, pp. 84–93, 2008.

simultaneously from left to right, reading a symbol from each of the input tapes, changes its state and writes a symbol on the output tape according to $\Delta$ (starting at $q_0$). If $w_i$ is shorter than $w_j$, then the automaton assumes the $\square$ symbol after the end of $w_i$. Once all input symbols are $\square$, the automaton concatenates the string $O(s)$, where $s$ is the current state, to the end of the string written to the output tape and then halts. The resulting string in the output tape, up to the first position where $\square$ is written, is the output of $\mathcal{M}$. We require that once $\square$ has been written to the output, all subsequent symbols written are $\square$ and the final output function is empty. Thus writing $\square$ can be thought of as terminating early. Each sequential $n$-variable Mealy automaton uniquely determines a function $f_{\mathcal{M}} : (\Sigma^\star)^n \to \Sigma^*$ called a **sequential automatic operation**.

**Definition 1.** *An algebra $\mathcal{A}=(A; f_1, \ldots, f_n)$ is* **sequential automatic** *if the domain $A$ is a regular language and operations $f_i$ on $A$ are sequential automatic.*

Operations computed by Mealy automata have been studied for many years. Mealy automata form a subclass of transducers, an area of active research in automata theory [3], and have also been studied by group theorists (see [9] for instance), initiated by Aleshin who used permutations computed by Mealy automata to solve the Burnside problem [1].

Some simple examples of sequential automatic algebras are finite algebras, the tree algebra $(\{0,1\}^\star; Left, Right)$ where $Left(x) = x0$ and $Right(x) = x1$, $(\omega; +)$, and $(\omega; S)$ where $S(n) + 1$ for $n \in \omega$.

If a sequential $n$-variable Mealy automaton $\mathcal{M}$ never writes the symbol $\square$, we call $\mathcal{M}$ a **strictly sequential $n$-variable Mealy automaton** and define a **strictly sequential automatic operation** and a **strictly sequential automatic algebra** accordingly. Strictly sequential automatic algebras form a subclass of sequential automatic algebras.

For the next definition we briefly explain finite automaton recognisable relations. An automaton $\mathcal{M}$ recognising a relation $R$ of arity $n$ behaves exactly as an $n$-variable Mealy automaton but with no outputs; instead $\mathcal{M}$ has a set $Q_f \subset Q$ of accepting states. $\mathcal{M}$ processes a tuple $(w_1, \ldots, w_n)$ in the same way as Mealy automata do, and accepts the tuple iff after processing the tuple, it is in one of the accepting states. Now we define automatic structures. These have been studied in [4], [5], [10], [12], [17].

**Definition 2.** *A relational structure $\mathcal{A}=(A; R_1, \ldots, R_m)$ is* **automatic** *if $A$, $R_1$, ..., $R_m$ are all finite automaton recognisable.*

We also use sequential automatic functions as mappings between equivalence classes. Let $f$ be a function computed by an $n$-variable sequential Mealy automaton. Let $E$ be an equivalence relation on $A$. We say that $f_{\mathcal{M}}$ **respects** $E$ if for all $(w_1, \ldots, w_n)$, $(w_1', \ldots, w_n')$ the condition $(w_1, w_1')$, ..., $(w_n, w_n') \in E$ implies that $(f_{\mathcal{M}}(w_1, \ldots, w_n), f_{\mathcal{M}}(w_1', \ldots, w_n')) \in E$. If every operation of an algebra respects $E$ then $E$ is called a **congruence** relation of the algebra.

**Definition 3.** *Let $\mathcal{A}=(A; f_1,\ldots,f_n)$ be a sequential automatic algebra. Let $E$ be a finite automaton recognisable congruence relation of $\mathcal{A}$. The quotient algebra $\mathcal{A}/E$ is called a* **generalised sequential automatic** *algebra. If $E$ additionally satisfies the property that for $x = x'\sigma^k$, where $x'$ does not end in $\sigma$, the equivalence class of $x$ is $\{x'\sigma^n | n \in \omega\} \cap A$, then we call the factor algebra $\mathcal{A}/E$* **continuous generalised sequential automatic***.*

Let $\mathcal{SSA}$, $\mathcal{SA}$, $\mathcal{CGSA}$ and $\mathcal{GSA}$ denote respectively the classes of strictly sequential automatic, sequential automatic, continuous generalised sequential automatic, and generalised sequential automatic algebras (all closed under isomorphisms). We have: $\mathcal{SSA} \subseteq \mathcal{SA} \subseteq \mathcal{CGSA} \subseteq \mathcal{GSA}$. We will provide separating examples for these later. This is unlike the case of automatic structures, where quotients by automatic congruence relations give no new structures.

The group $(\mathbb{Z}; +)$ is an example of a continuous generalised sequential automatic algebra. Here we represent numbers in base $-2$, allowing us to add integers without knowing their signs beforehand.

If $\mathcal{A}$ is (strictly, continuous generalised, generalised) sequential automatic and $\mathcal{B}$ is isomorphic to $\mathcal{A}$ then we call $\mathcal{A}$ a **(strictly, continuous generalised, generalised) sequential automatic presentation** of $\mathcal{B}$. Automatic presentations are defined similarly. Often we abuse our definitions and refer to algebras that have sequential automatic presentations as sequential automatic algebras, or structures with automatic presentations as automatic structures. When we describe an algebra as being automatic, it is to be taken as implicit that we are considering it as a relational structure.

A brief outline of the paper is as follows. Section 1 describes basic properties of the four classes of sequential automatic algebras and provides examples to separate them. The section also provides a classification theorem for generalised sequential algebras in the cases of finitely generated groups, Boolean algebras, and ordinals. Section 2 proves that if linearly ordered sets are defined as sequential automatic algebras then the order must be obtained from the lexicographic order on strings. This implies that the monadic second order theory of each sequential automatic linear order algebra is decidable. This also implies, from the result of Kuske [13], there exists an automatic linear order not isomorphic to a sequential automatic linear order algebra. Section 3 studies sequential automatic unary algebras and proves that the reachability problem for such algebras is decidable. This contrasts with automatic unary algebras, where the reachability problem is undecidable [17], [4]. An example is given of a (unary) permutation algebra that is automatic as a relational structure, but has no presentation as a sequential automatic algebra.

Finally, we stress that the goal is to show that sequential automatic algebras are more tame structures than their automatic counterparts. The ultimate goal in the study of sequential automatic algebras is to investigate whether or not natural problems asked about sequential automatic algebras, e.g. the isomorphism and the elementary equivalence problems, are decidable. This is the first paper devoted to this study.

# 1  General Properties, Separating Examples and Classification Results

The first part of following proposition implies that all decidability properties enjoyed by the class of automatic structures are present for automatic sequential algebras (see [10]). The second part states that sequential automatic algebras are closed under finite Cartesian products; the proof is straightforward:

**Proposition 1.** (1) *Every generalised sequential automatic algebra is automatic. In particular, there is an algorithm which, given a generalised sequential automatic algebra $A$, a first order formula $\phi(\bar{x})$, and a tuple $\bar{a} \in A$, decides if $A \models \phi(\bar{a})$. (2) If $A_1, \ldots, A_k$ are (continuous generalised, generalised or strictly) sequential automatic then so is $A_1 \times \ldots \times A_k$.*      $\square$

Now we separate $SSA$ from $SA$, $SA$ from $CGSA$, and $CGSA$ from $GSA$. The proof of the following proposition is easy:

**Proposition 2.** *If $A = (A; f_1, \ldots, f_n)$ is strictly sequential automatic then for each $a \in A$ and $f_i$ the set $f^{-1}(a) = \{\bar{x} \mid f_i(\bar{x}) = a\}$ is finite. Hence, algebras from the following classes are strictly sequential automatic iff they are finite: groups, rings, Boolean algebras, lattices with complements, vector spaces.*      $\square$

**Corollary 1.** $SSA$ *is a proper subset of* $SA$.

*Proof.* The algebra $(\omega; f)$, where $f(x) = 0$ for all $x \in \omega$, is a separator.      $\square$

**Proposition 3.** *A group is sequential automatic iff it is finite. Hence $SA$ is a proper subset of $CGSA$.*

*Proof.* Let $n$ be the length of the string for the unit element 1. When the automaton for multiplication reads a pair $(x, x^{-1})$ of length $> n$, it halts and outputs 1 based on prefixes of length $n$. If the group were infinite, there would be infinitely many such pairs and finitely many such prefixes, and so distinct $(x, x^{-1})$ and $(y, y^{-1})$ exist which share the same prefixes. Then $(x, y^{-1})$ therefore would output 1. A separating witness is the group $(\mathbb{Z}; +)$.      $\square$

**Definition 4.** *An algebra $A$ is **residually finite** if for all distinct $x, y \in A$ there is a homomorphism $\phi : A \to F$ onto a finite algebra $F$ such that $\phi(x) \neq \phi(y)$.*

**Proposition 4.** *All algebras in the class $CGSA$ are residually finite.*

*Proof.* We first show this for sequential automatic algebras $A = (A; f_1, \ldots, f_n)$. For $n \in \omega$, define $\phi_n$ to be the function such that for a string $w = \sigma_0 \ldots \sigma_k$, if $k \leq n$ then $\phi_n(w) = w$, otherwise $\phi_n(w) = \sigma_0 \ldots \sigma_n$. Define $\mathcal{F}_n = (F; f_1', \ldots, f_n')$, where $F$ is the image of $A$ under $\phi_n$ and $f_i' = \phi_n \circ f_i \circ \phi_n^{-1}$; $\phi_n$ is a homomorphism from $A$ to $\mathcal{F}_n$. Given two elements $x, y$ of $A$, let $n = \max(|x|, |y|)$ be the length of the longest string. Then $\phi_n : A \to \mathcal{F}_n$ is such that $\phi_n(x) \neq \phi_n(y)$. For a continuous generalised sequential algebra $A/E$, we construct $\mathcal{F}$ from $A$ as above, and then take $\mathcal{F}/E'$ where $E'$ is such that $x \cong_{E'} x'$ iff there exist $y, y' \in A$ such that $y \cong_E y'$ and $\phi(y) = x$, $\phi(y') = x'$.      $\square$

**Corollary 2.** *$\mathcal{CGSA}$ is a proper subclass of $\mathcal{GSA}$.*

*Proof.* The algebra $\mathcal{A} = (\omega; f)$, with $f(0) = 0$ and $f(n) = n-1$ is generalised sequential automatic but not residually finite and hence not continuous generalised sequential automatic.

This additive group of rational numbers is not residually finite. Hence:

**Corollary 3.** *The group $(\mathbb{Q}; +)$ does not belong to $\mathcal{CGSA}$.* $\square$

Now we give a classification result for the classes of finitely generated groups, Boolean algebras, and ordinals as continuous generalised sequential automatic algebras. These have been classified as automatic structures (see [12], [8]). We recall some definitions. For groups, a finitely generated group is **virtually abelian** if it contains an abelian subgroup of finite index. A finitely generated group is an automatic structure iff it is virtually abelian [14]. For Boolean algebras, $B_\omega$ denotes the Boolean algebra of finite and co-finite subsets of $\omega$. An infinite Boolean algebra is an automatic structure iff it is isomorphic to a finite Cartesian product of $B_\omega$ [12]. For ordinals, an ordinal is automatic iff it is strictly less than $\omega^\omega$ [17]. In order to treat ordinals as algebras we introduce the concept of a linear order algebra. Let $\mathcal{L} = (L; \leq)$ be a linearly ordered (lo) set. Define the function $f : L^2 \to L$ as $f(x, y) = \min(x, y)$. This function has the following properties for all $x, y, z \in L$: **1)** $f(x, y) = f(y, x)$; **2)** $f(x, y) = x$ or $f(x, y) = y$; **3)** if $f(x, y) = x$ and $f(y, z) = y$ then $f(x, z) = x$. Given any function $f : A \to A$ satisfying these conditions, we define a lo set by $x \leq_f y$ iff $f(x, y) = x$. We call algebras $(A; f)$, where $f$ satisfies the above conditions, **linear order (lo) algebras**. These transformations between lo sets and lo algebras preserve the isomorphism type, and a lo set is automatic iff the corresponding lo algebra is automatic (as relational structures).

**Definition 5.** *Let $\mathcal{L}_1, \mathcal{L}_2$ be two linear order algebras. The algebra $\mathcal{L}_1 +_\leq \mathcal{L}_2$ is over $L_1 \cup L_2$ such that if $x \in L_1$ and $y \in L_2$ then $\min(x, y) = \min(y, x) = x$, and otherwise $\min(x, y)$ follows from $L_1$ or $L_2$. The algebra $\mathcal{L}_1 \times_\leq \mathcal{L}_2$ is over the Cartesian product of $L_1$ and $L_2$ such that $\min((x_1, y_1), (x_2, y_2))$ is $(x_1, y_1)$ iff $\min(x_1, x_2) = x_1$, or $x_1 = x_2$ and $\min(y_1, y_2) = y_1$.*

**Proposition 5.** *Given two sequential automatic linear order algebras $\mathcal{L}_1, \mathcal{L}_2$, the algebras $\mathcal{L}_1 +_\leq \mathcal{L}_2$ and $\mathcal{L}_1 \times_\leq \mathcal{L}_2$ are sequential automatic. In particular, all ordinals less than $\omega^\omega$ are sequential automatic linear order algebras.* $\square$

**Theorem 1.**  *1. A finitely generated group is continuous generalised sequential automatic iff it is virtually abelian.*
   *2. An infinite Boolean algebra is continuous generalised sequential automatic iff it is isomorphic to a finite Cartesian product of $B_\omega$ to itself.*
   *3. An ordinal (as an lo algebra) is sequential automatic iff it is strictly below $\omega^\omega$.*

*Proof.* For part 1), the proof essentially follows [8]. Let $G$ be a finitely generated virtually abelian group. We can assume that that our group $G$ has a finitely generated free abelian normal subgroup $N$ of finite index. Then any element $g \in G$ can be given in the form $g = fn$ where $f \in F = G/N, n \in N$. Given two elements $g_1, g_2 \in G$, then $g_1 g_2 = f_1 n_1 f_2 n_2 = f_1 f_2 f_2^{-1} n_1 f_2 n_2$. Since $N$ is normal, $\phi_f(n) = f^{-1} n f$ gives an automorphism of $N$ for any $f$. Therefore when we multiply $f_1 n_1$ by $f_2 n_2$ we get $(f_1 f_2)n$ where $n = \phi_{f_2}(n_1) n_2 \in N$.

We describe in general terms the sequential automatic encoding of $G$. We represent group elements as a pair $(f, n)$ where $f \in F$ and $n \in N$. Because $F$ is finite and the $F$ part of the product depends only on the $F$ part of the two inputs, we can encode $f$ in the first digit. The rest of the string encodes $n$. When computing the product $(f_1, n_1)(f_2, n_2)$, the first state outputs $f_1 f_2$ and then branches to one of finitely many subautomata depending on $f_2$.

We need to show that a sequential automaton can compute $n_1 \phi_{f_2}(n_2)$. The subgroup $N$ is, by assumption, isomorphic to $\mathbb{Z}^r$ for some $r \in \omega$. Using the encoding of $\mathbb{Z}$ and Proposition 1, we can encode elements of $N$ and compute their sums. All automorphisms of $N \cong \mathbb{Z}^r$ can be determined from linear extension of their action on a minimal generating set, and therefore correspond to matrix multiplications. That means we can associate with each $f \in F$ an integer matrix $M_f$ which computes the automorphism $\phi_f$ on the vector representation of $N$. Multiplication by $M_f$ can be computed by a sequential automaton.

To compute $n_1 \phi_f(n_2)$, we combine the addition automaton for $N$ and the automaton for $M_f$, so that the first input string is given straight to the $N$ automaton, and the second input string is processed by the $M_f$ automaton, the output of which is given as the second input of the $N$ automaton. Now one uses the construction described to give a sequential presentation of $G$.

For part 2), by Proposition 1 we need to consider $B_\omega$. We encode subsets of $\omega$ in binary. The final character of a binary word is interpreted as being recurring (so when $\square$ is read following a $0(1)$ it is treated as a $0(1)$). Our language can code any finite or co-finite set. Given this encoding, the algebra $B_\omega$ is a continuous generalised sequential algebra.

The last part of the theorem follows from Proposition 5.                    $\square$

## 2   Linear Order Algebras

Here we investigate sequential automatic linear order (lo) algebras. The lexicographical order will play an essential role in describing these algebras.

**Example 1.** *Let $L$ be a regular language. Consider $\preceq_L$, the lexicographic order restricted to $L$. The algebra $(L; \min_\prec)$ is sequential automatic.*                    $\square$

As a corollary, one obtains that the dense order (the order of rational numbers) is a sequential automatic lo algebra; this is isomorphic to the $\preceq$ order on the set $\{w101 \in \{0,1\}^\star \mid w$ does not have $101$ as a substring$\}$.

**Example 2.** *An infinite sequential automatic lo algebra over a unary alphabet is isomorphic to $\omega + n$, $n + \omega^\star$, or $\omega + \omega^\star$, where $\omega^\star$ is $\omega$ reversed.*                    $\square$

**Theorem 2.** *A linear order algebra is sequential automatic iff it is isomorphic to the lexicographic order on a regular language.*

*Proof.* One direction is explained in Example 1. The idea for the other direction is that we have to decide which string is the minimum at the first position where they differ, because then we have to output the next digit from either one or the other. This is like the lexicographic order. Given a sequential automatic lo algebra $\mathcal{L} = (L; f)$ we explicitly construct a regular language and an isomorphism. Let $\mathcal{M}$ denote the automaton for recognising $L$. For the isomorphism, we use a slightly generalised form of sequential automata: instead of computing a function from $\Sigma^*$ to itself, we have two alphabets $\Sigma$ and $\Sigma'$ and compute a function from $\Sigma^*$ to $(\Sigma')^*$. Let $\Sigma$ be the alphabet for $\mathcal{L}$, and $\Sigma' = \{0, \ldots, |\Sigma|\}$.

For distinct strings $w_1, w_2$, let $i$ be the first position where they differ. If one is a prefix of the other, then this position is immediately after the end of this string, when a blank symbol $\square$ is first read. Let $u = f(w_1, w_2)$ be the output string. For $j < i$, $w_1(j) = w_2(j) = u(j)$. At position $i$ there is a choice: $u(i)$ must equal either $w_1(i)$ or $w_2(i)$, determining the rest of the string. The ordering of $w_1$ and $w_2$ gives an ordering on the pair of symbols $\{w_1(i), w_2(i)\}$; any pair of inputs that reach the current state of the automaton and read these symbols next must obey this ordering.

If we look at all pairs of symbols that can be read at this state, we get an ordering on each pair. We would like to combine these orderings to give an ordering on all of these symbols, and extend that to give an ordering on $\Sigma \cup \{\square\}$. However, it is possible to get orderings on pairs which are not consistent (that is, they cannot be simultaneously true in a linear ordering, for example $a < b$, $b < c$, $c < a$). We deal with this by keeping track of the state of $\mathcal{M}$. Given a state $q$ of $\mathcal{M}$, let $\Sigma_q$ be the set of all symbols (including $\square$) which can be read from $q$ as part of a path to a final state of $\mathcal{M}$. If we have a reachable state $(q, r)$ in the product of $\mathcal{M}$ and the automaton for $f$, the orderings given by $r$ for pairs in $\Sigma_q$ must be consistent and so we can combine them to give an ordering on $\Sigma_q$ and then extend this to an ordering on $\Sigma \cup \{\square\}$.

We construct a sequential automaton $\mathcal{A}$ mapping $L \subset \Sigma^*$ into $(\Sigma')^*$. Let the states be the product of $\mathcal{M}$ and the automaton for $f$. When a symbol $\sigma$ is read, we treat the $f$ part as being given the input $(\sigma, \sigma)$ and the $\mathcal{M}$ part as being given $\sigma$, and output the element of $\{0, \ldots, n\}$ corresponding to the position of $\sigma$ in the ordering on $\Sigma \cup \{\square\}$ associated with the pair of states. The final output function for each state outputs the element of $\{0, \ldots, n\}$ corresponding to the position of $\square$ in this ordering.

From the construction of $\mathcal{A}$, it preserves the order of $\mathcal{L}$ when we take the lexicographic ordering on its image, and the function computed by $\mathcal{A}$ is injective. It remains to show that the image of $L$ under this function is a regular language over $\Sigma'$. Construct another automaton by taking $\mathcal{A}$ and swapping the inputs and outputs on the transitions. We add a single final state $q_0$, and from each state $p = (q, r)$, where $q$ is a final state in $\mathcal{M}$, we add a transition to $q_0$ when $O(p)$ (the final output function in $\mathcal{A}$) is read. This gives an automaton which recognises exactly $\mathcal{A}(L)$. Therefore $\mathcal{A}$ is an isomorphism from $L$ to a regular language with the lexicographic order. $\square$

**Corollary 4.** *The monadic second order theory of any sequential linear order algebra is decidable.*

*Proof.* From the theorem above, the result follows from the fact that the lexicographical order is MSO definable in the binary tree [16]. $\square$

**Corollary 5.** *There exists an automatic linear order $(L; \leq)$ which is not isomorphic to any sequential automatic linear order algebra.*

*Proof.* In [13] Kuske constructed an automatic linear order that is not isomorphic to the lexicographical order on any regular set. $\square$

The third application is this. A *word* is $(\omega; \leq, P)$, where $P$ is a unary relation. A word is automatic if it is isomorphic to an automatic structure. The word $(\omega; \leq, P)$ can be transformed into the following algebra $(\omega; \min_{\leq}, f_P)$, where $f_P(x)$ is the maximum $y$ such that $y \leq x$ and $P(y)$. Call this a word algebra. Two words are isomorphic iff their corresponding word algebras are isomorphic.

**Corollary 6.** *If $(\omega; \min_{\leq}, f_P)$ is a sequential automatic word algebra then $(\omega; \leq, P)$ is a morphic word. Hence its monadic second order theory is decidable.*

*Proof.* The order can be replaced with the length-lexicographic order in which $P$ is still regular. From results in [2], it follows that $P$ is a characteristic of a morphic word. The rest follows from Thomas and Carton in [8]. $\square$

## 3   Sequential Automatic Unary Algebras

Here we consider algebras of the form $\mathcal{A} = (A; f_1, \ldots, f_n)$, where each $f_i$ is a unary function on $A$. These algebras are called **unary algebras**. A unary algebra can be transformed into a graph $(A, E)$ where $E(x, y)$ iff $f_1(x) = y \vee \ldots \vee f_n(x) = y$. The **reachability problem** for the unary algebra $\mathcal{A}$ is the set $\{(x, y) \mid$ in the graph $(A, E)$ there is a path from $x$ to $y\}$. For unary automatic algebras even with one unary operation the reachability problem is $\Sigma_1^0$-complete [17] [4]. In contrast, we have:

**Theorem 3.** *If $\mathcal{A} = (A; f_1, \ldots, f_n)$ is a sequential automatic unary algebra then the reachability problem for $\mathcal{A}$ is decidable in PSPACE.*

*Proof.* Suppose we want to check if $(x, y)$ is in the reachability relation, and that $y$ has length $n$. If $f(z)$ has length $|f(z)| \leq n$, then either $|z| \leq n$ or $f$ terminates early, after at most $n$ digits are read. In the second case, any string with the same first $n$ digits as $z$ would give the same output. Since all our functions $f_i$ are sequential, we need only consider the first $n+1$ digits when determining if $y$ is reachable. We need one extra digit to make sure we don't mistake a string of which $y$ is a prefix for $y$ itself. The algorithm is a straightforward search: begin at $x$, for each $f_i$ branch to search on the first $n+1$ digits of $f_i(x)$, terminate a branch when a string is reached that has already been visited, terminate the algorithm either positively when $y$ is reached, or negatively when all branches terminate. Since we only consider the first $n+1$ digits, the search space is finite, so the algorithm will always terminate. $\qquad\square$

**Proposition 6.** *There exists a sequential automatic unary algebra $\mathcal{A}$ in which the reachability relation is not recognised by pushdown automata.*

*Proof.* Consider the unary algebra $(A; f)$, where $A = 0^*1^*2^*$ and $f(0^i1^j2^k) = 0^{i+1}1^{j+1}2^{k+1}$. Let $x$ be the string 012. The set of elements reachable from 012 is $\{0^n1^n2^n \mid n \in \omega\}$, which is not recognised by pushdown automata. $\qquad\square$

An **algebra of permutations** is a unary algebra $(A; f_1, \dots, f_n)$ where each $f_i : A \to A$ is a bijection.

**Proposition 7.** *There is a PSPACE algorithm that, given a strictly sequential automatic unary algebra $\mathcal{A} = (A; f_1, \dots, f_n)$, where $A = \Sigma^\star$, decides if $\mathcal{A}$ is an algebra of permutations.*

*Proof.* Take a Mealy automaton $\mathcal{M}$ computing one of the operations of the algebra. Assume that all states are reachable. A strictly sequential Mealy automaton gives a permutation on $\Sigma^\star$ iff each state gives a permutation of its input and all of the final output strings are empty.

For each state $s \in \mathcal{M}$, we check whether the final output string $O(s)$ is empty, and whether for each pair of symbols $(\sigma_1, \sigma_2) \in \Sigma^2$, $\sigma_1 \neq \sigma_2$ the output when $\sigma_1$ is read in state $s$ is distinct from when $\sigma_2$ is read. The space required is a counter to run through the states which is proportional to $\log(|\mathcal{M}|)$. $\qquad\square$

**Theorem 4.** *(see also [9]): The length of a cycle on strings of length $n$ of a strictly sequential automatic permutation over $\Sigma$ is of the form $\prod_{i=1}^{n} a_i$ where $1 \leq a_i \leq |\Sigma|$.*

*Proof.* There are at most $|\Sigma|$ strings of length 1, and since these are permuted amongst themselves, cycles on strings of length 1 have lengths $1 \leq L \leq |\Sigma|$.

We proceed by induction. Given a cycle of length $L_n$ on strings of length $n$, we consider these strings truncated to the first $n-1$ digits. These strings, by the induction hypothesis, are in a cycle of length $L_{n-1} = \prod_{i=1}^{n-1} a_i$ where $1 \leq a_i \leq |\Sigma|$. Truncating to the first $n-1$ digits gives a homomorphism, so $L_{n-1}$ must divide $L_n$. Since $L_n/L_{n-1}$ cannot exceed $|\Sigma|$, the result follows. $\qquad\square$

**Corollary 7.** *There are algebras of permutations which are automatic but not strictly sequential automatic.*

*Proof.* Let $p : \omega \to \omega$ be any nonzero polynomial with positive coefficients. Let $L$ be a regular language over $\Sigma$ such that $growth_L(n) = |L \cap \Sigma^n| = p(n)$ for all $n$. Such languages exist (see [15] or [11]). Define the function $f : L \to L$ as follows: If $x \in L$ is not the largest length-lexicographically among all $y \in L$ with

$|x| = |y|$, then $f(x)$ is the length-lexicographically least element $z \in L$ greater than $x$ such that $|x| = |z|$; if $x \in L$ is the length-lexicographically greatest among all $y \in L$ with $|x| = |y|$, then $f(x)$ is the length-lexicographically least element $z \in L$ such that $|z| = |x|$. The structure $(L, f)$ is an automatic relational structure. The range of $p$ coincides with the set of all lengths of cycles of the permutation $f$. By the theorem above, $(L, f)$ cannot be isomorphic to a strictly sequential automatic permutation.                                          $\square$

It is an open question if there is an algorithm that, given two sequential automatic permutations, decides whether the permutations are isomorphic.

# References


1. Aleshin, M.: Finite automata and Burnside's problem for periodic groups. Mat. Notes 11, 199–203 (1972)
2. Bárány, V.: A Hierarchy of Automatic Words having a Decidable MSO Theory. In: Caucal, D. (ed.) Online Proceedings of the 11th Journées Montoises, Rennes (2006)
3. Béal, M., Carton, O., Prieur, C., Sakarovitch, J.: Squaring transducers: an efficient procedure for deciding functionality and sequentiality. Theor. Comput. Sci. 292(1), 45–63 (2003)
4. Blumensath, A.: Automatic Structures. Diploma Thesis, RWTH Aachen (1999)
5. Blumensath, A., Grädel, E.: Automatic Structures. In: 15th Annual IEEE Symposium on Logic in Computer Science, Santa Barbara, pp. 51–62 (2000)
6. Blumensath, A., Grädel, E.: Finite presentations of infinite structures: automata and interpretations. Theory of Computing Systems 37(6), 641–674 (2004)
7. Cannon, J., Epstein, D., Holt, D., Levy, S., Paterson, M., Thurston, W.: Word processing in groups. Jones and Bartlett (1992)
8. Carton, O., Thomas, W.: The monadic theory of morphic infinite words and generalizations. Information and Computation 176, 51–76 (2002)
9. Grigorchuk, R., Nekrashevich, V., Sushanski, V.: Automata, Dynamical systems, and groups. Tr. Mat. Inst. Steklova 231(Din. Sist. Avtom i Beskon. Gruppy), 134–214 (2000)
10. Khoussainov, B., Nerode, A.: Automatic Presentations of Structures. LNCS, vol. 960, pp. 367–392. Springer, Heidelberg (1995)
11. Khoussainov, B., Rubin, S.: Automatic Structures: Overview and Future Directions. Journal of Automata, Languages and Combinatorics 8, 287–301 (2003)
12. Khoussainov, B., Nies, A., Rubin, S., Stephan, F.: Automatic Structures: Richness and Limitations. In: LICS, pp. 44–53. IEEE Computer Society, Los Alamitos (2004)
13. Kuske, D.: Is Cantor's theorem automatic?. LNCS, vol. 2850, pp. 332–343. Springer, Heidelberg (2003)
14. Oliver, G., Thomas, R.: Automatic Presentations for Finitely Generated Groups. In: Diekert, V., Durand, B. (eds.) STACS 2005. LNCS, vol. 3404, pp. 693–704. Springer, Heidelberg (2005)
15. Saloma, A., Soittola, M.: Automata-theoretic Aspects of Formal Power Series. Springer, Heidelberg (1978)
16. Rabin, M.: Decidability of second order theories and automata on infinite trees. Trans. AMS 141, 1–35 (1969)
17. Rubin, S.: Automatic Structures. PhD Thesis, University of Auckland (2004)

# The Role of Classical Computation in Measurement-Based Quantum Computation

Dan Browne and Janet Anders

Department of Physics and Astronomy, University College London, Gower Street,
London WC1E 6BT
d.browne@ucl.ac.uk

**Abstract.** Measurement-based quantum computation (MBQC) is an important model of quantum computation, which has delivered both new insights and practical advantages. In measurement-based quantum computation, more specifically the "one-way model" of computation, the computation proceeds in the following way. A large number of quantum bits (qubits) are prepared in a special entangled state called a cluster state. The qubits are then measured in a particular basis and particular order, the measurements are adaptive, some bases depending on the outcome of previous measurements. A classical control computer processes the measurement results in order to feed forward the bases for future measurements. In this paper we focus on the role played by the classical computer in this model and investigate some of its properties.

**Keywords:** Quantum computation, computational models, measurement-based quantum computation.

## 1   Introduction

There remain many open questions in the relationship between quantum and classical computation. In particular, it is still unclear what precisely the features of quantum physics are which enable its (apparent) additional computational power? The development of equivalent but different models of computation has been a powerful tool in the evolution of classical computer science, and similarly, employing different descriptions of quantum computations, both as formal descriptions and physical recipes for its implementation, is proving very valuable.

One family of such models is known as measurement-based quantum computation (MBQC). The non-classical aspects of measurement are certainly a striking aspects of quantum physics. In MBQC, measurements play the central role in the computational model, a stark contrast to the reversibility of the unitary logic gates in the standard *circuit model* of quantum computation. In this paper, we shall focus on a model of MBQC introduced by Raussendorf and Briegel [1] and known as the *one-way quantum computer* [3,2,4,5,6]. In the one-way model, the computation consists of the preparation of a special entangled state called a *cluster state* on many qubits, and the measurement of single qubits in a particular order and particular basis. The bases for later measurements depends upon the

A. Beckmann, C. Dimitracopoulos, and B. Löwe (Eds.): CiE 2008, LNCS 5028, pp. 94–99, 2008.

results of earlier measurements which must be processed by a suitable classical *control computer*. Given a cluster state of sufficient size any quantum computation can be implemented, and the resources for this (number of entangled qubits) scales linearly in the number of gates in the equivalent classical circuit.

It is natural to ask whether cluster states are unique in enabling MBQC or whether other types of entangled states can accomplish a similar task. In fact, recently a much broader class of states have been discovered which share this property [7,8,9]. In this paper, we take a different approach and focus instead upon the classical control computer. This is an essential component and represents the only active information processing in the model. We will see here that focussing upon this aspect of the model reveals some interesting properties and motivates some intriguing new questions.

## 2 One-Way Model of Quantum Computation

The most common description of quantum computation is known as the circuit model. In that model, the computation is described as a sequence of quantum logic gates, drawn from a *universal gate set*. Several universal gate sets have been proposed, a common example consists of the CNOT gate: $|i\rangle|j\rangle \rightarrow |i\rangle|i \oplus j\rangle$ and arbitrary single qubit unitary transformations. The circuit model is familiar from classical computation, where {NAND,FANOUT} and {NOT,TOFFOLI} form irreversible and reversible universal gate sets.

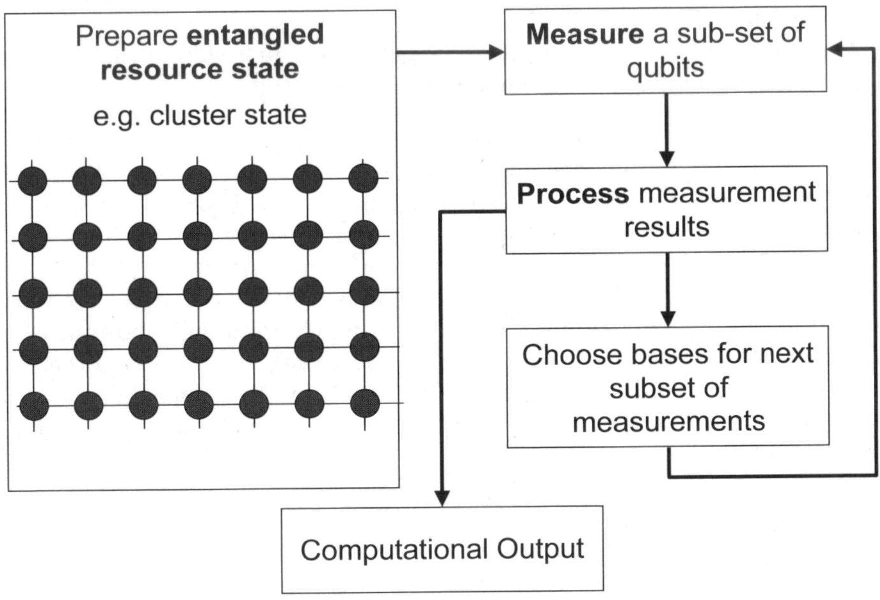


**Fig. 1.** A schematic outline of the one-way model of measurement-based quantum computation

The one-way model, on the other hand, is radically different from this circuit based approach, and has no natural classical analogue. In the one-way model, computation proceeds in the following steps (see figure 1):

1. A special multi-qubit entangled *resource state* is created on many qubits.
2. A subset of qubits are measured - these measurements a projective single qubit measurements in a particular set of bases.
3. The measurement outcomes are processed on a classical computer to determine the measurement bases for the next subset of qubits.
4. This next subset of qubits is measured.
5. The cycle of classical preprocessing on measurement outcomes and measurement is repeated until all qubits have been measured.
6. The final processing step returns the classical output of the computation.

The one-way model has a wealth of important features, including the following;

- Any quantum computation in the circuit model can be efficiently simulated in this model and vice versa - the models are thus of equal computational power.
- The measurement basis for each qubit is specified in advance up to a sign (i.e. one bit of information) - it is this sign which is dependent on prior measurements and is calculated by the control computer.
- The specification of the set of measurement-bases and their dependencies is an equivalent description of a quantum computation to the specification of a quantum circuit.
- The only entangling operations occur in the very first step of the computation, the construction of the entangled resource state.

A number of families of states have been identified which act as universal resource states for this model. In [1] the two-dimensional *cluster states* were shown to have this property. A cluster state is an entangled state on a square lattice of qubits. Each qubit is prepared in the superposition of basis states $(1/\sqrt{2})(|0\rangle+|1\rangle)$ and then entangling controlled-Z gates $(|x\rangle|y\rangle \rightarrow (-1)^{x \wedge y}|x\rangle|y\rangle)$ are applied between nearest neighbours in the lattice. Cluster states are represented graphically (as in figure 1) with graph nodes representing qubits and edges representing the entangling operations applied. Recently, a number of other universal families of resource states have been identified [7,10]. A striking feature of [10] is the significantly different way the randomness of measurement outcomes is accounted for.


## 3   CNOT Computers

**Definition.** A *CNOT computer* is a reversible classical computer, which implements circuits of logical gates drawn from the set {CNOT,NOT} on bits prepared in the 0 state.

The CNOT computer has a number of interesting properties.

- It is not a universal classical computer because CNOT and NOT do not form a universal set.
- Its computational power is described in terms of the complexity class Parity-L, often written $\oplus$L. $\oplus$L contains all problems which can be decided using a polynomial sized circuit on the CNOT computer.[11]
- In other words, $\oplus$L plays the role, for this model, played by the complexity class P for universal classical computation.
- The CNOT computer is believed to be less powerful than a universal classical computer since few of the problems in P have known efficient algorithms on it.
- The following problems, however, have been shown to lie in $\oplus$L, and have efficient CNOT constructions.
  - Calculating the parity of the bits in an $n$-bit string.
  - Inverting matrices over the Galois field GF(2) [12].
  - Simulating a restricted class of quantum computations known as the Clifford group [13].
- Any 2-bit reversible gate can be generated by CNOT and NOT, so it is reasonable to consider this the most powerful model of reversible computing in which three-bit gates are forbidden.

The main result of this paper is to show that for most variants of the one-way model, the classical control computer which calculates the measurement dependencies suffices to be a CNOT computer. We then use this observation to construct a new framework to describe one-way quantum computation and study its properties.

We shall not give a formal proof of this, as the technical aspects would go beyond the scope of this article, but instead will outline the reasoning which leads to this conclusion. The main idea is the following observation. When a cluster state is employed in the one-way model as described in [1], the functions which determine the sign of measurement bases can be reduced to calculating the *parity* of a subset of the previous measurement outcomes. As stated above, and as is simple to verify, the parity of bit strings can be computed efficiently on a CNOT computer.

A simple way to picture of the way the dependencies in measurements arise is due to Danos and Kashefi [14]. We label the outcomes of a single qubit measurement '0' and '1'. When a measurement is made on a qubit, which is part of an entangled state with other qubits, then the state of these qubits after the measurement depends upon the measurement outcome. Measurements in the one-way model are so chosen that the post-measurement states of other qubits are very simply related, as follows. Labelling these post-measurement state $|\psi_0\rangle$ and $|\psi_1\rangle$ respectively, according to whether '0' or '1' was measured,

$$|\psi_1\rangle = P|\psi_0\rangle$$

where P is a tensor product of Pauli operators ($X$, $Y$, or $Z$) and the identity, with support only upon qubits which have not yet been measured. This property is guaranteed for measurement-patterns on cluster states which satisfy a geometric criterion named "flow". The proof of this may be found in [14].

Measurement patterns in the one-way model are designed such that they will give the desired output on the final $n$ qubits to be measured if the preceding measurements all returned the '0' output. By applying the operator $P$ as a correction when '1' is obtained, one could ensure that the desired deterministic computation is obtained regardless of the measurement outcomes. In fact, the corrections never need be applied directly, they can simply be taken into account in the choice of measurement-basis for these later qubits.

When a qubit comes to be measured the cumulative correction operator arising from all previously made measurements must be applied. This cumulative operator will be a product of the correction operators from the previous measurements. Two important properties of Pauli operators are that they are self-inverse, $P^2 = \mathbb{1}$ and that pairs either commute or anti-commute. This means a product of single-qubit Pauli operators can always be reduced to the form $X^{p_x} Y^{p_y} Z^{p_z} = X^{p_x+p_y} Z^{p_y+p_z}$ (neglecting the physically irrelevant global phase). Here $p_x$ is the parity of the number of operators $X$ in the product, similarly for $Y$ and $Z$. The cumulative correction operator can be incorporated into the measurement in the following way. The effect of an $X$ correction is to flip the angle of the measurement basis, whereas the $Z$ correction means that the outcome of the measurement should be flipped. To establish whether these corrections must be applied, the parity of a subset of the previous measurements must be calculated. Since the parity of bit-strings can be calculated efficiently on the CNOT computer, it suffices as the control computer for cluster state based one-way quantum computation.

## 4   Discussion

In the final part of this paper, we shall discuss some of the consequences of this observation and some of the questions it motivates. The CNOT computer is a model of computation seemingly of lesser computational power than a universal classical computer and yet by allowing it access to measurement devices on a cluster state, one creates a device which is universal for quantum computation. Remarkably, the only data exchanged by the CNOT computer and each measurement device are a single bit in either direction.

There are a number of questions which naturally arise. Since the CNOT computer is not classically universal, what kind of resource states will enable classical universality in this model? It is clear that any resource state for universal quantum computation will also enable universal classical computation, but can one find quantum states which enable one but not the other? This would provide an interesting classical analogue of measurement based quantum computation, and a classification of resource states in terms of their computational resource power.

A further classification is also possible. One could consider whether there exist resource states which only enable measurement-based quantum computation

when the control computer is fully classically universal. There is already evidence
for the existence of such states. Recently, models of measurement-based quantum computation have been proposed [10] which utilise states with very different
properties to the cluster states. In particular, a different approach to attaining
deterministic computation is employed. These strategies are based upon buffer-
ing simulated logical gates with repeated attempts at the identity gate, sufficiently many times so that the errors cancel. The errors in this model are not
restricted to Pauli operators, but can be generators of any finite sub-group,
in which case the control computer must employ a counting strategy to identify when they have cancelled out. Since carrying digits for addition cannot be
straightforwardly achieved in the CNOT computer, it is possible that some of
the states described in [9,10] do indeed fall into this category, and require a
universal classical computer to enable MBQC.

In conclusion, we have shown that the classical computer required for the feedback control in one-way quantum computer need not be classically universal, but
a limited power CNOT computer suffices. The questions which this observation
motivates are fascinating and will form a rich subject for further research.

# References


1. Raussendorf, R., Briegel, H.: Phys. Rev. Lett. 86, 5188 (2001)
2. Raussendorf, R., Briegel, H.: Quantum. Inform. Compu. 2, 443 (2002)
3. Raussendorf, R., Browne, D., Briegel, H.: Phys. Rev. A 68, 022312 (2003)
4. Browne, D.E., Briegel, H.J.: One-way Quantum Computation - a tutorial introduction. In: Bruss, D., Leuchs, G. (eds.). Lectures on Quantum Information (2006)
5. Nielsen, M.: Rep. Math. Phys. 57, 147 (2006)
6. Jozsa, R.: Measurement-based quantum computation. In: Angelakis, D.G. (ed.)
   Quantum Information Processing - From Theory to Experiment (2006)
7. van den Nest, M., Miyake, A., Dür, W., Briegel, H.J.: Phys. Rev. Lett. 97, 150504
   (2006)
8. van den Nest, M., Dür, W., Miyake, A., Briegel, H.J.: New. J Phys. 9, 204 (2007)
9. Gross, D., Eisert, J., Schuch, N., Perez-Garcia, D.: Phys. Rev. A. 76 (2007)
10. Gross, D., Eisert, J.: Phys. Rev. Lett. 98, 220503 (2007)
11. Aaronson, S., Kuperberg, G.: The Complexity Zoo (accessed 10th March 2008),
    `http://qwiki.stanford.edu/wiki/Complexity_Zoo`
12. Damm, C.: Inform. Process. Lett. 36, 247 (1990)
13. Aaronson, S., Gottesman, D.: Phys. Rev. A 70, 052328 (2004)
14. Danos, V., Kashefi, E.: Phys. Rev. A 74, 052310 (2006)

# The Algebraic Counterpart of the Wagner Hierarchy


Jérémie Cabessa and Jacques Duparc

Université de Lausanne,
Faculty of Business and Economics HEC, Institute of Information Systems ISI,
CH-1015 Lausanne, Switzerland
Jeremie.Cabessa@unil.ch, Jacques.Duparc@unil.ch



**Abstract.** The algebraic study of formal languages shows that $\omega$-rational languages are exactly the sets recognizable by finite $\omega$-semigroups. Within this framework, we provide a construction of the algebraic counterpart of the Wagner hierarchy. We adopt a hierarchical game approach, by translating the Wadge theory from the $\omega$-rational language to the $\omega$-semigroup context.

More precisely, we first define a reduction relation on finite pointed $\omega$-semigroups by means of a Wadge-like infinite two-player game. The collection of these algebraic structures ordered by this reduction is then proven to be isomorphic to the Wagner hierarchy, namely a decidable and well-founded partial ordering of width 2 and height $\omega^\omega$.

**Keywords:** $\omega$-automata, $\omega$-rational languages, $\omega$-semigroups, infinite games, Wadge game, Wadge hierarchy, Wagner hierarchy.


## 1    Introduction

This paper stands at the crossroads of two mathematical fields, namely the algebraic theory of $\omega$-automata, and hierarchical games, in descriptive set theory.

The basic interest of the algebraic approach to automata theory consists in the equivalence between Büchi automata and finite $\omega$-semigroups [12] – an extension of the concept of a semigroup. These mathematical objects indeed satisfy several relevant properties. Firstly, given a finite Büchi automaton, one can effectively compute a finite $\omega$-semigroup recognizing the same $\omega$-language, and vice versa. Secondly, among all finite $\omega$-semigroups recognizing a given $\omega$-language, there exists a minimal one – called the syntactic $\omega$-semigroup –, whereas there is no convincing notion of Büchi (or Muller) minimal automaton. Thirdly, finite $\omega$-semigroups provide powerful characteristics towards the classification of $\omega$-rational languages; for instance, an $\omega$-language is first-order definable if and only if it is recognized by an aperiodic $\omega$-semigroup [7,10,18], a generalization to infinite words of Schützenberger, and McNaughton's and Papert famous results [9,16]. Even some topological properties (being open, closed, clopen, $\Sigma_2^0$, $\Pi_2^0$, $\Delta_2^0$) can be characterized by algebraic properties on $\omega$-semigroups (see [12,14]).

Hierarchical games, for their part, aim to classify subsets of topological spaces. In particular, the Wadge hierarchy [19] (defined via the Wadge games) appeared


A. Beckmann, C. Dimitracopoulos, and B. Löwe (Eds.): CiE 2008, LNCS 5028, pp. 100–109, 2008.

to be specially interesting to computer scientists, for it shed a light on the study of classifying $\omega$-rational languages. The famous Wagner hierarchy [20], known as the most refined classification of $\omega$-rational languages, was proven to be precisely the restriction of the Wadge hierarchy to these $\omega$-languages.

However, Wagner's original construction relies on a graph-theoretic analysis of Muller automata, away from the set theoretical and the algebraic frameworks. Olivier Carton and Dominique Perrin [2,3,4] investigated the algebraic reformulation of the Wagner hierarchy, a work carried on by Jacques Duparc and Mariane Riss [6]. But this new approach is not yet entirely satisfactory, for it fails to define precisely the algebraic counterpart of the Wadge (or Wagner) preorder on finite $\omega$-semigroups.

Our paper fill this gap. We define a reduction relation on subsets of finite $\omega$-semigroups by means of an infinite game, without any direct reference to the Wagner hierarchy. We then show that the resulting algebraic hierarchy is isomorphic to the Wagner hierarchy, and in this sense corresponds to the algebraic counterpart of the Wagner hierarchy. In particular, this classification is a refinement of the hierarchies of chains and superchains introduced in [2,4]. We finally prove that this algebraic hierarchy is also decidable.

## 2 Preliminaries

### 2.1 $\omega$-Languages

Given a finite set $A$, called the *alphabet*, then $A^*$, $A^+$, $A^\omega$, and $A^\infty$ denote respectively the sets of finite words, nonempty finite words, infinite words, and finite or infinite words, all of them over the alphabet $A$. Given a finite word $u$ and a finite or infinite word $v$, then $uv$ denotes the concatenation of $u$ and $v$. Given $X \subseteq A^*$ and $Y \subseteq A^\infty$, the concatenation of $X$ and $Y$ is denoted by $XY$.

We refer to [12, p.15] for the definition of *$\omega$-rational languages*. We recall that $\omega$-rational languages are exactly the ones recognized by finite Büchi, or equivalently, by finite Muller automata [12].

For any set $A$, the set $A^\omega$ can be equipped with the product topology of the discrete topology on $A$. The class of *Borel* subsets of $A^\omega$ is the smallest class containing the open sets, and closed under countable union and complementation.

### 2.2 $\omega$-Semigroups

The notion of an $\omega$-semigroup was first introduced by Pin as a generalization of semigroups [11,13]. In the case of finite structures, these objects represent a convincing algebraic counterpart to automata reading infinite words: given any finite Büchi automaton, one can build a finite $\omega$-semigroup recognizing (in an algebraic sense) the same language, and conversely, given any finite $\omega$-semigroup recognizing a certain language, one can build a finite Büchi automaton recognizing the same language.

**Definition 1 (see [12, p. 92]).** *An $\omega$-semigroup is an algebra consisting of two components, $S = (S_+, S_\omega)$, and equipped with the following operations:*

- *a binary operation on $S_+$, denoted multiplicatively, such that $S_+$ equipped with this operation is a semigroup;*
- *a mapping $S_+ \times S_\omega \longrightarrow S_\omega$, called mixed product, which associates with each pair $(s,t) \in S_+ \times S_\omega$ an element of $S_\omega$, denoted by $st$, and such that for every $s,t \in S_+$ and for every $u \in S_\omega$, then $s(tu) = (st)u$;*
- *a surjective mapping $\pi_S : S_+^\omega \longrightarrow S_\omega$, called infinite product, such that: for every strictly increasing sequence of integers $(k_n)_{n>0}$, for every sequence $(s_n)_{n\geq 0} \in S_+^\omega$, and for every $s \in S_+$, then*

$$\pi_S(s_0 s_1 \cdots s_{k_1-1}, s_{k_1} \cdots s_{k_2-1}, \ldots) = \pi_S(s_0, s_1, s_2, \ldots),$$

$$s\pi_S(s_0, s_1, s_2, \ldots) = \pi_S(s, s_0, s_1, s_2, \ldots).$$

Intuitively, an $\omega$-semigroup is a semigroup equipped with a suitable infinite product. The conditions on the infinite product ensure that one can replace the notation $\pi_S(s_0, s_1, s_2, \ldots)$ by the notation $s_0 s_1 s_2 \cdots$ without ambiguity. Since an $\omega$-semigroup is a pair $(S_+, S_\omega)$, it is convenient to call $+$-*subsets* and $\omega$-*subsets* the subsets of $S_+$ and $S_\omega$, respectively.

Given two $\omega$-semigroups $S = (S_+, S_\omega)$ and $T = (T_+, T_\omega)$, a *morphism of $\omega$-semigroups* from $S$ into $T$ is a pair $\varphi = (\varphi_+, \varphi_\omega)$, where $\varphi_+ : S_+ \longrightarrow T_+$ is a morphism of semigroups, and $\varphi_\omega : S_\omega \longrightarrow T_\omega$ is a mapping canonically induced by $\varphi_+$ in order to preserve the infinite product, that is, for every sequence $(s_n)_{n\geq 0}$ of elements of $S_+$, one has $\varphi_\omega(\pi_S(s_0, s_1, s_2, \ldots)) = \pi_T(\varphi_+(s_0), \varphi_+(s_1), \varphi_+(s_2), \ldots)$.

An $\omega$-semigroup $S$ is an $\omega$-*subsemigroup* of $T$ if there exists an injective morphism of $\omega$-semigroups from $S$ into $T$. An $\omega$-semigroup $S$ is a *quotient* of $T$ if there exists a surjective morphism of $\omega$-semigroups from $T$ onto $S$. An $\omega$-semigroup $S$ *divides* $T$ if $S$ is quotient of an $\omega$-subsemigroup of $T$.

The notion of pointed $\omega$-semigroup can be adapted from the notion of pointed semigroup introduced by Sakarovitch [15]. In this paper, a *pointed $\omega$-semigroup* denotes a pair $(S, X)$, where $S$ is an $\omega$-semigroup and $X$ is an $\omega$-subset of $S$. A mapping $\varphi : (S, X) \longrightarrow (T, Y)$ is a morphism of pointed $\omega$-semigroups if $\varphi : S \longrightarrow T$ is a morphism of $\omega$-semigroups such that $\varphi^{-1}(Y) = X$. The notions of $\omega$-subsemigroups, quotient, and division can then be easily adapted in this pointed context.

*Example 1.* Let $A$ be a finite set. The $\omega$-semigroup $A^\infty = (A^+, A^\omega)$ equipped with the usual concatenation is the *free $\omega$-semigroup* over the alphabet $A$ [2]. In addition, if $S = (S_+, S_\omega)$ is an $\omega$-semigroup with $S_+$ being finite, the morphism of $\omega$-semigroups $\varphi : S_+^\infty \longrightarrow S$ naturally induced by the identity over $S_+$ is called the *canonical morphism* associated with $S$.

In this paper, we strictly focus on *finite* $\omega$-semigroups, i.e. those whose first component is finite. It is proven in [12] that the infinite product $\pi_S$ of a finite $\omega$-semigroup $S$ is completely determined be the mixed products of the form $x\pi_S(s, s, s, \ldots)$ (denoted $xs^\omega$). We use this property in the next example.

*Example 2.* The pair $S = (\{0,1\}, \{0^\omega, 1^\omega\})$ equipped with the usual multiplication over $\{0,1\}$ and with the infinite product defined by the relations $00^\omega = 10^\omega = 0^\omega$ and $01^\omega = 11^\omega = 1^\omega$ is an $\omega$-semigroup.

Wilke was the first to give the appropriate algebraic counterpart to finite automata reading infinite words [21]. In addition, he established that the $\omega$-languages recognized by finite $\omega$-semigroups are exactly the ones recognized by Büchi automata, a proof that can be found in [21] or [12].

**Definition 2.** *Let $S$ and $T$ be two $\omega$-semigroups. One says that a surjective morphism of $\omega$-semigroups $\varphi : S \longrightarrow T$ recognizes a subset $X$ of $S$ if there exists a subset $Y$ of $T$ such that $\varphi^{-1}(Y) = X$. By extension, one also says that the $\omega$-semigroup $T$ recognizes $X$.*

**Proposition 1 (Wilke).** *An $\omega$-language is recognizable by a finite $\omega$-semigroup if and only if it is $\omega$-rational.*

*Example 3.* Let $A = \{a, b\}$, let $S$ be the $\omega$-semigroup given in Example 2, and let $\varphi : A^\infty \longrightarrow S$ be the morphism defined by $\varphi(a) = 0$ and $\varphi(b) = 1$. Then $\varphi^{-1}(0^\omega) = (A^*a)^\omega$ and $\varphi^{-1}(1^\omega) = A^*b^\omega$, and therefore these two languages are $\omega$-rational.

A *congruence of an $\omega$-semigroup* $S = (S_+, S_\omega)$ is a pair $(\sim_+, \sim_\omega)$, where $\sim_+$ is a semigroup congruence on $S_+$, $\sim_\omega$ is an equivalence relation on $S_\omega$, and these relations are stable for the infinite and the mixed products (see [12]). The quotient set $S/\sim = (S/\sim_+, S/\sim_\omega)$ is naturally equipped with a structure of $\omega$-semigroup. If $(\sim_i)_{i \in I}$ is a family of congruences on an $\omega$-semigroup, then the congruence $\sim$, defined by $s \sim t$ if and only if $s \sim_i t$, for all $i \in I$, is called the lower bound of the family $(\sim_i)_{i \in I}$. The upper bound of the family $(\sim_i)_{i \in I}$ is then the lower bound of the congruences that are coarser than all the $\sim_i$.

Given a subset $X$ of an $\omega$-semigroup $S$, the *syntactic congruence* of $X$, denoted by $\sim_X$, is the upper bound of the family of congruences whose associated quotient morphisms recognize $X$, if this upper bound still recognizes $X$, and is undefined otherwise. Whenever defined, the quotient $S(X) = S/\sim_X$ is called the *syntactic $\omega$-semigroup* of $X$, the surjective morphism $\mu : S \longrightarrow S(X)$ is the *syntactic morphism* of $X$, the set $\mu(X)$ is the *syntactic image* of $X$, and one has the property $\mu^{-1}(\mu(X)) = X$. The pointed $\omega$-semigroup $(S(X), \mu(X))$ will be denoted by $Synt(X)$. One can prove that the syntactic $\omega$-semigroup of an $\omega$-rational language is always defined, and is the unique (up to isomorphism) and minimal (for the division) pointed $\omega$-semigroup recognizing this language [12].

*Example 4.* Let $K = (A^*a)^\omega$ be an $\omega$-language over the alphabet $A = \{a, b\}$. The morphism $\varphi : A^\infty \longrightarrow S$ given in Example 3 is the syntactic morphism of $K$. The $\omega$-subset $X = \{0^\omega\}$ of $S$ is the syntactic image of $K$.

Finally, a pointed $\omega$-semigroup $(S, X)$ will be called *Borel* if the preimage $\pi_S^{-1}(X)$ is a Borel subset of $S_+^\omega$ (where $S_+^\omega$ is equipped with the product topology of the discrete topology on $S_+$). Notice that every finite pointed $\omega$-semigroup is Borel, since by Proposition 1, its preimage by the infinite product is $\omega$-rational, hence Borel (more precisely boolean combination of $\Sigma_2^0$) [12].

## 3    The Wadge and the Wagner Hierarchies

Let $A$ and $B$ be two alphabets, and let $X \subseteq A^\omega$ and $Y \subseteq B^\omega$. The *Wadge game* $\mathbb{W}((A, X), (B, Y))$ [19] is a two-player infinite game with perfect information, where Player I is in charge of the subset $X$ and Player II is in charge of the subset $Y$. Players I and II alternately play letters from the alphabets $A$ and $B$, respectively. Player I begins. Player II is allowed to skip her turn – formally denoted by the symbol "$-$" – provided she plays infinitely many letters, whereas Player I is not allowed to do so. After $\omega$ turns each, players I and II respectively produced two infinite words $\alpha \in A^\omega$ and $\beta \in B^\omega$. Player II wins $\mathbb{W}((A, X), (B, Y))$ if and only if ($\alpha \in X \Leftrightarrow \beta \in Y$). From this point onward, the Wadge game $\mathbb{W}((A, X), (B, Y))$ will be denoted $\mathbb{W}(X, Y)$ and the alphabets involved will always be clear from the context.

Along the play, the finite sequence of all previous moves of a given player is called the *current position* of this player. A *strategy* for Player I is a mapping from $(B \cup \{-\})^*$ into $A$. A *strategy* for Player II is a mapping from $A^+$ into $B \cup \{-\}$. A strategy is *winning* if the player following it must necessarily win, no matter what his opponent plays.

The *Wadge reduction* is defined via the Wadge game as follows: a set $X$ is said to be *Wadge reducible* to $Y$, denoted by $X \leq_W Y$, if and only if Player II has a winning strategy in $\mathbb{W}(X, Y)$. One then sets $X \equiv_W Y$ if and only if both $X \leq_W Y$ and $Y \leq_W X$, and also $X <_W Y$ if and only if $X \leq_W Y$ and $X \not\equiv_W Y$. The relation $\leq_W$ is reflexive and transitive, and $\equiv_W$ is an equivalence relation. A set $X$ is called *self-dual* if $X \equiv_W X^c$, and *non-self-dual* if $X \not\equiv_W X^c$. One can show [19] that the Wadge reduction coincides with the continuous reduction, that is $X \leq_W Y$ if and only if $f^{-1}(Y) = X$, for some continuous function $f : A^\omega \longrightarrow B^\omega$.

*The Wadge hierarchy* consists of the collection of all $\omega$-languages ordered by the Wadge reduction, and *the Borel Wadge hierarchy* is the restriction of the Wadge hierarchy to Borel $\omega$-languages. Martin's Borel determinacy [8] easily implies Borel Wadge determinacy, that is, whenever $X$ and $Y$ are Borel sets, then one of the two players has a winning strategy in $\mathbb{W}(X, Y)$. As a corollary, one can prove that, up to complementation and Wadge equivalence, the Borel Wadge hierarchy is a well ordering. Therefore, there exist a unique ordinal, called the *height* of the Borel Wadge hierarchy, and a mapping $d_W$ from the Borel Wadge hierarchy onto its height, called the *Wadge degree*, such that $d_W(X) < d_W(Y)$ if and only if $X <_W Y$, and $d_W(X) = d_W(Y)$ if and only if either $X \equiv_W Y$ or $X \equiv_W Y^c$, for every Borel $\omega$-languages $X$ and $Y$. The Borel Wadge hierarchy actually consists of an alternating succession of non-self-dual and self-dual sets with non-self-dual pairs at each limit level (as soon as finite alphabets are considered) [5,19].

*The Wagner hierarchy* is precisely the restriction of the Wadge hierarchy to $\omega$-rational languages, and hence corresponds to the most refined classification of such languages [6,12,20]. This hierarchy has a height of $\omega^\omega$, and it is decidable. The Wagner degree of an $\omega$-rational language can indeed be computed by analyzing the graph of a Muller automaton accepting this language [20].

Selivanov gave a complete set theoretical description of the Wagner hierarchy in terms of boolean expressions [17], and Carton, Perrin, Duparc, and Riss studied some algebraic properties of this hierarchy [2,4,6]. In this context, the present work provides a complete construction of the algebraic counterpart of the Wagner hierarchy.


## 4   The $\mathbb{SG}$-Hierarchy

We define a reduction relation on pointed $\omega$-semigroups by means of an infinite two-player game. This reduction induces a hierarchy of pointed $\omega$-semigroups. Many results of the Wadge theory [19] also apply in this framework, and provide a detailed description of this algebraic hierarchy.

Let $S = (S_+, S_\omega)$ and $T = (T_+, T_\omega)$ be two $\omega$-semigroups, and let $X \subseteq S_\omega$ and $Y \subseteq T_\omega$ be two $\omega$-subsets. The game $\mathbb{SG}((S, X), (T, Y))$ is an infinite two-player game with perfect information, where Player I is in charge of $X$, Player II is in charge of $Y$, and players I and II alternately play elements of $S_+$ and $T_+ \cup \{-\}$, respectively. Player I begins. Unlike Player I, Player II is allowed to skip her turn – denoted by the symbol "$-$" –, provided she plays infinitely many moves. After $\omega$ turns each, players I and II produced respectively two infinite sequences $(s_0, s_1, \ldots) \in S_+^\omega$ and $(t_0, t_1, \ldots) \in T_+^\omega$. Player II wins $\mathbb{SG}((S, X), (T, Y))$ if and only if $\pi_S(s_0, s_1, \ldots) \in X \Leftrightarrow \pi_T(t_0, t_1, \ldots) \in Y$. From this point onward, the game $\mathbb{SG}((S, X), (T, Y))$ will be denoted by $\mathbb{SG}(X, Y)$ and the $\omega$-semigroups involved will always be known from the context. A play in this game is illustrated below.

$$(X)\ \text{I}\ :\quad s_0 \qquad\qquad s_1 \cdots\cdots \quad\xrightarrow{\text{after }\omega\text{ moves}}\ (s_0, s_1, s_2, \ldots)$$

$$(Y)\ \text{II}\ :\qquad\qquad t_0 \qquad \cdots\cdots \quad\xrightarrow{\text{after }\omega\text{ moves}}\ (t_0, t_1, t_2, \ldots)$$

We now say that $X$ is $\mathbb{SG}$-*reducible* to $Y$, denoted by $X \leq_{SG} Y$, if and only if Player II has a winning strategy in $\mathbb{SG}(X, Y)$. We then naturally set $X \equiv_{SG} Y$ if and only if both $X \leq_{SG} Y$ and $Y \leq_{SG} X$, and also $X <_{SG} Y$ if and only if $X \leq_{SG} Y$ and $X \not\equiv_{SG} Y$. The relation $\leq_{SG}$ is reflexive and transitive, and $\equiv_{SG}$ is an equivalence relation.

Notice that if $(S, X)$ and $(T, Y)$ are two pointed $\omega$-semigroups, a given player has a winning strategy in the game $\mathbb{SG}(X, Y)$ if and only if this same player also has one in the Wadge game $\mathbb{W}(\pi_S^{-1}(X), \pi_T^{-1}(Y))$. Therefore Borel Wadge determinacy implies the determinacy of $\mathbb{SG}$-games involving Borel pointed $\omega$-semigroups.

The collection of Borel pointed $\omega$-semigroups ordered by the $\leq_{SG}$-relation is called *the $\mathbb{SG}$-hierarchy*, in order to underline the semigroup approach. Notice that the restriction of the $\mathbb{SG}$-hierarchy to Borel pointed free $\omega$-semigroups is exactly the Borel Wadge hierarchy. When restricted to finite pointed $\omega$-semigroups, this hierarchy will be called *the* $\mathbb{FSG}$-*hierarchy*, in order to underline the finiteness of the $\omega$-semigroups involved. As corollaries of the determinacy of Borel $\mathbb{SG}$-games, a straightforward generalization in this context of Martin and Wadge's results [8,19] shows that, up to complementation and $\leq_{SG}$-equivalence, the $\mathbb{SG}$-hierarchy is a well ordering. Therefore, there exist again a unique ordinal, called the *height* of the $\mathbb{SG}$-hierarchy, and a mapping $d_{SG}$ from the $\mathbb{SG}$-hierarchy onto its height, called the $\mathbb{SG}$-*degree*, such that $d_{SG}(X) < d_{SG}(Y)$ if and only if $X <_{SG} Y$, and $d_{SG}(X) = d_{SG}(Y)$ if and only if either $X \equiv_{SG} Y$ or $X \equiv_{SG} Y^c$, for every Borel $\omega$-subsets $X$ and $Y$. It directly follows from the Wadge analysis that the $\mathbb{SG}$-hierarchy has the same familiar "scaling shape" as the Borel or Wadge hierarchies: an increasing sequence of non-self-dual sets with self-dual sets in between, as illustrated in Figure 1, where circles represent the $\equiv_{SG}$-equivalence classes of pointed $\omega$-semigroups, and arrows stand for the $<_{SG}$-relation.

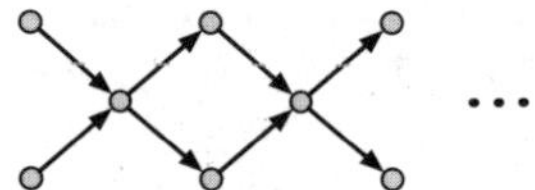

**Fig. 1.** The $\mathbb{SG}$-hierarchy

## 5   The $\mathbb{FSG}$ and the Wagner Hierarchies

This section shows that the $\mathbb{FSG}$-hierarchy is precisely the algebraic counterpart of the Wagner hierarchy. Consequently, this algebraic hierarchy has a height of $\omega^\omega$, and it is decidable.

Let $S = (S_+, S_\omega)$ be a finite $\omega$-semigroup, and let $\varphi : A^\infty \longrightarrow S$ be a surjective morphism of $\omega$-semigroups, for some finite alphabet $A$. Then every $\omega$-subset $X$ of $S_\omega$ can be lifted on an $\omega$-rational language $\varphi^{-1}(X)$ of $A^\omega$. The next proposition proves that this lifting induces an embedding from the $\mathbb{FSG}$-hierarchy into the Wagner hierarchy.

**Proposition 2.** *Let $(S, X)$ and $(T, Y)$ be two finite pointed $\omega$-semigroups, and let $\varphi : A^\infty \longrightarrow S$ and $\psi : B^\infty \longrightarrow T$ be two surjective morphisms of $\omega$-semigroups, where $A$ and $B$ are finite alphabets. Then $X \leq_{SG} Y$ if and only if $\varphi^{-1}(X) \leq_W \psi^{-1}(Y)$.*

*Proof (sketch).* A complete proof can be found in [1, pp. 86–88]. For the first direction, a given winning strategy for Player II in $\mathbb{SG}(X, Y)$ induces via $\varphi$ and $\psi^{-1}$ a winning strategy for this same player in the game $\mathbb{W}\left(\varphi^{-1}(X), \psi^{-1}(Y)\right)$. Conversely, a given winning strategy for Player II in $\mathbb{W}\left(\varphi^{-1}(X), \psi^{-1}(Y)\right)$ also induces via $\varphi^{-1}$ and $\psi$ a winning strategy for this same player in $\mathbb{SG}(X, Y)$.  $\square$

By the previous proposition, the Wadge reduction on $\omega$-rational languages and the $\mathbb{SG}$-reduction on $\omega$-subsets recognizing these languages coincide. The next corollary mentions that this property holds in particular for $\omega$-rational languages and their syntactic images, meaning that the $\mathbb{SG}$-reduction is the appropriate algebraic counterpart of the Wagner reduction. As a direct consequence, the Wagner degree is a *syntactic invariant*: if two $\omega$-rational languages have the same syntactic image, then they also have the same Wagner degree.

**Corollary 1.** *Let $K$ and $L$ be two $\omega$-rational languages and $\mu(K)$ and $\nu(L)$ be their syntactic images.*

(1) *$K \leq_W L$ if and only if $\mu(K) \leq_{SG} \nu(L)$.*

(2) *If $Synt(K) = Synt(L)$, then $K \equiv_W L$.*

*Proof.* Since $\mu$ and $\nu$ are syntactic morphisms, one has $\mu^{-1}(\mu(K)) = K$ and $\nu^{-1}(\nu(L)) = L$. Proposition 2 leads to the conclusion. For (2), if $Synt(K) = Synt(L)$, then $\mu(K) = \nu(L)$, and (1) leads to the conclusion. $\qquad\square$

As another consequence, the $\mathbb{SG}$-degree of an $\omega$-subset is invariant under surjective morphism, and in particular under syntactic morphism. Therefore, syntactic finite pointed $\omega$-semigroups are minimal representatives of their $\leq_{SG}$-equivalence class.

**Corollary 2.** *Let $\mu : S \longrightarrow T$ be a surjective morphism of finite $\omega$-semigroups, let $Y \subseteq T_\omega$, and let $X = \mu^{-1}(Y)$. Then $X \equiv_{SG} Y$.*

*Proof.* Let $\varphi : S^\infty_+ \longrightarrow S$ be the canonical morphism of $\omega$-semigroups associated with $S$, and let $\psi = \mu \circ \varphi : S^\infty_+ \longrightarrow T$. The mapping $\psi$ is a surjective morphism of $\omega$-semigroups. It satisfies $\psi^{-1}(Y) = \varphi^{-1} \circ \mu^{-1}(Y) = \varphi^{-1}(X)$, thus in particular, $\varphi^{-1}(X) \equiv_W \psi^{-1}(Y)$. Proposition 2 then shows that $X \equiv_{SG} Y$. $\qquad\square$

Finally, the following theorem proves that the Wagner hierarchy and the $\mathbb{FSG}$-hierarchy are isomorphic. The required isomorphism is the mapping which associates every $\omega$-rational language with its syntactic image. Therefore, the Wagner degree of an $\omega$-rational language and the $\mathbb{SG}$-degree of its syntactic image are the same.

**Theorem 1.** *The Wagner hierarchy and the $\mathbb{FSG}$-hierarchy are isomorphic.*

*Proof.* Consider the mapping from the Wagner hierarchy into the $\mathbb{SG}$-hierarchy which associates every $\omega$-rational language with its syntactic image. We prove that this mapping is an embedding. Let $K$ and $L$ be two $\omega$-rational languages, and let $X = \mu(K)$ and $Y = \nu(L)$ be their syntactic images. Corollary 1 ensures that $K \leq_W L$ if and only if $X \leq_{SG} Y$. We now show that, up to $\equiv_{SG}$-equivalence, this mapping is onto. Let $X$ be an $\omega$-subset of a finite $\omega$-semigroup $S = (S_+, S_\omega)$, let $\mu : S \longrightarrow S(X)$ be the syntactic morphism of $X$, and let $Y = \mu(X)$ be its syntactic image. Corollary 2 ensures that $X \equiv_{SG} Y$. Now, let also $\varphi : S^\infty_+ \longrightarrow S$ be the canonical morphism associated with $S_+$, and let $L = \varphi^{-1}(X)$. Then the morphism of $\omega$-semigroups $\psi = \mu \circ \varphi : S^\infty_+ \longrightarrow S(X)$ is the syntactic morphism of $L$ [12], and one has $\psi(L) = Y \equiv_{SG} X$. $\qquad\square$

As a corollary, we show that the $\mathbb{FSG}$-hierarchy is *decidable*: for every $\omega$-subset $X$ of the hierarchy, one can effectively compute the Cantor normal form of base $\omega$ of the ordinal $d_{SG}(X)$.

**Corollary 3.** *The $\mathbb{FSG}$-hierarchy has height $\omega^\omega$, and it is decidable.*

*Proof.* By the previous theorem, the $\mathbb{FSG}$-hierarchy and the Wagner hierarchy have the same height, namely $\omega^\omega$. In addition, given an $\omega$-subset $X$ of a finite $\omega$-semigroup $S = (S_+, S_\omega)$, one can effectively compute the $\mathbb{SG}$-degree of $X$ as follows. Let $\varphi : S_+^\infty \longrightarrow S$ be the canonical morphism associated with $S_+$, and let $L = \varphi^{-1}(X)$. Theorem 1 shows that the $\mathbb{SG}$-degree of $X$ is equal to the Wagner degree of $L$. Furthermore, the Wagner degree of $L$ can be effectively computed as follows. First, one can effectively compute an $\omega$-rational expression describing $L = \varphi^{-1}(X)$ [12, Corollary 7.4, p. 110]. Next, one can shift from this rational expression to some finite Muller automaton recognizing $L$, see [12, Chapter I, sections 10.1, 10.3, and 10.4]. Finally, the Wagner degree of the $\omega$-language recognized by a finite Muller automaton is effectively computable [20].    $\square$

*Example 5.* Consider the syntactic image $(S, X)$ of the $\omega$-language $K = (A^* a)^\omega$ given in example 4. We can prove that $d_{SG}((S, X)) = d_W(K) = \omega$.

## 6    Conclusion

This work is a first step towards the complete description of the algebraic counterpart of the Wagner hierarchy. Using a hierarchical game approach, we defined a reduction relation on finite pointed $\omega$-semigroups which was proven to be the algebraic counterpart of the Wadge (or Wagner) preorder on $\omega$-rational languages. As a direct consequence, the Wagner degree of $\omega$-rational languages is a syntactic invariant. The resulting algebraic hierarchy is then isomorphic to the Wagner hierarchy, namely a decidable partial order of width 2 and height $\omega^\omega$. But the decidability procedure presented in Corollary 3 relies on Wagner's naming procedure over Muller automata, and in this sense withdraws from the purely algebraic context.

The natural extension of this work would be to fill this gap, and hence describe an algorithm computing the Wagner degree of any $\omega$-rational set directly on its syntactic pointed $\omega$-semigroup, without any reference to some underlying Muller automata. This study is the purpose of a forthcoming paper.

We can also hope to extend this work to more sophisticated $\omega$-languages, like those recognized by deterministic counters, or even deterministic pushdown automata. This would obviously require to understand first the kind of infinite $\omega$-semigroups corresponding to such machines.

## References


1. Cabessa, J.: A Game Theoretical Approach to the Algebraic Counterpart of the Wagner Hierarchy. PhD thesis, Universities of Lausanne and Paris 7 (2007)
2. Carton, O., Perrin, D.: Chains and superchains for $\omega$-rational sets, automata and semigroups. Internat. J. Algebra Comput. 7(6), 673–695 (1997)

3. Carton, O., Perrin, D.: The Wadge-Wagner hierarchy of $\omega$-rational sets. In: Degano, P., Gorrieri, R., Marchetti-Spaccamela, A. (eds.) ICALP 1997. LNCS, vol. 1256, pp. 17–35. Springer, Heidelberg (1997)
4. Carton, O., Perrin, D.: The Wagner hierarchy. Internat. J. Algebra Comput. 9(5), 597–620 (1999)
5. Duparc, J.: Wadge hierarchy and Veblen hierarchy. I. Borel sets of finite rank. J. Symbolic Logic 66(1), 56–86 (2001)
6. Duparc, J., Riss, M.: The missing link for $\omega$-rational sets, automata, and semigroups. Internat. J. Algebra Comput. 16(1), 161–185 (2006)
7. Ladner, R.E.: Application of model theoretic games to discrete linear orders and finite automata. Information and Control 33(4), 281–303 (1977)
8. Martin, D.A.: Borel determinacy. Ann. of Math (2) 102(2), 363–371 (1975)
9. McNaughton, R., Papert, S.A.: Counter-Free Automata (M.I.T. research monograph no. 65). MIT Press, Cambridge (1971)
10. Perrin, D., Pin, J.-E.: First-order logic and star-free sets. J. Comput. System Sci. 32(3), 393–406 (1986)
11. Perrin, D., Pin, J.-É.: Semigroups and automata on infinite words. In: Semigroups, formal languages and groups (York, 1993), pp. 49–72. Kluwer Acad. Publ., Dordrecht (1995)
12. Perrin, D., Pin, J.-É.: Infinite Words. Pure and Applied Mathematics, vol. 141. Elsevier, Amsterdam (2004)
13. Pin, J.-É.: Logic, semigroups and automata on words. Annals of Mathematics and Artificial Intelligence 16, 343–384 (1996)
14. Pin, J.-E.: Positive varieties and infinite words. In: Lucchesi, C.L., Moura, A.V. (eds.) LATIN 1998. LNCS, vol. 1380. Springer, Heidelberg (1998)
15. Sakarovitch, J.: Monoïdes pointés. Semigroup Forum 18(3), 235–264 (1979)
16. Schützenberger, M.P.: On finite monoids having only trivial subgroups.. Inf. Control 8, 190–194 (1965)
17. Selivanov, V.: Fine hierarchy of regular $\omega$-languages. Theoret. Comput. Sci. 191(1-2), 37–59 (1998)
18. Thomas, W.: Star-free regular sets of $\omega$-sequences. Inform. and Control 42(2), 148–156 (1979)
19. Wadge, W.W.: Reducibility and determinateness on the Baire space. PhD thesis, University of California, Berkeley (1983)
20. Wagner, K.: On $\omega$-regular sets. Inform. and Control 43(2), 123–177 (1979)
21. Wilke, T.: An Eilenberg theorem for $\infty$-languages. In: Leach Albert, J., Monien, B., Rodríguez-Artalejo, M. (eds.) ICALP 1991. LNCS, vol. 510, pp. 588–599. Springer, Heidelberg (1991)

# Computing by Observing: A Brief Survey

Matteo Cavaliere

The Microsoft Research - University of Trento,
Centre for Computational and Systems Biology, CoSBi, Trento, Italy
`cavaliere@cosbi.eu`

**Abstract.** This paper is a brief survey of a computational paradigm
called *computing by observing* that stresses the role of an observer in
computation. The idea of the paradigm is that a computing device can
be obtained by combining a basic system and an observer that transforms
the trajectories of the basic system into a readable output. The paradigm
has been applied in several areas: natural computing (DNA computing
and membrane computing), automata and formal language theory. In
general, it has been shown that simple basic systems observed by simple
observers can produce that which only much more complex systems can
produce.

## 1   Introduction

A standard procedure in many fields of science is to conduct an experiment, observe the entire dynamics of a system and then take the result of this observation
as the final output. A momentary picture of the observed system is irrelevant,
and rather the entire dynamics of the system is evaluated.

Inspired by this, has been defined a framework where the computation is
obtained by observing the entire progress of a basic system. In other words, *the
entire observed progress constitutes the result of the computation.* The paradigm
(called *computing by observing*) stresses the role of the observer in computation
and defines a computation by balancing the power of the observed system and
the power of the external observer.

The paradigm, in the most general case, is sketched in Figure 1. Following a
set of rules the observer translates the trajectories of the observed basic system
into a readable output, usually strings.

In this abstract framework, the basic system has usually a dynamics specified
by some fixed rules (applied iteratively) and the observer is a mapping that
associates to each configuration reached by the system a label taken from a
finite set of labels. Here one considers observed systems and observers as formal
machines – the process of observation is actually a process used to filtrate and
interpret informations.

The framework was originally [5] applied to a membrane system. Following
the same idea, new bio-inspired models of computation have been obtained in [1]
(sticker systems) and in [2] (splicing systems). The generalization of the framework to formal language theory has been proposed in [7], to string-rewriting in

A. Beckmann, C. Dimitracopoulos, and B. Löwe (Eds.): CiE 2008, LNCS 5028, pp. 110–119, 2008.

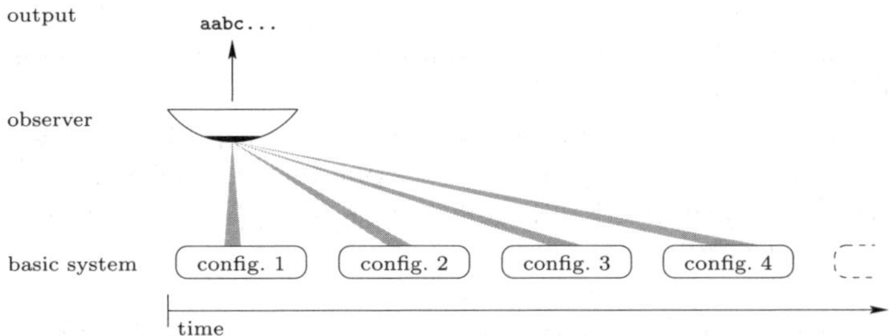


**Fig. 1.** Basic Idea of the Computing by Observing Framework

[4,6], where derivations of grammars and string-rewriting systems are observed by finite state automata. The passage from a basic system as concrete, observable, bio-physical thing to an abstract formal rewriting system can be imagined in the following manner: one can associate a set of (physical) objects to strings and can have some rewriting rules to evolve such objects.

In general, in all these mentioned works, already rather simple basic systems observed by simple observers were proved to be computationally universal (i.e., equivalent to Turing machines).

However, in most of these results it was necessary to design *both* a specific basic system and a specific observer to produce a specific result. On the other hand most basic systems cannot be (easily) programmed or modified (e.g., think to a basic system representing a biological system). However, their evolution can be monitored by using different observers. Then, it is natural to ask how much one can compute by fixing a basic system and choosing different observers (in other words, *computing by only observing*).

This question was (partially) answered in [3] where it was shown that every computational device can be obtained by observing in the "right" manner a *fixed* basic system (with both - observer and observed basic systems with limited computational power).

In this paper we briefly survey the results obtained by using such a paradigm in the formal languages area.

In what follows we use some basic notions and standard terminology from formal language theory (for details, the reader is invited to consult the corresponding chapters of the handbook of formal languages, [9]).

We only recall that $\mathcal{REG}$ and $\mathcal{CF}$ denotes the classes of regular and context-free grammars, respectively and that by REG, CF and RE we denote the classes of languages generated by regular, context-free, and type-0 grammars, respectively.

## 2   Computing by Observing

In this section we present *grammar/observer (G/O) systems* that are generative devices based on the framework discussed in the Introduction.

In this case, a formal grammar plays the role of the observed basic system and an automaton – be it a finite automaton or even a Turing machine – plays the role of the observer.

Initially, a formal definition of the observer is given, and then the definition of a G/O system is recalled. Three ways of working for a G/O system (*initial, free, always writing*) are presented. Examples of $G/O$ systems are presented later. This section is essentially based on [7].

### 2.1  An Example of Observer: Automata with Singular Output

As observer, we essentially use a device mapping arbitrarily long strings into just one singular symbol and we call it automata with singular output.

An automaton with singular output is a tuple $A = (Q, V, \Sigma, q_0, \delta, \sigma)$ with state set $Q$, input alphabet $V$, initial state $q_0 \in Q$, and a complete transition function $\delta$ as known from conventional finite state automata; further there is the output alphabet $\Sigma$ and a labeling function $\sigma : Q \mapsto \Sigma \cup \{\lambda\} \cup \{\bot\}$ where $\lambda$ denotes the empty string and $\bot \notin \Sigma$.

*The output of an automaton with singular output is the label of the state in which it halts/stops.* For a string $w \in V^*$ and an automaton $A$ we then write $A(w)$ for this output; for a sequence $w_1, \ldots, w_n$ of $n \geq 1$ strings over $V^*$ we write $A(w_1, \ldots, w_n)$ for the string $A(w_1) \cdots A(w_n)$.

The class of all (deterministic) automata with singular output will be denoted by $FA_O$. Clearly, in the same way observers can be obtained from other classes of automata such as pushdown automata, linear bounded automata or Turing machines.

### 2.2  G/O Systems

We combine a formal grammar and an observer obtaining the central notion of this section: $G/O$ *system*.

A $G/O$ *system* is a pair $\Omega = (G, A)$ constituted by a generative grammar $G = (N, T, S, P)$ and an automaton with singular output (observer) $A = (Q, V, \Sigma, q_0, \delta, \sigma)$ with output alphabet $\Sigma$, which then is also the output alphabet of the entire system $\Omega$. The automaton's input alphabet must be the union of $N$ and $T$ from the grammar to make the desired interaction possible, i.e., $V = N \cup T$.

We distinguish three different modes of generation that define three different models of $G/O$ *systems*:

1. writing a non-empty output in every step (*always writing G/O systems*),
2. writing a non-empty output in every step after an initialization phase of writing only $\lambda$ (*initial G/O systems*), and
3. changing between empty and non-empty output in an arbitrary manner (*free G/O systems*).

In the case of an *always writing G/O system* $\Omega$ the language generated by $\Omega$ is

$$L_a(\Omega) = \{A(w_1, w_2, \ldots, w_n) \mid S = w_0 \Rightarrow w_1 \Rightarrow \ldots \Rightarrow w_n,$$
$$w_n \in T^* \text{ and } A(w_i) \neq \lambda, \text{ for all } 1 \leq i \leq n\}.$$

Note that the very first sentential form, which is always the starting symbol, is excluded from the observation (otherwise, all words in $L(\Omega)$ would start with the same letter, if the observer was deterministic). The sentential forms $w_i, 1 \leq i \leq n$, are obtained by applying the productions of $G$, in the standard sequential way, as usually defined in grammars.

The best way to ensure the last condition, i.e., that $\lambda$ is never written as output, is of course to define the observer in such a way that it can never produces empty output.

For an *initial G/O system* the output is defined as

$$L_i(\Omega) = \{A(w_0, w_1, \ldots, w_n) \mid S = w_0 \Rightarrow w_1 \Rightarrow \ldots \Rightarrow w_n, \ w_n \in T^*,$$
$$\text{and for all } i \in \{1, \ldots, n\}, A(w_0, w_1, \ldots, w_{i-1}) \neq \lambda$$
$$\text{implies } A(w_i) \neq \lambda\}.$$

Finally, a *free G/O system* generates a language in the following non-restricted manner:

$$L_f(\Omega) = \{A(w_0, w_1, \ldots, w_n) \mid S = w_0 \Rightarrow w_1 \Rightarrow \ldots \Rightarrow w_n, \ w_n \in T^*\}.$$

Thus in all three cases the language contains all those words which the observer can produce during the possible terminating derivations of the underlying grammar. Derivations which do not terminate do not produce a valid output; this means that we only take into account finite words. Of course, by considering the other case of non-terminating derivations the G/O systems could also be used to generate languages of infinite words.

We will actually mainly present the variant where we consider as language produced by $\Omega$:
$$L_{\perp,a}(\Omega) = L_a(\Omega) \cap \Sigma^*.$$

In this way the strings in $L_a(\Omega)$ containing $\perp$ are filtered out and they are not present in $L_{\perp,a}(\Omega)$.

Analogously, the languages $L_{\perp,i}(\Omega)$ and $L_{\perp,f}(\Omega)$ are defined.

For a class $\mathcal{G}$ of grammars and a class $\mathcal{O}$ of observers, $\mathcal{L}_a(\mathcal{G}, \mathcal{O})$, $\mathcal{L}_i(\mathcal{G}, \mathcal{O})$, $\mathcal{L}_f(\mathcal{G}, \mathcal{O})$, $\mathcal{L}_{\perp,a}(\mathcal{G}, \mathcal{O})$, $\mathcal{L}_{\perp,i}(\mathcal{G}, \mathcal{O})$, and $\mathcal{L}_{\perp,f}(\mathcal{G}, \mathcal{O})$ denote the classes of all languages generated by G/O systems with grammars from $\mathcal{G}$ and observers from $\mathcal{O}$ in the respective modes. Quite obviously we obtain for fixed classes of grammars $\mathcal{G}$ and observers $\mathcal{O}$ the inclusions

$$\mathcal{L}_a(\mathcal{G}, \mathcal{O}) \subseteq \mathcal{L}_i(\mathcal{G}, \mathcal{O}) \subseteq \mathcal{L}_f(\mathcal{G}, \mathcal{O})$$

and the same for the variants where the special symbol $\perp$ can be written.

A description of a (free) G/O system $\Omega = (G, A)$ is presented in Figure 2.

For simplicity, in what follows, we present only the mappings that the observers define, without giving a real implementation (in terms of finite automata) for them.

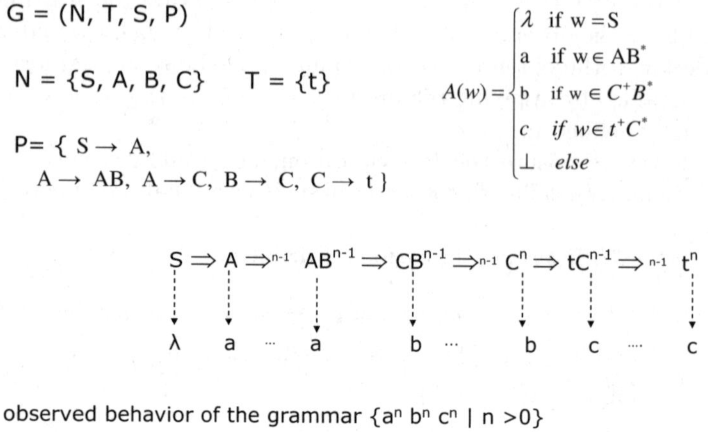

$$G = (N, T, S, P)$$

$$N = \{S, A, B, C\} \qquad T = \{t\}$$

$$P = \{\, S \rightarrow A,$$
$$A \rightarrow AB,\ A \rightarrow C,\ B \rightarrow C,\ C \rightarrow t\,\}$$

$$A(w) = \begin{cases} \lambda & \text{if } w = S \\ a & \text{if } w \in AB^* \\ b & \text{if } w \in C^+ B^* \\ c & \text{if } w \in t^+ C^* \\ \bot & \text{else} \end{cases}$$



**Fig. 2.** Consider $\Omega = (G, A)$. At each sentential form produced by the grammar $G$ the observer $A$ associates a symbol that can be $a, b, c, \bot$ or the empty string $\lambda$ (the vertical arrow is the observer mapping). The concatenation of these symbols is then an output string. For instance, in the figure the output string is $\lambda a \cdots ab \cdots bc \cdots c$. The mapping defined by the observer is specified by the regular expressions. The language $L_f(\Omega)$ is obtained by considering all possible halting derivations of $G$ and collecting all the output strings. Strings in $L_f(\Omega)$ containing $\bot$ are filtered out and not present in $L_{\bot,f}(\Omega)$ that in this case is $\{a^n b^n c^n \mid n > 0\}$.

## 2.3  Always Writing G/O Systems

The first mode of generation we present is the one of writing an output in every step, i.e., we consider the model of always writing G/O systems. This is maybe the most natural one, since in most cases the observation of an experiment should be complete, at least if about the outcome nothing is known beforehand.

We can see that every context-free language $L$ is in $\mathcal{L}_{\bot,a}(CF, FA_O)$. For this consider a context-free grammar in the Greibach normal form. There, all right sides of rules are elements of $TN^*$; this means that in every step exactly one terminal is produced. Since the grammar is still context-free, there is a leftmost derivation for every word in $L$. In this derivation all sentential forms except the initial $S$ are strings over $T^+N^*$. An observer can check that a derivation produces only sentential forms of this structure. Then it can output the rightmost terminal for each one of these sentential forms, and the result equals the string derived by the original grammar.

However, $\mathcal{L}_{\bot,a}(CF, FA_O)$ is bigger than only $CF$. As an example for a non-context-free language from this class we present $\{a^n b^n c^n \mid n > 0\}$. The grammar for this language is

$$G = (\{S, A, B, C\}, \{t\}, S, \{S \rightarrow A, A \rightarrow AB, A \rightarrow C, B \rightarrow C, C \rightarrow t\}).$$

The derivations whose observations will result in the output of words $a^n b^n c^n$ are the ones of the form

$$S \Rightarrow A \overset{n-1}{\Rightarrow} AB^{n-1} \Rightarrow CB^{n-1} \overset{n-1}{\Rightarrow} C^n \Rightarrow tC^{n-1} \overset{n-1}{\Rightarrow} t^n.$$

To produce the output and rule out all other derivations, the observing automaton $A$ will realize the following mapping from the set of sentential forms of $G$ into $\{a, b, c, \bot\}$:

$$A(w) = \begin{cases} a & \text{if } w \in AB^*, \\ b & \text{if } w \in C^+ B^*, \\ c & \text{if } w \in t^+ C^*, \\ \bot & \text{else.} \end{cases}$$

While $\{a^n b^n c^n \mid n > 0\}$ is still semilinear, also the language $\{a^{2^n} \mid n > 0\}$, which is not semilinear, lies in the class $\mathcal{L}_{\bot,a}(CF, FA_O)$. To show this, we first recall that a binary tree of depth $k$ has exactly $2^n - 1$ nodes, if the root is considered to already have depth one. Therefore a context-free grammar, which can have full binary trees as derivation trees, and an observer, which can check that only such derivations are made, can generate $\{a^{2^n} \mid n > 0\}$ when interacting in a G/O system. For this example our grammar is $G = (\{S, A, B, C, T_1, T_2, T_3\}, \{t\}, S, P)$, where the set $P$ of productions is

$$\{S \to A, A \to BB, B \to CC, C \to AA, A \to BT_1, B \to CT_2,$$
$$C \to AT_3, A \to t, B \to t, C \to t, T_1 \to t, T_2 \to t, T_3 \to t\}.$$

Now, for example, derivations resulting in the outputs $a^2$ and $a^8$ are

$$S \Rightarrow A \Rightarrow t$$

and

$$S \Rightarrow A \Rightarrow BB \Rightarrow CCB \Rightarrow CCCT_2 \Rightarrow tCCT_2 \Rightarrow ttCT_2 \Rightarrow tttT_2 \Rightarrow tttt,$$

respectively. The conditions that the observer has to check (for putting out $a$ in every step) are rather straightforward to see after the previous example.

A sentential form containing any $A$ must be completely changed to one from $B^*$, the same from $B$ to $C$, and finally from $C$ to $A$. This is done by use of the rules $A \to BB, B \to CC$ and $C \to AA$ respectively. To ensure that the entire sentential form is completely changed, the observer maps to $a$ only the sentential forms of the form $A^+ B^* \cup B^+ C^* \cup C^+ A^*$ – others result in the output $\bot$. We notice that there are never more than two different non-terminals present at the same time.

To stop the derivation the rightmost non-terminal of the sentential form must produce the corresponding $T_i$, with $i \in \{1, 2, 3\}$, by using one of the rules $A \to BT_1, B \to CT_2$, or $C \to AT_3$, and then the only possible further steps are to derive all non-terminals to $t$. In these cases, the sentential forms mapped to $a$ must be of the form $t^* A^* T_3 \cup t^* B^* T_1 \cup t^* C^* T_2 \cup t^+$.

Therefore the mapping of the observer is

$$A(w) = \begin{cases} a & \text{if} \quad w \in A^+B^* \cup B^+C^* \cup C^+A^* \cup \\ & \qquad t^*A^*T_3 \cup t^*B^*T_1 \cup t^*C^*T_2 \cup t^+, \\ \bot & \text{else.} \end{cases}$$

We note here one difference between the two examples: while in the first case for a word of length $n$ the workspace used by the grammar is $\frac{n}{3}$, in the second case the space is logarithmic in the length of the output.

Trying to characterize the class $\mathcal{L}_{\bot,a}(CF, FA_O)$ more closely, we can easily see that it is contained in the class of context-sensitive languages. In fact, the G/O system must write a symbol of output at each step and then the total space used by such a system for the generation of a word $w$ is bounded by a constant depending only on the context-free grammar. Therefore, every language generated by an always writing G/O systems is context-sensitive by the workspace theorem. It is an *open problem* whether or not the class $\mathcal{L}_{\bot,a}(CF, FA_O)$ corresponds exactly to the class of context-sensitive languages.

## 2.4   Initial G/O Systems

The second variant of G/O systems is called *initial G/O system* and in this model the sentential forms of an initial phase are mapped exclusively to $\lambda$. After the first non-empty output only non-empty outputs can be produced – looking back to the biochemical motivation of the concept of evolution and observer, this would correspond to a phase of initializing an experiment and then a phase of actual observation.

Such an initialization phase – which is not restricted, but can be much longer than the actually observed phase – greatly enhances the power of our G/O systems. Indeed, with the same classes of grammars and observers as in Section 2.3 we obtain computational universality in this case.

**Theorem 1.** $\mathcal{L}_{\bot,i}(CF, FA_O) = RE.$ *[7]*

## 2.5   Free G/O Systems

An immediate corollary of Theorem 1 is the fact that $\mathcal{L}_{\bot,f}(CF, FA_O) = RE$. However a $G/O$ system composed of even less powerful components, namely a locally commutative context-free grammar ($\mathcal{L}CCF$) and a finite state automaton is universal, if the output $\lambda$ can be used without any restriction. Notice that $LCCF$ languages are strictly included in $CF$ languages (e.g., [7]).

**Theorem 2.** $\mathcal{L}_{\bot,f}(\mathcal{L}CCF, FA_O) = RE.$ *[7]*

An *interesting fact* is that intuitively the observer's ability to produce $\bot$, i.e., to eliminate certain computations, seems a powerful and essential feature in all three models. However, we obtain all recursively enumerable languages over $\Sigma$ simply by intersection of a language over $\Sigma_\bot$ with the regular language $\Sigma^*$.

Now notice that recursive languages are closed under intersection with regular sets (can be obtained just by looking to the definition of recursive languages). Therefore, there must exist some grammar/observer systems $\Omega$ generating a *non-recursive* $\mathcal{L}_f(\Omega)$ (i.e., not using the filtering with $\perp$).

## 3   Computing by Only Observing

As discussed in the Introduction, it is interesting to understand how much one can compute by allowing only changes in the observer, keeping unchanged the observed basic system.

For this purpose, we say that a grammar is universal (modulo observation) for a family of languages if it can generate every language in the family when observed in an appropriate way.

Formally, as in [3]:

**Definition 3.** *A grammar $G$ is universal modulo observation (m.o.) for a family of languages $\mathcal{L}$ if $\mathcal{L} = \mathcal{L}_{\perp,f}(\{G\}, \mathrm{FA}_O)$.*

We show by means of an example that a G/O system can generate different classes of languages if the observer is changed while the grammar remains fixed.

Let us consider the following context-free grammar:

$$G = (\{S, A, B, C\}, \{t, p\}, S, \{S \to pS, S \to p,$$
$$S \to A, A \to AB, A \to C, B \to C, C \to t\}).$$

If $G$ is coupled with the observer $A'$ such that $A'(w) = a$ if $w \in \{S, A, B, C, t, p\}^+$, then $\Omega = (G, A')$ defines the language $L_{\perp,f}(\Omega) = \{a^i \mid i \geq 2\}$, a regular language. In fact, the derivation $S \to pS \overset{n-2}{\Rightarrow} p^{n-1}S \to p^n$ produces (when observed) the string $a^{n+1}$.

Keeping the same grammar $G$ we change the observer into $A''$ such that:

$$A''(w) = \begin{cases} \lambda & \text{if } w = S, \\ a & \text{if } w \in AB^*, \\ b & \text{if } w \in C^+B^*, \\ c & \text{if } w \in t^+C^*, \\ \perp & \text{else} \end{cases}$$

In this case, one can verifies that $\Omega = (G, A'')$ generates the language $L_{\perp,f}(\Omega) = \{a^n b^n c^n \mid n > 0\}$, a context-sensitive language.

This example suffices to underline that "part" of the computation can be done by choosing the right observer, keeping unchanged the underlying basic system.

Actually, this "part" can be really crucial: in fact, starting from universal type-0 grammars, one can construct an "universal" context-free grammar that can generate all recursively enumerable languages when observed in the correct manner. Formally,

**Theorem 4.** *There exists a context-free grammar that is universal m.o. for* RE. *[3].*

## 4    Restrictions on the Observed System

We can consider two restrictions on the observed systems and precisely on the considered context-free grammars: bounding the number of nonterminals and considering leftmost derivations only. In these cases we observe REG and CF, respectively.

The context-free grammar used as a universal grammar in Theorem 4 has no bound on the number of nonterminals in its sentential forms.

The next result shows that indeed this is a necessary property of context-free grammars that are observationally complete for type-0 grammars. In fact, when a bound is imposed, our observations constitute regular languages only. Recall that a context-free grammar is *nonterminal bounded* if there exists a constant $k$ such that all sentential forms generated by the grammar have at most $k$ nonterminals.

**Theorem 5.** *For every $G/O$ system $\Omega = (G, A)$ with $G$ nonterminal bounded context-free, $A \in FA_O$, $L_{\perp,f}(\Omega)$ is regular. [3]*

A *leftmost* generative grammar is a quadruple $G = (N, T, S, P)$ where $G$ has only leftmost derivations: we assume that $P \subseteq N^+ \times (N \cup T)^*$, and $\alpha \to \beta \in P$ implies $w\alpha x \Rightarrow w\beta x$ for $w \in T^*$ and $x \in (N \cup T)^*$.

A pushdown automaton corresponds to leftmost derivations in a type 0 grammar with productions of the form $pA \to aq\alpha$ for the instruction from state $p$ to $q$, reading $a$, popping $A$ and pushing $\alpha$ back to the stack. The following result implies that pushdown automata are less computationally powerful than context-free grammars when observed.

**Lemma 6.** *For every $G/O$ system $\Omega = (G, A)$ with $G$ a leftmost type-0 grammar, $A \in FA_O$, $L_{\perp,f}(\Omega)$ is context-free. [3]*

This result makes explicit the fact that, when considering this framework, the complexity of the produced output is influenced by the particular dynamics of the observed system.

## 5    Final Remarks

The survey has tried to show that a relevant parameter in defining the complexity of a computation can be the external observer. Theorem 4 proves that the observer can be crucial, if the observed system is sufficiently complex: it is possible to 'observe' every recursively enumerable language from a fixed context-free grammar.

It would be then extremely interesting to find grammars (in general systems) that can characterise (by changing the observer) families of languages with certain specified properties. Other directions of investigations are possible: for instance, what classes can be obtained without making use of the symbol $\perp$ (largely used in most of the proofs of the results presented here), or with $G/O$ systems using weaker observers (considering interesting restrictions on finite state automata). On this respect, which are the realistic limitations one should impose on the observer? An interesting suggestion, would be to impose a bound on the time (window) of observation. Another interesting new topic concerns the possibility of using such framework to learn in an easier way complex languages ([8]). Along these lines, we expect several interesting results and, maybe, being a new research area, new type of questions.

# References


1. Alhazov, A., Cavaliere, M.: Computing by observing bio-systems: The case of sticker systems. In: Ferretti, C., Mauri, G., Zandron, C. (eds.) DNA 2004. LNCS, vol. 3384. Springer, Heidelberg (2005)
2. Cavaliere, M., Jonoska, N., Leupold, P.: DNA splicing: Computing by observing. Natural Computing (to appear)
3. Cavaliere, M., Hoogeboom, H.J., Frisco, P.: Computing by only observing. In: H. Ibarra, O., Dang, Z. (eds.) DLT 2006. LNCS, vol. 4036. Springer, Heidelberg (2006)
4. Cavaliere, M., Leupold, P.: Observation of string-rewriting systems. Fundamenta Informaticae 74(4) (2006)
5. Cavaliere, M., Leupold, P.: Evolution and observation - A new way to look at membrane systems. In: Martín-Vide, C., Mauri, G., Păun, G., Rozenberg, G., Salomaa, A. (eds.) WMC 2003. LNCS, vol. 2933. Springer, Heidelberg (2004)
6. Cavaliere, M., Leupold, P.: Evolution and observation - A non-standard way to accept formal languages. In: Margenstern, M. (ed.) MCU 2004. LNCS, vol. 3354. Springer, Heidelberg (2005)
7. Cavaliere, M., Leupold, P.: Evolution and observation: A non-standard way to generate formal languages. Theoretical Computer Science 321 (2004)
8. Fernando, C., Cavaliere, M., Soyer, O., Goldstein, R.: Hebbian Learning of Context-Sensitive Languages (manuscript, 2008)
9. Rozenberg, G., Salomaa, A.: Handbook of Formal Languages. Springer, Heidelberg (1996)

# A Quantum Information-Theoretic Proof of the Relation between Horn's Problem and the Littlewood-Richardson Coefficients


Matthias Christandl

Centre for Quantum Computation, Department of Applied Mathematics and
Theoretical Physics, University of Cambridge, Wilberforce Road, Cambridge
CB3 0WA, United Kingdom
and
Arnold Sommerfeld Center for Theoretical Physics, Faculty of Physics,
Ludwig-Maximilians-Universität München, 80333 München, Germany
matthias.christandl@qubit.org



**Abstract.** Horn's problem asks for the conditions on sets of integers
$\mu$, $\nu$ and $\lambda$ that ensure the existence of Hermitian operators $A$, $B$ and
$A + B$ with spectra $\mu$, $\nu$ and $\lambda$, respectively. It has been shown that this
problem is equivalent to deciding whether $U_\lambda \subset U_\mu \otimes U_\nu$ for irreducible
representations of $\mathrm{GL}(d, \mathbb{C})$ with highest weights $\mu$, $\nu$ and $\lambda$. In this paper we present a quantum information-theoretic proof of the relation
between the two problems that is asymptotic in one direction. This result has previously been obtained by Klyachko using geometric invariant
theory [1]. The work presented in this paper does not, however, touch
upon the non-asymptotic equivalence between the two problems, a result
that rests on the recently proven *saturation conjecture* for $\mathrm{GL}(d, \mathbb{C})$ [2].


## 1 Introduction and Results

Given three spectra $\mu, \nu$ and $\lambda$, are there Hermitian operators $A$, $B$ with

$$(\mathrm{Spec}\, A, \mathrm{Spec}\, B, \mathrm{Spec}\, A + B) = (\mu, \nu, \lambda) \ \ ?$$

It is known as Horn's problem to characterise the set of triples $(\mu, \nu, \lambda)$ which
have an affirmative answer. Those form a convex polytope whose describing
inequalities have been conjectured by Horn in 1962 [3]. In this paper, we will
not be concerned with the characterisation of the polytope itself which has by
now been achieved [1] [2] but rather with the connection of Horn's problem
to the representation theory of $\mathrm{GL}(d, \mathbb{C})$. This connection was first noted by
B. V. Lidskii [4] and emerges as a natural twist to Klyachko's work on the
inequalities. More precisely, he proves the following two theorems relating the
admissible spectral triples to the Littlewood-Richardson coefficients $c_{\mu\nu}^\lambda$. $c_{\mu\nu}^\lambda$ is
the multiplicity of the irreducible representation $U_\lambda$ of $\mathrm{GL}(d, \mathbb{C})$ in the tensor
product representation $U_\mu \otimes U_\nu$ of $\mathrm{GL}(d, \mathbb{C})$. $\mu, \nu$ and $\lambda$ denote the highest weights
of the respective representations.


A. Beckmann, C. Dimitracopoulos, and B. Löwe (Eds.): CiE 2008, LNCS 5028, pp. 120–128, 2008.
© Springer-Verlag Berlin Heidelberg 2008

**Theorem 1.** *If $c^\lambda_{\mu\nu} \neq 0$, then there exist Hermitian operators $A$ and $B$ such that*

$$(Spec\ A,\ Spec\ B,\ Spec\ A + B) = (\mu, \nu, \lambda).$$

**Theorem 2.** *For Hermitian operators $A$, $B$ and $C := A + B$ with integral spectra $\mu$, $\nu$ and $\lambda$, there is an $N \in \mathbb{N}$ such that*

$$c^{N\lambda}_{N\mu, N\nu} \neq 0.$$

The original proofs of both theorems are based on deep results in geometric invariant theory. The contributions of this paper are elementary quantum-information-theoretic proofs of Theorem 1 and of the following variant of Theorem 2:

**Theorem 3.** *For all Hermitian operators $A$, $B$ and $C := A + B$ on $\mathbb{C}^d$ with spectra $\mu, \nu$ and $\lambda$, there is a sequence $(\mu^{(j)}, \nu^{(j)}, \lambda^{(j)})$, such that*

$$c^{\lambda^{(j)}}_{\mu^{(j)} \nu^{(j)}} \neq 0$$

*and*

$$\lim_{j \to \infty} \frac{\mu^{(j)}}{j} = Spec\ A$$

$$\lim_{j \to \infty} \frac{\nu^{(j)}}{j} = Spec\ B$$

$$\lim_{j \to \infty} \frac{\lambda^{(j)}}{j} = Spec\ A + B.$$

Theorems 2 and 3 can be shown to be equivalent with help of the fact that the triples $(\mu, \nu, \lambda)$ with nonvanishing Littlewood-Richardson coefficient form a finitely generated semigroup (see [5] for a similar equivalence in the context of the quantum marginal problem). Here, we choose to prove Theorem 3 since it more naturally fits our quantum information-theoretic approach. The basis of this approach is an estimation theorem for the spectrum of a density operator (Theorem 4) [6], which has recently been used [5] [7][8] to prove a relation analogous to the one presented in this paper between the Kronecker coefficient of the symmetric group and the spectra of a bipartite density operator and its margins.

In 1999, Knutson and Tao proved the *saturation conjecture* for $\mathrm{GL}(d, \mathbb{C})$, i.e. they proved that

$$c^{N\lambda}_{N\mu, N\nu} \neq 0 \text{ for some } N \in \mathbb{N} \text{ implies } c^\lambda_{\mu\nu} \neq 0.$$

This result implies that the $N$ in Theorem 2 can be taken to be one and the equivalence of the two problems is strict and not only asymptotic. The proof appeared in [2] and introduces the *honeycomb* model. A more compact version of this proof based on the *hive* model was given by [9], and a more accessible discussion can be found in [10].

We proceed with the introduction of the necessary group theory and quantum information theory before turning to the proofs.

# 2   Preliminaries

## 2.1   Spectrum Estimation

The tensor space $(\mathbb{C}^d)^{\otimes k}$ carries the action of the symmetric group $S_k$ which permutes the tensor factors and the diagonal action of $\mathrm{GL}(d, \mathbb{C})$: $g \mapsto g^{\otimes k}$. Since those actions commute in a maximal way, the tensor space decomposes in a form known as Schur-Weyl duality:

$$(\mathbb{C}^d)^{\otimes k} \cong \bigoplus_\lambda U_\lambda \otimes V_\lambda,$$

where $U_\lambda$ and $V_\lambda$ are irreducible representations of $\mathrm{GL}(d, \mathbb{C})$ and $S_k$, respectively. The sum extends over all labels $\lambda$ that are partitions of $k$ into $d$ parts, i.e. $\lambda = (\lambda_1, \ldots, \lambda_d)$ where the positive integers $\lambda_i$ obey $\lambda_i \geq \lambda_{i+1}$. As a label of an irreducible representation of $\mathrm{GL}(d, \mathbb{C})$, $\lambda$ is a *dominant weight* and as a label of an irreducible representation of $S_k$ it is a *Young frame*.

The following theorem has been discovered by Alicki, Rudnicki and Sadowski in the context of quantum optics [11] and independently by Keyl and Werner for use in quantum information theory [6]. In [7] a short account of Hayashi and Matsumoto's elegant proof [12] of this theorem is given.

**Theorem 4.** *Let $(\mathbb{C}^d)^{\otimes k} \cong \bigoplus_\lambda U_\lambda \otimes V_\lambda$ be the decomposition of tensor space according to Schur-Weyl duality and denote by $P_\lambda$ the projection onto $U_\lambda \otimes V_\lambda$. Then for any density operator $\rho$ with spectrum $r$ we have*

$$Tr\, P_\lambda \rho^{\otimes k} \leq (k+1)^{d(d-1)/2} \exp\left(-k D(\bar{\lambda}||r)\right) \tag{1}$$

*where $D(\cdot||\cdot)$ denotes the Kullback-Leibler distance of two probability distributions and $\bar{\lambda} = (\bar{\lambda}_1, \ldots, \bar{\lambda}_d) = (\frac{\lambda_1}{|\lambda|}, \ldots, \frac{\lambda_d}{|\lambda|})$. $|\lambda| = \sum_i \lambda_i = k$.*

This theorem can be interpreted as follows: The joint measurement of $k$ copies of the state $\rho$ by projection onto the spaces $U_{\lambda'} \otimes V_{\lambda'}$ will – with high probability – result in a measurement outcome $\lambda' = \lambda$ satisfying $\frac{\lambda}{k} \approx r$. $\frac{\lambda}{k}$ is therefore an estimate for the spectrum of $\rho$. Indeed the error exponent in eq.(1) is optimal [6].

## 2.2   Littlewood-Richardson Coefficients

Given two irreducible representations $U_\mu$ and $U_\nu$ of $\mathrm{GL}(d, \mathbb{C})$ with highest weights $\mu$ and $\nu$ we decompose the tensor product representation $U_\mu \otimes U_\nu$ of $\mathrm{GL}(d, \mathbb{C})$ (here, the group is represented simultaneously with $U_\mu$ and $U_\nu$) into irreducible representations of $\mathrm{GL}(d, \mathbb{C})$

$$U_\mu \otimes U_\nu \cong \bigoplus_\lambda c_{\mu\nu}^\lambda U_\lambda, \tag{2}$$

where $c_{\mu\nu}^\lambda$ denotes the multiplicity of $U_\lambda$ and is known as the *Littlewood - Richardson coefficient*. Since $\mathrm{GL}(d, \mathbb{C})$ is the complexification of $U(d)$, the unitary group in $d$ dimensions, we are allowed to – and will later on – regard all representations as representations of $U(d)$. The definition of the Littlewood-Richardson coefficient in eq. (2) is indeed the standard one. In the proofs below, however, we will work with a different definition given in terms of the symmetric group:

$$V_\lambda \downarrow^{S_n}_{S_k \times S_{n-k}} \cong \bigoplus_{\mu,\nu} c^\lambda_{\mu\nu} V_\mu \otimes V_\nu. \tag{3}$$

Here, we restricted the irreducible representation $V_\lambda$ of $S_n$ to the subgroup $S_k \times S_{n-k}$ and decomposed it into products of irreducible representations of $S_k$ and $S_{n-k}$. Observing that $S_n$ is self-dual, i.e. $V_\lambda^\star \cong V_\lambda$, this definition can be put into the following invariant-theoretic form

$$c^\lambda_{\mu\nu} = \dim(V_\lambda \otimes V_\mu \otimes V_\nu)^{S_k \times S_{n-k}}, \tag{4}$$

where $S_k \times S_{n-k}$ acts simultaneously on $V_\lambda$ and $V_\mu \otimes V_\nu$. Clearly, this characterisation only applies to Young frames, i.e. dominant weights with non-negative parts. The extension to the case of arbitrary dominant weights follows from the observation that $c^\lambda_{\mu\nu}$ is invariant under the transformation

$$\mu \mapsto \mu' := \mu + m(1^d)$$
$$\nu \mapsto \nu' := \nu + n(1^d) \tag{5}$$
$$\lambda \mapsto \lambda' := \lambda + (m+n)(1^d)$$

for integers $m$ and $n$, i.e. $c^\lambda_{\mu,\nu} = c^{\lambda'}_{\mu',\nu'}$. $(1^d)$ is short for $(1,\ldots,1)$ ($d$ ones).

# 3   Proofs

## 3.1   From Hermitian Operators to Density Operators

It will suffice to prove our results for nonnegative Hermitian operators and Young frames, i.e. dominant weights with nonnegative parts. In order to see this, assume that Theorem 1 holds for Young frames and consider an arbitrary triple of dominant weights $(\mu, \nu, \lambda)$ with $c^\lambda_{\mu,\nu} \neq 0$. Choose $m$ and $n$ large enough so that $(\mu', \nu', \lambda')$ defined above has no negative parts. Since $c^{\lambda'}_{\mu',\nu'} \neq 0$ there are positive Hermitian operators $A'$ and $B'$ with

$$(\operatorname{Spec} A', \operatorname{Spec} B', \operatorname{Spec} A' + B') = (\mu', \nu', \lambda').$$

This equation is equivalent to

$$(\operatorname{Spec} A, \operatorname{Spec} B, \operatorname{Spec} A + B) = (\mu, \nu, \lambda),$$

where $A := A' - m\mathbb{1}$ and $B := B' - n\mathbb{1}$. The latter is obtained by subtracting $(m(1^d), n(1^d), (m+n)(1^d))$ on both sides of the former observing that $\operatorname{Spec} (A' - m\mathbb{1}) = \operatorname{Spec} A' - m(1^d)$ (similarly for $B$). A similar argument can be carried out for Theorem 3.

Since we want to use intuition from quantum information theory, we define $p = \operatorname{Tr} A/(\operatorname{Tr} A + B)$, $\rho^A = A/\operatorname{Tr} A$ and $\rho^B = B/\operatorname{Tr} B$. The conditions on the spectra of $(A, B, A + B)$ are then equivalent to the conditions on the spectra of $(\rho^A, \rho^B, p\rho^A + (1-p)\rho^B)$, the convex mixture of density operators (i.e. trace one positive Hermitian operators).

We will therefore prove the following two theorems which are equivalent to Theorems 1 and 3 by the above discussion.

**Theorem 5.** *Let $(\mu, \nu, \lambda)$ be a triple of Young frames with $c_{\mu,\nu}^{\lambda} \neq 0$. Then there exist quantum states $\rho^A$ and $\rho^B$ such that*

$$Spec\ \rho^A = \bar{\mu}$$
$$Spec\ \rho^B = \bar{\nu}$$
$$Spec\ \rho^C = \bar{\lambda},$$

*where $p = \frac{|\mu|}{|\lambda|}$ and $\rho^C = p\rho^A + (1-p)\rho^B$.*

**Theorem 6.** *For all density operators $\rho^A$, $\rho^B$ and $\rho^C = p\rho^A + (1-p)\rho^B$ on $\mathbb{C}^d$ with spectra $\mu, \nu$ and $\lambda$ and $p \in [0,1]$, there is a sequence $(\mu^{(j)}, \nu^{(j)}, \lambda^{(j)})$, such that*

$$c_{\mu^{(j)},\nu^{(j)}}^{\lambda^{(j)}} \neq 0$$

*and*

$$\lim_{j\to\infty} \bar{\mu}^{(j)} = Spec\ \rho^A$$

$$\lim_{j\to\infty} \bar{\nu}^{(j)} = Spec\ \rho^B$$

$$\lim_{j\to\infty} \bar{\lambda}^{(j)} = Spec\ \rho^C.$$

### 3.2  Proof of Theorem 5

We assume without loss of generality that $0 < p \leq \frac{1}{2}$. It is well-known that the Littlewood-Richardson coefficients form a semigroup, i.e. $c_{\mu\nu}^{\lambda} \neq 0$ and $c_{\mu'\nu'}^{\lambda'} \neq 0$ implies $c_{\mu+\mu',\nu+\nu'}^{\lambda+\lambda'} \neq 0$[13][14]. As a consequence, $c_{\mu\nu}^{\lambda} \neq 0$ implies $c_{N\mu N\nu}^{N\lambda} \neq 0$ for all $N$. For every $N$ we will now construct density operators $\rho_N^A$ and $\rho_N^B$ whose limits $\rho^A := \lim_{N\to\infty} \rho_N^A$ and $\rho^B := \lim_{N\to\infty} \rho_N^B$ satisfy the claim of the theorem.

Fix a natural number $N$, set $n := N|\lambda|$ as well as $k := N|\mu|$ and let $p := \frac{k}{n}$. Since $c_{\mu\nu}^{\lambda}$ can only be nonzero if $|\mu| + |\nu| = |\lambda|$, we further have $n - k = N|\nu|$. As explained above, the nonnegativity of the parts of $\mu, \nu$ and $\lambda$ allows us to invoke the characterisation of the Littlewood-Richardson coefficient in terms of the symmetric group:

$$c_{N\mu,N\nu}^{N\lambda} = \dim(V_{N\mu} \otimes V_{N\nu} \otimes V_{N\lambda})^{S_k \times S_{n-k}},$$

where $S_k$ acts on $V_{N\mu}$, $S_{n-k}$ on $V_{N\nu}$ and $S_k \times S_{n-k} \subset S_n$ on $V_{N\lambda}$. Now pick a nonzero $|\Psi_N\rangle \in (V_{N\mu} \otimes V_{N\nu} \otimes V_{N\lambda})^{S_k \times S_{n-k}}$. Consider

$$\mathcal{H}^{(1)} \otimes \cdots \otimes \mathcal{H}^{(k)} \otimes \mathcal{H}^{(k+1)} \otimes \cdots \otimes \mathcal{H}^{(n)}$$
$$\otimes \mathcal{K}^{(1)} \otimes \cdots \otimes \mathcal{K}^{(k)} \otimes \mathcal{K}^{(k+1)} \otimes \cdots \otimes \mathcal{K}^{(n)}, \tag{6}$$

where $\mathcal{H}^{(i)}$ and $\mathcal{K}^{(j)}$ are isomorphic to $\mathbb{C}^d$. Embed the representation $V_{N\mu}$ in $\mathcal{H}^{(1)} \otimes \cdots \otimes \mathcal{H}^{(k)}$, $V_{N\nu}$ in $\mathcal{H}^{(k+1)} \otimes \cdots \otimes \mathcal{H}^{(n)}$ and $V_{N\lambda}$ in $\mathcal{K}^{(1)} \otimes \cdots \otimes \mathcal{K}^{(n)}$. The symmetric group $S_n$ permutes the pairs $\mathcal{H}^{(i)} \otimes \mathcal{K}^{(i)} \cong \mathbb{C}^{d^2}$ and its subgroup $S_k \times S_{n-k}$ permutes the first $k$ and the last $n-k$ pairs separately.

Any irreducible representation of the group $S_k \times S_{n-k}$ is isomorphic to a tensor product of irreducible representations of $S_k$ and $S_{n-k}$. $|\Psi_N\rangle$ is a trivial representation of $S_k \times S_{n-k}$ and can therefore only be isomorphic to the tensor product $V_k \otimes V_{n-k}$ of the trivial representations $V_k \equiv V_{(k,0,\ldots,0)}$ of $S_k$ and $V_{n-k} \equiv V_{(n-k,0,\ldots,0)}$ of $S_{n-k}$. On the first $k$ pairs the $k$-fold tensor product of $g \in \mathrm{U}(d^2)$ commutes with the action of $S_k$, and on the remaining pairs it is the $n-k$-fold tensor product of $g \in \mathrm{U}(d^2)$ which commutes with $S_{n-k}$. Schur-Weyl duality decomposes the space in (6) into

$$\bigoplus_{\tau,\tau'} U_\tau^{d^2} \otimes V_\tau \otimes U_{\tau'}^{d^2} \otimes V_{\tau'},$$

so that

$$|\Psi_N\rangle \in U_k^{d^2} \otimes V_k \otimes U_{n-k}^{d^2} \otimes V_{n-k},$$

and in terms of projectors onto those spaces

$$|\Psi_N\rangle\langle\Psi_N| \le P_k \otimes P_{n-k}$$
$$= [\dim U_k^{d^2} \int_{\mathbb{C}P^{d^2-1}} |\psi\rangle\langle\psi|^{\otimes k} d\psi] \otimes [\dim U_{n-k}^{d^2} \int_{\mathbb{C}P^{d^2-1}} |\phi\rangle\langle\phi|^{\otimes(n-k)} d\phi].$$

This directly implies

$$1 = \mathrm{Tr}\,|\Psi_N\rangle\langle\Psi_N| P_k \otimes P_{n-k}$$
$$\le \dim U_k^{d^2} \dim U_{n-k}^{d^2} \max_{\psi,\phi} \mathrm{Tr}\,|\Psi_N\rangle\langle\Psi_N||\psi\rangle\langle\psi|^{\otimes k} \otimes |\phi\rangle\langle\phi|^{\otimes(n-k)}$$

and therefore guarantees the existence of vectors $|\phi_N\rangle$ and $|\psi_N\rangle$ satisfying

$$\mathrm{Tr}\,[|\Psi_N\rangle\langle\Psi_N||\phi_N\rangle\langle\phi_N|^{\otimes k} \otimes |\psi_N\rangle\langle\psi_N|^{\otimes(n-k)}] \ge (\dim U_k^{d^2} \dim U_{n-k}^{d^2})^{-1}.$$

Since $|\Psi_N\rangle\langle\Psi_N| \le P_{N\mu} \otimes P_{N\nu} \otimes P_{N\lambda}$ we have

$$\mathrm{Tr}\,[P_{N\mu} \otimes P_{N\nu} \otimes P_{N\lambda}][|\phi_N\rangle\langle\phi_N|^{\otimes pn} \otimes |\psi_N\rangle\langle\psi_N|^{\otimes(1-p)n}]$$
$$\ge \mathrm{Tr}\,|\Psi_N\rangle\langle\Psi_N|[|\phi_N\rangle\langle\phi_N|^{\otimes pn} \otimes |\psi_N\rangle\langle\psi_N|^{\otimes(1-p)n}].$$

We define

$$\rho_N^A = \mathrm{Tr}\,_{\mathcal{K}^{(1)}} |\phi_N\rangle\langle\phi_N| = \mathrm{Tr}\,_{\mathcal{H}^{(1)}} |\phi_N\rangle\langle\phi_N| \tag{7}$$
$$\rho_N^B = \mathrm{Tr}\,_{\mathcal{K}^{(k+1)}} |\psi_N\rangle\langle\psi_N| = \mathrm{Tr}\,_{\mathcal{H}^{(k+1)}} |\psi_N\rangle\langle\psi_N| \tag{8}$$

and find, defining $\rho_N^C = p\rho_N^A + (1-p)\rho_N^B$ which satisfies

$$\operatorname{Tr} P_{N\lambda}(\rho_N^C)^{\otimes k} \geq \frac{1}{n+1}\operatorname{Tr} P_{N\lambda}(\rho_N^A)^{\otimes pn} \otimes (\rho_N^B)^{\otimes(1-p)n},$$

that

$$\operatorname{Tr} P_{N\mu}(\rho_N^A)^{\otimes pn} \geq (\dim U_k^{d^2} \dim U_{n-k}^{d^2})^{-1}$$
$$\operatorname{Tr} P_{N\nu}(\rho_N^B)^{\otimes(1-p)n} \geq (\dim U_k^{d^2} \dim U_{n-k}^{d^2})^{-1}$$
$$\operatorname{Tr} P_{N\lambda}(\rho_N^C)^{\otimes n} \geq (n+1)^{-1}(\dim U_k^{d^2} \dim U_{n-k}^{d^2})^{-1}.$$

Since $\dim U_n^{d^2} \leq n^{-d^2}$ these are inverse polynomial lower bounds, which, contrasted with the exponential upper bounds from Theorem 4,

$$\operatorname{Tr} P_\mu \rho_N^{A\,\otimes k} \leq (k+1)^{d(d-1)/2}\exp(-kD(\bar\mu||r^A)) \leq (k+1)^{d(d-1)/2}\exp(-k\epsilon^2/2)$$

and similarly for $\rho_N^B$ and $\rho_N^C$, imply

$$||\operatorname{Spec} \rho_N^A - \bar\mu|| \leq \epsilon$$
$$||\operatorname{Spec} \rho_N^B - \bar\nu|| \leq \epsilon$$
$$||\operatorname{Spec} \rho_N^C - \bar\lambda|| \leq \epsilon$$

for $\epsilon = O(d\sqrt{(\log N)/N})$. The proof is now completed, since $N$ was arbitrary and the existence of the limiting operators is guaranteed by the compactness of the set of density operators.

### 3.3   Proof of Theorem 6

We assume without loss of generality that $0 < p \leq \frac{1}{2}$. If $p$ is rational, consider a positive integer $n$ such that $k = pn$ (otherwise, approximate $p$ by a sequence of fractions $k/n$). Define purifications $|\psi\rangle^{AC}$ and $|\phi\rangle^{BC}$ of $\rho^A$ and $\rho^B$, respectively such that $p\operatorname{Tr}_A|\psi\rangle\langle\psi|^{AC} + (1-p)\operatorname{Tr}_B|\phi\rangle\langle\phi|^{BC} = \rho^C$. Consider the vector

$$|\tau\rangle = |\psi\rangle^{A_1 C_1} \otimes \cdots \otimes |\psi\rangle^{A_k C_k} \otimes |\phi\rangle^{B_1 C_{k+1}} \otimes \cdots \otimes |\phi\rangle^{B_{n-k} C_n},$$

where $|\psi\rangle^{A_j C_j} = |\psi\rangle^{AC}$ and $|\phi\rangle^{B_j C_{k+j}} = |\phi\rangle^{BC}$. This vector is invariant under the action of $S_k$ permuting systems $A_j C_j$ and $S_{n-k}$ permuting the systems $B_j C_{k+j}$ and is therefore of the form

$$|\tau\rangle = \sum_{\mu,\nu,\lambda} |\tau_{\mu\nu\lambda}\rangle,$$

for vectors $|\tau_{\mu\nu\lambda}\rangle \in U_\mu \otimes U_\nu \otimes U_\lambda \otimes \left(V_\mu \otimes V_\nu \otimes V_\lambda\right)^{S_k \times S_{n-k}}$.

By Theorem 4, for all $\epsilon > 0$ and $\mu$ with $\bar\mu \notin \mathcal{B}_\epsilon(r^A) = \{x := (x_1,\ldots,x_d) : ||x - r^A||_1 \leq \epsilon\}$ we have

$$\operatorname{Tr} P_\mu(\rho^A)^{\otimes k} \leq (k+1)^{d(d-1)/2}\exp(-kD(\bar\mu||r^A)) \leq (k+1)^{d(d-1)/2}e^{-\frac{k\epsilon^2}{2\ln 2}}$$

where Pinsker's inequality $D(\bar{\mu}\|r^A) \geq \frac{\|\bar{\mu}-r^A\|_1^2}{2\ln 2}$ was used in the last inequality. Similar statements hold for $\rho^B$ and $\rho^C$. Together with

$$\text{Tr}\, P_\lambda \text{Tr}_{A_1\cdots A_k B_1\cdots B_{n-k}} |\tau\rangle\langle\tau| = \text{Tr}\, P_\lambda (\text{Tr}_A |\psi\rangle\langle\psi|^{AC})^{\otimes k} \otimes (\text{Tr}_B |\phi\rangle\langle\phi|^{BC})^{\otimes n-k}$$

$$= \text{Tr}\, P_\lambda \frac{1}{n!} \sum_{\pi\in S_n} \pi (\text{Tr}_A |\psi\rangle\langle\psi|^{AC})^{\otimes k} \otimes (\text{Tr}_B |\phi\rangle\langle\phi|^{BC})^{\otimes n-k} \pi^{-1}$$

$$\leq (n+1)\text{Tr}\, P_\lambda (\rho^C)^{\otimes n}$$

we obtain (see [7])

$$\text{Tr}\, P_\mu \otimes P_\nu \otimes P_\lambda |\tau\rangle\langle\tau| \leq (n+1)^{d(d-1)/2}(n+3)e^{-\frac{pn\epsilon^2}{2}}.$$

This estimate can be turned around to give

$$\sum_{\substack{(\mu,\nu,\lambda):(\bar{\mu},\bar{\nu},\bar{\lambda})\in \\ (\mathcal{B}_\epsilon(r^A),\mathcal{B}_\epsilon(r^B),\mathcal{B}_\epsilon(r^C))}} \text{Tr}\, P_\mu \otimes P_\nu \otimes P_\lambda |\tau\rangle\langle\tau| \geq 1-\delta, \tag{9}$$

for $\delta := (n+1)^{d(d+8)/2}(n+3)e^{-\frac{pn\epsilon^2}{2\ln 2}}$, because the number of Young frames with $n$ boxes in $d$ rows is smaller than $(n+1)^d$.

For positive RHS of equation (9) the existence of a triple $(\mu,\nu,\lambda)$ with $\|\bar{\mu} - r^A\| \leq \epsilon$ (and for $\nu$ and $\lambda$ alike) and $|\tau_{\mu\nu\lambda}\rangle \neq 0$ is therefore guaranteed. In particular,

$$c_{\mu\nu}^\lambda = \dim(V_\mu \otimes V_\nu \otimes V_\lambda)^{S_k \times S_{n-k}} \neq 0$$

holds. The proof of the theorem is completed with the choice of an increasing sequence of appropriate integers $n$. The speed of convergence of the resulting sequence of normalised triples to $(r^A, r^B, r^C)$ can be estimated with $\epsilon = O(d\sqrt{(\log n)/n})$, a value for which the LHS of eq.(9) is bounded away from zero.

## Acknowledgment

The technique used in this paper was developed in collaboration with Graeme Mitchison and Aram Harrow in the context of the quantum marginal problem. I would like to thank both of them for many enlightening discussions. The hospitality of the *Accademia di Danimarca* in Rome, where part of this work was carried out, is gratefully acknowledged. This work was supported by the European Commission through the FP6-FET Integrated Project SCALA CT-015714, an EPSRC Postdoctoral Fellowship and a Nevile Research Fellowship of Magdalene College Cambridge.

## References

1. Klyachko, A.A.: Stable bundles, representation theory and Hermitian operators. Sel. math. New. ser. 4, 419–445 (1998)
2. Knutson, A., Tao, T.: The honeycomb model of $GL_n(C)$ tensor products I: Proof of the saturation conjecture. J. Am. Math. Soc. 12(4), 1055–1090 (1999)

3. Horn, A.: Eigenvalues of sums of Hermitian matrices. Pacif. J. Math. 12, 225–241 (1962)
4. Lidskii, B.V.: Spectral polyhedron of the sum of two Hermitian matrices. Func. Anal. Appl. 16, 139–140 (1982)
5. Christandl, M., Harrow, A.W., Mitchison, G.: On nonzero Kronecker coefficients and what they tell us about spectra. Comm. Math. Phys. 270(3), 575–585 (2007)
6. Keyl, M., Werner, R.F.: Estimating the spectrum of a density operator. Phys. Rev. A 64(5), 052311 (2001)
7. Christandl, M., Mitchison, G.: The spectra of density operators and the Kronecker coefficients of the symmetric group. Comm. Math. Phys. 261(3), 789–797 (2005)
8. Christandl, M.: The Structure of Bipartite Quantum States: Insights from Group Theory and Cryptography. PhD thesis, University of Cambridge (February 2006), quant-ph/0604183
9. Buch, A.: The saturation conjecture (after Knutson, A.,  Tao, T.:). Enseign. Math. 46, 43–60 (2000)
10. Knutson, A., Tao, T.: Honeycombs and sums of Hermitian matrices. Notices Amer. Math. Soc. 48, 175–186 (2001)
11. Alicki, R., Rudnicki, S., Sadowski, S.: Symmetry properties of product states for the system of $N$ $n$-level atoms. J. Math. Phys. 29(5), 1158–1162 (1988)
12. Hayashi, M., Matsumoto, K.: Quantum universal variable-length source coding. Phys. Rev. A 66(2), 022311 (2002)
13. Elashvili, A.G.: Invariant algebras. Advances in Soviet Math. 8, 57–64 (1992)
14. Zelevinsky, A.: Littlewood-Richardson semigroups, math.CO/9704228 (1997)

# Pell Equations
## and Weak Regularity Principles


Charalampos Cornaros[*]

Department of Mathematics, University of Aegean, GR-832 00 Karlovassi, Greece
kornaros@aegean.gr



**Abstract.** We study the strength of weak forms of the Regularity Principle in the presence of $IE_1$ (induction on bounded existential formulas) relative to other subsystems of $PA$. In particular, the Bounded Weak Regularity Principle is formulated, and it is shown that when applied to $E_1$ formulas, this principle is equivalent over $IE_1^-$ to $I\Delta_0 + exp$.


## 1 Introduction

The Regularity Principle asserts that if infinitely many elements $x$ are assigned at least one color from a bounded palette, then infinitely many $x$ are assigned the same color. In this paper, we study versions of this principle over weak subsystems of first-order PA (=Peano Arithmetic).

As usual (see, e.g., [5], [7]), $I\Gamma$ denotes the induction schema for formulas in $\Gamma$ (plus the base theory $PA^-$), $L\Gamma$ denotes the least number schema for formulas in $\Gamma$ (plus $PA^-$) and $B\Gamma$ denotes the collection schema for formulas in $\Gamma$ (plus $I\Sigma_0$), where $\Gamma$ is one of the formula classes $\Sigma_n, \Sigma_0(\Sigma_n), \Pi_n, E_n, U_n$. Also, $I\Delta_n$ denotes the induction schema for $\Delta_n$ formulas (plus $PA^-$) and $L\Delta_n$ denotes the induction schema for $\Delta_n$ formulas (plus $PA^-$). The parameter free counterpart of $I\Gamma$ is denoted by $I\Gamma^-$. We write $\phi(\boldsymbol{x}) \in \nabla_n(T)$, for some theory $T$ and some $\phi(\boldsymbol{x}) \in E_n$, if there exists some $\psi(\boldsymbol{x}) \in U_n$ such that $T$ proves $\forall \boldsymbol{x}(\phi(\boldsymbol{x}) \leftrightarrow \psi(\boldsymbol{x}))$.

Our aim in this section is to define the versions of the regularity principle that will be studied in the sequel and give a summary of our results, in view of results already known.

**Definition 1.** *(Regularity Principle)* $R\phi$ *denotes the following formula*

$$(\forall w)(\exists x > w)(\exists y \le u)\phi(x, y) \to (\exists y \le u)(\forall w)(\exists x > w)\phi(x, y).$$

*For a class of formulas* $\Gamma$, $\quad R\Gamma = \{R\phi | \phi \in \Gamma\}$.

For the formula classes $\Sigma_m$ and $\Pi_m$, the regularity axiom schema is a well studied combinatorial principle. Indeed, the following result is well-known (see Theorem 2.23 in [5]).

<hr>

[*] The author would like to thank S. Boughattas, for useful discussions concerning this paper and for drawing his attention to the contents of [2].


A. Beckmann, C. Dimitracopoulos, and B. Löwe (Eds.): CiE 2008, LNCS 5028, pp. 129–138, 2008.

**Theorem 1.** *For any $m \in \mathbb{N}$,*

$$I\Delta_0 + B\Sigma_{m+2} \Leftrightarrow I\Delta_0 + R\Sigma_{m+1} \Leftrightarrow I\Delta_0 + R\Pi_m \Leftrightarrow I\Delta_0 + B\Pi_{m+1}.$$

(Note that we do not include any induction principle in the formulation of the collection schemas.)

Although other formula classes will be discussed in passing, the primary focus of the present work is on the principle's application to formulas in the class $\Delta_0$ and especially in its subclass $E_1$.

The author's interest in this subject was aroused in the following way. In her paper [1], P. D'Aquino (building on earlier work of R. Kaye, see [8]) proved that $IE_1 + P \vdash I\Delta_0 + exp$, where $P$ is an axiom asserting that every Pell equation has a nontrivial solution, and $exp$ is the axiom $(\forall x)(\exists y)\, \varphi_e(x, y)$, where $\varphi_e(x, y)$ is one of the usual $\Delta_0$ formulas defining the graph of the function $y = 2^x$ in $\mathbb{N}$. As $I\Delta_0 + exp \vdash P$, it follows that, in any model of $I\Delta_0$, every Pell equation has a nontrivial solution if and only if the model has a well defined, and total, exponential function. Recall that a Pell equation is an equation of the form $Y^2 - dX^2 = 1$, where $d$ is not a square. It is a well known theorem of elementary number theory that any Pell equation has a non trivial solution. Obviously any proof of this fact must use a function exceeding what is available in $IE_1$ (and even $I\Delta_0$). The strategy used for the classical proof goes as follows.

Step 1. Using Dirichlet's Approximation lemma, one can prove that there are infinitely many pairs of positive integers of the form $(p, q)$ such that the norm $N(p, q) = p^2 - q^2 d$ is bounded above by the number $\lfloor 2\sqrt{d} \rfloor + 2$ where $\lfloor \; \rfloor$ is the floor function.

Step 2. Using a combinatorial principle (regularity will do), one concludes that there are infinitely many pairs $(p_i, q_i)$ that

(i) for each $i$, $p_i \geq 0, q_i > 0$,

(ii) have the same norm, let us say $N(N \neq 0$, as $d$ is not a square$)$

(iii) are congruent modulo $N$, i.e., $p_i \equiv p_j (mod N)$ and $q_i \equiv q_j (mod N)$ for each $i \neq j$.

It follows that $(p_1 p_2 - d q_1 q_2)^2 - d(p_1 q_2 - p_2 q_1)^2 = N^2$, for any two distinct pairs $(p_1, q_1), (p_2, q_2)$. Setting now $p_1 p_2 - d q_1 q_2 = NY$ and $p_1 q_2 - p_2 q_1 = NX$, for some integers $X, Y$, it clearly follows that $X \neq 0$ and $Y^2 - dX^2 = 1$.

Let us make some comments about this strategy. We know (see [2]) that Dirichlet's lemma can be proved in $IE_1$ without any use of combinatorial principles. So the strength needed to prove the existence of a nontrivial solution must come from the combinatorial principle employed in Step 2. This combinatorial principle cannot be $PHP\Delta_0$ (or even $PHP\Sigma_1$), where $PHP\Gamma$ denotes the pigeonhole principle for formulas in $\Gamma$, because $PHP\Sigma_1$ cannot prove $exp$ (see [4]). As we will check below, aside from $IE_1$ only an instance of $RE_1$ is needed to prove the following result.

**Theorem 2.** $IE_1 + RE_1 \vdash P$ *and, therefore,* $IE_1 + RE_1 \vdash I\Delta_0 + exp$.

This allows us to replace $I\Delta_0$ with $IE_1$ in Theorem 1.

**Corollary 3.** *For each* $m \in \mathbb{N}$,

$$I\Delta_0 + B\Sigma_{m+2} \Leftrightarrow IE_1 + R\Sigma_{m+1} \Leftrightarrow IE_1 + R\Pi_m \Leftrightarrow I\Delta_0 + B\Pi_{m+1}.$$

It would seem to be difficult to prove the converse of the second part of Theorem 2.

**Problem 1.** *Does* $I\Delta_0 + exp$ *prove* $RE_1$?

A. R. Woods (personal communication) has observed that, if we have a positive answer to Problem 1, then we would take a negative solution to the open problem of whether $I\Delta_0 + exp \vdash \Delta_0 \equiv E_1$. As we can show, $IE_1 + RU_1 \vdash I\Sigma_1$, without any additional assumptions (as usual, $U_1$ denotes the class of bounded universal formulas) and it follows that $RU_1$ is strong enough for $I\Delta_0 + exp$. In search for a "reversal" of the sort suggested by Problem 1, it is natural to try to reduce the strength of the regularity principle. One possibility is to restrict the rate of increase of $x$ in the definition of regularity down to polynomial growth.

**Definition 2.** *(Bounded Regularity Principle) For* $m \geq 1$, $R(\phi, m)$ *denotes the formula*

$$(\exists r > 1)(\forall w > 1)(\exists x > w)(\exists y \leq u)(rw^m \geq x \wedge \phi(x, y)) \rightarrow$$
$$(\exists y \leq u)(\forall w > 1)(\exists x > w)\phi(x, y).$$

As before, we can show that $IE_1 + R(E_1, 3) \vdash I\Delta_0 + exp$ (see Corollary 12). However, we have not managed to prove the converse implication. An alternative approach is to consider a new version of $R$, which we will call "weak regularity principle".

**Definition 3.** *(Weak Regularity Principle)* $WR\phi$ *denotes the formula*

$(\forall w)(\exists x > w)(\exists y \leq u)\phi(x, y) \rightarrow$
$(\forall w)(\exists y \leq u)(\exists x_1 > w)(\exists x_2 > w)[\phi(x_1, y) \wedge \phi(x_2, y) \wedge x_1 \neq x_2].$

In other words, this principle says: If it is possible to find infinitely many elements $x$, which can be assigned to some color $y$ from a bounded palette, then it is possible to find two arbitrarily large distinct elements, assigned to the same color of the palette. Actually this is the situation in Step 2. The "color" assigned to the pair $(p, q)$ can be considered to be (a code for) the triple $(N(p, q), p(mod|N(p, q)|), q(mod|N(p, q)|))$, where $|N(p, q)|$ is the absolute value of the norm of $(p, q)$. As we will see, it is enough to use regularity (bounded or not) for just $6D^3$ colors, where $D = \lfloor 2\sqrt{d} \rfloor + 2$ ($d$ not a square). Note that we need only two distinct pairs $(p_i, q_i)$ with the same "color", for the existence of a non trivial solution of the Pell equation $Y^2 - dX^2 = 1$. So weak regularity would suffice. Also note that the problem of finding infinitely many pairs $(p_i, q_i)$ with the above properties is obviously much harder than finding just two pairs with the same properties. But again, the last task is difficult because, otherwise, it would be easy to find a nontrivial solution of the Pell equation, which is not true, as it is generally believed that solving Pell equations appears to be harder than factoring.

Generally speaking, the phrase "infinitely many elements" could have a different meaning instead of the normal one, e.g. "arbitrarily large elements". For this reason, we have replaced $<$ with the symbol $\prec$, corresponding to a definable strict partial order. Then we can define the (weak) regularity principle in terms of $\prec$. We denote them by $R(\phi, \prec)$ and $WR(\phi, \prec)$ respectively. For example, $WR(\phi, \prec)$ is defined as follows:

$$(\forall w)(\exists x \succ w)(\exists y \leq u)\phi(x, y) \to$$
$$(\forall w)(\exists y \leq u)(\exists x_1 \succ w)(\exists x_2 \succ w)[\phi(x_1, y) \wedge \phi(x_2, y) \wedge x_1 \prec x_2],$$

where $u$ could be a free variable of $\phi$ but $w$ is not allowed to appear free in $\phi$. In the sequel we will also study some other interesting limit principles, namely $R(\Gamma_1, \Gamma_2)$ (respectively $WR(\Gamma_1, \Gamma_2)$), which denote the axiom schemas $R(\phi, \prec)$ (respectively $WR(\phi, \prec)$) for each $\phi \in \Gamma_1$, $\prec \in \Gamma_2$ plus the usual axioms describing that $\prec$ is a strict order.

Finally, we can also formulate a bounded version of weak regularity.

**Definition 4. (*Bounded Weak Regularity Principle*)** *By $WR(\phi, \prec, m)$ we denote the m-bounded version of the weak regularity principle, that is,*

$$(\exists r > 1)(\forall w > 1)(\exists x \leq rw^m)(\exists y \leq u)(\phi(x, y) \wedge x > 1 \wedge w \prec x) \to$$
$$(\forall w > 1)(\exists y \leq u)(\exists x_1, x_2)(w \prec x_1 \wedge w \prec x_2 \wedge x_1 \prec x_2 \wedge \phi(x_1, y) \wedge \phi(x_2, y)).$$

In the following sections, we will study the strength of the (bounded) weak regularity principle over various base theories. Our results can be summarized as follows.

- $WR(\Delta_{n+1}, \Delta_{n+1}) \leftrightarrow WR(E_1, \Sigma_0(\Sigma_n)) \leftrightarrow WR(U_1, \Sigma_0(\Sigma_n)) \leftrightarrow I\Sigma_{n+1}$.
- $IE_1^- + WR(\Sigma_0(\Sigma_n), <) \vdash B\Sigma_{n+1} + exp$.
- $WR(E_0, \Sigma_0(\Sigma_n), 1) \vdash I\Sigma_n$.
- For all $m \geq 3, n \geq 0$,
  $WR(\Delta_{n+1}, \Delta_{n+1}, m) \leftrightarrow WR(E_1, \Delta_{n+1}, m) \leftrightarrow B\Sigma_{n+1} + I\Delta_0 + exp$,
  $WR(\Delta_0, \Delta_0, m) \leftrightarrow WR(E_1, \Delta_0, m) \leftrightarrow I\Delta_0 + WR(\Delta_0, <, 1) \leftrightarrow I\Delta_0 + exp$.
- There exists a $\forall \exists U_1$-formula $W$, describing some particular instances of bounded weak regularity for $\nabla_1(IE_1^-)$ formulas and for the standard order $(<)$, such that $IE_1^- \vdash P \leftrightarrow W \leftrightarrow I\Delta_0 + exp$.

Let us make some comparison of the "infinite" principles $R$ and $WR$, with the well-known "finite" pigeonhole principles (we call them "finite" because they involve bounded sets). We know that (see [4]) for each $n \geq 1$,

- $PHP(\Sigma_0(\Sigma_n)) \Leftrightarrow I\Sigma_n$
- $I\Delta_0 + B\Sigma_{n+1} \Leftrightarrow PHP(\Sigma_{n+1})$.

But, unfortunately, $PHP(\Sigma_n)$ loses its strength for the case $n = 1$ or $n = 0$. Both cases of $PHP$ can be proved from (the strictly stronger) $I\Delta_0 + WR(\Delta_0, <)$ and $I\Delta_0 + WR(\Delta_0, <, 1)$ respectively. We could generally say that the strength of $WR$ (bounded or not) stands between those of $R$ and $PHP$. This is true for at least $\Sigma_0(\Sigma_n)$ formulas, because we can show that for all $n \geq 0, m \geq 1$,

$$IE_1 + R(\Sigma_0(\Sigma_n), m) \vdash WR(\Sigma_0(\Sigma_n), \Sigma_0(\Sigma_n)) \vdash WR(E_1, \Sigma_0(\Sigma_n)) \vdash$$
$$I\Delta_0 + WR(\Sigma_0(\Sigma_n), <) \vdash B\Sigma_{n+1} + exp \vdash PHP\Delta_{n+1} \vdash PHP(\Sigma_0(\Sigma_n)).$$

By the above results, it follows that there exist natural combinatorial principles equivalent to $I\Delta_0 + exp$ over $IE_1^-$. In view of this fact, one is naturally led to some interesting problems, such as the following.

**Problem 2.** *Are there any natural combinatorial principles with strength exactly $I\Delta_0 + \Omega_1$?*

**Problem 3.** *Are there "finite" combinatorial principles which have strength strictly greater than polynomial growth for models of $I\Delta_0$?*

In what follows, we presume that the reader is familiar with the relationships among different weak subsystems of PA (details can be found in [5], [7], [10]).

## 2   Limit Schemata and Weak Regularity

By modifying the proof of Theorem 1 in [3], one can obtain the following theorem.

**Theorem 4.** *If $f$ is a $\Sigma_0(\Sigma_n)$, $n \geq 0$, definable function (with parameters) over $I\Sigma_n$, then the following limit schemata are equivalent:*

*(i)  $I\Sigma_{n+1}$*
*(ii)  $(\exists x)(\forall m)(f(m) \leq x) \rightarrow (\exists m)(\forall n > m)(f(n) \leq f(m))$*
*(iii)  $(\exists x)(\forall m)(f(m) \leq x) \rightarrow (\exists m)(\forall n)(f(n) \leq f(m))$*
*(iv)  $(\forall n)(f(n) \leq f(n+1)) \wedge (\exists x)(\forall m)(f(m) \leq x) \rightarrow (\exists m)(\forall n > m)(f(n) = f(m)).$*

We now use this result to prove

**Theorem 5.** *For each $n \in \mathbb{N}$,*

*(i)  $I\Sigma_n + WR(E_1, \Sigma_0(\Sigma_n)) \vdash I\Sigma_{n+1}$*
*(ii)  $I\Sigma_n + WR(U_1, \Sigma_0(\Sigma_n)) \vdash I\Sigma_{n+1}.$*
*(iii)  $I\Sigma_n + WR(\Sigma_0(\Sigma_n), <) \vdash B\Sigma_{n+1}.$*

**Proof.** It suffices to prove part (ii) of Theorem 4. So let $f$ be a $\Sigma_0(\Sigma_n)$ definable function such that $(\exists x)(\forall m)(f(m) \leq x)$ but $(\forall m)(\exists n > m)(f(n) > f(m))$. Now define

$$n \succ m \Leftrightarrow (\exists n_1 \leq n)(\exists n_2 \leq n)(\exists m_1 \leq m)(\exists m_2 \leq m)[n = \langle n_1, n_2 \rangle \wedge$$
$$m = \langle m_1, m_2 \rangle \wedge n_2 = f(n_1) \wedge n_1 > m_1 \wedge n_2 > f(m_1)]$$

and let $\phi(x, y)$ denote the formula $y = (x)_2$, i.e., the $\langle\,\rangle$-second part of $x$, where $\langle\,\rangle$ is the standard coding of pairs. Clearly, $\phi$ is an $E_1$ (and $U_1$) formula and $\prec$

is $\Sigma_0(\Sigma_n)$ definable. So, using $WR(E_1, \Sigma_0(\Sigma_n))$, we can reach a contradiction. For (iii), let us suppose that some $\theta(x, y) \in \Pi_n$ does not satisfies collection, so

$$(\forall y < a)(\exists x)\theta(x, y) \text{ and } (\forall w)(\exists y < a)(\exists x > w)\theta(x, y).$$

Define $F(y) = x \leftrightarrow \theta(x, y) \wedge (\forall z < x)\neg\theta(z, y)$. Clearly, the range of $F$ is unbounded, thus $(\forall w)(\exists x > w)(\exists y < a)F(y) = x$. Using $WR(\Sigma_0(\Sigma_n), <)$, we reach a contradiction.    $\square$

As we will now see, $I\Sigma_n$ is redundant in parts (i) and (ii) of Theorem 5. Using Theorem 16, $I\Sigma_n$ can be replaced by $IE_1^-$ in (iii).

**Theorem 6.** *For any* $n \in \mathbb{N}$,

$$I\Sigma_{n+1} \vdash WR(\Delta_{n+1}, \Delta_{n+1}) \vdash WR(E_0, \Sigma_0(\Sigma_n), 1) \vdash I\Sigma_n.$$

**Proof.** To prove the first implication, let $M \models I\Sigma_{n+1}$, $u, w_0 \in M$ and suppose that $M \models (\forall w)(\exists x \succ w)(\exists y \leq u)\phi(x, y)$, for some $\phi \in \Delta_{n+1}$ and $\prec \in \Delta_{n+1}$. First, we define a $\Delta_{n+1}$-definable function $F : M \to M$ by

$$F(w) = x \Leftrightarrow x \succ w \wedge (\exists y \leq u)\phi(x, y) \wedge (\forall z < x)[(\exists y \leq u)\phi(z, y) \to z \not\succ w].$$

Using $I\Sigma_{n+1}$, we can easily prove, by induction on $w$, that

$$(\forall w)(\exists y \leq u)(\exists x)(x = F^{(w+1)}(w_0) \wedge \phi(x, y)). \tag{$\dagger$}$$

But the set $\{(w, y) : w \leq u + 1 \wedge y \leq u \wedge \exists x[x = F^{(w+1)}(w_0) \wedge \phi(x, y)]\}$ is $\Sigma_{n+1}$ definable, so it can be coded in $M$ by some element, i.e., there exists $t \in M$ such that $(\forall w \leq u + 1)(\exists y \leq u)(\langle w, y \rangle \in t)$. Using $PHP\Delta_0$ we can find $w_1, w_2$ with $w_1 < w_2$ and $y \leq u$ such that $\langle w_1, u \rangle \in t \wedge \langle w_2, u \rangle \in t$. So there exist two distinct $x_1 = F^{(w_1+1)}(w_0)$, $x_2 = F^{(w_2+1)}(w_0)$ such that

$$(\exists y \leq u)(\exists x_2 \succ w_0)(\exists x_1 \succ w_0)(\phi(x_1, y) \wedge \phi(x_2, y) \wedge x_1 \prec x_2).$$

To prove the second implication now, let us suppose that

$$M \models WR(E_0, \Sigma_0(\Sigma_n), 1) \text{ and } M \models \neg I\Sigma_n.$$

Then there exist $\theta(x, \boldsymbol{u}) \in \Sigma_n$ and $a, \boldsymbol{b} \in M$ such that

$$M \models \theta(0, \boldsymbol{b}) \wedge \forall x(\theta(x, \boldsymbol{b}) \to \theta(x + 1, \boldsymbol{b})) \wedge \neg\theta(a, \boldsymbol{b}).$$

Define $x \succ w \leftrightarrow (w > a \wedge x \leq a) \vee (w < a \wedge a \geq x > w \wedge \theta(x, \boldsymbol{b}) \wedge \theta(w, \boldsymbol{b})) \vee (w \leq a \wedge x < w \wedge \neg\theta(x, \boldsymbol{b}) \wedge \neg\theta(w, \boldsymbol{b}))$ and let $\phi(x, y)$ denote the formula $x = y$. Obviously, $(\forall w > 1)(\exists x \succ w)(\exists y \leq a)[\phi(x, y) \wedge 1 < x \leq aw]$, so, by the hypothesis, we obtain

$$(\forall w > 1)(\exists y \leq a)(\exists x_1 \succ w)(\exists x_2 \succ w)[\phi(x_1, y) \wedge \phi(x_2, y) \wedge x_1 \prec x_2].$$

But now taking $w = a$ we arrive at a contradiction.    $\square$

By repeating the above proof and noting that, over $I\Delta_0 + exp$, $I\Delta_{n+1}$ is equivalent to $B\Sigma_{n+1}$ (see [9]) and $x$ is bounded by a power of $w_0$ and $r$ in ($\dagger$) above, we obtain

**Corollary 7.** *For each $n \in \mathbb{N}$ and $m \geq 1$,*

$$I\Delta_0 + exp \vdash WR(\Delta_{n+1}, \Delta_{n+1}, m) \leftrightarrow WR(E_0, \Delta_{n+1}, m) \leftrightarrow B\Sigma_{n+1}.$$

In the next section we will prove the equivalence of $WR(\Delta_0, \Delta_0, 3)$ with $I\Delta_0 + exp$. More precisely, we will show the equivalence of the theories

$$IE_1^- + WR(E_1, <, m), \ \ I\Delta_0 + exp \text{ and } IE_1^- + WR(U_1, <, m),$$

for each $m \geq 3$. So we obtain also the following

**Corollary 8.** *For any $n \in \mathbb{N}$ and $m \geq 3$,*

$$WR(\Delta_{n+1}, \Delta_{n+1}, m) \leftrightarrow WR(E_1, \Delta_{n+1}, m) \leftrightarrow$$
$$WR(U_1, \Delta_{n+1}, m) \leftrightarrow I\Delta_0 + B\Sigma_{n+1} + exp.$$

## 3 Dirichlet's Approximation Lemma in $IE_1$

We start with some simple facts about $IE_1$. $IE_1$ can prove that $\sqrt{d}$ is irrational for every $d$ not a square, i.e.,

$$IE_1 \vdash \forall d [\forall a \leq d(a^2 \neq d) \rightarrow \forall x \forall y(x^2 = dy^2 \rightarrow x = y = 0)].$$

Note that, for any $M \models IE_1$ and $d \in M$ not a square, we can define in a natural way $=, +, \cdot$, and $<$ in the quadratic extension $M[\sqrt{d}]$, by means of $E_0$ formulas. For example, we define $p + \sqrt{d}q < r + \sqrt{d}s$ by

$$[p \geq r \rightarrow s > q \wedge (p-r)^2 < d(s-q)^2] \wedge [p < r \wedge s < q \rightarrow (p-r)^2 > d(s-q)^2].$$

Clearly, we can also extend $<$ for the case where $p, q, r, s$ are rational numbers, i.e., elements of $\mathbb{Q}(M)$. Note that this definition makes $<$ a total order on $\mathbb{Q}(M)^2$. We identify every $n \in M$ with $n + \sqrt{d}\, 0 \in M[\sqrt{d}]$. We also consider any element of $M[\sqrt{d}](= \mathbb{Q}(M)^2)$ of the form $r + \sqrt{d}\, 0$ as "rational" and any other element as "irrational". It is easy to prove the existence of $\lfloor x + \sqrt{d}y \rfloor$ (the integer part of $x + \sqrt{d}y$). It follows that $\lfloor \sqrt{d} \rfloor < \sqrt{d} < \lfloor \sqrt{d} \rfloor + 1$, so dividing by a sufficiently large element of $M$ we have proved that $M[\sqrt{d}]$ has arbitrarily small irrational elements between 0 and 1.

We continue now with a definition and a lemma that are necessary to prove Dirichlet's Approximation Lemma in $IE_1$ and come from [2].

**Definition 5.** *(**Farey series**) Let $I$ be a Bézout domain and $N > 0$ an arbitrary element of $I$. The Farey series $\mathfrak{F}_N$ of order $N$ is the ascending series of irreducible fractions between 0 and 1, whose denominators do not exceed $N$.*

**Lemma 9** *Suppose that $M \models IE_1$, $d \in M$ is not a square and $\alpha \in M[\sqrt{d}]$ irrational between 0 and 1. Then, for every $N > 0$ in $M$, there exist two successive elements of $\mathfrak{F}_N$ such that $\alpha$ lies between them.*

Using this Lemma, we can easily prove.

**Theorem 10.** *(Dirichlet's Approximation Lemma in $IE_1$)* *If $M, d$ are as in Lemma 9, then for every irrational $a$ in $M[\sqrt{d}]$ and every $Q > 0$, there exist $p, q \in M$, such that $(Q+1)^3 \geq q > Q$ and*

$$|a - \frac{p}{q}| < \frac{1}{q^2}.$$

**Theorem 11.** *There exists an $\nabla_1(IE_1^-)$ formula $\phi$, depending on a parameter $d$, such that*

$$IE_1 \vdash (\forall d)(d \neq \square \wedge WR(\phi_d, <, 3) \rightarrow (\exists x)(\exists y)(x > 1 \wedge x^2 - dy^2 = 1)),$$

*where $<$ denotes the usual ordering.*

**Proof.** The construction of $\phi_d$ follows the ideas described in the introduction. Let $\phi_d(x, y)$ be the formula

$$x = \langle p, q \rangle \wedge |p^2 - dq^2| < D \wedge q \neq 0 \wedge$$
$$y = (p^2 - dq^2 + D) + (p \bmod |p^2 - dq^2|)(2D) + (q \bmod |p^2 - dq^2|)(2D)^2,$$

where $D = \lfloor 2\sqrt{d} \rfloor + 2$, $q \bmod |p^2 - dq^2|$ and $p \bmod |p^2 - dq^2|$ are defined in the usual way. Thus $\phi_d$ can be described in an $E_1$ or an $U_1$ formula over $IE_1$, i.e., $\phi_d \in \nabla_1(IE_1)(= \nabla_1(IE_1^-))$. Using Dirichlet's Lemma we can easily prove that:

$$(\forall w > 1)(\exists x > w)(\exists y < 6D^3)(x \leq d^{12}w^3 \wedge \phi_d(x, y))$$

or, in other words,

$$\forall p \forall q \exists s \exists r \exists y < 6D^3(\langle p, q \rangle > 1 \rightarrow \langle p, q \rangle < \langle s, r \rangle \leq d^{12}\langle p, q \rangle^3) \wedge \phi_d(\langle x, y \rangle, y)),$$

where by $\langle \, , \, \rangle$ we denote the standard coding function (note that its basic properties are provable in $IE_1^-$).

Indeed, let $w = \langle p, q \rangle$ and consider $p - \sqrt{d} \, q \in M[\sqrt{d}]$. Applying Dirichlet's Lemma for $Q = \max\{d, q + p\}$ and $a = \sqrt{d}$, we get some $s - \sqrt{d} \, r$ such that $r > Q$ and $|s - \sqrt{d} \, r| < \frac{1}{r}$. From $s + r > p + q$ we deduce $\langle s, r \rangle > \langle p, q \rangle$ and $s + \sqrt{d} \, r = |s - \sqrt{d} \, r + 2\sqrt{d} \, r| < 2r\sqrt{d} + \frac{1}{r}$. By multiplying, we obtain

$$0 < |s^2 - dr^2| < 2\sqrt{d} + \frac{1}{r^2} < 2\sqrt{d} + 1 < D.$$

Hence $|s^2 - dr^2 + D| = s^2 - dr^2 + D < 2D$ and so $y < 6D^3$.

<u>Case 1.</u> $Q = d(\geq 2)$. From $s < dr$, it follows that $\langle s, r \rangle < \langle dr, r \rangle < 2d^2r^2 \leq 2d^2(d+1)^6 < 2^7 d^8 = 2^3 2^4 d^8 \leq w^3 d^{12}$.

<u>Case 2.</u> $Q = q + q \geq d$. Then $\langle s, r \rangle < \langle dr, r \rangle < 2d^2r^2 \leq 2d^2(Q+1)^6$. But from $\langle p, q \rangle > \frac{(Q+1)^2}{4}$ we take $4^3\langle p, q \rangle^3 > (Q+1)^6$ and thus $4^3 w^3 > (Q+1)^6$. It follows that $\langle s, r \rangle < 2d^2 4^3 w^3 \leq d^9 w^3$.

Hence, assuming $M \models WR(\phi_d, <, 3)$, we can take $r = d^{12}$ as a bound in any case to obtain two distinct pairs $(s_1, r_1), (s_2, r_2)$ such that $0 < r_1$, $0 < r_2$ and some $y < 6D^3$ such that $\phi_d(\langle s_1, r_1 \rangle, y) \wedge \phi_d(\langle s_2, r_2 \rangle, y)$. By the definition of $\phi_d$, it follows that

$$s_1^2 - dr_1^2 = s_2^2 - dr_2^2 \text{ and}$$
$$s_1 = s_2 \ mod|s_1^2 - dr_1^2| \wedge r_1 = r_2 \ mod|s_1^2 - dr_1^2|. \quad (*)$$

Furthermore, if we suppose that $\frac{s_1}{r_1} = \frac{s_2}{r_2}$, then for some $m \geq 0$ we have $s_1 = r_1 m$ and $s_2 = r_2 m$. Replacing $s_1$ by $r_1 m$ and $s_2$ by $r_2 m$ in $s_1^2 - dr_1^2 = s_2^2 - dr_2^2$, we get $r_2 = r_1$, i.e., a contradiction. From $(*)$, by multiplication, we take $(s_1 s_2 - dr_1 r_2)^2 - d(s_1 r_2 - s_2 r_1)^2 = k^2$, where $k = (s_1^2 - dr_1^2)$. Also $(*)$ implies that there exist $x, y, y \neq 0$, such that $s_1 s_2 - dr_1 r_2 = kx \wedge s_1 r_2 - s_2 r_1 = ky$. It follows that $x^2 - dy^2 = 1$ and $x > 1$, as required. $\qquad \square$

## 4  Bounded WR vs. *exp*

As already mentioned, it was proved in [1] that

$$IE_1^- + P \vdash\dashv I\Delta_0 + exp,$$

where $P$ is the axiom expressing "every Pell equation has a nontrivial solution". As a direct consequence of this result and Theorem 11, we obtain

**Corollary 12.**  $IE_1 + WR(E_1, <, 3) \vdash I\Delta_0 + exp$ *and so* $IE_1 + R(E_1, 3) \vdash I\Delta_0 + exp$.

A problem that naturally arises now is

**Problem 4.** *Does* $IE_1 + WR(E_1, <, n) \vdash I\Delta_0 + exp$ *for* $n = 1$ *or* $2$?

The converse of Corollary 12 can be easily seen to hold.

**Lemma 13** *Let* $M \models I\Delta_0 + exp$ *and* $\prec$ *be a strict partial order on* $M$ *defined by a* $\Delta_0$ *formula. Then, for every* $m > 0$ *and* $\phi \in \Delta_0$, $M \models WR(\phi, \prec, m)$.

**Corollary 14.** *For every* $m \geq 3$,

$$WR(E_1, \Delta_0, m) \leftrightarrow WR(U_1, \Delta_0, m) \leftrightarrow I\Delta_0 + exp.$$

Recall that if we take either $WR(E_1, \Delta_0)$ or $WR(U_1, \Delta_0)$ instead of their bounded versions, then we gain much power as Theorem 5 shows.

For the rest of the paper, by $W$ we denote the $\forall\exists U_1$-formula

$$(\forall d)(d \neq \square \to WR_d(\phi_d, <, 3)),$$

where $\phi_d(x, y)$ is the formula defined in Theorem 11 and $WR_d(\phi_d, <, 3)$ is

$$(\forall u)[u = 6(\lfloor 2\sqrt{d} \rfloor + 2)^3 \wedge (\forall w > 1)(\exists w < x \leq d^{12} w^3)(\exists y \leq u)\phi_d(x, y) \to$$
$$(\forall w > 1)(\exists y \leq u)(\exists x_1, x_2)(w < x_1 < x_2 \wedge \phi_d(x_1, y) \wedge \phi_d(x_2, y))].$$

Note that $P$ is a $\forall\exists E_0$-formula. Lemma 13 can be restated as $I\Delta_0 + exp \vdash W$ and so we have proved.

**Theorem 15.** $IE_1 \vdash P \leftrightarrow W$.

Note that we can improve Corollary 12 as follows.

**Theorem 16.** $IE_1^- + W \vdash I\Delta_0 + exp$.

**Proof.** As we saw in the proof of Theorem 11, $IE_1$ proves

$$(\forall d, p, q)(d \neq \square \rightarrow (\exists D \exists s \exists r \exists y < 6D^3)[\langle p, q \rangle > 1 \rightarrow$$
$$\langle p, q \rangle < \langle s, r \rangle < d^{12}\langle p, q \rangle^3 \wedge \phi_d(\langle s, r \rangle, y)]), \quad (*)$$

where $D = \lfloor 2\sqrt{d} \rfloor + 2$. By inspection, the formula above is $\forall E_1$. But, according to a result in [6], $IE_1$ is a $\exists \forall E_1$ conservative extension of $IE_1^-$ and so $(*)$ is provable in $IE_1^-$. Using this observation we are now able to prove $IE_1^- + W \vdash P$ as in the proof of Theorem 11. The result follows, since $IE_1^- + P \vdash I\Delta_0 + exp$ (see Corollary 4.10 in [1]). $\qquad\square$

# References


1. D'Aquino, P.: Pell equations and exponentiation in fragments of arithmetic, Ann. Pure Appl. Logic 77, 1–34 (1996)
2. Ayat, S.M.: The Skolem-Bang Theorems in Ordered Fields with an IP, arxiv.org/abs/0705.3356
3. Beklemishev, L.D., Visser, A.: On the limit existence principles in elementary arithmetic and $\Sigma_n^0$-consequences of theories. Ann. Pure Appl. Logic 136, 56–74 (2005)
4. Dimitracopoulos, C., Paris, J.: The pigeonhole principle and fragments of arithmetic. Z. Math. Logik Grundlag. Math. 32, 73–80 (1986)
5. Hájek, P., Pudlák, P.: Metamathematics of first-order arithmetic. Springer, Berlin (1993)
6. Kaye, R.: Diophantine induction and parameter-free induction, Ph. D. Dissertation, Manchester (1987)
7. Kaye, R.: Parameter-free universal induction. Z. Math. Logik Grundlag. Math. 35, 443–456 (1989)
8. Kaye, R.: Diophantine induction. Ann. Pure Appl. Logic 46, 1–40 (1990)
9. Slaman, T.: $\Sigma_n$-bounding and $\Delta_n$-induction. Proc. Amer. Math. Soc. 132, 2449–2456 (2004)
10. Wilmers, G.: Bounded existential induction. J. Symbolic Logic 50, 72–90 (1985)

# Computable Categoricity of Graphs with Finite Components

Barbara F. Csima[1,*], Bakhadyr Khoussainov[2,**], and Jiamou Liu[3,***]

[1] Department of Pure Mathematics, University of Waterloo
csima@math.uwaterloo.ca
www.math.uwaterloo.ca/~csima
[2] Department of Computer Science, University of Auckland
bmk@cs.auckland.ac.nz
www.cs.auckland.ac.nz/~bmk
[3] Department of Computer Science, University of Auckland
jliu036@ec.auckland.ac.nz
www.cs.auckland.ac.nz/~jliu036

**Abstract.** A computable graph is computably categorical if any two computable presentations of the graph are computably isomorphic. In this paper we investigate the class of computably categorical graphs. We restrict ourselves to strongly locally finite graphs; these are the graphs all of whose components are finite. We present a necessary and sufficient condition for certain classes of strongly locally finite graphs to be computably categorical. We prove that if there exists an infinite $\Delta_2^0$-set of components that can be properly embedded into infinitely many components of the graph then the graph is not computably categorical. We outline the construction of a strongly locally finite computably categorical graph with an infinite chain of properly embedded components.

## 1  Introduction

In this paper we are interested in computable graphs. A **computable graph** $\mathcal{G}$ is a pair $(V, E)$ where the set $V$ of vertices and the set $E$ of edges are both computable sets. All our graphs are undirected and infinite. If $\mathcal{G}$ is a computable graph isomorphic to a graph $\mathcal{G}'$ then $\mathcal{G}$ is called a **computable presentation of $\mathcal{G}'$** and $\mathcal{G}'$ is called **computably presentable**. For a computable graph $\mathcal{G}$ we can always assume that the set of vertices of $\mathcal{G}$ is $\omega$, the set of natural numbers.

The study of computable structures goes back to the late 1950s and finds its roots in the work of A. Malcev [15] and M. Rabin [16]. Later the theory has been developed by Yu. Ershov and A. Nerode and their colleagues (e.g. [3]). For the current state of the area see, for example, the book by Ershov and Goncharov [7], the Handbooks on computable models and algebra [5] [6]. See also [11].

* Partially supported by Canadian NSERC Discovery Grant 312501.
** B. Khoussainov has partially been supported by Marsden Fund of Royal New Zealand Society.
*** J. Liu is supported by NZIDRS of Education New Zealand.

A. Beckmann, C. Dimitracopoulos, and B. Löwe (Eds.): CiE 2008, LNCS 5028, pp. 139–148, 2008.

One of the central themes in the theory of computable structures is concerned with computable isomorphisms. We say that two computable graphs $G_1, G_2$ have the same **computable isomorphism type** if $G_1$ and $G_2$ are computably isomorphic.

**Definition 1.** *The number of computable isomorphism types of graph $\mathcal{G}$, denoted by $dim(\mathcal{G})$, is called the* **computable dimension** *of $\mathcal{G}$. If the computable dimension of $\mathcal{G}$ equals 1 then the graph $\mathcal{G}$ is called* **computably categorical**.

For example the graph $(\omega, E)$ where $E = \{\{i, i+1\} \mid i \in \omega\}$ is computably categorical. The graph consisting of $\omega$ many copies of $(\omega, E)$ is not computable categorical; in fact, it has computable dimension $\omega$. In general, providing examples of computably categorical graphs or graphs of computable dimension $\omega$ is easy. S. S. Goncharov in [9] was the first to provide examples of graphs of computable dimension $n$, where $n > 1$. In this paper we will be interested in the study of computably categorical graphs in a specific class of graphs called strongly locally finite graphs.

The study of computably categorical structures constitutes one of the major topics in the study of computable isomorphisms. Here the goal is to provide a characterization of computably categorical structures within specific classes of structures. This has been done for Boolean algebras [4], linearly ordered sets [17], trees [14], Abelian groups [8], ordered Abelian groups [12], etc. Hence, this paper fits the general program devoted to the study of computable isomorphisms.

Let $S$ be a sequence $\mathcal{G}_0, \mathcal{G}_1, \ldots$ of pairwise disjoint finite graphs. Define the new graph $\mathcal{G}_S$ as the disjoint union of these graphs. More formally, the set of vertices of $\mathcal{G}_S$ is $\bigcup_{i \in \omega} V_i$ and the set of edges is $\bigcup_{i \in \omega} E_i$.

Let $\mathcal{G}$ be a graph. We say that vertices $v$ and $w$ are **connected** if there is a path from $v$ to $w$. In this case we also say that $w$ is **reachable** from $v$. A **component** of $\mathcal{G}$ is a maximal subset of $\mathcal{G}$ in which any two vertices are connected. The component containing a vertex $v$ is denoted by $C(v)$.

We say that $\mathcal{G}$ is **strongly locally finite** if every component of $\mathcal{G}$ forms a finite graph. It is not hard to see that $\mathcal{G}$ is strongly locally finite if and only if $\mathcal{G}$ is $\mathcal{G}_S$ for some sequence $S$ of pairwise disjoint finite graphs. The following proposition gives a full description of computable dimensions for strongly locally finite graphs:

**Proposition 1.** *The computable dimension of any strongly locally finite graph is either $1$ or $\omega$. In particular, no strongly locally finite graph has a finite computable dimension $n$, where $n > 1$.*

*Proof.* We invoke the following well-known result of Goncharov [10]. If any two computable presentations of a structure $\mathcal{A}$ are isomorphic via a $\Delta_2^0$-function then the computable dimension of $\mathcal{A}$ is either 1 or $\omega$. Now, if $G$ is strongly locally finite then any two computable presentations of $G$ are isomorphic via a $\Delta_2^0$-function.   $\square$

By this proposition, it makes perfect sense to work towards a characterization of computably categorical strongly locally finite graphs. This is the subject of this paper.

Finally, *all* our graphs considered in this paper are strongly locally finite.

## 2   Computable Categoricity and the Size Function

Let $\mathcal{G}$ be a computable graph. Define the **size function** $size_{\mathcal{G}} : V \to \omega$ by $size_{\mathcal{G}}(v) = |C(v)|$, where $C(v)$ is the component of vertex $v$.

**Lemma 1.** *Let $\mathcal{G}_1, \mathcal{G}_2$ be computable presentations of $\mathcal{G}$ such that $size_{\mathcal{G}_1}, size_{\mathcal{G}_2}$ are computable. Then $\mathcal{G}_1$ and $\mathcal{G}_2$ are computably isomorphic.*

*Proof.* For $i \in \{1, 2\}$, we can effectively reveal $C(v)$ for any vertex $v$ in $\mathcal{G}_i$ by searching for the $size_{\mathcal{G}_i}(v)$ vertices that are connected to $v$. To construct a computable isomorphism between $\mathcal{G}_1$ and $\mathcal{G}_2$, map each $v$ to the corresponding vertex $v'$ in $\mathcal{G}_2$ such that $C(v) \cong C(v')$. In the construction, use the back and forth method of building the isomorphism. $\qquad\square$

The lemma implies that $\mathcal{G}$ is computably categorical if the size function is computable for all computable presentations of $\mathcal{G}$.

**Proposition 2.** *Suppose $size_{\mathcal{G}}$ is a computable function. The graph $\mathcal{G}$ is computably categorical if and only if the size function is computable for all computable presentations of $\mathcal{G}$.*

*Proof.* One direction is proved by Lemma 1. The other direction is straightforward since from $\mathcal{G}$ to any computable presentation $\mathcal{G}'$ of $\mathcal{G}$ there is a computable isomorphism $h$. Then $size_{\mathcal{G}'}(v) = size_{\mathcal{G}}(h(v))$. $\qquad\square$

In the rest of this section we suppose that $size_{\mathcal{G}}$ is computable. For any vertex $v \in V$, one effectively reveals the component of $v$ by using $size_{\mathcal{G}}(v)$. So, we effectively list (without repetition) $C_0, C_1, \ldots$ all components of $G$.

Given two finite graphs $\mathcal{H}_1 = (V_1, E_1)$ and $\mathcal{H}_2 = (V_2, E_2)$, we say $\mathcal{H}_1$ **properly embeds** into $\mathcal{H}_2$ if $V_1$ can be mapped injectively to a *proper* subset of $V_2$ that preserves the edge relation. We denote it by $\mathcal{H}_1 \prec \mathcal{H}_2$.

**Lemma 2.** *If there are infinitely many $i$ such that $\{j \mid C_i \prec C_j\}$ is an infinite set, then $\mathcal{G}$ is not computably categorical.*

*Proof.* Our goal is to build a graph $\mathcal{G}' = (\omega, E')$ such that $\mathcal{G}' \cong \mathcal{G}$ but $\mathcal{G}'$ is not computably isomorphic to $\mathcal{G}$. Let $\Phi_0, \Phi_1, \ldots$ be a standard enumeration of all partial computable functions from $\omega$ to $\omega$. We construct a graph $\mathcal{G}'$ that satisfies the following requirements:

$$P_e : \Phi_e \text{ is not an isomorphism from } G \text{ to } G'$$

The requirement $P_e$ has a higher **priority** than $P_t$ if $t > e$. We construct $\mathcal{G}'$ by stages. At stage $s$ we construct a finite graph $\mathcal{G}'_s$ so that $\mathcal{G}'_s$ is isomorphic to $\mathcal{G}$ restricted to $C_0 \cup \ldots \cup C_{s-1}$, $\mathcal{G}'_s \subset \mathcal{G}'_{s+1}$ for all $s$, and $f_s$ is the isomorphism constructed at stage $s$. Our desired graph will be $\mathcal{G}' = \cup_s \mathcal{G}'_s$. Set $\mathcal{G}'_0$ to be the empty graph. Set $f_0$ to be undefined.

At stage $s + 1$, consider $\mathcal{G}_s$ obtained by adding $C_s$ to $\mathcal{G}_{s-1}$. Let $C'_0, \ldots, C'_{s-1}$ be all components in $\mathcal{G}'_{s-1}$ such that each $C'_i$ is isomorphic to $C_i$ via the partial function $f_s$ for $i < s$. Find minimal $e \leq s + 1$ such that for some $i < s$ we have:

1. $\Phi_e$ has not been processed and $\Phi_{e,s+1}$ is defined on $C_i$.
2. $\Phi_{e,s+1}$ is a partial isomorphism.
3. The component $C'_j = \Phi_e(C_i)$ is free for $\Phi_e$, and $C_i \prec C_s$.

If such $e$ does not exist then go on to the next stage. Otherwise, act as follows:
(1) Extend $C'_j$ to a component, denoted by $C'_s$, such that $C'_s \cong C_s$; (2) Build
a new copy $C''_j$ isomorphic to $C_j$; (3) Redefine $f_s$ by mapping $C_j$ to $C''_j$ and
$C_s$ to $C'_s$. Declare $C'_s$ not free for all $\Phi_t$ with $t > e$, and declare $\Phi_e$ processed.
This completes the construction for $\mathcal{G}'_{s+1}$.

The correctness of the construction is now a standard proof. The proof is
based on the following two observations. First of all, one inductively shows that
each requirement $P_e$ is satisfied. Secondly, one proves that the function $f(v) =
\lim_s f_s(v)$ establishes an isomorphism (which is necessarily a $\Delta^0_2$-set). $\qquad\square$

For a computable graph $\mathcal{G}$ with a computable size function, let $C_0, C_1, \ldots$ be an
effective list of all components of $\mathcal{G}$. Define the **proper extension function**
$ext_{\mathcal{G}} : \omega \to \omega$ by $ext_{\mathcal{G}}(i) = |\{j \mid C_i \prec C_j\}|$.

**Lemma 3.** *Suppose there are finitely many $i$ such that the set $\{j \mid C_i \prec C_j\}$ is
infinite. If $ext_{\mathcal{G}}$ is not computable then $\mathcal{G}$ is not computable categorical.*

*Proof.* The construction of $\mathcal{G}'$ that is isomorphic but not computably isomorphic
to $\mathcal{G}$ is very similar to the construction for the previous lemma. The only difference is that we start with $\mathcal{G}_0$ as consisting of all (finitely many) components in
$\mathcal{G}$ that embed into infinitely many components. Therefore in this construction
let $C_0, C_1, \ldots$ list all *other* components in $\mathcal{G}$. The construction of the previous
lemma is then repeated.

Suppose $P_e$ is the requirement with the highest priority that is not satisfied.
Let $s$ be the stage when all requirements with higher priorities are satisfied. Since
$\Phi_e$ is an isomorphism, we can compute the the function $ext_{\mathcal{G}}$ as follows. Consider
$C_i$ for which $\Phi_e(C_i)$ is free for $\Phi_e$. Note that there are only finitely many $C_i$ that
are not free for $\Phi_e$. Let $t$ be the stage $> s$ such that $\Phi_{e,t}$ is defined on $C_i$. From
this stage on $C_i$ can not be properly embedded into $C_k$ for all $k > t$. Hence the
number of proper extensions of $C_i$ in $\mathcal{G}_t$ can be computed effectively. $\qquad\square$

We can now prove the following characterization theorem:

**Theorem 1.** *Let $\mathcal{G}$ be a graph such that $size_{\mathcal{G}}$ is a computable function. Then
the following are equivalent:*

1. *$\mathcal{G}$ is computably categorical.*
2. *The size function is computable in all computable presentations of $\mathcal{G}$.*
3. *There are finitely many $i$ such that the set $\{j \mid C_i \prec C_j\}$ is infinite and the
   function $ext_{\mathcal{G}}$ is computable.*

*Proof.* The equivalence of (1) and (2) follows from Proposition 2. The direction
(1) to (3) follows from the lemmas above. We prove the implication (3) $\to$ (1).
So, let $\mathcal{G}'$ be a computable presentation of $\mathcal{G}$. Take all components $C_i$ such that

$\{j \mid C_i \prec C_j\}$ is infinite. There are only finitely many such $C_i$; non-uniformly map them to isomorphic components in $\mathcal{G}'$.

Take $C_i$ such that $\{j \mid C_i \prec C_j\}$ is finite. Since $ext_{\mathcal{G}}$ is computable, we can list all components $X_1, \ldots, X_p$ in $\mathcal{G}$ that properly extend $C_i$. In $\mathcal{G}'$ find components $Y, Y_1, \ldots, Y_p$ such that $Y$ is isomorphic to $C_i$ and each $Y_i$ is isomorphic to $X_i$. Map $C_i$ isomorphically to $Y$. It is not hard to show, using the definition of the function $ext_{\mathcal{G}}$ and induction on the number of proper extensions of $C_i$, that $Y$ is a component of $\mathcal{G}'$ isomorphic to $C_i$. $\qquad\square$

## 3   A Sufficient Condition for Not Computably Categorical

In this section we do *not* assume computability of the size function $size_{\mathcal{G}}$. The theorem below gives us a version of Lemma 2 in this case.

**Theorem 2.** *Let $\mathcal{G}$ be a strongly locally finite graph on which the reachability relation is computable. If there exists an infinite $\Delta_2^0$ set of vertices $X$ such that $(\forall x \in X)(\exists^\infty v)[C(x) \prec C(v)]$, then $\mathcal{G}$ is not computably categorical.*

*Proof.* For each $s \in \omega$, let $\mathcal{G}_s$ be the restriction of the graph of $\mathcal{G}$ to vertices among $\{0, \ldots, s\}$. Since $\mathcal{G}$ is computable, we can uniformly compute $\mathcal{G}_s$. For each $v \in \{0, \ldots, s\}$, let $C_s(v)$ denote the connected component of $v$ in $\mathcal{G}_s$. Since the reachability relation on $\mathcal{G}$ is computable, we may assume without loss of generality that if $C_{\max(v,w)}(v) \neq C_{\max(v,w)}(w)$, then $C_s(v) \neq C_s(w)$ for all $s$. That is, when a new vertex is added to the graph of $\mathcal{G}$ it is immediately decided whether it is in the same component as any previously present vertices.

We will build a computable graph $\mathcal{H} \cong \mathcal{G}$ such that we meet for each $e \in \omega$ the requirement:

$$R_e : \Phi_e \text{ is not an isomorphism from } \mathcal{H} \text{ to } \mathcal{G}$$

We will construct $\mathcal{H}$ by stages. At each stage $s$ we will have a function $h_s : \mathcal{G}_s \cong \mathcal{H}_s$ and we will ensure that $h = \lim_s h_s$ exists.

If we declare that $h_s(v) = w$, then we will define $h_s$ such that $h_s : C_s(v) \cong C_s(w)$. If at a later stage $t$ the component of $v$ in $\mathcal{G}$ grows $(C_s(v) \subsetneq C_t(v))$, and we still have $h_t(v) = h_s(v)$, then we will add a new vertex to $\mathcal{H}_t$ and define $h_t$ to extend $h_s$ so that $h_t : C_t(v) \cong C_t(w)$.

To meet requirement $R_e$ we will find a vertex $v_e$ such that either $\Phi_e(v_e) \uparrow$ or $C(v_e) \prec C(\Phi_e(v_e))$.

Let $\{X_s\}_{s \in \omega}$ be a $\Delta_2^0$ approximation of $X$. For $n, s \in \omega$, let $x_{n,s} = \mu x [x \in X_s \wedge (\forall m < n)[x \notin C_s(x_{m,s})]]$. Note that since $X$ is $\Delta_2^0$ and since each component of $\mathcal{G}$ is finite, $x_n = \lim_s x_{n,s}$ exists for all $n$.

At each stage $s$ of the construction, we will have $v_{e,s} = x_{n,s}$ for some $n \geq e$. We will ensure that for each $e \in \omega$, $v_e = \lim_s v_{e,s}$ exists and provides the witness for meeting requirement $R_e$.

The basic idea for meeting a single requirement $R_0$ is as follows. We let $v_{0,s} = x_{0,s}$ at every stage $s$. If we ever see that $\Phi_{0,s}(v_{0,s}) \downarrow$, and if $\Phi_0$ appears to be an isomorphism in the sense that the component of $v_{0,s}$ in $\mathcal{G}_s$ is isomorphic to the component of $\Phi_0(v_{0,s})$ in $\mathcal{H}_s$, then we begin to search for a new component to appear in $\mathcal{G}$ that properly extends the component of $v_{0,s}$. If $v_{0,s} \in X$, then we will find such a component. So, at the same time as searching for the component, we also run the approximation of $X$ to see if $v_{0,t} \neq v_{0,s}$ at some later stage $t$. If we first find out that $v_0$ changes, then we continue to wait for $\Phi_0$ to converge on this new $v_0$. If we are provided with a new component extending that of $v_{0,s}$ then we re-define our map $h$ and extend the graph $\mathcal{H}$ so that the component of $\Phi_0(v_{0,s})$ in $\mathcal{H}$ is now isomorphic to the new large component, and we include a new component in $\mathcal{H}$ that is isomorphic to the component of $v_{0,s}$ in $\mathcal{G}$. Thus at the end of stage $s + 1$, we will have $C_s(v_{0,s}) \prec C_s(\Phi_e(v_{0,s}))$. This will have us meet requirement $R_0$ unless the component of $v_{0,s}$ in $\mathcal{G}$ grows at some later stage. If this happens, we again search for a proper extension of the component of $v_0$ in $\mathcal{G}$ to complete the diagonalization. Note that after a certain stage, $v_{0,s}$ will never change, and will always be a member of $X$. Since the component of $v_0$ in $\mathcal{G}$ is finite, it can grow only finitely often. If after the component of $v_0$ in $\mathcal{G}$ has fully appeared we see that $\Phi_0(v_0) \downarrow$, then we will at that point succeed in meeting requirement $R_0$.

The only extra complication for multiple requirements is that we want to ensure that $h : \mathcal{H} \cong \mathcal{G}$, so we must make sure that if some $w \in \text{range}(h_s)$, then $h^{-1}(w)$ exists. That is, we only re-define $h_s^{-1}(w)$ finitely often. This is where we will use the $v_e$ instead of just $x_e$ as witnesses. If we find that $\Phi_e(v_{e,s}) \downarrow$, but is mapped to some component where we have already redefined $h$ for the sake of higher priority requirements, then instead of proceeding with the diagonalization, we will change $v_e$ to be the next member of $X$ (i.e., if $v_{e,s} = x_{n,s}$, we would let $v_{e,s+1} = x_{n+1,s+1}$). Since each requirement only causes $h$ to be re-defined finitely often, $v_e$ will only be re-defined finitely often for this reason. If we notice that we were wrong about our guess for $x_n$ (i.e., $x_{n,s} \neq x_{n,s+1}$), then we will drop back down all the $v_{e,s} \geq x_{n,s}$ to be as small as possible.

We now give the formal construction.

We may assume without loss of generality that if $C_s(v) \neq C_s(v')$, and if $\Phi_e(v) \downarrow$ and $\Phi_e(v') \downarrow$ then $C_s(\Phi_e(v)) \neq C_s(\Phi_e(v'))$. This is because since $\mathcal{G}$ has the computable reachability relation, $C_s(v) \neq C_s(v') \Rightarrow C(v) \neq C(v')$, so if $\Phi_e$ maps $v$ and $v'$ to the same component in $\mathcal{H}$ then we immediately have $R_e$ satisfied. We also assume that $\Phi_{e,s}(x) \downarrow \Rightarrow (\forall y < x)[\Phi_{e,s}(y) \downarrow]$.

*Stage 0:* Let $v_{e,0} = x_{e,0}$ for all $e \in \omega$. Let $h_0(0) = 0$. Let $\mathcal{H}_0$ have the single vertex 0 and no edges.

*Stage $s + 1$:*

*Step 1:* Choose the least $e$ such that $\Phi_{e,s+1}(v_{e,s}) \downarrow$ and $C_{s+1}(v_{e,s}) \cong C_{s+1}(\Phi_{e,s+1}(v_{e,s}))$, and such that $x_{n,s+1} = x_{n,s}$, where $n$ is such that $x_{n,s} = v_{e,s}$. If no such $e$ exists, move to Step 2. If $h^{-1}$ or $h$ have already been re-defined at earlier stages by higher priority requirements on $\Phi_{e,s+1}(v_{e,s})$ or $h^{-1}(\Phi_{e,s+1}(v_{e,s}))$, respectively, then set $v_{e,s+1} = x_{n+1,s+1}$. For $i > e$, let $v_{i,s+1} = x_{n+1+i-e,s+1}$. For $i < e$, let $v_{i,s+1} = x_{m,s+1}$, where $m$ is such that $x_{m,s} = v_{i,s}$. Move to stage $s + 2$.

Otherwise, speed up the enumeration of $\mathcal{G}$ and the approximation of $X$ until we either find some $t > s$ such that $v_{e,t} \neq v_{e,s}$ (more precisely, $x_{n,t} \neq x_{n,s}$, where $v_{e,s} = x_{n,s}$), or we find some $t > s$ such that there exists $v \in \mathcal{G}_t$, $v \notin \mathrm{dom}(h_s)$, and $C_t(v_{e,s+1}) \prec C_t(v)$. In the first case, move to step 2. In the second case, re-define $\mathcal{H}$ setting $h_{s+1}(v) = \Phi_{e,s+1}(v_{e,s})$ and expand the component of $\Phi_{e,s+1}(v_{e,s})$ to be isomorphic to $C_t(v)$. Also introduce a new component isomorphic to $C_t(h_s^{-1}(\Phi_e(v_{e,s})))$ into $\mathcal{H}_{s+1}$, and define $h_{s+1}$ on $C_t(h_s^{-1}(\Phi_e(v_{e,s})))$ accordingly.

*Step 2:* Let $n$ be least such that $x_{n,s+1} \neq x_{n,s}$. For $e$ such that $v_{e,s} = x_{m,s}$ with $m < n$, let $v_{e,s+1} = v_{e,s}$. Let $e$ be least such that $v_{e,s} = x_{m,s}$ with $m \geq n$. For $i \geq e$, let $v_{i,s+1} = x_{n+i-e,s+1}$.

*Step 3:* For all new vertices $v$ introduced into $\mathcal{G}_{s+1}$ (there may be more than 1 since we sped up the enumeration in step 1), if not already done so in step 1, introduce corresponding new vertices into $\mathcal{H}_{s+1}$. Extend $h_{s+1}$ accordingly.

This completes the construction.

The correctness of the construction is based on the following observations. Firstly, for each $e$, $v_e = \lim_s v_{e,s}$ exists; this tells us that each requirement $R_e$ is met and is eventually satisfied. Secondly, for each $v \in \mathcal{G}$, $h(v) = \lim_s h_s(v)$ exists, and that for each $w \in \mathcal{H}$, $h^{-1}(w) = \lim_s h_s^{-1}(w)$ exists. These together with the fact that at each stage $s$, $h_s : \mathcal{G}_s \cong \mathcal{H}_s$ show that $h$ establishes an isomorphism between $\mathcal{G}$ and $\mathcal{H}$. Thus $\mathcal{G} \cong \mathcal{H}$, but $\mathcal{G}$ is not computable isomorphic to $\mathcal{H}$, and hence $\mathcal{G}$ is not computably categorical. □

We note that with essentially the same proof Theorem 2 can be strengthened by removing the assumption that the reachability relation is computable. We also note that since every infinite $\Sigma_2^0$ set has an infinite $\Delta_2^0$ subset, we need only assume there exists an infinite $\Sigma_2^0$ set of such vertices.

## 4   Infinite Chains of Embedded Components

From the two theorems above, one may suggest that the existence of an infinite chain of properly embedded components in a graph may imply that the graph is not computably categorical. One may also suggest that the $\Delta_2^0$-bound in Theorem 2 could be replaced with a $\Delta_3^0$-bound. The main result of this section is to refute these two suggestions and outline of a proof for the following result:

**Theorem 3.** *There is a strongly locally finite computably categorical graph that possesses an infinite chain of properly embedded components. In fact, the set $\{v \mid C(v)$ is properly embedded into $\omega$ many components$\}$ is computable in $0''$.*

*Proof.* Let $\Phi_0, \Phi_1, \ldots$ be a standard enumeration of all partial computable functions from $\omega^2$ to $\{0, 1\}$. Based on this, one builds an effective enumeration of all computable graphs $\mathcal{G}_0, \mathcal{G}_1, \mathcal{G}_2, \ldots$ uniformly. On $i$ at stage $t$ we have:   (1)

$V_{i,t} \subseteq V_{i,t+1}$, $E_{i,t} \subseteq E_{i,t+1}$ for $t \in \omega$; (2) $\bigcup_t \mathcal{G}_{i,t} = \mathcal{G}_i$, where $\mathcal{G}_{i,t} = (V_{i,t}, E_{i,t})$; (3) $V_{i,t} = \{0, \ldots, k_{i,t}\}$ where $k_{i,t}$ is the maximal $j \leq t$ such that for all $n, m \leq j$ the values $\Phi_{i,t}(n, m)$ are defined.

Our goal is to construct a graph $\mathcal{G} = (\omega, E)$ such that $\mathcal{G}$ has an infinite sequence $C_0 \prec C_1 \prec C_2 \prec \ldots$ of properly embedded components, and the construction of $\mathcal{G}$ meets the following requirements:

$$R_e : \text{If } \mathcal{G}_e \cong \mathcal{G} \text{ then } \mathcal{G}_e \text{ is computably isomorphic to } \mathcal{G}$$

Here we show how to satisfy just one requirement $R_e$. The general construction (that we omit in this paper due to space limitations) is based on putting all strategies for $R_e$ on a priority tree. The general construction produces a true path through the tree, the true path is computable in $0''$, and all the requirements $R_e$ are satisfied along the path. The general construction is somewhat similar to and simpler than the constructions in [1], [2], and [13] .

The rest of the proof will handle one requirement $R_e$. We need some notation and definitions. We use cycles as defined in the previous section.

Let $H$ be a graph and $v$ be its vertex. To **attach** a cycle $C_n$ to $v$ means to extend the graph $H$ by adjoining to $H$ the graph $C_n$ and adding the edge $\{v, 1\}$.

The graph $\mathcal{G}$ that we construct will be strongly locally finite such that each component of $\mathcal{G}$ will consist of a vertex $v$ together with finitely many cycles attached to $v$. We call such components **special-cyclic components**.

We approximate $\mathcal{G}_e$ as $\mathcal{G}_{e,0} \subseteq \mathcal{G}_{e,1} \subseteq \mathcal{G}_{e,2} \subseteq \ldots$ such that every component of $\mathcal{G}_{e,t}$ is special-cyclic. During the construction we guarantee the following. If $\mathcal{G}_{e,t}$ provides a component $C$ that cannot be embedded into $\mathcal{G}_t$ then $C$ will never be embedded into $\mathcal{G}$. In this case $R_e$ is satisfied, and we ensure that $\mathcal{G}$ has an infinite sequence of properly embedded components. Thus, we can always assume that $\mathcal{G}_{e,t}$ is embedded into $\mathcal{G}$ currently built. During the construction we also guarantee that no two components of $\mathcal{G}$ are isomorphic.

The graph $\mathcal{G}_t$ denotes approximation to $\mathcal{G}$ at stage $t$. Components of $\mathcal{G}_t$ are denoted by $H_{j,t}$, and we assume a natural order between the components (e.g. $H_1 < H_2$ if the minimum vertex in $H_1$ is less than the minimum vertex in $H_2$).

At stage $t$, the function $f_t$ denotes a partial isomorphism from $\mathcal{G}_{e,t}$ into $\mathcal{G}_t$ that we build. We will also have finitely many selected components in $\mathcal{G}_t$. We say $H_{j,t}$ (a component of $\mathcal{G}_t$) is **covered** if there is a component $H_{e,j,t}$ of $\mathcal{G}_e$ such that $f_t$ maps $H_{e,j,t}$ into $H_{j,t}$. We say that $R_e$ is in the **waiting state** (at stage $t$) if there are selected and uncovered components of $\mathcal{G}_t$. We say that $R_e$ **recovers** (at stage $t$) if for every selected component $H$ in $\mathcal{G}_t$ there is a (necessarily) unique component $H'$ in $\mathcal{G}_{e,t}$ such that $H'$ embeds into $H$ and $H'$ can not be embedded into any other component of $\mathcal{G}_t$ and $f$ has not been defined on $H'$. Now we describe our stagewise construction of $\mathcal{G}$ against one $R_e$.

*Stage 0.* Set $\mathcal{G}_0$ to contain two special-cyclic components such that one component has a cycle of length 3 attached and the other has a cycle of length 4 attached. Select both components. *Mark* the first component. $R_e$ is now in the waiting state. The function $f_0$ is empty.

*Stage $t + 1$.* Compute $\mathcal{G}_{e,t+1}$. Assume $R_e$ is in the waiting state. Build a new component $H$ in $\mathcal{G}_t$ such that the unselected components built in the previous stage properly embedded into $H$, and such that none of the selected components embed into $H$. This builds $\mathcal{G}_{t+1}$.

Assume $R_e$ has recovered. Let $H_1$ be the marked component. Let $H_2$ be the first selected component such that $H_2$ is not marked. For every selected component $H$, consider $H'$ in $\mathcal{G}_{e,t+1}$ such that $H'$ embeds into $H$, $f_t$ is not defined on $H'$, and $H'$ does not embed into any other component of $\mathcal{G}_t$. Extend $f_t$ to $f_{t+1}$ by mapping all such $H'$ into $H$. Extend $\mathcal{G}_t$ to $\mathcal{G}_{t+1}$ as follows:

1. Let $t'$ be the last recovery stage before stage $t + 1$. To all components built between stages $t' + 1$ and $t + 1$ attach new and distinct cycles of distinct unused lengths. This makes these components non-embeddable into each other. Declare these components newly selected.
2. Declare all the components selected at stage $t'$ unselected.
3. Consider $H_1$ and $H_2$. To $H_2$ attach a cycle of length $n$ if $H_1$ has a cycle of length $n$ attached to it and $H_2$ has no cycle of length $n$ attached. Remove the mark from $H_1$ and mark the newly extended $H_2$. Declare $H_2$ selected.
4. Construct a new component with a new cycle of unused length in $\mathcal{G}_{t+1}$.

This finishes the description of stage $t + 1$. Set $\mathcal{G} = \bigcup_t \mathcal{G}_t$. Now we show that $\mathcal{G}$ is a desired graph.

**Lemma 4.** *Suppose there is a stage $t$ after which $R_e$ never recovers. Then $\mathcal{G}$ has an infinite chain of properly embedded components and $R_e$ is satisfied.*

*Proof.* After stage $t$, the construction builds an infinite chain of properly embedded components. Also, $\mathcal{G}_e \not\cong \mathcal{G}'$ and hence $R_e$ is satisfied.     $\square$

**Lemma 5.** *If $\mathcal{G}_e \cong \mathcal{G}$ then $\bigcup_t f_t$ effectively extends to an isomorphism.*

*Proof.* It must be the case that $\Phi_e$ is total. Let $H_1$ be a component of $\mathcal{G}$.

Case 1. The component $H_1$ is never marked. In this case, by construction, $H_1$ must contain a cycle of length $n$ such that no other component of $\mathcal{G}$ has a cycle attached of the same length. Assume $H_1$ is selected at stage $t'$. In the next recovery stage $t + 1$, $f_{t+1}$ maps $H'_{1,t+1}$ into $H_{1,t+1}$. Since $H'_1$ is the only component that contains a cycle of length $n$, we will have $H_1 \cong H'_1$.

Case 2. Assume $H_1$ is marked at stage $t'$ and let $t + 1$ be the next recovery stage after $t'$. We can assume that $f_{t'}$ maps $H'_1$ to $H_1$.

At stage $t + 1$ we have $H_2$ (see stage $t + 1$). $H_2$ contains a cycle of length $m$ such that no other component has a cycle of length $m$. At stage $t + 1$ we also have a mapping $f_{t+1}$ such that $f_{t+1}$ maps $H'_{1,t+1}$ to $H_{1,t+1}$ and $H'_{2,t+1}$ to $H_{2,t+1}$ and $f_{t'} \subseteq f_{t+1}$. Let $t_1$ be the next recovery stage after $t + 1$. Again it must be the case that $f_{t+1} \subseteq f_{t_1}$ as otherwise $\mathcal{G}_e$ contains two components containing cycles of length $m$ (after which the construction guarantees that $\mathcal{G}$ contains no two components with cycles of length $m$).     $\square$

**Lemma 6.** *Assume $\mathcal{G}_e \cong \mathcal{G}$. Then $\mathcal{G}$ contains an infinite chain of properly embedded components.*

*Proof.* The components marked at recovery stages form the desired chain.     $\square$

These lemmas prove that the construction is correct to satisfy one $R_e$. In the general construction our priority $T$ will be the binary tree over the alphabet $r, w$ with the order $r < w$, where $r$ represents recovery and $w$ represents the waiting state. The nodes of length $e$ in $T$ will be devoted to satisfy $R_e$.    □

## References


1. Cholak, P., Goncharov, S., Khoussainov, B., Shore, R.A.: Computably categorical structures and expansions by constants. J. Symbolic Logic 64(1), 13–37 (1999)
2. Cholak, P., Shore, R.A., Solomon, R.: A computably stable structure with no Scott family of finitary formulas. Arch. Math. Logic 45(5), 519–538 (2006)
3. Crossley, J.N. (ed.): Aspects of effective algebra, Vic., 1981. Upside Down A Book Co. Yarra Glen.
4. Dzgoev, V.D., Gončarov, S.S.: Autostability of models. Algebra i Logika 19, 45–58 (1980)
5. Yu., L., Ershov, S.S., Goncharov, A., Nerode, J.B.: Handbook of recursive mathematics. Studies in Logic and the Foundations of Mathematics, vol. 1, 138. North-Holland, Amsterdam (1998); Recursive model theory
6. Ershov, Y.L., Goncharov, S.S., Nerode, A., Remmel, J.B., Marek, V.W. (eds.): Handbook of recursive mathematics. Studies in Logic and the Foundations of Mathematics, vol. 2, 139, pp. 621–1372. North-Holland, Amsterdam (1998); Recursive algebra, analysis and combinatorics
7. Ershov, Y.L., Goncharov, S.S.: Constructive models. Siberian School of Algebra and Logic. Consultants Bureau, New York (2000)
8. Gončarov, S.S.: Autostability of models and abelian groups. Algebra i Logika 19(1), 23–44 (1980)
9. Gončarov, S.S.: The problem of the number of nonautoequivalent constructivizations. Algebra i Logika 19(6), 621–639, 745 (1980)
10. Goncharov, S.S.: Limit equivalent constructivizations. In: Mathematical logic and the theory of algorithms, "Nauka" Sibirsk. Otdel., Novosibirsk. Trudy Inst. Mat., vol. 2, pp. 4–12 (1982)
11. Goncharov, S.S.: Computability and computable models. In: Mathematical problems from applied logic. II. Int. Math. Ser (N. Y.), vol. 5, pp. 99–216. Springer, New York (2007)
12. Goncharov, S.S., Lempp, S., Solomon, R.: The computable dimension of ordered abelian groups. Adv. Math. 175(1), 102–143 (2003)
13. Hirschfeldt, D.R.: Degree spectra of relations on structures of finite computable dimension. Ann. Pure Appl. Logic 115(1-3), 233–277 (2002)
14. Lempp, S., McCoy, C., Miller, R., Solomon, R.: Computable categoricity of trees of finite height. J. Symbolic Logic 70(1), 151–215 (2005)
15. Malćev, A.I.: Constructive algebras. I. Uspehi Mat. Nauk, 16(3 (99)), 3–60 (1961)
16. Rabin, M.O.: Computable algebra, general theory and theory of computable fields. Trans. Amer. Math. Soc. 95, 341–360 (1960)
17. Remmel, J.B.: Recursively categorical linear orderings. Proc. Amer. Math. Soc. 83, 387–391 (1981)

# P Automata:
# Membrane Systems as Acceptors

Erzsébet Csuhaj-Varjú

Computer and Automation Research Institute
Hungarian Academy of Sciences
Kende utca 13-17, 1111 Budapest, Hungary
csuhaj@sztaki.hu

The concept of a membrane system (a P system) was introduced by Gheorghe
Păun in 1998 [9,10], with the aim of formulating a computational device abstracted from the architecture and the functioning of the living cell. Since that
time, the theory of membrane systems has proved to be a successful area in
bio-inspired computing.

The main ingredient of a P system is a hierarchically embedded structure of
membranes. Each membrane encloses a region that contains objects and might
also contain other membranes. The outmost membrane is called the skin membrane. There are rules associated to the regions describing the evolution of the
objects which represent chemical substances. The evolution rules correspond
to chemical reactions, and the evolution of the system to a computation. The
main features of P systems include the transformation (rewriting) of objects,
their moving among the different regions (communication), and possibly other
additional capabilities such as, for example, a dynamically changing membrane
structure, or special constraints added to the sets of rules. At any moment in
time, the membrane system can be described by its configuration (its state)
which consists of the actual membrane structure and the contents of the regions.

Membrane systems can be considered as computational devices: starting from
an initial configuration, the system evolves by passing from one configuration
to another one and if it halts, i.e., no rule can be applied anymore, then the
computation is successful. When changing the configuration, the rules can be
applied in a sequential manner or in a maximally parallel manner. When they
are applied in the sequential manner, one rule is applied in each region in each
derivation step, and when they are applied in the maximally parallel manner, as
many rules are applied simultaneously in each region as it is possible. For more
information on P systems, the reader may consult the monograph [11] and the
publications listed at the P systems webpage [12].

The generic model of P systems, briefly described above, restricts its functioning to the dynamics of the contents of the regions and the membrane structure: no feature is provided for modeling the interaction of the cell and its
surrounding environment. Since membrane systems attempt to provide a formal
framework for describing living cells and the cells are in interaction with their

A. Beckmann, C. Dimitracopoulos, and B. Löwe (Eds.): CiE 2008, LNCS 5028, pp. 149–151, 2008.

biological environments, those variants of membrane systems where the P system communicates with its environment are subjects of a well-motivated, important research area.

One approach to modelling P systems which communicate with their environments is the framework called P automata. In this case, the environment is represented by an infinite (or finite) supply of objects from which multisets of objects are imported by the skin membrane in the system during the computation. These P systems resemble acceptors, since the P system consumes input, and the change of its state (configuration) depends on both the imported objects and its actual configuration. Furthermore, the sequences of imported multisets of objects can be distinguished as accepted or rejected input sequences. The language accepted by the P automaton is determined by the set of multiset sequences that are imported by an accepting computation, i.e., a computation starting from the initial configuration and ending in an accepting state. (Accepting states are configurations specified together with the system.)

The first variant of P automata, called one-way P automaton, having only communication rules, was introduced in [4,5]. Almost at the same time, a closely related notion, the analyzing $P$ system was defined in [7]. Both one-way P automata and analyzing P systems are able to describe any recursively enumerable language, even having a very small number of membranes. In [4], it was shown that any recursively enumerable language can be obtained as a mapping of the language of a one-way P automaton, where the P automaton has seven membranes. Later, the result was improved in [6] by reducing the number of necessary membranes to two.

Since the introduction of the generic notion, several variants of P automata have been defined and studied, different from each other in the following features: the way of defining the acceptance/rejection of the input (defined by accepting states or by halting), the type of communication with the environment (one-way or two-way communication), the types of communication rules used by the regions, the way of functioning of the membrane system (whether or not it has evolution rules), and whether or not the membrane structure changes in the course of the computation. For more information the reader is referred to the summary [1], the PhD dissertation [8], and the articles listed in the on-line bibliography at the P systems webpage [12].

P automata can be studied from many points of view. In addition to investigating them as computational models of cells interacting with their environments, comparisons of models from classical automata theory and from P automata theory are also of particular interest, since these accepting P systems combine properties of both conventional automata and unconventional computational devices. Notice, for example, that unlike a conventional automaton, a P automaton does not have a separate set of states, its actual state is identified by its actual configuration. Similarly, while in the case of conventional automata the whole input can be found on the input tape at the beginning of the computation, P automata variants allow the input to be determined step by step in the course of the computation.

In this talk we give an overview on the computational power and size of the main variants of P automata, and demonstrate how some well-known language classes as, for example, the class of context-sensitive languages can be represented by these constructs in a very natural manner. We also discuss similarities and differences between P automata and some well-known types of conventional automata and close the talk by proposing some open problems for future research.

To give a flavor of the results, in [2,3] it was shown that if the rules of the P automata (using so-called antiport rules with promoters and inhibitors) are applied sequentially, then the accepted language class is strictly included in the class of languages accepted by one-way Turing machines with a logarithmically bounded workspace, while if the rules are applied in the maximally parallel manner, then these P automata determine the class of context-sensitive languages.

# References


1. Csuhaj-Varjú, E.: P automata. In: Mauri, G., Păun, G., Pérez-Jímenez, M.J., Rozenberg, G., Salomaa, A. (eds.) WMC 2004. LNCS, vol. 3365, pp. 19–35. Springer, Heidelberg (2005)
2. Csuhaj-Varjú, E., Ibarra, O.H., Vaszil, G.: On the Computational Complexity of P Automata. In: Ferretti, C., et al. (eds.) Preliminary Proceedings of DNA10. Tenth International Meeting on DNA Computing, Milan, June 7-10, 2004, pp. 97–106. University of Milano-Bicocca, Italy (2004)
3. Csuhaj-Varjú, E., Ibarra, O.H., Vaszil, G.: On the Complexity of P Automata. Natural Computing 5(2), 109–126 (2006)
4. Csuhaj-Varjú, E., Vaszil, G.: P automata. In: Păun, G., Zandron, C. (eds.) Pre-Proceedings of the Workshop on Membrane Computing, Curtea de Arges, Romania, August 19-23, 2002, pp. 117–192 (2002); MolCoNet-IST-2001-32008
5. Csuhaj-Varjú, E., Vaszil, G.: P Automata or Purely Communicating Accepting P systems. In: Păun, G., Rozenberg, G., Salomaa, A., Zandron, C. (eds.) WMC 2002. LNCS, vol. 2597, pp. 219–233. Springer, Heidelberg (2003)
6. Freund, R., Martín-Vide, C., Obtułowicz, A., Păun, G.: On Three Classes of Automata-like P Systems. In: Ésik, Z., Fülöp, Z. (eds.) DLT 2003. LNCS, vol. 2710, pp. 292–303. Springer, Heidelberg (2003)
7. Freund, R., Oswald, M.: A Short Note on Analysing P Systems. Bulletin of the EATCS 78, 231–236 (2002)
8. Oswald, M.: P Automata. PhD dissertation, Technical University of Vienna (2003)
9. Păun, G.: Computing with membranes. TUCS Report 208, Turku Centre for Computer Science (1998)
10. Păun, G.: Computing with membranes. Journal of Computer and System Sciences 61(1), 108–143 (2000)
11. Păun, G.: Membrane Computing. An Introduction. Springer, Heidelberg (2002)
12. The P systems web page at http://psystems.disco.unimib.it

# On the Processing Power of Protozoa

Mark Daley

Departments of Computer Science and Biology
The University of Western Ontario
daley@csd.uwo.ca

The ciliated protozoa are a diverse, and ubiquitously occuring, group of unicellular eukaryotic organisms of striking complexity. Almost all ciliates are binucleated; that is, unlike other eukaryotes, they have two types of nuclei: micronuclei and macronuclei. The macronuclei behave functionally as one would typically expect of a eukaryotic nucleus: they express proteins and take care of the general "housekeeping" functions of the cell. Ciliate macronuclei are ampliploid[1] and contain very little non-coding DNA. The diploid micronuclei, arranged much more like a typical eukaryotic nucleus, are inert during the "day-to-day" functioning of the cell and serve instead as a storehouse for germline information.

Ciliates reproduce asexually, through fission, but also engage in the sexual process of conjugation. During conjugation, haploid micronuclear genomes are exchanged between ciliates, and post-conjugation, both ciliates destroy their old macronuclei and generate new macronuclei from the new micronucleus. Comparing the micronuclear version of a gene to the macronuclear version one notes several segments of the micronuclear gene that have been eliminated in the macronucleus. Segments of micronuclear genes that are so eliminated are called Internal Eliminated Sequences, or IESs, while the segments which remain are called Macronuclear Destined Sequences or MDSs. The process of IES elimination during macronuclear development is quite complex and has been studied extensively in several ciliate genera [6].

In most ciliates studied thus far, MDSs occur in the micronucleus in the same order (denoted the "orthodox order") as they do in the macronucleus, though they may be separated by IESs. In the stichotrichous ciliates, however, several genes have been observed in which the MDSs occur in a *scrambled* order in the micronucleus relative to the orthodox macronuclear order [9]. In order to produce a functional macronucleus, the ciliate must have some form of information-processing mechanism that allows scrambled micronuclear genes to be descrambled into functional macronuclear genes.

The computational nature of this descrambling process has attracted the interest of computer scientists who, together with ciliate biologists, have proposed a series of models for gene descrambling. The two earliest proposed models were based on simple sets of intramolecular [4] and intermolecular [7] recombination events guided by so-called pointer sequences flanking IESs and MDSs.

---

[1] Macronuclei contain many copies of each gene, ranging from tens of copies (e.g. 45 for most genes in *Tetrahymena thermophila*) to tens of thousands (e.g., $\beta$-tubulin in *Euplotes crassus*) or more.

A. Beckmann, C. Dimitracopoulos, and B. Löwe (Eds.): CiE 2008, LNCS 5028, pp. 152–153, 2008.
© Springer-Verlag Berlin Heidelberg 2008

Although recent experimental results have not proven consistent with these models, they still provide higher-level metrics for assessing the complexity of scrambled genes [5].

More recently, descrambling models based on the template-guided recombination of DNA molecules [10] or DNA-RNA hybrids [1] have been proposed. These models are strongly biologically motivated and are not only consistent with observed biological data, but have very recently been successfully directly tested [8]. Given that it has now been demonstrated that we can co-opt the descrambling mechanism to perform somewhat arbitrary descramblings, the question of the computational power of this process becomes natural.

Viewing these biological models as formal computing systems, a wide range of computational powers have been demonstrated varying from very weak (e.g., simple template-guided recombination is no more powerful than a finite-state machine [3]) to universal (e.g., the intermolecular model [7], TGR with context [2]).

In this talk we will give a high-level overview of models for gene descrambling along with an exposition of computability results. We will conclude with the presentation of a new knot-theoretic extension to the TGR model which addresses the issue of the thermodynamic unfavourability of the current model through an appeal to the topological properties of the molecules undergoing recombination.

# References


1. Angeleska, A., Jonoska, N., Saito, M., Landweber, L.F.: RNA-guided DNA assembly. Journal of Theoretical Biology (in press, corrected proof, 2007)
2. Daley, M., McQuillan, I.: On computational properties of templated-guided DNA recombination. In: Carbone, A., Pierce, N.A. (eds.) DNA 2005. LNCS, vol. 3892, pp. 27–37. Springer, Heidelberg (2006)
3. Daley, M., McQuillan, I.: Template-guided DNA recombination. Theor. Comput. Sci. 330(2), 237–250 (2005)
4. Ehrenfeucht, A., Harju, T., Petre, I., Prescott, D.M., Rozenberg, G.: Computation in Living Cells: Gene Asssembly in Ciliates. Springer, Heidelberg (2003)
5. Harju, T., Li, C., Petre, I., Rozenberg, G.: Complexity measures for gene assembly. Knowledge Discovery and Emergent Complexity in Bioinformatics, 42–60 (2007)
6. Jahn, C.L., Klobutcher, L.A.: Genome remodeling in ciliated protozoa. Annual Review of Microbiology 56(1), 489–520 (2002)
7. Landweber, L.F., Kari, L.: Universal molecular computation in ciliates. In: Winfree, E., Landweber, L.F. (eds.) Evolution as Computation. Springer, Heidelberg (1999)
8. Nowacki, M., Vijayan, V., Zhou, Y., Schotanus, K., Doak, T.G., Landweber, L.F.: RNA-mediated epigenetic programming of a genome-rearrangement pathway. Nature (advanced online publication) (November 2007)
9. Prescott, D.M.: Genome gymnastics: unique modes of DNA evolution and processing in ciliates. Nat. Rev. Genet. 1(3), 191–198 (2000)
10. Prescott, D.M., Ehrenfeucht, A., Rozenberg, G.: Template-guided recombination for IES elimination and unscrambling of genes in stichotrichous ciliates. Journal of Theoretical Biology 222(3), 323–330 (2003)

# Computing Equilibria in Large Games We Play

Constantinos Daskalakis[*]

Computer Science, U. C. Berkeley
costis@cs.berkeley.edu

**Abstract.** Describing a game using the standard representation requires information exponential in the number of players. This description complexity is impractical for modeling large games with thousands or millions of players and may also be wasteful when the information required to populate the payoff matrices of the game is unknown, hard to determine, or enjoy high redundancy which would allow for a much more succinct representation. Indeed, to model large games, succinct representations, such as graphical games, have been suggested which save in description complexity by specifying the graph of player-interactions. However, computing Nash equilibria in such games has been shown to be an intractable problem by Daskalakis, Goldberg and Papadimitriou, and whether approximate equilibria can be computed remains an important open problem. We consider instead a different class of succinct games, called *anonymous games*, in which the payoff of each player is a symmetric function of the actions of the other players; that is, every player is oblivious of the identities of the other players. We argue that many large games of practical interest, such as congestion games, several auction settings, and social phenomena, are anonymous and provide a polynomial time approximation scheme for computing Nash equilibria in these games.

Game Theory is important for the study of large competitive environments, such as the Internet, the market, and even social and biological systems. However, due to the scale of such systems, many natural computational problems within game theory, e.g. computing a Nash equilibrium, require complexity considerations. In fact, these considerations originate at the game-representation itself: Describing a game using the standard normal form requires information exponential in the number of players, since one needs to specify the payoff of each player for every selection of strategies by the players of the game. This exponential description complexity is forbidding for games of many players, and, in any case, the required details to fill in the payoff tables might be either hard to determine, or exhibit high redundancy which would allow for more succinct representations.

An important class of succinct representations is that of *graphical games*, in which a graph of player-interactions is specified, and a player's payoff only depends on the strategies of her neighbors in the graph. If the strategic environment

[*] Supported by a Microsoft Research Fellowship, NSF grant CCF − 0635319, a gift from Yahoo! Research, and a MICRO grant.

A. Beckmann, C. Dimitracopoulos, and B. Löwe (Eds.): CiE 2008, LNCS 5028, pp. 154–157, 2008.

that is being considered exhibits a sparse structure of player-interactions the graphical game representation results in large savings in description complexity; in particular, an interaction-graph of constant degree results in polynomial (in the number of players) description complexity, which renders computational questions for these games practically meaningful. However, computing Nash equilibria in sparse graphical games has been shown to be computationally intractable [4], and interest has been shifted recently towards computing approximate Nash equilibria, that is randomized strategies for the players of the game such that no player has more than some small incentive to change her strategy if the other players do not. Approximate equilibria of any desired approximation can be computed efficiently provided that the treewidth of the player-interactions is bounded [7]; but, whether there exists a polynomial time approximation scheme for the general case remains an important open problem.

In this paper, we propose the study of another class of succinct games, called *anonymous games* [2,3]. In an anonymous game, every player is different, that is, she has her own payoff function, but this function is symmetric with respect to the strategies of the other players; in other words, the players are oblivious of the identities of the other players. Several important classes of multi-player games fall in the class of anonymous games, e.g., auction settings in which the payoff of a bidder is only affected by the distribution of the other bids and not on the identities of the other bidders. Anonymous games also capture congestion games where the delay incurred by a driver for choosing a route only depends on the number of other cars on her route, but not on the identities of the drivers. Finally, anonymous games have been suggested for modeling social phenomena, e.g., social pressure related to veiling in Islamic societies [2].

We also consider generalizations of the model, e.g., *typed anonymous games* in which the players are partitioned into *types* and the payoff function of every player is symmetric with respect to the strategies of the other players within each type. A typed anonymous game with $n$ players, $s$ strategies per player, and $t$ types of players requires description size of the order of $n^{st}$, which is polynomial in $n$, if $s$ and $t$ are constants. We present positive approximation results for anonymous games, both for pure strategy equilibria and mixed strategy equilibria; in particular, we show the following:

**Theorem 1 ([5]).** *In any anonymous game with $s$ strategies per player there exists an $O(s^2\lambda)$-approximate pure Nash equilibrium, where $\lambda$ is the Lipschitz constant of the utility functions of the players (assumed to be such that, for any partitions $x$ and $y$ of the players into the $s$ strategies, every player's utility function $u$ satisfies $|u(x) - u(y)| \leq \lambda||x - y||_1$). Such an approximate equilibrium can be found in linear time.*

**Theorem 2 ([5,6]).** *There is a Polynomial Time Approximation Scheme (PTAS) for the mixed Nash equilibrium problem for anonymous games with a constant number of strategies per player.*

Our pure Nash equilibrium result follows from a discretized version of Brouwer's fixed point theorem. A discrete function is defined over the *signatures* of the pure strategy profiles of the game, that is the set of partitions of $n-1$ players (without identities) into $s$ strategies. The function maps a partition into another if the latter is derived from the former after a best response move by each of the first $n-1$ players of the game conditioned on their environment being the former partition. This discrete function of Nash dynamics is interpolated into a continuous function and Brouwer's fixed point theorem guarantees the existence of a fixed point of that function. The challenge is to show that, close to the fixed point of the continuous function, there exists a discrete point which gives rise to an approximate pure Nash equilibrium of the game.

Our polynomial time approximation scheme for mixed Nash equilibria is based on deep results in probability theory regarding the approximation of sums of multinomial distributions by other distributions [1]. To discretize the search space for mixed Nash equilibria, we consider mixed strategy profiles where every player's mixed strategy is restricted so that each pure strategy is played with probability which is an integer multiple of some small constant $\delta < 1$. Searching the space of discretized mixed strategy profiles can be done efficiently in time $O\left(n^{(1/\delta)^s}\right)$; hence the question arises of what $\delta$ is required to guarantee that at least one of the discretized mixed strategy profiles is an $\epsilon$-approximate mixed Nash equilibrium. We show that the required $\delta$ is a function of $\epsilon$ and $s$ only which implies a polynomial time approximation scheme for the mixed Nash equilibrium problem in anonymous games. To establish the bound on $\delta$ we need to bound the total variation distance between a sum of multinomial distributions and a sum of (appropriately rounded) discretized multinomial distributions. In particular, we show the following:

**Theorem 3 ([6]).** *Let $\{p_i\}_{i\in[n]}$ be a set of probability vectors, where $p_i \in \Delta_1^s$, $\Delta_1^s$ is the standard $(s-1)$-simplex, and let $\{\mathcal{X}_i\}_{i\in[n]}$ be a set of independent $s$-dimensional random unit vectors, such that $\Pr[\mathcal{X}_i = e_\ell] = p_{i,\ell}$, where $e_\ell$ is the unit vector along dimension $\ell$, $\ell \in [s]$. Let $z > z_0$ be a positive integer, where $z_0$ is some absolute constant. Then there exists another set of probability vectors $\{\widehat{p}_i\}_{i\in[n]}$, where $\widehat{p}_i \in \Delta_1^s$, such that*

*1. $|\widehat{p}_{i,\ell} - p_{i,\ell}| = O(\frac{1}{z})$, for all $i \in [n], \ell \in [s]$;*
*2. $\widehat{p}_{i,\ell}$ is an integer multiple of $\frac{1}{2^s}\frac{1}{z}$, for all $i \in [n], \ell \in [s]$;*
*3. if $p_{i,\ell} = 0$, then $\widehat{p}_{i,\ell} = 0$, for all $i \in [n], \ell \in [s]$;*
*4. if $\{\widehat{\mathcal{X}}_i\}_{i\in[n]}$ are independent random vectors such that $\Pr[\widehat{\mathcal{X}}_i = e_\ell] = \widehat{p}_{i,\ell}$, $\ell \in [s]$, then*

$$\left\|\sum_i \mathcal{X}_i - \sum_i \widehat{\mathcal{X}}_i\right\|_{TV} = O\left(f(s)\frac{\log z}{z^{1/5}}\right)$$

*where $f(s)$ is a function of $s$ only.*

The above theorem states, roughly, that, for any constant $c > 5$, there is a way to quantize any set of $n$ mixed strategies into another set of mixed strategies, whose probabilities of choosing each pure strategy are integer multiples of $\delta$,

$\delta \in [0,1]$, so that the total variation distance between the (random) number of players playing each pure strategy before and after the quantization is bounded by $O(f(s)2^{s/c}\delta^{1/c})$, which does not depend on $n$. This bound, which is of more general interest, immediately implies a polynomial time approximation scheme for finding mixed Nash equilibria in anonymous games with a constant number of strategies per player. The result can be generalized for typed anonymous games and several extensions of the model, e.g., settings where the players are partitioned into *neighborhoods* and the payoff function of a player is symmetric with respect to the actions of her neighbors and symmetric with respect to the actions of her non-neighbors. As no other polynomial time approximation scheme is known for finding mixed Nash equilibria in large games (except for graphical games with bounded treewidth [7]), this algorithm furthers our understanding of mixed Nash equilibrium computation. It also offers a surprising structural property of the space of mixed Nash equilibria with regards to approximation.

# References


1. Barbour, A.D., Chen, L.H.Y.: An Introduction to Stein's Method. In: Barbour, A.D., Chen, L.H.Y. (eds.). Lecture Notes Series, vol. 4, Institute for Mathematical Sciences, National University of Singapore, Singapore University Press and World Scientific, Singapore (2005)
2. Blonski, M.: The women of Cairo: Equilibria in large anonymous games. Journal of Mathematical Economics 41(3), 253–264 (2005)
3. Blonski, M.: Anonymous Games with Binary Actions. Games and Economic Behavior 28(2), 171–180 (1999)
4. Daskalakis, C., Goldberg, P.W., Papadimitriou, C.H.: The Complexity of Computing a Nash Equilibrium. In: The 38th ACM Symposium on Theory of Computing, STOC 2006 (2006); SIAM Journal on Computing special issue for STOC 2006 (to appear, 2006)
5. Daskalakis, C., Papadimitriou, C.H.: Computing Equilibria in Anonymous Games. In: the 48th Annual IEEE Symposium on Foundations of Computer Science, FOCS 2007 (2007)
6. Daskalakis, C., Papadimitriou, C.H.: Discretizing the Multinomial Distribution and Nash Equilibria in Anonymous Games. ArXiv Report (2008)
7. Daskalakis, C., Papadimitriou, C.H.: Computing Pure Nash Equilibria via Markov Random Fields. In: The 7th ACM Conference on Electronic Commerce, EC 2006 (2006)

# A Week-End Off:
# The First Extensive Number-Theoretical Computation on the ENIAC


Liesbeth De Mol[1,*] and Maarten Bullynck[2,**]

[1] Center for Logic and Philosophy of Science, University of Ghent, Blandijnberg 2,
9000 Gent, Belgium
elizabeth.demol@ugent.be
[2] IZWT, Gaussstrasse 20, 42119 Wuppertal, Germany and REHSEIS, Paris, France
bullynck@uni-wuppertal.de



**Abstract.** The first extensive number-theoretical computation run on the ENIAC, is reconstructed. The problem, computing the exponent of 2 modulo a prime, was set up on the ENIAC during a week-end in July 1946 by the number-theorist D.H. Lehmer, with help from his wife Emma and John Mauchly. Important aspects of the ENIAC's design are presented and the reconstruction of the implementation of the problem on the ENIAC is discussed in its salient points.

**Keywords:** ENIAC, Derrick H. Lehmer, number theory, Fermat's little theorem, early programming, parallelism, prime sieve.


## 1   Introduction

Just too late to help win the Second World War, just in time to start off the computer age, the first digital electronic general-purpose computer (See e.g. [2]), the ENIAC, was presented to the public February 15, 1946 at Penn University. The Ballistic Research Laboratories (Aberdeen Proving Ground) had "assembled a 'Computations Committee' to prepare for utilizing the machine after its completion" [1, p. 693], and the ENIAC was extensively test-run during its first months.

One of the members of this committee was the number-theorist Derrick H. Lehmer who spent a Fourth-of-July weekend testing the ENIAC. The Lehmer family – Derrick, Emma and two teenage kids – arrived at the Moore school on Friday 5 p.m. where they met John Mauchly, who was, together with Presper Jr. Eckert, the father of the ENIAC-design. He helped them set up the ENIAC for the implementation of an interesting number-theoretical problem and stayed on as an operator through the week-end [10, p. 451]. Lehmer's computation was important to the post-war reputation of electronic computers among mathematicians.


---

* Postdoctoral Fellow of the Fund for Scientific Research – Flanders (FWO).
** Postdoctoral Research Grant, Alexander-von-Humboldt-Stiftung.

A. Beckmann, C. Dimitracopoulos, and B. Löwe (Eds.): CiE 2008, LNCS 5028, pp. 158–167, 2008.
© Springer-Verlag Berlin Heidelberg 2008

The ENIAC was an electronic and highly parallel machine. As a consequence it was revolutionary fast in its time. However, setting up a program on the ENIAC was time-consuming. The original control system of the ENIAC was decentralized, its elements were distributed over the different units of the machine. The programming had to be done "directly", i.e., there was no converter code, let alone a programming language available to set up a 'program'. Instead, one had to connect the different parts of the machine through cables and adaptors. Only in 1948 was the ENIAC rewired into a serial computer that could be set up using a converter code that selected subroutines from a function table, thus simulating a stored-program computer [11]. The original "local programming" method can best be described "as analogous to the design and development of a special-purpose computer out of ENIAC component parts for each new application." [3, p. 31] Or as Jean Bartik, one of the ENIAC's female programmers, put it: 'The ENIAC was a son-of-a-bitch to program.'

The particularities of programming the ENIAC make it interesting for present-day computer scientists. As W.B. Fritz, one of the ENIAC-operators remarks: "Anyone now doing research in parallel computing might take a look at ENIAC during this first time period, for indeed ENIAC was a parallel computer with all of the problems and opportunities this entails." [3, p. 31] More importantly even, 'programming' the ENIAC invites one to question many things taken for granted nowadays, amongst them the relation between the user and the machine.

In this paper, we will present Lehmer's problem and show how to set it up on the ENIAC. Given the complexity of 'programming' (actually wiring) the ENIAC, we will not be able to provide all details here. Instead we will merely outline our reconstruction, providing details only for certain aspects of the set-up.[1]

## 1.1   How a Number-Theorist Got Involved with Computers

Derrick H. Lehmer (1905-1991) was born into number theory. His father, Derrick N. Lehmer, was a reputed number-theorist, best known for his factor table up to 10,000,000 and his stencil sheets to find factors of large numbers. Already as a young boy D.H. devised a way, using bicycle chains, to mechanize these stencils, building the first electro-mechanical number sieve in 1926, the first photo-electric sieve in 1932. Given this lasting interest in making mathematical tables (always a time-consuming job) and particularly in mechanizing sieves, it is clear that D.H. Lehmer perfectly fitted into the 'Computations Committee'.

Lehmer also contributed in other ways to the advance of the computer age. He helped R.C. Archibald' to publish the important journal *Mathematical Tables and other Aids to Computation* (known as MTAC) from 1943 to 1950, and served as the editor-in-chief himself until 1959. This journal was an important channel of publication during the early years of computing, offering a fast way of

---

[1] We want to thank Martin Carlé (Humboldt University Berlin) for his stimulation to get involved into the ENIAC's technical details and wirings. This reconstruction is intended as a contribution to the ENIAC NOMOI project.

getting new results in print, and it reads like the diary-like records of advances in hardware and computing techniques from 1943 to 1959. Lehmer also often ventilated his strong opinions on computing in this journal, making a case for seeing mathematics (especially number theory) as an observational science for which data had to be gathered through computation. Lehmer argued strongly and convincingly that the computer would change the world of mathematics.

## 1.2   The Structure of the ENIAC

The Electrical Numerical Integrator And Computer (ENIAC) was first described in a proposal John Mauchly submitted to Penn University, and was ultimately built with U.S. Army money by a team of engineers under the direction of Presper Eckert Jr. It used about 18,000 vacuum tubes and 1,500 relays. From 1945 to 1947 this was a highly parallel computer, though its parallelism was hardly ever used [2, p. 376] one of the notable exceptions being Lehmer's program.

The ENIAC had a modular architecture. It comprised 20 accumulators, a multiplier, a divider and square rooter, a constant transmitter, 3 function tables, a master programmer, a cycling unit, an initiating unit and also a card reader and a printer. The constant transmitter and the function tables are the ENIAC's main memory storage units. Function tables had to be set manually with switches, the constant transmitter could store upto 16 numbers read from a punched card (before or during the computation) and 4 numbers that had to be set manually with switches.

The ENIAC had two kinds of electronic circuits: the *numerical circuits* for storing and processing electric signals representing numbers and *programming circuits* for controlling the communication between the different parts of the machine. Most of the units had both kind of circuits, except for the master programmer and the initiating unit which consisted only of programming circuits. All units had to be programmed locally. The program switches, located on the front face of the units, had to be set before a computation started to specify which operations had to be performed. It will become clearer in the remainder of the paper how to control the order of the operations.

Of crucial importance in the ENIAC was the central programming pulse (CPP), emitted by the cycling unit once every 1/5000th of a second, marking the beginning and end of a computation cycle. These pulses synchronized the operations of the units. When a unit completed an operation it emitted one of these as a program output pulse, stimulating the next operation. All the units required an integral number of cycles. E.g. an accumulator required 1/5000th of a second for an addition, so we speak also of an addition time instead of a cycle.

We will now detail the design of the accumulator and the master programmer and explain how to set up some basic processes such as branching and iteration. For more details we refer to [2,5] and especially the detailed *Report on the ENIAC* by Adèle Goldstine [4].[2]

---

[2] We would also like to refer to Till Zopke's Java-applet ENIAC simulation [12], that proved handy for testing some easy cablings.

**The accumulator.** The accumulators were the main arithmetic units of the ENIAC and could be used to add or subtract. Each accumulator held a 10-place decimal number and a sign, stored in ten decade ring counters and a PM counter. It had 5 input channels ( $\alpha$ to $\epsilon$) to receive a number. It had two output channels ($A$ and $S$) to transmit a number $n$ (through A) or its complement $10^{10} - n$ (through S). In one addition time, the accumulator could either receive a number $n$ adding (if $n \geq 0$) or subtracting (if $n < 0$) it to/from its content, or transmit the number it stored through one or both of its outputs. The program part of the accumulator consisted of 16 program controls: 4 receivers and 12 transceivers. A transceiver had a program pulse input and output terminal, a clear-correct switch (to clear or not clear its content after a cycle; it could also be used to round off numerical results), an operation switch (to be set to $\alpha$ to $\epsilon$, A, S, AS or 0, determining whether the accumulator should receive or transmit a number, or do nothing) and a repeat switch (with which it could either receive or transmit up to 9 times). When a transceiver received a program pulse through a program cable at addition time $r$, the operations set on the program switch associated with that transceiver were executed. When these had been finished after $n$ ($1 \leq n \leq 9$) addition times, a program pulse was transmitted through the output of the transceiver at addition time $r + n$. A receiver differs from a transceiver in that it has no output terminal and no repeater switch.

**The master programmer.** The master programmer provided a certain amount of centralized programming memory. It consisted of 10 independently functioning units, each having a 6-stage counter (called the stepper), 3 input terminals (the stepper input, direct input and clear input), and 6 output terminals for each stage of the stepper. Each such stage $s$ was associated with a fixed number $d_s$ by manually setting decade switches, and with 1 to 5 decade counters.

If a pulse arrived at the stepper input (SI) of a stepper, one was added to the counter of stage $s$. If this number equaled the preset number $d_s$, it cleared the counter of stage $s$ and cycled to the next stage $s + 1$. In both cases, a program pulse was emitted through the output terminal of stage $s$. A pulse at the direct input (DI) cycled *immediately* to the next stage, and a pulse at the clear input (CI) reset the stepper to its initial configuration. In neither case a program pulse was emitted. In this way the master programmer could be used, among other things, to sequence operations and to iterate a given subroutine.

**Branching.** The ENIAC was capable of discriminating between program sequences by examining the magnitude of some numerical result. This "magnitude discrimination" or "branching" was possible because 9 digit pulses were transmitted for sign indication M and none for sign indication P. The fact that digit pulses were transmitted for every digit except for 0 could be exploited in a similar manner. The digit pulse corresponding to the sign (or a digit) could then be converted into a program pulse by connecting the PM lead (or another digit lead) of the A and/or S output terminal of an accumulator to the program pulse input terminal of an otherwise unused 'dummy (program) control' by using a special adaptor [4, Sec. 4.5].

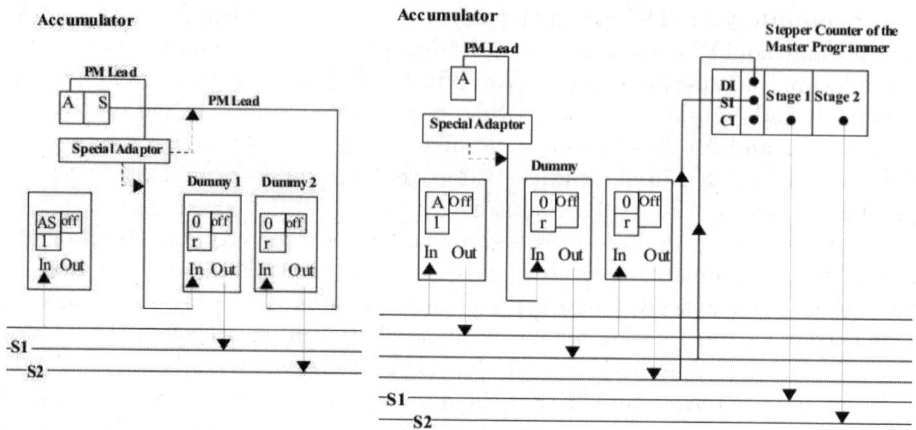


**Fig. 1.** Wiring schemes for each of the branching methods. The if-routine is started when a program pulse is received at the first program cable. Depending on the sign of the number in the accumulator, either subroutine $S_1$ or $S_2$ will be executed.

There were basically two ways to do this branching: either by using the two output channels A and S of an accumulator and two dummy program controls, or, by using only one output channel, A or S, one dummy control and the master programmer. Fig. 1 shows *possible* implementations of these two 'if' methods. They were reconstructed on the basis of the (limited) information on branching in [2,4].

## 2    Lehmer's ENIAC 'Program'

### 2.1    The Mathematical Problem

Primality tests and factoring methods belong to the most important topics in number theory. One class of primality tests are (partial) converses of Fermat's little theorem. Unfortunately, the direct converse of the theorem is false in general: If $a^{p-1} \equiv 1 \bmod p$ for an arbitrary $a$ (not dividing $p$), then $p$ is often though not always a prime. A way out of the failing general converse is to list all exceptions, i.e., all composite $p$ for which $a^{p-1} \equiv 1 \bmod p$ is true. Taking $a$ equal to 2 is computationally the most advantageous choice. To compute the composite $p$'s for which $2^{p-1} \equiv 1 \bmod p$ one needs a table of all exponents $e$ of 2 modulo the prime $p$, $e$ being the least value $n$ such that $2^n \equiv 1 \bmod p$ and $e$ is some divisor of $p - 1 = ef$. Kraïtchik published such an exponent table in 1924 for $p < 300000$, but it contained quite some errors [6]. D.H. Lehmer now proposed to use the ENIAC to compute a table of exponents correcting and extending Kraïtchik's table to $p < 4.5 \cdot 10^6$. The results of Lehmer's calculation on the ENIAC were published as a list of corrigenda to Kraïtchik's table in 1947 (MTAC 2 (19) p. 313). In 1949 an article discussing some computational details of setting up this 'program' on the ENIAC appeared [7].

## 2.2   Description of the Main Steps of the Computation

As Lehmer remarks right at the beginning of his description of the ENIAC set-up: "The method used by the ENIAC to find the exponent of 2 modulo $p$ differs greatly from the one used by human computers" [7, p. 301]. A number-theorist knows that the exponent $e$ of 2 so that $2^e \equiv 1 \bmod p$ ($p$ prime) is either a divisor of or equal to $p-1$, and can thus restrict himself to doing trial divisions with suitable divisors of $p-1$ only. On the ENIAC, however, it is more expeditive to compute $2^t$ for all $t < p-1$. This 'idiot approach' takes, in the worst case, "less than 2.4 seconds, less time than it takes to copy down the value of $p$", whereas the sophisticated method requires "much outside information via punched cards [...] to be prepared by hand in advance." [7, p. 302]

Lehmer's flow-diagram [7, p. 303] gives the following main steps of the ENIAC computation:

**Step 1.** Initiation and preliminary set-up, go to Step 2.
**Step 2.** Increase $p$ by 2, goto Step 3.
**Step 3.** Sieve on $p$: Is $p$ divisible by a prime $\leq 47$? Yes/No, goto Step 2/Step 4.
**Step 4.** Exponent routine to find $e$. Is $e > 2,000$? Yes/No, goto Step 7/Step 5.
**Step 5.** Does $e$ divide $p-1$? Yes/No, goto Step 6/Step 7.
**Step 6.** Punch $p, e$ and $f$ ($p-1 = ef$), goto Step 7.
**Step 7.** Erase exponent calculation, goto Step 2.

The sieve set-up (step 3) can take advantage of the parallelism of the ENIAC's hardware. Thus implemented, this becomes a fast method for minimizing the chance that $p = 2n+1$ is not a prime. The sieve implemented on the ENIAC sieved out all numbers $p = 2n+1$ having prime factors $\leq 47$. As Lehmer notes [7, p. 302], about 86 percent of the composites were eliminated after step 3 (sieve). The remaining 24 percent were required to pass a further test: namely $p-1$ must be divisible by $e$ (step 5). This requirement is so strict that the remaining number of composites is very small. Finally, these were eliminated by hand through comparison with D.N. Lehmer's list of primes.

The exponent $e$ is calculated by a simple recursive routine (step 4), building up a sequence of positive integers $r_k$, where $r_k$ is nothing but the remainder on division of $2^k$ by $p$. The start value $r_1$ is set to 2. Given $r_k$, the next value $r_{k+1}$ is set to $2r_k - p$. If $2r_k - p < 0$, $r_{k+1}$ is set to $2r_k$. If $2r_k - p > 0$ then the ENIAC checks whether $r_{k+1}$ - 2 is negative. If the reply is yes, then $r_{k+1}$ is 1 and $e = k+1$. If no, then the ENIAC checks whether $k+1 = 2001$. If not, then $r_{k+2}$ is the next value computed. If yes, then the ENIAC gives up the search for $e$ and tries the next value of $p$.

## 2.3   Outline of the Set-Up of the Computation on the ENIAC

We will now give an outline of how to wire the ENIAC to perform Lehmer's computation. Due to lack of space not all details can be given. We will focus on the parallel set-up of the sieve and then give a sketch of the complete reconstruction. The reconstruction is based upon the 7-step-diagram (sec. 2.2) and some implementational details given in Lehmer's [7,8,9]. Important to note: Lehmer's description does not completely determine the actual machine implementation, the design of the ENIAC, however, seriously limits the possibilities.

**Wiring a sieve.** Sieves check for every subsequent number of a specific sequence if the number fulfills $j$ conditions/congruences. If it fulfills one or more congruences, it is sieved out, if not, the number is let through. The best known sieve, Eratosthenes's, checks for every natural number $n$ if it is divisible by a prime ($n \equiv 0 \bmod p_j$?), the numbers that pass are relative prime to the primes $p_j$.

Our sieve implementation uses an accumulator $A_{p_j}$ for each prime $p_j \leq 47$, ($1 \leq j \leq 14$ ) except for 2 (See sec. 2.2). To begin (step 1), each $A_{p_j}$ is set to the complement of $p_j - 1$. E.g. $A_{p_{14}}$ will contain M 9999999954. An initiating pulse is sent to the first transceiver (T1) of all $A_{p_j}$'s, its operation switch (OS) is set to $\alpha$ and the repeat switch (RS) to 1, as well as to the constant transmitter (CT) such that it will transmit the number 2. This results in the addition of 2 in all 14 accumulators. The CT then transmits a program pulse (PP), finishing the first cycle of the computation. The next steps of the computation check for each of the $A_{p_j}$ *in parallel* whether the number $P = 2r + 1$ (the first $P$ being 3) is or is not divisible by one of the $p_j$. This is done with the second branching method (See Sec. 1.2), by connecting the PM lead of the S output of each of the $A_{p_j}$ to 14 dummy controls (T2). This works because if $P$ is divisible by $p_j$, the number contained in $A_{p_j}$ will be P 0000000000 and thus positive, while it will be negative in all other cases (this is why we use complements). If a given $A_{p_j}$ stores P 0000000000, and $P$ is thus divisible by $p_j$, $A_{p_j}$ has to be reset to the complement of $2p_j$.[3] This was a difficult problem to solve, because only those accumulators that store P 0000000000 should receive a value, and each of these must receive a different number. The problem for the ENIAC to decide which accumulators should receive and which should not, was solved by *directly* connecting the program pulse output terminal of each of the dummy controls of the $A_{p_j}$ to the program pulse input terminal of one of the transceivers (T3) of each of the $A_{p_j}$. This could be done by using a *loaded program jumper* [4, 11.6.1]. Each T3 of an $A_{p_j}$ is set to receive once through input channel $\alpha$, $\beta$ or $\gamma$ depending on the group $A_{p_j}$ belongs to. The transmission of 14 different numbers to the 14 $A_{p_j}$'s is done by using the three function tables and special digit adaptors. The 14 $A_{p_j}$'s are divided into three groups: $A_{p_1} - A_{p_5}$, $A_{p_6} - A_{p_{10}}$, $A_{p_{11}} - A_{p_{14}}$. In each group, the PP output terminal of T1 of rsp. $A_{p_1}$, $A_{p_6}$ and $A_{p_{11}}$ is connected to three different program cables. The first of these cables sends a PP to function table 1, the second to function table 2 and the last to function table 3. The argument clear switch of each of the tables is set to O. Without going into the details of this setting, it is important to know that in this specific wiring, the switch is set to O so that the function table will transmit the value $f(0)$ to the input channel it is connected to. Each of the function tables contains rsp. one of the following values: M 610142226, M 3438465862 and M 64828694 at place 0 (function value $f(0)$). These numbers are nothing but the

---

[3] We use $2p_j$ since only numbers of the form $2r + 1$ are sent through the sieve.

concatenation of the values $2p_j$ which have to be sent to those $A_{p_j}$ for which $p_j$ divides $2r + 1$ ($A_{p_j}$ stores P 0000000000). Five addition times after each of the function tables has received a program pulse, each of these values will be sent through the respective input channels $\alpha$, $\beta$ and $\gamma$.

Now, if e.g. accumulator $A_{p_1}$ has been set to receive through $\alpha$ by its dummy control it will receive the value M 610142226 through $\alpha$. A special adaptor is inserted at the input terminal $\alpha$ of $A_{p_1}$. It is used to combine a shifter adaptor – which is used to shift the digit lines a certain number of times to the left or to the right – and a deleter adaptor – which makes it possible to select only those digits needed. Setting both deleter and shifter in the correct way for $A_{p_1}$, the number M 0000000006 (instead of M 610142226) will be subtracted from the content of $A_{p_1}$. After this, $A_{p_1}$ will contain M 9999999994 which is the value needed for the sieve to work properly.

Figure 2 shows the details of the wiring of this sieve for $A_{p_1}$. Note that two stepper counters plus two extra transceivers are used to let ENIAC decide whether $2r+1$ passes the sieve test. The first transceiver, situated in Accumulator $A_{p_1}$ is needed to delay this decision through the second stepper counter (waiting until the function tables have sent their values $f(0)$). The second transceiver (situated right next to one of the stepper counters) is used to send a program pulse to the direct input of a second stepper, thus "making" the decision. The whole sieve process takes 12 addition times and is thus very fast. Fundamental in this wiring was the synchronization of all the different steps.

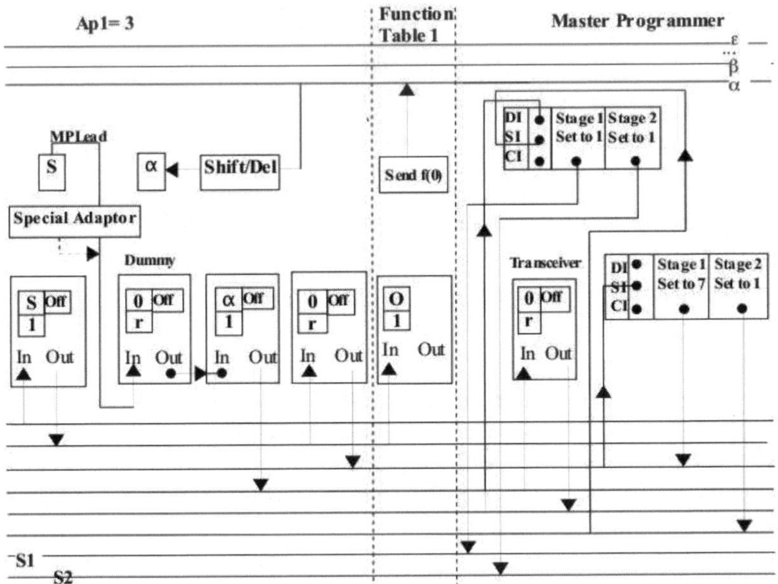


**Fig. 2.** Parallel Implementation of the sieve for $A_{p_1}$, $p_1 = 3$

**Outline of the complete program.** We will now provide an outline of our reconstruction of the complete computing-the exponent-of-2-modulo-$p$-program (Sec. 2.2). Steps 2 and 3 were already discussed in the previous paragraph.

**Step 1.** The setting of the $A_{p_j}$'s as well as the three function tables has already been discussed. The numbers $+ 2$, $-2$ and $-1$ are set manually on the constant transmitter (CT). An accumulator $A_P$ used to store the number $P = 2r + 1$ being processed, should be set to store the number 1. The computation is started by an initiating pulse sent to the 14 accumulators and the CT.

**Step 2.** This was already discussed. Besides the $A_{p_j}$'s also the transceiver of $A_P$, as well as a transceiver of one other accumulator used in step 4 ($A_{e_1}$), should be set to receive through $\alpha$.

**Step 3.** This was already discussed.

**Step 4.** In this subroutine 5 accumulators are used, i.e. $A_{e_1}$, $A_{e_2}$, $A_{e_3}$, $A_P$ and $A_E$ (which is used to count the number of iterations done before $2_{r_k} - p = 1$ or $k > 2000$ and thus to determine the exponent). Three stepper counters are used. One is used to check whether $k > 2000$ and receives a PP with each iteration of the subroutine. The others are used in combination with dummy controls: one to check whether $2_{r_k} - p > 0$ and one to check whether $2_{r_k} - p = 1$. We do not have the space here to give the details of the wiring.

**Step 5.** For this routine 4 accumulators are used, including $A_P$, $A_E$ and the cleared $A_{e_1}$ (which is used to store $f$). Besides these three the last unused accumulator $A_{20}$ is introduced into the computation. Again we cannot provide details here. It is important that at the start of the subroutine, $A_{20}$ receives $P$ from $A_P$ and next $-1$ from the CT. At the end of the computation, $A_{e_1}$ will contain $f$, the number of times $e$ can be subtracted from $p - 1$.

**Step 6.** There are several ways to wire this routine. One way is to use for $A_P$, $A_{e_1}$ and $A_E$ accumulators for which there is a static output to the printer.

**Step 7.** This can be done by using the clear-correct switch for the accumulators involved (except for $A_P$).


## 3     Discussion

Our proposed reconstruction of one of the first extensive computer 'programs', executed 1946 on the ENIAC, cannot be said to be complete nor definitive yet, but we project to present the complete wiring with discussion of difficult points in another, more extended paper in the near future. Already our outline brings out some salient points.

It is clear that 'programming' the ENIAC is perhaps sometimes nearer to engineering than to programming as we know it today. The 'if's are intricate digit-pulse-to-program-pulse conversions, which, in combination with the steppers of the master programmer, can be used to sequence the computation; different adaptors have to be inserted at the appropriate places, and the parallelism has to handled subtly by synchronizing the numerical and the program parts of the wiring. This polymorphy of combining the units, cables and adaptors of the ENIAC allows for many small tricks to be implemented in the 'program', but forbids a general approach to setting up the ENIAC. Such a general approach only arrived with the ENIAC's rewiring into a serial machine, using the function tables for 'indexing' the subroutines.

Concluding, we would like to encourage research on early computer programs on the ENIAC or on other early computers. Their study and analysis might contribute to understand the beginnings and achievements of the computer age. It might especially clarify the development of programming techniques and computational methods in correlation with the development of the hardware, as well as the evolution of the interaction and interface between the operator/programmer and the computer.

# References


1. Alt, F.: Archaeology of computers – reminiscences, 1945–1947. Comm. ACM 15(7), 693–694 (1972)
2. Burks, A.W., Burks, A.R.: The ENIAC: First general-purpose electronic computer. IEEE Ann. Hist. Comp. 3(4), 310–399 (1981)
3. Fritz, W.B.: Eniac – A problem solver. IEEE Ann. Hist. Comp. 16(1), 25–45 (1994)
4. Adele, K.: Goldstine. Report on the ENIAC, Technical report I. Technical report, Moore School of Electrical Engineering, University of Pennsylvania, Philadelphia (June 1946), Published in 2 vols
5. Goldstine, H.H., Goldstine, A.: The electronic numerical integrator and computer (ENIAC). Math. Tables Aids Comp. 2(15), 97–110 (1946)
6. Lehmer, D.H.: On the converse of Fermat's theorem. Amer. Math. Monthly 43(6), 347–354 (1936)
7. Lehmer, D.H.: On the converse of Fermat's theorem II. Amer. Math. Monthly 56(5), 300–309 (1949)
8. Lehmer, D.H.: The sieve problem for all-purpose computers. Math. Tables Aids Comp. 7(41), 6–14 (1953)
9. Lehmer, D.H.: The influence of computing on research in number theory. In: Proc. Sympos. Appl. Math., vol. 20, pp. 3–12. Amer. Math. Soc. (1974)
10. Lehmer, D.H.: A history of the sieve process. In: Howlett, J., Metropolis, N., Rota, G.-C. (eds.) A History of Computing in the Twentieth Century, pp. 445–456. Academia Press, New York (1980)
11. Neukom, H.: The second life of ENIAC. IEEE Ann. Hist. Comp. 28(2), 4–16 (2006)
12. Zoppke, T., Rojas, R.: The virtual life of ENIAC: Simulating the operation of the first electronic computer. IEEE Ann. Hist. Comp. 28(2), 18–25 (2006)

# Phase Transitions for Weakly Increasing Sequences

Michiel De Smet[*] and Andreas Weiermann

Ghent University
Building S22 - Krijgslaan 281
9000 Gent - Belgium
{mmdesmet,weierman}@cage.ugent.be

**Abstract.** Motivated by the classical Ramsey for pairs problem in reverse mathematics we investigate the recursion-theoretic complexity of certain assertions which are related to the Erdös-Szekeres theorem. We show that resulting density principles give rise to Ackermannian growth. We then parameterize these assertions with respect to a number-theoretic function $f$ and investigate for which functions $f$ Ackermannian growth is still preserved. We show that this is the case for $f(i) = \sqrt[d]{i}$ but not for $f(i) = \log(i)$.

**Keywords:** Ackermann function, weakly increasing sequences, Erdös-Szekeres, Dilworth, Ramsey theory, phase transitions.

## 1 Introduction

It is well known that every infinite sequence of natural numbers contains an infinite subsequence which is weakly increasing. It is quite natural to ask, which strength can be generated from this principle (which we call ISP) and for this purpose we consider a miniaturization of ISP in terms of densities. This density approaches Paris's original independence result for PA in terms of Ramsey-densities, which one can find in [6]. As usual the density statement for ISP follows from an application of König's Lemma to ISP.

We consider phase transitions related to ISP. This contributes to a general research program of the second author about phase transitions in logic and combinatorics. For more information, see [8], [9], [10], [11], and [12].

Let $f$ be a number-theoretic function and $X$ a set of natural numbers. Then a function $F : X \to \mathbb{N}$ is called $f$-regressive if $F(x) \leq f(x)$, for all $x \in X$. We define $X$ to be 0-ISP-dense($f$) if $|X| > f(\min(X))$ and to be $(n+1)$-ISP-dense($f$) if for all $f$-regressive $F : X \to \mathbb{N}$ there exists a $Y \subseteq X$ such that $F \restriction Y$ is weakly increasing and such that $Y$ is $n$-ISP-dense($f$). Then, due to König's Lemma, for any fixed $f$, every natural number $n$ and every natural number $a$ there exists a natural number $b := \mathrm{ISP}_f(n, a)$ such that the interval $[a, b]$ is $n$-ISP-dense($f$).

It is easy to see that for a constant function $f$ the function $n \mapsto \mathrm{ISP}_f(n, n)$ is primitive recursive. Moreover, one could prove that the function $n \mapsto \mathrm{ISP}_f(n, n)$

[*] Aspirant Fonds Wetenschappelijk Onderzoek (FWO) - Flanders.

A. Beckmann, C. Dimitracopoulos, and B. Löwe (Eds.): CiE 2008, LNCS 5028, pp. 168–174, 2008.

is Ackermannian for $f(i) = i$. So in-between constant functions and the identity function there will be a threshold region for $f$ where the function $n \mapsto \mathrm{ISP}_f(n, n)$ switches from being primitive recursive to being Ackermannian. We show that for $f(i) = \log(i)$ the function $n \mapsto \mathrm{ISP}_f(n, n)$ remains primitive, even elementary, recursive whereas for every fixed $d$ and $f(i) = \sqrt[d]{i}$ the function $n \mapsto \mathrm{ISP}_f(n, n)$ becomes Ackermannian. [Cf. Figure 1]

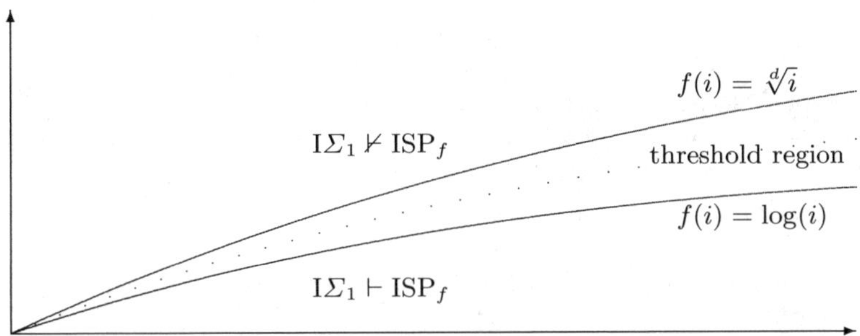


**Fig. 1.** Phase transitions for $\mathrm{ISP}_f$

Our results are intended to contribute partly to the $\mathrm{RT}_2^2$ problem in reverse mathematics. One can find information on this problem in, for example, [3] and [7]. We will need the well known Erdös-Szekeres theorem, which states that a given sequence $a_0, \ldots, a_{n^2}$ of real numbers contains a weakly increasing subsequence of length $n + 1$ or a strictly decreasing subsequence of length $n + 1$. CAC is the assertion that a given infinite partial order has either an infinite chain or an infinite antichain.

It is obvious that over $\mathrm{RCA}_0$, $\mathrm{RT}_2^2$ yields ISP and the related Erdös-Szekeres- or CAC-principles. Moreover, over $\mathrm{RCA}_0$ $\mathrm{RT}_2^2$ also yields the infinitary Erdös-Moser principle EM stating that every complete infinite directed graph has an infinite transitive subgraph. Now EM and CAC are particularly interesting for studying $\mathrm{RT}_2^2$ since $\mathrm{RCA}_0 + \mathrm{EM} + \mathrm{CAC}$ prove $\mathrm{RT}_2^2$. Therefore classifying the strength of EM and CAC may yield progress in classifying the strength of $\mathrm{RT}_2^2$. It is somewhat surprising that even ISP generates all primitive recursive functions with its miniaturization. But this should not be seen as an indication that $\mathrm{RCA}_0 + \mathrm{RT}_2^2$ prove the totality of the Ackermann function.

## 2 Classifying Phase Transitions for Weakly Increasing Sequences

Let us define the notion of ISP-density in a formal way.

**Definition 1.** *Let $f$ be a number-theoretic function. Then $X$ is called 0-ISP-dense$(f)$ if $|X| > f(\min(X))$. $X$ is defined to be $(n + 1)$-ISP-dense$(f)$ if for all $f$-regressive $F : X \to \mathbb{N}$, there exists $Y \subseteq X$ such that $Y$ is $n$-ISP-dense$(f)$ and*

$$y < y' \rightarrow F(y) \leq F(y'),$$

*for all $y, y' \in Y$.*

Given $f : \mathbb{N} \rightarrow \mathbb{N}$, we define

$$F_0(i) := i + 1;$$
$$F_{k+1}(i) := F_k^{f(i)}(i);$$
$$F(i) := F_i(i).$$

We have a look at the notion of ISP-density in the case of $f(i) = \lfloor \log(i) \rfloor$, where we set $\log(0) = 0$. Thus, if $\log(i)$ is not a natural number, we consider the greatest integer less than $\log(i)$, as $f$ needs to be a number-theoretic function. Henceforth we leave out the left and right floor symbols and write down $f(i) = \log(i)$, for the sake of clarity.

In the following proof we need the Erdös-Szekeres theorem, which is given in the introduction.

**Theorem 1.** $I\Sigma_1 \vdash (\forall n)(\forall a)(\exists b)([a, b] \text{ is } n\text{-}ISP\text{-}dense(\log))$.

*Proof.* If $a = 0$ and $n \leq 3$, the $\Sigma_1$-completeness of $I\Sigma_1$ yields that one can prove the existence of an appropriate $b$. Now assume that $a > 0$ and $n > 3$. Put $b := 2^{2^{(n+1)\cdot(n+a+1)}}$. We claim that any $Y \subseteq [a, b]$ with $|Y| > 2^{2^{(k+1)\cdot(n+a+1)}}$ is $k$-ISP-dense(log). Proof by induction on $k$. Assume the assertion holds for $k - 1$. Assume that $F : Y \rightarrow \mathbb{N}$, such that $F(y) \leq \log(y)$ for every $y \in Y$. Then there exists by the Erdös-Szekeres theorem a set $Z \subseteq Y$ such that $F \upharpoonright Z$ is weakly increasing or strictly decreasing and $|Z| > 2^{2^{(k+1-1)\cdot(n+a+1)}}$. Remark that

$$
\begin{aligned}
F(\min(Z)) &\leq \log(\min(Z)) \\
&\leq \log(2^{2^{(n+1)\cdot(n+a+1)}} - 2^{2^{k\cdot(n+a+1)}}) \\
&\leq \log(2^{2^{k\cdot(n+a+1)}} \cdot (2^{2^{(n+1)\cdot(n+a+1)}-2^{k\cdot(n+a+1)}} - 1)) \\
&\leq 2^{k\cdot(n+a+1)} + 2^{(n+1)\cdot(n+a+1)} - 2^{k\cdot(n+a+1)} \\
&= 2^{(n+1)\cdot(n+a+1)} \\
&< 2^{2^{k\cdot(n+a+1)}} \\
&< |Z|.
\end{aligned}
$$

So if $F$ is strictly decreasing on $Z$, then $F(\max(Z)) < 0$, a contradiction. Thus $F$ is weakly increasing on $Z$. Since $|Z| > 2^{2^{(k+1-1)\cdot(n+a+1)}}$ the induction hypothesis yields that $Z$ is $(k-1)$-ISP-dense(log), which implies that $Y$ is $k$-ISP-dense(log). If $k = 0$ then $|Y| > 2^{2^{(n+a+1)}} > \log(b) \geq \log(\min(Y))$. $\qquad\square$

Thus $f(i) = \log(i)$ yields a lower bound. We now consider this result with regard to the function $ISP_f$, mentioned in the introduction. Let us first give a formal definition.

**Definition 2.** *Let $f$ be a number-theoretic function. Define $\mathrm{ISP}_f : \mathbb{N} \to \mathbb{N}$ by $\mathrm{ISP}_f(n) := \mathrm{ISP}_f(n, n)$, where $\mathrm{ISP}_f(n, a)$ is the least natural number $b$, such that $[a, b]$ is $n$-ISP-dense($f$).*

Theorem 1 states that $\mathrm{ISP}_{\log}$ is a provably total function of $I\Sigma_1$. As explained in [7], this is equivalent with $\mathrm{ISP}_{\log}$ being primitive recursive.

To obtain an upper bound, we consider an arbitrary natural number $d$ and the number-theoretic function $f(i) = \lfloor \sqrt[d]{i} \rfloor$. Thus, if $\sqrt[d]{i}$ is not a natural number, we consider the greatest integer less than $\sqrt[d]{i}$, since $f$ needs to be a number-theoretic function. As done in the previous case, we will leave out the left and right floor symbols, for the sake of clarity. Using this $f$, the function $F$ is Ackermannian, which is demonstrated in [5].

**Lemma 1.** *Assume that $[a, b]$ is $n$-ISP-dense($\sqrt[d]{\ }$). Then there exists $Y \subseteq [a, b]$ such that $Y$ is $(n-1)$-ISP-dense($\sqrt[d]{\ }$) and such that for all $i$ such that $F_1^{i+1}(a) \leq b$ we have that $Y \cap [F_1^i(a), F_1^{i+1}(a)[$ contains exactly one element.*

*Proof.* Define $G_0 : [a, b] \to \mathbb{N}$ as follows. $G_0(F_1^i(a) + j) := \sqrt[d]{F_1^i(a)} - j$ where $j \in \mathbb{N}$, such that $F_1^i(a) \leq F_1^i(a) + j < F_1^{i+1}(a)$. Remark that

$$
\begin{aligned}
F_1^i(a) + j &< F_1^{i+1}(a) \\
&= F_1(F_1^i(a)) \\
&= F_0^{\sqrt[d]{F_1^i(a)}}(F_1^i(a)) \\
&= F_1^i(a) + \sqrt[d]{F_1^i(a)},
\end{aligned}
$$

which implies $j < \sqrt[d]{F_1^i(a)}$, and so $G_0(x) > 0$, for every $x \in [a, b]$. Then $G_0(x) \leq \sqrt[d]{x}$ for every $x \in [a, b]$. Since $[a, b]$ is $n$-ISP-dense($\sqrt[d]{\ }$), we can choose $Y \subseteq [a, b]$ which is $(n-1)$-ISP-dense($\sqrt[d]{\ }$) and on which $G_0$ is weakly increasing. On every interval $[F_1^i(a), F_1^{i+1}(a)[$ the function $G_0$ is strictly decreasing. Hence $Y \cap [F_1^i(a), F_1^{i+1}(a)[$ contains at most one element. If

$$
Y \cap [F_1^i(a), F_1^{i+1}(a)[ = \emptyset,
$$

then add $F_1^i(a)$ to $Y$. Since supersets of $(n-1)$-ISP-dense($\sqrt[d]{\ }$) sets are again $(n-1)$-ISP-dense($\sqrt[d]{\ }$), we are done. $\qquad\square$

**Lemma 2.** *Suppose $n \geq k > 0$. Assume that a nonempty $Y \subseteq [a, b]$ is $(n - k)$-ISP-dense($\sqrt[d]{\ }$) and that $Y \cap [F_k^i(a), F_k^{i+1}(a)[$ contains exactly one element for all $i$ such that $F_k^{i+1}(a) \leq b$. Then there exists a nonempty $Z \subseteq Y$ such that $Z$ is $(n - k - 1)$-ISP-dense($\sqrt[d]{\ }$) and for all $i$ such that $F_{k+1}^{i+1}(a) \leq b$ we have that $Z \cap [F_{k+1}^i(a), F_{k+1}^{i+1}(a)[$ is a singleton.*

*Proof.* Define $G_k : Y \to \mathbb{N}$ by $G_k(y_j^i) := \sqrt[d]{F_{k+1}^{i-1}(a)} - j$ for that $y_j^i \in Y$ with

$$
y_j^i \in [F_{k+1}^{i-1}(a), F_{k+1}^i(a)[
$$

172     M. De Smet and A. Weiermann

and
$$y_j^i \in [F_k^j(F_{k+1}^{i-1}(a)), F_k^{j+1}(F_{k+1}^{i-1}(a))[.$$

Remark that
$$F_k^j(F_{k+1}^{i-1}(a)) < F_{k+1}^i(a)$$
$$= F_{k+1}(F_{k+1}^{i-1}(a))$$
$$= F_k^{\sqrt[d]{F_{k+1}^{i-1}(a)}}(F_{k+1}^{i-1}(a)),$$

which yields $j < \sqrt[d]{F_{k+1}^{i-1}(a)}$, and so $G_k(y) > 0$, for every $y \in Y$. Furthermore, it is obvious that $G_k(y) \leq \sqrt[d]{y}$, for every $y \in Y$. Since Y is $(n-k)$-ISP-dense($\sqrt[d]{}$), we can choose a nonempty $Z \subseteq Y$ which is $(n-k-1)$-ISP-dense($\sqrt[d]{}$) and on which $G_k$ is weakly increasing. On every interval $Y \cap [F_{k+1}^{i-1}(a), F_{k+1}^i(a)[$ the function $G_k$ is strictly decreasing. Hence $Z \cap [F_{k+1}^{i-1}(a), F_{k+1}^i(a)[$ contains at most one element. If
$$Z \cap [F_{k+1}^{i-1}(a), F_{k+1}^i(a)[ = \emptyset,$$

then add $F_{k+1}^{i-1}(a)$ to $Z$. Since supersets of $(n-k-1)$-ISP-dense($\sqrt[d]{}$) sets are $(n-k-1)$-ISP-dense($\sqrt[d]{}$), we are done. $\qquad\square$

**Lemma 3.** *If* $Y \subseteq [a,b]$ *is* $n$-*ISP-dense*($\sqrt[d]{}$) *and* $a \geq 1$, *then* $\max (Y) \geq F_{n+1}(a)$.

*Proof.* Assume $Y \subseteq [a,b]$ is $n$-ISP-dense($\sqrt[d]{}$). Then by the preceding two lemmas there exists a nonempty $Z \subseteq [a,b]$ which is 0-ISP-dense($\sqrt[d]{}$) and such that $[F_{n+1}^i(a), F_{n+1}^{i+1}(a)[ \cap Z$ is a singleton for all $i$ with $F_{n+1}^{i+1}(a) \leq b$. Since $a \geq 1$ and $Z$ is 0-ISP-dense($\sqrt[d]{}$), $Z$ contains at least two elements and the second one is greater than $F_{n+1}(a)$. $\qquad\square$

If we put $a = n+1$, then Lemma 3 yields that $\text{ISP}_{\sqrt[d]{}}(n) \geq F(n+1)$, for every $n \in \mathbb{N}$. Since $F$ is Ackermannian, so is $\text{ISP}_{\sqrt[d]{}}$.

**Theorem 2.** *Let* $d \in \mathbb{N}$. *Then*
$$I\Sigma_1 \nvdash (\forall n)(\forall a)(\exists b)([a,b] \text{ is } n\text{-}ISP\text{-}dense(\sqrt[d]{})).$$

*Proof.* The provably total functions of $I\Sigma_1$ are exactly the primitive recursive functions. Thus, if $I\Sigma_1$ would prove the existence of an appropriate $b$ for all $n$ and all $a$, it would prove the existence of such $b$ for all $n$ and all $a \geq 1$. Hence $\text{ISP}_{\sqrt[d]{}}$ would be primitive recursive, which is a contradiction. $\qquad\square$

Combining Theorem 1 and Theorem 2, we can sharpen the range of the threshold region in comparison with our first estimation. Indeed, instead of claiming that the phase transition occurs for a function $f$ in-between constant functions and the identity function, we now can say that the threshold region for $f$ is in-between log and $\sqrt[d]{}$. Moreover, as we have recently found out, one can do even better. Let $A_d$ be the $d$th approximation of the Ackermann function Ack. Following claims will be proved in a future paper.

**Claim 1.** *Let $d$ be a natural number and $f(i) = i^{\frac{1}{A_d^{-1}(i)}}$. Then*

$$I\Sigma_1 \vdash (\forall n)(\forall a)(\exists b)([a,b] \text{ is } n\text{-}ISP\text{-}dense(f)).$$

**Claim 2.** *Let $f(i) = i^{\frac{1}{\mathrm{Ack}^{-1}(i)}}$. Then*

$$I\Sigma_1 \nvdash (\forall n)(\forall a)(\exists b)([a,b] \text{ is } n\text{-}ISP\text{-}dense(f)).$$

Hence we will have obtained a precise threshold region. One can also investigate ISP-density, allowing small ordinals in the range of the functions $F$ under consideration. To do so, it is necessary to extend the notion of ISP-density. Assume $l$ is a natural number and $\omega^l$-$n$-ISP-density-$(f)$ is defined in an appropriate way, then we can prove similar results for the threshold region as obtained above.

## 3 The Erdös-Szekeres Theorem and the Dilworth Decomposition Theorem

As we have already seen in the introduction, the Erdös-Szekeres theorem considers only a finite amount of elements. However, there also exists a corresponding infinitary statement, which states that a given infinite sequence $a_0, \ldots, a_n, \ldots$ of real numbers contains an infinite weakly increasing subsequence or an infinite strictly decreasing subsequence.

These two results are related via an appropriate density notion which we call ES-density (Erdös-Szekeres-density). In fact we consider ES-density with respect to a parameter function $f : \mathbb{N} \to \mathbb{N}$.

The well known Dilworth theorem states that a given partial order with distinct elements $a_0, \ldots, a_{n^2}$ contains a chain of length $n + 1$ or an antichain of length $n + 1$. The corresponding infinitary statement, which is called CAC, has already been given in the introduction. It states that a given infinite partial order contains an infinite chain or an infinite antichain.

Also these two results are related via an appropriate density notion which we call D-density (Dilworth-density) and which depends on a number-theoretic parameter function $f$.

Considering ES-density and D-density, one can prove similar results about the according phase transitions, as obtained in the case of the weakly increasing sequences.

The following problem arises. If we combine $I\Sigma_1$ with D-density in an appropriate way, then do we obtain the same provably recursive functions as given by $RCA_0 + CAC$?

The idea is to approximate Ramsey for pairs here. CAC would be a first step. Studying CAC we encounter the Erdös-Moser (EM) result, which is given in the introduction. EM states that every complete infinite directed graph has an infinite transitive subgraph. It is easy to see that $RT_2^2$ proves CAC and EM. But we also have a reversal, namely $RCA_0 + EM + CAC \vdash RT_2^2$. See, for example,

[1] and [7] for more information. We notice that EM can not be proved within $RCA_0$, since there exists a recursive complete infinite directed graph for which there is no recursive infinite transitive subgraph.

# References


1. Bovykin, A., Weiermann, A.: The strength of infinitary ramseyan principles can be accessed by their densities. In: Annals of Pure and Applied Logic (preprint, 2005), http://logic.pdmi.ras.ru/~andrey/research.html
2. Cholak, P., Marcone, A., Solomon, R.: Reverse mathematics and the equivalence of definitions for well and better quasi-orders. J. Symbolic Logic 69(3), 683–712 (2004)
3. Cholak, P.A., Jockusch, C.G., Slaman, T.A.: On the strength of Ramsey's theorem for pairs. J. Symbolic Logic 66(1), 1–55 (2001)
4. Kojman, M., Lee, G., Omri, E., Weiermann, A.: Sharp thresholds for the phase transition between primitive recursive and Ackermannian Ramsey numbers. Journal of Combinatorial Theory, Series A (in press); available online March 7, 2008
5. Omri, E., Weiermann, A.: Classifying the phase transition threshold for Ackermannian functions. Ann. Pure Appl. Logic (to appear)
6. Paris, J.B.: Some independence results for Peano arithmetic. J. Symbolic Logic 43(4), 725–731 (1978)
7. Simpson, S.G.: Subsystems of second order arithmetic. In: Perspectives in Mathematical Logic, Springer, Berlin (1999)
8. Weiermann, A.: Analytic combinatorics, proof-theoretic ordinals, and phase transitions for independence results. Ann. Pure Appl. Logic 136(1-2), 189–218 (2005)
9. Weiermann, A.: Phasenübergänge in Logik und Kombinatorik. Mitt. Dtsch. Math.-Ver. 13(3), 152–156 (2005)
10. Weiermann, A.: An extremely sharp phase transition threshold for the slow growing hierarchy. Math. Structures Comput. Sci. 16(5), 925–946 (2006)
11. Weiermann, A.: Phase transition thresholds for some natural subclasses of the recursive functions. In: Beckmann, A., Berger, U., Löwe, B., Tucker, J.V. (eds.) CiE 2006. LNCS, vol. 3988, pp. 556–570. Springer, Heidelberg (2006)
12. Weiermann, A.: Phase transition thresholds for some Friedman-style independence results. MLQ Math. Log. Q. 53(1), 4–18 (2007)

# Succinct NP Proofs from an Extractability Assumption

Giovanni Di Crescenzo[1] and Helger Lipmaa[2]

[1] Telcordia Technologies, Piscataway, NJ, USA
giovanni@research.telcordia.com
[2] University College London, London, UK

**Abstract.** We prove, *using a non-standard complexity assumption*, that any language in $\mathcal{NP}$ has a *1-round* (that is, the verifier sends a message to the prover, and the prover sends a message to the verifier) argument system (that is, a proof system where soundness holds against polynomial-time provers) with communication complexity *only polylogarithmic in the size of the $\mathcal{NP}$ instance*. We also show formal evidence that the nature of the non-standard complexity assumption we use is analogous to previous assumptions proposed in the cryptographic literature. The question of whether complexity assumptions of this nature can be considered acceptable or not remains of independent interest in complexity-theoretic cryptography as well as complexity theory.

## 1 Introduction

A conventional $\mathcal{NP}$ proof system requires a single message from prover to verifier and communication at most polynomial in the length of the instance to the $\mathcal{NP}$ language. Several variants of this proof system have been proposed in the literature, motivated by various theoretical and practical applications. In particular, interactive proof systems [4,12] added two major ingredients: interaction between prover and verifier, and randomization in messages exchanged, so that the traditional completeness and soundness were allowed not to hold with some very small probability. In this paper we focus on simultaneously minimizing the round complexity and the communication complexity of the messages exchanged between prover and verifier.

*History and previous work.* In [15], the author proposed a 2-round (or, 4-message) argument system for $\mathcal{NP}$ with polylogarithmic communication complexity under standard intractability assumptions; here, an argument system [7] is a proof system where soundness holds against all polynomial-time provers. (For related constructions see also universal arguments of [5] and CS proofs of [20].) It has long been an open question whether one can build a 1-round argument system for $\mathcal{NP}$ with even sublinear communication complexity. The impressive communication-efficiency property of both PCP schemes and PIR protocols motivated the authors of [1,6] to propose some combination of both tools to obtain such an argument system. As an example of how to combine these two tools, define a database equal to the PCP-transformed witness for the statement $x \in L$; then, the verifier's message contains multiple PIR queries for random and independent positions in this database, and the prover's message contains answers to these queries according to the PIR protocol; finally, the verifier can compute the

A. Beckmann, C. Dimitracopoulos, and B. Löwe (Eds.): CiE 2008, LNCS 5028, pp. 175–185, 2008.

database content with respect to the queried positions, and can apply the PCP verifier to accept or not the common input. (This is the protocol proposed in [6]; the protocol proposed in [1] is a variation of it, in that it adds some randomization steps to the verifier's message). As described, this protocol satisfies the completeness property, and would seem to only require a polylogarithmic (in $n$) number of queries to satisfy soundness, under an appropriate assumption about the privacy property of the PIR protocol used. (The intuition here being that the PIR privacy property implies that the prover cannot guess the random indices queried by the verifier, and therefore can only meet the verifier's checks with probability only slightly larger than the PCP error probability for inputs not in the language.) Unfortunately, this protocol was proved to be *not* sound in [11], one main technical reason being that a prover can use different databases to answer to each query by the verifier. Furthermore, the authors in [11] suggest the intuitive reasoning that the PIR privacy alone might be too weak to imply the soundness for a resulting argument system along these lines.

*Our result and assumption.* In this paper we break through the linear-size barrier for 1-round argument systems for NP languages, and in fact, achieve polylogarithmic communication complexity. Our departure point is again the 1-round protocol from [6] based on PIR schemes and PCP systems, but we add to it some crucial modifications based on hash function families and Merkle trees, bearing some similarities to the 2-round (4-message) protocol from [15] and to another 2-round protocol from [17] proposed for 2-party private computation. As we cannot use similar proofs as for 2-round protocols, we prove the soundness of our 1-round argument system using a non-black-box but *non-standard* assumption. However, we give formal evidence that our assumption is of similar nature to certain assumptions that were recently introduced in the cryptographic literature. Very roughly speaking, our assumption can be described as follows: the only way for a prover to send a 'correct' PIR answer to a PIR query from the verifier, along with a compression of the database and a proof that the answer belongs to the compressed database, is to actually know a "correct" database. Assumptions of this nature have been used to solve various problems, including long-standing ones, in cryptography (starting with [9]) and zero-knowledge (starting with [14]). While using these assumptions, researchers warned that they are very strong and suggested that their validity is further studied. Recently, in [10], these assumptions were abstracted into a class of "extractability-assumptions", and several variants of this class were defined, the strongest variants being proved to be false (assuming intractability assumptions often used in the cryptography literature and believed to be true), and the weakest variants (including the one we use here) remaining unsettled. All proofs are omitted due to space restriction.

## 2   Definitions and Tools

We recall formal definitions for the main tools and protocols used in our main construction: 1-round argument systems, PCP proofs, private information retrieval schemes, collision-resistant function families, and Merkle trees. We assume some familiarity with properties of these tools and with various other notions of proof systems for $\mathcal{NP}$ languages. We use the notation $y \leftarrow A(x)$ to denote the probabilistic experiment of running (the possibly probabilistic) algorithm $A$ on input $x$, and denoting its output as $y$.

**Definition 1 (One-round argument systems).** Let $L$ be a language in $\mathcal{NP}$. Let $(P, V)$ be a pair where $V = (V_1, V_2)$, and $P$, $V_1$, $V_2$ are probabilistic algorithms running in time polynomial in their first input. We say that $(P, V)$ is a *1-round argument system with parameters* $(\delta_c, \delta_s)$ for $L$ if the following properties hold:

*Completeness.* $\forall x \in L$, and all witnesses $w$ for $x$, it holds that $V_2(x, s, vmes, pmes) =$ accept with probability $\geq 1 - \delta_c(n)$, where $(vmes, s) \leftarrow V_1(x)$, $pmes \leftarrow P(x, w, vmes)$.
*Soundness.* $\forall x \notin L$, for all probabilistic polynomial time algorithms $P'$, it holds that $V_2(x, s, vmes, pmes) =$ accept with probability $\leq \delta_s(n)$, where $(vmes, s) \leftarrow V_1(x)$ and $pmes \leftarrow P'(x, vmes)$.

$(P, V)$ is a *1-round argument system* for $L$ if there exists negligible functions $\delta_c, \delta_s$ such that $(P, V)$ is a 1-round argument system with parameters $(\delta_c, \delta_s)$ for $L$.

**Probabilistically Checkable Proof (PCP) Systems.** In this paper we consider PCP systems [2,3] that are non-adaptive and have efficient generation (as essentially all such systems in the literature). In the formal definition of PCP systems, by $\pi_i$ we denote the bit at location $i$ of $m$-bit string $\pi$.

**Definition 2.** Let $L$ be a language in $\mathcal{NP}$. Let $(P, V)$ be a pair where $V = (V_1, V_2)$, and $P$, $V_1$, $V_2$ are probabilistic algorithms running in time polynomial in their first input. We say that $(P, V)$ is a *(non-adaptive) probabilistically checkable proof system with parameters* $(q, \delta_c, \delta_s)$ for $L$ if for some polynomial $p$ the next properties hold:

*Completeness.* $\forall x \in L$ such that $|x| = n$, for all witnesses $w$ certifying that $x \in L$, $P(x, w)$ returns with probability $\geq 1 - \delta_c(n)$ a proof tape $\pi = \pi_1 | \cdots | \pi_{p(n)}$ such that $V_2(x, (i_1, \pi_{i_1}), \ldots, (i_q, \pi_{i_q})) =$ accept, where $(i_1, \ldots, i_q) \leftarrow V_1(x)$.
*Soundness.* $\forall x \notin L$, for all proof tapes $\pi'$, $V_2(x, (i_1, \pi'_{i_1}), \ldots, (i_q, \pi'_{i_q})) =$ accept holds with probability $\leq \delta_s(n)$, where $(i_1, \ldots, i_q) \leftarrow V_1(x)$.

We say that $(P, V)$ is a *(non-adaptive) probabilistically checkable proof system* for $L$ if there exists a polynomial $q$ and negligible functions $\delta_c, \delta_s$ such that $(P, V)$ is a (non-adaptive) probabilistically checkable proof system with parameters $(q, \delta_c, \delta_s)$ for $L$.

In this paper we use PCP systems with slightly superlogarithmic query complexity and negligible soundess error; that is, having parameters $(q, 1 - o(1/\mathrm{poly}(n)), o(1/\mathrm{poly}(n)))$, for $q = (\log n)^{1+\epsilon}$, for some constant $\epsilon > 0$. PCP systems in the literature having such parameters include [22].

**Private Information Retrieval (PIR) Schemes.** We review the formal definition of PIR schemes in the single-database model [16].

**Definition 3.** Let $D$, $Q$, $R$ be algorithms running in polynomial time in the length of their first input ($Q$, $R$ may be probabilistic). Let $n$ be a security parameter, $m$ be the length of the database and $\ell$ the bit length of the database elements. We say that $(D, Q, R)$ is a *(single-database) PIR scheme with parameters* $(n, m, \ell, \delta_c, \delta_p)$ if:

*Correctness:* For any $m$-element database $db$ of $\ell$-bit strings and any location $i \in [m]$, with probability $\geq 1 - \delta_c(n)$ it holds that $R(1^n, 1^m, 1^\ell, i, (q, s), a) = db[i]$, where $(q, s) \leftarrow Q(1^n, 1^m, 1^\ell, i)$, and $a \leftarrow D(1^n, 1^\ell, db, q)$.
*Privacy:* For any family of probabilistic circuits $\{A_n\}_{n \in \mathbb{N}}$ running in time $t(n)$ and any $i, j \in [m]$, it holds that $|p_i - p_j| \leq \delta_p(n)$, where for $h = i, j$, it holds that $p_h = \text{Prob}\left[ (q, s) \leftarrow Q(1^n, 1^m, 1^\ell, h) \ : \ A_n(1^n, q) = 1 \right]$,

PIR schemes with communication complexity only polylogarithmic in the size $m$ of the database, have been proposed, under hardness assumptions about number-theoretic problems, in [8,13,18]. We can use any of these schemes in our main result.

**Collision-resistant (CR) hash function families.** This popular cryptographic primitive can be informally described as a family of compression functions for which it is hard to compute two preimages of the same output. We recall the formal definition of a CR hash function secure against adversaries that run in time superpolynomial in the hash function's security parameter. This will allow us to use CR hash functions with a security parameter polylogarithmic in another, global, security parameter.

**Definition 4.** Let $\mathcal{H} = \{H_u\}$ be a family of functions $H_u : \{0, 1\}^{2v} \to \{0, 1\}^v$, where $u$ is a function index satisfying $|u| = n$. We say that $\mathcal{H}$ is a *collision-resistant function family with parameters* $(n, 2v, v, t, \epsilon)$ if for any algorithm $A$ running in time $t(n)$, it holds that $\text{Prob}\left[ u \leftarrow \{0, 1\}^n; (x_1, x_2) \leftarrow A(u) \ : \ H_u(x_1) = H_u(x_2) \right] \leq \epsilon(n)$. We say that $\mathcal{H}$ is a *superpolynomial-time collision-resistant function family* if it is a collision-resistant function family with parameters $(n, 2v, v, t, \epsilon)$, for some $v$ polylogarithmic in $n$, some $\epsilon$ negligible in $n$ and some $t$ superpolynomial in $n$.

**Merkle trees.** Starting from any collision-resistant hash function family, with hash functions $H_u$ mapping $2\ell$-bit inputs to $\ell$-bit outputs, Merkle defined in [19] the following tree-like construction to compress a polynomial number $m = 2^t$ of $\ell$-bit strings $x_0, \dots, x_{m-1}$ into a single $\ell$-bit string $y$. The output $\text{MTree}(H_u; x_0, \dots, x_{m-1})$ is recursively defined as $H_u(\text{MTree}(H_u; x_0, \dots, x_{m/2-1}), \text{MTree}(H_u; x_{m/2}, \dots, x_{m-1}))$, where $\text{MTree}(H_u; x) = x$, for any $\ell$-bit string $x$. In the rest of the paper, we will use the following notation: the output computed as $y = \text{MTree}(H_u; x_0, \dots, x_{m-1})$ is also denoted as *root*; for any $\ell$-bit string $x_i$ associated to the $i$-th leaf of the tree, we define the *$i$-th certification path* as the sequence of values that are necessary to certify that $x_i$ is a leaf of the Merkle tree with root $y$. This construction has been often used in several results in complexity theory and interactive proofs, including [5,15,20].

## 3   An Extractability Assumption

We now present the assumption that will be used in the rest of the paper, and prove that it is an extractability assumption, a notion recently introduced in [10] which generalizes assumptions studied in various papers (starting with [9]). We start by recalling definitions from [10].

**Extractability Hardness Assumptions.** Informally speaking, an extractability assumption considers any probabilistic polynomial time algorithm $A$ that, on input a security parameter in unary and an index, returns a secret output and a public output. Then, the assumption states that if $A$ satisfies certain efficiency or hardness properties (to be defined later), then for any adversary algorithm $Adv$ trying to simulate $A$, there exists an efficient algorithm $Ext$ that, given the security parameter, the index, $Adv$'s public output and random bits, can compute a matching secret output. Actually, it is more appropriate to talk about a class of extractability assumptions, varying over the specific algorithms $A$, and the algorithms that generate the index taken as input by $A$. Towards formal definitions, we first note that the problem of generating an extractability assumption may not be well-defined for all probabilistic polynomial-time algorithms (for instance, some algorithms may not have a secret output at all), but, instead, has to be defined for algorithms with very specific properties. This motivates the following definition of an extractable-algorithm candidate.

**Definition 5.** Let $Ind$ be a set samplable in polynomial time whose elements we also call *indices*. Let $A$ be a probabilistic polynomial time algorithm (or, alternatively, $A$ is a deterministic algorithm that takes as input a sufficiently long and uniformly distributed string $R$). On input a security parameter $1^n$, random string $R$ and an index $ind \in Ind$, $A$ returns a triple $(s, m, h)$ in time polynomial in $n$. Let Setup $=$ (Sample, Verify) be a pair of probabilistic polynomial time algorithms such that Sample, on input $1^n$, generates pairs $(ind, sind)$, where $ind \in Ind$, and Verify, on input $(1^n, ind, sind, m, h)$, returns *accept* with probability 1 if $\exists R, s$ such that $A(1^n, R, ind) = (s, m, h)$ or with probability negligible in $n$ otherwise. We say that $(A, Ind, \text{Setup})$ is an *extractable-algorithm candidate* if there exists a polynomial $r$ such that:

*Efficient public output computation.* There exists an efficient algorithm Eval that, on input $ind \in Ind$ and $s$, returns values $(R, m, h)$ such that $(s, m, h) = A(1^n, R, ind)$
*Secret output hardness.* For any efficient algorithm $Adv$ the probability $p_s$ that $\exists R'$ such that $A(1^n, R', ind) = (s', m, h)$ is negligible in $n$. Here, $R \leftarrow \{0, 1\}^{r(n)}$, $(ind, sind) \leftarrow$ Sample$(1^n)$, $(s, m, h) \leftarrow A(1^n, R, ind)$, and $s' \leftarrow Adv(ind, m, h)$.
*Hard-core output hardness.* For any efficient algorithm $Adv$, the probability $p_h$ that $\exists R'$ such that $A(1^n, R', ind) = (s, m, h')$ is negligible in $n$. Here, $R \leftarrow \{0, 1\}^{r(n)}$, $(ind, sind) \leftarrow$ Sample$(1^n)$, $(s, m, h) \leftarrow A(1^n, R, ind)$, and $h' \leftarrow Adv(ind, m)$.

Towards recalling the formal definition of the class of extractability assumptions, we note that even if an adversary $A$ succeeds in generating $m$ without any knowledge of $s$, then the hard-core output hardness requirement would make it hard for this specific adversary to generate $h$. The latter fact of course does not imply a proof that any $A$ returning $m, h$ actually knows $s$, but this class of assumptions postulates that this is indeed the case, by allowing $s$ to be efficiently extracted from $A$, given the randomness used in computing 'valid' outputs $m, h$ (for any algorithm $A$ that satisfies the above three properties). In the formalization, we also need a pair of algorithms Setup = (Sample, Verify), where Sample generates the index $ind$ taken as input by $A$, and Verify checks whether the output $(m, h)$ returned by $A$ has the correct form, by returning $accept$ with probability 1 if $\exists R, s$ such that $A(1^n, R, ind) = (s, m, h)$ or with probability negligible in $n$ otherwise.

**Assumption 1 [EA assumption].** Let $(A, Ind, \text{Setup})$, be an extractability assumption candidate, where Setup = (Sample, Verify) and $Ind$ is a set that is samplable in polynomial time. For any polynomial-time algorithm $Adv$, there exists a polynomial time algorithm $Ext$, called the $A$-*extractor*, such that, denoting by $aux$ a polynomial-length auxiliary-input (modeling $A$'s history), the probability that $\text{Verify}(1^n, ind, sind, m, h) = accept$ and $\nexists R' \in \{0, 1\}^{r(n)}$ such that $A(1^n, R', ind) = (s', m, h)$, is negligible in $n$, where $R \leftarrow \{0, 1\}^{r(n)}$, $(ind, sind) \leftarrow \text{Sample}(1^n)$, $(m, h) \leftarrow Adv(1^n, R, ind, aux)$ and $s' \leftarrow Ext(1^n, R, ind, m, h, aux)$.

## 3.1   Our EA Assumption

We now formally describe the EA assumption that we will use in this paper, which is based on PIR schemes and Merkle trees based on CR hash functions. Informally speaking, we consider an extractable-algorithm candidate that randomly chooses a large string $r$ and compresses into a much shorter string $root$ using Merkle trees, and defines a database $db$ as follows: the $i$-th record $db_i$ is set equal to the $i$-th bit $r_i$ of string $r$, concatenated with the logarithmic number of strings that are used to compute $root$ from $r_i$ within the Merkle tree computation. The index $ind$ is generated as a PIR query to a random index $j \in \{1, \ldots, |r|\}$, and the algorithm $A$ returns, on input the security parameter $1^n$ and $ind$, a main output, computed as the string $root$ associated with the root of the Merkle tree, a hard-core output, computed as the PIR answer using $ind$ as a query and $db$ as a database, and a secret output, computed as the string $r$.

Before proceeding more formally, we sketch why this construction of an extractable-algorithm candidate satisfies Definition 5. First, it satisfies the efficient output computation requirement as the efficient computability of the main output follows from the analogous property of the Merkle tree, and the efficient computability of the secret and hard-core outputs follow from the analogous property of the answers in a PIR scheme. Second, it satisfies the secret output hardness as an adversary able to compute the secret output from the main output and the index can be used to contradict the collision-resistance property of the hash function family. Third, it satisfies the hard-core output unpredictability requirement as it is hard to compute a valid PIR answer only from the PIR query and the root of the Merkle tree, as for database $db$, this would imply a way to break the collision-resistance property of the hash function family used.

**Formal description.** We formally define set $Ind$, algorithm $A$ and the pair of algorithms Setup = (Sample, Verify), and then prove that $(A, Ind, \text{Setup})$ is an extractable-algorithm candidate under appropriate assumptions. We will consider databases with $m = \text{poly}(n)$ records (the actual polynomial not being important for our result to hold).

*Set Ind:* This is the set of CR hash functions indices and PIR queries on databases with $m$ records; formally: $Ind = \{(u, query) \mid u \leftarrow \{0,1\}^k; (query, s_q) \leftarrow Q(1^n, 1^m, 1^\ell, i)$ for some $i \in \{1, \ldots, m\}$ and some random string used by $Q\}$.

*Algorithm* Sample*:* This is the querying algorithm in the PIR scheme; on input $1^n$, algorithm Sample randomly chooses $i \in \{1, \ldots, m\}$, $u \in \{0,1\}^k$, computes $(query, s_q) \leftarrow Q(1^n, 1^m, 1^\ell, i)$ and returns: $ind = (1^m, u, query)$ and $sind = (i, s_q)$.

*Algorithm* $A$*:* On input $1^n, ind$, algorithm $A$ first randomly chooses an $m$-bit string $r$ and computes a Merkle tree compression of $r$, thus obtaining $root$ and the $i$-th certification path $path_i$ from the $i$-th bit $r_i$ in $r$ to $root$, for $i = 1, \ldots, m$. Then $A$ defines an $m$-record database $db$ as follows: for $i = 1, \ldots, m$, the $i$-th record of $db$ contains a unique $v$-bit representation of bit $r_i$, concatenated with the $i$-th certification path $path_i$. Then, $A$ computes

1. the main output as $main = root$;
2. the hard-core output $h$ equal to the PIR answer to the query from $ind$ using $db$ as a database; that is, $h = D(1^n, 1^m, 1^\ell, db, query)$, where $ind = (1^m, u, query)$;
3. the secret output $s$ equal to the $m$-bit string $r$.

*Algorithm* Verify*:* This is the retrieving algorithm in the PIR scheme; formally, algorithm Verify$(1^n, ind, sind, main, h)$ is defined as follows:

1. rewrite $ind$ as $ind = (1^m, query)$, $sind$ as $sind = (i, s_q)$, $main$ as $main = root$;
2. compute $db[i] = R(1^n, 1^m, 1^\ell, i, (query, s_q), h)$ and rewrite $db[i]$ as $r_i | path_i$;
3. check that $path_i$ is a valid $i$-th certification path from $r_i$ to $root$ using hash function $H_u$; if yes, then return: $accept$ otherwise return: $reject$.

We obtain the following theorem.

**Theorem 1.** Let $n$ be a security parameter. Assume the existence of a family of CR hash functions with parameters $(n', 2v, v, t, \epsilon)$, such that $n', v$ are polylogarithmic in $n$ and $t$ is superpolynomial in $n$ but subexponential in $v$. Also, assume there exists a (single-database) PIR scheme having parameters $(n, m, \ell, \delta_c, \delta_p)$, with communication complexity polylogarithmic in $n$, where $m$ is polynomial in $n$, $\ell$ is polylogarithmic in $n$ and $\delta_c$ is negligible in $n$. Then the above triple $(A, Ind, \text{Setup})$ is an extractable-algorithm candidate with parameters $(n, p_s, p_h)$, where: (1) $p_s$ is negligible in $n$; (2) if $\delta_p$ is negligible in $n$ then so is $p_h$; (3) if $(s, h, m)$ denotes $A$'s output on input $(1^n, ind)$, then $|s| + |h| + |m|$ is at most polylogarithmic in $n$.

## 4   A Low-Communication 1-Round Argument for $\mathcal{NP}$

We are now ready to present the main result of the paper.

**Theorem 2.** Let $L$ be a language in $\mathcal{NP}$ and let $(A, Ind, \text{Setup})$ be the triple proposed in Section 3, and proved to be an extractable-algorithm candidate assuming the existence of PIR schemes and CR hash functions. If $(A, Ind, \text{Setup})$ satisfies the EA assumption, then there exists a 1-round argument system $(P, V)$ for $L$ such that: if the assumed PIR scheme has communication complexity polylogarithmic in the database size then $(P, V)$ has communication complexity polylogarithmic in $n$, the length of the common input to the argument system.

We once again caution the reader that this result is based on a quite non-standard hardness assumption. Preliminary studies on variants of this assumption [10] indicate that the strongest variants are actually false (under intractability assumptions that are often used in the cryptography literature and believed to be true) no matter what is the specific extractable-algorithm candidate. Luckily, the variant used here seems significantly different and it is still open whether it can be proved to be false for all extractable-algorithm candidates (under some conventional intractability assumption) or can be considered a reasonable assumption for at least one of them.

**Informal Description of** $(P, V)$**.** Our argument system is obtained as an appropriate combination of the following tools: PIR schemes with efficient communication complexity, PCP systems, CR hash function families and Merkle trees. As in [1], the starting point is the protocol from [6]: the verifier asks to receive some random entries in the PCP-transformed witness through some random PIR queries; the prover computes the PCP-transformed witness and uses it as a database from which to compute and send the PIR answers to the verifier; finally, the latter can check that the indices retrieved from the database corresponding to entries that would be accepted by the PCP verifier. As this protocol was shown to be not sound from [11], we attempt to modify it so that it achieves soundness under the EA assumption for the extractable-algorithm candidate presented in Section 3. Consequently, we modify the prover so that for every PIR query, it also computes a Merkle-tree compression of the PCP-transformed witness and defines each database record to contain not only a bit of the PCP-transformed witness but also the certification path to the Merkle-tree root. We then note that this modification, while enabling us to use the EA assumption, may still not be very helpful as a cheating prover might choose to apply the Merkle-tree compression algorithm to a new string $\pi^j$ for every query $query^j$ made by the verifier. (In essence, this is a variant of the main objection raised by [11] about the protocol in [1].) In our protocol such attacks are avoided by modifying prover and verifier so that the prover only computes a single Merkle-tree root and the verifier can efficiently check that, for each of the verifier's PIR queries, the prover uses certification paths that refer to the same root (and thus to the same single database containing them).

**Formal Description.** By $x$ we denote the $n$-bit common input to our argument system $(P, V)$. Protocol $(P, V)$, formally described in Figure 1, uses the following tools:

1. A collision-resistant hash function family $\mathcal{H} = \{H_u\}$, such that $H_u : \{0, 1\}^{2v} \rightarrow \{0, 1\}^v$, where $|u|, v$ are polylogarithmic in $n$.
2. The Merkle tree construction $Mtree$ defined in Section 2, based on the collision-resistant hash function family $\mathcal{H}$.

3. A (non-adaptive) PCP system (pcpP,pcpV), where pcpV=(pcpV$_1$,pcpV$_2$), with parameters $(q, \delta_c, \delta_s)$, where $q$ is polylogarithmic in $n$; $\delta_c, \delta_s$ are negligible in $n$.
4. A (single-database) PIR scheme $(D, Q, R)$ with parameters $(n, m, \ell, \delta_c, \delta_p)$ with communication complexity polylogarithmic in $m$, and where $m$ is polynomial in $n$, $\ell$ is polylogarithmic in $n$ and $\delta_c, \delta_p$ are negligible in $n$.

We now prove that $(P, V)$ (formally described at the end of the section) satisfies Theorem 2. We start by noting that $V$ runs in polynomial time. This follows since algorithms $Q, R$ from the assumed PIR scheme and algorithm pcpV from the assumed PCP scheme run in polynomial time; and, furthermore, since checking whether a given string is an $i$-th certification path in a Merkle tree can be done in polynomial time.

**Communication complexity:** The communication complexity of $(P, V)$ is polylog$(n)$ as: both the value $u$ and each PIR query sent by $V$ have length polylogarithmic in $n$; the number $q$ of PIR queries sent to $P$ is also polylogarithmic in $n$; since the database record length $\ell$ is $O(v \log m)$, and $v$ is chosen to be polylogarithmic in $n$, then so is $\ell$ and so is the length of each of the PIR answers sent by $P$.

**Completeness.** Assume $x \in L$. Then the completeness (with $\delta_c$ negligible in $n$) follows from the correctness property of the PIR scheme used and the completeness of the PCP proof system used.

**Soundness (main ideas).** Assume that $x \notin L$ and that there exists a cheating prover making $V$ accept with non-negligible probability. Then with the same probability this prover produces a main output $main$ and $q$ corresponding hard-core outputs $h_1, \ldots, h_q$ of algorithm $A$, in correspondence of the PIR queries from $V$, from which one can obtain indices $ind_1, \ldots, ind_q$ for $A$. Now, we distinguish two cases, according to whether, after applying the EA assumption to triple $(ind_i, main, h_i)$ and thus extracting string $W_i$, for $i = 1, \ldots, q$, the extracted strings $W_1, \ldots, W_q$ are all equal or not.

*Case (a):* there exists $a, b \in \{1, \ldots, q\}$ such that $W_a \neq W_b$. In this case we can derive an efficient algorithm that breaks the collision-resistance of the hash function family $\mathcal{H}$. Even in this case, as while proving the secret output hardness in the proof of Theorem 1, we need to use the extractable algorithm assumption to extract two different strings $W_a, W_b$ such that $MTree(H_u; W_a) = MTree(H_u; W_b)$.

*Case (b):* $W_1 = \cdots = W_q$. In this case, we can derive an efficient algorithm that distinguishes which among two $q$-tuples of random values in $\{1, \ldots, m\}$ was used to compute $V$'s PIR queries (by a simple hybrid argument, this is then used to efficiently break the privacy of the PIR scheme used). Very roughly speaking, this is done by observing the following. First, for the $q$-tuple actually used by $V$'s PIR queries, the prover is able to provide entries from the PCP-transformed witness that would be accepted by the PCP verifier. Instead, the $q$-tuple not used by $V$'s PIR queries has distribution uniform and independent from the exchanged communication. Then the probability that the string $W_1$ used by the prover contains entries from the PCP-transformed witness that would be accepted by the PCP verifier in correspondence with this $q$-tuple can be showed to be negligible using the soundness of the PCP proof system used.

---

**Common input:** $n$-bit instance $x$
$P$'s **private input:** a witness $w$ certifying that $x \in L$.

$V$ **(message 1):**
1. Randomly choose an index $u$ for a hash function $H_u$ from $\mathcal{H}$;
2. for $j = 1, \ldots, q$,
   randomly and independently choose database index $i_j \in \{1, \ldots, m\}$;
   compute PIR query $(query_j, aux_j) = Q(1^n, 1^m, 1^\ell, i_j)$;
3. send $u, query_1, \ldots, query_q$ to $P$.

$P$**(message 2):**
1. Run the PCP prover on input instance $x$ and witness $w$ and let
   $\pi =$pcpP$(x, w)$;
2. compute $root = Mtree(H_u; \pi)$ and send $root$ to $V$;
3. for $i = 1, \ldots, m$;
   let $path_i$ be the $i$-th certification path for the $i$-th bit $\rho_i$ of $\pi$,
   define the content of the $i$-th record of database $db$ as $(\pi_i | path_i)$;
4. for $j = 1, \ldots, q$,
   compute $ans_j = D(1^n, 1^m, 1^\ell, db, query_j)$ and send $ans_j$ to $V$.

$V$ **(decision):**
1. For $j = 1, \ldots, q$,
   compute $db_{i_j} = R(1^n, 1^m, 1^\ell, i_j, (query_j, aux_j), ans_j)$;
   rewrite $db_{i_j}$ as $db_{i_j} = (path_{i_j} | \pi_{i_j})$;
   check that $path_{i_j}$ is an $i_j$-th certification path for $\pi_{i_j}$ and $root$;
2. check that pcp$V_2(x, (i_1, \pi_{i_1}), \ldots, (i_q, \pi_{i_q})) =$accept;
3. if all verifications are satisfied then accept else reject.

---

**Acknowledgements.** The second author was partially supported by the the Estonian Science Foundation, grant 6848.

# References


1. Aiello, W., Bhatt, S.N., Ostrovsky, R., Rajagopalan, S.R.: Fast Verification of Remote procedure Calls: Short Witness-Indistinguishable One-Round Proofs for $\mathcal{NP}$. In: Welzl, E., Montanari, U., Rolim, J.D.P. (eds.) ICALP 2000. LNCS, vol. 1853. Springer, Heidelberg (2000)
2. Arora, S., Safra, S.: Probabilistic Checking of Proofs: A New Characterization of NP. Journal of the ACM 45(1), 70–122 (1998)
3. Arora, S., Lund, C., Motwani, R., Sudan, M., Szegedy, M.: Proof verification and the hardness of approximation problems. Journal of the ACM 45(3), 501–555 (1998)
4. Babai, L., Moran, S.: Arthur-Merlin Games: a Randomized Proof System, and a Hierarchy of Complexity Classes. Journal of Computer and System Sciences 36, 254–276 (1988)
5. Barak, B., Goldreich, O.: Universal Arguments and Their Applications. In: Proc. of IEEE Conference on Computational Complexity (2002)
6. Biehl, I., Meyer, B., Wetzel, S.: Ensuring the Integrity of Agent-Based Computation by Short Proofs. In: Proc. of Mobile Agents 1998. LNCS. Springer, Heidelberg (1998)

7. Brassard, G., Chaum, D., Crépeau, C.: Minimum Disclosure Proofs of Knowledge. Journal of Computer and System Sciences 37(2), 156–189 (1988)
8. Cachin, C., Micali, S., Stadler, M.: Computationally Private Information Retrieval with Polylogarithmic Communication. In: Stern, J. (ed.) EUROCRYPT 1999. LNCS, vol. 1592. Springer, Heidelberg (1999)
9. Damgård, I.: Towards Practical Public-key Systems Secure against Chosen Ciphertext Attack. In: Feigenbaum, J. (ed.) CRYPTO 1991. LNCS, vol. 576. Springer, Heidelberg (1992)
10. Di Crescenzo, G.: Extractability Complexity Assumptions (August 2006) (unpublished manuscript)
11. Dwork, C., Langberg, M., Naor, M., Nissim, K., Reingold, O.: Succinct NP Proofs and Spooky Interactions (December 2004) (unpublished manuscript)
12. Goldwasser, S., Micali, S., Rackoff, C.: The Knowledge Complexity of Interactive Proof-Systems. SIAM Journal on Computing 18(1) (1989)
13. Gentry, C., Ramzan, Z.: Single-Database Private Information Retrieval with Constant Communication Rate. In: Caires, L., Italiano, G.F., Monteiro, L., Palamidessi, C., Yung, M. (eds.) ICALP 2005. LNCS, vol. 3580. Springer, Heidelberg (2005)
14. Hada, S., Tanaka, T.: On the existence of 3-round Zero-Knowledge Protocols. In: Krawczyk, H. (ed.) CRYPTO 1998. LNCS, vol. 1462. Springer, Heidelberg (1998)
15. Kilian, J.: A note on Efficient Zero-knowledge Proofs and Arguments. In: Proc. of ACM STOC 1991 (1991)
16. Kushilevitz, E., Ostrovsky, R.: Replication is not needed: Single Database, computationally-private information retrieval. In: Proc. of 38th IEEE FOCS 1997 (1997)
17. Laur, S., Lipmaa, H.: Consistent Adaptive Two-Party Computations, Cryptology ePrint Archive, Report 2006/088 (2006)
18. Lipmaa, H.: An Oblivious Transfer Protocol with Log-Squared Communication. In: Zhou, J., López, J., Deng, R.H., Bao, F. (eds.) ISC 2005. LNCS, vol. 3650. Springer, Heidelberg (2005)
19. Merkle, R.: A Certified Digital Signature. In: Brassard, G. (ed.) CRYPTO 1989. LNCS, vol. 435. Springer, Heidelberg (1990)
20. Micali, S.: CS proofs. In: Proc. of 35th IEEE FOCS 1994 (1994)
21. Russell, A.: Necessary and Sufficient Conditions for Collision-Free Hashing. J. Cryptology 8(2), 87–100 (1995)
22. Samorodnitsky, A., Trevisan, L.: A PCP characterization of NP with Optimal Amortized Query Complexity. In: Proc. of the 32nd ACM STOC 2000 (2000)

# Describing the Wadge Hierarchy for the Alternation Free Fragment of $\mu$-Calculus (I)
## The Levels Below $\omega_1$


Jacques Duparc[1] and Alessandro Facchini[1,2,*]

[1] University of Lausanne, Faculty of Business and Economics - ISI,
University of Lausanne, CH-1015 Lausanne
[2] LaBRI, University of Bordeaux 1, 351 cours de la Libération,
FR-33405 Talence cedex
{jacques.duparc,alessandro.facchini}@unil.ch



**Abstract.** The height of the Wadge Hierarchy for the Alternation Free
Fragment of $\mu$-calculus is known to be at least $\epsilon_0$. It was conjectured
that the height is exactly $\epsilon_0$. We make a first step towards the proof of
this conjecture by showing that there is no $\Delta_2^\mu$ definable set in between
the levels $\omega^\omega$ and $\omega_1$ of the Wadge Hierarchy of Borel Sets.

**Keywords:** $\mu$-calculus, Wadge games, topological complexity, parity
games, weakly alternating automata.


## 1 Introduction

Propositional modal $\mu$-calculus extends basic modal logic by adding two fixpoint
operators. It is a very expressive and interesting formal system. Indeed, when
considering binary trees, it is as expressive as $S2S$, monadic second order logic
of two successors, and therefore it is very well suited for specifying properties of
transition systems. However, altough much studied, the $\mu$-calculus remains very
difficult to understand. This can already be seen when considering the alternation
free fragment. Indeed, the languages definable by a formula of this sublogic are
exactly the ones recognized by a weakly alternating automaton. Since the class
of weakly recognizable languages corresponds to the intersection of Büchi and
co-Büchi languages, this fragment captures some languages not recognized by a
deterministic automaton.

We are interested in the analysis of the topological complexity of non-deterministic tree languages. Very little is known from this perspective, but what concerns the rather small class of weakly recognizable languages. In particular, in
[DM07] the authors show that the Wadge hierarchy of this class has height at
least $\epsilon_0$, and they conjecture that the bound is tight. By showing that the degree of a tree language definable by an alternation free formula is either below


* Research supported by a grant from the Swiss National Science Foundation, n.
200021-116508: Project "Topological Complexity, Games, Logic and Automata".



A. Beckmann, C. Dimitracopoulos, and B. Löwe (Eds.): CiE 2008, LNCS 5028, pp. 186–195, 2008.

$\omega^\omega$ or above $\omega_1$, we make a first step towards the proof of this fundamental conjecture. Indeed, we think that by adapting the very same strategy used in this paper, it would be possible to prove a generalization of our main result: if a set definable by an alternation free formula is Wadge equivalent to a set of the form $\sum_{i=k}^{0} B_i \bullet \beta_i + E \bullet \alpha$, with every $B_i$ and $E$ initializable sets, if $\alpha < \omega_1$ then $\alpha < \omega^\omega$. The next, harder, final step towards the complete answer to the conjecture would then be to prove that if the degree of a set corresponding to the exponentiation of the class of models of an alternation free formula is $\omega_1^\alpha$, where $\alpha < \omega_1$, then in this case too $\alpha < \omega^\omega$ holds.

Due to space limitations, we omit some of the proofs. They can be found in [DF$\infty$].

## 2   The Propositional Modal $\mu$-Calculus

### 2.1   Syntax and Semantics

The $\mu$-calculus $\mathcal{L}_\mu$ is the logic resulting of the addition of greatest and least fixpoint operators to propositional modal logic. More precisely, given a set P, the collection $\mathcal{L}_\mu$ of fixpoint formulas is defined as follows:

$$\varphi ::= p \mid \neg p \mid \top \mid \bot \mid \varphi \wedge \varphi \mid \varphi \vee \varphi \mid \Diamond\varphi \mid \Box\varphi \mid \mu x.\varphi \mid \nu x.\varphi$$

where $p, x \in \mathsf{P}$ and $x$ occurs only positively in $\eta x.\varphi$ ($\eta = \nu, \mu$).

Let $W$ be a non empty alphabet. A tree over $\wp(\mathsf{P})$ is a partial function $t :$ $W^* \to \wp(\mathsf{P})$ with a prefix closed domain. Those trees can have infinite and finite branches. A tree is called binary if $W = \{0, 1\}$. In the sequel we only consider binary trees over $\wp(\mathsf{P})$.

Let $T^\omega_{\wp(\mathsf{P})}$ denote the set of full binary trees over $\wp(\mathsf{P})$. Given $v \in \mathsf{dom}(t)$, by $t \downarrow_v$ we denote the subtree of $t$ rooted in $v$. By $^n\{0, 1\}$ we denote the set of words over $\{0, 1\}$ of length $n$, and by $t[n]$ we denote the finite initial binary tree of height $n + 1$ given by the restriction of $t$ over $\bigcup_{0 \leq i \leq n} {}^i\{0, 1\}$.

Throughout the paper, $\mu$-formulas will be interpreted over full binary trees over $\wp(\mathsf{P})$.

Given a binary tree $t$, the set $||\varphi||_t$ of nodes satisfying a formula $\varphi$ is defined inductively by the usual conditions for propositional modal logic, plus the conditions for the fixpoint operators specified by:

$$||\mu x.\varphi(x)||_t = \bigcap \{A \subseteq \{0, 1\}^* : ||\varphi(A)||_t \subseteq A\}$$

$$||\nu x.\varphi(x)||_t = \bigcup \{A \subseteq \{0, 1\}^* : A \subseteq ||\varphi(A)||_t\}$$

We say that a tree $t$ is a model of a $\mu$-formula iff the root $\varepsilon$ of the tree is such that $\varepsilon \in ||\varphi||_t$. Clearly if $v \in ||\varphi||_t$, then $t \downarrow_v$ is isomorphic to a model of $\varphi$. In this case we simply say that $t \downarrow_v$ is a model of $\varphi$. We denote by $||\varphi||$ the class of all models of $\varphi$.

Fixpoint operators can be viewed as quantifiers. Therefore we use the standard terminology and notations as for quantifiers. For instance $\mathsf{free}(\varphi)$ denotes the class of all propositional variables occurring free in $\varphi$.

If $\psi$ is a subformula of $\varphi$, we write $\psi \leq \varphi$. We write $\psi < \varphi$ when $\psi$ is a proper subformula. We denote by $\mathsf{sub}(\varphi)$ the set of all subformulas of $\varphi$.

A formula $\varphi$ of $\mathcal{L}_\mu$ is said to be *well-named* if no two distincts occurrences of fixpoint operators in $\varphi$ bind the same variable, and no variable has both free and bound occurrences in $\varphi$.

A propositional variable $p$ is guarded in a formula $\varphi$ of $\mathcal{L}_\mu$ if every occurrence of $p$ in $\varphi$ is in the scope of a modal operator. A formula $\psi$ of $\mathcal{L}_\mu$ is said to be *guarded* iff for every subformula of $\psi$ of the form $\eta x.\delta$, $x$ is guarded in $\delta$.

It is easy to check that every fixpoint formula is equivalent to a well-named and guarded $\mu$-formula. From now on we consider that every $\mu$-formula is well-named and guarded.

Because $\varphi$ is well named, if $x$ is not free in $\varphi$ then there is exactly one subformula $\eta x.\delta \leq \varphi$ which bounds $x$. This formula is denoted by $\varphi_x$.

Suppose $\psi < \phi$ and that $\{x_1, \ldots, x_n\}$ is the set of free variables of $\psi$ which are bounded in $\phi$. Then, following [ES89], we say that the *closure of $\psi$ with respect to $\phi$*, denoted by $\mathsf{cl}_\phi(\psi)$, is the formula obtained by substituing in $\psi$ every $x_i$ with its binding formula $\phi_{x_i}$ in $\phi$. The *closure of a $\mu$-formula $\varphi$*, denoted by $\mathsf{cl}(\varphi)$, is given by

$$\mathsf{cl}(\varphi) = \bigcup_{\phi \leq \varphi} \mathsf{cl}_\varphi(\phi)$$

Note that $\mathsf{cl}(\varphi)$ is finite.

Given $\varphi \in \mathcal{L}_\mu$ , we say that $\varphi$ is *reduced* if there is no $\psi < \varphi$ such that $\psi \equiv \bot$ or $\psi \equiv \top$. Since every $\mu$-formula is equivalent to a reduced $\mu$-formula, we also suppose that every $\mu$-formula is reduced.


## 2.2  The Alternation Free Fragment

Let $\Phi \subseteq \mathcal{L}_\mu$. For $\eta \in \{\nu, \mu\}$, $\eta(\Phi)$ is the smallest class of formulas such that:

- $\Phi, \neg\Phi \subset \eta(\Phi)$;
- If $\psi(x) \in \eta(\Phi)$ and $x$ occurs only positively, then $\eta x.\psi \in \eta(\Phi)$;
- If $\psi, \varphi \in \eta(\Phi)$, then $\psi \wedge \varphi, \psi \vee \varphi, \Diamond\psi, \Box\psi \in \eta(\Phi)$;
- If $\psi, \varphi \in \eta(\Phi)$ and $x$ is bound in $\psi$, then $\varphi[x/\psi] \in \eta(\Phi)$

With the help of this definition, we introduce the syntactical hierarchy for the modal $\mu$-calculus. For all $n \in \mathbb{N}$, we define the class of $\mu$-formulas $\Sigma_n^\mu$ and $\Pi_n^\mu$ inductively as follows:

- $\Sigma_0^\mu = \Pi_0^\mu$ - formulas without fixpoints;
- $\Sigma_{n+1}^\mu = \mu(\Pi_n^\mu)$;
- $\Pi_{n+1}^\mu = \nu(\Sigma_n^\mu)$.

$$\Delta_n^\mu := \Sigma_n^\mu \cap \Pi_n^\mu$$

The fixpoint alternation depth ($ad$) of a formula is the number of non-trivial nestings of alternating least and greatest fixpoints. Formally, the alternation depth of $\varphi \in \mathcal{L}_\mu$ is given by

$$\mathsf{ad}(\varphi) = n \text{ iff } n = \inf\{k : \varphi \in \Delta_{k+1}^\mu\}$$

All $\Sigma_n^\mu$ and $\Pi_n^\mu$ form the *syntactical modal $\mu$-calculus hierarchy*, which is strict. The fixpoint alternation free fragment corresponds to the class $\Delta_2^\mu$.

## 3   Evaluation Game for the $\mu$-Calculus

In this section, given $\varphi \in \mathcal{L}_\mu$ and a binary tree $t$, we characterize the corresponding parity game, keeping in mind that by $\mathsf{ad}(\psi)$ we denote the alternation depth of the $\mu$-formula $\psi$.

We define the game's arena $\langle V_0, V_1, E \rangle$, and the corresponding parity game as follows. The set $V$ of vertices corresponds to pairs of $\mu$-formulas of the closure of $\varphi$ and members of $\mathsf{dom}(t)$. Then we can split $V$ into $V_0$ and $V_1$ in order to define the locations of Player 0 and Player 1. We say that some vertex $\langle \psi, v \rangle \in V$ belongs to $V_0$ iff one of the following conditions holds:

1. $\psi = \bot$;
2. $\psi = p$ and $p \notin t(v)$;
3. $\psi = \neg p$ and $p \in t(v)$;
4. $\psi = \delta_1 \vee \delta_2$;
5. $\psi = \Diamond \delta$.

Every other vertex belongs to $V_1$.

The moves $E$ are defined as follows:

1. if $\psi = \bot, \top, p, \neg p$ no moves are possible ;
2. from $\langle \phi_1 \vee \phi_2, v \rangle$ player 0 can move to $\langle \phi_i, v \rangle$, and from $\langle \phi_1 \wedge \phi_2, v \rangle$ player 1 can move to $\langle \phi_i, v \rangle$, $i = 1, 2$;
3. from $\langle \Diamond \phi, v \rangle$ player 0 can move to $\langle \phi, w \rangle$, with $w = v0, v01$, and from $\langle \Box \phi, v \rangle$ player 1 can move to $\langle \phi, w \rangle$, with $w = v0, v1$;
4. from $\langle \eta x.\phi(x), v \rangle$, the only available move leads to $\langle \phi(x/\eta x.\phi(x)), v \rangle$, with $\eta = \mu, \nu$.

In order to complete the definition, we define the (partial) priority function $\Omega : V \to \omega$. Let $\langle \psi, s \rangle \in V$. Then, if $\psi = \eta X.\delta$, $\eta = \mu, \nu$, we have that:

$$\Omega(\langle \psi, s \rangle) = \begin{cases} \mathsf{ad}(\eta X.\delta) & \text{if } \eta = \mu \text{ and } \mathsf{ad}(\psi) \text{ is odd, or} \\ & \eta = \nu \text{ and } \mathsf{ad}(\psi) \text{ is even;} \\ \mathsf{ad}(\eta X.\delta) - 1 & \text{if } \eta = \mu \text{ and } \mathsf{ad}(\psi) \text{ is even, or} \\ & \eta = \nu \text{ and } \mathsf{ad}(\psi) \text{ is odd.} \end{cases}$$

Thus, if the play $\pi$ is finite, Player 0 wins iff the last vertex of the play belongs to $V_1$, and if the play $\pi$ is infinite, Player 0 wins iff the greatest priority appearing infinitely often is even.

We denote by $\mathcal{E}(\varphi, t)$ the parity game associated to $\varphi \in \mathcal{L}_\mu$ and the tree $t$, starting at $\langle \varphi, \varepsilon \rangle$.

**Proposition 1 ([ES89]).** $t \in ||\varphi||$ *iff Player 0 has a winning strategy for* $\mathcal{E}(\varphi, t)$. ∎

This result can be seen as the "game-theoretical version" (restricted to full binary trees) of what is usually called the *fundamental theorem* of the semantic of the modal $\mu$-calculus.

We will be mainly interested on the "infinitary" part of a parity game $\mathcal{E}(\varphi, t)$. Given a parity game $\mathcal{E}(\varphi, t)$, its *infinitary reduction*, denoted by $\mathcal{E}^\infty(\varphi, t)$, is the parity game defined as follows:

if there is no leaf in the game tree of $\mathcal{E}(\varphi, t)$, then $\mathcal{E}^\infty(\varphi, t) = \mathcal{E}(\varphi, t)$, otherwise we modify the game tree of $\mathcal{E}(\varphi, t)$ recursively from the leaves to the root as follows:

1. for every leaf $\langle \phi, v \rangle$ of the game tree of $\mathcal{E}(\varphi, t)$, if $v \in ||\phi||_t$, then we substitute $\langle \top, v \rangle$ to the node $\langle \phi, v \rangle$ into the tree. Otherwise we substitute $\langle \bot, v \rangle$ to the node $\langle \phi, v \rangle$ into the tree,

2. if in the resulting tree, the node we are considering is a conjuctive node, that is of the form $\langle \phi \wedge \psi, v \rangle$, then we erase the subtree starting from the conjuctive node and then:

(a) we change $\langle \phi \wedge \psi, v \rangle$ into $\langle \varphi, v \rangle$ iff the node is such that

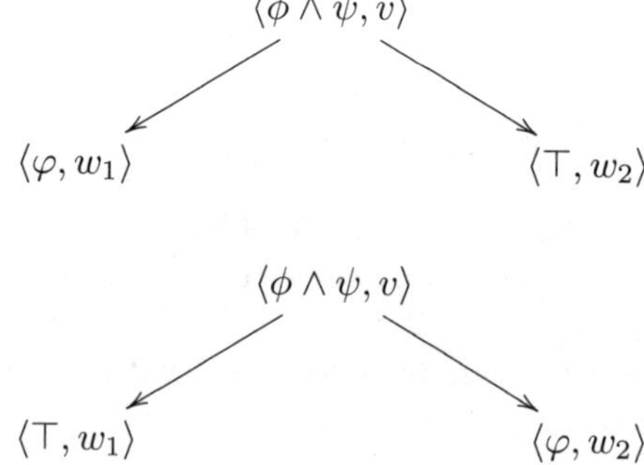

or

Dually in the case of a disjunctive node, that is of a node of the form $\langle \phi \vee \psi, v \rangle$;

(b) we change $\langle \phi \wedge \psi, v \rangle$ into $\langle \bot, v \rangle$ iff the node is such that

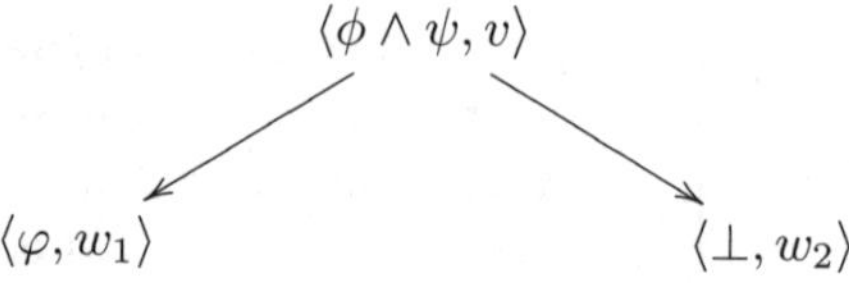

or

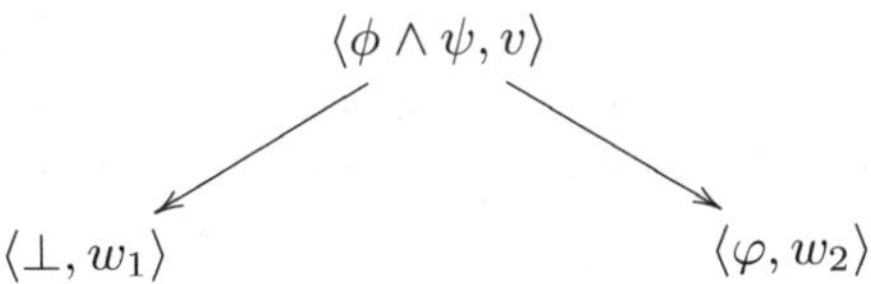

Dually in the case of a disjunctive node;

3. if in the resulting tree, the node we are considering is an existential node, that is of the form $\langle \Diamond \phi, v \rangle$, then we erase the subtree starting from the existential node and then

   (a) we change $\langle \Diamond \phi, v \rangle$ into $\langle \bot, v \rangle$ iff the existential node is such that every successor is of the form $\langle \bot, w \rangle$. Dually in the case of a universal node, that is of the form $\langle \Box \phi, v \rangle$;

   (b) we change $\langle \Diamond \phi, v \rangle$ into $\langle \top, v \rangle$ iff the node is such that there is a successor of the form $\langle \top, w \rangle$. Dually in the case of a universal node;

4. if in the resulting tree the considered node is a fixpoint node, that is of the form $\langle \eta x.\phi, v \rangle$, with $\eta = \mu, \nu$, then we erase the subtree starting from the fixpoint node and we change the fixpoint node into $\langle \psi, v \rangle$ iff the only successor $\langle \psi, w \rangle$ of $\langle \eta x.\phi, v \rangle$ is such that $\psi = \top, \bot$ or $\eta x.\phi \not\leq \psi$.

At the end of the process, the game tree reduces to a single node $\langle \phi, \varepsilon \rangle$. Then we set

$$\mathcal{E}^\infty(\varphi, t) = \mathcal{E}(\phi, t)$$

By construction, Player 0 has a winning strategy for $\mathcal{E}(\varphi, t)$ iff Player 0 has a winning strategy for $\mathcal{E}^\infty(\varphi, t)$.

## 4   The Wadge Hierarchy

Consider the space $T_{\mathsf{B}}^\omega$ equipped with the standard Cantor topology. Then, if $T, U \subseteq T_{\mathsf{B}}^\omega$, we say that $T$ is continuously reducible to $U$, if there exists a continuous function $f$ such that $T = f^{-1}(U)$. We write $T \leq_w U$ iff $T$ is continuously reducible to $U$. This particular order is called the *Wadge ordering*. If $T \leq_w U$ and $U \leq_w T$, then we write $T \equiv_w U$. If $T \leq_w U$ but not $U \leq_w T$, then we write $T <_w U$. Thus, the Wadge hierarchy is the partial order induced by $<_w$ on the equivalence classes given by $\equiv_w$.

Let $T$ and $U$ be two arbitrary sets of full binary trees. The Wadge game $\mathcal{G}_w(T, U)$ is played by two players, player I and player II. Both player build a tree, say $t_I$ and $t_{II}$. At every round, player I plays first, and both players add a finite number of children to the terminal nodes of their corresponding tree. Player II is allowed to skip its turn, but not forever.

We say that player II wins the game iff $t_I \in T \Leftrightarrow t_{II} \in U$. This game was designed precisely in order to obtain:

**Lemma 1 (Wadge).** *Let $T, U \subseteq T_{\mathsf{P}}^\omega$. Then $T \leq_w U$ iff Player I has a winning strategy in the game $\mathcal{G}_W(T, U)$.* ∎

Recall that a language $L$ is called *self dual* if it is equivalent to its complement, otherwise it is called *non-self dual*.

By Borel determinacy, we have that, if $T, U \subseteq T_B^\omega$ are Borel, then $\mathcal{G}_W(T, U)$ is determined. As a consequence, a variant of Martin-Monk's result shows that $<_w$ is well-founded. Thus, we can define by induction the *Wadge degree* for sets of finite Borel rank:

- $d_w(\emptyset) = d_w(\emptyset^C) = 1$
- $d_w(L) = \sup\{d_w(M) + 1 : M \text{ is non self dual}, M <_w L\}$ for $L >_W \emptyset$.

Weakly recognizable tree languages are all Borel. Moreover, in [DM07], by showing that the family of tree languages recognized by weak alternating automata is closed under the set-theoretical counterpart of ordinal sum $(+)$, multiplication $(\bullet)$ by ordinals $< \omega^\omega$, and pseudo-exponentiation with base $\omega_1$. It was proved that the Wadge hierarchy of this class of tree languages has height at least $\epsilon_0$. The authors conjecture that, in fact, the height *is* $\epsilon_0$. Because the weakly recognizable languages are the $\Delta_2^\mu$ definable tree languages, the preceding results and conjecture apply also to the fixpoint alternation free fragment of the $\mu$-calculus.

# 5  Main Result

In this section we show that there is no $\Delta_2^\mu$ definable set whose level is between $\omega^\omega$ and $\omega_1$. This means that if $L$ is $\Delta_2^\mu$ definable, then $d_w(L) < \omega^\omega$ or $d_w(L) \geq \omega_1$.

Recall that we suppose that models of $\mu$-formulas are infinite binary trees. First of all, we have to modify the construction of the arena of the *infinitary reduction* of the evaluation game of a $\mu$-formula to the case of a finite initial segment of a model. More precisely, apply the procedure previously described in section 3 to the subset $\{\langle \phi, v \rangle : v \in \bigcup_{0 \leq i \leq n} {}^i\{0, 1\} \wedge \phi \in \mathsf{cl}(\varphi)\}$, by considering that a node $\langle \phi, v \rangle$ is a leaf of the restriction iff it is a leaf in the original game graph. Suppose that the results of the procedure is the node $\langle \xi, \varepsilon \rangle$ and that $\xi \notin \{\top, \bot\}$, otherwise the case is trivial. Then, the *reduction of $\varphi$ to $t[n]$* is the bipartite graph $\langle V_0^*, V_1^*, E^* \rangle$ defined as follows. The set $V^*$ of vertices corresponds to pairs of $\mu$-formulas of the closure of $\xi$ and members of $\bigcup_{0 \leq i \leq n} {}^i\{0, 1\}$. The moves $E^*$ are defined as expected. However, we now modify the splitting of $V^*$ into $V_0^*$ and $V_1^*$. We say that some vertex $\langle \psi, v \rangle \in V^*$ belongs to $V_0^*$ iff it is a disjunctive or an existential node *and* $E(\langle \psi, v \rangle) \neq \emptyset$.

We are now ready to define the class of reaching games of $\varphi$ over $t[n]$. Such a reaching game is played by two players, player 0 and player 1, over the arena $\langle V_0^*, V_1^*, E^* \rangle$. Let $\{S_1, \ldots, S_n\}$ be an enumeration of all the subsets of $\bigcup_{0 \leq i \leq n} {}^i\{0, 1\}$. Given a subset $S_j$, we say that *Player 0 wins the reaching game of $\varphi$ over $t[n]$ with respect to $S_j$* iff a position $\langle \psi, v \rangle \in V_1^*$ can be reached where Player 1 cannot move and $v \in S_j$. This game is denoted by $\mathcal{R}(\varphi, t[n], S_j)$.

Whitout loss of generality, we can suppose that for every $v \in \mathsf{dom}(t)$, $t(v) \setminus \bigcup_{w \in \mathsf{dom}(t) \setminus \{v\}} t(w) \neq \emptyset$. Therefore, for every $v \in t[n]$, there is a $\mu$-formula $\phi_v$ without fixpoint and $p \in \mathsf{free}(\phi_v) \setminus t(v)$ such that for every $\psi \in \mathcal{L}_\mu$, $t \in ||\phi_v(p/\psi)||$

iff $t \downarrow_v \in \|\psi\|$. We can read $\phi_v$ as the formula describing the only way to reach the node $v$ starting from the root of the tree $t$. Moreover, for every $n$, there is a formula $\phi_{t[n]}$ which completely describes this initial tree of height $n+1$, that is, for every tree $s$, if $s \in \|\phi_{t[n]}\|$, then $s[n]$ and $t[n]$ are bisimilar.

Note that, for very $S_j$, the number of winninig strategies for Player 0 in the reaching game $\mathcal{R}(\varphi, t[n], S_j)$ is finite. Let $\{f_1, \ldots, f_k\}$ be an enumeration of the winning strategies for player 0 in $\mathcal{R}(\varphi, t[n], S_j)$. Fix a winning strategy $f_i$. Let $F_i \subseteq S_j$ be the set of nodes such that, $v \in F_i$ iff there exists a final winning position having as second component $v$ if player 0 plays according to $f_i$. Clearly $F_i$ is finite. Suppose $F_i = \{v_{i,1}, \ldots, v_{i,k}\}$. Then, for every $v_{i,l} \in F_i$, consider the finite set $\Phi_{i,l} \subseteq \mathsf{cl}(\varphi)$ such that $\psi \in \Phi_{i,l}$ iff $\langle \psi, v_{i,j} \rangle$ is a possible final winning position when player 0 plays according to $f_i$. Suppose $\Phi_{i,l} = \{\psi_{(i,1_l)}, \ldots, \psi_{(i,n_l)}\}$, for $v_{i,l} \in F_i$. Thus, the formula

$$\tau_i := \phi_{v_{i,1}}\big(p/(\psi_{(i,1_1)} \wedge \ldots \wedge \psi_{(i,n_1)})\big) \wedge \ldots \wedge \phi_{v_{i,k}}\big(p/(\psi_{(i,1_k)} \wedge \ldots \wedge \psi_{(i,n_k)})\big)$$

can be seen as describing the tree corresponding to the winning strategy $f_i$ for player 0 in the reaching game $\mathcal{R}(\varphi, t[n], S_j)$.

More generally, given

$$\sigma_j := \bigvee_{1 \leq i \leq l} \tau_i$$

and assuming that $t \in \|\varphi\|$, Player 0 has a winning strategy in the reaching game $\mathcal{R}(\varphi, t[n], S_j)$ iff $t \in \|\sigma_j\|$.

Consider now the set $M_{\mathcal{R}(\varphi, t[n])} \subseteq \{S_1, \ldots, S_n\}$ given by the following conditions:

for every $S_i \in M_{\mathcal{R}(\varphi, t[n])}$:

1. Player 0 has a winning strategy in $\mathcal{R}(\varphi, t[n], S_i)$;
2. There is no $S_k \subset S_i$ such that Player 0 has a winning strategy in $\mathcal{R}(\varphi, t[n]S_k)$

By construction, for every tree $s$ we obtain that:

$$s \in \|\varphi \wedge \phi_{t[n]}\| \text{ iff } s \in \Big\| \bigvee_{S_j \in M_{\mathcal{R}(\varphi, t[n])}} \sigma_j \Big\|$$

This terminates the "$\mu$-calculus" part of the proof of the main result, the other half of it being the following descriptive set theoretical result[1]:

**Lemma 2 ([D03]).** *Let $A \subseteq T_A^\omega$ be an initializable set of trees; $\lambda, \delta$ be some countable ordinals and $B \subseteq T_B^\omega$, if $A \bullet (\delta + 1) \leq_w B \leq_w A \bullet \lambda$ then there is a finite tree $t$ such that*

$$\begin{cases} B_t \equiv_w A \bullet (\delta + 1), \\ \qquad \text{or} \\ B_t \equiv_w {}^{\neg}(A \bullet (\delta + 1)). \end{cases}$$

*where $B_t$ denotes the set of members of $B$ extending $t$.* ∎

---

[1] Note that originally this lemma was stated for infinite words. We adapt it for full binary trees in a straightforward manner.

The core of the proof of our main result relies on finding a "reasonable" upper bound for the degrees of conjunctive and disjunctive formulas. It is not difficult to verify that:

**Lemma 3.** *Let* $\varphi_1, \ldots \varphi_n \in \Delta_2^\mu$. *Suppose for every* $1 \leq i \leq n$, $||\varphi_i||$ *non self dual,* $d_w(||\varphi_i||) < \omega_1$. *Then:*

1. $d_w(||\varphi_1 \wedge \cdots \wedge \varphi_n||) \leq \sup\{d_w(||\varphi_i||) : 1 \leq i \leq n\} \cdot n;$
2. $d_w(||\varphi_1 \vee \cdots \vee \varphi_n||) \leq \sup\{d_w(||\varphi_i||) : 1 \leq i \leq n\} \cdot n.$ ∎

From this lemma, it is then possible to prove that:

**Proposition 2.** *Let* $\varphi \in \Delta_2^\mu$, $||\varphi||$ *non self dual and* $1 < d_w(||\varphi||) < \omega_1$. *Let* $t$ *be a full binary tree such that for a certain* $n$ *it holds that* $||\varphi \wedge \phi_{t[n]}|| <_w ||\varphi||$. *Then there exists* $k \in \mathbb{N}$:

$$d_w(||\varphi \wedge \phi_{t[n]}||) = d_w(|| \bigvee_{S_j \in M_{\mathcal{R}(\varphi, t[n])}} \sigma_j ||) \leq \lambda \cdot k$$

*where* $\lambda = \sup\{d_w \xi : \xi \in \bigcup_{S_j \in M_{\mathcal{R}(\varphi, t[n])}} \bigcup_i \bigcup_{v_{i,j} \in F_i} \Phi_{i,j}\}$. ∎

Note that in the last proposition, $k$ is given by considering the number of final winning positions for player 0 in the reaching games $\mathcal{R}(\varphi, t[n], S_j)$.

By proposition 2 we can prove that:

**Corollary 1.** *Assume* $\varphi \in \Delta_2^\mu$ *non self dual,* $1 < d_w||\varphi|| < \omega_1$ *and* $B \subset T_B^\omega$ *satisfying both* $B \leq_w ||\varphi||$ *and* $\mathrm{cof}(d_w B) = \omega$, *then, for every* $n$, *there is* $\lambda < d_w B$ *such that for every tree* $t$ *and every* $n$:

$$if ||\varphi \wedge \phi_{t[n]}|| <_w B \text{ then } d_w(||\varphi \wedge \phi_{t[n]}||) < \lambda \cdot \omega$$ ∎

This almost immediately leads to:

**Theorem 1.** *Let* $\varphi \in \Delta_2^\mu$. *Let* $\alpha > 1$ *be a countable ordinal, and suppose* $d_w||\varphi|| = \alpha$. *Then* $\alpha < \omega^\omega$.

*Proof of theorem 1:* Towards a contradiction, we assume that $\alpha \geq \omega^\omega$, and apply corollary 1. Consider $B$ a canonical set of Wadge degree $\omega^\omega$. By corollary 1, there exists $\lambda < d_w(B)$ such that for every tree $t$ and every $n$:

$$if ||\varphi \wedge \phi_{t[n]}|| <_w B \text{ then } d_w(||\varphi \wedge \phi_{t[n]}||) < \lambda \cdot \omega$$

Fix such a $\lambda$. Since $d_w(B) = \omega^\omega$, we have that $\lambda < \omega^\omega$. Therefore, there is $n < \omega$ such that $\lambda < \omega^n$. Hence $\lambda \cdot \omega < \omega^{n+1}$ holds.

By lemma 2, there is a tree $t$, an integer $n$ and $\varrho \in \{-, +\}$ such that $||\varphi||_t \equiv_w \varrho(\emptyset \bullet (\omega^{n+1} + 1))$. Finally, since $\omega^{n+1} + 1 < \omega^\omega$, $d_w(||\varphi||_t) < \lambda \cdot \omega$ holds, we obtain the following contradiction:

$$\lambda \cdot \omega < \omega^{n+1} < \omega^{n+1} + 1 = d_w(||\varphi||_t) < \lambda \cdot \omega$$ ∎

This theorem proves that there is no $\Delta_2^\mu$ definable set in between the levels $\omega^\omega$ and $\omega_1$ of the Wadge Hierarchy of Borel Sets, an enthusiastic first step in proving the whole conjecture about the Wadge Hierarchy of weakly recognizable tree languages.

## 6   Conclusion

We have seen how, with the help of reaching games of a $\mu$-formula over an initial segment of a model, it is possible to determine upper bounds for the Wadge degree of disjunctive and conjunctive alternation free formulas. This result is one of the key step in order to verify that in between levels $\omega^\omega$ and $\omega_1$ of the Wadge hierarchy of Borel sets there is no tree language definable by an alternating free formula.

The main result of our paper constitute a first step towards the proof of the fundamental conjecture stating that the height of Wadge hierarchy of weakly recognizable tree languages is exactly $\epsilon_0$ and that the set of ordinals involved is the smallest one that contains 1 and is closed under ordinal sum, multiplication by $\omega$ and exponentiation of base $\omega_1$.

## References


[AN01]    Arnold, A., Niwinski, D.: Rudiments of $\mu$-Calculus. Studies in Logic, vol. 146. Elsevier, Amsterdam (2001)

[BGL07]   Berwanger, D., Grädel, E., Lenzi, G.: The variable hierarchy of the $\mu$-calculus is strict. Theory of Computing Systems 40, 437–466 (2007)

[D01]     Duparc, J.: Hierarchy and Veblen Hierarchy Part 1: Borel Sets of Finite Rank. Journal of Symbolic Logic 66(1), 56–86 (2001)

[D03]     Duparc, J.: A Hierarchy of Deterministic Context-Free $\omega$-languages. Theoretical Computer Science 290, 1253–1300 (2003)

[DF$\infty$]   Duparc, J., Facchini, A.: Describing the Wadge Hierarchy for the Alternation Free Fragment of $\mu$-Calculus (I) - The Levels Below $\omega_1$. Draft version, http://www.hec.unil.ch/logique/recherche/travaux/

[DM07]    Duparc, J., Murlak, F.: On the Topological Complexity of Weakly Recognizable Tree Languages. In: Csuhaj-Varjú, E., Ésik, Z. (eds.) FCT 2007. LNCS, vol. 4639, pp. 261–273. Springer, Heidelberg (2007)

[EJ91]    Emerson, E.A., Jutla, C.S.: Tree Automata, $\mu$-Calculus and Determinacy (Extended Abstract). In: FOCS 1991, pp. 368–377 (1991)

[ES89]    Streett, R.S., Emerson, E.A.: An automata theoretic decision procedure for the propositional $\mu$-calculus. Information and Computation 81(3), 249–264 (1989)

# Subrecursive Complexity of Identifying the Ramsey Structure of Posets


Willem L. Fouché

Department of Decision Sciences,
University of South Africa, P.O. Box 392, 0003 Pretoria, South Africa
fouchwl@unisa.ac.za



**Abstract.** We show that finite ordinal sums of finite antichains are Ramsey objects in the category of finite posets and height-preserving embeddings. Our proof yields a primitive recursive algorithm for constructing the finite posets which contain the required homogeneities. We also find, in terms of the classical Ramsey numbers, best possible upper bounds for the heights of the posets in which the homogeneous structures can be found.


**Mathematics Subject Classification (2000):** 03D20, 68Q17, 06A07, 05D10.

**Keywords:** subrecursive hierarchy, partially ordered sets (posets), structural Ramsey theory.

## 1 Introduction

In [3] the author studied the structural Ramsey theory of finite posets. The paper [3] makes substantial use of the Graham-Rothschild theorem [6]. The proof of the latter result requires an Ackermann-type recursion and is consequently not a proof in primitive recursive arithmetic.

In this paper we establish, by means of primitive recursive constructions, Ramsey properties of posets with respect to height-preserving embeddings. Our main theorem is a generalisation of the results in [2]. The methods of the present paper can also be used to find a primitive recursive proof of the main result in [3]. This will be discussed in a sequel to this paper. A proof of Theorem 2 in this paper for $h = H = 2$ is established in [4] and applied to determining the Ramsey structure of bipartite graphs. In this connection see also [5]. Results of this type have important applications to model theory [1] and the topological dynamics of automorphism groups of countably categorical structures as was demonstrated in the decisive recent paper by Kechris, Pestov and Todorcevic [9].

In the seventies, largely because of the work by Nešetřil, Rödl [11], [12] and, independently by Abramson, Harrington [1], remarkable progress was made with the problem of determining the Ramsey objects in various classes of combinatorial configurations. During the second half of the eighties the proofs were simplified by Nešetřil and Rödl [14] by introducing new amalgamation techniques.


A. Beckmann, C. Dimitracopoulos, and B. Löwe (Eds.): CiE 2008, LNCS 5028, pp. 196–205, 2008.

(See also the survey [10] for a clear exposition of these techniques.) A variation of these techniques will play an important role in the proof of main theorem of this paper. (See Theorem 2.)

In this paper, unless otherwise stated, all the structures referred to will be finite. If $\mathbf{P}$ is a poset, then the *height* of an element $x$ of $\mathbf{P}$, denoted by ht $x$, is the largest size of any chain in $\mathbf{P}$ having $x$ as the maximum element. The height of $\mathbf{P}$, denoted by ht $\mathbf{P}$, is the maximum value of ht $x$ as $x$ varies over the elements of $\mathbf{P}$. If $\mathbf{P}$ and $\mathbf{Q}$ are posets, then an *embedding* of $\mathbf{P}$ into $\mathbf{Q}$ is an injective mapping $\lambda$ from the underlying set of $\mathbf{P}$ into the underlying set of $\mathbf{Q}$, such that for elements $x, y$ of $\mathbf{P}$, it is the case that $x < y$ iff $\lambda(x) < \lambda(y)$. If, moreover, ht $\lambda(x) =$ ht $\lambda(y)$ whenever ht $x =$ ht $y$, we say that the embedding is *height-preserving*. We write $[\mathbf{Q}, \mathbf{P}]_h$ for the set of copies of $\mathbf{P}$ under height-preserving embeddings of $\mathbf{P}$ in $\mathbf{Q}$. A poset is said to be *complete* if it is the ordinal sum of finitely many antichains. Hence a poset $\mathbf{P}$ is complete iff for elements $x, y$ of $\mathbf{P}$, it is the case that $x < y$, whenever ht $x <$ ht $y$.

For natural numbers $s, h, H$, we write $R(s, h, H)$ (the associated Ramsey number [17]) for the smallest natural number $N$, such that for any $s$-colouring of the $h$-subsets of a set $X$ consisting of $N$ elements, there is an $H$-subset $Y$ of $X$ such that all the $h$-subsets of $Y$ are of the same colour. In [2] the author proved.

**Theorem 1.** *Suppose $\mathbf{C}$ is a chain of size $h$ and $\mathbf{Q}$ is a poset of height $H$. Then, for every natural number $s$, there is a poset $\mathbf{R}$ of height $R(s, h, H)$ such that, for any $s$-colouring $\chi$ of the copies of $\mathbf{C}$ in $\mathbf{R}$, there is an embedding $\lambda$ of $\mathbf{Q}$ into $\mathbf{R}$ such that $\chi$ is monochromatic on the set of copies of $\mathbf{C}$ in $\mathbf{Q}'$, where $\mathbf{Q}'$ is the image of $\mathbf{Q}$ under $\lambda$.*

In this paper we generalise this theorem to arbitrary complete posets.

**Theorem 2.** *Suppose $\mathbf{P}$ is a complete poset of height $h$ and $\mathbf{Q}$ is a poset of height $H$. Then, for every natural number $s$, there is a poset $\mathbf{R}$ of height $R(s, h, H)$ such that, for any colouring $\chi : [\mathbf{R}, \mathbf{P}]_h \to [s]$, there is a height-preserving embedding $\lambda$ of $\mathbf{Q}$ into $\mathbf{R}$ such that $\chi$ is monochromatic on $[\mathbf{Q}', \mathbf{P}]_h$ where $\mathbf{Q}'$ is the image of $\mathbf{Q}$ under $\lambda$.*

The proof will show that one can find a poset $\mathbf{R}$ that satisfies the conclusion of the theorem such that its size depends primitive recursively on $s, h, H$ and the sizes of $\mathbf{P}$ and $\mathbf{Q}$, respectively. Moreover, the problem of constructing $\mathbf{R}$ from $\mathbf{P}, \mathbf{Q}$ and $s$ has a time complexity which also depends primitive recursively on the parameters $\mathbf{P}, \mathbf{Q}$ and $s$. It follows that there is a primitive recursive algorithm for yielding $\mathbf{R}$ from $s, h, H, \mathbf{P}$ and $\mathbf{Q}$.

In the standard Erdös-notation we can express the theorem as follows: For posets $\mathbf{P}$ and $\mathbf{Q}$ with $\mathbf{P}$ *complete* and of heights $h, H$, respectively, there is, for any $s \geq 1$, a poset $\mathbf{R}$ of height $R(s, h, H)$ such that

$$\mathbf{R} \longrightarrow (\mathbf{Q})_s^{\mathbf{P}}. \tag{1}$$

We shall show that the theorem is the best posssible as far as the height of $\mathbf{R}$ in (1) is concerned. Indeed, for any complete poset $\mathbf{P}$ of height $h$, say, there is, for any $H \geq h$, a poset $\mathbf{Q}$ of height $H$, such that, if $\mathbf{R}$ satisfies (1), then the height of $\mathbf{R}$ will be at least the Ramsey number $R(s, h, H)$.

For given natural numbers $s, h, H$ and for posets $\mathbf{P}, \mathbf{Q}$ as above, let $P(s, \mathbf{P}, \mathbf{Q})$ be the smallest natural number $r$, such that, for some poset $\mathbf{R}$ of height $R(s, h, H)$ and size $r$, the relation (1) holds. The size of $P(s, \mathbf{P}, \mathbf{Q})$ can be estimated in terms of the so-called *loop complexity* $l(P)$ of the primitive recursive function $P$. With every primitive recursive function $g$, one can associate a natural number, $L(g)$, the loop complexity of $g$. (See, for example,[19] for a discussion). Roughly speaking, one says $L(g) \leq n$, if, starting with addition, one can find a program for computing $g$ which involves at most $n$ nested loops. For example, the loop complexity of the function computing the well-known Ramsey numbers is at most three while, as was shown by Shelah [18], for the Jales-Hewitt function, the loop complexity is at most four. Our proof will yield a loop complexity of at most seven for the function $P$.

## 2   Preliminaries

If $\mathbf{P}$ is a poset of height $h$ and if $\sigma \subset [h]$, we write $\mathbf{P}(\sigma)$ for the induced subposet spanned by the subset $\{x \in \mathbf{P} : \text{ht } x \in \sigma\}$. We shall make frequent use of the following result, a proof of which can be found on pp 256-259 of [2].

**Theorem 3.** *If $\mathbf{P}$ is any poset of height $h$, say, then for any $H \geq h$, there is a graded poset $\mathbf{Q}$ of height $H$, such that, for each $h$-subset $\sigma$ of $[H]$, there is a copy of $\mathbf{P}$ in $\mathbf{Q}(\sigma)$.*

An easy consequence of this result is that Theorem 2 is the best possible as far as the height of $\mathbf{R}$ is concerned. For given $\mathbf{P}$ of height $h$, let $\mathbf{Q}$ be a poset of height $H$ satisfying the conclusion of Theorem 3. Suppose $\mathbf{R}$ is any poset such that (1) holds. If $R = \text{ht } \mathbf{R}$, consider any $s$-colouring of the $h$-subsets of $[R]$. This induces an $s$-colouring of $[\mathbf{R}, \mathbf{P}]_h$ by giving each copy $P'$ the colour of the set of heights of the elements of $P'$. To find a monochromatic $Q'$ , is a stronger requirement than that of finding a monochromatic $H$-subset of $[R]$. It follows that the inequality $R \geq R(s, h, H)$ must hold.

If $A$ is set and $n \geq 1$, we write $A^{[n]}$ for the set of $n$-subsets of $A$. For integers $n_1, \ldots, n_h \geq 1$ and $m, n, s \geq 1$, we write

$$n \implies [m, (n_1, \ldots, n_h)]_s \qquad (2)$$

when the following holds:

> For sets $Y_1, \ldots, Y_h$ with $|Y_i| \geq n$, and an $s$-colouring $\chi : \prod_{i=1}^{h} Y_i^{[n_i]} \to [s]$, there is, for each $i \in [h]$, some subset $Z_i$ of $Y_i$, each set $Z_i$ having $m$ elements, such that $\chi$ is monochromatic on $\prod_{i=1}^{h} Z_i^{[n_i]}$.

The existence of $n$, for given $n_i, m$ and $s$ is a well-known result (the so-called product Ramsey theorem, see for example [7]).

Let $A$ be a finite alphabet and let $\lambda$ be a symbol which does not occur in $A$. A 1-parameter word over $A$ is a word $w = w(\lambda)$ over the alphabet $A \cup \{\lambda\}$ such that there is at least one occurrence of $\lambda$ in $w$. For $a \in A$ we denote by $w(a)$ the word that we obtain from $w$ by replacing all occurences of $\lambda$ by $a$. The following theorem is well-known and is due to Hales-Jewitt [8].

**Theorem 4.** *For $r, s \geq 1$, there is a number $HJ(s, r)$ such that for any $s$-colouring of the words of length $HJ(s, r)$ over an alphabet $A$ consisting of $r$ letters, there is a 1-parameter word $w(\lambda)$ over $A$ such that all the words $w(a)$, $a \in A$ are of the same colour.*

A primitive recursive bound for the number $HJ(s, r)$ was found by Shelah [18]. For other expositions of Shelah's proof, see for example [7] and [16].

## 3   Layered Posets

In this section we prove a technical lemma on partitions of a class of structures which we shall call layered posets. We begin by introducing the following notation: Let $A$ be a finite set of natural numbers and let

$$\sigma = \sigma_1 | \ldots | \sigma_h |$$

be a partition into $h$ blocks of a finite set $A$ of natural numbers such that, if $1 \leq i < j \leq h$, each element of $\sigma_i$ is less than each element of $\sigma_j$. We call such a partition an *ordered* partition of $A$. For a given ordered partition $\sigma$ of the finite set $A$ into $h$ blocks, a $\sigma$-*layered poset* is a pair $\mathbf{M} = (\mathbf{P}, \ell)$ where $\mathbf{P}$ is a poset of height $h$ and $\ell$ is a surjection from the underlying set of $\mathbf{P}$ onto $A$ such that the following condition is satisfied:

- If $x < y(\mathbf{P})$ and $i, j \in [h]$ are such that $\ell(x) \in \sigma_i$ and $\ell(y) \in \sigma_j$, then $i < j$.

We refer to the sets $L_i := \ell^{-1}(\{i\})$ with $i \in A$, as the *layers* of the poset. The layers are partitioned into $h$ blocks $\Sigma_1, \cdots, \Sigma_h$ which are "controlled" by the partition $\sigma$, i.e., each block of layers is of the form $\Sigma_k = \{L_j : j \in \sigma_k\}$, for some $1 \leq k \leq h$. Finally, it is required that if $x < y$ in $\mathbf{P}$, then the layer containing $x$ will be in a block that precedes the block in which the layer containing $y$ lies. If $x$ is in some layer in $\Sigma_k$ we say that $x$ is in the $k$th block of the $\sigma$-layered poset. Note that in this case ht $x \leq k$ for if $C$ is any chain in $\mathbf{P}$ having $x$ as the maximum element, the elements of $C$ will be in different blocks of $(\mathbf{P}, \ell)$.

If $\mathbf{M} = (\mathbf{P}, \ell)$ and $\mathbf{N} = (\mathbf{Q}, k)$ are $\sigma$-layered posets, then an embedding $\mu$ from $\mathbf{M}$ into $\mathbf{N}$ is a poset embedding $\mu : \mathbf{P} \to \mathbf{Q}$ which is layer-preserving in the sense that $\ell(x)) = k(\mu(x))$ for all $x$ in the underlying set of $\mathbf{M}$. (We do not require that $\mu$ be height-preserving.) A $\sigma$-layered poset is said to be *transversal* if each of its layers consists of exactly one point and, moreover, if an element has height $j$, say, then it will be in the $j$th block of the $\sigma$-layered set. Note that an embedding of a transversal $\sigma$-layered set into an arbitrary $\sigma$-layered set will always be height-preserving since all elements of a given height will be in the same block, a property which must be preserved by any embedding.

Let $\mathbf{P}$ be a poset and let $L$ be total order on its underlying set such that

$$\text{ht } x < \text{ht } y \Rightarrow x < y(L). \tag{3}$$

With $(\mathbf{P}, L)$ we can associate the following transversal layered poset structure: Let $a$ be the size and $h$ the height of $\mathbf{P}$. Set $A = [a]$ and let $\sigma = \sigma_1| \cdots |\sigma_h|$ be the ordered partition such that the number of elements in $\sigma_j$ is the number of elements of $\mathbf{P}$ of height $j$, for $j = 1, \ldots, h$. We define $\ell : \mathbf{P} \to [a]$ by the requirement that, if $x \in \mathbf{P}$ is the $k$th element of the underlying set of $\mathbf{P}$ with respect to the total ordering $L$, then $\ell(x) = k$. Note that, if $x < y(\mathbf{P}$, then ht $x <$ ht $y$ and, therefore, $\ell(x) < \ell(y)$ on account of (3). It follows that the layers in $\Sigma_i$ are the singleton sets corresponding with the elements of $\mathbf{P}$ of height $i$. The poset structure remains unchanged. We denote this associated $\sigma$-layered poset also by $(\mathbf{P}, L)$.

**Example.** Let $\mathbf{P}$ be the poset on $\{a, b, c\}$ such that $a < c$ $(\mathbf{P})$ and $b$ is an isolated point. Let $L$ be the total order $a < b < c$ $(L)$. Then the corresponding layers $L_1, L_2, L_3$ are $\{a\}, \{b\}, \{c\}$, respectively and the blocks $\Sigma_1, \Sigma_2$ are $\{L_1, L_2\}$ and $\{L_3\}$, respectively. Only the first and third layers are connected by a relation in $\mathbf{P}$. If, however, the total order is given by $b < a < c$ $(L)$, then the sequence of layers are given by $\{b\}, \{a\}, \{c\}$ and in this case only the second and third layers are connected by a relation in $\mathbf{P}$.

In general, one can associate many transversal layered structures with a poset $\mathbf{P}$ (by varying over $L$). The complete posets, however, can carry exactly one transversal layered poset structure. In this case, any two elements in distinct blocks are connected by a relation in the underlying poset. This observation plays a crucially important role in what follows.

**Proposition 1.** *Let $\sigma$ be an ordered partition of some finite set $A$ of natural numbers. For $s \geq 1$ and $\sigma$-layered posets $\mathbf{M}$ and $\mathbf{A}$, where $\mathbf{A}$ is the transversal $\sigma$-layered poset corresponding to a complete poset, there is a $\sigma$-layered poset $\mathbf{N}$ such that*

$$\mathbf{N} \longrightarrow (\mathbf{M})_s^{\mathbf{A}}.$$

*Proof.* Let $a$ denote the number of layers of $\mathbf{M}$. Let $L_1, \ldots, L_a$ be the layers of $\mathbf{M}$ and let $A_1, \ldots, A_r$ be an enumeration (without repetition) of the copies of $\mathbf{A}$ in $\mathbf{M}$. Set

$$m := HJ(s, r),$$

the corresponding Hales-Jewitt number (see Theorem 4). We next define the $\sigma$-layered poset $\mathbf{N}$ which will satisfy the conclusion of the proposition. The underlying set of $\mathbf{N}$ is $N := \cup_{i=1}^{a} N_i$, where each

$$N_i = \underbrace{L_i \times \cdots \times L_i}_{m}.$$

The $N_i$ will be the layers of $\mathbf{N}$ and are arranged into blocks by $\sigma$. Write $\mathbf{P}$ for the poset structure of $\mathbf{M}$. We define the a poset structure $\mathbf{Q}$ on $\mathbf{N}$ as follows: If $x = (x_1, \ldots, x_m) \in N_i$ and $y = (y_1, \ldots, y_m) \in N_j$, then

$$x < y(\mathbf{Q}) \Leftrightarrow x_k < y_k(\mathbf{P}), \quad k = 1, \ldots, m.$$

It is clear that if $x < y(\mathbf{Q})$ then the index $i$, respectively $j$, of the block in $\mathbf{N}$ containing $x$, respectively $y$, are the same as of $x_1$, respectively $y_1$, in $\mathbf{M}$ so that we can conclude that $i < j$. Hence $\mathbf{N}$ is indeed a $\sigma$-layered poset.

Let $w(\lambda)$ be a 1-parameter word of length $m$ over the alphaber $[r]$. (Recall that $r$ is the number of copies of $\mathbf{A}$ in $\mathbf{M}$.) For $1 \leq i \leq m$ and $1 \leq k \leq r$ we denote by $a(i,k)$ the element of $A_k$ on the layer $L_i$ of $\mathbf{M}$. For the given parameter word $w(\lambda)$ we now define a mapping $\nu : \mathbf{M} \to \mathbf{N}$ as follows: Suppose $w = w_1 \cdots w_m$. If $b \in \mathbf{M}$ then $\nu(b) = (y_1, \ldots, y_m)$ where, if $b$ belongs to the $i$th layer $L_i$, we have, $y_j = b$ when $\lambda$ appears in the $j$th position of $w(\lambda)$ and $y_j = a(i, w_j)$, otherwise. We now show that

$$b_1 < b_2(\mathbf{P}) \Leftrightarrow \nu(b_1) < \nu(b_2)(\mathbf{Q}).$$

If the left-hand inequality holds and $b_1$, respectively $b_2$, belongs to the block $\Sigma_u$, respectively $\Sigma_v$, then $u < v$. Since each $A_k$ is complete, each element of $A_k$ belonging to $\Sigma_u$ is less that every element of $A_k$ belonging to $\Sigma_v$. We conclude that $\nu(b_1) < \nu(b_2)$. The converse follows directly from the fact that there is at least one occurrence of $\lambda$ in $w(\lambda)$. It is also clear that the mapping $\nu$ is injective and is layer-preserving. We conclude that it is an embedding. In this way we can associate with a 1-parameter word of length $m$ over $[r]$ an embedding of $\mathbf{M}$ into $\mathbf{N}$.

Write $[\mathbf{N}, \mathbf{A}]$ for the set of copies of $\mathbf{A}$ in $\mathbf{N}$. We next define a mapping

$$\phi : [r]^m \longrightarrow [\mathbf{N}, \mathbf{A}].$$

Informally, $\phi(w_1 \cdots w_m)$ is the $a \times m$ array with columns $A_{w_1}, \ldots, A_{w_m}$. That is, its intersection with the $i$th layer $N_i$ is the element $(a(i, w_1), \ldots, a(i, w_m))$. It follows from our constructions that if $w(\lambda)$ is a 1-parameter word of length $m$ over $[r]$, then, for $a \in [r]$:

$$\nu(A_a) = \phi(w(a)),$$

where $\nu$ is the embedding corresponding to $w(\lambda)$.

An $s$-colouring $\chi$ of $[\mathbf{N}, \mathbf{A}]$ induces via $\phi$ an $s$-colouring $\chi_1$ of $[r]^m$ and hence, by the Hales-Jewitt theorem, there is a 1- parameter word $w(\lambda)$ such that $\chi_1$ assumes a constant value on $w(1), \ldots, w(r)$. By our preceding constructions, this means that there is a copy $\mathbf{M}'$ of $\mathbf{M}$, namely $\mathbf{M}' = \nu(\mathbf{M})$, where $\nu$ is the embedding associated with $w(\lambda)$, such that $\chi$ assumes a constant value on the copies $\phi(w(1)), \ldots, \phi(w(r))$ of $\mathbf{A}$ in $\mathbf{M}'$. Since $\nu(A_1), \cdots, \nu(A_r)$ are pairwise distinct and since $[\mathbf{M}', \mathbf{A}]$ has exactly $r$ elements, we conclude that $\chi$ is monochromatic on $[\mathbf{M}', \mathbf{A}]$. This concludes the proof of the proposition.

## 4   The First Amalgamation

In this section, we prove Theorem 2 for the case when both $\mathbf{P}$ and $\mathbf{Q}$ are of the same height.

**Lemma 1.** *Suppose $\mathbf{P}$ and $\mathbf{Q}$ are posets of the same height $h$ where $\mathbf{P}$ is complete. Then, for every $s \geq 1$, there is a poset $\mathbf{R}$ of height $h$ such that*

$$\mathbf{R} \longrightarrow (\mathbf{Q})_s^{\mathbf{P}}.$$

*Proof.* We first introduce the following notation: If $\tau$ and $\sigma$ are ordered set partitions into $h$ blocks, we write $\sigma \prec \tau$ if the $j$th block of $\sigma$ is a subset of the the $j$th block of $\tau$ for $j = 1, \ldots, h$. If $\mathbf{M}$ is a $\tau$-layered poset and if $\sigma \prec \tau$, a $\sigma$-layered poset carried by $\sigma$ is by definition a $\sigma$-layered poset spanned by some of the elements of $\mathbf{M}$, each such element being on a level belonging to a block which appears in $\sigma$.

For $j = 1, \ldots, h$, let $\alpha_j$, respectively $\beta_j$, be the number of elements of height $j$ of $\mathbf{P}$, respectively $\mathbf{Q}$. Set

$$b = \max\{\beta_j : j = 1, \ldots, h\}.$$

Let $N$ be so large that

$$N \implies [b, (\alpha_1, \ldots, \alpha_h)]_s.$$

(See (2).) We fix total orders $L_1, L_2$ on $\mathbf{P}$ and $\mathbf{Q}$ respectively, such that (3) holds when $L = L_1$ and when $L = L_2$. In the sequel we shall view these two posets as transversal layered posets, with the layered structures induced by $L_1$ and $L_2$.

Let $\tau = \tau_1 | \cdots | \tau_h |$ be the ordered partition of $[Nh]$ having $h$ blocks each of length $N$. Let $\mathbf{Q}^0$ be a $\tau$-layered poset of height $h$ such that, if $\sigma$ is an ordered partition with $\sigma \prec \tau$ where the blocks of $\sigma$ have sizes $\beta_1, \ldots, \beta_h$, respectively, then $\sigma$ will carry a copy of the transversal $(\mathbf{Q}, L_2)$. (This can be easily arranged by taking suitably many disjoint copies of $\mathbf{Q}$.)

Let $s_1, \ldots, s_t$ be an enumeration (without repetition) of the elements of

$$\tau_1^{[\alpha_1]} \times \cdots \times \tau_h^{[\alpha_h]}.$$

Note that we can view each $s_j$ as an ordered partition of an appropriate set in such a way that $s_j \prec \tau$. We shall inductively construct a sequence $\mathbf{Q}^1, \ldots, \mathbf{Q}^t$ of $\tau$-layered posets in terms of the $s_j$ in such a way that the underlying poset $\mathbf{R}$, say, of $\mathbf{Q}^t$ will satisfy the conclusion of the lemma.

Suppose $0 \le k < t$ and $\mathbf{Q}^k$ has been constructed. Let $D^{k+1}$ be the $s_{k+1}$-layered subposet of $\mathbf{Q}^k$ carried by $s_{k+1}$. Apply Proposition 1 to find an $s_{k+1}$-layered poset $E^{k+1}$ such that

$$E^{k+1} \longrightarrow (D^{k+1})_s^{(\mathbf{P}, L_1)}.$$

Here, we view $(\mathbf{P}, L_1)$ as an $s_{k+1}$-layered poset. Attach a copy of $\mathbf{Q}^k$ to each copy of $D^{k+1}$ in $E^{k+1}$ in such a way that any two distinct copies meet only in $E^{k+1}$. At this stage we have a structure with a layering and an ordering which need not be transitive. Now take the transitive closure of this ordering and let $\mathbf{Q}^{k+1}$ be the resulting structure. This is a $\tau$-layered poset for if $y$ covers $x$ in $\mathbf{Q}^{k+1}$, then $x < y$ in some copy of $\mathbf{Q}^k$ with its original structure, so that, if $x$, respectively $y$, belongs to block $\tau_i$, respectively $\tau_j$, then $i < j$. It follows that for any $x, y \in \mathbf{Q}^{k+1}$, if $x < y$ in $\mathbf{Q}^{k+1}$ and if $x, y$ belongs to $\tau_i, \tau_j$, then $i < j$.

We now show that every copy $Q^k$ of $\mathbf{Q}^k$ attached to a copy of $D^{k+1}$ will retain the original structure of $\mathbf{Q}^k$ even after the transitive closure has been effected. Suppose $x < y$ in $\mathbf{Q}^{k+1}$, where both $x$ and $y$ are in $Q^k$. Let $C$ be a chain in $\mathbf{Q}^{k+1}$ of the form $x = x_0 < x_1 < \ldots < x_f = y$, where for all $i < f$, we have

$x_i < x_{i+1}$ in the structure $\mathbf{Q}^{k+1}$ prior to when the the transitive dclosure has been effected. If this chain ever leaves $Q^k$, it must do so via $E^{k+1}$ whereafter it must again return to $E^{k+1}$. Denote the intersection points $x_{i_1}, \cdots, x_{i_p}$, say, where $i_1 < \ldots < i_p$, of $C$ with $E^{k+1}$ by $y_1, \cdots, y_p$, respectively. If $1 \leq j < p$, then $y_j < y_{j+1}$ in some copy of $Q^k$ with its original structure. Since both $y_j$ and $y_{j+1}$ are in $E^{k+1}$, we conclude that $y_j < y_{j+1}$ in $E^{k+1}$ with its original structure. The final intersection point $y_p$ must be in $E^{k+1} \cap Q^k$ because $y_p \leq y$ and one can finally reach $y$ only via elements of $Q^k$ in $E^{k+1}$. We conclude that $x \leq y_1 \leq y_p \leq y$ in $Q^k$ with its original structure.

Note that the height of $\mathbf{Q}^{k+1}$ is again $h$: Firstly, we may inductively assume that ht $\mathbf{Q}^k = h$ and since $\mathbf{Q}^{k+1}$ contains a copy of $\mathbf{Q}^k$, the former must have a height which is at least $h$. Secondly, if $x < y$ in $\mathbf{Q}^{k+1}$, then $x$ and $y$ will be on layers belonging to different blocks. Since there are $h$ blocks, we conclude that the height of $\mathbf{Q}^{k+1}$ cannot be more than $h$.

Write $\mathbf{R}$ for the poset structure of $\mathbf{Q}^t$. Let $\chi$ be an $s$-colouring of $[\mathbf{R}, \mathbf{P}]_h$. By means of a downward induction one can now prove the following statement:

> $(\dagger)_k$: For $0 \leq k \leq t$ there is a copy $Q^k$ of $\mathbf{Q}^k$ in $\mathbf{R}$ such that the copies of $\mathbf{P}$ carried by $s_j$ depends on $j$ only, for every $j = k+1, \ldots, t$.

Suppose $(\dagger)_{k+1}$ holds, where $k+1 \leq t$ and let $Q^{k+1}$ witness the truth of $(\dagger)_{k+1}$. By our construction, there is a copy of $D^{k+1}$ in $Q^{k+1}$ in which all the copies of $\mathbf{P}$ are of the same colour. Write $Q^k$ for the copy of $\mathbf{Q}^k$ that is attached to this $D^{k+1}$. Then $Q^k$ witnesses the truth of $(\dagger)_k$.

Since $(\dagger)_0$ holds, there is a copy $Q^0$ of $\mathbf{Q}^0$ such that any copy of $\mathbf{P}$ in $Q^0$ will have a colour that depends only on the ordered partition that carries the copy. In this way we find that we have an induced $s$-colouring $\chi_1$ of

$$\tau_1^{[\alpha_1]} \times \cdots \times \tau_h^{[\alpha_h]}.$$

By our choice of $N$, there is an ordered partition $\sigma = \sigma_1|\sigma_2\ldots|\sigma_h$ having blocks of sizes $\beta_1, \ldots, \beta_h$ such that $\chi_1$ is monochromatic on

$$\sigma_1^{[\alpha_1]} \times \cdots \sigma_h^{[\alpha_h]}.$$

Take a copy of $\mathbf{Q}$ in $Q^0$ carried by $\sigma$ and we are done.

## 5   The Second Amalgamation

In this section we prove Theorem 2. Let $\mathbf{P}$ and $\mathbf{Q}$ be posets of heights $h$ and $H$, respectively, and let $s \geq 1$. Set $R = R(s, h, H)$. Apply Theorem 3 to find a poset $\mathbf{Q}^0$ of height $R$ such that for each $H$-subset $\sigma$ of $[R]$, the poset $\mathbf{Q}^0(\sigma)$ contains a copy of $\mathbf{Q}$. Let $s_1, \ldots, s_t$ be an enumeration of the $h$-subsets of $[R]$. We now inductively define a sequence of posets $\mathbf{Q}^k$, $k = 1, \ldots, t$, each of height $R$, as follows: Suppose $\mathbf{Q}^k$ has been defined where $k \geq 0$. Set $D^{k+1} = \mathbf{P}^k(s_{k+1})$. This is a poset of height $h$. Apply Lemma 1 to find a poset $E^{k+1}$ of height $h$ such that

$$E^{k+1} \longrightarrow (D^{k+1})_s^{\mathbf{P}}.$$

Attach to each copy (image under a *height-preserving* embedding) of $D^{k+1}$ in $E^{k+1}$ a copy of $\mathbf{Q}^k$ such that, again, any two copies meet only in $E^{k+1}$ and take the transitive closure. Call the resulting poset $\mathbf{Q}^{k+1}$. We note that an element of $E^{k+1}$ will have the same height in each attached copy of $\mathbf{Q}^k$ with the original structure. Moreover, a criss-crossing argument as in the proof of Lemma 1 shows that each attached copy will retain its original poset structure in $\mathbf{Q}^{k+1}$.

We now show that each $y \in \mathbf{Q}^{k+1}$ will have a height in $\mathbf{Q}^{k+1}$ which is the same as its height in any copy $Q'$ of $\mathbf{Q}^k$ which contains $y$: Let $Q'$ be a copy of $\mathbf{Q}^k$ and, for a fixed $y \in Q'$, let $C$ be a saturated chain in $\mathbf{Q}^{k+1}$ of the form $x_0 < \ldots < x_y = y$. Denote the intersection points of $C$ with $E^{k+1}$ by $x_{i_1}, \cdots, x_{i_p}$, where $i_1 < \ldots < i_p$. Set $i_0 = 0$ and, for $j < p$, let $Q_j$ denote the copy of $\mathbf{Q}^k$ such that the subinterval $[x_{i_j}, x_{i_{j+1}}]$ of $C$ lies in $Q_j$. Since the height of $x_{i_1}$ in $Q_0$ and $Q_1$ (with their original structures) is the same, their is a chain of length $i_1$ in $Q_1$ having $x_{i_1}$ as maximum element which can be continued to a chain in $Q_1$ of length $i_2$ up to the point $x_{i_2}$. Continuing in this way, we eventually find a chain in $Q_{p-1}$ of size $i_p$ up to the point $x_{i_p}$. But the height of the latter point is the same in $Q'$ and $Q_{p-1}$. It follows that there is a chain in $Q'$ of size $i_p$ up to $x_{i_p}$ which can be extended to a chain in $Q'$ of length $f$ having $y = x_f$ as maximum element.

We conclude that each attached copy $Q'$ is the image of $\mathbf{Q}^k$ under a height-preserving embedding of $\mathbf{Q}^k$ into $\mathbf{Q}^{k+1}$. Another consequence is that ht $\mathbf{Q}^{k+1}=R$.

Set $\mathbf{R} = \mathbf{Q}^t$. It again follows that for an $s$-colouring of $[\mathbf{R}, \mathbf{P}]_h$, there will be a copy $Q^0$ of $\mathbf{Q}^0$ such that the colour of a copy of $\mathbf{P}$ in $Q^0$ will only depend on its set of heights. By our choice of $R$ there will be an $H$-subset $\sigma$ of $[R]$ such that all the copies in $Q^0$ whose set of heights is contained in $\sigma$ will be of the same colour. Let $Q$ be a copy of $\mathbf{Q}$ in $Q^0(\sigma)$. Then all the copies of $\mathbf{P}$ in $Q$ will be of the same colour. This concludes the proof of the theorem.

# References


1. Abramson, F.G., Harrington, L.O.: Models without indiscernables. J. Symb. Logic 43, 572–600 (1978)
2. Fouché, W.L.: Chain partitions of ordered sets. Order 13, 255–266 (1996)
3. Fouché, W.L.: Symmetry and the Ramsey degree of posets. Discrete Math. 167/168, 309–315 (1997)
4. Fouché, W.L.: Symmetry and the Ramsey degree of finite relational structures. J. Comb. Theory Ser A 85, 135–147 (1999)
5. Fouché, W.L., Pretorius, L.M., Swanepoel, C.J.: The Ramsey degrees of bipartite graphs: A primitive recursive proof. Discrete Math. 293, 111–119 (2005)
6. Graham, R.L., Rothschild, B.L.: Ramsey's theorem for $n$-parameter sets. Trans. Amer. Math. Soc. 159, 257–292 (1971)
7. Graham, R.L., Rothschild, B.L., Spencer, J.L.: Ramsey Theory. Wiley, New York (1990)
8. Hales, A.W., Jewett, R.I.: Regularity and positional games. Trans. Amer. Math. Soc. 106, 222–229 (1963)
9. Kechris, A.S., Pestov, V., Todorcevic, S.: Fraïssé limits, Ramsey theory, and topological dynamics of automorphism groups. GAFA 15, 106–189 (2005)

10. Nešetřil, J.: Ramsey theory. In: Graham, R.L., Grötschel, M., Lovász, L. (eds.) Handbook of Combinatorics, vol. 2, pp. 1331–1403. North Holland, Amsterdam (1995)
11. Nešetřil, J., Rödl, V.: Partitions of relational and set systems. J. Comb. Theory Ser A 83, 289–312 (1977)
12. Nešetřil, J., Rödl, V.: Ramsey classes of set systems. J. Comb. Theory Ser A 34, 183–201 (1983)
13. Nešetřil, J., Rödl, V.: Combinatorial partitions of finite posets and lattices– Ramsey lattices. Algebra Universalis 19, 106–119 (1984)
14. Nešetřil, J., Rödl, V.: The partite construction and Ramsey set systems. Discrete Math. 75, 327–334 (1989)
15. Paoli, M., Trotter, W.T., Walker, J.W.: Graphs and orders in Ramsey theory and in dimension theory. In: Rival, I. (ed.) Graphs and Order, Reidel (1984)
16. Prömel, H.J., Voigt, B.: Graham-Rothschild parameter sets. In: Nešetřil, J., Rödl, V. (eds.) Mathematics of Ramsey Theory, pp. 113–149. Springer, Berlin (1990)
17. Ramsey, F.P.: On a problem of formal logic. Proc. London Math. Soc. 30, 264–286 (1930)
18. Shelah, S.: Primitive recursive bounds for van der Waerden numbers. J. Amer. Math. Soc. 1, 683–697 (1988)
19. Sommerhalder, R., van Westhenen, S.C.: The Theory of Computability. Addison Wesley, Reading (1988)

# Solving Simple Stochastic Games[*]

Hugo Gimbert[1] and Florian Horn[2]

[1] LaBRI, Université Bordeaux 1, France
`hugo.gimbert@labri.fr`
[2] LIAFA, Université Paris 7, France
`florian.horn@liafa.jussieu.fr`

**Abstract.** We present a new algorithm for solving Simple Stochastic Games (SSGs), which is fixed parameter tractable when parametrized with the number of random vertices. This algorithm is based on an exhaustive search of a special kind of positional optimal strategies, the *f-strategies*. The running time is $\mathcal{O}(\ |V_{\mathrm{R}}|! \cdot (\log(|V|)|E| + |p|)\ )$, where $|V|, |V_{\mathrm{R}}|, |E|$ and $|p|$ are respectively the number of vertices, random vertices and edges, and the maximum bit-length of a transition probability. Our algorithm improves existing algorithms for solving SSGs in three aspects. First, our algorithm performs well on SSGs with few random vertices, second it does not rely on linear or quadratic programming, third it applies to all SSGs, not only stopping SSGs.

## Introduction

Simple Stochastic Games (SSGs for short) are played by two players Max and Min in a sequence of steps. Players move a pebble along edges of a directed graph $(V, E)$. There are three type of vertices: $V_{\mathrm{Max}}$ is the set of vertices of player Max, $V_{\mathrm{Min}}$ the set of vertices of player Min and $V_{\mathrm{R}}$ the set of random vertices. When the pebble is on a vertex of $V_{\mathrm{Max}}$ or $V_{\mathrm{Min}}$, the corresponding player chooses an outgoing edge and moves the pebble along it. When the pebble is on a random vertex, the successor is chosen randomly according to some fixed probability distribution: from vertex $v \in V_{\mathrm{R}}$ the pebble moves towards vertex $w \in V$ with some probability $p(w|v)$ and the probability that the game stops is $0$, *i.e.* $\sum_{w \in V} p(w|v) = 1$. An SSG is depicted on Figure 1, with vertices of $V_{\mathrm{Max}}$ represented as $\bigcirc$, vertices of $V_{\mathrm{Min}}$ represented as $\square$, and vertices of $V_{\mathrm{R}}$ represented as $\Diamond$.

Player Max and Min have opposite goals, indeed player Max wants the pebble to reach a special vertex $t \in V$ called the *target vertex*, if this happens *the play is won by player* Max. In the opposite case, the play proceeds forever without reaching $t$ and is *won by player* Min. For technical reasons, we assume that $t$ is a vertex of player Max and is absorbing. *Strategies* are used by players to choose their moves, a strategy tells where to move the pebble depending on the sequence of previous vertices, *i.e.* the finite path followed by the pebble from the beginning of the play. The *value* of a vertex $v$ is the maximal probability

[*] This research was partially supported by french project ANR "DOTS".

A. Beckmann, C. Dimitracopoulos, and B. Löwe (Eds.): CiE 2008, LNCS 5028, pp. 206–209, 2008.

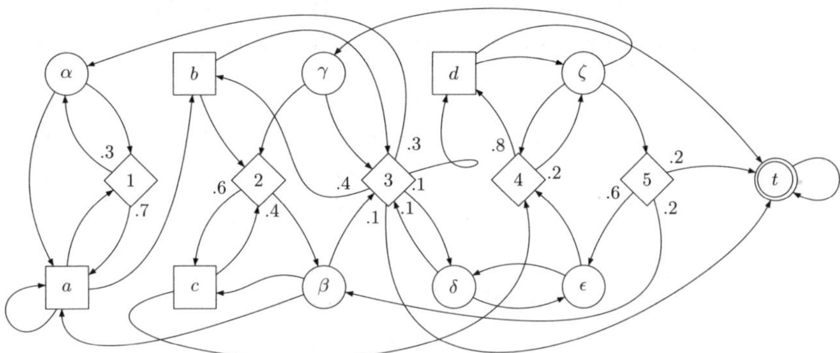

**Fig. 1.** A Simple Stochastic Game

with which player Max can enforce the play to reach the target vertex. When player Max, respectively player Min, uses an *optimal strategy* he ensures reaching the target with a probability greater, respectively smaller, than the value of the initial vertex.

We are interesting in *solving* SSGs, that is computing values and optimal strategies.

*Existing algorithms for solving SSGs.* The complexity of solving SSGs was first considered by Condon [Con92], who proved that deciding whether the value of an SSG is greater than $\frac{1}{2}$ is in NP ∩ co-NP. The algorithm provided in [Con92] consists in first transforming the input SSG in a *stopping SSG* where the probability to reach a sink vertex is 1. The transformation keeps unchanged the fact that the initial vertex has value strictly greater than $\frac{1}{2}$ but induces a quadratic blowup of the size of the SSG. The algorithm then non-deterministically guesses the values of vertices, which are rational numbers of linear size, and checks that these values are the unique solutions of some *local optimality equations*.

Three other kinds of algorithms for solving SSGs are presented in [Con93]. These algorithms require transformation of the initial SSG into an equivalent *stopping* SSG and are based on local optimality equations. First algorithm computes values of vertices using a quadratic program with linear constraints. Second algorithm computes iteratively from below the values of the SSGs, and the third is a strategy improvement algorithm *à la* Hoffman-Karp. These two last algorithms require solving an exponential number of linear programs, as it is the case for the algorithm recently proposed in [Som05].

Finally, these four algorithms suffer three main drawbacks.

First, these algorithms rely on solving either an exponential number of linear programs or a quadratic program, which may have prohibitive algorithmic cost and makes the implementation tedious.

Second, these algorithms only apply to the special class of *stopping* SSGs. Although it is possible to transform any SSG into a stopping SSG with arbitrarily small change of values, computing exact values this way requires to modify drastically the original SSG, introducing either $|V|^2$ new random vertices or new transition probabilities of bit-length quadratic in the original bit-length. This also makes the implementation tedious.

Third, the running time of these algorithms may *a priori* be exponential whatever be the number of random vertices of the input SSG, including the case of SSGs with no random vertices at all, also known as reachability games on graphs. However it is well-known that reachability games on graphs are solvable in linear time.

Notice that randomized algorithms do not perform much better since the best randomized algorithms [Lud95, Hal07] known so far run in sub-exponential expected time $e^{O(\sqrt{n})}$.

*Our results.* In this paper we present an algorithm that computes values and optimal strategies of an SSG in time $\mathcal{O}(\ |V_\mathrm{R}|! \cdot (\log(|V|)|E| + |p|)\ )$, where $|V_\mathrm{R}|$ is the number of random vertices, $|V|$ is the number of vertices and $|p|$ is the maximal bit-length of transition probabilities.

The key point of our algorithm is the fact that optimal strategies may be looked for in a strict subset of positional strategies, called the class of **f**-strategies. The **f**-strategies are in correspondence with permutations of random vertices. Our algorithm does an exhaustive search of optimal **f**-strategies among the $|V_\mathrm{R}|!$ available ones and check their optimality. Optimality is easy to check, it consists in computing a reachability probability in a Markov Chain with $V_\mathrm{R}$ states, which amounts to solving a linear system with at most $|V_\mathrm{R}|$ equations.

*Comparison with existing work.* We improve existing results by three aspects: our algorithm performs better on SSGs with few random vertices, it is arguably much simpler, and we provide new insight about the structure of optimal strategies.

Our algorithm performs much better on SSGs with few random vertices than previously known algorithms. Indeed, its complexity is $\mathcal{O}(\ |V_\mathrm{R}|! \cdot (\log(|V|)|E| + |p|)\ )$, hence when there are no random vertices at all, our algorithm matches the usual quadratic complexity for solving reachability games on graphs. When the number of random vertices is fixed, our algorithm performs in polynomial time, and on the class of SSGs such that $|V_\mathrm{R}| \leq \sqrt{|V_\mathrm{Max}| + |V_\mathrm{Min}|}$ our algorithm is sub-exponential.

Whereas the complexity is optimal when there are no random vertices, this is no more the case when there are no vertices for player Max or Min. In that case, there exists polynomial time algorithm, whereas the complexity of our algorithm remains exponential in the number of random vertices.

Our algorithm is arguably simpler than previously known algorithms. Indeed, it does not require use of linear or quadratic programming. Although linear programs can be solved in polynomial time [Kac79, Ren88], this requires high-precision arithmetic. By contrast, our algorithm is very elementary: it enumerates permutations of the random vertices and for each permutation, it solves a linear system of equations.

Our algorithm is also simpler because it applies directly to any kind of SSGs, whereas previously known algorithms require the transformation of the input

SSG into a stopping SSG of quadratic size. As a consequence, we can use the same algorithm for solving other types of infinite duration game; this is ongoing work.

The full paper is available online [GH07].

*Acknowledgments.* We thank an anonymous referee for his very insightful and helpful comments.

# References

[Con92]  Condon, A.: The complexity of stochastic games. Information and Computation 96, 203–224 (1992)

[Con93]  Condon, A.: On algorithms for simple stochastic games. In: Advances in computational complexity theory. DIMACS series in discrete mathematics and theoretical computer science, vol. 13, pp. 51–73 (1993)

[Der72]  Derman, C.: Finite State Markov Decision Processes. Academic Press, London (1972)

[Dix82]  Dixon, J.D.: Exact solution of linear equations using $p$-adic expansions. Numerische Mathematik 40, 137–141 (1982)

[GH07]  Gimbert, H., Horn, F.: Solving simple stochastic games with few random vertices (2007), http://hal.archives-ouvertes.fr/hal-00195914/fr/

[Hal07]  Halman, N.: Simple stochastic games, parity games, mean payoff games and discounted payoff games are all LP-type problems. Algorithmica 49, 37–50 (2007)

[Kac79]  Kachiyan, L.G.: A polynomial time algorithm for linear programming. Soviet Math. Dokl. 20, 191–194 (1979)

[Lud95]  Ludwig, W.: A subexponential randomized algorithm for the simple stochastic game problem. Information and Computation 117, 151–155 (1995)

[Ren88]  Renegar, J.: A polynomial-time algorithm, based on newton's method, for linear programming. Mathematical Programming 40, 59–93 (1988)

[Sha53]  Shapley, L.S.: Stochastic games. Proceedings of the National Academy of Science USA 39, 1095–1100 (1953)

[Som05]  Somla, R.: New algorithms for solving simple stochastic games. Electr. Notes Theor. Comput. Sci. 119(1), 51–65 (2005)

# The Shrinking Property for NP and coNP


Christian Glaßer[1], Christian Reitwießner[2], and Victor Selivanov[2,*]

[1] Julius-Maximilians-Universität Würzburg, Germany
`{glasser,reitwiessner}@informatik.uni-wuerzburg.de`
[2] A.P. Ershov Institute of Informatics Systems, Siberian Division of the Russian
Academy of Sciences, Russia
`vseliv@nspu.ru`



**Abstract.** We study the shrinking and separation properties (two notions well-known in descriptive set theory) for NP and coNP and show that under reasonable complexity-theoretic assumptions, both properties do not hold for NP and the shrinking property does not hold for coNP. In particular we obtain the following results.

1. NP and coNP do not have the shrinking property, unless PH is finite. In general, $\Sigma_n^P$ and $\Pi_n^P$ do not have the shrinking property, unless PH is finite. This solves an open question from [25].
2. The separation property does not hold for NP, unless UP $\subseteq$ coNP.
3. The shrinking property does not hold for NP, unless there exist NP-hard disjoint NP-pairs (existence of such pairs would contradict a conjecture by Even, Selman, and Yacobi [6]).
4. The shrinking property does not hold for NP, unless there exist complete disjoint NP-pairs.

Moreover, we prove that the assumption NP $\neq$ coNP is too weak to refute the shrinking property for NP in a relativizable way. For this we construct an oracle relative to which P $=$ NP $\cap$ coNP, NP $\neq$ coNP, and NP has the shrinking property. This solves an open question by Blass and Gurevich [2] who explicitly ask for such an oracle.


## 1 Introduction

The shrinking property and the separation property are well-known notions from descriptive set theory. In this paper we study these notions with respect to complexity classes like NP.

**Definition 1.** *1. A class $\mathcal{C}$ has the* shrinking property, *if for all $A, B \in \mathcal{C}$ there exist disjoint sets $A', B' \in \mathcal{C}$ such that $A' \subseteq A$, $B' \subseteq B$, and $A' \cup B' = A \cup B$.*
*2. A class $\mathcal{C}$ has the* separation property, *if for all disjoint $A, B \in \mathcal{C}$ there exists an $S \in \mathcal{C} \cap \text{co}\mathcal{C}$ that separates $A$ and $B$.*

Both properties were introduced long ago in descriptive set theory (see e.g. [16]) where they play an important role. A simple result states that the class

---


$^\star$ Supported by RFBR grant 07-01-00543a.

O of open subsets of the Baire space has the shrinking property but does not have the separation property. Later the properties were studied in computability theory (see e.g. [24]) and again it turned out that they are very important, in particular due to their close relation to undecidability of first-order theories. In particular, for many natural theories $T$ the set of the sentences provable in $T$ and the set of the sentences false in a finite model of $T$ are recursively (even effectively) inseparable (see e.g. the survey [5] for additional details). Another simple result states that the class RE of computably enumerable sets has the shrinking property, but does not have the separation property. It turned out (see e.g. [18]) that there is a deep and fruitful analogy between O (and more general classes as e.g. levels of the Borel hierarchy) and RE (and more general classes as e.g. levels of the arithmetical hierarchy). More recently it was shown [27,28] that the shrinking and separation properties are also interesting for the theory of finite automata on infinite words.

Note that the shrinking property is also known as the reduction property (see e.g. [18,24]). We follow Blass and Gurevich [2] and use the first name in this paper, because the word "reduction" has also a quite different meaning.

Since there is an analogy between NP and RE, complexity theorists started to study the separation and shrinking properties for NP and coNP. While the separation property was investigated rather comprehensively (see e.g. [10,11]), the shrinking property was not considered systematically so far. In this respect, Blass and Gurevich [2] and Selivanov [25] show some first results and identify open questions. As one might expect, the status of both properties in the context of complexity theory is not as clear as in computability theory or descriptive set theory: they turn out to be closely related to some well-known conjectures.

In this paper we continue the study of the separation and shrinking properties in complexity theory, and we give evidence that NP does not have these properties. We show that under reasonable complexity-theoretic assumptions (like an infinite PH and UP $\not\subseteq$ coNP) both properties do not hold for NP and the shrinking property does not hold for coNP. Moreover, $\Sigma_n^P$ and $\Pi_n^P$ do not have the shrinking property, unless the PH is finite. This solves an open question from [25]. We relate the shrinking and separation properties for NP to other well-known notions. For example, we show that the shrinking property does not hold, unless there exist NP-hard disjoint NP-pairs. The existence of such pairs contradicts a conjecture that is related to security aspects of public-key cryptosystems [6,12]. Moreover, the shrinking property does not hold for NP, unless there exist complete disjoint NP-pairs. Such complete pairs are studied because of their relations to the theory of propositional proof systems [22,23]. We will also see that the shrinking property for NP is closely related to selectivity, nondeterministic function classes and inverting polynomial-time computable functions [14].

Along with the above-mentioned oracle-independent results, we establish some oracle separations. In particular, we prove that the assumption NP $\neq$ coNP is too weak to refute the shrinking property for NP in a relativizable way. For this we construct an oracle relative to which NP has the shrinking property and $(NP \cap coNP) = P \neq NP$. It follows that relative to this oracle, NP $\subseteq$ NPSV-sel and $NP \neq coNP$. Moreover, with our construction we solve an open problem by Blass and Gurevich [2] who explicitly ask for such an oracle.

## 2    Preliminaries

### 2.1    Disjoint NP-Pairs

Even, Selman, and Yacobi [6,7] showed that the security of public-key cryptosystems depends on the computational complexity of certain promise problems. Such problems can be written as pairs of disjoint sets, and it turned out that pairs of disjoint NP-sets are crucially important for the analysis of the cracking problem for public-key cryptosystems.

A *disjoint* NP-*pair* is a pair of nonempty sets $A$ and $B$ such that $A, B \in NP$ and $A \cap B = \emptyset$. Let DisjNP denote the class of all disjoint NP-pairs. Given a disjoint NP-pair $(A, B)$, a *separator* is a set $S$ such that $A \subseteq S$ and $B \subseteq \overline{S}$ (i.e., $S$ *separates* $(A, B)$). Let $Sep(A, B)$ denote the class of all separators of $(A, B)$.

Fortnow and Rogers [9] investigated the existence of disjoint sets in NP (resp., coNP) that are P-inseparable. Grollman and Selman [12] showed that certain one-way functions exist if and only if there exists a disjoint NP-pair $(A, B)$ that is P-inseparable (i.e., $Sep(A, B) \cap P = \emptyset$).

**Definition 2 ([12,23,17]).** *Let $(A, B)$ and $(C, D)$ be disjoint pairs.*

1. *$(A, B)$ is* many-one reducible in polynomial-time *to $(C, D)$, $(A, B) \leq_{\mathrm{m}}^{\mathrm{PP}} (C, D)$, if for every separator $T \in Sep(C, D)$, there exists a separator $S \in Sep(A, B)$ such that $S \leq_{\mathrm{m}}^{\mathrm{P}} T$.*
2. *$(A, B)$ is* uniformly many-one reducible in polynomial-time *to $(C, D)$, $(A, B) \leq_{\mathrm{um}}^{\mathrm{PP}} (C, D)$, if there exists a polynomial-time computable function $f$ such that for every separator $T \in Sep(C, D)$, there exists a separator $S \in Sep(A, B)$ such that $S \leq_{\mathrm{m}}^{\mathrm{P}} T$ via $f$.*
3. *Analogously one defines Turing reducibility and uniform Turing reducibility.*

**Theorem 1 ([12,23,11]).** *For all disjoint pairs $(A, B)$ and $(C, D)$,*

$$(A, B) \leq_{\mathrm{T}}^{\mathrm{PP}} (C, D) \ \Leftrightarrow \ (A, B) \leq_{\mathrm{uT}}^{\mathrm{PP}} (C, D)$$

$$(A, B) \leq_{\mathrm{m}}^{\mathrm{PP}} (C, D) \ \Leftrightarrow \ (A, B) \leq_{\mathrm{um}}^{\mathrm{PP}} (C, D)$$

$$\Leftrightarrow \ there \ is \ a \ polynomial\text{-}time\text{-}computable \ total \ function \ f \ such \ that \ f(A) \subseteq C \ and \ f(B) \subseteq D.$$

Razborov [23] and Pudlák [22] showed that disjoint NP-pairs are closely related to propositional proof systems. For example, if optimal propositional proof systems exist, then there exist complete disjoint NP-pairs. A disjoint pair $(A, B)$ is $\leq_{\mathrm{m}}^{\mathrm{PP}}$-*complete* (resp., $\leq_{\mathrm{T}}^{\mathrm{PP}}$-*complete*) for the class DisjNP if $(A, B) \in$ DisjNP and for all disjoint pairs $(C, D) \in$ DisjNP, $(C, D) \leq_{\mathrm{m}}^{\mathrm{PP}} (A, B)$ (resp., $(C, D) \leq_{\mathrm{T}}^{\mathrm{PP}} (A, B)$).

**Definition 3.** *Let $(A, B) \in$ DisjNP and let $\leq_r$ be from $\{\leq_{\mathrm{m}}^{\mathrm{PP}}, \leq_{\mathrm{T}}^{\mathrm{PP}}, \leq_{\mathrm{um}}^{\mathrm{PP}}, \leq_{\mathrm{uT}}^{\mathrm{PP}}\}$.*

1. *$X \leq_r (A, B) \overset{df}{\Longleftrightarrow} (X, \overline{X}) \leq_r (A, B)$.*

2. $(A, B)$ *is* $\leq_{\mathrm{m}}^{\mathrm{PP}}$*-hard for NP* $\stackrel{df}{\Longleftrightarrow}$ SAT$\leq_{\mathrm{m}}^{\mathrm{PP}}(A, B)$.

3. $(A, B)$ *is* $\leq_{\mathrm{T}}^{\mathrm{PP}}$*-hard for NP (NP-hard for short)* $\stackrel{df}{\Longleftrightarrow}$ SAT$\leq_{\mathrm{T}}^{\mathrm{PP}}(A, B)$.

The following conjecture is due to Even, Selman, and Yacobi.

**Conjecture 1 ([6]).** There is no NP-hard disjoint NP-pair.

If this conjecture is true, then no public-key cryptosystem is NP-hard to crack (see Theorem 2 for more consequences). Homer and Selman [15] construct a relativized world where P $\neq$ NP, but all disjoint NP-pairs are P-separable. In particular, Conjecture 1 holds in this world.

## 2.2   Function Classes

The study of NP search problems and the difficulty of inverting polynomial-time computable functions led to the notion of partial, multivalued functions that are computable by NP-machines. For each partial, multivalued function $f$, set-$f(x)$ denotes the *set of values* of $f$ on input $x$. If $f(x)$ is undefined, then set-$f(x) = \emptyset$.

**Definition 4 ([3]).** *We define some function classes:*

1. NPMV *is the class of partial, multivalued functions $f$ for which there is a nondet. polynomial-time machine $N$ such that for every $x$, it holds that* set-$f(x) = \{y \mid N(x)\,has\,an\,accepting\,path\,that\,outputs\,y\}$.
2. NP$k$V $\stackrel{df}{=} \{f \in$ NPMV $\mid \forall x, |$set-$f(x)| \leq k\}$ *where $k \geq 1$*
3. NPSV $\stackrel{df}{=}$ NP1V *(the class of partial, singlevalued NPMV-functions)*
4. NPbV $\stackrel{df}{=} \{f \in$ NP2V $\mid \forall x,$ set-$f(x) \subseteq \{0, 1\}\}$
5. PF *is the class of partial functions computable in (det.) polynomial time.*
6. *For any class of functions $\mathcal{F}$, let $\mathcal{F}_t \stackrel{df}{=} \{f \in \mathcal{F} \mid f$ is total$\}$.*

For partial, multivalued functions $f$ and $g$, we say that $g$ is a *refinement* of $f$, if for all $x$, $g(x)$ is defined if and only if $f(x)$ is defined, and set-$g(x) \subseteq$ set-$f(x)$. For function classes $\mathcal{F}$ and $\mathcal{G}$ we write $\mathcal{F} \subseteq_c \mathcal{G}$, if for every $f \in \mathcal{F}$ there exists a $g \in \mathcal{G}$ such that $g$ is a refinement of $f$.

Selman [30] gives a systematic comparison of classes of functions that are computed by nondeterministic polynomial-time transducers. Moreover, this paper identifies relations between these function classes and disjoint NP-pairs. A comprehensive overview of function classes can be found in [31].

Fenner et al. [8] introduced and studied the class NPbV. In particular, the paper investigates and gives several equivalent formulations of the hypotheses NPMV$_t \subseteq_c$ PF and NPbV$_t \subseteq_c$ PF. For example, the latter is equivalent to the hypothesis that all disjoint pairs of coNP-sets are P-separable.

**Definition 5 ([29,13,14]).** *Let $\mathcal{F}$ be any class of functions (possibly multivalued and/or partial). A set $A$ is $\mathcal{F}$-selective if there is a function $f \in \mathcal{F}$ such that for every $x$ and $y$ it holds that* set-$f(x, y) \subseteq \{x, y\}$ *and* $(\{x, y\} \cap A \neq \emptyset \Rightarrow \emptyset \neq$ set-$f(x, y) \subseteq A)$. *By $\mathcal{F}$-sel we denote the class of sets that are $\mathcal{F}$-selective.*

**Theorem 2 ([6,12,30]).** *If Conjecture 1 holds, then* $\mathrm{NP} \neq \mathrm{coNP}$, $\mathrm{NP} \neq \mathrm{UP}$, $\mathrm{NPMV} \not\subseteq_c \mathrm{NPSV}$, *and no public-key cryptosystem is* $\mathrm{NP}$-*hard to crack.*

A polynomial-time computable function $f$ is *honest*, if there is a polynomial $q$ such that for every $y$ in the range of $f$ there exists an $x$ in the domain of $f$ such that $f(x) = y$ and $|x| \leq q(|y|)$.

There are several equivalent formulations of the hypothesis $\mathrm{NPMV} \subseteq_c \mathrm{NPSV}$. Later we will show that one can add the shrinking property for NP to this list of equivalent formulations.

**Theorem 3 ([30,14]).** *The following are equivalent:*

1. $\mathrm{NPMV} \subseteq_c \mathrm{NPSV}$
2. $\mathrm{NP2V} \subseteq_c \mathrm{NPSV}$
3. $\mathrm{SAT} \in \mathrm{NPSV}$-sel
4. $\mathrm{NP} \subseteq \mathrm{NPSV}$-sel
5. *The inverses of honest functions in* $\mathrm{PF}$ *have refinements in* $\mathrm{NPSV}$.

Hemaspaandra et al. [14] use selectivity to show that the assertions above (e.g. $\mathrm{NPMV} \subseteq_c \mathrm{NPSV}$) imply a collapse of the polynomial-time hierarchy. Their proof relativizes to all oracles [14]. Naik et al. [20] improve this result and show that the output-multiplicity hierarchy $\{\mathrm{NP}k\mathrm{V}\}_{k \geq 1}$ is infinite unless the polynomial-time hierarchy collapses.

**Theorem 4 ([14]).** *If* $\mathrm{NP} \subseteq \mathrm{NPSV}$-sel *then* $\mathrm{ZPP}^{\mathrm{NP}} = \mathrm{PH}$.

We are going to study the relationships between the following assertions.

**Definition 6.** *Define the following assertions.*

> **A1:** DisjNP *does not have a* $\leq_{\mathrm{T}}^{\mathrm{PP}}$-*complete disjoint pair.*
> **A2:** DisjNP *does not have a* $\leq_{\mathrm{m}}^{\mathrm{PP}}$-*complete disjoint pair.*
> **A3:** *There is no* NP-*hard disjoint* NP-*pair.*
> **A4:** *The separation property does not hold for* coNP.
> **A4′:** $\mathrm{NPbV}_t \not\subseteq_c \mathrm{NPSV}$.
> **A5:** $\mathrm{UP} \not\subseteq \mathrm{coNP}$.
> **A6:** *The* PH *is infinite.*
> **A7:** *The separation property does not hold for* NP.
> **A8:** *The shrinking property does not hold for* NP.
> **A8′:** $\mathrm{NPMV} \not\subseteq_c \mathrm{NPSV}$.
> **A9:** *The shrinking property does not hold for* coNP.
> **A9′:** *There is no disjoint* NP-*pair that is* $\leq_{\mathrm{m}}^{\mathrm{PP}}$-*hard for* NP.
> **A9″:** $\mathrm{NP} \neq \mathrm{coNP}$.
> **A10:** *There is a* P-*inseparable, disjoint* NP-*pair.*

**Proposition 1.** $\mathrm{A4} \Rightarrow \mathrm{A8}$,    $\mathrm{A7} \Rightarrow \mathrm{A9}$,    $\neg\mathrm{A9''} \Rightarrow (\neg\mathrm{A4} \wedge \neg\mathrm{A7} \wedge \neg\mathrm{A8} \wedge \neg\mathrm{A9})$.

**Theorem 5 ([11]).** $\mathrm{A1} \Rightarrow \mathrm{A2} \Rightarrow \mathrm{A9'}$,    $\mathrm{A1} \Rightarrow \mathrm{A3} \Rightarrow \mathrm{A9'}$,    $\mathrm{A9'} \Leftrightarrow \mathrm{A9''}$.

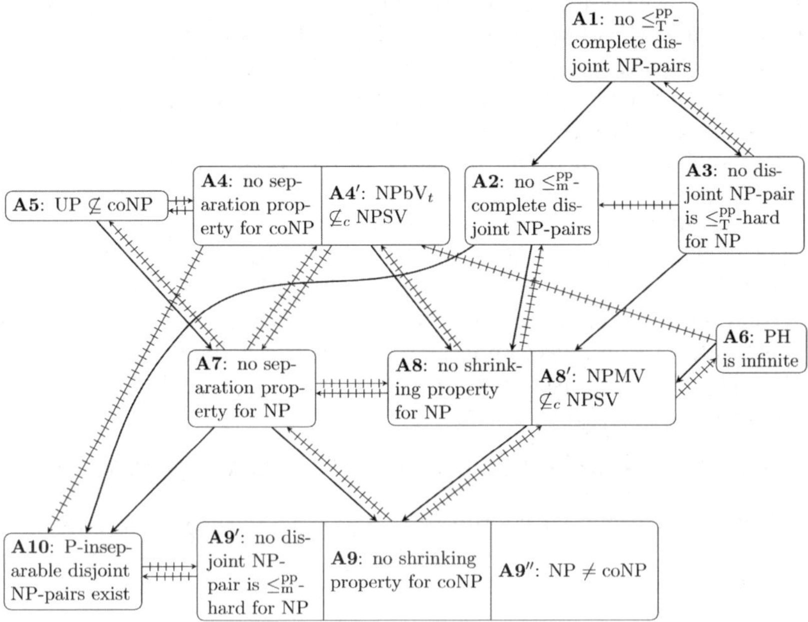

**Fig. 1.** Summary of the obtained results. Normal arrows denote relativizable implications, crossed-out arrows denote implications that do not hold relative to some oracle. Assertions that share a box are equivalent.

In the following we establish new relationships between the assertions given in Definition 6. Fig. 1 gives a summary of the relationships and their relativizability. In particular, we will answer the following open questions:

**Open Problem 1 [2, problem 3]:** *Find an oracle relative to which the shrinking property holds for* NP *but* NP $\neq$ coNP. *Better yet, find an oracle relative to which the shrinking property holds for* NP *and* (NP $\cap$ coNP) = P $\neq$ NP.

**Open Problem 2 [25]:** *Is there an oracle relative to which the polynomial-time hierarchy does not collapse and for all* $n \geq 1$, $\Sigma_n^P$ *(resp.,* $\Pi_n^P$*) has the shrinking property?*

In Theorem 12 we construct the oracle that is asked for in the first problem. The Corollaries 1 and 2 will give negative answers to the second question.

## 3   Connections to Reasonable Assumptions

In this section we establish implication relationships between the introduced notions. Our results imply that, under reasonable complexity-theoretic assumptions like an infinite PH and UP $\not\subseteq$ coNP, the shrinking and separation properties do not hold for NP and the shrinking property does not hold for coNP. In particular, we will relate the shrinking and separation properties to well-known notions like the classes $\Sigma_n^P$ and $\Pi_n^P$ of the PH, the function classes NPMV and NPSV, NP-hard disjoint NP-pairs, and complete disjoint NP-pairs.

Moreover, with the Corollaries 1 and 2 we will solve the open problem in [25] that is mentioned at the end of Section 2.

First let us relate the shrinking property for NP to known and well-understood classes of functions that are computable in nondeterministic, polynomial time. As a consequence, the shrinking property for NP is equivalent to all hypotheses mentioned in Theorem 3. From a result by Hemaspaandra et al. [14] it then follows that the shrinking property does not hold for NP, unless the PH collapses to its second level. Later we will see that this evidence is optimal in the sense that relativizable techniques cannot strengthen the collapse to the first level. Moreover, the collapse consequence of the shrinking property for NP solves an open question from [25].

**Theorem 6.** NP *has the shrinking property* $\Leftrightarrow$ NPMV $\subseteq_c$ NPSV.

The last theorem together with a result by Hemaspaandra et al. [14] immediately implies a collapse of the PH, if NP has the shrinking property. This solves the $\Sigma_n^P$-part of the open problem in [25] (the second open problem that is mentioned at the end of Section 2).

**Corollary 1.** *For $n \geq 1$, the shrinking property for $\Sigma_n^P$ implies* $\mathrm{ZPP}^{\Sigma_n^P} = \mathrm{PH}$. *In particular, the shrinking property for* NP *implies* $\mathrm{ZPP}^{\mathrm{NP}} = \mathrm{PH}$.

**Theorem 7.** NP *has the shrinking prop.* $\Rightarrow$ NP-*hard disjoint* NP-*pairs exist.*

*Remark 1.* Interestingly, the analog of the last theorem in computability theory is false, i.e., RE has the shrinking property and the analog of Conjecture 1 holds: There exists an m-complete disjoint pair (equivalently, an effectively inseparable pair) $(A, B)$ of computably enumerable sets (see e.g. [24]), but every disjoint pair of computably enumerable sets can be separated by some set whose degree is strictly less than $0'$ [32].

**Theorem 8.** coNP *has the shrinking property* $\Leftrightarrow$ NP = coNP.

Since Theorem 8 is relativizable, we can solve the $\Pi_n^P$-part of the open problem in [25]. Together with Corollary 1 this completely solves that problem.

**Corollary 2.** *For $n \geq 1$, $\Pi_n^P$ has the shrinking property* $\Leftrightarrow$ $\Sigma_n^P = \Pi_n^P$.

**Theorem 9.** NP *has the shrinking prop.* $\Rightarrow$ DisjNP *has* $\leq_m^{\mathrm{pp}}$-*complete pairs.*

**Theorem 10.** UP $\not\subseteq$ coNP $\Rightarrow$ *the separation property does not hold for* NP.

**Theorem 11.** coNP *has the separation property* $\Leftrightarrow$ $\mathrm{NPbV}_t \subseteq_c$ NPSV.

We remark that all results in this section are relativizable.

## 4   Oracle Separations

We now concentrate on those implications between the assertions A1–A10 that were left open in Section 3. For most of them we can find oracle constructions showing that the implication cannot be established by relativizable techniques.

The main result in this section is the construction of an oracle relative to which NP has the shrinking property and $(NP \cap coNP) = P \neq NP$. This oracle has three applications: First, it provides a relativized world in which some of the open implications do not hold. Second, it shows that the assumption $NP \neq coNP$ is too weak to refute the shrinking property for NP with relativizable techniques. Third, with this oracle we solve an open problem by Blass and Gurevich [2] who explicitly ask for the existence of such an oracle.

**Theorem 12.** *There exists an oracle $O$ relative to which the following holds:*

1. NP *has the shrinking property.*
2. $P = NP \cap coNP$.
3. $UP \not\subseteq coNP$.

**Corollary 3.** *Relative to the oracle $O$ in Theorem 12, all of the following holds:* $\neg A1$, $\neg A2$, $\neg A3$, $\neg A4$, A5, A7, $\neg A8$, A9, A9', A9'', A10, $ZPP^{NP} = PH$.

Recall that Hemaspaandra et al. showed that $NP \subseteq NPSV$-sel implies a collapse of the polynomial hierarchy to $ZPP^{NP}$. Our oracle constructed in Theorem 12 shows that relativizable techniques cannot strengthen this collapse to NP.

**Corollary 4.** *There is an oracle relative to which* $NP \subseteq NPSV$-sel *and* $NP \neq coNP$.

Reversely, Corollary 1 shows that there is no oracle relative to which NP has the shrinking property and $\Sigma_2^P \neq \Pi_2^P$. In this sense the oracle constructed in Theorem 12 is nearly optimal.

**Theorem 13 ([15]).** *There exists an oracle $O$ relative to which all of the following holds:* $\neg A1$, $\neg A2$, A3, $\neg A5$, $\neg A7$, A8, A9, A9', A9'', $\neg A10$.

**Theorem 14.** *We summarize known oracle constructions.*

1. *There exists an oracle relative to which all NP-pairs are P-separable, but coNP does not have the separation property. (A4 $\not\Rightarrow$ A10) [9, Theorem 3.3].*
2. *There exists an oracle relative to which coNP has the separation property, but NP does not have the separation property. (A7 $\not\Rightarrow$ A4) [9, Theorem 3.5]*
3. *There exists an oracle relative to which NP and coNP do not have the separation property and $UP \subseteq coNP$. (A7 $\wedge$ A4 $\not\Rightarrow$ A5) [9, Theorem 3.4]*
4. *There exists an oracle relative to which there are no NP-hard disjoint NP-pairs, but there exist $\leq_m^{pp}$-complete disjoint NP-pairs. (A3 $\not\Rightarrow$ A2) [11, Theorem 6.1]*
5. *There exists an oracle relative to which $NP2V \not\subseteq_c NPSV$, but $NP2V_t \subseteq_c PF \subseteq NPSV$. In particular, $NPMV \not\subseteq_c NPSV$ and $NPbV_t \subseteq_c NPSV$. (A8' $\not\Rightarrow$ A4') [20, Corollary 6]*

6. *There exists an oracle relative to which* NP2V $\not\subseteq_c$ NPSV *and* PH $= \Delta_2^P$. *In particular,* NPMV $\not\subseteq_c$ NPSV *and* PH *is finite.* (A8' $\not\Rightarrow$ A6) [20, Theorem 5]
7. *There exists an oracle relative to which* UP $\neq$ NP $=$ PSPACE. *In particular,* P $\neq$ NP $\cap$ coNP *and hence* P-*inseparable disjoint* NP-*pairs exist.* (A10 $\not\Rightarrow$ A9'') [21, Lemma 4.7]
8. *There exists an oracle relative to which* P $=$ NP $\cap$ coNP $\neq$ UP, NPMV$_t$ $\subseteq_c$ NPSV$_t$, *and* PH *is infinite. Here* coNP *has the separation property and* UP $\not\subseteq$ coNP. (A5 $\wedge$ A6 $\not\Rightarrow$ A4) [4]
9. *Relative to a random oracle,* NP $\neq$ coNP, NP $\neq$ UP, *and* NPMV $\not\subseteq_c$ NPSV. *In particular, neither* NP *nor* coNP *have the shrinking property.* [1,19,20]

## 5  Conclusions and Open Questions

The results of this paper show that, similar to descriptive set theory and computability theory, the separation and shrinking properties are important also for complexity theory, because they are closely related to many other fundamental notions. In contrast to descriptive set theory and computability theory, these properties are probably false for complexity classes like NP or coNP, because they contradict widely believed conjectures. The negative solution to the problem in [25] gives a clear evidence that the refinements of the PH studied in [25] behave probably much worse than the analogous refinements of the Borel hierarchy in descriptive set theory and of the arithmetical hierarchy in computability theory (see [26] for additional details).

Our summary in Fig. 1 motivates several open problems. In particular, we would like to know answers for the following questions:

1. Does an infinite PH imply that the separation property does not hold for NP (resp., coNP)?
2. Is there an oracle relative to which A8 $\not\Rightarrow$ A3? By Theorem 2 and Corollary 1 it suffices to construct an oracle relative to which UP $=$ NP and $\Sigma_2^P \neq \Pi_2^P$. However, it is a known open question whether such a relativized world exists. An even stronger result would be an oracle relative to which A2 $\not\Rightarrow$ A3.

## References

1. Bennett, C., Gill, J.: Relative to a random oracle $P^A \neq NP^A \neq coNP^A$ with probability 1. SIAM Journal on Computing 10, 96–113 (1981)
2. Blass, A., Gurevich, Y.: Equivalence relations, invariants, and normal forms. SIAM Journal on Computing 13(4), 682–689 (1984)
3. Book, R.V., Long, T., Selman, A.L.: Quantitative relativizations of complexity classes. SIAM Journal on Computing 13, 461–487 (1984)
4. Buhrman, H., Fortnow, L., Koucký, M., Rogers, J.D., Vereshchagin, N.K.: Inverting onto functions and polynomial hierarchy. In: Diekert, V., Volkov, M.V., Voronkov, A. (eds.) CSR 2007. LNCS, vol. 4649, pp. 92–103. Springer, Heidelberg (2007)
5. Ershov, Y.L., Lavrov, I.A., Taimanov, A.D., Taitslin, M.A.: Elementary theories. Uspechi Matematicheskikh Nauk. 20(4), 37–108 (1965); In Russian, English translation: Russian Mathematical Surveys, 20(4), 35–105 (1965)

6. Even, S., Selman, A.L., Yacobi, J.: The complexity of promise problems with applications to public-key cryptography. Information and Control 61, 159–173 (1984)
7. Even, S., Yacobi, Y.: Cryptocomplexity and NP-completeness. In: de Bakker, J.W., van Leeuwen, J. (eds.) ICALP 1980. LNCS, vol. 85, pp. 195–207. Springer, Heidelberg (1980)
8. Fenner, S., Fortnow, L., Naik, A., Rogers, J.: On inverting onto functions. In: Proceedings 11th Conference on Computational Complexity, pp. 213–223. IEEE Computer Society Press (1996)
9. Fortnow, L., Rogers, J.: Separability and one-way functions. In: Du, D.-Z., Zhang, X.-S. (eds.) ISAAC 1994. LNCS, vol. 834. Springer, Heidelberg (1994)
10. Glaßer, C., Selman, A.L., Sengupta, S.: Reductions between disjoint NP-pairs. Information and Computation 200, 247–267 (2005)
11. Glaßer, C., Selman, A.L., Sengupta, S., Zhang, L.: Disjoint NP-pairs. SIAM Journal on Computing 33(6), 1369–1416 (2004)
12. Grollmann, J., Selman, A.L.: Complexity measures for public-key cryptosystems. SIAM Journal on Computing 17(2), 309–335 (1988)
13. Hemachandra, L.A., Hoene, A., Ogiwara, M., Selman, A.L., Thierauf, T., Wang, J.: Selectivity. In: Proceedings 5th International Conference on Computing and Information, pp. 55–59. IEEE Computer Society (1993)
14. Hemaspaandra, L., Naik, A., Ogihara, M., Selman, A.L.: Computing solutions uniquely collapses the polynomial hierarchy. SIAM Journal on Computing 25, 697–708 (1996)
15. Homer, S., Selman, A.L.: Oracles for structural properties: The isomorphism problem and public-key cryptography. Journal of Computer and System Sciences 44(2), 287–301 (1992)
16. Kechris, A.S.: Classical Descriptive Set Theory. Springer, New York (1994)
17. Köbler, J., Messner, J., Torán, J.: Optimal proof systems imply complete sets for promise classes. Information and Computation 184(1), 71–92 (2003)
18. Moschovakis, Y.N.: Descriptive Set Theory. North Holland, Amsterdam (1980)
19. Naik, A.: The structural complexity of intractable search functions. PhD thesis, State University of New York, Buffalo (1994)
20. Naik, A., Rogers, J., Royer, J., Selman, A.L.: A hierarchy based on output multiplicity. Theoretical Computer Science 207, 131–157 (1998)
21. Ogiwara, M., Hemachandra, L.: A complexity theory of feasible closure properties. Journal of Computer and System Sciences 46, 295–325 (1993)
22. Pudlák, P.: On reducibility and symmetry of disjoint NP-pairs. In: Sgall, J., Pultr, A., Kolman, P. (eds.) MFCS 2001. LNCS, vol. 2136, pp. 621–632. Springer, Heidelberg (2001)
23. Razborov, A.: On provably disjoint NP-pairs. Technical Report TR94-006, Electronic Colloquium on Computational Complexity (1994)
24. Rogers Jr., H.: Theory of Recursive Functions and Effective Computability. McGraw-Hill, New York (1967)
25. Selivanov, V.L.: Two refinements of the polynomial hierarchy. In: Enjalbert, P., Mayr, E.W., Wagner, K.W. (eds.) STACS 1994. LNCS, vol. 775, pp. 439–448. Springer, Heidelberg (1994)
26. Selivanov, V.L.: Fine hierarchies and boolean terms. Journal of Symbolic Logic 60, 289–317 (1995)
27. Selivanov, V.L.: Fine hierarchy of regular omega-languages. Theoretical Computer Science 191(1-2), 37–59 (1998)

28. Selivanov, V.L.: Fine hierarchy of regular aperiodic omega-languages. In: Harju, T., Karhumäki, J., Lepistö, A. (eds.) DLT 2007. LNCS, vol. 4588, pp. 399–410. Springer, Heidelberg (2007)
29. Selman, A.L.: P-selective sets, tally languages, and the behavior of polynomial-time reducibilities on NP. Mathematical Systems Theory 13, 55–65 (1979)
30. Selman, A.L.: A taxonomy on complexity classes of functions. Journal of Computer and System Sciences 48, 357–381 (1994)
31. Selman, A.L.: Much ado about functions. In: Proceedings 11th Conference on Computational Complexity, pp. 198–212. IEEE Computer Society Press (1996)
32. Shoenfield, J.R.: Degrees of models. Journal of Symbolic Logic 25(3), 233–237 (1960)

# On the Hardness of Truthful Online Auctions with Multidimensional Constraints

Rica Gonen

Yahoo! Research Labs
701 First Street
Sunnyvale, CA 94089
gonenr@yahoo-inc.com

**Abstract.** This paper assess the prospect of creating truthful mechanisms for sponsored search auctions where advertisers have budget and time constraints. While the existing impossibility in this area by [4] addresses the situation where advertisers have budget limitations and static prices but not time limitations; our result applies to the common setting in practice where advertisers have time and budget limitations, prices are dynamic and advertisers act strategically on their time limitation as well as their budget.

We show that in cases where advertisers' arrival and departure times are private information, no truthful deterministic mechanism for budgeted sponsored search with time constrained advertisers can perform well with respect to social welfare maximization. Essentially, to protect itself from advertisers' time manipulation a truthful mechanism must give up significant social welfare.

It is also shown that even in cases where advertisers' arrival and departure times are known to the mechanism, the existence of advertiser budgets is itself a problem. In this case a budgeted sponsored search mechanism with time constrained advertisers can not achieve non-trivial welfare approximation when using a local pricing scheme (a player pricing history is not taken into account). Also, it is shown that for a dynamic global pricing scheme no such truthful mechanism exists.

## 1   Introduction and Related Work

In the last three years sponsored search auctions have gained much attention (e.g., [5,14]) due to their implications on the profitability of the three major search engines, i.e., Yahoo!, Google, and MSN. In a general sponsored search auction advertisers bid on a set of keywords (sometimes with budget constraints as in [3,14,15]) and a search engine allocates each keyword to the highest $K$ bidding advertisers' ads and charges every allocated advertiser a price for presenting his ad if the ad is clicked on by the user.

When approaching the design of a sponsored search auction one may consider several challenging problems; budget constraints, incentives, and the online nature of the problem.

A. Beckmann, C. Dimitracopoulos, and B. Löwe (Eds.): CiE 2008, LNCS 5028, pp. 221–230, 2008.

Budget constraints: As advertisers are charged only for their clicked ads, search engines look to maximize advertisers' expected value. As such, for every keyword the advertisers are ranked according to their reported value per click multiplied by their click-through rate. The click-through rate is an estimated parameter by nature and since the demand for keywords can be large, advertisers find themselves in a situation where they have no control over the overall payment the search engine will require them to pay for services rendered. For this reason many advertisers prefer to mitigate their risk by reporting a budget limitation in addition to their value per click. Consequently the budgeted sponsored search auction can be understood as modeling a risk-neutral advertiser who enters a multi-unit auction for click-throughs and is interested in a finite number of click-throughs. Of course, it would be easier to explicitly state the number of click-throughs desired but a budget is a good proxy for this expressiveness when lacking complete information on the cost of clicks.

The online nature of the problem: Sponsored search auctions are traditionally viewed as online auctions due to the online demand presented by user queries for keywords. Another online aspect less considered in the literature is the online fashion in which advertisers can submit their advertising requests. An advertiser my submit a bid for a keyword and a budget constraint for the next half day and then return and submit his bid a week later when a new advertising campaign is launched. It is well known that the search engines companies are facing strategic behavior of advertisers with regards to timing their participation in the system.

Incentives: As advertisers are utility maximizers and would like to minimize their cost for the ad referrals they receive they may misreport their value per click, their budget constraints, or even manipulate the times at which they request the search engine to consider their bid.

## 1.1  Previous Work

The sponsored search auctions currently in use charge every allocated advertiser a "generalized second price" for presenting the ad (if it was clicked on). Unfortunately as was shown in [5,7] the prices charged by the "generalized second price" payment scheme do not incent advertisers to reveal their true value for getting a click on a keyword when more then a single ad is presented on a page (which in practice is usually the case). For the basic model where advertisers have no budget constraints and are present in a single page auction, some schemes are suggested to overcome the incentive problem when designing an incentive-compatible sponsored search auction mechanism (e.g., [2,9]).

The budget-constrained bidding format imposes challenges on welfare maximization and revenue maximization regardless of the incentive issues. Several attempts were made to investigate the welfare/revenue maximization problem in sponsored search auctions with budget-constrained advertisers. [3,14] assumed that advertisers report their true value and budget while [1,4,15] investigated revenue maximization while guaranteeing the truthful behavior of advertisers.

Unlike the non budgeted basic model that was shown to be truthful and to maximize welfare, the budget constrained sponsored search auction can not maximize welfare while maintaining truthfulness, as was show by [4]. [4]'s model studies a multi-unit auction with multiple players each of whom has a private valuation and budget. In the context of sponsored search this model assumes that all advertisers are constantly present in the auction. Essentially [4]'s impossibility shows that any such truthful auction will allocate all units (keyword impressions) to a single player (advertiser) while the remaining advertisers will not be allocated any keyword. Such an allocation is imposed by a truthful mechanism as the single allocated player induces high prices for all other players, who end up unable to win any of the units. The winning player on the other hand enjoys lower prices induced by the other players. As the winning player is budget constrained he will not be able to buy all the units, which is the source of the inefficiency. When taking into consideration the [4] model in the sponsored search auction context one must note an interesting subtlety: Units (keywords) are allocated sequentially after the winning advertiser expends his budget and he leaves the auction, at this point in time the mechanism's inefficiency might be saved if prices are recalculated using only the remaining advertisers. Another way in which the inefficiency presented by [4] can potentially be saved is with the introduction of a time constraint under which the budget can be used. Our model extends [4] as it consider an online sponsored search as well as time constrained advertisers.

## 1.2   Our Results

This paper investigates the complete model in which incentive compatibility, budget constraints, and time constraints are taken into consideration in the context of a sponsored search auction. Two types of pricing schemes are investigated. The first is the local pricing scheme where prices of a keyword auction in time period $t$ are determined independent of other same-keyword auctions in time periods $\tau \neq t$. The second is the global pricing scheme where prices of a keyword auction in time period $t$ can be dependent on local prices of other same-keyword auctions in time periods $\tau \neq t$. The two types of pricing schemes are examined in two forms: the dynamic form and the static form. Unfortunately the picture revealed by this paper is not very encouraging with respect to designing truthful (dominant strategy) time-constrained and budget-constrained sponsored search mechanisms as all of the results presented are impossibility results.

We divide the paper into two parts. In the first part the model assumes that every advertiser has four private parameters; value per click, budget, arrival time, and departure time. The main theorem in this part shows that no truthful mechanism with any pricing scheme (local or global) can achieve a non-trivial social welfare in a time-constrained budget-constrained sponsored search mechanism. The above theorem is shown for local pricing schemes. We then show that no global pricing scheme in this setting can be truthful. In the second part we relax the assumption that advertisers' time constraints are private values and assume that the mechanism knows the advertisers' true time constraints.

The results in this section do not lead to more encouraging conclusions. The main theorem in the second part, similar to the first section, shows that no truthful mechanism with any pricing scheme (local or global) can achieve a non-trivial social welfare in time-constrained budget-constrained sponsored search mechanisms. The later theorem is shown for local pricing schemes. The global pricing scheme divides into a dynamic and a static form. The dynamic global pricing scheme is shown not to be truthful in this case while the static global pricing scheme although truthful can not always guarantee a non-trivial social welfare. Our results extends [4] paper even in the case of static global pricing scheme where all advertisers has identical time constraints as [4] only prove their impossibility result for two players and two goods while our result can handel $N$ advertisers and $M$ goods.

There are two positive results in the existing literature that achieve non-trivial welfare maximization and truthfulness. Let us explain these mechanisms in the context of our impossibility result. The first mechanism is a positive result designing a $\delta$-gain truthful time-constrained budget-constraint sponsored search mechanism and was given in [10]. The result uses a global pricing scheme and allows for dominant strategy $\delta$-gain truthful behavior. Essentially the $\delta$-gain concept allows a player a small bounded gain from lying and by that discourages advertisers from manipulating the system. Since the mechanism is not strictly truthful it is not covered by our result. The second positive result is from [15]. [15]'s result achieves a truthful welfare maximizing budget-constrained sponsored search mechanism. The mechanism presented is a global static pricing scheme in a setting where advertisers are not time constrained at all (whether the mechanism knows their time parameters or not). Therefore one might think that [15]'s result should fall under the impossibility result of [4], but this is not the case as impressions of keywords are allocated randomly to advertisers, thus creating a randomized algorithm while [4] and our impossibility results refer to deterministic designs.

Finally our main impossibility result is inspired by [11]. This paper presented an impossibility result for online ascending auctions with gradually expiring items. [11]'s setting and context is of a scheduling problem. As such, our setting differs from theirs by introducing demand that results from a budget constraint and not a unit demand.

Some other literature [6] considered online mechanisms with time parameter. [6] work differ significantly from ours as they assume only arrival time as a parameter and no departure time. In addition unlike our setting they do not consider the lies of reporting an earlier arrival time than the true one.

There are also several previous works in the economic literature that address Bayesian budget constraints for single-item settings e.g., [13]. Our work, like [4] and unlike the existing economic literature, considers dominant strategy implementations.

This paper is organized in the following way: in the next section we formally introduce our model and give the necessary definitions. In section 3 we present our impossibility results that assume advertisers have time constrained private parameters. In section 4 we relax the former assumption and conclude in section 5.

## 2  Model and Definitions

In our model $N$ risk-neutral, utility-maximizing advertisers bid for advertising slots based on a keyword. This paper focuses on the bidding process for a single keyword, though the results easily extend to multiple keywords. It is therefore supposed w.l.o.g. that the keyword appears at every time $t$. Whenever that keyword appears in the search at time $t$, $K_t$ slots of advertisements appear in the search results.

The mechanism used to derive the impossibility results are assumed to run from time starting at $t = 1$ and end at $t = T$. Each time period is called a round. During each round the mechanism allocates advertisers to the $K_t$ slots or to some portion of the slots if there are too many slots. The mechanism also assumes that all slots offered in a single time round are of identical "quality". However, the impossibility easily extends to the case where slots are not of equal quality.

Advertisers arrive and depart the system in an online manner and may arrive and depart several times. Each advertiser $i$ has a private value per click (independent of the slot the ad originally appeared in) which is denoted by $v_i$. For every arrival and departure each advertiser $i$ also has an arrival and departure time, denoted $a_i$ and $l_i$ respectively, and a privately known budget denoted $b_i$. In section 3 we assume that arrival and departure times are privately known parameters to each advertiser. In 4 we relax this assumption and assume that the advertisers true arrival and departure times are known to the mechanism, which is used to further elucidate impossibilities around this setting. The mechanism objective function is to maximize social welfare, i.e., $\sum_{i \in N} \omega_i v_i$ where $\omega_i$ is the number of clicks resulting from the allocation constructed by the mechanism. We also assume advertisers have quasi-linear utilities (as long as their budget is not expended, in which case an advertiser's utility is unboundedly small) and that every advertiser will act rationally in order to maximize his own utility: his obtained overall value from clicks minus his price. An advertiser may arrive before at or after his true arrival time and declare any value or budget and any deadline[1].

Since our advertisers are budget constrained and are charged for the clicks they receive, we denote $i$'s remaining budget at time $t$ as $B_i^t \geq 0$.

Our work focuses on exploring the design possibilities of a truthful budget-constrained, time-constrained sponsored search mechanisms and therefore we start by defining a truthful in dominant strategy budget-constrained, time-constrained sponsored search mechanism.

Let $S_i$ be the domain of all advertiser $i$'s vectors of types $s_i = (a_i, l_i, v_i, b_i)$, and let $S_{-i} = \times_{j \neq i} S_j$. By the revelation principle it is enough to consider direct revelation mechanisms.

Consider the number of clicks resulting from the allocation constructed by the mechanism upon receiving the type $s_i \in S_i$ from player $i$ and $s_{-i} \in S_{-i}$ from the other players, and denote that number $\omega_i$.

---

[1] The true arrival time can be think of in this context as a start time of an advertising campaign.

**Definition 1. *Truthfulness (in dominant strategies)*** *An online budget constrained, time constrained, sponsored search mechanism is truthful if there exist price functions $p_i : S_1 \times \ldots \times S_n \to R$ such that for any $i$, any $s_{-i} \in S_{-i}$, any true type $s_i \in S_i$, and any $\bar{s}_i \neq s_i$*

$$\omega_i v_i - p_i(s_i, s_{-i}) \geq \bar{\omega}_i v_i - p_i(\bar{s}_i, s_{-i})$$

This paper presents a number of impossibility results for sponsored search mechanisms, some of which depend on the pricing scheme structure.

The following definitions presents two types of pricing schemes in the sponsored search auction mechanism: one in which prices are determined locally and the other in which prices are determined globally. The locally determined pricing scheme computes the price of a click for advertiser $i$ at time $t$ by considering only advertisers that are present at time $t$ (other than $i$). The globally determined pricing scheme computes the price of a click for advertiser $i$ at time $t$ by considering all local prices until time $t$. The reason to consider the local prices until time $t$ and not the local prices of all times is that the global pricing scheme is an online pricing scheme and therefore only has a look back. For an example of a global pricing scheme used in an online mechanism see [10]. We proceed by the formal definitions:

Let $S^t_{-i}$ be the domain of all advertisers' (except for $i$) vectors of types such that for all $s_{-i} \in S_{-i}$, $s_{-i} = (a_{-i}, l_{-i}, v_{-i}, b_{-i})$ and $t \in [a_{-i}, l_{-i}]$.

**Definition 2.** *A pricing scheme is called a* local *pricing scheme if for every advertiser $i$ and every time $t$ there exists a price function $p^L_i : S^t_{-i} \to R$.*

**Definition 3.** *A pricing scheme is called a* global *pricing scheme if for every advertiser $i$ and every time $t$ there exists a price function $p^G_i : p^L_i(S^1_{-i}), .., p^L_i(S^t_{-i}) \to R$.*

## 3 First Impossibility: With Private Time Parameters

In this section we focus our attention on showing our main theorem. The main theorem proves that no truthful deterministic mechanism for budget-constrained time-constrained sponsored search with private time parameters for advertisers can perform well with respect to social welfare maximization. Essentially the proof shows that in order to protect itself from advertisers' time manipulation a truthful mechanism must give up most of the social welfare.

To prove the theorem we utilize a claim which shows that in order to maintain advertisers' truthful behavior a mechanism that uses a local pricing scheme must allocate the advertiser in the lowest priced period where he is available.

Naturally the theorem also applies to the simple setting where advertisers are only interested in a single click and their budget is their value per click.

As the our claim discuss social welfare performance of the mechanism we will use the following definition of $c$-approximation mechanism.

**Definition 4.** *A mechanism is a $c$-approximation to the social welfare if for any $s \in S$ and any allocation of the mechanism, the social welfare obtained in the allocation is at least a $1/c$ fraction of the optimal social welfare with respect to $s$.*

In the theorem below we benchmark the performance of the algorithm's social welfare maximization to the number of web pages presented to the algorithm. The number of web pages presented is assumed to be $\delta M$ where $\delta$ is some positive constant. To ease the understanding of the theorem's proof we fist demonstrate it assuming $M$ web pages presented to the algorithm with advertisers' budget which is small (equal to their value) and then extend the proof to show the result for advertiser with large budget (equal to $\delta$ times their value) while assuming $\delta M$ web pages presented.

The proofs of theorem 1 and claim 3 can be found in the full paper [8].

**Theorem 1.** *Any truthful deterministic mechanism for an online budget constrained, time constrained, sponsored search auction cannot always obtain more than $O(1/(\delta M))$ fraction of the optimal welfare where advertisers have private time parameters.*

*Claim.* Set some $c$-approximation truthful deterministic mechanism. Then, for any player $i$ and any time $t \in [a_i, l_i]$ there exists a price function $p_i^t : S_{-i}^t \to R$ such that, for any combination of players $s_{-i}^t$, if $v_i \leq b_i$ then:

- If $v_i > p_i^t(s_{-i}^t)$ and $p_i^t(s_{-i}^t) < p_i^{t'}(s_{-i}^{t'})$ for all $t' \neq t, a_i \leq t' \leq l_i$ then $i$ is allocated at time $t$ and pays $p_i^t(s_{-i}^t)$ if he is clicked on.
- If $v_i < p_i^t(s_{-i}^t)$ then $i$ is not allocated a slot.

Although theorem 1 shows that any truthful deterministic mechanism for an online budget-constrained time-constrained sponsored search auction cannot always obtain more than $O(1/(\delta M))$ fraction of the optimal welfare it is implied by the proof's construction that there exists a truthful deterministic mechanism that always obtains at least $1/(\delta M)$ fraction of the optimal welfare.

Careful examination of theorem 1 may lead one to suspect that theorem 1 only applies for truthful mechanisms with local pricing schemes, as our proof of claim 3 utilizes such prices to prove theorem 1. However, we can show that theorem 1 holds for any pricing scheme, local or global. To prove the above we show that a global pricing scheme does not yield a truthful mechanism for online budget-constrained time-constrained sponsored search auctions where the time parameters are the advertisers' private information.

**Lemma 1.** *There does not exist a truthful budget-constrained time-constrained sponsored search mechanism where time parameters are the advertisers' private information that has a global pricing scheme.*

The proof of lemma 1 can be found in the full paper [8].

In this section we focused our attention on impossibility results in a setting where advertisers have four privately known parameters, including their arrival and departure, in the auction and their budget. The truthful mechanisms are proven to be under performing in terms of social welfare maximization as they "protect" themselves from time manipulations by advertisers.

A natural question to ask is whether the truthful budget-constrained time-constrained sponsored search mechanisms can perform better with respect to welfare maximization where time manipulations are not possible, i.e., the algorithm knows the true time parameters of the advertisers.

Recall [4]' model which assumes that all advertisers are constantly present in the auction. Essentially their impossibility shows that any such truthful auction will allocate all units (keyword impressions) to a single player (advertiser) while the rest of the advertisers will not be allocated any keyword. Such allocation is imposed by a truthful mechanism as the single allocated player induces high prices for all other players, who end up unable to win any of the units. The winning player on the other hand enjoys lower prices induced by the other players. As the winning player is budget constrained he will not be able to buy all the units, which is the source of the inefficiency. When taking into consideration the [4] model in the sponsored search auction context one must notice an interesting subtlety. As the units (keywords) are allocated sequentially once the winning advertiser fulfills his budget, he leaves the auction. At that point if prices are recalculated to include only the remaining advertisers the inefficient solution might be come efficient.

To better understand the impact of budget-constrained time-constrained advertisers on the impossibility of designing a truthful sponsored search mechanisms, in the next section we relax the assumption that advertisers have privately known arrival and departure times. We assume that the mechanism knows all of the advertisers' true arrival and departure times such that advertisers can not act as if they have a different arrival or departure time. By relaxing the time manipulation of advertisers we are able to show impossibilities that are encountered when attempting to remove budget manipulations.

## 4  Second Impossibility: Relaxing Private Time Parameters

In the previous section we investigated the effect of the advertisers' private time parameters on the truthful online budget-constrained time-constrained sponsored search mechanism. In this section we relax the assumption that advertisers have privately known arrival and departure times and assume that the mechanism knows all advertisers' true arrivals and departures. Relaxing the privately known time parameters assumption allows us to capture the difficulties imposed by the time-constrained budgeted environment.

The budgeted nature of the problem creates two types of pricing scheme formats. The first format, static pricing schemes, is where prices are predetermined by the mechanism given the knowledge of all advertisers' arrival and departure times and the report of their budget and value. In the second format, dynamic pricing schemes, prices keep updating depending on the advertisers currently available to the mechanism. As players are budget constrained different allocations will lead to different prices for the same player in the dynamic format.

In section 3 we examined two types of pricing schemes and their affect on the truthful time-constrained budget-constrained sponsored search mechanism. For local pricing schemes we showed that the truthful time-constrained budget-constrained sponsored search mechanism can not achieve non trivial welfare approximation and showed that the global pricing scheme does not maintain truthfulness at all. In this section we refine our examination of different pricing scheme types by examining the dynamic and static characteristics of the local and global pricing schemes.

We start by showing that even when time constraints are not private to advertisers, the budget-constrained time-constrained sponsored search mechanism can not achieve non-trivial welfare approximation when using a local pricing scheme.

We then continue by distinguishing between the two forms of prices in the global pricing scheme case. For dynamic global pricing schemes we can show that such schemes will not maintain truthfulness and for static global pricing schemes we can show that although they maintain truthfulness they can not guarantee a non-trivial welfare approximation.

Due to lack of space we defer the technical details of the section to the full paper [8].

## 5   Conclusions

This paper assess the prospect of creating truthful mechanisms for sponsored search auctions where advertisers have budget and time constraints. In cases where advertisers' arrival and departure times are private information, no truthful deterministic mechanism for sponsored search with budget constrained and time constrained advertisers can perform well with respect to social welfare maximization. Even in cases where advertisers' true arrival and departure times are known to the mechanism, a sponsored search mechanism with budget constrained and time constrained advertisers can not achieve non-trivial welfare approximation when using a local pricing scheme nor can such a truthful mechanism exist for a dynamic global pricing scheme.

Given the impossibility results presented in this paper there are several ways in which one may approach the design of a budget-constrained time-constrained sponsored search mechanism. One possible solution in to pursue mechanisms that are $\delta$-Gain truthful as was done in [10]. If the $\delta$-Gain truthful approach is applied with a global pricing scheme as in [10] the global pricing scheme may yield a non trivial welfare approximation. Another possible approach is to design a randomized algorithm for solving the problem as was done in [15] for the budget constrained but not time constrained advertisers. One may also take the approach followed in [11] of a set-Nash equilibrium solution concept and design a semi-myopic mechanism that achieves a non trivial welfare approximation for the budget-constrained time-constrained sponsored search auction.

# References


1. Abrams, Z.: Revenue Maximization when Bidders have budgets. In: Proc. Symposium on Discrete Algorithms, pp. 1074–1082 (2006)
2. Aggarwal, G., Goel, A., Motwani, R.: Truthful Auctions for Pricing Search Keywords. In: Proceding of EC 2006 (2006)
3. Abrams, Z., Mendelevitch, O., Tomlin, J.: Optimal Delivery of Sponsored Search Advertisements Subject to Budget Constraints. In: Proc. EC 2007 (2007)
4. Borgs, C., Chayes, J., Immorlica, N., Mahdian, M., Saberi, A.: Multi-unit auctions with budget-constrained bidders. In: Proc. 6th ACM Conference on Electronic Commerce, pp. 44–51 (2005)
5. Edelman, B., Ostrovsky, M., Schwarz, M.: Internet Advertising and the Generalized Second Price Auction: Selling Billions of Dollars Worth of Keywords. American Economic Review 97 (2007)
6. Friedman, E., Parkes, D.C.: Pricing WiFi at Starbucks- Issues in Online Mechanism Design. In: Proc. 4th ACM Conf. on Electronic Commerce (EC 2003) (2003)
7. Gonen, R.: Untruthful Behavior in Google Slot Auctions (unpublished manuscript, 2004)
8. Gonen, R.: On the Hardness of Truthful Online Auctions with Multidimensional Constraints, http://www.ricagonen.com
9. Gonen, R., Pavlov, E.: An Incentive-Compatible Multi Armed Bandit Mechanism. In: Third Workshop on Sponsored Search Auctions WWW 2007, PODC 2007 (2007)
10. Gonen, R., Pavlov, E.: An Adaptive Sponsored Search Mechanism $\delta$-Gain Truthful in Valuation, Time, and Budget (submited for review, 2007)
11. Lavi, R., Nisan, N.: Online Ascending Auctions for Gradually Expiring Items. In: SODA 2005 (2005)
12. Myerson, R.: Optimal Auction Design. Mathematics of Operations Research, 58–73 (1981)
13. Maskin, E.S.: Auctions, development and privatization: Ecient auctions with liquidity- constrained buyers. European Economic Review 44, 667–681 (2000)
14. Mehta, A., Saberi, A., Vazirani, U., Vazirani, V.: Adwords and the Generalized Bipartite Matching Problem. In: Proceedings of the Symposium on the Foundations of Computer Science, pp. 264–273 (2005)
15. Pavlov, E.: Truthful Polynomial Time Optimal Welfare Keywords Auctions with Budget Constraints. In: Joint workshop NetEcon+IBC (2007)

# Effective Dimensions and Relative Frequencies


Xiaoyang Gu[*] and Jack H. Lutz[*,**]

Department of Computer Science, Iowa State University, Ames, IA 50011 USA
{xiaoyang,lutz}@cs.iastate.edu



**Abstract.** Consider the problem of calculating the fractal dimension of a set $X$ consisting of all infinite sequences $S$ over a finite alphabet $\Sigma$ that satisfy some given condition $P$ on the asymptotic frequencies with which various symbols from $\Sigma$ appear in $S$. Solutions to this problem are known in cases where

(i) the fractal dimension is classical (Hausdorff or packing dimension), or

(ii) the fractal dimension is effective (even finite-state) and the condition $P$ *completely* specifies an empirical distribution $\pi$ over $\Sigma$, i.e., a limiting frequency of occurrence for *every* symbol in $\Sigma$.

In this paper we show how to calculate the finite-state dimension (equivalently, the finite-state compressibility) of such a set $X$ when the condition $P$ only imposes *partial* constraints on the limiting frequencies of symbols. Our results automatically extend to less restrictive effective fractal dimensions (e.g., polynomial-time, computable, and constructive dimensions), and they have the classical results (i) as immediate corollaries. Our methods are nevertheless elementary and, in most cases, simpler than those by which the classical results were obtained.

**Keywords:** effective fractal dimensions, empirical frequencies, finite-state dimension, randomness, saturated sets.


## 1   Introduction

The most fundamental statistics used in the analysis of data for purposes of compression or prediction are the empirical frequencies with which various symbols appear. When *every* symbol has a frequency that is known and stable throughout the data, the problems of compression and prediction are well understood, with the main insights now over a half-century old [5,15,16]. However, when only *partial* constraints on the empirical frequencies–e.g., the *relative* frequencies of some of the symbols–are known, these problems become more challenging.


[*] This author's research was supported in part by National Science Foundation Grants 0344187, 0652569 and 0728806, and by Spanish Government MEC Projects TIC 2002-04019-C03-03 and TIN 2005-08832-C03-02. Part of the results were announced (without proceedings) at the special session on Randomness in Computation in Fall 2005 Central Section Meeting of the American Mathematical Society.

[**] Corresponding author.



A. Beckmann, C. Dimitracopoulos, and B. Löwe (Eds.): CiE 2008, LNCS 5028, pp. 231–240, 2008.

This paper shows how to calculate the finite-state dimension (equivalently, the compressibility or predictability by finite-state machines [6,14]) of a set $X$ of infinite sequences over a finite alphabet $\Sigma$ when membership of a sequence $S$ in $X$ is determined by some given condition $P$ on the asymptotic frequencies with which various symbols from $\Sigma$ appear in $S$. Our results hold even when $P$ only imposes partial constraints on the limiting frequencies of symbols, and they automatically extend to less restrictive effective dimensions, such as polynomial-time, computable, and constructive dimensions. In order to explain our results and their significance, we briefly review four lines of research that are precursors of our work.

## 1.1   Classical Fractal Dimensions

In 1919, Hausdorff [13] developed a rigorous way of assigning a dimension to every subset of an arbitrary metric space. His definition agrees with the intuitive notion of dimension for "smooth" sets (e.g., smooth curves have dimension 1; smooth surfaces have dimension 2), but assigns non-integer dimensions to some more exotic sets, and hence came to be called a "fractal" dimension. In 1949, Eggleston [7], building on work of Besicovitch [3] and Good [10], proved that, for any probability measure $\pi$ on a finite alphabet $\Sigma$, the set of all sequences in which each symbol $a \in \Sigma$ has asymptotic frequency $\pi(a)$ has Hausdorff dimension $\mathcal{H}_{|\Sigma|}(\pi)$, the Shannon entropy of $\pi$, normalized to range over $[0, 1]$. In retrospect, Hausdorff dimension is an information-theoretic concept [27], but these developments essentially all took place prior to Shannon's development of information theory [28].

In the early 1980's, another fractal dimension called packing dimension, was introduced [29,30]. Packing dimension agrees with Hausdorff on "regular" sets, but is larger on some sets [9].

## 1.2   Shannon Information Theory

In 1948, Shannon [28] developed a probabilistic theory of information (Shannon entropy) that has been enormously productive and is the setting in which most work on compression and prediction has been carried out [5].

## 1.3   Effective Fractal Dimensions

In 2000, Lutz [18,19] proved a new characterization of Hausdorff dimension in terms of betting strategies and used this characterization to formulate effective fractal dimensions ranging from polynomial-time and polynomial-space dimensions to computable and constructive dimensions. Pushing this effort further, finite-state dimension was introduced the following year [6], and is now known to characterize the compressibility [6] and predictability [14] of sequences over finite alphabets. In [1], packing dimension was shown to have a betting-strategy characterization that is exactly dual to that of Hausdorff dimension, thereby giving dual "strong dimensions" at each of the levels of effectivity for which dimensions had been defined. Each of the papers mentioned here extended the above-mentioned result of Eggleston to the effective dimension(s) introduced. Hence, as indicated in our first paragraph, the compression and prediction problems are well understood, even at the finite-state level, when a set $X$ of sequences is defined in terms of given, well-defined, asymptotic frequencies of all symbols.

### 1.4   Classical Dimensions of Saturated Sets

In 2002, Barreira, Saussol, and Schmeling [2] considered the classical fractal dimensions of sets of sequences defined in terms of conditions placing (typically partial) constraints on the frequencies and relative frequencies of symbols. The example by which they introduced their work was the set $X$ of all sequences over the alphabet $\{0, 1, 2, 3\}$ in which there are asymptotically five times as many 0's as 1's. (No constraint is placed on the frequency of any individual symbol.) Using sophisticated techniques (multifractal analysis and ergodic theory), they showed how to compute the classical Hausdorff dimensions of sets of this kind. As it turns out, Volkmann [31] and his student Cajar [4] had previously defined a set $X$ of sequences to be *saturated* if membership in it is completely determined by the asymptotic behaviors (not necessarily convergent) of the frequencies of symbols and investigated the Hausdorff dimensions of many saturated sets. Olsen [21,22,23,24] and Olsen and Winter [25,26] also used multifractal analysis to study such sets.

### 1.5   Our Results

We show how to calculate the finite-state dimensions of saturated sets. We give a pointwise characterization of the dimensions of such sets, and we prove a general correspondence principle stating that, if $X$ is any saturated set, then the finite-state dimension of $X$ is exactly its classical Hausdorff dimension, and the finite-state strong dimension of $X$ is exactly its classical packing dimension. We also give completely elementary methods (no multifractal analysis or ergodic theory) for computing the finite-state dimensions of various types of saturated sets. By our correspondence principle, this yields elementary proofs that these results also hold for classical fractal dimensions and less restrictive effective fractal dimensions.

The rest of this paper is organized as follows. Section 2 lists the basic definitions and conventions we use in this paper. Section 3 reviews the definitions of Hausdorff dimension, packing dimension, finite-state dimension, and finite-state strong dimension. We give a few example of calculating the dimensions of exotic saturated sets in Section 4. In Section 5, we discuss finite-state dimensions of saturated sets in detail and give insight into why a maximum entropy principle holds.

## 2   Preliminaries

Let $m \geq 2$ be an integer. We work with the $m$-ary alphabet $\Sigma_m = \{0, 1, \ldots, m-1\}$. $\Sigma_m^*$ is the set of all (finite) *strings* on $\Sigma_m$ including the empty string $\lambda$.

$\mathbf{C}_m = \Sigma_m^\infty$ is the set of all (infinite) $m$-ary *sequences*. $\mathbf{C} = \mathbf{C}_2$ is the Cantor space. $\Delta(\Sigma_m)$ is the set of all probability measures on $\Sigma_m$.

Let $i$ be an integer such that $0 \leq i \leq m - 1$. The symbol counting function $\#_i : (\mathbf{C}_m \cup \Sigma_m^*) \times \mathbb{N} \to \mathbb{N}$ is defined such that for every string or sequence $S$ and $n \in \mathbb{N}$, $\#_i(S, n)$ is the number of occurrences of $i$ in the first $n$ bits of $S$. The symbol frequency function $\pi_i : (\mathbf{C}_m \cup \Sigma_m^*) \times \mathbb{N} \to [0, 1]$ is defined such that $\pi_i(S, n) = \#_i(S, n)/n$. The empirical measure function $\vec{\pi} : (\mathbf{C}_m \cup \Sigma_m^*) \times \mathbb{N} \to \Delta(\Sigma_m)$ is defined such that $\vec{\pi}(S, n) = (\pi_0(S, n), \ldots, \pi_{m-1}(S, n))$. Intuitively, $\vec{\pi}$ extracts empirical probability measures from the first $n$ bits of a string or a sequence based on the actual frequencies of digits.

## 3    The Four Dimensions

Hausdorff dimension and packing dimension are important tools in mathematics used to study the size of sets and the properties of dynamic systems. All countable sets have 0 for both of these dimensions. In order to study relative size of countable sets from the eyes of computers with different resources, Lutz generalized Hausdorff dimension to effective dimensions by using his gale characterization of Hausdorff dimension [18]. Athreya, Hitchcock, Lutz, and Mayordomo then gave a dual gale characterization of packing dimension, with which, they generalized packing dimension to effective strong dimensions [1]. We first review the definitions related to gales. Note that $\Sigma_m$ is an alphabet with $m$ symbols and $m \geq 2$.

**Definition.** Let $s \in [0, \infty)$. An *s-supergale* is a function $d : \Sigma_m^* \to [0, \infty)$ such that for all $w \in \Sigma_m^*$ $m^s d(w) \geq \sum_{a \in \Sigma_m} d(wa)$. The *success set* of an $s$-supergale $d$ is $S^\infty[d] = \{S \in \mathbf{C} \mid \limsup_{n \to \infty} d(S[0..n - 1]) = \infty\}$. The *strong success set* of $d$ is $S_{\mathrm{str}}^\infty[d] = \{S \in \mathbf{C} \mid \liminf_{n \to \infty} d(S[0..n - 1]) = \infty\}$.

Now we conveniently give the gale characterizations of Hausdorff and packing dimensions as definitions. Please refer to Falconer [8] for classical definitions.

**Definition.** ([18,1]). Let $X \subseteq \mathbf{C}_m$. The *Hausdorff dimension* of $X$ is

$$\dim_{\mathrm{H}}(X) = \inf \{s \in [0, \infty) \mid X \subseteq S^\infty[d] \text{ for some } s\text{-supergale } d \}.$$

The *packing dimension* of $X$ is

$$\dim_{\mathrm{P}}(X) = \inf \{s \in [0, \infty) \mid X \subseteq S_{\mathrm{str}}^\infty[d] \text{ for some } s\text{-supergale } d\}.$$

Finite-state dimension and strong dimension are finite-state counterparts of classical Hausdorff dimension [13] and packing dimension [20,29] introduced by Dai, Lathrop, Lutz, and Mayordomo [6] and Athreya, Hitchcock, Lutz, and Mayordomo [1] in the Cantor space $\mathbf{C}$. Finite-state dimensions are defined by using the gale characterizations of the Hausdorff dimension [18] and the packing dimension [1] and restricting the gales to the ones whose underlying betting strategies can be carried out by finite-state gamblers. In this section, we give the definitions of the finite-state dimensions for space $\mathbf{C}_m$ and review their basic properties. Now, we define finite-state gamblers on alphabet $\Sigma_m$.

**Definition.** ([6]) A *finite-state gambler* (*FSG*) is a 5-tuple $G = (Q, \Sigma_m, \delta, \vec{\beta}, q_0)$ such that $Q$ is a non-empty finite set of *states*; $\Sigma_m$ is the input alphabet; $\delta : Q \times \Sigma_m \to Q$ is the *state transition function*; $\vec{\beta} : Q \to \Delta(\Sigma_m)$ is the *betting function*; $q_0 \in Q$ is the *initial state*.

The extended transition function $\delta^* : Q \times \Sigma_m^* \to Q$ is defined such that

$$\delta^*(q, wa) = \begin{cases} q & \text{if } w = a = \lambda, \\ \delta(\delta^*(q, w), a) & \text{if } w \neq \lambda. \end{cases}$$

We use $\delta$ for $\delta^*$ and $\delta(w)$ for $\delta(q_0, w)$ for convenience.

The betting function $\beta_i : Q \to \Delta(\Sigma_m)$ specifies the bets the FSG places on each input symbol in $\Sigma_m$ with respect to a state $q \in Q$.

**Definition.** ([6]). Let $G = (Q, \Sigma_m, \delta, \vec{\beta}, q_0)$ be an FSG. The *s-gale* of $G$ is the function $d_G : \Sigma_m^* \to [0, \infty)$ defined by the recursion

$$d_G(wb) = \begin{cases} 1 & \text{if } w = b = \lambda, \\ m^s d_G(w) \beta_i(\delta(w))(b) & \text{if } b \neq \lambda, \end{cases}$$

for all $w \in \Sigma_m^*$ and $b \in \Sigma_m \cup \{\lambda\}$. For $s \in [0, \infty)$, a function $d : \Sigma_m^* \to [0, \infty)$ is a *finite-state s-gale* if it is the $s$-gale of some finite-state gambler.

Note that in the original definition of a finite-state gambler the range of the betting function $\vec{\beta}$ is $\Delta(\{0, 1\}) \cap \mathbb{Q}^2$ [6,1]. In the following observation, we show that allowing the range of $\vec{\beta}$ to have irrational probability measures does not change the notions of finite-state dimension and strong dimension.

**Observation 3.1.** *Let* $G = (Q, \Sigma_m, \delta, \vec{\beta}, q_0)$ *be an FSG. For each* $\epsilon > 0$, *there exists an FSG* $G' = (Q, \Sigma_m, \delta, \vec{\beta'}, q_0)$ *with* $\vec{\beta'} : Q \to \Delta(\Sigma_m) \cap \mathbb{Q}^m$ *such that for all* $s \in [0, \infty)$, $S^\infty[d_G^{(s)}] \subseteq S^\infty[d_{G'}^{(s+\epsilon)}]$ *and* $S_{\mathrm{str}}^\infty[d_G^{(s)}] \subseteq S_{\mathrm{str}}^\infty[d_{G'}^{(s+\epsilon)}]$.

In this paper, we allow the finite-state gamblers to place irrational bets.

**Definition.** ([6,1]). Let $X \subseteq \mathbf{C}_m$. The *finite-state dimension* of $X$ is

$$\dim_{\mathrm{FS}}(X) = \inf \left\{ s \in [0, \infty) \mid X \subseteq S^\infty[d] \text{ for some finite-state } s\text{-gale } d \right\}$$

and the *finite-state strong dimension* of $X$ is

$$\mathrm{Dim}_{\mathrm{FS}}(X) = \inf \left\{ s \in [0, \infty) \mid X \subseteq S_{\mathrm{str}}^\infty[d] \text{ for some finite-state } s\text{-gale } d \right\}.$$

We will use the following basic properties of the Hausdorff, packing, finite-state, strong finite-state dimensions.

**Theorem 3.2.** *([6,1]). Let $X, Y, X_i \subseteq \Sigma_m^\infty$ for $i \in \mathbb{N}$.*

1. $0 \leq \dim_{\mathrm{H}}(X) \leq \dim_{\mathrm{FS}}(X) \leq 1,\ 0 \leq \dim_{\mathrm{P}}(X) \leq \mathrm{Dim}_{\mathrm{FS}}(X) \leq 1.$
2. $\dim_{\mathrm{H}}(X) \leq \dim_{\mathrm{P}}(X),\ \dim_{\mathrm{FS}}(X) \leq \mathrm{Dim}_{\mathrm{FS}}(X).$
3. *If $X \subseteq Y$, then the dimension of $X$ is at most the dimension of $Y$.*
4. $\dim_{\mathrm{FS}}(X \cup Y) = \max\{\dim_{\mathrm{FS}}(X), \dim_{\mathrm{FS}}(Y)\}$ *and* $\mathrm{Dim}_{\mathrm{FS}}(X \cup Y) = \max\{\mathrm{Dim}_{\mathrm{FS}}(X), \mathrm{Dim}_{\mathrm{FS}}(Y)\}.$
5. $\dim_{\mathrm{H}}\left(\bigcup_{i=0}^\infty X_i\right) = \sup_{i \in \mathbb{N}} \dim_{\mathrm{H}}(X_i),\ \dim_{\mathrm{P}}\left(\bigcup_{i=0}^\infty X_i\right) = \sup_{i \in \mathbb{N}} \dim_{\mathrm{P}}(X_i).$

## 4    Relative Frequencies of Digits

As we have mentioned in Section 1, Besicovitch in 1934 and Eggleston in 1949 proved the following two identities respectively.

**Theorem 4.1.** $\dim_{\mathrm{H}}(\mathrm{FREQ}^{\leq \beta}) = \mathcal{H}_2((\beta, 1 - \beta))$ *[3] and* $\dim_{\mathrm{H}}(\mathrm{FREQ}_\beta) = \mathcal{H}_2((\beta, 1 - \beta))$ *[7], where* $\beta \in [0, \frac{1}{2}]$, $\mathrm{FREQ}^{\leq \beta} = \{S \in \mathbf{C} \mid \limsup_{n \to \infty} \pi_0(S, n) \leq \beta\}$, *and* $\mathrm{FREQ}_\beta = \{S \in \mathbf{C} \mid \lim_{n \to \infty} \pi_0(S, n) = \beta\}$.

In this section, we will calculate the finite-state dimension of some more exotic sets that contain $m$-adic sequences that satisfy certain conditions placed on the frequencies of digits. The proofs in this section use straightforward constructions of finite-state gamblers. Both the constructions and analysis use completely elementary techniques.

Let $\mathcal{H}_{\beta,m}(\alpha) = -(\alpha \log_m \alpha + \beta\alpha \log_m \beta\alpha + (1 - \alpha - \beta\alpha) \log_m \frac{1 - \alpha - \beta\alpha}{m-2})$. Let

$$\alpha^*(x) = \begin{cases} \frac{1}{m} & x < 1 \\ \frac{1}{1 + x + (m-2)x^{\frac{x}{x+1}}} & \text{otherwise.} \end{cases}$$

Note that

$$\mathcal{H}_{\beta,m}(\alpha^*(\beta)) = \sup_{\alpha \in [0, \frac{1}{1+\beta}]} \mathcal{H}_{\beta,m}(\alpha) = \begin{cases} 1 & \text{if } \beta < 1, \\ \log_m(m - 2 + \frac{1+\beta}{\beta^{\frac{\beta}{\beta+1}}}) & \text{otherwise.} \end{cases}$$

**Theorem 4.2.** *Let $\beta' \geq \beta \geq 0$. Let*

$$X = \left\{ S \ \middle|\ \liminf_{n \to \infty} \frac{\pi_1(S, n)}{\pi_0(S, n)} \geq \beta \ \text{and} \ \limsup_{n \to \infty} \frac{\pi_1(S, n)}{\pi_0(S, n)} \geq \beta' \right\}.$$

*Then* $\dim_{\mathrm{H}}(X) = \dim_{\mathrm{FS}}(X) = \mathcal{H}_{\beta',m}(\alpha^*(\beta'))$ *and* $\dim_{\mathrm{P}}(X) = \mathrm{Dim}_{\mathrm{FS}}(X) = \mathcal{H}_{\beta,m}(\alpha^*(\beta)).$

**Corollary 4.3.** *(Theorem 2 [2]). Let $\beta \geq 0$. Let*

$$X = \left\{ S \ \middle|\ \lim_{n \to \infty} \frac{\pi_1(S, n)}{\pi_0(S, n)} = \beta \right\}.$$

*Let $\beta' = \max\{\beta, 1/\beta\}$. Then*

$$\dim_{\mathrm{H}}(X) = \mathcal{H}_{\beta,m}(\alpha^*(\beta')) = \log_m\left( m - 2 + \frac{1 + \beta'}{\beta^{\frac{\beta'}{\beta'+1}}} \right)$$

Note that $\dim_{\mathrm{P}}(X)$, $\dim_{\mathrm{FS}}(X)$, and $\mathrm{Dim}_{\mathrm{FS}}(X)$ all takes the value of $\dim_{\mathrm{H}}(X)$, which were not proven in [2].

*Proof.* We prove the case where $\beta' = \beta$. The other case is similar by switching 0's and 1's in the sequences. Let $Y = \left\{ S \;\middle|\; \liminf_{n \to \infty} \frac{\pi_1(S,n)}{\pi_0(S,n)} \geq \beta \right\}$. Let

$$Z = \left\{ S \;\middle|\; \begin{array}{l} \lim_{n \to \infty} \pi_0(S,n) = \alpha^*(\beta),\; \lim_{n \to \infty} \pi_1(S,n) = \beta\alpha^*(\beta), \\ \text{and } (\forall i > 1) \lim_{n \to \infty} \pi_i(S,n) = \frac{1-\alpha^*(\beta)-\beta\alpha^*(\beta)}{m-2} \end{array} \right\}.$$

By Eggleston's theorem, $\dim_{\mathrm{H}}(Z) = \mathcal{H}_{\beta,m}(\alpha^*(\beta))$. Since $Z \subseteq X \subseteq Y$, it follows immediately from Theorem 4.2 that $\dim_{\mathrm{H}}(X) = \mathcal{H}_{\beta,m}(\alpha^*(\beta))$. $\qquad\square$

## 5   Saturated Sets and Maximum Entropy Principle

In Section 4, we calculated the finite-state dimensions of many sets defined using properties on asymptotic frequencies of digits. They are all saturated sets. Now we formally define saturated sets and investigate their collective properties.

Let $\Pi_n(S) = \{\vec{\pi}(S,k) \mid k \geq n\}$ for all $n \in \mathbb{N}$. Let $\bar{\Pi}_n(S) = \overline{\Pi_n(S)}$, i.e., $\bar{\Pi}_n(S)$ is the closure of $\Pi_n(S)$. Define $\Pi : \mathbf{C}_m \to \mathcal{P}(\Delta(\Sigma_m))$ such that for all $S \in \mathbf{C}_m$, $\Pi(S) = \bigcap_{n \in \mathbb{N}} \bar{\Pi}_n(S)$.

**Definition.** Let $X \subseteq \mathbf{C}_m$. We say that $X$ *is saturated* if for all $S, S' \in \mathbf{C}_m$,

$$\Pi(S) = \Pi(S') \Rightarrow [S \in X \iff S' \in X].$$

When we determine an upper bound on the finite-state dimensions of a set $X \subseteq \mathbf{C}_m$, it is in general not possible to use a single probability measure as the betting strategy even when $X$ is saturated. However, when certain conditions are true, a simple 1-state finite-state gambler may win on a huge set of sequences with different empirical digit distribution probability measures.

In the following, we formalize such a condition and reveal some relationship between betting and the Kullback-Leibler distance (relative entropy) [5]. Note that $m$-dimensional Kullback-Leibler distance $\mathcal{D}_m(\vec{\beta} \parallel \vec{\alpha})$ is defined as

$$\mathcal{D}_m(\vec{\beta} \parallel \vec{\alpha}) = \mathrm{E}_{\vec{\beta}} \log_m \frac{\vec{\beta}}{\vec{\alpha}}.$$

**Definition.** Let $\vec{\alpha}, \vec{\beta} \in \Delta(\Sigma_m)$. We say that $\vec{\alpha}$ $\epsilon$-*dominates* $\vec{\beta}$, denoted as $\vec{\alpha} >>^\epsilon \vec{\beta}$, if $\mathcal{H}_m(\vec{\alpha}) \geq \mathcal{H}_m(\vec{\beta}) + \mathcal{D}_m(\vec{\beta} \parallel \vec{\alpha}) - \epsilon$. We say that $\vec{\alpha}$ *dominates* $\vec{\beta}$, denoted as $\vec{\alpha} >> \vec{\beta}$, if $\vec{\alpha} >>^0 \vec{\beta}$.

Note that $\mathcal{H}_m(\vec{\beta}) + \mathcal{D}_m(\vec{\beta} \parallel \vec{\alpha}) = \mathrm{E}_{\vec{\beta}} \log_m \frac{1}{\vec{\beta}} + \mathrm{E}_{\vec{\beta}} \log_m \frac{\vec{\beta}}{\vec{\alpha}} = \mathrm{E}_{\vec{\beta}} \log_m \frac{1}{\vec{\alpha}}$, where $\mathrm{E}_{\vec{\beta}} \log_m \frac{\vec{\beta}}{\vec{\alpha}} = \sum_{i=0}^{m-1} \beta_i \log_m \frac{\beta_i}{\alpha_i}$. It is very easy to see that the uniform probability measure dominates all probability measures.

**Observation 5.1.** *Let* $\vec{\alpha} = (\frac{1}{m}, \ldots, \frac{1}{m})$. *Let* $\vec{\beta} \in \Delta(\Sigma_m)$. *Then* $\vec{\alpha} >> \vec{\beta}$.

Here, we give a few interesting properties of the domination relation.

**Theorem 5.2.** *Let $\vec{\alpha} = (\alpha_0, \ldots, \alpha_{m-1}) \in \Delta(\Sigma_m)$. Let $\vec{\beta} = (\beta_0, \ldots, \beta_{m-1}) \in \Delta(\Sigma_m)$ be such that $\beta_j = 1$, where $j = \arg\max\{\alpha_0, \ldots, \alpha_{m-1}\}$. Then $\vec{\alpha} >> \vec{\beta}$ and $\mathcal{H}_m(\vec{\beta}) = 0$.*

**Theorem 5.3.** *Let $\vec{\alpha}, \vec{\beta} \in \Delta(\Sigma_m)$, $\epsilon \geq 0$, and $r \in [0,1]$. If $\vec{\alpha} >>^\epsilon \vec{\beta}$, then $\vec{\alpha} >>^\epsilon r\vec{\alpha} + (1-r)\vec{\beta}$.*

**Theorem 5.4.** *Let $\vec{\mu} = (\frac{1}{m}, \ldots, \frac{1}{m}) \in \Delta(\Sigma_m)$ be the uniform probability measure. Let $\vec{\beta} \in \Delta(\Sigma_m)$. Let $s \in [0,1]$. Let $\vec{\alpha} = s\vec{\mu} + (1-s)\vec{\beta}$. Then $\vec{\alpha} >> \vec{\beta}$.*

The following theorem relates the domination relation to finite-state dimensions.

**Theorem 5.5.** *Let $\vec{\alpha} \in \Delta(\Sigma_m)$ and $X \subseteq \Sigma_m^\infty$.*

1. *If $\vec{\alpha} >>^\epsilon \vec{\pi}(S,n)$ for infinitely many $n$ for every $\epsilon > 0$ and every $S \in X$, then $\dim_{\mathrm{FS}}(X) \leq \mathcal{H}_m(\vec{\alpha})$.*
2. *If $\vec{\alpha} >>^\epsilon \vec{\pi}(S,n)$ for all but finitely many $n$ for every $\epsilon > 0$ and every $S \in X$, then $\mathrm{Dim}_{\mathrm{FS}}(X) \leq \mathcal{H}_m(\vec{\alpha})$.*

Theorem 5.5 tells us that if we can find a single dominating probability measure for $X \subseteq \mathbf{C}_m$, then a simple 1-state FSG may be used to assess the dimension of $X$. However, in the following, we will see that the domination relationship is not even transitive.

**Theorem 5.6.** *Domination relation defined above is not transitive.*

Fix $\vec{\alpha} \in \Delta(\Sigma_m)$ with $\mathcal{H}_m(\vec{\alpha}) \neq 1$, the hyperplane $H$ in $\mathbb{R}^m$ defined by $\mathcal{H}_m(\vec{\alpha}) = \sum_{i=0}^{m-1} x_i \log_m \frac{1}{\alpha_i}$ divides the simplex $\Delta(\Sigma_m)$ into two halves $A$ and $B$ with $A \cap B \subseteq H$. Suppose $(\frac{1}{m}, \ldots, \frac{1}{m}) \in B$, then $A = \{\vec{\beta} \in \Delta(\Sigma_m) \mid \vec{\alpha} >> \vec{\beta}\}$.

So it is not always possible to find a single probability measure that dominates all the empirical probability measures of sequences in $X \subseteq \mathbf{C}_m$. Nevertheless, we take advantage of the compactness of $\Delta(\Sigma_m)$ and give a general solution for finding the dimensions of $X \subseteq \mathbf{C}_m$, when $X$ is saturated. The following theorem is our pointwise maximum entropy principle for saturated sets. It says that the dimension of a saturated set is the maximum pointwise asymptotic entropy of the empirical digit distribution measure.

**Theorem 5.7.** *Let $X \subseteq \mathbf{C}_m$ be saturated. Let $H = \sup_{S \in X} \liminf_{n \to \infty} \mathcal{H}_m(\vec{\pi}(S,n))$ and $P = \sup_{S \in X} \limsup_{n \to \infty} \mathcal{H}_m(\vec{\pi}(S,n))$. Then $\dim_{\mathrm{FS}}(X) = \dim_{\mathrm{H}}(X) = H$ and $\mathrm{Dim}_{\mathrm{FS}}(X) = \dim_{\mathrm{P}}(X) = P$.*

This theorem automatically gives a solution for finding an upper bounds for dimensions of arbitrary $X$.

**Corollary 5.8.** *Let $X \subseteq \mathbf{C}_m$ and let $H$ and $P$ be defined as in Theorem 5.7. Then $\dim_{\mathrm{FS}}(X) \leq H$ and $\mathrm{Dim}_{\mathrm{FS}}(X) \leq P$.*

Due to the space limit of the proceeding, we omit the calculation of the dimensions of some interesting saturated sets using Theorem 5.7 here. Interested readers are encouraged to read the full version of this paper.

# 6   Conclusion

A general saturated set usually has an uncountable decomposition in which, the dimension of each element is easy to determine, while the dimension of the whole set, which is the uncountable union of all the element sets, is very difficult to determine and requires advanced techniques in multifractal analysis and ergodic theory. By using finite-state gamblers and gale characterizations of dimensions, we are able to obtain very general results calculating the classical dimensions and finite-state dimensions of saturated sets using completely elementary analysis. This indicates that gale characterizations will play a more important role in dimension-theoretic analysis and that finite-state gamblers are very powerful.

**Acknowledgments.** We thank anonymous referees for helpful comments.

# References


1. Athreya, K.B., Hitchcock, J.M., Lutz, J.H., Mayordomo, E.: Effective strong dimension, algorithmic information, and computational complexity. SIAM Journal on Computing 37, 671–705 (2007)
2. Barreira, L., Saussol, B., Schmeling, J.: Distribution of frequencies of digits via multifractal analyais. Journal of Number Theory 97(2), 410–438 (2002)
3. Besicovitch, A.S.: On the sum of digits of real numbers represented in the dyadic system. Mathematische Annalen 110, 321–330 (1934)
4. Cajar, H.: Billingsley dimension in probability spaces. Lecture notes in mathematics, vol. 892 (1981)
5. Cover, T.M., Thomas, J.A.: Elements of Information Theory. John Wiley & Sons, Inc., New York (1991)
6. Dai, J.J., Lathrop, J.I., Lutz, J.H., Mayordomo, E.: Finite-state dimension. Theoretical Computer Science 310, 1–33 (2004)
7. Eggleston, H.: The fractional dimension of a set defined by decimal properties. Quarterly Journal of Mathematics 20, 31–36 (1949)
8. Falconer, K.: The Geometry of Fractal Sets. Cambridge University Press, Cambridge (1985)
9. Falconer, K.: Fractal Geometry: Mathematical Foundations and Applications, 2nd edn. Wiley, Chichester (2003)
10. Good, I.J.: The fractional dimensional theory of continued fractions. In: Proceedings of the Cambridge Philosophical Society, vol. 37, pp. 199–228 (1941)
11. Gu, X.: A note on dimensions of polynomial size circuits. Theoretical Computer Science 359(1-3), 176–187 (2006)
12. Gu, X., Lutz, J.H., Moser, P.: Dimensions of Copeland-Erdős sequences. Information and Computation 205(9), 1317–1333 (2007)
13. Hausdorff, F.: Dimension und äusseres Mass. Mathematische Annalen 79, 157–179 (1919)
14. Hitchcock, J.M.: Fractal dimension and logarithmic loss unpredictability. Theoretical Computer Science 304(1–3), 431–441 (2003)
15. Huffman, D.A.: A method for the construction of minimum redundancy codes. In: Proc. IRE, vol. 40, pp. 1098–1101 (1952)
16. Kelly, J.: A new interpretation of information rate. Bell Systems Technical Journal 35, 917–926 (1956)

17. Lutz, J.H.: Gales and the constructive dimension of individual sequences. In: Proceedings of the 27th International Colloquium on Automata, Languages, and Programming, pp. 902–913 (2000); Revised as [19]
18. Lutz, J.H.: Dimension in complexity classes. SIAM Journal on Computing 32, 1236–1259 (2003); Preliminary version appeared In: Proceedings of the Fifteenth Annual IEEE Conference on Computational Complexity, pp. 158–169 (2000)
19. Lutz, J.H.: The dimensions of individual strings and sequences. Information and Computation 187, 49–79 (2003); Preliminary version appeared as [17]
20. McMullen, C.T.: Hausdorff dimension of general Sierpinski carpets. Nagoya Mathematical Journal 96, 1–9 (1984)
21. Olsen, L.: Multifractal analysis of divergence points of deformed measure theoretical Birkhoff averages. Journal de Mathématiques Pures et Appliquées. Neuvième Série 82(12), 1591–1649 (2003)
22. Olsen, L.: Applications of multifractal divergence points to some sets of $d$-tuples of numbers defined by their $n$-adic expansion. Bulletin des Sciences Mathématiques 128(4), 265–289 (2004)
23. Olsen, L.: Applications of divergence points to local dimension functions of subsets of $\mathbb{R}^d$. Proceedings of the Edinburgh Mathematical Society 48, 213–218 (2005)
24. Olsen, L.: Multifractal analysis of divergence points of the deformed measure theoretical Birkhoff averages. III. Aequationes Mathematicae 71(1-2), 29–53 (2006)
25. Olsen, L., Winter, S.: Multifractal analysis of divergence points of the deformed measure theoretical Birkhoff averages II (preprint, 2001)
26. Olsen, L., Winter, S.: Normal and non-normal points of self-similar sets and divergence points of self-similar measures. Journal of the London Mathematical Society (Second Series) 67(1), 103–122 (2003)
27. Ryabko, B., Suzuki, J., Topsoe, F.: Hausdorff dimension as a new dimension in source coding and predicting. In: 1999 IEEE Information Theory Workshop, pp. 66–68 (1999)
28. Shannon, C.E.: A mathematical theory of communication. Bell System Technical Journal 27, 379–423, 623–656 (1948)
29. Sullivan, D.: Entropy, Hausdorff measures old and new, and limit sets of geometrically finite Kleinian groups. Acta Mathematica 153, 259–277 (1984)
30. Tricot, C.: Two definitions of fractional dimension. Mathematical Proceedings of the Cambridge Philosophical Society 91, 57–74 (1982)
31. Volkmann, B.: Über Hausdorffsche Dimensionen von Mengen, die durch Zifferneigenschaften charakterisiert sind. VI. Mathematische Zeitschrift 68, 439–449 (1958)

# Reachability in Linear Dynamical Systems

Emmanuel Hainry

LORIA, Université Henri Poincaré
Campus scientifique, BP 239 - 54506 Vandœuvre-lès-Nancy, France
Emmanuel.Hainry@loria.fr

**Abstract.** Dynamical systems allow to modelize various phenomena or processes by only describing their local behaviour. It is however useful to understand the behaviour in a more global way. Checking the reachability of a point for example is a fundamental problem. In this document we will show that this problem that is undecidable in the general case is in fact decidable for a natural class of continuous-time dynamical systems: linear systems. For this, we will use results from the algebraic numbers theory such as Gelfond-Schneider's theorem.

**Keywords:** Dynamical Systems, Reachability, Skolem-Pisot problem, Gelfond-Schneider Theorem.

## 1 Introduction

A dynamical system is described by a function (the dynamics of the system) and a space on which this function is defined and in which the system will evolve. The evolution of a dynamical system is hence described in a very simple way but it can be hard to grasp where a point that undergoes the dynamics will go. Hence the problem of deciding whether given a certain point, the system will eventually reach another given point is fundamental.

Indeed, many natural phenomena can be described using dynamical systems. Examples come from mathematics [1], physics, biology [2]; the famous Lorenz' attractor [3] is an example of a dynamical system describing a meteorological phenomenon. However, as standard as those systems are, and as simple as the description of their dynamics may be, many important problems such as limit and reachability are undecidable.

Some positive results are known for some very specific classes but on the whole, it is very difficult to know much about such systems. Even considering polynomial systems yields many undecidable problems: [4] shows that it is possible to simulate a Turing machine using a polynomial dynamical system. It is hence undecidable whether or not a trajectory will reach the region corresponding to the halting state of the machine. This particular problem can be seen as a continuous version of the Skolem-Pisot problem [5,6,7] which studies whether a component of a discrete linear system will reach 0. This problem is not different from deciding if this system reaches a hyperplane of the space, described by $y_k = 0$ where $k$ is the number of the component considered. The Skolem-Pisot problem is equivalent to deciding whether a linear recurrent sequence reaches 0.

A. Beckmann, C. Dimitracopoulos, and B. Löwe (Eds.): CiE 2008, LNCS 5028, pp. 241–250, 2008.

It is still open whether the Skolem-Pisot problem is decidable. Some results are known but they don't yet enlighten the whole decision problem. As an example of recent developments, [7] shows that in small dimensions, the problem is decidable, and [8] shows that this problem is NP-hard. As this problem also arises in a continuous context it would be interesting to study the continuous Skolem-Pisot problem for continuous-time linear dynamical systems. Considering a continuous space may make the study of this problem easier than in a discrete space, indeed if two points on the two different sides of the aimed hyperplane are reached, continuity (and the intermediate values theorem) implies that the hyperplane will also be reached. Even if the discrete version of this problem had many possible interpretations, no natural interpretation appears in the continuous case.

The (point to point) reachability problem, which is undecidable in the general case, has been shown undecidable for various restricted classes of dynamical systems, such as Piecewise Constant Derivative systems [9] where the dynamics are really simple as it consists of a sharing of the space into regions where the derivative will be constant. Other results on the subject of reachability and undecidability of problems in hybrid systems are studied in [10,11,12,13].

It has been shown [14] that in discrete-time linear dynamical systems, the reachability problem is decidable. The class of linear dynamical systems in the continuous field is hence a good candidate for a class of dynamical systems where reachability might be decidable. It is however not trivial to extend the result on discrete dynamical systems to continuous dynamical systems, indeed, it uses algebraic properties of the orbit that are not preserved in a continuous setting. In this paper, we will hence focus on linear continuous-time dynamical systems and show that reachability is decidable for those systems. This result is a necessary step if we want to study the continuous Skolem-Pisot problem that also deals with linear dynamical systems.

The section 2 presents the problems we are going and mathematical notions that will be useful in the following. The section 3 contains results of undecidability: for polynomial dynamical systems, the Skolem-Pisot problem and the reachability problem are undecidable.The next section is the core of this paper: it contains the theorem 4 which is the core of this paper and proves the decidability of reachability in linear dynamical systems. The proof of this result details in fact the algorithm used to decide the question. It is composed of two parts: the part 4.2 shows how to solve the problem in the specific case where the matrix is in Jordan form; the part 4.1 recalls that putting the matrix into Jordan form is doable.

## 2    Prerequisites

### 2.1    Linear Continuous-Time Dynamical Systems

The dynamics of a linear dynamical system are described by a linear differential equation. To describe such a system, we take a matrix of real numbers which will represent the dynamics and a vector of reals that is the initial point. We use here classical definitions and notations that can be found in [15].

**Definition 1 (Linear continuous-time dynamical system).** *Given a matrix $A \in \mathbb{R}^{n \times n}$ and a vector $X_0 \in \mathbb{R}^n$. We define $X$ as the solution of the following Cauchy problem:*

$$\begin{cases} X' & = AX \\ X(0) & = X_0. \end{cases}$$

*$X$ is called a trajectory of the system.*

**Definition 2 (Reachability).** *Given $A \in \mathbb{R}^{n \times n}$, $X_0 \in \mathbb{R}^n$, $Y \in \mathbb{R}^n$, the system is said to reach $Y$ from $X_0$ if there exists $t \in \mathbb{R}$ such that $X(t) = Y$ with $X$ the trajectory defined with the dynamics $A$ and the initial point $X_0$.*

**Definition 3 ($\omega$-limit points).** *Given a trajectory $X$, a point $Y$ is an $\omega$-limit point of $X$ if there is an diverging increasing sequence $(t_n) \in \mathbb{R}^{\mathbb{N}}$ such that $Y = \lim_{n \to +\infty} X(t_n)$.*

**Definition 4 ($\omega$-limit sets).** *The $\omega$-limit set of a dynamical system is the set of its $\omega$-limit points: $\omega(X) = \cap_n \overline{\cup_{t>n} X(t)}$, where $\overline{A}$ is the closure of the set $A$.*

The problems we are interested in are the reacability problem (which we will prove decidable in Linear Dynamical Systems) and the Skolem-Pisot problem.

*Problem 1 (Reachability problem).* Given a trajectory $X$ defined from $A \in \mathbb{K}^{n \times n}$ and $X_0 \in \mathbb{K}^n$, a point $Y \in \mathbb{K}^n$, decide whether $Y$ can be reached from $X_0$.

The classical Skolem-Pisot problem originally consists in determining if a linear recurrent sequence has a zero. It can however be defined as a hyperplane reachability problem.

*Problem 2 (Skolem-Pisot problem).* Given a trajectory $X$, given $C \in \mathbb{K}^n$ defining an hyperplane[1] of $\mathbb{K}^n$, decide if $\exists t \in \mathbb{R}$ such that $C^T X(t) = 0$? In other words, does the trajectory $X$ intersect the hyperplane defined by $C$?

The problems we will consider will be those for which the field $\mathbb{K}$ is in fact the set of rational numbers $\mathbb{Q}$.

## 2.2   Polynomials

Let us now recall a few notations, mathematical tools and algorithms on polynomials. In the following, we use a field $\mathbb{K}$ that is a subfield of $\mathbb{C}$.

**Definition 5 (Ring of polynomials).** *We denote $\mathbb{K}[X]$ the ring of one variable polynomials with coefficients in $\mathbb{K}$. A polynomial can be written as $P(X) = \sum_{i=1}^n a_i X^i$, with $a_i \in \mathbb{K}$ and $a_n \neq 0$. The integer $n$ is the degree of $P$.*

---

[1] The hyperplane defined by $C$ is the set of points $Y$ such that $C^T Y = [0]$.

**Definition 6 (Roots of a polynomial).** *The set $Z(P)$ of roots of a polynomial $P$ is defined as $Z(P) = \{x \in \mathbb{C}; P(x) = 0\}$*

**Definition 7 (Algebraic numbers).** *The set of roots of polynomials with coefficients in $\mathbb{Q}$ is the set of algebraic numbers.*

An algebraic number can be represented uniquely by the minimal polynomial it nulls (minimal in $\mathbb{Q}[X]$ for the division) and a ball containing only one root of the polynomial. Note that the size of the ball can be chosen using only the values of the coefficients of the polynomial as [16] shows a bound on the distance between roots of a polynomial from its coefficient.

**Definition 8 (Representation of an algebraic number).** *An algebraic number $\alpha$ will be represented by $(P, (a, b), \rho)$ where $P$ is the minimal polynomial of $\alpha$, $a + ib$ is an approximation of $\alpha$ such that $|\alpha - (a + ib)| < \rho$ and $\alpha$ is the only root of $P$ in the open ball $\mathcal{B}(a + ib, \rho)$.*

It can be shown that given the representations of two algebraic numbers $\alpha$ and $\beta$, the representations of $\alpha + \beta$, $\alpha - \beta$, $\alpha\beta$ and $\alpha/\beta$ can be computed. See [17,18] for details.

We will also need specific results on algebraic numbers that come from [19,20].

**Proposition 1 (Baker).** *Given $\alpha \in \mathbb{C} - \{0\}$, $\alpha$ and $e^{\alpha}$ are not both algebraic numbers.*

**Theorem 1 (Gelfond-Schneider).** *Let $\alpha$ and $\beta$ be two algebraic numbers. If $\alpha \notin \{0, 1\}$ and $\beta \notin \mathbb{Q}$, then $\alpha^{\beta}$ is not algebraic*

### 2.3 Matrices

**Definition 9 (Characteristic polynomial).** *Given a matrix $A \in \mathbb{K}^{n \times n}$, its characteristic polynomial is $\chi_A(X) = \det(X I_n - A)$*

**Definition 10 (Exponential of a matrix).** *Given a matrix $A$, its exponential denoted $\exp(A)$ is the matrix*

$$\sum_{i=1}^{+\infty} \frac{1}{i!} A^i.$$

Note that the exponential is well defined for all real matrices.

All matrices can be put in Jordan form, which allows to compute easily the exponential. To find more about Jordan matrices and blocks, the reader may consult [15] or [21].

**Definition 11 (Jordan block).** *A Jordan block is a square matrix of one of the two following forms*

$$\begin{bmatrix} \lambda & & & \\ 1 & \lambda & & \\ & \ddots & \ddots & \\ & & 1 & \lambda \end{bmatrix} \quad ; \quad \begin{bmatrix} B & & & \\ I_2 & B & & \\ & \ddots & \ddots & \\ & & I_2 & B \end{bmatrix} \quad \text{with } B = \begin{bmatrix} a & -b \\ b & a \end{bmatrix} \text{ and } I_2 = \begin{bmatrix} 1 & 0 \\ 0 & 1 \end{bmatrix}$$

**Definition 12 (Jordan form).** *A matrix that contains Jordan blocks on its diagonal is said to be in Jordan form.*

$$\begin{bmatrix} D_1 & 0 & \cdots & 0 \\ 0 & D_2 & \ddots & \vdots \\ \vdots & \ddots & \ddots & 0 \\ 0 & \cdots & 0 & D_n \end{bmatrix}$$

**Proposition 2 ([21]).** *Any matrix $A \in \mathbb{R}^{n \times n}$ is similar to a matrix in Jordan form. In other words,*

$$\exists P \in GL(\mathbb{R}^{n \times n}) \text{ and } J \text{ in Jordan form such that } A = P^{-1}JP.$$

## 3  Undecidability for Polynomial Dynamical Systems

Many biological phenomena can be modelised using polynomial dynamical systems rather than linear dynamical systems. A famous example comes from meteorological systems which were described by Lorenz in [3]. Lorenz' attractor has a quite chaotic behaviour which gives the intuition that the reachability problem in polynomial dynamical systems is not decidable. Other polynomial differential systems yields fractal basins of attraction. In other words, this dynamical systems has exactly two $\omega$-limit points depending on the initial point and, the set of starting points that will lead to the first of those attractors is a fractal, for example a Julia set.

In those systems, from already known results, we can infer that the Skolem-Pisot problem and the reachability problem are undecidable.

**Theorem 2.** *The Skolem-Pisot problem is undecidable for polynomial dynamical systems.*

*Proof.* From [4], we know that it is possible to simulate a Turing machine using a polynomial differential system. The halt of the Turing machine is then equivalent to the system reaching the hyperplane $z = q_f$ which stands for the halting state. This is an instance of the Skolem-Pisot problem.

**Theorem 3.** *Reachability is undecidable for polynomial dynamical systems.*

*Proof.* Let us modify the Turing machine of the previous proof so that from the halting state, the machine erases its tape then enters a special state. Simulating this machine by the same mechanism from [4], the dynamical system reaches the point representing blank tapes and special state if and only if the original machine halts. This means we can translate any instance of the halting problem into a reachability in polynomial differential systems problem.

## 4    Decidability for Linear Dynamical Systems

This section is devoted to proving the main theorem of this article: theorem 4.

**Theorem 4.** *The reachability problem for continuous time linear dynamical systems with rational coefficients is decidable.*

To decide whether a point is reachable we will try to obtain an expression of the trajectory $X$ that is usable and with this expression search for the different $t$ that could be solution. We will first consider the case where the matrix is in Jordan form: this case will be studied in section 4.2. The section 4.1 will show how to put the matrix in Jordan form. Note that the Jordan matrix will have algebraic coefficients and not only rational ones.

### 4.1    To Put the Matrix in Jordan Form

To be able to do what we have done in the previous section, we will want to find a Jordan matrix similar to the one considered. Building the Jordan form of a matrix implies knowing its eigenvalues, for that we need to compute the roots of the characteristic polynomial of the matrix.

This consist in the following steps that are classical: computing the characteristic polynomial; factorizing the polynomial in $\mathbb{Q}[X]$; computing the roots; jordanizing the matrix.

### 4.2    If the Matrix Is in Jordan Form

Let us suppose that the matrix $A$ is in Jordan form with algebraic coefficients and that the $X_0$ and $Y$ vectors are also composed of algebraic elements. This

means $A = \begin{bmatrix} D_1 & 0 & \cdots & 0 \\ 0 & D_2 & \ddots & \vdots \\ \vdots & \ddots & \ddots & 0 \\ 0 & \cdots & 0 & D_k \end{bmatrix}$ with the $D_i$ being Jordan blocks.

The solution of the Cauchy system $\begin{cases} X' = AX \\ X(0) = X_0 \end{cases}$ is $X(t) = \exp(tA)X_0$.

We then need to compute the exponential of $tA$. It is easy to check that

$$\exp(tA) = \begin{bmatrix} \exp(tD_1) & & & \\ & \exp(tD_2) & & \\ & & \ddots & \\ & & & \exp(tD_k) \end{bmatrix}$$

Finding a $t \in \mathbb{R}$ such that $X(t) = Y$ is equivalent to finding such a $t$ for each component $i$ and ensuring this is always the same $t$. We are going to solve the equation Jordan block by Jordan block. It means we choose an $i$ such that the corresponding part of $X_0$ is not null (in the other case it is easy to decide if either all $t \in \mathbb{R}$ will be solutions or no $t$ will be solution) and search for a $t$ such that $\exp(tD_i) \begin{bmatrix} x_1 \\ \vdots \\ x_{n_i} \end{bmatrix} = \begin{bmatrix} y_1 \\ \vdots \\ y_{n_i} \end{bmatrix}$ where the $x_j$ and $y_j$ are the elements of $X_0$ and $Y$ corresponding to the block $i$. To simplify the notations, we will forget $i$ and just consider the problem as being $\exp(tD)X_0 = Y$ and $k$ being the size of this block.

There are two cases to consider: the two different forms of Jordan blocks. For each of those cases, a few sub cases are to be considered which revolve around the nullity of the real part of the eigenvalue. Let us note that as we deal with algebraic numbers, it is possible to verify if the real part or the imaginary part is null.

**First form: A real eigenvalue.** The first form of Jordan blocks corresponds to a real eigenvalue $\lambda$. Two cases need to be dealt with: $\lambda = 0$ and $\lambda \neq 0$

*If $\lambda \neq 0$.* The exponential is $\exp(tD) = e^{t\lambda} \begin{bmatrix} 1 & & & & \\ t & 1 & & & \\ \frac{t^2}{2} & t & 1 & & \\ \vdots & \ddots & \ddots & \ddots & \\ \frac{t^k}{k!} & \cdots & \frac{t^2}{2} & t & 1 \end{bmatrix}$. If $X_{0_1}$ is not 0,

then there is at most one $t \in \mathbb{R}$ solution. Indeed, let us consider $x_i$, the first non null element of $\{x_1, x_k\}$. The only possible $t$ is then $\frac{1}{\lambda} \ln\left(\frac{y_i}{x_i}\right)$.

We want to verify that this $t$ is coherent with the rest of the block. Let us remark that $e^{t\lambda} = \frac{y_i}{x_i}$ is an algebraic number. If the block has size more than 1, then $t$ verifies some algebraic equations hence the proposition 1 says $\lambda t = 0$, it is easy to verify if $t = 0$ is the solution of the block.

*If $\lambda = 0$.* The case with $\lambda = 0$ means we are searching for a $t$ such that

$$\begin{bmatrix} 1 & & & & \\ t & 1 & & & \\ t^2/2 & t & 1 & & \\ \vdots & \ddots & \ddots & \ddots & \\ \frac{t^k}{k!} & \cdots & t^2/2 & t & 1 \end{bmatrix} \begin{bmatrix} x_1 \\ \vdots \\ x_k \end{bmatrix} = \begin{bmatrix} y_1 \\ \vdots \\ y_k \end{bmatrix}$$

For such a $t$ to exist, we need to have $x_1 = y_1$, $x_2 + tx_1 = y_2$, ... Let us say that $x_i$ is the first non-null element of $X$. Then the only candidate for $t$ is $\frac{y_{i+1} - x_{i+1}}{x_i}$. Since this candidate is algebraic, it is easy to check whether this $t$ is a solution for the block.

**Second form.** The second form corresponds to complex eigenvalues. The Jordan block is $D = \begin{bmatrix} B & & & \\ I_2 & B & & \\ & \ddots & \ddots & \\ & & I_2 & B \end{bmatrix}$ with $B = \begin{bmatrix} a & -b \\ b & a \end{bmatrix}$ and $I_2 = \begin{bmatrix} 1 & 0 \\ 0 & 1 \end{bmatrix}$. The expo-

nential is $\exp(D) = \mathrm{e}^{ta} \begin{bmatrix} B_2 & & & & \\ tB_2 & B_2 & & & \\ \frac{t^2}{2} B_2^2 & tB_2 & B_2 & & \\ \vdots & \ddots & \ddots & \ddots & \\ \frac{t^k}{k!} B_2^k & \cdots & \frac{t^2}{2} B_2^2 & tB_2 & B_2 \end{bmatrix}$ with $B_2 = \begin{bmatrix} \cos(tb) & -\sin(tb) \\ \sin(tb) & \cos(tb) \end{bmatrix}$.

There are two cases to consider, whether $a$ is null or not.

*If $a = 0$.* In the case where the eigenvalue has a null real part, the $\exp(ta)$ term disappears. Let us suppose $c$ is the smallest odd number such that $x_j \neq 0$ or $x_{j+1} \neq 0$. We first want to solve $\begin{bmatrix} y_j \\ y_{j+1} \end{bmatrix} = B_2 \begin{bmatrix} x_j \\ x_{j+1} \end{bmatrix}$. Let us remark that, since $B_2$ is a rotation, if $\sqrt{x_j^2 + x_{j+1}^2} \neq \sqrt{y_j^2 + y_{j+1}^2}$, there is no solution and in the other case, there is an infinity of solutions. We can express the solution of this system $t \in \alpha + \frac{2\pi}{b}\mathbb{Z}$ where $\alpha$ is not explicitly algebraic as its expression uses $\tan^{-1}$. Let us remark that for all those candidate $t$, the matrix $B_2$ is the same, namely $B_2 = \begin{bmatrix} \cos(\alpha) & -\sin(\alpha) \\ \sin(\alpha) & \cos(\alpha) \end{bmatrix}$. Those $\cos(\alpha)$ and $\sin(\alpha)$ are algebraic numbers that can be computed: we can write an expression in $x_j$, $x_{j+1}$, $y_j$ and $y_{j+1}$ for each combination of signs for those numbers.[2]

We then have to verify whether the following components of $X$ and $Y$ are compatible with those $t$. We have $\begin{bmatrix} y_{j+2} \\ y_{j+3} \end{bmatrix} = t \begin{bmatrix} y_j \\ y_{j+1} \end{bmatrix} + \begin{bmatrix} \cos(\alpha) & -\sin(\alpha) \\ \sin(\alpha) & \cos(\alpha) \end{bmatrix} \begin{bmatrix} x_{j+2} \\ x_{j+3} \end{bmatrix}$. Since $y_j$ or $y_{j+1}$ is non null (as $\sqrt{y_j^2 + y_{j+1}^2} = \sqrt{x_j^2 + x_{j+1}^2} \neq 0$), there is then at most one solution and we can express it as an algebraic number.

**Conclusion for $a = 0$.** We are able to discriminate 3 possible cases: either there is no solution, either there is exactly one candidate $t$ (defined with a fraction and a few subtractions of elements of $X$ and $Y$) either there is an infinity of candidate $t$ (defined as $\pm\alpha + \frac{2\pi}{b}\mathbb{Z}$ with the $\alpha$ being fractions of elements of $X$ and $Y$). This last case will need to be compared with the results for the other Jordan blocks to decide whether there will be solutions or not for the whole system.

*If $a \neq 0$.* In the case where $a \neq 0$, the term $\exp(ta)$ makes the solution not simply turn around the origin but describe a spiral. If $a > 0$, this spiral is diverging, if $a < 0$ it is converging to the origin. We just have to study the norm of $Y$.

---

[2] For example, if $x_j > 0$, $x_{j+1} > 0$, $y_j > 0$ and $y_{j+1} > 0$, we have $\sin(\alpha) = \sqrt{\frac{y_j^2}{y_j^2 + y_{j+1}^2} \frac{x_j^2 + x_{j+1}^2}{(x_j + x_{j+1})^2}}$ and $\cos(\alpha)$ satisfies a similar expression.

We want to solve the system $e^{ta} \begin{bmatrix} \cos(tb) & -\sin(tb) \\ \sin(tb) & \cos(tb) \end{bmatrix} \begin{bmatrix} x_1 \\ x_2 \end{bmatrix} = \begin{bmatrix} y_1 \\ y_2 \end{bmatrix}$ with $x_1$ or $x_2$ not null (if they are, we will choose another $x_j$). Let us consider the norms of the two sides of this equation: $e^{ta} \sqrt{x_1^2 + x_2^2} = \sqrt{y_1^2 + y_2^2}$. As we have chosen $x_1$ or $x_2$ to be non null, we can write $e^{ta} = \sqrt{\frac{y_1^2 + y_2^2}{x_1^2 + x_2^2}}$. We hence have exactly one $t$ candidate to be the solution. This $t$ is the logarithm of an algebraic number and we can check whether $tb$ is the correct angle (this is the combination of a non algebraic solution with an infinity of solutions).

**Putting together the solutions.** As we have seen, for one block, we may have no solution, one solution or an infinity of solutions. We must then bring the blocks together. In the case where one block has no solution, the problem is solved. In the case where there is exactly one solution, it can be algebraic (if $\lambda = 0$, or $\lambda > 0$ and there is more than one component to check), in which case it is easy to compute formally $\exp(tA)X_0$ and compare it with $Y$.

If we only have non explicitly algebraic solutions, we know that the solution must verify $\forall i, \exp(a_i t) = z_i$ with $a_i$ and $z_i$ algebraic numbers. We must then have $e^{\frac{a_1}{a_2} \ln(z_1)} = z_2$. From theorem 1, it implies that $a_1/a_2 \in \mathbb{Q}$ or $z_1 \in \{0, 1\}$. $z_1 = 0$ is not compatible, $z_1 = 1$ means that $t$ is rational and does not belong to this case. $a_1/a_2 \in \mathbb{Q}$ can be checked easily (it means the degree of the minimal polynomial is at most 1). Then we must check that $z_1^{a_1/a_2} = z_2$ which is possible for a rational exponent. This verification must be done for all pairs of $a_i$.

If we have several infinities of candidates, we have to decide whether those infinities have a common point. To decide whether the $\alpha_i + \frac{2\pi}{b_i}\mathbb{Z}$ intersect, we need to know whether the $b_i$ have an integer common multiple. If they don't, then there will exist an infinity of $t$ belonging to all those sets; if they do, only a finite number of $t$ need to be tested.

The last case is if we have on one hand a non algebraic solution and on the other hand an infinity of solutions. We can summarize this case as the simultaneous resolution of two constraints: $\begin{cases} e^{at} = z \\ \begin{bmatrix} \cos(bt) & -\sin(bt) \\ \sin(bt) & \cos(bt) \end{bmatrix} \begin{bmatrix} x_1 \\ x_2 \end{bmatrix} = \begin{bmatrix} y_1 \\ y_2 \end{bmatrix} \end{cases}$. We will rephrase the second part as $\begin{bmatrix} e^{ibt} & 0 \\ 0 & e^{-ibt} \end{bmatrix} \begin{bmatrix} 1 & -i \\ 1 & i \end{bmatrix} \begin{bmatrix} x_1 \\ x_2 \end{bmatrix} = \begin{bmatrix} 1 & -i \\ 1 & i \end{bmatrix} \begin{bmatrix} y_1 \\ y_2 \end{bmatrix}$.

And we can write the whole system as the following: $\begin{cases} e^{at} = z \\ e^{ibt} = z_2 \\ e^{-ibt} = z_3 \end{cases}$, where $a$, $b$, $z$, $z_2$, and $z_3$ are algebraic numbers (some are complex). We have already been confronted with such a system (but it had only two components) and we know that from theorem 1 it means that $i\frac{b}{a}$ belongs to $\mathbb{Q}$ or $z \in \{0, 1\}$. $i\frac{b}{a} \in \mathbb{Q}$ can be verified easily as it is an algebraic number; $z = 0$ is impossible, 1 means $e^{at} = 1$ hence $a = 0$ (which belongs to another case) or $t = 0$ hence $z_2 = z_3 = 1$ in which case, $t = 0$ is a solution to the problem.

# References


1. Hirsch, M.W., Smale, S., Devaney, R.: Differential Equations, Dynamical Systems, and an Introduction to Chaos. Elsevier Academic Press (2003)
2. Murray, J.D.: Mathematical Biology, 2nd edn. Biomathematics, vol. 19. Springer, Berlin (1993)
3. Lorenz, E.N.: Deterministic non-periodic flow. Journal of the Atmospheric Sciences 20, 130–141 (1963)
4. Graça, D.S., Campagnolo, M.L., Buescu, J.: Robust simulations of Turing machines with analytic maps and flows. In: Cooper, S.B., Löwe, B., Torenvliet, L. (eds.) CiE 2005. LNCS, vol. 3526, pp. 169–179. Springer, Heidelberg (2005)
5. Mignotte, M.: Suites récurrentes linéaires. Séminaire Delange-Pisot-Poitou. Théorie des nombres 15, G14–1–G14–9 (1974)
6. Berstel, J., Mignotte, M.: Deux propriétés décidables des suites récurrentes linéaires. Bulletin de la Société Mathématique de France 104, 175–184 (1976)
7. Halava, V., Harju, T., Hirvensalo, M., Karhumäki, J.: Skolem's problem - on the border between decidability and undecidability. Technical Report 683, Turku Center for Computer Science (2005)
8. Blondel, V., Portier, N.: The presence of a zero in an integer linear recurrent sequence is NP-hard to decide. Linear algebra and its Applications 351–352, 91–98 (2002)
9. Bournez, O.: Complexité algorithmique des systèmes dynamiques continus et hybrides. PhD thesis, École Normale Supérieure de Lyon (1999)
10. Asarin, E., Maler, O., Pnueli, A.: Reachability analysis of dynamical systems having piecewise-constant derivatives. Theoretical Computer Science 138, 35–65 (1995)
11. Asarin, E., Schneider, G.: Widening the boundary between decidable and undecidable hybrid systems. In: Brim, L., Jančar, P., Křetínský, M., Kucera, A. (eds.) CONCUR 2002. LNCS, vol. 2421, pp. 193–208. Springer, Heidelberg (2002)
12. Asarin, E., Schneider, G., Yovine, S.: On the decidability of the reachability problem for planar differential inclusions. In: Di Benedetto, M.D., Sangiovanni-Vincentelli, A.L. (eds.) HSCC 2001. LNCS, vol. 2034, pp. 89–104. Springer, Heidelberg (2001)
13. Blondel, V., Tsitsiklis, J.N.: A survey of computational complexity results in systems and control. Automatica 36(9), 1249–1274 (2000)
14. Kannan, R., Lipton, R.J.: Polynomial-time algorithm for the orbit problem. Journal of the ACM 33(4), 808–821 (1986)
15. Hirsch, M.W., Smale, S.: Differential Equations, Dynamical Systems, and Linear Algebra. Academic Press (1974)
16. Mignotte, M.: An inequality about factors of polynomials. Mathematics of Computation 28(128), 1153–1157 (1974)
17. Bostan, A.: Algorithmique efficace pour des opérations de base en calcul formel. PhD thesis, École polytechnique (2003)
18. Brawley, J.V., Carlitz, L.: Irreducibles and the composed product for polynomials over a finite field. Discrete Mathematics 65(2), 115–139 (1987)
19. Baker, A.: Transcendental Number Theory. Cambridge University Press (1990)
20. Gelfond, A.O.: Transcendental and Algebraic Numbers. Dover Publications (2003)
21. Lelong-Ferrand, J., Arnaudiès, J.M.: Cours de mathématiques, tome 1: algèbre. Dunod (1971)

# Hybrid Functional Interpretations

Mircea-Dan Hernest[1] and Paulo Oliva[2,*]

[1] Informatics Institute, University of Innsbruck, Austria
`dan.hernest@uibk.ac.at`
[2] Department of Computer Science, Queen Mary, University of London
`pbo@dcs.qmul.ac.uk`

**Abstract.** We show how different functional interpretations can be combined via a multi-modal linear logic. A concrete hybrid of Kreisel's modified realizability and Gödel's Dialectica is presented, and several small applications are given. We also discuss how the hybrid interpretation relates to variants of Dialectica and modified realizability with non-computational quantifiers.

**Keywords:** Functional interpretations, modified realizability, Dialectica interpretation, linear logic, program extraction from proofs, uniform quantifiers.

## 1  Introduction

The second author recently devised a unified presentation [1] of different functional interpretations of intuitionistic logic, including Kreisel's modified realizability [2] and Gödel's Dialectica interpretation [3]. As it turns out, these distinct interpretations diverge only in the treatment of the structural rule of contraction. Therefore, due to its finer handling of contractions, linear logic [4] gives us the optimal setting for further analysing and comparing these interpretations [5,6].

In this article we show that functional interpretations not only can be better understood modulo linear logic, but can also be successfully combined into what we term *hybrid interpretations*, where features of different interpretations can coexist. Consider for instance the handling of extensionality when working in the language of all finite types. In the case of modified realizability we can safely adopt a fully extensional setting with primitive equality for basic types (say $n = m$ for numbers $n, m \in \mathbb{N}$) and higher-type equality defined as $f \stackrel{\rho \to \tau}{=} g :\equiv \forall x^\rho (fx \stackrel{\tau}{=} gx)$ together with the axiom schema of extensionality

$$x \stackrel{\rho}{=} y \;\to\; fx \stackrel{\tau}{=} fy \tag{1}$$

for all finite types $\rho, \tau$. However, when it comes to Dialectica interpretation, the translation of (1) requires witnesses for the universal quantifiers within $x \stackrel{\rho}{=} y$, which cannot be majorised in general [7] and hence cannot be expressed inside Gödel's system T. But recall that intuitionistic proofs can be embedded into linear logic ones, with intuitionistic implications $A \to B$ translated as linear implications $!A \multimap B$. The difficulty

---

* The second author gratefully acknowledges support of the Royal Society of the UK under grant 516002.K501/RH/kk.

A. Beckmann, C. Dimitracopoulos, and B. Löwe (Eds.): CiE 2008, LNCS 5028, pp. 251–260, 2008.

of Dialectica in dealing with full extensionality is that the "negative information" in the assumption $!A \equiv !(x \overset{\rho}{=} y)$ of (1) should not (and cannot) be witnessed. In other words, the modality "!" should in this case rather be treated as by Kreisel's modified realizability, also when carrying out a Dialectica interpretation.

This distinguished treatment of the modalities is possible because, as pointed out by Girard (cf. [8] and [9], p84), the modalities are not canonical, thus different modalities can coexist into a single system. Therefore, we intend to make use of what we name a *multi-modal linear logic*, which includes distinct modalities corresponding to each of the various functional interpretations. E.g., in the extensionality example (1) we rather use Kreisel's modified realizability modality ($!_k A$) in order to express that the information in the premise of the axiom schema should not be witnessed:

$$!_k (x \overset{\rho}{=} y) \;\; \multimap \;\; fx \overset{\tau}{=} fy \, . \tag{2}$$

This generalises Spector's quantifier-free rule of extensionality (see [10]) since it allows us to derive $rs \overset{\tau}{=} rt$ from $s \overset{\rho}{=} t$ in any context of the form $!_k \Delta$, and has the advantage that visibly (2) requires no realizer.

In contrast to modified realizability, the Dialectica interpretation is well-suited to deal with classical proofs via negative translation, as it interprets the Markov principle

$$\neg \forall x A_{\mathsf{qf}}(x) \;\; \rightarrow \;\; \exists x \neg A_{\mathsf{qf}}(x) \, . \tag{3}$$

Since the premise of (3) corresponds in linear logic to $? \exists x A_{\mathsf{qf}}^{\perp}(x)$, the modality "?" should rather be treated as by the Dialectica interpretation, even when attempting to do a modified realizability, hence (3) should be replaced by

$$!_g \, ?_g \, \exists x \, A_{\mathsf{qf}}(x) \;\; \multimap \;\; \exists x \, ?_g \, A_{\mathsf{qf}}(x) \, . \tag{4}$$

For proofs which use both extensionality (2) and Markov principle (4), constructive information will be extracted whenever such a labelling of the modalities is possible.

The setting of multi-modal linear logic also allows for a unified study of the non-computational ("nc" for short) quantifiers introduced by Berger in the context of modified realizability [11] and adapted by the first author to Dialectica interpretation [12].

The paper is organised as follows. In the next section we present the formal system of multi-modal linear logic. In Section 3 we introduce the hybrid functional interpretation of the multi-modal system. In Section 4 we present a few illustrative applications of the hybrid interpretation. A comparison between the use of nc quantifiers and of our hybrid logic is given in Section 5. Section 6 discusses possible extensions of the hybrid interpretation to include other modalities. In Section 6.1 we present some ideas for an algorithm which decorates a (linear translation of a) given intuitionistic proof with different modalities so as to achieve a desired outcome of the extraction program.

## 2  Multi-modal Linear Logic $\mathsf{LL}_{\mathsf{h}}^{\omega}$

We build upon an extension of classical linear logic to the language of all finite types, introduced by the second author in [13]. The set of *finite types* $\mathcal{T}$ is inductively defined by: $i, b \in \mathcal{T}$ and if $\rho, \sigma \in \mathcal{T}$ then $\rho \to \sigma \in \mathcal{T}$. For simplicity, we deal with only two basic finite types $i$ (integers) and $b$ (booleans). We use no linear types nor linear terms.

We assume that the terms of $\mathsf{LL}_\mathsf{h}^\omega$ contain all typed $\lambda$-terms, i.e. variables $x^\rho$ for each finite type $\rho$, $\lambda$-abstractions $(\lambda x^\rho.t^\sigma)^{\rho \to \sigma}$, term applications $(t^{\rho \to \sigma} s^\rho)^\sigma$, and conditionals $(s^b)(t^\rho, r^\rho)$. The atomic formulas of $\mathsf{LL}_\mathsf{h}^\omega$ are $A_{\mathsf{at}}, B_{\mathsf{at}}, \ldots$ and $A_{\mathsf{at}}^\perp, B_{\mathsf{at}}^\perp, \ldots$. For simplicity, the standard propositional constants $0, 1, \perp, \top$ of linear logic have been omitted, since the hybrid interpretation of atomic formulas is trivial (see Definition 1).

Formulas are built from atomic formulas via connectives $A \,\invamp\, B$ (par), $A \otimes B$ (tensor), $A \diamond_z B$ (if-then-else) and quantifiers $\forall x A$ and $\exists x A$. Exponentials will be treated in Section 2.1 and nc-quantifiers in Section 5. The *linear implication* $A \multimap B$ abbreviates $A^\perp \,\invamp\, B$ where the *linear negation* $(\cdot)^\perp$ is an abbreviation so that $(A^\perp)^\perp$ is syntactically equal to $A$ (see [4, 13]). Recall that the structural rules of linear logic do not include the usual rules of weakening and contraction. These are added separately, in a controlled manner via the use of modalities (cf. Section 2.1). The rules for the multiplicative connectives and quantifiers are the usual ones for (one-sided) classical linear logic (see [4, 13]). Following [5], we deviate from the standard formulation of linear logic and use the if-then-else logical constructor $A \diamond_z B$ instead of standard additive conjunction and disjunction[1]. The rules for $A \diamond_z B$ are given in [13] (Table 3). In terms of quantification over booleans, the standard additives can be defined as

$$A \wedge B :\equiv \forall z^b (A \diamond_z B) \qquad A \vee B :\equiv \exists z^b (A \diamond_z B) .$$

**Notation for tuples.** We use bold face variables $\boldsymbol{f}, \boldsymbol{g}, \ldots, \boldsymbol{x}, \boldsymbol{y}, \ldots$ for tuples of variables, and bold face terms $\boldsymbol{a}, \boldsymbol{b}, \ldots, \boldsymbol{\gamma}, \boldsymbol{\delta}, \ldots$ for tuples of terms. Given the sequences of terms $\boldsymbol{a}$ and $\boldsymbol{b}$, by $\boldsymbol{a}(\boldsymbol{b})$ we mean the sequence of terms $a_0(\boldsymbol{b}), \ldots, a_n(\boldsymbol{b})$. Similarly for the multiple simultaneous substitution $\boldsymbol{a}[\boldsymbol{b}/\boldsymbol{x}]$.

## 2.1  Kreisel and Gödel Modalities

The second author [5, 13] has recently studied possible different interpretations for the exponentials ! and ?, and how these correspond to well-known functional interpretations of intuitionistic logic. We here introduce syntactically distinct exponentials (see Table 1) and show how these different interpretations can coexist (whence the "hybrid" denomination). For simplicity we have considered only the so-called "Kreisel" and "Gödel" modalities, denoted $!_k$ and respectively $!_g$, together with their duals $?_k$

**Table 1.** Rules for the exponentials $* \in \{k, g\}$

$$\dfrac{?_*\Gamma, A}{?_*\Gamma, !_*A}\,(!_*) \qquad \dfrac{\Gamma, A}{\Gamma, ?_*A}\,(?_*) \qquad \dfrac{\Gamma, ?_*A, ?_*A}{\Gamma, ?_*A}\,(\mathrm{con}_*) \qquad \dfrac{\Gamma}{\Gamma, ?_*A}\,(\mathrm{wkn}_*)$$

---

[1] See Girard's comments in [4] (p13) and [9] (p73) on the relation between the additive connectives and the if-then-else construct.

and $?_g$. This will correspond to a combination of modified realizability and Dialectica interpretation into a single functional interpretation which supersedes both of them.

Moreover, a partial order of information is put on the distinct modalities in the form of the following "relaxing" rules

$$\frac{\Gamma, ?_g A}{\Gamma, ?_k A} \ (\text{?-relax}) \qquad \text{and} \qquad \frac{\Gamma, !_k A}{\Gamma, !_g A} \ (\text{!-relax})$$

meaning that at anytime we can choose to "forget" some information we had. This is because, as will be reflected in the hybrid interpretation given below, the Gödel "whynot" is meant to carry a finer information than $?_k$, whereas the Kreisel "bang" is more general than $!_g$. The usual rules for both kinds of exponentials are presented in Table 1.

In mixing both Kreisel's and Gödel's interpretations, we must add also the following restriction on the "Gödel" contraction rule $\mathrm{con}_g$ (for terminology see Section 3 below):

$(*)$ if the contraction formula $A$ in $\mathrm{con}_g$ is *computationally relevant*, then it must not contain any Kreisel whynot $?_k$ in front of a computationally relevant subformula, and also no Kreisel bang $!_k$ in front of a *refutation relevant* subformula.

As we will see, $(*)$ ensures that the interpretation of such formulas $A$ is quantifier-free (hence decidable); $(*)$ is necessary and sufficient for attaining Theorem 1.


## 3   A Hybrid Functional Interpretation

To each formula $A$ of $\mathsf{LL}_\mathsf{h}^\omega$ we associate a not necessarily quantifier-free formula $|A|_y^x$ of linear logic $\mathsf{LL}^\omega$ (defined in [13]) where $x, y$ are fresh variables not appearing in $A$. The length and types of $x, y$ are inductively determined by the formula $A$. The variables $x$ in the superscript are called the *witnessing variables*, while the subscript variables $y$ are called the *challenge variables*. Intuitively, the interpretation of $A$ is a two-player (Eloise and Abelard) one-move game, where $|A|_y^x$ is the adjudication relation. We want that Eloise has a winning move whenever $A$ is provable in $\mathsf{LL}_\mathsf{h}^\omega$. Moreover, the hybrid linear logic proof of $A$ will provide Eloise's winning move $a$, i.e., $\forall y |A|_y^a$ will hold in $\mathsf{LL}^\omega$, where $a$ is a tuple of terms of corresponding types.

Formulas for which the tuple of witnessing variables is not empty are considered *computationally relevant*, and formulas for which the sequence of challenge variables is not empty are considered *refutation relevant*. An $?_k$ (respectively $!_k$) in front of a computationally (respectively refutation) irrelevant formula will be called *redundant*.

**Definition 1 (Hybrid Interpretation).** *The interpretation of atomic formulas are the atomic formulas themselves, with empty sets of witnessing and challenge variables, i.e.* $|A_\mathrm{at}| :\equiv A_\mathrm{at}$ *and* $|A_\mathrm{at}^\perp| :\equiv A_\mathrm{at}^\perp$. *Assuming* $|A|_y^x$ *and* $|B|_w^v$ *already defined, we define*

$$|A \,\invamp\, B|_{y,w}^{f,g} :\equiv |A|_y^{fw} \,\invamp\, |B|_w^{gy} \qquad\qquad |\exists z A(z)|_f^{x,z} :\equiv |A(z)|_{fz}^x$$

$$|A \otimes B|_{f,g}^{x,v} :\equiv |A|_{fv}^x \otimes |B|_{gx}^v \qquad\qquad |\forall z A(z)|_{y,z}^f :\equiv |A(z)|_y^{fz}$$

$$|A \,\Diamond_z\, B|_{y,w}^{x,v} :\equiv |A|_y^x \,\Diamond_z\, |B|_w^v \,.$$

*Finally, we can give different interpretations to the modalities as:*

$$|!_k A|^{\boldsymbol{x}} :\equiv !\forall \boldsymbol{y}|A|^{\boldsymbol{x}}_{\boldsymbol{y}} \qquad |!_g A|^{\boldsymbol{x}}_{\boldsymbol{f}} :\equiv !|A|^{\boldsymbol{x}}_{\boldsymbol{f}\boldsymbol{x}}$$

$$|?_k A|_{\boldsymbol{y}} :\equiv ?\exists \boldsymbol{x}|A|^{\boldsymbol{x}}_{\boldsymbol{y}} \qquad |?_g A|^{\boldsymbol{f}}_{\boldsymbol{y}} :\equiv ?|A|^{\boldsymbol{f}\boldsymbol{y}}_{\boldsymbol{y}} .$$

*It is easy to see that* $|A^{\perp}|^{\boldsymbol{y}}_{\boldsymbol{x}} \equiv (|A|^{\boldsymbol{x}}_{\boldsymbol{y}})^{\perp}$ *and thus* $|A \multimap B|^{\boldsymbol{f},\boldsymbol{g}}_{\boldsymbol{x},\boldsymbol{w}} \equiv |A|^{\boldsymbol{x}}_{\boldsymbol{f}\boldsymbol{w}} \multimap |B|^{\boldsymbol{g}\boldsymbol{x}}_{\boldsymbol{w}} .$

We prove the soundness of our interpretation, i.e., we show how Eloise's winning move in the game $|A|^{\boldsymbol{x}}_{\boldsymbol{y}}$ can be algorithmically extracted from a proof of $A$ in $\mathsf{LL}^{\omega}_{\mathsf{h}}$.

**Theorem 1 (Soundness of Hybrid Interpretation).** *Let* $A_0, \ldots, A_n$ *be a sequence of formulas of* $\mathsf{LL}^{\omega}_{\mathsf{h}}$*, with* $\boldsymbol{z}$ *as the only free-variables. If the sequent* $A_0, \ldots, A_n$ *is provable in* $\mathsf{LL}^{\omega}_{\mathsf{h}}$*, then terms* $\boldsymbol{a}_0, \ldots, \boldsymbol{a}_n$ *can be automatically synthesised from its formal proof, such that the translated sequent* $|A_0|^{\boldsymbol{a}_0}_{\boldsymbol{x}_0}, \ldots, |A_n|^{\boldsymbol{a}_n}_{\boldsymbol{x}_n}$ *is provable in* $\mathsf{LL}^{\omega}$*, where* $\mathsf{FV}(\boldsymbol{a}_i) \in \{\boldsymbol{z}, \boldsymbol{x}_0, \ldots, \boldsymbol{x}_n\}\backslash\{\boldsymbol{x}_i\}.$

**Proof:** Ignoring the rules for Gödel exponentials, the proof is given in [5], with $! :\equiv !_k$ and $? :\equiv ?_k$. The addition of Gödel exponentials to the language (together with their interpretation) does not alter the facts. Also the rules for Gödel exponentials are treated in [13], but independent of the Kreisel exponentials. Hence all we need to prove is that the proofs in [13] still hold after adding Kreisel exponentials to the language (together with their interpretation). It is easy to notice that, due to the restriction $(*)$ we added on $\mathrm{con}_g$, the interpretation of $?_g A$ is quantifier-free, hence decidable. This is because (so far) only non-redundant $!_k$ or $?_k$ could introduce quantifiers in the translated formula. $\square$

## 4   Simple Applications to Program Extraction

In this section we present some examples where it pays off to analyse proofs using both Kreisel and Gödel modalities. Some information might not be relevant while some other might be. One can thus use $!_k A$ and $?_k A$ to ignore the computationally irrelevant parts of the proof, in a way very similar in effect with light Dialectica [12].

### 4.1   Example 1

Consider theorems of the form

$$\forall x A \rightarrow \forall y B \rightarrow \forall z C \tag{5}$$

possibly with parameters, where the negative information on $x$ is irrelevant, while the one on $y$ is of our interest. In this case, we would rather view this theorem as

$$!_k \forall x A \multimap !_g \forall y B \multimap \forall z C . \tag{6}$$

For instance, consider the simple intuitionistic theorem

$$\forall f \left( \forall n(f(n) \leq 1) \rightarrow \forall m(f(m) \neq f(m+1)) \rightarrow \forall l(f(l) = f(l+2)) \right) . \tag{7}$$

From a proof of this, using labelling (6), our hybrid interpretation extracts a realizer $\Phi(f, l)$ s.t.

$$\forall f, l \left( \forall n(f(n) \leq 1) \rightarrow (f(\Phi(f,l)) \neq f(\Phi(f,l)+1)) \rightarrow (f(l) = f(l+2)) \right) .$$

Indeed, one such witness is $\Phi(f, l) :=$ if $(f(l+1) = f(l+2))$ then $l+1$ else $l$.

## 4.2  Example 2

More concretely, we consider the well-known example of extracting the Fibonacci numbers from a minimal logic proof of their weak existence. The example was first used in [14] to illustrate the so-called "refined A-translation" and then in [15] to illustrate the light Dialectica (see also Section 4.3 of [12]). The semi-classical Fibonacci proof is a minimal-logic proof of $\forall n \exists^{\mathrm{d}} m\, G(n,m)$ , where

$$\exists^{\mathrm{d}} m\, G(n,m) \quad :\equiv \quad \forall m\,(G(n,m) \to \bot) \to \bot \tag{8}$$

from assumptions expressing that $G$ is the graph of the Fibonacci function ($G$ is viewed as a predicate constant without computational content), i.e., $G(0,0)$, $G(1,1)$ and

$$\forall l_1, l_2, l_3 \left(G(l_1, l_2) \to G(l_1 + 1, l_3) \to G(l_1 + 2, l_2 + l_3)\right). \tag{9}$$

Note that such a specification fits into the form (5) (with $C:\equiv\bot$). As was noticed by the first author in [15], the negative universally quantified $l_1$, $l_2$ and $l_3$ do not need to be witnessed in order to extract an algorithm for computing the Fibonacci numbers as a witness for $m$ as function of $n$. The proof in [14] can thus be translated to a hybrid linear logic proof such that, in the pattern of (6), statement (8) becomes

$$?_g\, \exists m\, G(n,m)$$

and (9) becomes (we tacitly removed a number of redundant $!_g$ from the front of $G$'s)

$$!_k \forall l_1, l_2, l_3 \left(G(l_1, l_2) \multimap G(l_1 + 1, l_3) \multimap G(l_1 + 2, l_2 + l_3)\right)$$

and therefore only $m$ is witnessed, by the usual Fibonacci algorithm defined as $F_n := F_{n-1} + F_{n-2}$ and $F_1 := 1$ and $F_0 := 0$.

## 4.3  Example 3

The Dialectica interpretation and modified realizability also treat the induction rule[2]

$$\frac{A(0) \quad A(n) \to A(n+1)}{A(l)}\ (\mathrm{IND})$$

in slightly different ways. In both cases, the proofs of $A(0)$ and $A(n) \to A(n+1)$ provide a realiser $t[l]$ for the witnessing variables of $A(l)$, i.e., $|A|_y^t$. However, only during the extraction of $t$ via Dialectica interpretation a functional which refutes $A(n)$ when given a refutation for $A(n+1)$ will also be extracted. Such realizer is nonetheless not used in the construction of the desired term $t$. Therefore we rather always treat induction in the way modified realizability does, even when constructing a Dialectica witness. In our multi-modal setting, this can be achieved by formulating induction as

$$\frac{A(0) \quad !_k A(n) \multimap A(n+1)}{A(l)}\ (\mathrm{IND})$$

since the Kreisel modality blocks the witnessing of counter-example flows.

---

[2] The induction stated here corresponds to the induction rule with no open assumption in natural deduction systems.

### 4.4   Example 4

Consider the representation of real numbers as Cauchy sequences of rationals with a fixed rate of convergence. A real number being positive carries the extra information of a lower bound on how far from zero the limit of the sequence can be (cf. [10]). In order to avoid going into the representation level, when analysing the proof that a certain real function $f$ is positive at $x$, i.e. $f(x) >_{\mathbb{R}} 0$, it is often useful to view this as $\exists l(f(x) >_{\mathbb{R}} 2^{-l})$. Although witnessing $l$ gives us some lower bound on the value of $f(x)$, the formula $f(x) >_{\mathbb{R}} 2^{-l}$ still carries information on how far above $2^{-l}$ the value of $f(x)$ is. This extra information is usually irrelevant in practice and the purely existential matrix can be treated as quantifier-free, given that we can always forget these witnesses later. When automatising program extraction, it thus proves to be useful to make sure that the interpretation will not witness the innermost existential quantifier at all. This can be achieved by viewing the statement $f(x) >_{\mathbb{R}} 0$ as $\exists l ?_k (f(x) >_{\mathbb{R}} 2^{-l})$.

## 5   Comparison to Light Dialectica

As we noticed above, the effect of applying the hybrid functional interpretation on the semi-classical Fibonacci proof is equivalent to that of light Dialectica. This is not unexpected, since the two are related by a shared feature: the occultation of certain non-relevant quantifiers. Whereas light Dialectica needs a stronger restriction on the introduction rule for the nc-universal quantifier, a direct correspondence exists between the so-called $\mathrm{ncm-FC}$ condition of [12] and the present hybrid-interpretation restriction $(*)$ on contraction formulas of $\mathrm{con}_g$. Both have the purpose of ensuring that the translated contraction formula is decidable. We can thus see the hybrid interpretation as a simplification of the light Dialectica. On the other hand, there are situations which the latter can handle, whereas the former cannot. The reason is that a $!_k A$ discards all challenge terms of $|A|$ and symmetrically a $?_k A$ discards all witness terms of $|A|$. In contrast, by means of nc-quantifiers one can exactly "pick" which variables of $A$ do not need to be witnessed or challenged. In this sense, light Dialectica appears to be finer than the hybrid functional interpretation. Nonetheless, optimal is to have both techniques available in a single interpretation, combining their different syntactic natures. One can then easily choose to use either of them, or even both when necessary. For this reason we designed the following "light" hybrid interpretation, which supersedes both the hybrid and the light interpretations (light Dialectica [12] and light modified realizability [11]). Moreover, the nc-quantifiers could be useful in a purely linear context already.

### 5.1   The Light Hybrid Interpretation

To the language of our system we add the symbols $\overline{\forall}$ and $\overline{\exists}$ for non-computational universal and existential quantifiers respectively, usually abbreviated nc-forall, nc-exists. The hybrid interpretation of the nc quantifiers is

$$|\overline{\exists}z A(z)|_{\boldsymbol{y}}^{\boldsymbol{x}} := \exists z |A(z)|_{\boldsymbol{y}}^{\boldsymbol{x}} \qquad\qquad |\overline{\forall}z A(z)|_{\boldsymbol{y}}^{\boldsymbol{x}} := \forall z |A(z)|_{\boldsymbol{y}}^{\boldsymbol{x}} \qquad ,$$

hence the translated formula includes the "regular" quantifiers corresponding to the nc quantifiers - a further reason, besides the interpretation of non-redundant $!_k$, $?_k$, that

$|B|_w^v$ is not quantifier-free for general $B$. Therefore, the restriction on the contraction rule $\mathrm{con}_g$ is enhanced so that computationally relevant contraction formulas must not contain any nc-quantifier, besides satisfying condition $(\ast)$. Moreover, corresponding rules must be devised for the introduction of the new $\overline{\forall}$ and $\overline{\exists}$. These are just copies of the rules $(\forall)$ and $(\exists)$:

$$\frac{\Gamma, A}{\Gamma, \overline{\forall}zA} \; (\overline{\forall}) \qquad \frac{\Gamma, A[t/z]}{\Gamma, \overline{\exists}zA} \; (\overline{\exists})$$

but in the case of $(\overline{\forall})$ with an extension of the restriction that $z$ is not free in $\Gamma$: further $z$ must not be free in the terms $t$ of the $(\exists)$ instances in the proof of the premise $\Gamma, A$, nor in the computationally relevant contraction formulas of this proof. Notice the context-dependency of the above restriction, which is better expressed as "$z$ must not be free in the witnessing terms of the translations of $\Gamma, A$, after mining the proof of this premise sequent". In fact, the latter form is both necessary and sufficient, whereas the former is largely sufficient but can be optimised to become necessary as well (just as in [12]).

**Theorem 2 (Soundness of light hybrid interpretation).** *Theorem 1 still holds after the addition of nc-quantifiers with their introduction rules and hybrid interpretation.*

**Proof:** Notice that the presence of nc-quantifiers does not modify the set of free variables of the translated formula (since nc-quantified variables are regular-quantified in the translation). The interpretation of $(\overline{\exists})$ is just an instance of $(\exists)$. Similarly, the interpretation of $(\overline{\forall})$ is an instance of $(\forall)$, but we must check the restriction that $z$ is not free in the witnessed translation $|\Gamma|_v^\gamma$ of $\Gamma$. The extra restrictions we set on $(\overline{\forall})$ ensure that $z$ is not free in $\gamma$ and by the usual restriction $z$ is not among the free variables of $\Gamma$, which appear free in $|\Gamma|_v^\gamma$ as well. Since $z$ does not appear in the list of challenge variables for $|\overline{\forall}zA|$, essential is also that $z$ cannot be free in the witnesses $a$ from $|A|_x^a$. $\qquad\square$

# 6  Future Work: Extension and Automation

We can also consider other modalities, e.g., Howard ($!_h$) and Diller-Nahm ($!_d$), together with their duals $?_h$ and $?_d$. We assume a general ordering on all four modalities as $k > h > d > g$ and add the following weakening rules w.r.t. this partial order:

$$\frac{\Gamma, ?_i A}{\Gamma, ?_j A} \qquad \frac{\Gamma, !_j A}{\Gamma, !_i A} \qquad (j > i \text{ and } i, j \in \{k, h, d, g\})$$

meaning that anytime we can choose to "forget" some information we had. The interpretation of the new exponentials should be, following [13], as follows:

$$|!_d A|_f^x := !\forall y \in fx\, |A|_y^x \qquad |?_d A|_y^f := ?\exists x \in fy\, |A|_y^x$$

$$|!_h A|_f^x := !\forall y \leq^* fx\, |A|_y^x \qquad |?_h A|_y^f := ?\exists x \leq^* fy\, |A|_y^x$$

In some cases, when decidability of formulas is an issue, we might need to use the Diller-Nahm interpretation instead of Dialectica. Consider the following example (also (7) could serve as an example, if we assume that $f(m) = f(m+1)$ is undecidable)

$$\forall f^{\mathbb{N}\to\mathbb{R}}\left(\forall m(f(m) <_\mathbb{R} f(m+1)) \to \forall n(f(n) <_\mathbb{R} f(n+2))\right). \qquad (10)$$

Note that $<_\mathbb{R}$ is an undecidable relation. The best we can do is to collect a finite set of witnesses for $m$ (as functions of $n$). Like in Example 4 above, also here are we not interested in the redundant information hidden within $f(m) <_\mathbb{R} f(m+1)$. For the sake of program-extraction, formula (10) is thus better labelled as

$$\forall f^{\mathbb{N}\to\mathbb{R}} \left( !_d \forall m\, ?_k \left( f(m) <_\mathbb{R} f(m+1) \right) \;\multimap\; \forall n\, ?_k \left( f(n) <_\mathbb{R} f(n+2) \right) \right).$$

We can produce a finite collection of witnesses for $m$ as $\Phi(f,n) := \{n, n+1\}$ so that

$$\forall f^{\mathbb{N}\to\mathbb{R}}, n \left( \forall m \in \Phi(f,n)\, (f(m) <_\mathbb{R} f(m+1)) \;\to\; (f(n) <_\mathbb{R} f(n+2)) \right).$$

Note that the same effect can be achieved via a light Diller-Nahm interpretation, using an $\overline{\exists}$ for the existential quantifier hidden within $<_\mathbb{R}$, rather than the Kreisel whynot $?_k$.

## 6.1  Automated Decoration of Modalities

Given a proof of a mathematical theorem, once the desired information (i.e., quantified variables to be realized) is selected, we can automatically view the intuitionistic proof as a hybrid linear logic proof, with the modalities decorated in such way that the proof analysis will give us the information requested. For instance, in a theorem of the form

$$\forall x\, A \;\to\; \forall y\, \exists z\, B \tag{11}$$

it could be that we are interested only in the negative universal information $x$, and not in the positive existential information $z$. Hence we rather present (11) as a specification in multi-modal linear logic decorated like

$$!_g \forall x\, A \multimap \forall y\, ?_k \exists z\, B \qquad \text{or} \qquad !_g \forall x\, A \multimap \forall y\, \overline{\exists} z\, B.$$

An automated tool can try to figure out if such a labelling of the given proof is possible. If it is, the hybrid interpretation will then return the realizer $t$ and a linear logic proof of

$$\forall y\, (!A[ty/x] \multimap ?\exists z B) \qquad \text{or} \qquad \forall y\, (!A[ty/x] \multimap \exists z B)$$

which can finally be translated back to an intuitionistic proof of $\forall y\, (A[ty/x] \to \exists z B)$.

The input to such a "decorating algorithm" is the intuitionistic proof of an intuitionistic formula $A$ and $A^\star$, a (light) hybrid decoration of the linear logic translation of $A$. We would like to transform the proof of $A$ into a (light) hybrid linear proof of $A^\star$. For this we should establish how the rules of intuitionistic logic could be translated to proofs in hybrid linear logic. In general, an intuitionistic proof of $B$ from uncancelled assumptions $A_0, \ldots, A_n$ gets canonically translated to a linear proof of $B$ from $!_0 A_0, \ldots, !_n A_n$, where $!_i$ is one of the possible modalities, hence a proof of the sequent $?_0 A_0^\perp, \ldots, ?_n A_n^\perp, B$. Whenever a linear cut rule is to be applied, one has to make sure that the exponential flavours in the cut formula from the left sub-tree are isomorphically corresponding to the exponential flavours in its linear negation from the right sub-tree. Moreover, this correspondence has to be coordinated recursively down into the sub-trees. It is nevertheless not enough to simply ensure that the exponential flavours propagate soundly from conclusion to axioms and assumptions. One has to also verify the various restrictions: those involved by contraction rules like $\mathrm{con}_g$ and, if the nc-quantifiers are used as well, those involved by the $(\overline{\forall})$ introduction rule.

Sometimes, hybrid decorated specifications $A^\star$ simply fail to have a hybrid proof.

## 7    Conclusion

Hybrid interpretations successfully combine peculiar features of different functional interpretations. A few restrictions need to be satisfied when mixing the corresponding distinct modalities. The possibility of "colouring" the exponentials in a linear proof translation of the given specification with the desired flavours can be investigated by an algorithm. The non-computational quantifiers smoothly add to the picture in a way that uniformly explicates the structure of both light Dialectica and light modified realizability. Illustrative applications of the hybrid interpretations were here presented.

Example 3 brings an important optimisation of the usual treatment of induction by Dialectica. Full extensionality and the Markov principle are simultaneously treated under certain restrictions. Although not previously noticed, similar effects could be achieved via the nc quantifiers, but using the hybrid modalities appears to be smoother.

The user of hybrid interpretations thus has a large choice of techniques. The hunt for new applications has now just opened and the reader is warmly invited!

## References


1. Oliva, P.: Unifying functional interpretations. Notre Dame Journal of Formal Logic 47(2), 263–290 (2006)
2. Kreisel, G.: Interpretation of analysis by means of constructive functionals of finite types. In: Heyting, A. (ed.) Constructivity in Mathematics, pp. 101–128. North Holland, Amsterdam (1959)
3. Gödel, K.: Über eine bisher noch nicht benützte Erweiterung des finiten Standpunktes. Dialectica 12, 280–287 (1958)
4. Girard, J.Y.: Linear logic. Theoretical Computer Science 50(1), 1–102 (1987)
5. Oliva, P.: Modified realizability interpretation of classical linear logic. In: Proc. of the 22nd Annual IEEE Symposium on Logic in Computer Science (LICS 2007). IEEE Press (2007)
6. Oliva, P.: An analysis of Gödel's Dialectica interpretation via linear logic. Dialectica (2008), http://www.dcs.qmul.ac.uk/~pbo
7. Howard, W.A.: Hereditarily majorizable functionals of finite type. In: Troelstra, A.S. (ed.) WG 1988. Lecture Notes in Mathematics, vol. 344, pp. 454–461. Springer, Berlin (1973)
8. Danos, V., Joinet, J.B., Schellinx, H.: The structure of exponentials: Uncovering the dynamics of linear logic proofs. In: Mundici, D., Gottlob, G., Leitsch, A. (eds.) KGC 1993. LNCS, vol. 713, pp. 159–171. Springer, Heidelberg (1993)
9. Girard, J.Y.: Towards a geometry of interaction. Contemporary Mathematics 92 (1989)
10. Kohlenbach, U., Oliva, P.: Proof mining: a systematic way of analysing proofs in mathematics. Proceedings of the Steklov Institute of Mathematics 242, 136–164 (2003)
11. Berger, U.: Uniform Heyting Arithmetic. Annals Pure Applied Logic 133, 125–148 (2005)
12. Hernest, M.D.: Optimized programs from (non-constructive) proofs by the light (monotone) Dialectica interpretation. PhD Thesis, École Polytechnique and Universität München (2006), http://www.brics.dk/~danher/teza/
13. Oliva, P.: Computational interpretations of classical linear logic. In: Leivant, D., de Queiroz, R. (eds.) WoLLIC 2007. LNCS, vol. 4576, pp. 285–296. Springer, Heidelberg (2007)
14. Berger, U., Schwichtenberg, H., Buchholz, W.: Refined program extraction from classical proofs. Annals of Pure and Applied Logic 114, 3–25 (2002)
15. Hernest, M.D.: Light Dialectica program extraction from a classical Fibonacci proof. Electronic Notes in Theoretical Computer Science 171(3), 43–53 (2007)

# The Algorithm Concept – Tool for Historiographic Interpretation or Red Herring?

Jens Høyrup
jensh@ruc.dk
http://www.akira.ruc.dk/~jensh/

**Abstract.** With starting point in Donald Knuth's paper "Ancient Babylonian Algorithms", and using the algebraic reading of pre-Modern mathematical texts as a parallel, the paper discusses the relevance of the algorithm concept, on one hand as an analytical tool for the understanding and comparison of mathematical procedures, on the other as a possible key to how pre-Modern reckoners *thought* their mathematics and to how they thought *about* it.

**Keywords:** pre-Modern mathematics; algorithm concept as an historiographic analytical tool; algebra as a historiographic analytical tool.

A "red herring": the smoked herring drawn
across the trail of the fox in order to distract
 the hounds and make the hunt last longer

To August Ziggelaar
*on occasion of his eighty years'
birthday, 17 January 2008*

## 1  A Parallel but Preceding Issue

When the Rhind Mathematical Papyrus – the most important single source for ancient Egyptian mathematics – was first published by August Eisenlohr in 1877, he interpreted some of the calculations of the text by means of that kind of equation algebra which in his times was currently taught in school. In 1880, Moritz Cantor followed him in the first edition of the first volume of his *Vorlesungen über Geschichte der Mathematik*,[1] making it thereby (if any specific excuse was needed) the canonical way to read the text. It remained so in spite of the well-argued objections formulated by Léon Rodet already in 1881 (with the conclusion [24: 205] that "when studying the history of a science, exactly as when one

---

[1] Still in the third edition from 1907 [3: 74].

A. Beckmann, C. Dimitracopoulos, and B. Löwe (Eds.): CiE 2008, LNCS 5028, pp. 261–272, 2008.
© Springer-Verlag Berlin Heidelberg 2008

wants to obtain something, one should `rather ask God himself than his saints'"[2]). In his third edition, Cantor [3: 76] refers to Rodet's objections and alternative interpretation through the method of a "single false position" (p. 76),[3] but sees no genuine difference. Eric Peet, in his new edition of the Rhind Papyrus [20: 60], characterizes the matter as "not one of essence but of form".

Egyptian mathematics was not alone in this situation. In 1886, H. G. Zeuthen published *Die Lehre von den Kegelschnitten im Altertum*, arguing that in *Elements* II.1–10 the ancient Greek geometers possessed "what one may call a *geometric algebra*, since on one hand, like algebra, its deals with general magnitudes, irrational as well as rational, on the other uses other means than ordinary language in order to make its procedures intelligible and impress them on memory" [29: 7]. What Zeuthen had in mind was obviously a much more modern kind of algebra than what Eisenlohr had thought of, and the assertion is rather unobjectionable *if* Zeuthen's whole explanation is taken into account.[4] But it was not, and the resulting conventional wisdom of twentieth-century historiography was that the ancient Greek mathematicians had algebra *without qualification*, "dressed up" as geometry but algebra in "mathematical essence".

Algebra was also the obvious interpretational tool when "Babylonian algebra" was discovered and deciphered in the years around 1930 – and in subsequent decades it was taken for the very truth, as historians and historically interested mathematicians read the commentary and popularizations of the "saints" (i.e., of Neugebauer, van der Waerden and others) – see [7].

Cautious objections against the existence of a "Babylonian algebra" were raised by Michael Mahoney in 1971 [17], based however on a definition of algebra which excluded everything written before Viète, and therefore perhaps not very relevant for historians interested, e.g., in al-Khwārizmī's or Fibonacci's algebra. A famous clash in 1975–1978 between Sabetai Unguru [26], B. L. van der Waerden [27], Hans Freudenthal [5] and André Weil [28] at least made it clear that the status of Greek "geometric algebra" was under discussion.

In mild form, the association of large areas of pre-Modern mathematics to algebra is reflected in the characterization of problems as "equations". An illustrative example chosen at random (that is, from a book which I happened to review recently) is the statement that a twelfth-century *Liber augmentis et diminutionis* shows "how linear

---

[2] My translation, as everywhere in the following when no translator is identified.

[3] The method may be illustrated on the problem by which Fibonacci introduces the method in the *Liber abbaci* [2: 173]: $^1/_4$ and $^1/_3$ of a tree are underground, and this part is 21 palms. We posit a length for the tree, of which the fractions can be taken conveniently – most obviously 12. $^1/_4 + ^1/_3$ of 12 palms are 7 palms – but we need 21 palms. Therefore the initial guess should be multiplied by $^{21}/_7 = 3$.

The method can also be used for homogeneous problems of (for instance) the second degree; then the scaling factor is the square root of the error factor.

The Rhind Papyrus only uses the method for first-degree problems, but elsewhere in the Pharaonic mathematical corpus it is applied to homogeneous problems of the second degree (to find the sides of a rectangle from their ratio and the area).

[4] Admittedly, soon afterwards Zeuthen [29: 12] expresses *Elements* II.1–10 as algebraic equations dealing with $a, b, c, \ldots$ – but then he explains that these must be understood as statements about lines and rectangles.

equations with one unknown or systems of linear equations with two unknowns may be solved with the help of the rule of double false position"[5] [4: I, 5]. This also illustrates why some historians object to the automatic algebraic reading. One problem of the treatise runs as follows [16: I, 326]:

> Somebody traded with a quantity of money, and this quantity was doubled for him. From this he gave away two dragmas, and traded with the rest, and it was doubled for him. From this he gave away four dragmas, after which he traded with the rest, and it was doubled. But from this he gave away six dragmas, and nothing remained for him.

Actually, the treatise solves this problem (and many others) not only through application of the "double false position" but also by stepwise reverse calculation and by means of what the treatise calls its *regula*, the formulation and solution of a first-degree equation in which the unknown initial quantity is called a *thing* and treated exactly as an *x*. Seeing *the problem itself* simply as "an equation" misses the need for what Viète following Pappos called "zetetics", the *formulation* of the problem as an equation – and, in the present case, masks that zetetics is no automatic process, since the problem may as well be translated for instance into a system of three equations with three unknowns (the successive amounts traded with).

A translation of a literary text always identifies that which the reader is supposed *not* to know – the words of the foreign language – within a framework which the reader is supposed to know. In cases where the semantic structures of the two languages are different, it is sometimes possible for the translator to make a choice depending on local semantics without telling the reader – in a classic example, translating English "wood" into German "Holz" if the material is thought of, and into "Wald" if the "wood" refers to many trees growing together. If an English pun is involved, an explanatory note is needed for the German reader.

Such a note is, *mutatis mutandis*, what Zeuthen gave. His reference to "algebra" was a tool for making his readers understand how the theorems from *Elements* II were used. *Applied thus*, the reference to the reader's notion of algebra was hence a fruitful as well as legitimate explanatory tool – and even a way to make the reader reflect upon his own notion of algebra.

Zeuthen's followers forgot the note, and many of those who explained Egyptian and Babylonian mathematics as "algebra" never thought of making similar notes. Thereby "algebra" became a red herring, distracting from analysis of what goes on in the ancient texts and what went on in the mind of its carriers instead of elucidating it.

## 2  Seeing Historical Texts through Algorithms

In recent decades, it has become customary to appeal to the algorithm concept, mostly as an alternative, at times as a supplement to "algebra". The precedent invites us to ask whether this is a new and better tool or another red herring?

The first publication to use the notion of algorithms as a tool to understand what goes on in historical texts was probably Donald Knuth's "Ancient Babylonian Algorithms"

---

[5] That is, making two guesses and finding the correct value from the two errors that arise by means of a calculation which follows the principle of the "alligation rule" (though the latter link is never made). In mathematical principle, we may see the method as a linear interpolation, and some medieval mathematicians indeed provided a corresponding geometric proof.

from 1972 [15]. He did not see algorithms as an alternative way to explain Babylonian mathematics but states indeed (p. 622) that the

> Babylonian mathematicians [...] were adept at solving many types of algebraic equations. But they did not have an algebraic notation that is quite as transparent as ours; they represented each formula by a step-by-step list of rules for its evaluation, i.e. by an algorithm for computing that formula. In effect, they worked with a "machine language" representation of formulas instead of a symbolic language.

There are at least three layers in this. *Firstly*, that the algorithm is a prescription for finding a result – it provides neither the idea behind the procedure nor any proof of its correctness, and cannot do that (on this level of mathematics) as long as everything is understood as a prescribed sequence of abstract numerical operations – as, *secondly*, was Knuth's understanding of the mathematical texts, based on the translations and the interpretation of the time [7]. Only the abstract understanding of the numbers of the texts as devoid of ontological reference allows us to consider them as elements of a "machine language". *Thirdly*, that an "algorithm" is a "step-by-step list of rules"; this may seem uncontroversial – but see below, note 15.

Knuth gives this illustration (from the tablet BM 85200+VAT 6599 #24[6]). What is at stake is to find the length and the width of the base of a cistern, whose volume is given (in the usual transcription of sexagesimal numbers) as 27;46,40 (meaning $27 + {}^{46}/_{60} + {}^{40}/_{3600}$), and whose depth is 3;20, given that the length exceeds the width by 0;50. I conserve Knuth's parenthetical explanations:

A (rectangular) cistern.
The height is 3,20, and a volume of 27,46,40 has been excavated.
The length exceeds the width by 50. (The object is to find the length and the width.)
You should take the reciprocal of the height, 3,20, obtaining 18.
Multiply this by the volume, 27,46,40, obtaining 8,20. (This is the length times the width; the problem has been reduced to finding x and y, given that x–y = 50 and xy = 8,20. A standard procedure for solving such equations, which occurs repeatedly in Babylonian manuscripts, is now used.)
Take half of 50 and square it, obtaining 10,25.
Add 8,20, and you get 8,30,25. (Remember that the radix point position always needs to be supplied. In this case, 50 stands for $^5/_6$ and 8,20 stands for $8\,^1/_3$, taking into account the sizes of typical cisterns!)
The square root is 2,55.
Make two copies of this, adding (25) to the one and subtracting from the other.
You find that 3,20 (namely $3\,^1/_3$) is the length and 2,30 (namely $2\,^1/_2$) is the width.
This is the procedure.

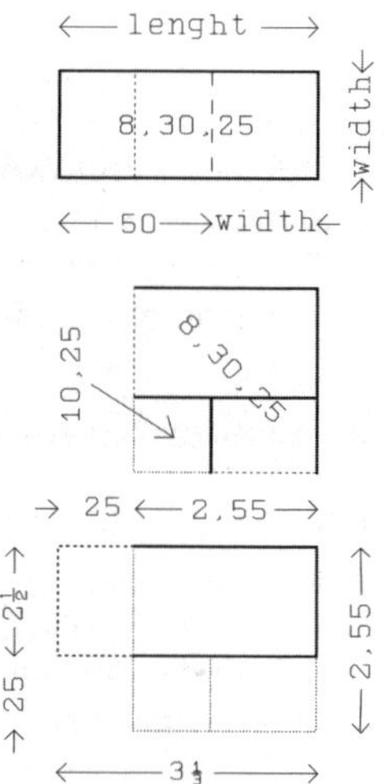


---



[6] Knuth translates freely from the translation in [19: I, 198, 205]. Revised transliteration and re-translation in agreement with recent insights in [9: 146].

We observe that until the beginning of the "standard procedure", the numbers are not ontologically abstract (in other words, deprived of semantics), not "machine language" but intrinsically also an explanation – knowing that the volume is the product of base and height, we understand that division of the volume by the height (which the Babylonians performed as a multiplication by its reciprocal) must give the base.

What Knuth could not know in 1972 is that the "standard procedure" refers to a sequence of geometric cut-and-paste operations – shown here alongside the prescription. His "square root" is thus the side of a square, and the "two copies" (the text actually says "posit it twice") are the two sides which meet in a corner. What Knuth renders "adding (25) to one and subtracting from the other" (actually "join to one, remove from one") is a recurrent ellipsis for a sub-sub-procedure in which the half-excess is joined to one side and removed from the other – often *first* removed and *only afterwards* – because *the same* line segment is involved and therefore has to be at disposition – joined to the other side. Even this part therefore is not written in "machine language" but semantically loaded; the inherent references to the geometric diagram[7] which is manipulated provides a justification of the procedure which is just as adequate as the one that follows from our manipulations of an algebraic equation.[8]

Removal of the reference to the "machine language", a misunderstanding induced by the translation into modern arithmetical language, does not prevent us from speaking of the prescription as an "algorithm": it still consists of a "step-by-step list of rules". However, as Knuth points out (p. 674), he only finds "straight-line calculations, without any branching or decision-making involved. In order to construct algorithms that are really non-trivial from a computer-scientist's point of view, we need to have some operations that affect the flow of control". The closest he gets is the reading of a text with repetition as an expanded macro-iteration.

He *might* have pointed to that use of an embedded sub-routine which he observes in the text he quotes. This feature of the Babylonian texts was explored in some depth by Jim Ritter [23]. Ritter centred the discussion on the tablet Str. 368,[9] which has the same embedded sub-routine as the example discussed by Knuth – with one small difference. Instead of performing the bisection within the subroutine, the main procedure omits a previous doubling that should produce the number to be bisected. The same pairwise cancellation of operations, one inside and the other outside the sub-routine, is found in other texts. The algorithmic interpretation can of course be saved (we may just speak of two related but different sub-routines) – but the two-level algorithmic interpretation can still be seen to be only a formalization of the sequence of operations, and not to

---

[7] These references are visible in the terminology, which is only rendered inadequately by Knuth. The Old Babylonian mathematical terminology (that is, the terminology of the earlier second millennium BCE, the period from which most mathematical texts stem) distinguishes two different "additive operations", two different "subtractions", two different "halves", and no less than four "multiplications" (one of which is not a genuine multiplication but a rectangle construction).

[8] Karine Chemla has repeatedly used the formulation that the text is "algorithm and proof in one". For the whole geometric interpretation of the procedure, see for instance [9].

[9] Transliteration and translation in [19: I, 311*f*].

cover that *insight* from which the sequence of steps is planned – which would not astonish Knuth, cf. above.

In what Knuth regarded as the trivial sense, Babylonian mathematical texts – more precisely, the "procedure texts"[10] – can certainly be understood as consisting of algorithms. The texts teach by means of paradigmatic examples, that is, by means of steps in sequence; the ontological identifications of the entities which are operated on ("the height", "the volume", etc.) just show that the algorithm is not a purely numerical one; occasional explanatory remarks ("because he has said that ...", referring to the statement) we may understand as "comment fields".

In this sense, however, even a Euclidean construction (say, *Elements* I,1, "On a given line segment to construct an equilateral triangle", ed. [6: I, 10]) can be read as an algorithm, with the only difference that the comments field (here a proof) follows after the completion of the algorithmic prescription ("With centre A and distance AB to draw the circle BΓΔ ..."; and with centre B and distance BA to draw ..." ). Even this is a trivial linear algorithm, even though it may be applied as a sub-routine in other constructions (thus already in *Elements* I.2).[11] We may legitimately ask whether a conceptual tool which can be applied so widely is really informative (but the answer will probably depend on taste rather than on arguments).

Greek mathematics is certainly more than geometrical construction, and the "comment fields" of constructional propositions attach these to the general endeavours of theory and demonstration. On the other hand, the concentration on paradigmatic examples was not a Babylonian monopoly. Knuth (p. 676) already refers to the ancient Egyptians and to Indian and Chinese mathematics (rightly, indeed, with the only difference that Indian and Chinese sources regularly state their "algorithms" in the abstract before giving the paradigmatic examples); and the list need not stop there. If the preponderant use of (branch-free) algorithms characterizes these types of mathematics, should we not expect it also to characterize the way their carriers understood mathematics?

Old Babylonian and late medieval texts allow us to reach at least a partial answer to this question. Before we turn to the carriers' perspective, however, we shall take up a final aspect of the use of the algorithm notion as a historiographical tool.

## 3  Algorithmic Analysis

In order to distil from a text problem its "mathematical substance" (and thus to decide if and why the procedure is adequate), some kind of formalization is often needed. To take a simple example, the "rule of three":[12]

---

[10] Beyond these texts, which describe the procedure to be followed in problem solutions, the corpus of mathematical texts encompasses "catalogues" listing only problem statements (at times with indication of the solution), mathematical tables and tablets containing only numerical calculations.

[11] More interesting embedding is present in ancient Greek geometry at the level of the formulaic language, as discussed by Germaine Aujac [1] and particularly by Reviel Netz [18: 127–167]. But this has hardly anything to do with algorithms, it only shows that the notion of embedding is interesting on its own – cf. [8].

[12] I quote from Jacopo da Firenze's *Tractatus algorismi* from 1307, ed., trans. [12: 237].

> 7 *tornesi* are worth 9 *parigini*.[13] Say me, how much will 20 *tornesi* be worth? Do thus, the thing that you want to know is that which 20 *tornesi* will be worth. And the not similar (thing) is that which 7 *tornesi* are worth, that is, they are worth 9 *parigini*. And therefore we should multiply 9 *parigini* times 20, they make 180 *parigini*, and divide in 7, which is the third thing. Divide 180, from which results 25 and $^5/_7$. And 25 *parigini* and $^5/_7$ will 20 *tornesi* be worth. And thus the similar computations are done.

At first we replace 7 by $a$, 9 by $b$ and 20 by $c$, and then we say that $c$ *tornesi* are worth $^{(cb)}/_a$ *parigini*. We might also have argued that 20 is $^{20}/_7$ times as much as 7, and the value of the 20 *tornesi* hence $(^{20}/_7) \cdot 9$ *parigini*, that is, $(^c/_a) \cdot b$ (this method was called "by ratio" by the Arabic mathematician Ibn Thabāt [21: 43] around 1200 and preferred by some Arabic mathematicians).

If read as computational prescriptions (that is, as straight-line algorithms, "first multiply ...then divide ..." respectively "first divide ... then multiply"), these formulae are quite adequate. The danger is, however, that they are read as algebraic formulae, in which case the reader might believe that the two methods are identical – which is clearly a bad approach to historical texts, since it conflates an opaque procedure (the rule of three) and a transparent one. The frequent references in general histories of mathematics to the presence of the rule of three in Babylonian and Egyptian mathematics shows that the mere possibility of translating into algebraic formulae suffices to produce the mistake.

At times, moreover, even a literal reading of an algebraic formula does not allow an unambiguous reconstruction of the computational procedure which it expresses – and thus not to decide whether two texts actually use the same procedure. For instance, we may look at this problem from the Late Babylonian tablet BM 34568:[14]

> The diagonal and the length I have accumulated: 9. 3 the width. What the length and the diagonal. Since you do not know,
> 9 steps of 9, 81, and 3 steps of 3, 9. 9 from 81 you lift:
> remaining 72. 72 steps of ½ you go: 36. 9 steps of what
> may I go so that 36 (is produced)? 9 steps of 4 you go: 36. 4 the length.
> 4 from 9 you lift: remaining 5. 5 the diagonal.

To render this procedure by the line

$$l \text{ is found as } \frac{\frac{1}{2} \cdot ([d+l]^2 - w^2)}{d+l}, \ d \text{ as } (d+l)-l$$

(as done in [10: 13] apart from a missing fraction line in the print) is only adequate because $d+l$ is a given number; if it had been calculated, the formula would not tell whether it was calculated twice in the formula for $l$ or once, and saved.

---

[13] "Tornesi" are minted in Tours, "parigini" in Paris.

[14] I use the translation in [9: 393], but replace the sexagesimal place value numbers with decimal ones.

**pMoscow, no. 19**

| Translation | Numerical Algorithm | Symbolical Algorithm |
|---|---|---|
| [1] Method of calculating a quantity, calculated $1\,\overline{2}$ times | $1\,\overline{2}$ | $D_1$ |
| together with [2] 4 | 4 | $D_2$ |
| and it has come to 10. | 10 | $D_3$ |
| What is the quantity that says it? | | |
| [3] Then you calculate the difference of these 10 to these 4. Then 6 results. | $10 - 4 = 6$ | (1) $D_3 - D_2$ |
| [4] Then you divide 1 by $1\,\overline{2}$. Then $\overline{\overline{3}}$ results. | $1 : 1\,\overline{2} = \overline{\overline{3}}$ | (2) $1 : D_1$ |
| [5] Then you calculate $\overline{\overline{3}}$ of these 6 Then 4 results. | $\overline{\overline{3}} \cdot 6 = 4$ | (3) (2) · (1) |
| Behold it is 4 (the quantity that) [6] says it. What has been found by you is correct. | | |

Annette Imhausen's algorithmic representation of an Egyptian problem [13: 165]

An alternative formalism, able to better grasp the structure and details of complicated calculations for analysis, was proposed by Jim Ritter [23][15] and amply used in adapted shape by Annette Imhausen first in 2002 [13] and next in her dissertation from 2003 [14]: In a three-column scheme, the single steps of the text (in translation), the numerical steps and their explanation in symbols stand in parallel. In the symbol column, the outcome of each computation is given a *new* name,[16] which produces an unambiguous trail.

## 4  The Participants' Point of View

So much about the algorithm concept as a tool for historiographic text analysis. We should now return to its possible adequacy as a mirror for the original reckoners' understanding of what they were doing.

---

[15] The paper circulated for long before its final publication in 2004. I read it myself in 1997; a preprint [22] appeared in 1998.

It should be added that Jim Ritter's notion of an algorithm is much broader than Knuth's "step-by-step list of rules". He introduces "another, more general level of the algorithm, more general than that of the calculational techniques or that of the arithmetical operations, the level of method of solution, the choice of strategy of solution", and exemplifies this by the method of a single false position which can be seen to underlie several of his examples. This has the apparent advantage of making the carriers' understanding part of the algorithm. As far as I can see, however, the algorithm concept is dissolved by the inclusion of a level which is not linked to the steps of the algorithm (as are the "comments", be they Babylonian or Euclidean) but which is on the other hand common to many algorithms that differ in their steps. Instead of seeing the algorithms used in weather prediction as encompassing the physical theories and differential equations on which they are based, it seems to me to leave more room for analysis to separate the physical and mathematical theories from their implementation in computer algorithms. I shall therefore go on using the usual ("Knuthian") understanding of the term.

[16] It is noteworthy that the same principle was followed by Jordanus of Nemore in the earlier part of the thirteenth century, when he introduced a letter formalism with the purpose of proving the correctness of arithmetical and algebraic theorems (and *not* of making symbolic algebraic calculations).

Many Old Babylonian procedure texts start the prescription by a phrase "You, by your doing". Is it adequate to read this as a reference to a specific algorithm individualized as such? If so, we might perhaps expect to find occasional references to such algorithms by name.

We do indeed find a few references by name to particular methods. What is striking, however, is that the occurrences of the names show them to point to methods that can be varied, *not* to precise algorithms (not even to what can naturally be interpreted as branched algorithms). One, *makṣarum*/"bundling", refers to the division of a surface (in the actual case, a triangle) or a volume (in the actual case, a cube) into a bundle of smaller surfaces or volumes of the same shape [9: 66, 254]; the other, "the Akkadian [method]" refers to the quadratic completion which we have encountered in the sub-routine discussed above – but it turns up in a procedure of a different and quite peculiar character [9: 194].

This corresponds well to the flexible use of the sub-routine which we discussed above; the Old Babylonian reckoners hence appear to have conceived of their methods as procedures which could be applied flexibly as required by varying contexts, not as a tool-box of fixed algorithms. Only the very standardized set of problems occurring in the texts cause us to find *exactly* the same procedure time and again, and thus giving *us* the impression that fixed algorithms are involved.

I am not aware of the presence of elements in Egyptian mathematical discourse which allow a similar analysis; however, the actual algorithms constructed by Annette Imhausen are often so varied in their details that even they are likely to represent the modern analysis only, not the way the Egyptians understood their mathematical practice.

As far as the Indian and Chinese material is concerned, my inability to read the texts in the original language prevents me from forming a definite opinion; however, the initial abstract formulation of rules which are then followed by examples may suggest that ("trivial") algorithmic thinking was closer to the way Chinese and Indian reckoners thought.

I am much more familiar with the culture of practical arithmetic represented by Leonardo Fibonacci and the Italian and Provencal abbacus treatises of the fourteenth and fifteenth centuries. [17] Within this culture, the word which *might* represent something close to an algorithm is *regula* (*regola*, *reghola*, etc.). It is still reflected in our modern notion of the "rule of three" (the *regola delle tre cose* of the abbacus masters) referred to above. This really looks like an algorithm, and indeed a quite trivial one – but trivial only until we start reading the texts closely. Indeed, if we look at for instance the presentation in Jacopo da Firenze's *Tractatus algorismi* [12: 236–240] we find that it is divided into several cases: all three numbers are integers, one of the first two numbers contains fractions, or both of these do. But the three cases are not treated in parallel – the second and third only tell to multiply adequately by the denominators, leaving it tacitly understood that the rest is as in the first case; although it is not said (and perhaps not precisely conceptualized) it is obvious that the substructure is a less trivial algorithm:

---

[17] Høyrup 2005 gives the reasons that Fibonacci must be seen as an early representative of the same broad mathematical culture as the later abbacus writings and not as the "father" of the abbacus school.

**IF** all three numbers are integer **GO L**;
**IF** only one of the former numbers contains a fraction with denominator $p$, multiply both of these by $p$;
**GO L**;
**IF** both of them contain fractions, with respective denominators $p$ and $q$, multiply both by a common multiple of $p$ and $q$;
**GO L**;
**L:**
(multiply and divide)

In the case of the presentation of the "rule of double false position" (see note 5), this structure is even more explicit. Some of the abbacus books, and also Fibonacci, occasionally operate with negative numbers conceptualized as "debts"; but they never do so in the rule of double false. Therefore, the formula to be used depends on whether both guesses turn out to result in an excess (or both in a deficit), or one in an excess, the other in a deficit. The algorithm may not be presented in full – in Barthélemy de Romans' *Compendy de la praticque des nombres* [25: 390] all that is said is thus *plus et plus, meins et meins, sustrayons. Plus et meins, adjoustons* ("excess and excess, deficit and deficit, we subtract. Excess and deficit, we add"). This only describes the initial branching structure, and leaves out the linear part as already known.

However, a *regula* is mostly not an algorithm, neither straight-line nor branched. For instance, the *regula* of the *Liber augmentis et diminutionis* (see text around note 5), reappearing as *regula recta* in Fibonacci's *Liber abbaci*, refers to a general and very flexible method: the application of first-degree equation algebra. Several other *regulae* are similarly open-ended; actually, even the rule of three may be adequately but tacitly adapted to problems of inverse proportionality. Application of the algorithm concept thus allows to trace a substructure *in statu nascendi* in the thinking of the abbacus masters; but if they had been asked what they meant by *regola*, the answer would most likely not have made us think of an algorithm.

All in all, we may conclude that the algorithm concept, when applied to pre-modern mathematical texts, *may* represent a valid mapping of their procedures – at times useful, at times as trivial as the algorithms which it digs out of the sources. If believed to correspond to the way the early reckoners thought *about* their activity, it is likely to be a red herring (barring perhaps Chinese and Sanskrit texts). If used to trace emerging substructures in the way they *thought* their mathematics it is mostly also misleading – but not always; used with delicacy it may sometimes offer a valuable tool.

# References


[1] Aujac, G.: Le langage formulaire dans la géométrie grecque. Revue d'Histoire des Sciences 37(2), 97–109 (1984)

[2] Boncompagni, B. (ed.): Scritti di Leonardo Pisano matematico del secolo decimoterzo. I. Il Liber abbaci di Leonardo Pisano. Tipografia delle Scienze Matematiche e Fisiche, Roma (1857)

[3] Cantor, M.: Vorlesungen über Geschichte der Mathematik. Erster Band, von den ältesten Zeiten bis zum Jahre 1200 n. Chr. <Superscript>3</Superscript>Teubner, Leipzig (1907)

[4]  Folkerts, M.: The Development of Mathematics in Medieval Europe: The Arabs, Euclid, Regiomontanus. Variorum Collected Studies Series, vol. CS811. Ashgate, Aldershot (2006)

[5]  Freudenthal, H.: What Is Algebra and What Has It Been in History? Archive for History of Exact Sciences 16, 189–200 (1977)

[6]  Heiberg, J.L.: Euclidis Elementa. 5 vols. Euclidis Opera omnia, vol. I-V. Teubner, Leipzig (1883–1888)

[7]  Høyrup, J.: Changing Trends in the Historiography of Mesopotamian Mathematics: An Insider's View. History of Science 34, 1–32 (1996)

[8]  Høyrup, J.: Embedding: Multi-purpose Device for Understanding Mathematics and Its Development, or Empty Generalization? Filosofi og Videnskabsteori på Roskilde Universitetscenter. 3. Række: Preprints og Reprints, Nr. 8 (2000)

[9]  Høyrup, J.: Lengths, Widths, Surfaces: A Portrait of Old Babylonian Algebra and Its Kin. In: Studies and Sources in the History of Mathematics and Physical Sciences. Springer, New York (2002)

[10]  Høyrup, J.: Seleucid Innovations in the Babylonian 'Algebraic' Tradition and Their Kin Abroad. In: Dold-Samplonius, et al. (eds.) From China to Paris: 2000 Years Transmission of Mathematical Ideas. Boethius, vol. 46, pp. 9–29. Steiner, Stuttgart (2002)

[11]  Høyrup, J.: Leonardo Fibonacci and Abbaco Culture: a Proposal to Invert the Roles. Revue d'Histoire des Mathématiques 11, 23–56 (2005)

[12]  Høyrup, J.: Jacopo da Firenze's Tractatus Algorismi and Early Italian Abbacus Culture. In: Science Networks. Historical Studies, Birkhäuser, Basel etc, vol. 34 (2007)

[13]  Imhausen, A.: The Algorithmic Structure of the Egyptian Mathematical Problem Texts. In: Steele, J.M., Imhausen, A. (eds.) Under One Sky. Astronomy and Mathematics in the Ancient Near East. Alter Orient und Altes Testament, vol. 297, pp. 147–166. Ugarit-Verlag (2002)

[14]  Imhausen, A.: Ägyptische Algorithmen. Eine Untersuchung zu den mittelägyptischen mathematischen Aufgabentexten. Ägyptologische Abhandlungen, vol. 65. Harrassowitz, Wiesbaden (2003)

[15]  Knuth, D.: Ancient Babylonian Algorithms. Communications of the Association of Computing Machinery 15, 671–677 (1972); A correction of an erratum in 19, 108 (1976) is of no importance here

[16]  Libri, G.: Histoire des mathématiques en Italie. 4 vols. Jules Renouard. Paris (1838–1841)

[17]  Mahoney, M.S.: Babylonian Algebra: Form vs. Content. Studies in History and Philosophy of Science 1, 369–380 (1972)

[18]  Netz, R.: The Shaping of Deduction in Greek Mathematics: A Study in Cognitive History. Ideas in Context, vol. 51. Cambridge University Press, Cambridge (1999)

[19]  Neugebauer, O. (ed.) Mathematische Keilschrift-Texte. 3 vols. Quellen und Studien zur Geschichte der Mathematik, Astronomie und Physik. Abteilung A: Quellen. 3. Band, erster-dritter Teil. Julius. Springer, Berlin (1935, 1935, 1937)

[20]  Peet, T.E. (ed.): The Rhind Mathematical Papyrus, British Museum 10057 and 10058. Introduction, Transcription, Translation and Commentar. University Press of Liverpool, London (1923)

[21]  Rebstock, U. (ed.): Die Reichtümer der Rechner (Ġunyat al-Ḥussāb) von Aḥmad b.Ṭabāt (gest. 631/1234). Die Araber – Vorläufer der Rechenkunst. Beiträge zur Sprach- und Kulturgeschichte des Orients, vol. 32. Verlag für Orientkunde Dr. H. Vorndran, Walldorf-Hessen (1993)

[22]  Ritter, J.: Reading Strasbourg 368: A Thrice-Told Tale. Max-Planck-Institut für Wissenschaftsgeschichte. Preprint 103 (1998)

[23] Ritter, J.: Reading Strasbourg 368: A Thrice-Told Tale. In: Chemla, K. (ed.) History of Science, History of Text. Boston Studies in the Philosophy of Science, vol. 238, pp. 177–200. Kluwer, Dordrecht (2004); (I used an electronic manuscript version kindly put at my disposition by Jim Ritter and therefore have to omit page references)

[24] Rodet, L.: Les prétendus problèmes d'algèbre du manuel du calculateur égyptien (Papyrus Rhind). Journal asiatique. septième série 18, 184–232, 390–559 (1881)

[25] Spiesser, M. (ed.) Une arithmétique commerciale du XV<Superscript>e</Superscript> siècle. Le Compendy de la praticque des nombres de Barthélemy de Romans. De Diversis artibus, Brepols, Turnhout, vol. 70 (2003)

[26] Unguru, S.: On the Need to Rewrite the History of Greek Mathematics. Archive for History of Exact Sciences 15, 67–114 (1975)

[27] van der Waerden, B.L.: Defence of a "Shocking" Point of View. Archive for History of Exact Sciences 15, 199–210 (1976)

[28] Weil, A.: Who Betrayed Euclid? Archive for History of Exact Sciences 19, 91–93 (1978)

[29] Zeuthen, H.G.: Die Lehre von den Kegelschnitten im Altertum. Höst & Sohn, København (1886)

# Adversarial Scheduling Analysis of Game-Theoretic Models of Norm Diffusion


Gabriel Istrate[1],[*], Madhav V. Marathe[2], and S.S. Ravi[3]

[1] e-Austria Institute, V.Pârvan 4, cam. 045B, Timişoara RO-300223, Romania
`gabrielistrate@acm.org`
[2] Network Dynamics and Simulation Science Laboratory, and Dept. of Computer Science Virginia Tech.
`mmarathe@vbi.vt.edu`
[3] Computer Science Dept., S.U.N.Y. at Albany, Albany, NY 12222, U.S.A.
`ravi@cs.albany.edu`



**Abstract.** In [IMR01] we advocated the investigation of robustness of results in the theory of learning in games under adversarial scheduling models. We provide evidence that such an analysis is feasible and can lead to nontrivial results by investigating, in an adversarial scheduling setting, Peyton Young's model of diffusion of norms [You98]. In particular, our main result incorporates *contagion* into Peyton Young's model.

**Keywords:** evolutionary games, adversarial scheduling, Markov chains.


## 1 Introduction

Game-theoretic equilibria are *steady-state properties*; that is, given that all the players' actions correspond to an equilibrium point, it would be irrational for any of them to deviate from this behavior when the others stick to their strategy. The fundamental problem facing this type of concept is that it does not predict *how players arrive at this equilibrium in the first place*, or how they "choose" one such equilibrium, if several such points exist. The theory of equilibrium selection of Harsányi and Selten assumes some form of prior coordination between players, in the form of a *tracing procedure*. This strong prerequisite is often unrealistic.

*Evolutionary game theory* (EGT) [Wei95] attempts to explain the emergence of equilibria as the result of an evolutionary "learning" process. Models of this type assume one (or several) populations of *agents*, that interact by playing a game, and updating their behavior based on the outcome of this interaction.

Results in EGT are important not as realistic models of strategic behavior. Rather, they provide possible explanations for experimentally observed features of real-world social dynamics. Similar issues apply in the area of *agent-based social simulation* [GT05, BEM05]. In [Eps07, AE96], Epstein et al. advocated a generative approach to social science: in order to better understand a given phenomenon one should be able to generate it via simulations.

---


[*] Corresponding author.



A. Beckmann, C. Dimitracopoulos, and B. Löwe (Eds.): CiE 2008, LNCS 5028, pp. 273–282, 2008.

Given that such mathematical models or simulations are emerging as tools for policy-making (see e.g. [NBB99, ECC$^+$04, EG+04, FC+06]), how can we be sure that the conclusions that we derive from the output of the simulation do not crucially depend on the particular assumptions and features we embed in it? Part of the answer is that these results have to display "robustness" with respect to the various idealizations inherent in the model, be it mathematical or computational.

In this paper we are only concerned with one such issue: *scheduling*, i.e., the order in which agents get to update their strategies. Three alternatives have been studied extensively, both in the mathematical and the computer simulation literature: in the *synchronous mode* every player updates at every step. A popular alternative is *uniform matching*. Models of the latter type assume an underlying (hyper)graph topology (describing the sets of players allowed to simultaneously update in one step as a result of game playing) and choose a (hyper)edge uniformly at random from the available ones. Employing uniform matching in multiagent models of social systems is unrealistic for it assumes *perfect* and *global* randomness. Another alternative is to assume that the schedule is given as a a permutation of players that is repeated [BH+06, BH+07].

We investigate in an adversarial setting Peyton Young's model of evolution of norms [You98] (see also [You03]). The dynamics models an important aspect of social networks, the emergence of *conventions*, and has been proposed as an evolutionary justification for the emergence of certain rules in the pragmatics of natural language [Roo04]. Our results can be summarized as follows: results on selection of strict-dominant equilibria under random noise extend (Theorem 1) to a class of nonadaptive schedulers. However, such an extension fails for adaptive schedulers, even those with fairness properties similar to those of a random scheduler. Our main result (Theorem 2) extends the convergence to the strictly-dominant equilibrium to a class of "nonmalicious" adaptive schedulers that models *contagion* and has a certain *reversibility* property (the class of such schedulers includes the random scheduler as a special case). However for this class of schedulers the *convergence time* is *not* necessarily the one from the case of random scheduling.

Besides the relevance of our results to evolutionary game-theory, we hope that the concepts and techniques relevant to this paper can contribute to the theory of *rapidly mixing Markov chains*.

## 2   Preliminaries

We consider adversarial analysis of *population games* [Blu01]. Systems of interest in this class consist of a number of *agents*, defined as the vertices of a hypergraph $H = (V, E)$. Each edge of this hypergraph represents a particular choice of all players who can play (one or more simultaneous instances of) a game $G$ that defines the dynamics. Each player has a *state* (generally a mixed strategy of $G$) chosen from a certain set $S$. The global state of the system is an element of $\overline{S} = S^V$. The dynamics proceeds by choosing one edge $e$ of $H$ (according to a scheduling mechanism that is specified by the scheduler), letting the agents in $e$ play the game, and updating their states as a result of game playing.

## 2.1 Schedulers

Denote by $X^*$ the set of finite words over alphabet $X$.

**Definition 1.** *A (probabilistic) scheduler assigns a probability distribution $p_{w,s}$ on $E$ to each triple $(w, s, s_0)$ consisting of initial prefixes $w \in E^*, s \in \overline{S}^*$ with $|w| = |s|$ and starting state $s_0 \in \overline{S}$. The next element $e \in E$ to be scheduled, given prefixes $w, s$ and initial state $s_0$, is sampled from $E$ according to $p_{w,s,s_0}$.*

*A non-adaptive probabilistic scheduler is specified by (a) a collection (multi-set) $\Sigma = \{\mathcal{D}_1, \ldots, \mathcal{D}_m\}$ of probability distributions on the set $E$ such that every $x \in E$ belongs to the support of some distribution $\mathcal{D}_i$ and (b) a fixed permutation $\pi$ of $\Sigma$. The scheduler proceeds by (possibly concurrently) scheduling elements of $E$ sampled from a distribution from $\Sigma$ chosen according to (consecutive repetitions of) permutation $\pi$. For $C > 0$, a non-adaptive probabilistic scheduler is $C$ individually-fair if for every $x \in E$, the probability that $x$ is scheduled during one round of $\pi$ is at least $C/|E|$.*

One can define, for any given triple $(w, s, s_0)$, where $w \in E^*, s \in \overline{S}^*$ and $s_0 \in \overline{S}$, a probability $\pi_{w,s,s_0}$ that, starting from state $s_0$ the scheduler uses $w$ as the initial prefix of its schedule and evolves its global state according to string $s$. Let $\Omega$ denote the resulting probability space. We divide each trajectory of a probabilistic scheduler into *rounds*: the first round is the smallest initial segment that schedules each element of $E$ at least once, the second round is the smallest segment starting at the end of the first round that schedules each element at least once, and so on.

**Definition 2.** *If $f(\cdot)$ is a function on integers, we say that a family of probabilistic schedulers, indexed by $n$, the cardinality of the set $E$, is $O(f(n))$-fair w.h.p. if there exists a monotonically decreasing function $g : (0, \infty) \to (0, 1)$, with $\lim_{\epsilon \to \infty} g(\epsilon) = 0$ such that for every state $s \in \overline{S}$, the random variable $l_i$ measuring the length of the $i$'th round, satisfies the condition $\underline{\lim}_{n \to \infty} \mathrm{Prob}[l_i > \epsilon \cdot f(n)] < g(\epsilon)$.*

Random scheduling can be specified by a non-adaptive probabilistic scheduler whose set $\Sigma$ consists of just one distribution, namely the uniform distribution on $E$. This scheduler is 1-individually fair and, by the well-known Coupon Collector Lemma it is also $O(n \log(n))$-fair w.h.p.

## 2.2 Peyton Young's Model of Norm Diffusion

The setup of this model is the following: agents located at the vertices of a graph $G$ interact by playing a two-person symmetric game with payoff matrix $M = (m_{i,j})_{i,j \in \{\mathbf{A}, \mathbf{B}\}}$ displayed in Figure 1. It is assumed that strategy $\mathbf{A}$ is a so called *strict risk-dominant equilibrium*. That is, we have $a - d > b - c > 0$. Each undirected edge $\{i, j\}$ has a positive weight $w_{ij} = w_{ji}$ that measures its "importance". When scheduled, agents play (using the same strategy, identified as *the agent's state*) against each of their neighbors. If agent $i$ is the one to

| strategies | **A** | **B** |
|:---:|:---:|:---:|
| **A** | a,a | c,d |
| **B** | d,c | b,b |

**Fig. 1.** Payoff matrix

update, $\overline{x}$ is the joint profile of agents' strategies, and $z \in \{\mathbf{A}, \mathbf{B}\}$ is the candidate new state, $p^{\beta}(x_i \rightarrow z|\overline{x}) \sim e^{\beta \cdot \nu_i(z, \overline{x}_{-i})}$, where $\nu_i(z, \overline{x}_{-i})$, the payoff of the $i$'th agent should he play strategy $z$ while the others' profile remains the same is given by $\nu_i(z, \overline{x}_{-i}) = \sum_{(i,j) \in E} w_{ij} m_{z,x_j}$. Under random scheduling, the process we defined is a variant of the best-response dynamics. This latter process (viewed as a Markov chain) is not ergodic. Indeed, the since in game $G$ it is always better to play the same strategy as your partner, the dynamics has at least two fixed points, states "all $\mathbf{A}$" and "all $\mathbf{B}$", defined as the two states where all labels have a common value.

An important property of Peyton Young's dynamics is that it corresponds to a *potential game*: there exists a function $\rho : \overline{V} \rightarrow \mathbf{R}$ such that, for any player $i$, any possible actions $a_1, a_2$ of player $i$, and any action profile $\overline{a}$ of the other players, $u_i(a_1, a) - u_i(a_2, a) = \rho(a_1, a) - \rho(a_2, a)$ (where $u_i$ is the utility function of player $i$). In other words changes in utility as a result of strategy update correspond to changes in a global potential function. An explicit potential is given by $\rho^*(x) = \sum_{(h,k) \in E} w_{h,k} m_{x_h, x_k}$.

## 2.3 Stochastic Stability

A fundamental concept we are dealing with is that of a *stochastically stable state* for dynamics described by a Markov chain.

**Definition 3.** *Consider a Markov process $P^0$ defined on a finite state space $\Omega$. For each $\epsilon > 0$, define a Markov process $P^{\epsilon}$ on $\Omega$. $P^{\epsilon}$ is a* regular perturbed *Markov process if all of the following conditions hold.*

- *$P^{\epsilon}$ is irreducible for every $\epsilon > 0$.*
- *For every $x, y \in \Omega$, $\lim_{\epsilon > 0} P^{\epsilon}_{xy} = P^0_{xy}$.*
- *If $P_{xy} > 0$ then there exists $r(m) > 0$, the resistance of transition $m = (x \rightarrow y)$, such that as $\epsilon \rightarrow 0$, $P^{\epsilon}_{xy} = \Theta(\epsilon^{r(m)})$.*

*Let $\mu^{\epsilon}$ be the (unique) stationary distribution of $P^{\epsilon}$. A state $s$ is* stochastically stable *if $\underline{\lim}_{\epsilon \rightarrow 0} \mu^{\epsilon}(s) > 0$.*

**Definition 4.** *A tree rooted at node $j$ is a set $T$ of edges such that for any state $w \neq j$ there exists a unique (directed) path from $w$ to $j$. The* resistance *of a rooted tree $T$ is the sum of resistances of all edges in $T$.*

The following characterization of stochastically stable states is presented as Lemma 3.2 in the Appendix of [You98]:

**Proposition 1.** *Let $P^\epsilon$ be a regular perturbed Markov process, and for each $\epsilon > 0$ let $\mu^\epsilon$ be the unique stationary distribution of $P^\epsilon$. Then $\lim_{\epsilon \to 0} \mu^\epsilon = \mu^0$ exists, and $\mu^0$ is a stationary distribution of $P^0$. The stochastically stable states are precisely those states $z$ such that there exists a tree rooted at $z$ of minimal resistance (among all rooted trees).*

**Definition 5.** *Given a graph $G$, a nonempty subset $S$ of vertices and a real number $0 \le r \le 1/2$, we say that $S$ is $r$-close-knit if $\forall S' \subseteq S, S' \ne \emptyset$, $\frac{e(S',S)}{\sum_{i \in S'} deg(i)} \ge r$, where $e(S',S)$ is the number of edges with one endpoint in $S'$ and the other in $S$, and $deg(i)$ is the degree of vertex $i$. A graph $G$ is $(r,k)$-close-knit if every vertex is part of a $r$-close-knit set $S$, with $|S| = k$.*

**Definition 6.** *Given $p \in [0,1]$, the $p$-inertia of the process is the maximum, over all states $x_0 \in S$, of $W(\beta, p, x_0)$, the expected waiting time until at least $1 - p$ of the population is playing action $A$ conditional on starting in state $x_0$.*

The model in [You03, You98] assumes independent individual updates, arriving at random times governed (for each agent) by a Poisson arrival process with rate one. Since we are, however, interested in adversarial models that do not have an easy description in continuous time we will assume that the process proceeds in discrete steps. At each such step a random node is scheduled. It is a simple exercise to translate the result in [You03, You98] to an equivalent one for global, discrete-time scheduling. The conclusions of this translation are: (1) The stationary distribution of the process is *the Gibbs distribution*, $\mu_\beta(x) = \frac{e^{\beta \rho(x)}}{\sum_z e^{\beta \rho(x)}}$, where $\rho$ is the potential function of the dynamics. (2) "All **A**" is the unique stochastically-stable state of the dynamics. (3) Let $r^* = \frac{b-c}{a-d+b-c}$, and let $r > r^*$, $k > 0$. On a family of $(r,k)$-close-knit graphs the convergence time is $O(n)$.

## 3   Results

First we note that Peyton Young's results easily extend to non-adaptive schedulers. *Adaptive* schedulers on the other hand, even those of fairness no higher than that of the random scheduler, can preclude the system from ever entering a state where a proportion higher than $r$ of agents plays the risk-dominant strategy:

**Theorem 1.** *The following hold:*

*(i) For all non-adaptive schedulers, the state "all **A**" is the unique stochastically stable state of the system.*

*(ii) Let $\mathcal{G}$ be a class of graphs that are $(r,k)$-close-knit for some fixed $r > r^*$. Let $f = f(n)$ be a class of non-adaptive $\Theta(1)$ individually fair schedulers. Given any $p \in (0,1)$ there exists a $\beta_p$ such that for all $\beta > \beta_p$ there exists a constant $C$ such that the $p$-inertia of the process (under scheduling given by $f$) is at most $C \cdot m \cdot n$, where $m = m(n)$ is the number of rounds of $f$.*

*(iii) For every $r \in (0,1)$ there exists an adaptive scheduler, $O(n \log(n))$-fair w.h.p. (where the constant hidden in the "O" notation depends on $r$), that can*

*forever prevent the system, started on the "all **B**'s" configuration, from ever having more than a fraction of r of the agents playing **A**.*

The proof of Theorem 1 is fairly simple and is left to [IMR08] because of space reasons.

### 3.1   Main Result: Diffusion of Norms by Contagion

Adaptive schedulers can display two very different notions of adaptiveness: (i) The next node depends only on the set of previously scheduled nodes, or (ii) It crucially depends on the *states* of the system so far.

The adaptive schedulers in Theorem 1(iii) crucially use the second, stronger, kind of adaptiveness. In the sequel we study a model that displays adaptiveness of type (i) but not of type (ii). The model is specified as follows: To each node $v$ we associate a probability distribution $D_v$ on the vertices of $G$. We then choose the next scheduled node according to the following process. If $t_i$ is the node scheduled at stage $i$, we chose $t_{i+1}$, the next scheduled node, by sampling from $D_{t_i}$. In other words, the scheduled node performs a (non-uniform) random walk on the vertices of graph $G$. To exclude technical problems such as the periodicity of this random walk, we assume that it is always the case that $v \in supp(D_v)$. Also, let $H$ be the directed graph with edges defined as follows: $(x, y) \in E[H] \iff (y \in supp(D_x))$. This dynamics generalizes both the class of non-adaptive schedulers from previous result and the random scheduler (for the case when $H$ is the complete graph). In the context of van Rooy's evolutionary analysis of signalling games in natural language [Roo04], it functions as a simplified model for an essential aspect of emergence of linguistic conventions: transmission via *contagion*.

It is easy to see that the dynamics can be described by an aperiodic Markov chain $M$ on the set on $V^{\{\mathbf{A},\mathbf{B}\}} \times V$, where a state $(\overline{w}, x)$ is described as follows:

- $\overline{w}$ is the set of strategies chosen by the agents.
- $x$ is the label of the last agent that was given the chance to update its state.

If the directed graph $H$ is strongly connected then the Markov chain $M$ is irreducible, hence it has a stationary distribution $\Pi$. We will, therefore, limit ourselves in the sequel to settings with strongly connected $H$. We will, further, assume that the dynamics is *weakly reversible*, i.e. $(x \in supp(D_y))$ if and only if $(y \in supp(D_x))$. This, of course, means that the graph $H$ is undirected. Note that since we do not constrain otherwise the transition probabilities of distributions $D_i$, the stationary distribution $\Pi$ of the Markov chain does *not*, in general, decompose as a product of component distributions. That is, one cannot generally write $\Pi(w, x)$ as $\Pi(w, x) = \pi(w) \cdot \rho(x)$, for some distributions $\pi, \rho$.

**Theorem 2.** *The set $Q = \{(w, x)|w = V^{\mathbf{A}}\}$ is the set of stochastically stable states for the diffusion of norms by contagion.*

*Proof.* States in $Q$ are obviously reachable from one another by zero-resistance moves, so it is enough to consider one state $y \in Q$ and prove that it is stochastically stable. To do so, by Proposition 1, all we need to do is to show that $y$ is the root of a tree of minimal resistance. Indeed, consider another state $x \in Q$ and let $T$ be a minimum potential tree rooted at $x$.

*Claim.* There exists a tree $\overline{T}$ rooted at $y$ having potential less or equal to the potential of the tree $T$, strictly smaller in case $x$ is not a state having all its first-component labels equal to $A$.

Let $\pi_{y,x} = (x_0, i_0) \rightarrow (x_1, i_1) \rightarrow \ldots \rightarrow (x_k, i_k) \rightarrow (x_{k+1}, i_{k+1}) \rightarrow \ldots \rightarrow (x_r, i_r)$ be the path from $y$ to $x$ in $T$ (that is $(x_0, i_0) = y$, $(x_r, i_r) = x$).

We will define $\overline{T}$ by viewing the set of edges of $T$ as partitioned into subsets of edges corresponding to paths as follows (see Figure 2 (a)):

  (i)   The set of edges of path $\pi_{y,x}$.
 (ii)   The set of edges of the subtree rooted at $y$.
(iii)   Edges of tree components (perhaps consisting of a single node) rooted at a node of $\pi_{y,x}$, other than $y$ (but possibly being $x$).

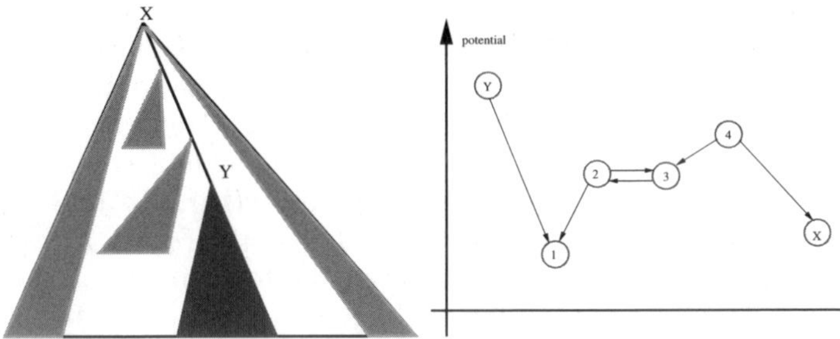


**Fig. 2.** (a). Decomposition of edges of tree $T$ (b). Resistance of edges on a path between two nodes $X$ and $Y$.

To obtain $\overline{T}$ we will transform each tree (path) in the above decomposition of $T$ into one that will be added to $\overline{T}$. The transformation goes as follows:

  (i)   Instead of path $\pi_{y,x}$ we add path $\varPi_{x,y}$ from $x$ to $y$ defined by: $\varPi_{x,y} = (x_r, i_r) \rightarrow (x_{r-1}, i_r) \rightarrow (x_{r-2}, i_{r-1}) \rightarrow \ldots \rightarrow (x_0, i_1) \rightarrow (x_0, i_0) = y$.
 (ii)   Rooted trees of type (2) are included into tree $\overline{T}$ as well.
(iii)   The transformation is more complicated for the third type of edges, and we explain it in detail. Let $W_k$ be a tree component of $T$, connected to path $\pi_{y,x}$ at connection point $(x_k, i_k)$.
        **Case 1:** $x_k = x_{k-1}$. Then the point $(x_k, i_k) = (x_{k-1}, i_k)$ belongs to path $\varPi_{x,y}$ as well, so one can just add the rooted tree $W_k$ to $\overline{T}$ as well.
        **Case 2:** $x_k \neq x_{k-1}$ *and the move* $(x_{k-1}, i_{k-1}) \rightarrow (x_k, i_k)$ *has positive resistance.* In this case, since in configuration $x_{k-1}$ and scheduled node $i_k$ we have a choice of either moving to $x_k$ or staying in $x_{k-1}$, it follows that the move $(x_k, i_k) \rightarrow (x_{k-1}, i_k)$ has zero resistance.
        Therefore we can add to $\overline{T}$ the tree $\overline{W_k} = W_k \cup \{(x_k, i_k) \rightarrow (x_{k-1}, i_k)\}$. The tree has the same resistance as the one of tree $W_k$.

**Case 3:** $x_{k-1} \neq x_k$ and the move $(x_{k-1}, i_{k-1}) \to (x_k, i_k)$ has zero resistance. Let $j$ be the smallest integer such that either $x_{k+j+1} = x_{k+j}$ or $x_{k+j+1} \neq x_{k+j}$ and the move $(x_{k+j}, i_{k+j}) \to (x_{k+j+1}, i_{k+j+1})$ has positive resistance. In this case, one can first replace $W_k$ by $W_k \cup \{(x_k, i_k) \to (x_{k+1}, i_{k+1}), (x_{k+1}, i_{k+1}) \to \ldots \to (x_{k+j}, i_{k+j})\}$ without increasing its total resistance. Then we apply one of the techniques from Case 1 or Case 2.

**Case 4:** $x_{k-1} \neq x_k$, the move $(x_{k-1}, i_{k-1}) \to (x_k, i_k)$ has zero resistance, and all moves on $\pi_{y,x}$, from $x_k$ up to $x$ have zero resistance. Then define $\overline{W_k} = W_k \cup (x_k, i_k) \to (x_{k+1}, i_{k+1}) \to \ldots \to x$.

It is easy to see that no two sets $W_k$ intersect on an edge having positive resistance. The union of the paths of all the sets is a directed associated graph $W$ rooted at $y$, that contains a rooted tree $\overline{T}$ of potential no larger than the potential of $W$. Since transformations in cases (i),(iii) do not increase tree resistance, to compare the potentials of $T$ and $W$ it is enough to compare the resistances of paths $\pi_{y,x}$ and $\Pi_{x,y}$.

We come now to a fundamental property of the game $G$: since it is a potential game, the resistance $r(m)$ of a move $m = (a_1, j_1) \to (a_2, j_2)$ only depends on the values of the potential function at three points: $a_1, a_2$ and $a_3$, where $a_3$ is the state obtained by assigning node $j_2$ the value not assigned by move to $a_2$. Specifically, $r(m) > 0$ if either $\rho^*(a_2) < \rho^*(a_1)$, in which case $r(m) = \rho^*(a_1) - \rho^*(a_2)$, or $a_2 = a_1$ and $\rho^*(a_3) > \rho^*(a_1)$, in which case $r(m) = \rho^*(a_3) - \rho^*(a_1)$. In other words, the resistance of a move is positive in the following two cases: (1) The move leads to a decrease of the value of the potential function. In this case the resistance is equal to the difference of potentials. (2) The move corresponds to keeping the current state (thus not modifying the value of the potential function), but the alternate move would have increased the potential. In this case the resistance is equal to the value of this increase.

Let us now compare the resistances of paths $\pi_{y,x}$ and $\Pi_{x,y}$. First, the two paths contain no edges of infinite resistance, since they correspond to possible moves under Markov chain dynamics $P^\epsilon$. If we discount second components, the two paths correspond to a single sequence of states $Z$ connecting $x_0$ to $x_r$, more precisely to *traversing $Z$ in opposite directions*. (The last move in $\Pi_{x,y}$ has zero resistance and can thus be discounted). Resistant moves of type (2) are taken into account by both traversals, and contribute the same resistance value to both paths. So, to compare the resistances of the two paths it is enough to compare resistance of moves of type (1). Moves of type (1) of positive resistance are those that lead to a decrease in the potential function. Decreasing potential in one direction corresponds to increasing it in the other (therefore such moves have zero resistance in the opposite direction).

An illustration of the two types of moves is given in Figure 2 (b), where the path between $X$ and $Y$ goes through four other nodes, labeled 1 to 4. The relative height of each node corresponds to the value of the potential function at that node. Nodes 2 and 3 have equal potential, so the transition between 2 and 3 contributes an equal amount to the resistance of paths in both directions (which may be positive or not). Other than that only transitions of positive resistance are pictured.

The conclusion of this argument is that $r(\pi_{y,x}) - r(\Pi_{x,y}) = \rho^*(x) - \rho^*(y) \geq 0$, and $r(\pi_{y,x}) - r(\Pi_{x,y}) > 0$ unless $x$ is an "all $A$" state.

### 3.2  The Inertia of Diffusion of Norms with Contagion

Theorem 2 shows that random scheduling is not essential in ensuring that stochastically stable states in Peyton Young's model correspond to all players playing $A$: the same result holds in the model with contagion. On the other hand, the result on the $p$-inertia of the process on families of close-knit graphs is *not* robust to such an extension. Indeed, consider the *line graph* $L_{2n+1}$ on $2n+1$ nodes labelled $-n, \ldots, -1, 0, 1 \ldots n$. Consider a random walk model such that: (a) the origin of the random walk is node 0, and (b) the walk goes left, goes right or stays in place, each with probability $1/3$. It is a well-known property of the random walk that it takes $\Omega(n^2)$ time to reach nodes at distance $\Omega(n)$ from the origin. Therefore, the $p$-inertia of this random walk dynamics is $\Omega(n^2)$ even though for every $r > 0$ there exists a constant $k$ such that the family $\{L_{2n+1}\}$ is $(r, k)$-close-knit for large enough $n$.

In the journal version of the paper we will present an upper bound on the $p$-inertia for the diffusion of norms with contagion based on concepts similar to the *blanket time* of a random walk [WZ96].

## 4   Conclusions and Acknowledgments

Our results have made the original statement by Peyton Young more robust, and have highlighted the (lack of) importance of various properties of the random scheduler in the results from [You98]: the *reversibility* of the random scheduler, as well as its inability to use the global system state are important in an adversarial setting, while its fairness properties are not crucial for convergence, only influencing convergence time. Also, the fact that the stationary distribution of the perturbed process is the Gibbs distribution (true for the random scheduler) does not necessarily extend to the adversarial setting.

This work has been supported by the Romanian CNCSIS under a PN-II "Idei" Grant, by the U.S. Department of Energy under contract W-705-ENG-36 and NSF Grants CCR-97-34936, CNS-062694, SES-0729441 and NIH-NIGMS MIDAS project 5U01GM070694-06.

## References

[AE96]     Axtell, R., Epstein, J.: Growing Artificial Societies: Social Science from the Bottom Up. The MIT Press (1996)

[Blu01]    Blume, L.: Population games. In: Durlauf, S., Peyton Young, H. (eds.) Social Dynamics: Economic Learning and Social Evolution. MIT Press (2001)

[BEM05]   Barrett, C., Eubank, S., Marathe, M.: Modeling and Simulation of Large Biological, Information and Socio-Technical Systems: An Interaction Based Approach. In: Goldin, D., Smolka, S., Wegner, P. (eds.) Interactive Computation: The New Paradigm. Springer, Heidelberg ( to appear, 2005)

[BH+06]   Barrett, C., Hunt III, H., Marathe, M., Ravi, S.S., Rosenkrantz, D., Stearns, R.: Dichotomy Theorems for Reachability Problems in Sequential Dynamical Systems. Journal of Computer and System Sciences 72, 1317–1345 (2006)

[BH+07]   Barrett, C., Hunt III, H., Marathe, M., Ravi, S.S., Rosenkrantz, D., Stearns, R., Thakur, M.: Computational Aspects of Analyzing Social Network Dynamics. In: Proc. International Joint Conference on Artificial Intelligence (IJCAI 2007), Hyderabad, India (January 2007)

[ECC+04]  Epstein, J.M., Cummings, D., Chakravarty, S., Singa, R., Burke, D.: Toward a Containment Strategy for Smallpox Bioterror. An Individual-Based Computational Approach. Brookings Institution Press (2004)

[Eps07]   Epstein, J.: Generative Social Science: Studies in Agent-based Computational Modeling. Princeton University Press (2007)

[EG+04]   Eubank, S., Guclu, H., Anil Kumar, V.S., Marathe, M., Srinivasan, A., Toroczkai, Z., Wang, N.: Modeling Disease Outbreaks in Realistic Urban Social Networks. Nature 429, 180–184 (2004)

[FC+06]   Ferguson, N.L., Cummings, D., Fraser, C., Cajka, J., Cooley, P., Burke, D.: Strategies for mitigating an influenza pandemic. Nature (April 2006)

[GT05]    Gilbert, N., Troizch, K.: Simulation for social scientists, 2nd edn. Open University Press (2005)

[IMR01]   Istrate, G., Marathe, M.V., Ravi, S.S.: Adversarial models in evolutionary game dynamics. In: Proceedings of the 13th ACM-SIAM Symposium on Discrete Algorithms (SODA 2001) (2001) (journal version in preparation)

[IMR08]   Istrate, G., Marathe, M.V., Ravi, S.S.: Adversarial scheduling analysis of game-theoretic models of norm diffusion. Technical report 0803.2495, arXiv.org (2008)

[NBB99]   Nagel, K., Beckmann, R., Barrett, C.: TRANSIMS for transportation planning. In: Bar-Yam, Y., Minai, A. (eds.) Proceedings of the Second International Conference on Complex Systems, Westview Press (1999)

[Roo04]   Van Rooy, R.: Signalling games select Horn strategies. Linguistics and Philosophy 27, 423–497 (2004)

[Wei95]   Weibull, J.: Evolutionary Game Theory. M.I.T. Press (1995)

[WZ96]    Winkler, P., Zuckerman, D.: Multiple cover time. Random Structures and Algorithms 9(4), 403–411 (1996)

[You93]   Young, H.P.: The evolution of conventions. Econometrica 61(1), 57–84 (1993)

[You98]   Young, H.P.: Individual Strategy and Social Structure: an Evolutionary Theory of Institutions. Princeton University Press (1998)

[You03]   Young, H.P.: The diffusion of innovations in social networks. In: Blume, L., Durlauf, S. (eds.) The Economy as a Complex System III, Oxford University Press (2003)

# A Simple P-Matrix Linear Complementarity Problem for Discounted Games*

Marcin Jurdziński and Rahul Savani**

Department of Computer Science, University of Warwick, UK
{mju,rahul}@dcs.warwick.ac.uk

**Abstract.** The values of a two-player zero-sum binary discounted game
are characterized by a P-matrix linear complementarity problem (LCP).
Simple formulas are given to describe the data of the LCP in terms of
the game graph, discount factor, and rewards. Hence it is shown that the
unique sink orientation (USO) associated with this LCP coincides with
the strategy valuation USO associated with the discounted game. As an
application of this fact, it is shown that Murty's least-index method for
P-matrix LCPs corresponds to both known and new variants of strategy
improvement algorithms for discounted games.

**Keywords:** Discounted game, linear complementarity problem, P-
matrix, strategy improvement algorithm, unique sink orientation, zero-
sum game.

## 1 Introduction

Discounted (stochastic) games were introduced by Shapley [15]. The monograph
of Filar and Vrieze [6] discusses discounted (stochastic) games in detail. For
clarity, we only consider non-stochastic discounted games in this paper. One
motivation for studying these games is that there is a polynomial time reduction
to discounted games from parity games (via mean-payoff games) [14,21], which
are equivalent to model checking for the modal mu-calculus. A polynomial-time
algorithm for parity games is a long-standing open question.

Our contribution is a transparent reduction from binary discounted games to
the P-matrix linear complementarity problem (LCP). The simple formulas for
the LCP data allow us to show that the unique sink orientation of the cube
associated with the P-matrix LCP [16] is the same as the strategy valuation
USO for the game. As an application of this fact, it is shown that Murty's least-
index method for P-matrix LCPs corresponds to both known and new variants of
strategy improvement algorithms for binary discounted games. For games with
outdegree greater than two, one gets *generalized* LCPs. Discounted games can
be reduced in polynomial time to simple stochastic games [21]. Recently (non-
binary) simple stochastic games have been reduced to P-matrix (generalized)

* This research was supported in part by EPSRC projects EP/D067170/1,
EP/E022030/1, and EP/D063191/1 (DIMAP).
** Corresponding author.

A. Beckmann, C. Dimitracopoulos, and B. Löwe (Eds.): CiE 2008, LNCS 5028, pp. 283–293, 2008.

LCPs [17,8]. The monograph of Cottle et. al. [4] is the authoritative source on the linear complementarity problem.

## 2   Discounted Games

A (perfect-information binary) discounted game $\Gamma=(S,\lambda,\rho,r_\lambda,r_\rho,\beta,S_{\mathrm{Min}},S_{\mathrm{Max}})$ consists of: a set of states $S = \{\,1,2,\dots,n\,\}$; *left* and *right successor* functions $\lambda,\rho : S \to S$, respectively; *reward* functions $r_\lambda, r_\rho : S \to \mathbb{R}$ for left and right edges respectively, with $r_\lambda(s) = r_\rho(s)$ if $\lambda(s) = \rho(s)$; a *discount factor* $\beta \in [0,1)$; and a partition $(S_{\mathrm{Min}}, S_{\mathrm{Max}})$ of the set of states. A sequence $\langle s_0, s_1, s_2, \dots \rangle \in S^\omega$ is a *play* if for all $i \in \mathbb{N}$, we have that $\lambda(s_i) = s_{i+1}$ or $\rho(s_i) = s_{i+1}$. We define the *($\beta$-)discounted payoff* $\mathcal{D}(\pi,\beta)$ of a play $\pi = \langle s_0, s_1, s_2, \dots \rangle$ by $\mathcal{D}(\pi,\beta) = \sum_{i=0}^\infty \beta^i r(s_i, s_{i+1})$, with $r(s_i, s_{i+1})$ denoting $r_\lambda(s_i)$ or $r_\rho(s_i)$ as appropriate.

A function $\mu : S_{\mathrm{Min}} \to S$ is a positional strategy for player Min if for every $s \in S_{\mathrm{Min}}$, we have that $\mu(s) = \lambda(s)$ or $\mu(s) = \rho(s)$. Strategies $\chi : S_{\mathrm{Max}} \to S$ for player Max are defined analogously. We write $\Pi_{\mathrm{Min}}$ and $\Pi_{\mathrm{Max}}$ for the sets of positional strategies for player Min and Max, respectively. For strategies $\mu \in \Pi_{\mathrm{Min}}$ and $\chi \in \Pi_{\mathrm{Max}}$, and a state $s \in S$, we write $\mathrm{Play}(s,\mu,\chi)$ for the play $\langle s_0, s_1, s_2, \dots \rangle$, such that $s_0 = s$, and for all $i \in \mathbb{N}$, we have that $s_i \in S_{\mathrm{Min}}$ implies $\mu(s_i) = s_{i+1}$, and $s_i \in S_{\mathrm{Max}}$ implies $\chi(s_i) = s_{i+1}$. A function $\sigma : S \to S$ is a (combined) positional strategy. For a combined positional strategy $\sigma : S \to S$, we write $\mathrm{Play}(s,\sigma)$ for the play $\mathrm{Play}(S, \sigma{\restriction}S_{\mathrm{Min}}, \sigma{\restriction}S_{\mathrm{Max}})$.

For every $s \in S$, we define the *lower value* $\mathrm{Val}_*(s,\beta)$ and the *upper value* $\mathrm{Val}^*(s,\beta)$ by

$$\mathrm{Val}_*(s,\beta) = \max_{\chi \in \Pi_{\mathrm{Max}}} \min_{\mu \in \Pi_{\mathrm{Min}}} \mathcal{D}(\mathrm{Play}(s,\mu,\chi),\beta),$$

$$\mathrm{Val}^*(s,\beta) = \min_{\mu \in \Pi_{\mathrm{Min}}} \max_{\chi \in \Pi_{\mathrm{Max}}} \mathcal{D}(\mathrm{Play}(s,\mu,\chi),\beta).$$

The inequality $\mathrm{Val}_*(s,\beta) \le \mathrm{Val}^*(s,\beta)$ always holds. We say that the *value* exists in a state $s \in S$, if we have $\mathrm{Val}_*(s,\beta) = \mathrm{Val}^*(s,\beta)$; we then write $\mathrm{Val}(s,\beta)$ for $\mathrm{Val}_*(s,\beta) = \mathrm{Val}^*(s,\beta)$. We say that the discounted game is *positionally determined* if for all $s \in S$, the value exists in $s$.

We identify functions $\overline{v} : S \to \mathbb{R}$ and $n$-vectors $\overline{v} \in \mathbb{R}^n$. For $s \in S$, depending on which interpretation is more natural in context, we write either $\overline{v}(s)$ or $\overline{v}_s$. We do the same for $n$-vectors of variables, for which Latin letters $v, w$, and $z$ are typically used. We say that $\overline{v} : S \to \mathbb{R}$ is a solution of the optimality equations $\mathrm{Opt}(\Gamma)$ if for all $s \in S$, we have

$$\overline{v}(s) = \begin{cases} \min\{\, r_\lambda(s) + \beta \cdot \overline{v}(\lambda(s)),\ r_\rho(s) + \beta \cdot \overline{v}(\rho(s))\,\} & \text{if } s \in S_{\mathrm{Min}}, \\ \max\{\, r_\lambda(s) + \beta \cdot \overline{v}(\lambda(s)),\ r_\rho(s) + \beta \cdot \overline{v}(\rho(s))\,\} & \text{if } s \in S_{\mathrm{Max}}. \end{cases} \tag{1}$$

**Theorem 1 ([15]).** *Every discounted game is positionally determined. Moreover, the optimality equations $Opt(\Gamma)$ have a unique solution $\overline{v} : S \to \mathbb{R}$, and for every $s \in S$, we have $Val(s,\beta) = \overline{v}$.*

It follows from the existence of a solution to the optimality equations that there exist optimal *pure* positional strategies [21]. Hence, without loss of generality, we consider only pure positional strategies.

## 3   A P-Matrix LCP for Discounted Games

### 3.1   An LCP for Discounted Games

Consider the following set of constraints over variables $v(s), w(s), z(s)$, for all $s \in S$:

$$v(s) + w(s) = r_\lambda(s) + \beta v(\lambda(s)), \text{ if } s \in S_{\text{Min}}, \tag{2}$$
$$v(s) - w(s) = r_\lambda(s) + \beta v(\lambda(s)), \text{ if } s \in S_{\text{Max}}, \tag{3}$$
$$v(s) + z(s) = r_\rho(s) + \beta v(\rho(s)), \text{ if } s \in S_{\text{Min}}, \tag{4}$$
$$v(s) - z(s) = r_\rho(s) + \beta v(\rho(s)), \text{ if } s \in S_{\text{Max}}, \tag{5}$$
$$w(s), z(s) \geq 0 \tag{6}$$
$$w(s) \cdot z(s) = 0. \tag{7}$$

Non-negative variables $w(s)$ and $z(s)$ should be thought of as slack variables which turn inequalities such as $v(s) \leq r_\lambda(s) + \beta v(\lambda(s))$ if $s \in S_{\text{Min}}$, or $v(s) \geq r_\rho(s) + \beta v(\rho(s))$ if $s \in S_{\text{Max}}$, into equations. Note that variables $w$ are slacks for left successors, and variables $z$ are slacks for right successors. The natural inequalities for left and right successors, turned into equations (2)–(5) using non-negative slack variables (6), together with the *complementarity* condition (7), for all $s \in S$, yield the following characterization.

**Proposition 1.** *There is a unique solution* $\overline{v}, \overline{w}, \overline{z} : S \to \mathbb{R}$ *of constraints (2)–(7), and* $\overline{v}$ *is the unique solution of Opt($\Gamma$).*

A linear complementarity problem [4] LCP($M, q$) is the following set of constraints:

$$w = Mz + q, \tag{8}$$
$$w, z \geq 0, \tag{9}$$
$$w_s \cdot z_s = 0, \text{ for every } s \in S, \tag{10}$$

where $M$ is an $n \times n$ real matrix, $q \in \mathbb{R}^n$, and $w$ and $z$ are $n$-vectors of real variables. In order to turn constraints (2)–(7) into a linear complementarity problem LCP($M, q$), we rewrite equations (2)–(5) in matrix notation and eliminate variables $v(s)$, for all $s \in S$.

For a predicate $p$, we define $[p] = 1$ if $p$ holds, and $[p] = 0$ if $p$ does not hold. For $\sigma : S \to S$, define the $n \times n$ matrix $T_\sigma$ by $(T_\sigma)_{st} = [\sigma(s) = t]$, for all $s, t \in S$. For every $n \times n$ matrix $A$, we define the matrix $\widehat{A}$ by setting $(\widehat{A})_{st} = (-1)^{[s \in S_{\text{Min}}]} A_{st}$, for every $s, t \in S$. Observe that $\widehat{A}$ is obtained from $A$ by multiplying all entries in every row $s$, such that $s \in S_{\text{Min}}$, by $-1$.

Equations (2)–(3) and (4)–(5) can be written as

$$\widehat{I}v = w + \widehat{I}r_\lambda + \beta\widehat{T_\lambda}v,$$
$$\widehat{I}v = z + \widehat{I}r_\rho + \beta\widehat{T_\rho}v,$$

respectively, where $v, w,$ and $z$ are $n$-vectors of real variables, and $r_\lambda, r_\rho \in \mathbb{R}^n$ are the vectors of rewards. By eliminating $v$ we get

$$w + \widehat{I}r_\lambda = (\widehat{I} - \beta\widehat{T_\lambda})(\widehat{I} - \beta\widehat{T_\rho})^{-1}(z + \widehat{I}r_\rho),$$

and hence we obtain an $\mathrm{LCP}(M, q)$ equivalent to constraints (2)–(7), where

$$M = (\widehat{I} - \beta\widehat{T_\lambda})(\widehat{I} - \beta\widehat{T_\rho})^{-1}, \tag{11}$$
$$q = M\widehat{I}r_\rho - \widehat{I}r_\lambda. \tag{12}$$

**Proposition 2.** *There is a unique solution $\overline{w}, \overline{z} \in \mathbb{R}^n$ of the $\mathrm{LCP}(M, q)$, and $(\widehat{I} - \beta\widehat{T_\lambda})^{-1}(\overline{w} + \widehat{I}r_\lambda) = (\widehat{I} - \beta\widehat{T_\rho})^{-1}(\overline{z} + \widehat{I}r_\rho)$ is the unique solution of $\mathrm{Opt}(\Gamma)$.*

Invertibility of $(\widehat{I} - \beta\widehat{T_\lambda})$ and $(\widehat{I} - \beta\widehat{T_\rho})$ is guaranteed by Theorem 4.

## 3.2    The P-Matrix Property

For an $n \times n$ matrix $A$ and $\alpha \subseteq S$, such that $\alpha \neq \emptyset$, the *principal submatrix* $A_{\alpha\alpha}$ of $A$ is the matrix obtained from $A$ by removing all rows and columns in $S \setminus \alpha$. A principal minor of $A$ is the determinant of a principal submatrix of $A$. An $n \times n$ matrix is a P-matrix [4] if all of its principal minors are positive. The importance of P-matrices for LCPs is captured by the following theorem.

**Theorem 2 (Theorem 3.3.7, [4]).** *A matrix $M \in \mathbb{R}^{n \times n}$ is a P-matrix if and only if the $\mathrm{LCP}(M, q)$ has a unique solution for every $q \in \mathbb{R}^n$.*

There are many algortithms for LCPs that work for P-matrices, but not in general. As stated by the following theorem, the matrices that arise from discounted games are P-matrices.

**Theorem 3.** *The matrix $M = (\widehat{I} - \beta\widehat{T_\lambda})(\widehat{I} - \beta\widehat{T_\rho})^{-1}$ is a P-matrix.*

*Proof.* By Proposition 2, every LCP $(M, q)$ arising from a discounted game has a unique solution. Given $M$, every $q \in \mathbb{R}^n$ can arise from a game (to see this, set $r_\lambda = 0$ in (12) and note that $M$ is invertible), hence $M$ is a P-matrix by Theorem 2. $\qquad\square$

We give an alternative proof of Theorem 3 that does not rely on the fixed point theorem underlying Theorem 1 and Proposition 2. For this we recall the following two theorems from linear algebra. An $n \times n$ matrix $A$ is *strictly row-diagonally dominant* if for every $i$, $1 \leq i \leq n$, we have $|A_{ii}| > \sum_{j \neq i} |A_{ij}|$.

**Theorem 4 (Levy-Desplanques [9]).** *Every strictly row-diagonally dominant square matrix is invertible.*

A *convex combination* of $n \times n$ matrices $B$ and $C$ is a matrix $QB + (I - Q)C$, where $Q$ is a diagonal matrix with diagonal entries $q_1, q_2, \ldots, q_n \in [0, 1]$.

**Theorem 5 (Johnson-Tsatsomeros [10]).** *Let $A = BC^{-1}$, where $B$ and $C$ are square real matrices. Then $A$ is a P-matrix iff every convex combination of $B$ and $C$ is invertible.*

*Proof (alternative proof of Theorem 3).* For every $\beta \in [0, 1)$, both $(\widehat{I} - \beta\widehat{T_\lambda})$ and $(\widehat{I} - \beta\widehat{T_\rho})$ are strictly row-diagonally dominant, and so is every convex combination of them. By Theorem 4, every such convex combination is invertible, and hence by Theorem 5, the matrix $M = (\widehat{I} - \beta\widehat{T_\lambda})(\widehat{I} - \beta\widehat{T_\rho})^{-1}$ is a P-matrix.  $\square$

It is well-known that one-player discounted games, where $S = S_{\mathrm{Min}}$ or $S = S_{\mathrm{Max}}$, can be solved in polynomial time via a simple linear program [5]. We briefly note that in this case the matrix $M$ is hidden-K, giving another proof that the LCP $(M, q)$ is solvable via a linear program [13]. A matrix $X$ is a Z-matrix if all off-diagonal entries are non-positive. A P-matrix $M$ is hidden-K if and only if there exist Z-matrices $X$ and $Y$ such that $MX = Y$ and $Xe > 0$, where $e$ is the all-one vector (see page 212 of [4]). Without loss of generality, suppose $S = S_{\mathrm{Max}}$, so $\widehat{I} = I$, $\widehat{T_\lambda} = T_\lambda$, and $\widehat{T_\rho} = T_\rho$. Then, by (11), we have $M(I - \beta T_\rho) = (I - \beta T_\lambda)$, which gives the hidden-K property.

## 3.3  Understanding $q$ and $M$

For every $\sigma : S \to S$ with $\sigma(s) \in \{\lambda(s), \rho(s)\}$, let $\overline{v}^\sigma \in \mathbb{R}^n$ be the vector of discounted payoffs of $\sigma$-plays $\langle \mathcal{D}(\mathrm{Play}(s, \sigma), \beta)\rangle_{s \in S}$. We define $r_\sigma \in \mathbb{R}^n$ as follows. For $s \in S$,

$$r_\sigma(s) = \begin{cases} r_\lambda(s) & \text{if } \sigma(s) = \lambda(s), \\ r_\rho(s) & \text{if } \sigma(s) = \rho(s). \end{cases}$$

**Proposition 3.** *For $\sigma : S \to S$, we have $\overline{v}^\sigma = (\widehat{I} - \beta\widehat{T_\sigma})^{-1}\widehat{I}r_\sigma$.*

*Proof.* The discounted payoff of the play $\mathrm{Play}(s, \sigma)$ is the unique solution of the system of equations $v = r_\sigma + \beta T_\sigma v$, which is equivalent to $\widehat{I}v = \widehat{I}r_\sigma + \beta\widehat{T_\sigma}v$, and hence $\overline{v}^\sigma = (\widehat{I} - \beta\widehat{T_\sigma})^{-1}\widehat{I}r_\sigma$.  $\square$

**Proposition 4.** *If $q \in \mathbb{R}^n$ is as defined in (12), then $q = \widehat{I}\big(\overline{v}^\rho - (r_\lambda + \beta T_\lambda \overline{v}^\rho)\big)$.*

*Proof.* By Proposition 3, we have

$$q = M\widehat{I}r_\rho - \widehat{I}r_\lambda = (\widehat{I} - \beta\widehat{T_\lambda})(\widehat{I} - \beta\widehat{T_\rho})^{-1}\widehat{I}r_\rho - \widehat{I}r_\lambda = (\widehat{I} - \beta\widehat{T_\lambda})\overline{v}^\rho - \widehat{I}r_\lambda \ .  \qquad \square$$

For $\sigma : S \to S$, define the $n \times n$ matrix $D_\sigma$ in the following way. For $s \in S$, let $\mathrm{Play}(s, \sigma) = \langle s_0, s_1, \ldots, s_{k-1}, \langle t_0, t_1, \ldots, t_{\ell-1}\rangle^\omega\rangle$. Then for $t \in S$, we define

$$(D_\sigma)_{st} = \begin{cases} \beta^i & \text{if } t = s_i \text{ for some } i, \ 0 \le i < k, \\ \frac{\beta^{k+i}}{1 - \beta^\ell} & \text{if } t = t_i \text{ for some } i, \ 0 \le i < \ell, \\ 0 & \text{otherwise.} \end{cases}$$

**Proposition 5.** *For $\sigma : S \to S$, we have $\overline{v}^\sigma = D_\sigma r_\sigma$.*

*Proof.* Let $s \in S$ and $\mathrm{Play}(s, \sigma) = \langle s_0, s_1, \ldots, s_{k-1}, \langle t_0, t_1, \ldots, t_{\ell-1} \rangle^\omega \rangle$. Then we have

$$
\begin{aligned}
\mathcal{D}(\mathrm{Play}(s,\sigma), \beta) &= \sum_{i=0}^{k-1} \beta^i r_\sigma(s_i) + \beta^k \sum_{j=0}^{\infty} \sum_{i=0}^{\ell-1} \beta^{j\ell+i} r_\sigma(t_i) \\
&= \sum_{k-1} \beta^i r_\sigma(s_i) + \sum_{i=0}^{\ell-1} \left( \beta^{k+i} \sum_{j=0}^{\infty} \beta^{j\ell} \right) r_\sigma(t_i) \\
&= \sum_{k-1} \beta^i r_\sigma(s_i) + \sum_{i=0}^{\ell-1} \frac{\beta^{k+i}}{1 - \beta^\ell} \cdot r_\sigma(t_i) \\
&= (D_\sigma r_\sigma)_s. \qquad\qquad \square
\end{aligned}
$$

By Proposition 5, the discounted payoff of the play $\mathrm{Play}(s,\sigma)$ is equal to $(D_\sigma r_\sigma)_s = \sum_{t \in S} (D_\sigma)_{st} \cdot r_\sigma(t)$. Therefore, we can think of $(D_\sigma)_{st}$ as the coefficient of the contribution of the reward $r_\sigma(t)$ on the edge that leaves state $t \in S$, towards the total discounted payoff of the play, which is starting from state $s$, and that is following strategy $\sigma$ onwards.

**Lemma 1.** *Let $M$ be the $n \times n$ matrix as defined in (11). Then for every $s, t \in S$, we have $M_{st} = (-1)^{[s \in S_{\mathrm{Min}}] + [t \in S_{\mathrm{Min}}]} \left( (D_\rho)_{st} - \beta (D_\rho)_{\lambda(s)t} \right)$.*

*Proof.* The following follows from Propositions 3 and 5, and from $\widehat{I}^{-1} = \widehat{I}$:

$$
\begin{aligned}
M &= (\widehat{I} - \beta \widehat{T_\lambda})(\widehat{I} - \beta \widehat{T_\rho})^{-1} \widehat{I} I^{-1} \\
&= (\widehat{I} - \beta \widehat{T_\lambda}) D_\rho \widehat{I}.
\end{aligned}
$$

Therefore, for all $s, t \in S$, we have

$$
\begin{aligned}
M_{st} &= (-1)^{[s \in S_{\mathrm{Min}}]} \cdot (-1)^{[t \in S_{\mathrm{Min}}]} (D_\rho)_{st} - (-1)^{[s \in S_{\mathrm{Min}}]} \beta \cdot (-1)^{[t \in S_{\mathrm{Min}}]} (D_\rho)_{\lambda(s)t} \\
&= (-1)^{[s \in S_{\mathrm{Min}}] + [t \in S_{\mathrm{Min}}]} \left( (D_\rho)_{st} - \beta (D_\rho)_{\lambda(s)t} \right). \qquad\qquad \square
\end{aligned}
$$

## 4   Algorithms

### 4.1   Unique Sink Orientations of Cubes

A unique sink orientation (USO) of an $n$-dimensional hypercube is an orientation of its edges such that every face has a unique sink. The USO problem is to find the unique sink of the $n$-cube, using calls to an oracle that gives the orientation of edges adjacent to a vertex. For more details about USOs see [19].

For an LCP $(M, q)$, the vector $q$ is *nondegenerate* if it is not a linear combination of any $n-1$ columns of $(I, -M)$. Every P-matrix LCP $(M, q)$ of dimension $n$ with nondegenerate $q$ corresponds to a USO $\psi(M, q)$ of the $n$-cube [16].

A *principal pivot transform* (PPT) of the LCP $(M, q)$ is a related LCP with the role of $w_i$ and $z_i$ exchanged for all $i \in \alpha$ for some $\alpha \subseteq \{1, \ldots, n\}$. We denote by $M_i$ the $i$-th column of $M$ and by $e_i = I_i$ the $i$-th unit vector. For each $\alpha \subseteq \{1, \ldots, n\}$, define the $n \times n$ matrix $B^\alpha$ as,

$$(B^\alpha)_i = \begin{cases} -M_i, & \text{if } i \in \alpha \,, \\ e_i, & \text{if } i \notin \alpha \,. \end{cases}$$

The $\alpha$-PPT of $(M, q)$, written $(M^\alpha, q^\alpha)$, is found as follows. Start with the matrix $A = [I, -M, q]$, which comes from the equation $Iw - Mz = q$, see (8). Obtain $A'$ from $A$ by exchanging $I_i$ with $-M_i$ for all $i \in \alpha$. Then $(B^\alpha)^{-1} A' = [I, -M^\alpha, q^\alpha]$.

The vertices of $\psi(M, q)$ correspond to the subsets $\alpha \subseteq \{1, \ldots, n\}$. At vertex $\alpha$, the $n$ adjacent edges are oriented according to the sign of $q^\alpha = ((B^\alpha)^{-1}q)$. For exactly one $\alpha$, we have $q^\alpha \geq 0$, so that $z = 0$ is a trivial solution of the LCP $(M^\alpha, q^\alpha)$; this is the sink of $\psi(M, q)$.

For a binary discounted game $\Gamma$, each subset $\alpha \subseteq \{1, \ldots, n\}$ corresponds to a choice of right-successor function, $\rho^\alpha$, with

$$\rho^\alpha(s) = \begin{cases} \lambda(s) & \text{if } s \in \alpha, \\ \rho(s) & \text{if } s \notin \alpha, \end{cases}$$

for all $s \in S$. The sink of $\psi(M, q)$ is an $\alpha$ such that $\rho^\alpha$ is an optimal (combined) strategy.

## 4.2   Strategy Improvement and the Strategy Valuation USO

In this section we outline strategy improvement algorithms for solving discounted games. Such algorithms also exist for other classes of zero-sum games, such as parity games, mean-payoff games, and simple stochastic games [1,20]. For the *all-switching* variant of strategy improvement, no super-linear examples are known for any of these classes of games.

Underlying strategy improvement algorithms are corresponding USOs. For binary games, as considered here, these are USOs of cubes, for games with out-degree larger than two, USOs of grids; see [7].

**Definition 1.** *For a pair of strategies, a state $s \in S$ is switchable if the optimality equation $s$, given by (1), does not hold. That is, for a right successor function $\rho$, used to denote a strategy pair, and a left successor function $\lambda$, used to denote the alternative choices to $\rho$, a state $s \in S_{Max}$ is switchable under strategy pair $\rho$ if*

$$r_\lambda(s) + \beta \cdot \overline{v}^\rho(\lambda(s)) > r_\rho(s) + \beta \cdot \overline{v}^\rho(\rho(s)), \tag{13}$$

*and state $s \in S_{Min}$ is switchable under $\rho$ if*

$$r_\lambda(s) + \beta \cdot \overline{v}^\rho(\lambda(s)) < r_\rho(s) + \beta \cdot \overline{v}^\rho(\rho(s)). \tag{14}$$

The rewards of the game are nondegenerate if there is no strategy pair $\sigma$ such that for some state $s \in S$ we have $r_\lambda(s) + \beta \cdot \overline{v}^\sigma(\lambda(s)) = r_\rho(s) + \beta \cdot \overline{v}^\sigma(\rho(s))$. For the purpose of defining the strategy valuation USO $\tau(\Gamma)$, we only consider nondegenerate rewards. We associate a vertex of $\tau(\Gamma)$ with the strategy pair $\sigma$.

**Definition 2.** *For a game $\Gamma$, the strategy valuation USO $\tau(\Gamma)$ is defined as follows. At vertex $\sigma$ an edge is outgoing if and only if the corresponding state is switchable.*

**Proposition 6.** *For a game $\Gamma$ and the corresponding $(M, q)$ defined by (11)-(12), we have $\tau(\Gamma) = \psi(M, q)$ .*

*Proof.* For $s \in S$, we have $q_s = (-1)^{[s \in S_{\mathrm{Min}}]}\big((r_\rho + \beta \overline{v}^\rho(\rho(s))) - (r_\lambda + \beta \overline{v}^\rho(\lambda(s)))\big)$, by Proposition 6. Thus, if $s \in S_{\mathrm{Max}}$, then $q_s < 0$ if and only if (13) is satisfied, and if $s \in S_{\mathrm{Min}}$, then $q_s < 0$ if and only if (14) is satisfied. $\qquad\square$

For a fixed strategy $\chi$ of Max, a *best response* of Min, $BR(\chi)$, is a strategy that for all $s \in S_{\mathrm{Min}}$ does not satisfy (14). For a state $s \in S$, there are two opposite facets, i.e., $(n-1)$-dimensional faces, of $\tau(\Gamma)$ such that in one all strategies are consistent with $\lambda(s)$, and in the other all are consistent with $\rho(s)$. Thus, the strategy $\chi$ of Max defines a subcube of $\tau(\Gamma)$ as the intersection of the facets that are consistent with $\chi$. $BR(\chi)$ is the sink in this subcube.

**Algorithm 1.** [Strategy Improvement for Max]

Input: Discounted game $\Gamma$

---

**repeat:**

1. $\rho' \leftarrow \big(BR(\rho{\restriction}S_{\mathrm{Max}}), \rho{\restriction}S_{\mathrm{Max}}\big)$.
2. Obtain $\rho''$ from $\rho'$ by switching at a nonempty subset of switchable $s \in S_{\mathrm{Max}}$ under $\rho'$.
3. $\rho \leftarrow \rho''$

**until** $\rho'' = \rho'$.

The proof of correctness of strategy improvement for simple stochastic games in Section 3.3 of [2] can be easily adapted to discounted games using the fact that, for every strategy $\sigma$, the matrix $D_\sigma = (I - \beta T_\sigma)^{-1}$ is nonnegative and has positive diagonal.

Algorithm 1 has the following interpretation in terms of $\tau(\Gamma)$. In Step 1., find the best response of Min as the sink in the subcube of $\tau(\Gamma)$ consistent with $\chi$. In Step 2., from this sink, jump to the antipodal vertex in the subcube spanned by the chosen set of outgoing edges (switchable states). The algorithm can be seen as repeating Step 2. in the *strategy-improvement* USO $\tau_{\mathrm{Max}}(\Gamma)$, which is an *inherited* USO where the vertices correspond to the strategies of Max only. To obtain $\tau_{\mathrm{Max}}(\Gamma)$ from $\tau(\Gamma)$, we "drop" the dimensions corresponding to Min: at vertex $\gamma$ in $\tau_{\mathrm{Max}}(\Gamma)$, the orientation is consistent with that at $(BR(\gamma), \gamma)$ in $\tau(\Gamma)$, which is the sink in the subcube of $\tau(\Gamma)$ consistent with $\gamma$. For more details on inherited USOs, see Section 3 of [19]. It is a long-standing open question whether the all-switching variant of strategy improvement is polynomial.

With degenerate rewards, there is at least one edge that does not have a well-defined orientation. Strategy improvement still works, by considering any such edge as incoming to the current vertex.

### 4.3   Murty's Least-Index Method

In this section we outline Murty's least-index method for P-matrix LCPs. We show that, applied to the LCP $(M, q)$ derived from a discounted game $\Gamma$ according to (11) and (12), the least-index method can be considered as a strategy improvement algorithm.

**Algorithm 2.** [Murty's least-index method]

Input: LCP $(M, q)$. Initialization: Set $\alpha := \emptyset$, $\bar{q} := q$.

---

**while** $\bar{q} \not\geq 0$ **do:**

$\quad s \leftarrow \min_{\{1,\ldots,n\}} \{i \mid \bar{q}_i < 0\};$

$\quad \alpha \leftarrow \alpha \oplus \{s\};$

$\quad \bar{q} \leftarrow (B^\alpha)^{-1} q.$

For a proof of the correctness of this Murty's least-index method, see [16]. Given Lemma 6, we see that, applied to the LCP derived from $\Gamma$, in each iteration Algorithm 2 makes a single switch in a switchable state with the lowest index.

**Proposition 7.** *Suppose Murty's least-index method is applied to the LCP arising from a discounted game $\Gamma$. If $S_{Max} = \{1, \ldots, k\}$ $(S_{Min} = \{1, \ldots, k\})$ for some $k \in \{1, \ldots, n\}$, then the algorithm corresponds to a single-switch variant of strategy improvement for Min (Max).*

*Proof.* Suppose $S_{\mathrm{Max}} = \{1, \ldots, k\}$. Then before any states of Min are switched, we have $q_1, \ldots, q_k \geq 0$, i.e., Max is playing a best response. Then, if possible, a single switch for Min with lowest index is made. $\qquad\square$

Murty's least-index method gives a new algorithm for binary discounted games, and hence also for binary mean-payoff and parity games. For a given game, the method depends on an initial strategy pair, and an ordering of the states. As described by Proposition 7, for certain orderings of the states the method corresponds to a single-switch variant of strategy improvement in which the subroutine of computing best responses is also done via single-switch strategy improvement; for general orderings however it is a different algorithm.

## 5  Further Research

There are several algorithms for P-matrix LCPs that should be investigated in the context of discounted and simple stochastic games. For example, there is the Cottle-Dantzig prinicpal pivoting method [3] and Lemke's algorithm [12], which are pivoting methods. There are also interior point methods known for P-matrix LCPs [11].

The reduction from mean-payoff games to discounted games requires "large" discount factors. Can we design efficient algorithms for smaller discount factors? For small enough discount factor, the matrix $M$ is close to the identity matrix and hence hidden-K, so the LCP can be solved as a linear program.

Whether all-switching strategy improvement is a polynomial-time algorithm is a long-standing open question. An exponential lower bound has been given for USOs in [18], but so far games that give rise to these example have not been constructed. What about upper bounds for strategy improvement for one-player discounted games? Are the inherited (strategy improvement) USOs, which we know to be acyclic, linearly inducible? Do they at least satisfy the Holt-Klee condition, which is known to hold for P-matrix LCPs, but is not necessarily preserved by inheritance [7]?

*Acknowledgements.* We thank Hugo Gimbert for stimulating us to formulate and study an LCP for solving discounted games.

## References


1. Condon, A.: The complexity of stochastic games. Information and Computation 96, 203–224 (1992)
2. Condon, A.: On algorithms for simple stochastic games. In: Advances in Computational Complexity Theory, pp. 51–73. American Mathematical Society (1993)
3. Cottle, R.W., Dantzig, G.B.: Complementary pivot theory of mathematical programming. Linear Algebra and Its Applications 1, 103–125 (1968)
4. Cottle, R.W., Pang, J.-S., Stone, R.E.: The Linear Complementarity Problem. Academic Press (1992)
5. Derman, C.: Finite State Markov Decision Processes. Academic Press (1972)
6. Filar, J., Vrieze, K.: Competitive Markov Decision Processes. Springer, Heidelberg (1997)
7. Gärtner, B., Morris, W.D., Rüst, L.: Unique sink orientations of grids. In: Jünger, M., Kaibel, V. (eds.) IPCO 2005. LNCS, vol. 3509, pp. 210–224. Springer, Heidelberg (2005)
8. Gärtner, B., Rüst, L.: Simple stochastic games and P-matrix generalized linear complementarity problems. In: Liśkiewicz, M., Reischuk, R. (eds.) FCT 2005. LNCS, vol. 3623, pp. 209–220. Springer, Heidelberg (2005)
9. Horn, R.A., Johnson, C.R.: Matrix Analysis. Cambridge University Press (1985)
10. Johnson, C.R., Tsatsomeros, M.J.: Convex sets of nonsingular and P-matrices. Linear and Multilinear Algebra 38, 233–239 (1995)
11. Kojima, M., Noma, T., Megiddo, N., Yoshise, A. (eds.): A Unified Approach to Interior Point Algorithms for Linear Complementarity Problems. LNCS, vol. 538. Springer, Heidelberg (1991)

12. Lemke, C.E.: Bimatrix equilibrium points and mathematical programming. Management Science 11, 681–689 (1965)
13. Mangasarian, O.L.: Linear complementarity problems solvable by a single linear program. Mathematical Programming 10, 263–270 (1976)
14. Puri, A.: Theory of Hybrid Systems and Discrete Event Systems. PhD thesis, University of California, Berkeley (1995)
15. Shapley, L.S.: Stochastic games. Proc. Nat. Acad. Sci. U.S.A. 39, 1095–1100 (1953)
16. Stickney, A., Watson, L.: Digraph models of Bard-type algorithms for the linear complementarity problem. Mathematics of Operations Research 3, 322–333 (1978)
17. Svensson, O., Vorobyov, S.: Linear complementarity and P-matrices for stochastic games. In: Virbitskaite, I., Voronkov, A. (eds.) PSI 2006. LNCS, vol. 4378, pp. 408–421. Springer, Heidelberg (2007)
18. Szabó, T., Schurr, I.: Jumping doesn't help in abstract cubes. In: Jünger, M., Kaibel, V. (eds.) IPCO 2005. LNCS, vol. 3509, pp. 225–235. Springer, Heidelberg (2005)
19. Szabó, T., Welzl, E.: Unique sink orientations of cubes. In: IEEE Symposium on Foundations of Computer Science (FOCS), pp. 547–555 (2001)
20. Vöge, J., Jurdziński, M.: A discrete strategy improvement algorithm for solving parity games (Extended abstract). In: Emerson, E.A., Sistla, A.P. (eds.) CAV 2000. LNCS, vol. 1855, pp. 202–215. Springer, Heidelberg (2000)
21. Zwick, U., Paterson, M.: The complexity of mean payoff games on graphs. Theoretical Computer Science 158, 343–359 (1996)

# Implementing Spi Calculus
# Using Nominal Techniques


Temesghen Kahsai[1,*] and Marino Miculan[2]

[1] Department of Computer Science, Swansea University, UK
csteme@swan.ac.uk
[2] DiMI, University of Udine, Italy
miculan@dimi.uniud.it



**Abstract.** The aim of this work is to obtain an interactive proof environment based on Isabelle/HOL for reasoning formally about cryptographic protocols, expressed as processes of the spi calculus (a $\pi$-calculus with cryptographic primitives). To this end, we formalise syntax, semantics, and *hedged bisimulation*, an environment-sensitive bisimulation which can be used for proving security properties of protocols. In order to deal smoothly with binding operators and reason up-to $\alpha$-equivalence of bound names, we adopt the new *Nominal datatype* package. This simplifies both the encoding, and the formal proofs, which turn out to correspond closely to "manual proofs".


## 1 Introduction

It is well known that proving security properties of communication protocols is difficult and error-prone. Since Paulson's seminal work [19], *interactive (semi-automatised) proof assistants* (such as Isabelle [18]) have been recognised as valid aid tools to this end. In principle, (a model of) the system can be formalised in a proof assistant, the security property can be formally stated as a "theorem", and a formal proof can be carried out interactively by the user, and checked by the environment. Sophisticated semi-automatic proof search tactics may simplify some derivation steps. This approach should be seen as complementary, and not in contrast, to the many fully automatised tools, such as those based on (symbolic) model checking, SAT solvers, etc.; see e.g. [3,6,10,16]. Actually, both approaches have strong and weak points: automatic tools are quite successful on particular finite-state systems, but they naturally suffer of state-explosion problems and usually cannot be used for proving general properties. On the other hand, interactive tools may be tedious and slow to use, but in principle can be used for proving any valid property. In fact, the best solution would be *integrated proof assistants*, that is semi-automatic interactive tools where the verification of decidable subgoals can be left to automatic proof search tools; see e.g. [17,24] for some application of this approach to model checking problems.


* The author is supported by EPSRC under the grant EP/D037212/1.



A. Beckmann, C. Dimitracopoulos, and B. Löwe (Eds.): CiE 2008, LNCS 5028, pp. 294–305, 2008.

Among the many formal models for security, a particularly successful one is the *spi calculus* [2], a process calculus intended to describe and reason about the behaviour of cryptographic protocols; security properties can be expressed rigorously as statements of *behavioural equivalence* between processes. These equivalences can be characterised by *environmental (i.e. context-sensitive) bisimulations*, where the environment keeps track of the knowledge accumulated by an attacker that observes the evolution of the protocols. Some fully automatised approaches for deciding these equivalences have been proposed (notably [6,8,14]), but their applicability is limited to small decidable fragments.

In this paper, we intend to complement these tools with an interactive proof assistant for spi calculus. More precisely, we give a formalisation of the spi calculus in the proof assistant Isabelle/HOL, using the recently developed *Nominal package*. We formalise syntax, operational semantics, and two environmental bisimulations. In our opinion this work is useful for many reasons. First, we readily obtain an interactive environment which can be effectively used for proving (formally and error-free) security properties of protocols expressed as spi calculus processes, as well as meta-theoretic results about the spi calculus itself. Secondly, the rigorous encoding of a calculus in the metalanguage of a logical framework is normative, since it forces to spell out in full detail all aspects of the calculus, thus giving the possibility of identifying and fixing problematic issues which may be overlooked on paper. Thirdly, this formalisation can be used for integrating automatised tactics and proof search strategies, as described above. Fourthly, this is the first application of these techniques to environmental bisimulations, and this study can be ported to other environmental bisimulations such as those recently studied for higher-order languages in [21]. Finally, extensive case studies like this are useful test-beds for state-of-art and still under development logical frameworks and proof assistants.

Regarding this last aspect, one feature of our work is that we use Isabelle/HOL extended with the *Nominal package* [22] (or Isabelle/Nominal for short). The Nominal package implements in Isabelle/HOL the ideas of Nominal Logic, introduced in the seminal works by Gabbay and Pitts [11,20]. These techniques aim to simplify the manipulation and reasoning of data with binding operators, by automatically identifying terms up to $\alpha$-equivalence of bound names. This aspect is fundamental in the spi calculus, where (bound) variables and names are crucially used for representing channels, keys, nonces, etc. Of course other encoding methodologies are possible in principle, but the hassle of dealing explicitly with different representations of the same process would hinder the usability of the resulting encoding in interactive proofs.

Although it is still under development, the Nominal package is already quite usable; see [4] for an extensive implementation of the theory of $\pi$-calculus in Isabelle/Nominal, which have been inspiration for the present work. In fact, we think that case studies like the present work can give useful insights for further improvements of the Nominal package, thanks to the several distinguishing features of spi calculus with respect to $\pi$-calculus (such as message passing and peculiar context-sensitive bisimulations).

*Synopsis.* In Section 2 we recall the spi calculus: its syntax, semantics and hedged bisimilarity. In Section 3 we give a brief introduction to the Nominal package. In Section 4 we describe the implementation of the spi calculus in Isabelle/Nominal, together with an example of a formal proof. Section 5 concludes the paper with some future and related work.

The complete Isabelle source code, with further examples and an encoding of "framed bisimulation" (which we cannot report in this paper due to lack of space), can be found at `http://www.cs.swan.ac.uk/~csteme/SpiInIsabelle/`.

## 2   Spi Calculus

The spi calculus [2] is a process calculus extending $\pi$-calculus [15] with primitives for describing and reasoning about the behaviour of cryptographic protocols. Security properties, such as secrecy, authenticity (and also authentication via a reduction to secrecy [5]), can be expressed as statements of behavioral equivalence. In this section we recall the syntax and the semantics of the shared-key spi-calculus, following [9].[1]

*Syntax.* We assume an infinite set of *names* $\mathcal{N}$, ranging from $a, b, c, \ldots k, l, m, n$ and $x, y, z$. The set of *expressions* is defined by the following grammar:

$$\begin{aligned} \zeta, \eta &::= a \mid \mathsf{E}_\eta(\zeta) \mid \mathsf{D}_\eta(\zeta) & &\text{expressions } \mathcal{E} \\ \delta &::= a \mid \mathsf{E}_\delta(\delta) & &\text{decryption-free expressions } \mathcal{D} \\ M, N &::= a \mid \mathsf{E}_k(M) & &\text{messages } \mathcal{M} \end{aligned}$$

$\mathsf{E}_\eta(\zeta)$ represents the cipher-text obtained by encrypting the expression $\zeta$ with the expression $\eta$ as key, using some given (perfect) shared-key cryptosystem. $\mathsf{D}_\eta(\zeta)$ represents the decryption of $\zeta$ using the key $\eta$, if successful.

The *guards* $\mathcal{G}$ are defined by the following grammar:

$$\phi, \psi ::= tt \mid \phi \wedge \psi \mid \neg\phi \mid let\ z = \zeta\ in\ \phi \mid is\_Name(\delta) \mid [\delta = \delta]$$

Decryption constructors can occur only in the $\zeta$ of the "let" construct: the formula $let\ z = \zeta\ in\ \phi$ evaluates the expression $\zeta$, and if evaluation succeeds (i.e., $\zeta$ contains no encrypted expressions which cannot be decrypted), binds its value to $z$ and evaluates $\phi$. Equality and name tests can be performed only on decryption-free expressions; this means that before comparing two expressions, or checking whether an expression is a name, all pending decryptions have to be solved.

Finally, the set of *processes* is defined as follows:

$$\begin{aligned} P, Q, R ::= &\mathbf{0} \mid \overline{\delta}\langle N \rangle.P \mid \delta(x).P \mid P + P \mid P \mid Q \\ &\mid (\nu n)P \mid {!P} \mid \phi P \mid let\ x = \zeta\ in\ P \end{aligned}$$

---

[1] Of course more expressive languages, e.g. with primitives for public-key cryptography, can be dealt with easily; for the sake of simplicity, in this paper we prefer to consider a simpler language.

The name $n$ is bound in $(\nu n)P$, and $x$ is bound in $P$ by *let* $x = \zeta$ *in* $P$ and $M(x).P$. For an intuitive description of spi calculus processes, see [1]. Some syntactic conventions: $fn(P)$ indicates the sets of names free in process $P$. A *concretion* is an expression of the form $\nu m_1, \ldots, m_k \langle M \rangle P$, where $M$ is a term, $P$ is a process, $k \geq 0$, and the names $m_1, \ldots, m_k$ are bound in $M$ and $P$. An *agent* is a process or a concretion. The meta-variables $A$ and $B$ range over arbitrary agents, and $fn(A)$ stands for the sets of free names of an agent $A$. An *action* is either a name $a$ with a message $M$ (representing input), or a co-name $\overline{m}$ (representing output) or the distinguished *silent action* $\tau$.

*Operational semantics.* Expressions and boolean guards are evaluated by two functions $[\![.]\!] : \mathcal{E} \to \mathcal{M} \cup \{\bot\}$, $[\![.]\!] : \mathcal{G} \to \{tt, ff\}$. The behaviour of processes is described by the *commitment relation* $(P \xrightarrow{\alpha} A)$, where $P$ is a process, $\alpha$ is an action, and $A$ is an agent. See [9] for a complete description of these two notions.

*Environment-sensitive bisimulations.* Bisimulations for spi calculus are based on the idea of an environment observing a pair of processes, trying to distinguish one from the other using the knowledge accumulated during the evolution of these processes. The environment typically observes the transitions derived from the operational semantics of the processes. *Hedged bisimulation* is an improved form of environment-sensitive bisimulation. It has been introduced in [9], following ideas of [7], in order to highlight the differences between different environment-sensitive bisimulation. The main idea of hedges is to keep track of the correspondence of different names which play the same role.

**Definition 1.** *A* hedge *is a finite subset of* $\mathcal{M} \times \mathcal{M}$. $\mathcal{H}$ *denotes the set of all hedges.*

*A hedge $h$ is consistent iff for $(M, N) \in h$:*

1. $M \in \mathcal{N}$ *iff* $N \in \mathcal{N}$;
2. *if* $(M', N') \in h$ *then* $M = M'$ *iff* $N = N'$
3. *if* $M = \mathsf{E}_{M_2}(M_1)$ *and* $N = \mathsf{E}_{N_2}(N_1)$ *then* $M_2 \notin \pi_1(h)$ *and* $N \notin \pi_2(h)$, *where* $M_1, M_2, N_1$ *and* $N_2$ *are expressions (and* $\pi_1(h)$ *is the first projection).*

The *synthesis* $S(.)$ of a hedge is defined inductively. We write $h \vdash M \leftrightarrow N$ for $(M, N) \in S(h)$. Intuitively, $h \vdash M \leftrightarrow N$ means that, using the knowledge $h$, the environment is unable to distinguish two processes $P$ and $Q$, if the first emits $M$ and the second $N$. The rules for synthesis are the following:

$$(\textsc{Synth. hedge}) \frac{(m, n) \in h}{h \vdash m \leftrightarrow n} \qquad (\textsc{Synth. enc}) \frac{h \vdash M \leftrightarrow M' \quad h \vdash N \leftrightarrow N'}{h \vdash \mathsf{E}_N(M) \leftrightarrow \mathsf{E}_{N'}(M')}$$

The *analysis* $A(h)$ is the smallest subset of $\mathcal{M} \times \mathcal{M}$ containing $h$ and satisfying the following rule:

$$\frac{(\mathsf{E}_a(M), \mathsf{E}_b(N)) \in A(h) \quad (a, b) \in A(h)}{(M, N) \in A(h)}$$

The *irreducibles* $I(h)$ of a hedge $h$ is $I(h) \triangleq A(h) \backslash \{(\mathsf{E}_a(M), \mathsf{E}_b(N)) \mid (a, b) \in A(h) \wedge M, N \in \mathcal{N}\}$. Intuitively, the analysis decrypts as much as possible pairs of messages using pairs of names that are considered equivalent by the environment; irreducibles terms are those which cannot be decrypted further.

**Definition 2 (Hedged simulation).** *A hedged relation $\mathcal{R}$ is a subset of $\mathcal{H} \times P \times P$. We say that $\mathcal{R}$ is consistent if $h \vdash P\mathcal{R}Q$ implies that $h$ is consistent.*

*A consistent hedged relation $\mathcal{R}$ is a hedged simulation if whenever $h \vdash P\mathcal{R}Q$ the following conditions hold:*

1. *If $P \xrightarrow{\tau} P'$ then there exists a process $Q'$ such that $Q \Longrightarrow Q'$ and $h \vdash P'\mathcal{R}Q'$.*
2. *If $P \xrightarrow{a\,M} P'$, $h \vdash a \leftrightarrow b$, $B$ is a finite set of names disjoint from $fn(P) \cup fn(Q) \cup fn(h)$, and $N$ is a message such that $h \cup Id_B \vdash M \leftrightarrow N$, then there exists $Q'$ such that $Q \xrightarrow{b\,N} Q'$ and $h \cup ID_B \vdash P'\mathcal{R}Q'$*
3. *If $P \xrightarrow{\bar{a}} (\nu\vec{c})\langle M\rangle P'$, $h \vdash a \leftrightarrow b$ and $\vec{c}$ is disjoint from $fn(P) \cup fn(\pi_1(h))$, then there exists $Q', N, \vec{d}$ with $\vec{d}$ disjoint from $fn(Q) \cup fn(\pi_2(h))$ such that $Q \xRightarrow{\bar{b}} (\nu\vec{d})\langle N\rangle Q'$ and $I(h \cup \{(M, N)\}) \vdash P'\mathcal{R}Q'$.*

$\mathcal{R}$ is a *hedged bisimulation* if both $\mathcal{R}$ and $\mathcal{R}^{-1}$ are *hedged simulations*. The *hedged bisimilarity* is the greatest hedged bisimulation, and it is denoted by $\sim_h$.

It turns out that hedged bisimilarity corresponds to barbed equivalence [9].


# 3   Isabelle/Nominal

The *Nominal package* [22] for Isabelle/HOL aims to provide a framework for reasoning about process calculi and programming languages with binding operators in a convenient way, so that formal proofs should be easy to carry out as informal "pencil-and-paper" proofs. The work is based on the nominal logic [20]; the main technical novelty introduced by Urban et al. [22] is that the construction for $\alpha$-equivalent terms is done without adding any axiom to the Isabelle/HOL logic; therefore the theory is implemented just as a package of Isabelle/HOL, without the need of changing the underlying proof assistant.

A *nominal datatype* definition is like an ordinary datatype, but it explicitly tags the binding occurrences of names. For instance, in the syntax of the usual untyped $\lambda$-calculus the notation `<<name>>term` stands for "a `term` abstracted over `name`", that is, with a name bound in `term`. The package automatically provide the $\alpha$-equivalence between terms; e.g., `(lam x (var x))` and `(lam y (var y))` are equal. Moreover, the package generates automatically powerful induction rules over terms up-to $\alpha$-equivalence (among other useful properties). This saves the user much hassle in large proofs.

The core of the nominal logic relies on the notion of *name swapping*. Atoms (i.e., names) are manipulated not by renaming substitutions but by permutations (bijective mappings from atoms to atoms). In the Nominal package, permutations are represented as finite lists of atom swappings (*i.e*, pairs of atoms). The operation of permutation applies to all names in a term, including the binding and bound occurrences: if $T$ be a term, and $a$ and $b$ are names then $(a\ b) \bullet T$ denotes the term where all instances of $a$ in $T$ becomes $b$ and vice versa. For further details, the reader can refer to `http://isabelle.in.tum.de/nominal/`.

# 4   Encoding Spi Calculus in Isabelle/Nominal

In this section, we describe the implementation of spi calculus in the general purpose proof assistant Isabelle [18], using its instantiation HOL-Nominal implementing higher-order intuitionistic logic and including the Nominal package. For improving readability of theories and proofs, we use *Isar* [23] (Intelligible semi-automatized reasoning), with occasionally some syntactic sugar.

## 4.1   Implementation of Syntax and Semantics

We introduce one type of nominal atoms *name*, which will be used in binders. *Expressions, decryption free expressions, messages* and *processes* are declared as nominal datatypes; actually, only processes have binders, but in the current version of the Nominal package, building nominal datatypes over normal datatypes is not easy.[2]

Due to the use of Isabelle syntax, there may be a slight change of notation from Section 2, but the rationale will be clear.

**nominal_datatype** *expr = Name name     | Sk_enc expr expr     | Sk_dec expr expr*
**nominal_datatype** *dfexpr = Df_Name name      | Df_Sk_enc dfexpr dfexpr*
**nominal_datatype** *mess = M_Name name      | M_Sk_enc name mess*
**nominal_datatype** *Proc = Pnil                    | in_pref dfexpr ≪name≫Proc*
     *| out_pref dfexpr dfexpr Proc     | par Proc Proc*
     *| res ≪name≫Proc                 | bang Proc*
     *| boolean_guard guard Proc       | letp expr ≪name≫Proc*

*Sk_enc* and *Sk_dec* represent $\mathsf{E}_\eta(\zeta)$ and $\mathsf{D}_\eta(\zeta)$ respectively. By declaring those datatypes as nominal datatype, the nominal package generates a powerful induction rule, where bound names occurring in the inductive cases will be automatically chosen to be different from any name (or variable) already used in a proof. Thus, naming clashing are avoided automatically. The Nominal package derives several proofs so that Isabelle's type system can in most circumstances automatically infer when a type is a permutation type. Nominal datatypes are always permutation types and their elements are finitely supported.

Substitution operators for the different datatypes are implemented as functions by using the recursion combinator that is automatically generated by the Nominal package for the datatypes terms defined above; this allows us to define recursively functions that respect $\alpha$-equivalence classes.

*subst_mess_Name* :: *mess* $\Rightarrow$ *name* $\Rightarrow$ *name* $\Rightarrow$ *mess*
*subst_mess_Name m n* $\equiv$ $\lambda a.$ (*M_Name a*)

*subst_mess_enc* :: *mess* $\Rightarrow$ *name* $\Rightarrow$ *name* $\Rightarrow$ *mess* $\Rightarrow$ *mess* $\Rightarrow$ *mess*
*subst_mess_enc m n* $\equiv$ $\lambda t\_ m2.$ (*M_Sk_enc t m2*)

---

[2] The problem is that datatypes used within nominal datatypes must satisfy several "equivariance properties". These properties are automatically proved for nominal datatypes, but are left to the user for normal datatypes.

$$subst\_mess :: mess \Rightarrow mess \Rightarrow name \Rightarrow mess \quad (\_[\_ \sim\sim \_])$$
$$e[m \sim\sim n] \equiv (mess\_rec \ (subst\_mess\_Name \ m \ n) \ (subst\_mess\_enc \ m \ n)) \ e$$

$mess\_rec$ denote the recursion combinator for $mess$. The notation $e[m \sim\sim n]$ is a syntactic sugar, and it can be read as $e$ with $m$ for $n$, and represent the substitution of all occurances of $n$ of type $name$ in the message $e$ with $m$. Evaluation operators for expressions and boolean guards $[\![.]\!]$ are implemented as a nominal partial recursive function(**nominal-primrec**). The output of the function that evaluates expressions is of type $mess\ option$, i.e. if the evaluation is succesfull we expect a message otherwise an error. In Isabelle/HOL the type $t\ option$ models the result of a computation that may terminate with an error (represented by $None$) or return the value $v$ (represented by $Some\ v$).

**consts**
    $eval\_expr :: expr \Rightarrow mess\ option$
**nominal-primrec**
    $eval\_expr\ (Name\ a) = \ Some\ (M\_Name\ a)$
    $eval\_expr\ (Sk\_enc\ e1\ e2) = (case\ eval\_expr(e2)\ of$
                $None \Rightarrow None$
                $|\ Some\ M \Rightarrow (case\ eval\_expr(e1)\ of$
                $None \Rightarrow None$
                $|\ Some\ k \Rightarrow mess\_case\ k\ (\%l.\ Some(M\_Sk\_enc\ l\ M))\ (\%x\ y.\ None)))$

$eval\_expr\ (Sk\text{-}dec\ e1\ e2) = (case\ eval\text{-}expr(e2)\ of$
                $None \Rightarrow None$
                $|Some\ X \Rightarrow (mess\_case\ X\ (\%l.\ None)$
                        $(\%k\ M.\ (case\ eval\_expr(e1)\ of$
                $None \Rightarrow None$
                $|Some\ N \Rightarrow mess\_case\ N\ (\%k'.(if\ k=k'\ then\ (Some\ M)\ else\ None))$
                        $(\%x\ y.\ None)))))$

$eval\_expr\ (Sk\_enc\ e1\ e2)$ asserts if the evaluation of $e2$ is $None$ then the evaluation of $(Sk\_enc\ e1\ e2)$ is $None$; otherwise if the evaluation is of type $(Some\ M)$ we make a case distinction: if the evaluation of $e2$ is $None$ the function return $None$ otherwise if the evaluation is of type $(Some\ k)$ it returns $M\_Sk\_enc$ $l\ M$ otherwise it returns $None$. $eval\_expr\ (Sk\_dec\ e1\ e2)$ follows the same kind of evaluation. In Isabelle "case" expressions are just sugared syntax for a special case combinator which is automatically defined whenever we define a datatype. For nominal datatypes, however, this is not yet supported, hence we define a case combinator ($mess\_case$) for this purpose.

**consts** $mess\_case :: mess \Rightarrow (name \Rightarrow \ 'a) \Rightarrow (name \Rightarrow mess \Rightarrow \ 'a) \Rightarrow \ 'a$
**nominal-primrec**
    $mess\_case\ (M\_Name\ n)\ c1\ c2 = c1\ n$
    $mess\_case\ (M\_Sk\_enc\ n\ m)\ c1\ c2 = c2\ n\ m$

The commitment relation is defined by induction; as an example, we report the commitment of (GUARD) rule.

$$comm\_guard: \ [\![\ eval\_guard(g);\ \ P - \alpha \ \longmapsto P'\ ]\!] \Longrightarrow (g\ \gamma\ P) - \alpha \ \longmapsto P'$$

In $[\![\ldots]\!]$ we have both the precondition and the side condition of the rule. $P - \alpha \longmapsto P'$ stands for the commitment of $P$ doing an action $\alpha$ and then behaves like $P'$. Guards are evaluated to a boolean value by *eval_guard*, which is the (encoding of) evaluation of guards defined in Section 2. *eval_guard* is defined by recursion on the syntax of guards, much like *eval_dfexpr*; notice that the guards have a binder. ($g\gamma P$ is the guarded process, denoted by $\phi P$ in Section 2.)

## 4.2 Implementation of Hedged Bisimulation

In Isabelle/Nominal, hedges are just sets of term pairs. The consistency of hedges (*Cons H*) is defined by a predicate formed by three clauses corresponding to Definition 1. The notions of analysis (*Analysis H*) and irreducibles (*Irreducibles H*) are implemented as inductive sets, via the *least fixed point*.

Let us focus now on hedged bisimilarity, which is defined by coinduction. We describe in detail the second condition (about input transitions):

**consts** *HedgedBisim* :: (*hedge* × *Proc* × *Proc*) *set*
**coinductive** *HedgedBisim*
**intros**
 *HedgedBisim_Def*: $[\![H \in$ *Finites*; (*Cons H*);
*(clauses for $\tau$, omitted)*
     $\forall P'\ a\ b\ M\ N\ B.(P - (a\ M) \longmapsto P') \wedge (H\vdash(a\leftrightarrow b)) \wedge (B\in$ *Finites*$) \wedge (B\ \natural$ $(H,P,Q)) \wedge (((H\cup(ID\ B)))\vdash(M\leftrightarrow N)) \longrightarrow (\exists\,Q'.(Q,b,N,Q')\in$ *commIn* $\wedge\ (((H\cup(ID$ $B)),P',Q') \in$ *HedgedBisim*$))$;
    *(symmetric clause for input, omitted)*
     $\forall P'\ (c::name\ list)\ a\ b\ M.(c\ \natural\ (H,P)) \longrightarrow (P - cobarb(a) \longmapsto$ (*concAgent* (*ConcChain c M P'*))) $\wedge$ $((H)\vdash((M\_Name\ a)\leftrightarrow(M\_Name\ b))) \longrightarrow (\exists\,Q'\ (d::name$ *list*$)\ N.(d\ \natural\ (H,Q)) \wedge ((Q,b,(ConcChain\ d\ N\ Q'))\in$ *commOut*$) \wedge (((Irreducibles(H$ $\cup\{(M,N)\})),P',\ Q') \in$ *HedgedBisim*$))$;
    *(symmetric clause for output, omitted)* $]\!] \Longrightarrow (H,P,Q) \in$ *HedgedBisim*

$(H \in$ *Finites*$)$ and $(Cons\ H)$ require the hedge $H$ to be a consistent finite set of message pairs. $(Q, b, N, Q') \in$ *commIn*, where $b$ is of type *name* and $N$ is of type *mess*, stands for the commitment relation of the process $Q$ and abstraction $Q'$ under the input $b\ N$, possibly preceded by $\tau$ transitions. Notice that the freshness of the names in $B$ is easily ensured by the hypothesis $(B\#(H,P,Q))$.

The clauses for the output transitions are represented similarly; here again, we use the freshness predicate from Nominal package, to encode the freshness of locally scoped (i.e., newly created) names $c, d$.

## 4.3 Example: "Perfect Encryption"

In order to explain how the implementation of spi calculus presented above can be used, we give an example proof by proving a simple bisimilarity, that is the *simple encryption* property taken from [1]. We want to prove that for all $M, M'$, the processes

$$(\nu k)\bar{c}\langle \mathsf{E}_k(M)\rangle \quad \text{and} \quad (\nu k)\bar{c}\langle \mathsf{E}_k(M')\rangle \tag{1}$$

are hedged bisimilar. This means that there is no way to distinguish the cleartext message if the encryption key is kept secret.

In the proof given in [1], a bisimulation $S$ is defined, such that

$$(\{c, n\}, \{\}) \vdash (\nu\ k)\bar{c}\langle E_k(M)\rangle\ S\ (\nu\ k)\bar{c}\langle E_k(M')\rangle$$

However, instead of explicitly define such $S$ and prove that it is a bisimulation, we can take advantage of the coinductive support provided by the Isabelle/Nominal environment to directly prove that

$$(\{c, n\}, (\nu\ k)\bar{c}\langle E_k(M)\rangle, (\nu\ k)\bar{c}\langle E_k(M')\rangle) \in HedgedBisim$$

and the Isar proof sketch is the following:

```
lemma perfectEncyption:
  shows ({c, n},((⟨ν(a)⟩( c<(a∫ M∫ )>.PNil)), ((⟨ν(a)⟩(c<(a∫ N∫ )>.PNil))) ∈ Hedged-
Bisim   (is ?x ∈ _ )
proof -
  have ?x : {?x} by simp
  then show ?thesis
  proof coinduct
    case (HedgedBisim z)
      have ?HedgedBisim_Def sorry
    then show ?case by blast
  qed
qed
```

Let us analyze how the proof proceeds. Basically, the proof is done by using forward reasoning, through the *coinduct* proof method; then the proof splits into several sub-cases, before working towards one of the disjuncts. We also need to fill in a sensible starting point $?x : \{?x\}$ and take special care of the types here. $?x : \{?x\}$ is solved by *simp* (an Isabelle method that solves the goal using simplification rules); *?thesis* stands for the current goal to be proved. At this point the proofs is done by case analysis. Applying the rule *HedgedBisim_Def*, we are left with eight subgoals (the 6 clauses of coinductive definition, plus finiteness and consistency of hedge). Each case is then proved on its own (quite tediously); in the proof sketch above, this part is replaced by the command **sorry** (which proves anything but it is very convenient for top down proof development; this command can be replaced by the actual proofs later on). In particular, the proof of these cases exploits the features provided by the Nominal package for handling freshness of bound names. We are able to finish the coinduction step, working from the case assumptions to the conclusion *?case*, via the *blast* method (a classical reasoner which tries to solve automatically the current goal).

## 5   Conclusions and Future Work

In this paper we have presented a formalization of the spi calculus, a calculus of cryptographic processes, in the Isabelle/HOL proof assistant using the new

Nominal package. The proof environment so obtained allows for formal proofs which closely correspond to the (traditional) manual proofs, and in some sense are even simpler because we don't have to figure out and define explicitly bisimulations beforehand, thanks to the support provided by Isabelle to coinductive proofs. The Nominal package really played an important role to this end: it allows for a smooth handling of binding operators, thus reducing the overhead in encoding the system and conducting a formal proof.

However, although the Nominal package is already usable and fruitful, in our opinion some details need to be improved, in particular the support for recursively defined functions and case analysis over nominal datatypes. Moreover, at the moment is not possible to do *reflection*, i.e. implementing computations inside the logic rather than in the meta-language, and this due to the fact that the current version does not support co-generation.

*Related work.* Theorem provers and proof assistants have been widely used to model process algebra and reason about correctness. Paulson's work [19] is arguably the first application of Isabelle to the verification of cryptographic protocols. Actually, the $\pi$-calculus is a paradigmatic example for proof environments and encoding techniques designed to handle binding operators. The closest development to ours is [4], where Bengtson and Parrow have formalised the $\pi$-calculus in Isabelle/HOL using the Nominal package, providing a library which allows users to carry proofs about $\pi$-calculus. Other approaches to binding management are de Bruijn indexes [12], and *(weak) higher order abstract syntax* [13]. De Bruijn indexes are quite cumbersome to use in interactive proofs, because names disappear completely. On the other hand, the (weak) HOAS approach needs to postulate some key properties (the so called *Theory of Contexts*) as axioms. Although proved to be consistent with (classical) higher order logic, the Theory of Contexts is inconsistent with the Axiom of Choice; therefore, its portability to the Isabelle environment is still under discussion.

*Future work.* A first research stemming from the present work is to prove some general meta-theoretic results about spi-calculus, similarly to the work done by Briais in Coq [8]. Another interesting possibility is to implement special tactics to be used during the proof developments for proving decidable equivalences. These new commands can be written completely inside Isabelle; a possible way, following ProVerif approach, is to translate the protocol and the goal into classical logic, and then take advantage of the powerful support provided by Isabelle/HOL to classical reasoning. Alternatively (and more efficiently), the new commands can call auxiliary tools, external to Isabelle. To this end, the recent works about symbolic bisimulation of spi calculus [8] may be useful.

# References

1. Abadi, M., Gordon, A.D.: A bisimulation method for cryptographic protocols. Nord. J. Comput. 5(4), 267 (1998)
2. Abadi, M., Gordon, A.D.: A calculus for cryptographic protocols: The Spi calculus. Journal of Information and Computation 148(1), 1–70 (1999)

3. Armando, A., Basin, D.A., Boichut, Y., Chevalier, Y., Compagna, L., Cuéllar, J., Drielsma, P.H., Héam, P.-C., Kouchnarenko, O., Mantovani, J., Mödersheim, S., von Oheimb, D., Rusinowitch, M., Santiago, J., Turuani, M., Viganò, L., Vigneron, L.: The AVISPA tool for the automated validation of internet security protocols and applications. In: Etessami, K., Rajamani, S.K. (eds.) CAV 2005. LNCS, vol. 3576, pp. 281–285. Springer, Heidelberg (2005)
4. Bengtson, J., Parrow, J.: Formalising the $\pi$-calculus using nominal logic. In: Seidl, H. (ed.) FOSSACS 2007. LNCS, vol. 4423, pp. 63–77. Springer, Heidelberg (2007)
5. Blanchet, B.: From secrecy to authenticity in security protocols. In: Hermenegildo, M.V., Puebla, G. (eds.) SAS 2002. LNCS, vol. 2477, pp. 342–359. Springer, Heidelberg (2002)
6. Blanchet, B., Abadi, M., Fournet, C.: Automated verification of selected equivalences for security protocols. In: Proc. 20th LICS, pp. 331–340. IEEE (2005)
7. Boreale, M., Nicola, R.D., Pugliese, R.: Proof techniques for cryptographic processes. SIAM J. Comput. 31(3), 947–986 (2001)
8. Borgström, J., Briais, S., Nestmann, U.: Symbolic bisimulation in the spi calculus. In: Gardner, P., Yoshida, N. (eds.) CONCUR 2004. LNCS, vol. 3170, pp. 161–176. Springer, Heidelberg (2004)
9. Borgström, J., Nestmann, U.: On bisimulations for the spi calculus. Mathematical Structures in Computer Science 15(3), 487–552 (2005)
10. Clarke, E.M., Jha, S., Marrero, W.: Verifying security protocols with brutus. ACM Trans. Softw. Eng. Methodol. 9(4), 443–487 (2000)
11. Gabbay, M.J., Pitts, A.M.: A new approach to abstract syntax involving binders. In: Proc. 14th LICS, pp. 214–224. IEEE (1999)
12. Hirschkoff, D.: Bisimulation proofs for the $\pi$-calculus in the Calculus of Constructions. In: Gunter, E.L., Felty, A.P. (eds.) TPHOLs 1997. LNCS, vol. 1275. Springer, Heidelberg (1997)
13. Honsell, F., Miculan, M., Scagnetto, I.: $\pi$-calculus in (co)inductive type theory. Theoretical Computer Science 253(2), 239–285 (2001)
14. Hüttel, H.: Deciding framed bisimilarity. In: Proceedings of Infinity 2002. Electronic Notes in Theoretical Computer Science, vol. 68, pp. 1–18 (2003)
15. Milner, R., Parrow, J., Walker, D.: A calculus of mobile processes. Inform. and Comput. 100(1), 1–77 (1992)
16. Mitchell, J.C., Mitchell, M., Stern, U.: Automated analysis of cryptographic protocols using Mur$\phi$. In: IEEE Symposium on Security and Privacy, pp. 141–151. IEEE Computer Society (1997)
17. Namjoshi, K.S.: Certifying model checkers. In: Berry, G., Comon, H., Finkel, A. (eds.) CAV 2001. LNCS, vol. 2102, pp. 2–13. Springer, Heidelberg (2001)
18. Nipkow, T., Paulson, L.C.: Isabelle-91. In: Kapur, D. (ed.) CADE 1992. LNCS, vol. 607, pp. 673–676. Springer, Heidelberg (1992)
19. Paulson, L.C.: The inductive approach to verifying cryptographic protocols. Journal of Computer Security 6(1-2), 85–128 (1998)
20. Pitts, A.M.: Nominal logic, a first order theory of names and binding. Information and Computation 186, 165–193 (2003)
21. Sangiorgi, D., Kobayashi, N., Sumii, E.: Environmental bisimulations for higher-order languages. In: Proc. LICS, pp. 293–302. IEEE Computer Society (2007)

22. Urban, C., Tasson, C.: Nominal techniques in Isabelle/HOL. In: Nieuwenhuis, R. (ed.) CADE 2005. LNCS (LNAI), vol. 3632, pp. 38–53. Springer, Heidelberg (2005)
23. Wenzel, M.: Isar - a generic interpretative approach to readable formal proof documents. In: Bertot, Y., Dowek, G., Hirschowitz, A., Paulin, C., Théry, L. (eds.) TPHOLs 1999. LNCS, vol. 1690, pp. 167–184. Springer, Heidelberg (1999)
24. Yu, S., Luo, Z.: Implementing a model checker for LEGO. In: Fitzgerald, J.S., Jones, C.B., Lucas, P. (eds.) FME 1997. LNCS, vol. 1313, pp. 442–458. Springer, Heidelberg (1997)

# An Enhanced Theory of Infinite Time Register Machines


Peter Koepke[1] and Russell Miller[2]

[1] Mathematisches Institut, Universität Bonn, Germany
koepke@math.uni-bonn.de
[2] Queens College and The Graduate Center, City University of New York, USA
Russell.Miller@qc.cuny.edu



**Abstract.** *Infinite time register machines* (ITRMs) are register machines which act on natural numbers and which are allowed to run for arbitrarily many ordinal steps. Successor steps are determined by standard register machine commands. At limit times a register content is defined as a lim inf of previous register contents, if that limit is finite; otherwise the register is *reset* to 0. (A previous weaker version of infinitary register machines, in [6], would halt without a result in case of such an overflow.) The theory of infinite time register machines has similarities to the infinite time TURING machines (ITTMs) of HAMKINS and LEWIS. Indeed ITRMs can decide all $\Pi_1^1$ sets, yet they are strictly weaker than ITTMs.

**Keywords:** ordinal computability, hypercomputation, infinitary computation, register machine.


## 1 Introduction

JOEL D. HAMKINS and ANDY LEWIS [3] defined *infinite time* TURING *machines* (ITTMs) by letting an ordinary TURING machine run for arbitrarily many ordinal steps, taking appropriate limits at limit times. An ITTM can compute considerably more functions than a standard TURING machine. In this paper we introduce *infinite time register machines* (ITRMs) which may be seen as ordinary *register* machines running for arbitrarily many ordinal steps. Successor steps are determined by standard register machine commands. At limit times the register contents are defined as lim inf's of the previous register contents, if that limit is finite; otherwise the register is *reset* to 0.

Our ITRMs may be viewed as a specialization of the *ordinal register machines* (ORMs) examined in [8]. The stages are still allowed to range over all ordinals, but we now have a space bound of $\omega$ on the contents of the registers. Of course, this requires a rule for the action of the machine when a register overflows. In previous versions (see for example [6]), the machines halted or crashed when encountering an overflow; those machines exactly corresponded to hyperarithmetic definitions. Our machines reset a register to 0 whenever it overflows. We view this as a more natural rule, defining richer descriptive classes which are more


A. Beckmann, C. Dimitracopoulos, and B. Löwe (Eds.): CiE 2008, LNCS 5028, pp. 306–315, 2008.

in analogy with the ITTM-definable classes, and we believe that the theorems in this paper support our view. Therefore, we propose to use the name *infinite time register machine* to refer to our machines in this paper. The machines defined in [6] by the first author could be called *non-resetting infinite time register machines*.

Our ITRMs are to ORMs as the ITTMs of Hamkins and Lewis are to *ordinal Turing machines*, or OTMs, as defined in [7]. In both cases the ordinal machines have unbounded time and space, whereas in the operation of traditional finite time machines, both time and space were bounded by $\omega$. ITTMs and ITRMs fall in between, with $\omega$-much space but unbounded time, hence are denoted as "infinite time" machines. With the bound on space, of course, the ITRMs necessarily follow different procedures than the ORMs at limit stages. (For ITTMs and OTMs the corresponding difference concerns head location, not cell contents.)

Many results for ITRMs in this paper reflect this connection with ITTMs. Notably, we show that ITRMs are $\Pi_1^1$-complete in the sense that for any lightface $\Pi_1^1$-set $A$ of reals there is an ITRM such that a given real $x$ is accepted by the machine iff $x \in A$ (Theorem 2). In particular the class WO of codes for wellorders is ITRM-decidable (Theorem 1), and likewise ITTM-decidable (by results in [3]). On the other hand ITRMs are strictly weaker than ITTMs because the latter are able to solve the halting problem for the former (Theorem 3). Moreover, for a given number $N$ of registers, the halting problem for ITRMs with $N$ registers is ITRM-decidable, using of course more registers (Theorem 4). In further research we plan to develop the theory of ITRMs along the lines of the ITTMs in [3].

## 2 Infinite Time Register Machines

We base our presentation of infinite time machines on the *unlimited register machines* of [1].

**Definition 1.** *An* unlimited register machine *URM has registers* $R_0, R_1, \ldots$ *which can hold* natural numbers. *A register program consists of commands to increase or to reset a register. The program may jump on condition of equality of two registers.*

*An URM* program *is a finite list* $P = I_0, I_1, \ldots, I_{s-1}$ *of instructions, each of which may be of one of five kinds:*

a) *the* zero instruction $Z(n)$ *changes the contents of* $R_n$ *to* 0, *leaving all other registers unaltered;*

b) *the* successor instruction $S(n)$ *increases the natural number contained in* $R_n$ *by* 1, *leaving all other registers unaltered;*

c) *the* oracle *instruction* $O(n)$ *replaces the content of the register* $R_n$ *by the number* 1 *if the content is an element of the oracle, and by* 0 *otherwise;*

d) *the* transfer instruction $T(m, n)$ *replaces the contents of* $R_n$ *by the natural number contained in* $R_m$, *leaving all other registers unaltered;*

e) *the* jump instruction $J(m, n, q)$ *is carried out within the program $P$ as follows: the contents $r_m$ and $r_n$ of the registers $R_m$ and $R_n$ are compared, all registers are left unaltered; then, if $r_m = r_n$, the URM proceeds to the $q$th instruction of $P$; if $r_m \neq r_n$, the URM proceeds to the next instruction in $P$.*

*Since the program is finite, it can use only finitely many of the registers, and the exact number of registers used will often be important. The instructions of the program can be addressed by their indices which are called* program states. *At each ordinal time $\tau$ the machine will be in a* configuration *consisting of a program state $I(\tau) \in \omega$ and the register contents which can be viewed as a function $R(\tau) : \omega \to \omega$. $R(\tau)(n)$ is the content of the register $R_n$ at time $\tau$. We also write $R_n(\tau)$ instead of $R(\tau)(n)$.*

**Definition 2.** *Let $P = I_0, I_1, \ldots, I_{s-1}$ be an URM program. Let $Z \subseteq \omega$, which will serve as an oracle. A pair*

$$I : \theta \to \omega, R : \theta \to (^{\omega}\omega)$$

*is an* (infinite time register) computation *by $P$ if the following hold:*

a) *$\theta$ is an ordinal or $\theta = \mathrm{Ord}$; $\theta$ is the* length *of the computation;*
b) *$I(0) = 0$; the machine starts in state 0;*
c) *If $\tau < \theta$ and $I(\tau) \notin s = \{0, 1, \ldots, s-1\}$ then $\theta = \tau + 1$; the machine halts if the machine state is not a program state of $P$;*
d) *If $\tau < \theta$ and $I(\tau) \in s$ then $\tau + 1 < \theta$; the next configuration is determined by the instruction $I_{I(\tau)}$, with $I(\tau + 1) = I(\tau) + 1$ unless otherwise specified:*
   i. *if $I_{I(\tau)}$ is the zero instruction $Z(n)$ then define $R(\tau + 1) : \omega \to \mathrm{Ord}$ by setting $R_k(\tau + 1)$ to be 0 (if $k = n$) or $R_k(\tau)$ (if not).*
   ii. *if $I_{I(\tau)}$ is the successor instruction $S(n)$ then define $R_k(\tau + 1)$ to be $R_k(\tau) + 1$ (if $k = n$) or $R_k(\tau)$ (if not).*
   iii. *if $I_{I(\tau)}$ is the oracle instruction $O(n)$ then define $R_k(\tau+1)$ to be $R_k(\tau)$ (if $k \neq n$); or 1 (if $k = n$ and $R_k(\tau) \in Z$); or 0 (if $k = n$ and $R_k(\tau) \notin Z$).*
   iv. *if $I_{I(\tau)}$ is the transfer instruction $T(m, n)$ then define $R_k(\tau + 1)$ to be $R_m(\tau)$ (if $k = n$) or $R_k(\tau)$ (if not).*
   v. *if $I_{I(\tau)}$ is the jump instruction $J(m, n, q)$ then let $R(\tau + 1) = R(\tau)$, and set $I(\tau + 1) = q$ (if $R_m(\tau) = R_n(\tau)$) or $I(\tau + 1) = I(\tau) + 1$ (if not).*
e) *If $\tau < \theta$ is a limit ordinal, then $I(\tau) = \liminf_{\sigma \to \tau} I(\sigma)$ and*

$$\forall k \in \omega \ R_k(\tau) = \begin{cases} \liminf_{\sigma \to \tau} R_k(\sigma), & \text{if } \liminf_{\sigma \to \tau} R_k(\sigma) < \omega \\ 0, & \text{if } \liminf_{\sigma \to \tau} R_k(\sigma) = \omega. \end{cases}$$

*By the second clause in the definition of $R_k(\tau)$ the register is* reset *in case $\liminf_{\sigma \to \tau} R_k(\sigma) = \omega$.*

*The computation is obviously determined recursively by the initial register contents $R(0)$, the oracle $Z$ and the program $P$. We call it the* (infinite time register) computation *by $P$ with input $R(0)$ and oracle $Z$. If the computation halts then $\theta = \beta + 1$ is a successor* ordinal *and $R(\beta)$ is the final register content. In this case we say that $P$ computes $R(\beta)(0)$ from $R(0)$ and the oracle $Z$, and we write $P : R(0), Z \mapsto R(\beta)(0)$.*

**Definition 3.** *An n-ary partial function $F : \omega^n \rightharpoonup \omega$ is computable if there is a register program $P$ such that for every n-tuple $(a_0, \ldots, a_{n-1}) \in \mathrm{dom}(F)$ holds*

$$P : (a_0, \ldots, a_{n-1}, 0, 0, \ldots), \emptyset \mapsto F(a_0, \ldots, a_{n-1}).$$

*Here the oracle instruction is not needed.*

Obviously any standard recursive function is computable.

**Definition 4.** *A subset $x \subseteq \omega$, i.e., a (single) real number, is computable if its characteristic function $\chi_x$ is computable.*

*A subset $A \subseteq \mathcal{P}(\omega)$ is computable if there is a register program $P$, and an oracle $Y \subseteq \omega$ such that for all $Z \subseteq \omega$:*

$$Z \in A \text{ iff } P : (0, 0, \ldots), Y \times Z \mapsto 1, \text{ and } Z \notin A \text{ iff } P : (0, 0, \ldots), Y \times Z \mapsto 0$$

*where $Y \times Z$ is the cartesian product of $Y$ and $Z$ with respect to the pairing function*

$$(y, z) \mapsto \frac{(y + z)(y + z + 1)}{2} + z.$$

*Here we allow a single real parameter $Y$ (or equivalently, finitely many such parameters), mirroring the approach in [6].*

## 3 Computing $\Pi_1^1$-Sets

We describe an ITRM-program to check the oracle $Z$ for illfoundedness. Illfoundedness will be witnessed by some infinite descending chain. Initial segments of such a chain will be kept on a finite *stack* of natural numbers. Code a stack $(r_0, \ldots, r_{m-1})$ by $r = 2^m \cdot 3^{r_0} \cdot 5^{r_1} \cdots p_m^{r_{m-1}}$ where $p_i$ is the $i$-th prime number. In the subsequent program we shall treat one register as a stack, with content `stack` equal to $r$ above, and with associated operations `push`, `pop`, `length-stack`, `stack-is-decreasing`; this last predicate checks that the elements of the stack, except possibly the bottom element, form a decreasing sequence in the oracle $Z$. All of these are computable by an ITRM. The specific coding of stack contents leads to a controlled limit behaviour:

**Proposition 1.** *Let $\alpha < \tau$ where $\tau$ is a limit ordinal. Assume that in some ITRM-computation using a stack, the stack contains $r = (r_0, \ldots, r_{m-1})$ for cofinally many times below $\tau$ and that all contents in the time interval $(\alpha, \tau)$ are endextensions of $r = (r_0, \ldots, r_{m-1})$. Then at time $\tau$ the stack contents are $r = (r_0, \ldots, r_{m-1})$.*

The following program $P$ on an ITRM outputs `yes`/`no` depending on whether the oracle $Z$ codes a wellfounded relation. The program is a backtracking algorithm which searches for a "leftmost" infinite descending chain in $Z$. A stack is used to organize the backtracking. We present the program in simple pseudo-code and assume that it is translated into a register program according to Definition 1 so that the order of commands is kept. Also the stack commands like push are understood as *macros* which are inserted into the code with appropriate renaming of variables and statement numbers. The ensuing Lemma explains the operation of the program and proves its correctness.

```
        push 1; %% marker to make stack non-empty
        push 0; %% try 0 as first element of descending sequence
        FLAG=1; %% flag that fresh element is put on stack
Loop: Case1: if FLAG=0 and stack=0 %% inf descending seq found
            then begin; output 'no'; stop; end;
        Case2: if FLAG=0 and stack=1 %% inf descending seq not found
            then begin; output 'yes'; stop; end;
        Case3: if FLAG=0 and length-stack > 1
        %% top element cannot be continued infinitely descendingly
            then begin; %% try next
            pop N;
            push N+1;
            FLAG:=1; %% flag that fresh element is put on stack
            goto Loop;
            end;
        Case4: if FLAG=1 and stack-is-decreasing
            then begin;
            push 0; %% try to continue sequence with 0
            FLAG:=0; FLAG:=1; %% flash the flag
            goto Loop;
            end;
        Case5: if FLAG=1 and not stack-is-decreasing
            then begin;
            pop N;
            push N+1; %% try next
            FLAG:=0; FLAG:=1; %% flash the flag
            goto Loop;
            end;
```

Notice that the program will always loop back to Loop until it halts.

**Lemma 1.** *Let $I : \theta \to \omega$, $R : \theta \to (^{\omega}\omega)$ be the computation by $P$ with oracle $Z$ and trivial input $(0, 0, \ldots)$. Then the computation satisfies:*

a) *Suppose the machine is in state* Loop *with stack contents $(1, r_0, \ldots, r_{m-1})$ so that $(r_0, \ldots, r_{m-1})$ descend strictly in $Z$. Moreover suppose that* Flag=1 *and that $Z$ is wellfounded below $r_{m-1}$. Then the machine will reach the state* Loop *with the same stack contents and* Flag=0 *after a certain interval of time; during that interval, $(1, r_0, \ldots, r_{m-1})$ will always be an initial segment of the stack.*

b) *Suppose the machine is in state* Loop *with stack contents $(1, r_0, \ldots, r_{m-1})$ so that $(r_0, \ldots, r_{m-1})$ descend strictly in $Z$. Moreover suppose that* Flag=1 *and that $Z$ is illfounded below $r_{m-1}$. Let $r_m$ be the smallest integer such that $r_m Z r_{m-1}$ and $Z$ is illfounded below $r_m$ . Then the machine will reach the state* Loop *with stack contents* $(1, r_0, \ldots, r_{m-1}, r_m)$ *and* Flag=1 *after a certain interval of time; during that interval,* $(1, r_0, \ldots, r_{m-1})$ *will always be an initial segment of the stack.*

*c) If $Z$ is wellfounded then the computation stops with output* '**yes**'.

*d) If $Z$ is illfounded then the computation stops with output* '**no**'.

*Proof.* a) is proved by induction on $r_{m-1}$ in the wellfounded part of $Z$. So consider a situation $(1, r_0, \ldots, r_{m-1})$ as in a) and assume that a) already holds for all appropriate stacks $(1, r'_0, \ldots, r'_{m'-1})$ with $r'_{m'-1} Z r_{m-1}$. By Case4, Case3, and the inductive assumption, the machine will check through all extensions $(1, r_0, \ldots, r_{m-1}, N)$ with $N \in \omega$ of the stack and always get to state Loop with Flag=0. The limit of these checks is a stack $(1, r_0, \ldots, r_{m-1})$ with Flag=0, as required.

b) Consider the situation described in b). The program checks through all extensions $(1, r_0, \ldots, r_{m-1}, N)$ with $N < r_m$ of the stack. Case5 rejects those $N$ which fail $N Z r_{m-1}$, and part (a) shows that the others are also rejected. So Case3 finally puts $r_m$ on the stack, with Flag=1.

c) and d) follow from a) and b) resp.

Parts c) and d) of the Lemma imply immediately:

**Theorem 1.** *The set* WO $= \{Z \subseteq \omega \mid Z$ *codes a wellorder*$\}$ *is computable by an ITRM.*

**Theorem 2.** *Every $\Pi^1_1$ set $A \subseteq \mathcal{P}(\omega)$ is ITRM-computable.*

*Proof.* Let $f$ be a recursive function so that $Y \in A \leftrightarrow f(Y) \in$ WO. Given a real $Y$ an ITRM can decide whether $Y \in A$ by letting the above WO-algorithm run on $f(Y)$. Note that the algorithm needs to decide whether certain integers stand in the relation $f(Y)$. This can be reduced to computing a certain digit of $f(Y)$ which is possible using the oracle $Y$ and a fixed algorithm for computing $f$.

Since $\Pi^1_1$-sets can be decided, it is also possible to decide Boolean combinations of $\Pi^1_1$-sets by ITRMs. By induction on ordertypes one can prove a running time estimate for the WO-algorithm:

**Lemma 2.** *For an oracle $Z$ coding a well order of ordertype $\alpha$ the WO-program runs at least $\alpha$ steps before it halts.*

## 4   ITRMs, ITTMs, and Halting Problems

A computation by an ITRM can be simulated by an ITTM. If the register $R_m$ contains the number $i$ this can be represented as an initial segment of $i$ 1's on the $m$-th tape of an ITTM. If $\lambda$ is a limit ordinal and the contents of the register $R_m$ yield $\liminf_{\tau \to \lambda} R_m(\tau) = i^* \leqslant \omega$ then the $m$-th tape will hold an initial segment of $i^*$ 1's at time $\lambda$. If $i^*$ is finite, this is the correct simulation of the ITRM. If $i^* = \omega$ this may be checked by an auxiliary program which then *resets* the register to 0. Thus every class of reals which is computable by an ITRM is computable by an ITTM, and hence must be $\Delta_2^1$, by Theorem 2.5 in [3].

In fact ITRMs are strictly weaker than ITTMs. A *configuration* is a tuple $(I, R)$ of a program state $I$ and register contents $R : \omega \to \omega$ where $R(n) = 0$ for almost all $n < \omega$. The following halting criterion for ITRMs uses a wellfounded pointwise partial order of configurations:

$$(I_0, R_0) \leqslant (I_1, R_1) \text{ iff } I_0 \leqslant I_1 \text{ and } \forall n < \omega \ R_0(n) \leqslant R_1(n).$$

**Lemma 3.** *Let*

$$I : \theta \to \omega, R : \theta \to (^\omega\omega)$$

*be the infinite time register computation by $P$ with input $(0, 0, \ldots)$ and oracle $Z$. Then this computation does* not *halt iff there are $\tau_0 < \tau_1 < \theta$ such that $(I(\tau_0), R(\tau_0)) = (I(\tau_1), R(\tau_1))$ and*

$$\forall \tau \in [\tau_0, \tau_1] \ (I(\tau_0), R(\tau_0)) \leqslant (I(\tau), R(\tau)).$$

*Proof.* Assume that the computation does not halt. Let $A$ be the set of all configurations which occur class-many times in this computation, and fix a stage $\tau^-$ after which only configurations in $A$ occur. We claim that $A$ is downwards directed in the partial order of configurations: for $(I_0, R_0), (I_1, R_1) \in A$ choose an ascending $\omega$-sequence $\tau^- < \tau_0 < \tau_1 < \cdots$ of stages such that each $(I_i, R_i)$ occurs at all stages of the form $\tau_{2 \cdot k + i}$ with $i < 2$. Then the configuration $(I, R)$ occuring at stage $\tau = \sup_n \tau_n$ has $(I, R) \leqslant (I_0, R_0)$ and $(I, R) \leqslant (I_1, R_1)$, by the rules for limit stages.

Let $(I_0, R_0)$ be the unique $\leqslant$-minimal element of $A$. Choose stages $\tau_0, \tau_1$ such that $t^- < t_0 < t_1 < \theta$ and $(I(\tau_0), R(\tau_0)) = (I(\tau_1), R(\tau_1)) = (I_0, R_0)$. This is the situation required by the lemma.

For the converse assume that there are $\tau_0 < \tau_1 < \theta$ such that $(I(\tau_0), R(\tau_0)) = (I(\tau_1), R(\tau_1))$ and

$$\forall \tau \in [\tau_0, \tau_1] \ (I(\tau_0), R(\tau_0)) \leqslant (I(\tau), R(\tau)).$$

Then one can easily show by induction, using the $\lim\inf$ rules:

If $\sigma \geqslant \tau_0$ is of the form $\sigma = \tau_0 + (\tau_1 - \tau_0) \cdot \alpha + \beta$ with $\beta < \tau_1 - \tau_0$ then

$$(I(\sigma), R(\sigma)) = (I(\tau_0 + \beta), R(\tau_0 + \beta)).$$

In particular the computation will not stop.

**Theorem 3.** *The* halting problem *for ITRMs*

$$\{(P, Z) \mid P \text{ is a register program, } Z \subseteq \omega, \text{ and the computation by } P$$
$$\text{with input } (0, 0, \ldots) \text{ and oracle } Z \text{ halts}\}$$

*is decidable by an ITTM with oracle $Z$.*

*Proof.* The criterion of Lemma 3 can be implemented on an ITTM with an auxiliary tape on which we have one cell for each possible configuration of the ITRM. We use the ITTM to simulate the ITRM computation by a program $P$ with input $(0, 0, \ldots)$ and oracle $Z$. At stage $\tau$ of the simulation we erase from the auxiliary tape all 1's for configurations which are not $\leqslant (I(\tau), R(\tau))$, and put a 1 in the cell for the configuration $(I(\tau), R(\tau))$. If there was already a 1 in this cell, then we conclude from Lemma 3 that the computation never halts. At limit stages the same procedure applies. (There may be infinitely many 1's on the auxiliary tape at a limit stage, of which cofinitely many will immediately be erased. For an ITTM, this poses no difficulty.) These two processes continue until either the simulated ITRM computation halts or we conclude as above that it will never halt. By Lemma 3, one of these alternatives must happen.

For a fixed number of registers these ideas can be transfered to an ITRM (with more registers).

**Theorem 4.** *The restricted halting problem*

$$\{(P, Z) \mid P \text{ is a register program using at most } N \text{ registers, } Z \subseteq \omega, \text{ and}$$
$$\text{the computation by } P \text{ with input } (0, 0, \ldots) \text{ and oracle } Z \text{ halts}\}$$

*is decidable by an ITRM with oracle $Z$, for every $N < \omega$.*

*Proof.* We introduce some notation to handle configurations of the $N$ register machine. View a configuration $(I, R)$ as the $(N + 1)$-sequence

$$(R(0), \ldots, R(N - 1), I)$$

and use letters $c, c', \ldots$ to denote configurations. Write $c \leqslant c'$ iff $\forall m \leqslant N \; c(m) \leqslant c'(m)$. Let $I : \theta \to \omega$, $R : \theta \to (^\omega \omega)$ be the infinite time resetting register computation by $P$ with input $(0, 0, \ldots)$ and oracle $Z$. The computation is a sequence $(c(\tau) | \tau < \theta)$ of configurations.

By Lemma 3, the computation does not stop $(\theta = \infty)$ iff

$$\exists \sigma < \tau < \theta \; (c(\sigma) = c(\tau) \wedge \forall \sigma' \in [\sigma, \tau] \; c(\sigma) \leqslant c(\sigma')).$$

This motivates the definition

$$C(\tau) = \{c(\sigma) | \sigma < \tau \wedge \forall \sigma' \in [\sigma, \tau] \; c(\sigma) \leqslant c(\sigma')\}.$$

Then the halting criterion is simply

$$\exists \tau (c(\tau) \in C(\tau)).$$

Note that the initial configuration $(0, \ldots, 0)$ is an element of $C(\tau)$ for all $\tau > 0$.

The Theorem will be proved by showing that (a code for) $C(\tau)$ can be easily computed, and indeed by an ITRM. For technical reasons we introduce some auxiliary sequences of configuration sets. For $m \leqslant N$ let $C_m(\tau)$ be the *finite* set

$$C_m(\tau) = \{c(\sigma) | \sigma < \tau \wedge \forall \sigma' \in [\sigma, \tau] \; c(\sigma) \leqslant c(\sigma') \wedge \forall i \leqslant N \; c(\sigma)(i) \leqslant c(\tau)(m)\}.$$

Obviously $C(\tau) = C_{m_0}(\tau)$ where $c(\tau)(m_0) = \max_{i \leqslant N} c(\tau)(i)$. To handle sets of the form $C_m(\tau)$ as natural numbers and register contents we assume that we have a recursive enumeration or Gödelization $c_0, c_1, \ldots$ of configurations with $N$ registers. Finite sets $C$ of configurations can be coded by the natural number

$$C^* = \prod_{c_k \in C} p_k \,,$$

which can be stored in a machine register.

Consider a simulation of the computation $(c(\tau)|\tau < \theta)$ on some register machine with sufficiently many registers. We argue that the sequence $(C(\tau)^*|\tau < \theta)$ can be uniformly computed alongside the simulation, which solves the halting problem. We proceed by induction on $\tau < \theta$.

$C(0) = \{(0, \ldots, 0)\}$ only contains the initial configuration.

If $C(\tau)$, $c(\tau)$ and $c(\tau + 1)$ are given, then

$$C(\tau + 1) = \begin{cases} \{c \in C(\tau)|c \leqslant c(\tau + 1)\} \cup \{c(\tau)\}, \text{ if } c(\tau) \leqslant c(\tau + 1); \\ \{c \in C(\tau)|c \leqslant c(\tau + 1)\}, \text{ else.} \end{cases}$$

Hence $C(\tau + 1)^*$ can be computed by an ordinary register machine from $C(\tau)^*$, $c(\tau)$, and $c(\tau + 1)$.

Finally consider the limit time $\lambda < \theta$.

In case that $c(\lambda) = (0, \ldots, 0)$ then $C(\lambda) = \{(0, \ldots, 0)\}$. $C(\lambda)^*$ is easily computable, and moreover the criterion for divergence of the computation is fulfilled. So consider the case that $c(\lambda) \neq (0, \ldots, 0)$. Choose $m_0$ such that

$$c(\lambda)(m_0) = \max_i c(\lambda)(i) > 0.$$

Then $C_{m_0}(\lambda) = C(\lambda)$ and

(1) $c(\lambda)(m_0) = \liminf_{\tau \to \lambda} c(\tau)(m_0)$.
(2) $\liminf_{\tau \to \lambda} C_{m_0}(\tau)^*$ exists and is finite.

*Proof*. By (1) there is a cofinal subset $T \subseteq \lambda$ such that

$$\forall \tau \in T \ \ c(\tau)(m_0) = c(\lambda)(m_0).$$

For $\tau \in T$ we have

$$C_{m_0}(\tau) \subseteq \{c \mid \forall i \leqslant N \ c(i) \leqslant c(\lambda)(m_0)\}.$$

The right-hand side is a fixed finite set. So for $\tau \in T$, $C_{m_0}(\tau)^*$ is bounded by some fixed integer. Thus the liminf is finite. $qed(2)$
(3) Let $c_k \leqslant c(\lambda)$. Then $p_k|C_{m_0}(\lambda)^*$ iff $p_k|\liminf_{\tau \to \lambda} C_{m_0}(\tau)^*$.

*Proof*. Let $p_k|C_{m_0}(\lambda)^*$. Then $c_k \in C_{m_0}(\lambda)$. Take $\sigma < \lambda$ with $c_k = c(\sigma)$ such that for all $\sigma' \in [\sigma, \lambda]$, both $c_k \leqslant c(\sigma')$ and $c(\sigma')(m_0) \geq c(\lambda)(m_0)$. Then for all $\tau \in (\sigma, \lambda)$ we have $c_k \in C_{m_0}(\tau)$ and $p_k|C_{m_0}(\tau)^*$. Since $\liminf_{\tau \to \lambda} C_{m_0}(\tau)^*$ will be equal to one of those $C_{m_0}(\tau)^*$ we get that $p_k|\liminf_{\tau \to \lambda} C_{m_0}(\tau)^*$.

Conversely assume $p_k | \liminf_{\tau \to \lambda} C_{m_0}(\tau)^*$. Take $\tau_0 < \lambda$ such that $\forall \tau \in [\tau_0, \lambda]$ $c(\tau) \geqslant c(\lambda)$. Take $\tau_1 \in [\tau_0, \lambda)$ such that $\liminf_{\tau \to \lambda} C_{m_0}(\tau)^* = C_{m_0}(\tau_1)^*$. Then $p_k | C_{m_0}(\tau_1)^*$, $c_k \in C_{m_0}(\tau_1)$ and by the choice of $\tau_0$ also $c_k \in C_{m_0}(\tau)$ for all $\tau \in [\tau_1, \lambda)$. Since $c_k \leqslant c(\lambda)$ we have $c_k \in C_{m_0}(\lambda)$ and $p_k | C_{m_0}(\lambda)^*$. $qed(3)$

This means that $C_{m_0}(\lambda)^* = C(\lambda)^*$ can be computed from $(C_{m_0}(\tau)^* | \tau < \lambda)$ by a $\liminf$-operation. To compute the sequence $(C(\tau)^* | \tau < \theta)$ alongside $(c(\tau) | \tau < \lambda)$ we can use $N + 1$ new registers $R_0, \ldots, R_N$ to store the values $C_0(\tau)^*, \ldots, C_N(\tau)^*$. Initially these registers are set to $\{(0, \ldots, 0)\}^*$. Given $C_0(\tau)^*, \ldots, C_N(\tau)^*, c(\tau)$, and $c(\tau+1)$ one can compute $C_0(\tau+1)^*, \ldots, C_N(\tau+1)^*$ by an ordinary register program on some extra registers and transfer these values to $R_0, \ldots, R_N$. For limit $\lambda < \theta$ the $\liminf$-rule sets $R_0, \ldots, R_N$ to

$$\liminf_{\tau \to \lambda} C_0(\tau)^*, \ldots, \liminf_{\tau \to \lambda} C_N(\tau)^*.$$

By (3), an ordinary register program on further extra registers can compute the value $C(\lambda)^* = C_{m_0}(\lambda)^*$, from which it can then compute

$$C_0(\lambda)^*, \ldots, C_N(\lambda)^*$$

and transfer them to $R_0, \ldots, R_N$.

This means for all $\tau \in [\omega, \theta)$, $C_m(\tau)$ will be the $(\tau+1)$-st value transferred to the register $R_m$, concluding the proof of Theorem 4.

The Theorem implies that the machines get eventually stronger by increasing the number of registers. As a consequence there cannot be a universal ITRM.

# References


1. Cutland, N.J.: Computability: An Introduction to Recursive Function Theory. In: Perspectives in Mathematical Logic. Cambridge University Press (1980)
2. Dimitriou, I., Hamkins, J.D., Koepke, P.(eds.): BIWOC – Bonn International Workshop on Ordinal Computability. Bonn Logic Reports (2007)
3. Hamkins, J.D., Lewis, A.: Infinite Time Turing Machines. J. Symbolic Logic 65(2), 567–604 (2000)
4. Hamkins, J.D., Linetsky, D., Miller, R.: The complexity of quickly ORM-decidable sets. In: Cooper, S.B., Löwe, B., Sorbi, A. (eds.) CiE 2007. LNCS, vol. 4497. Springer, Heidelberg (2007)
5. Hamkins, J.D., Miller, R.: Post's problem for ordinal register machines. In: Cooper, S.B., Löwe, B., Sorbi, A. (eds.) CiE 2007. LNCS, vol. 4497, pp. 358–367. Springer, Heidelberg (2007)
6. Koepke, P.: Infinite time register machines. In: Beckmann, A., Berger, U., Löwe, B., Tucker, J.V. (eds.) CiE 2006. LNCS, vol. 3988, pp. 257–266. Springer, Heidelberg (2006)
7. Koepke, P.: Turing computations on ordinals. B. Symbolic Logic 11, 377–397 (2005)
8. Koepke, P., Siders, R.: Computing the recursive truth predicate on ordinal register machines. In: Beckmann, A., et al. (eds.) Logical approaches to computational barriers. Computer Science Report Series, vol. 7, pp. 160–169 (2006)

# Many Facets of Complexity in Logic

Antonina Kolokolova

Memorial University of Newfoundland
`kol@cs.mun.ca`

**Abstract.** There are many ways to define complexity in logic. In finite model theory, it is the complexity of describing properties, whereas in proof complexity it is the complexity of proving properties in a proof system. Here we consider several notions of complexity in logic, the connections among them, and their relationship with computational complexity. In particular, we show how the complexity of logics in the setting of finite model theory is used to obtain results in bounded arithmetic, stating which functions are provably total in certain weak systems of arithmetic. For example, the transitive closure function (testing reachability between two given points in a directed graph) is definable using only NL-concepts (where NL is non-deterministic log-space complexity class), and its totality is provable within NL-reasoning.

## 1   Introduction

How do we talk about the concept of hardness in the context of mathematical logic? Historically, there are several approaches using different notions of hardness, including the following:

1. The hardness of *describing* an object (by some formula)
2. The hardness of *proving* properties of an object in a formal system.
3. The hardness of *solving* a problem (e.g., when the problem is expressed in the language of logic. )

These notions of hardness loosely correspond to the following fields

1. Finite model theory (in particular, descriptive complexity).
2. Bounded arithmetic and proof complexity
3. Complexity theory and some areas of computational logic

Each of these fields has tight and well-studied connections to computational complexity theory. However, the direct relationships among different notions of hardness in logic are only now becoming a focus of attention. Here we will present, as main example of a such a direct relationship, results in bounded arithmetic we obtain using the notion of hardness from descriptive complexity.

We start with the canonical notion of hardness as defined in complexity theory.

A. Beckmann, C. Dimitracopoulos, and B. Löwe (Eds.): CiE 2008, LNCS 5028, pp. 316–325, 2008.

## 2   The Computational Complexity Setting

The computational complexity of a problem is measured in terms of resources used by an algorithm to solve the problem. The standard resources are space (memory) and time. Algorithms can be deterministic and non-deterministic. For example, the famous class NP consists of problems solvable by the non-deterministic polynomial-time algorithms. Similarly, the problems solvable by non-deterministic *log-space* algorithms comprise the class NL.

A complexity class that has very robust definitions in the context of logic is uniform $AC^0$, the class of problems solvable by a uniform family of constant-depth polynomial-size boolean circuits. In bounded arithmetic, this class corresponds to the theory $V^0$ or, in the first-order setting, $I\Delta_0$ (where $I\Delta_0$ is Peano Arithmetic with induction restricted to bounded formulae). In the descriptive complexity setting $AC^0$ corresponds to the first-order logic (data complexity of first-order logic is $AC^0$ when the language contains arithmetic). We will discuss it in more detail later.

A classical example of a problem complete for NP is 3-colorability: given a graph as an input, determine if it can be colored with three colors so that no edge connects vertices of the same color. A problem complete for NL is graph reachability (determining if there is a path between two vertices). The class $AC^0$ is one of the few for which there are non-trivial lower bounds: the Parity problem of determining if an input string has an even number of 1s is not in $AC^0$. That is, there is no first-order formula which would be true on all and only structures with even number of elements (in fact, since this result holds in a non-uniform setting, adding to the languages any kind of numerical predicates, even undecidable ones, would still not help a first-order formula to express a parity of a string). Thus binary addition can be done in $AC^0$, but multiplication cannot.

## 3   Finite Model Theory and Descriptive Complexity

Finite model theory developed as a subfield of model theory with emphasis on finite structures. In this new setting many of the standard techniques of model theory, most notably compactness, do not apply. Since testing if a first-order formula has a finite model is undecidable [Tra50], our focus will be on the complexity of model checking: given a finite structure and formula of some logic, decide if this structure is a model of this formula. Considering both the formula and the structure as inputs gives fairly high complexity: checking if a first-order formula holds on a Boolean (two-element) structure is complete for PSPACE (class of problems solvable in polynomial space). Here we consider *data complexity* of the model checking, where a formula is fixed and the only input is the structure. For example, a formula might encode the conditions for a graph to be 3-colourable, and structures are graphs.

This leads us to descriptive complexity, the area studying the direct correspondence between data complexity of logics of different power and complexity classes. In this setting a logic is said to *capture* a complexity class over a class of structures if, informally, the model checking problem for the logic is solvable in the complexity class and every problem in the class is representable in the logic. A classical reference on the subject is the book "Descriptive complexity" by Immerman [Imm99].

**Definition 1 (Capture by a logic).** *Let $C$ be a complexity class, $L$ a logic and $K$ a class of finite structures. Then $L$ captures $C$ on $K$ if*

1. *For every $L$-sentence $\phi$ and every $\mathcal{A} \in K$, testing if $\mathcal{A} \models \phi$ with $\phi$ fixed and an encoding of $\mathcal{A}$ as an input can be done in $C$.*
2. *For every collection $K'$ of structures closed under isomorphism, if this collection is decidable in $C$ then there is a sentence $\phi_{K'}$ of $L$ such that $\mathcal{A} \models \phi_{K'}$ iff $\mathcal{A} \in K'$, for every $\mathcal{A} \in K$.*

In descriptive complexity the class of structures is often fixed to be arithmetic structures, that is, structures with $\min, \max, +, \times, \leq, =$ in the language which receive standard interpretations. In particular, the universe of a structure is always considered to be $\{0, \ldots, n - 1\}$.

*Example 1 (*PARITY$(X)$*).* This is a formula over successor structures (with $\min, \max, S \in \tau$), models of which have interpretations of $X$ as sets with an odd number of 1's. It encodes a dynamic-programming algorithm for computing parity of $X$: $P_{odd}(i)$ is true (and $P_{even}(i)$ is false) iff the prefix of $X$ of length $i$ contains an odd number of 1's.

$$\exists P_{even} \exists P_{odd} \forall i P_{even}(\min) \wedge \neg P_{odd}(\min)$$
$$\wedge (P_{odd}(\max) \leftrightarrow \neg X(max)) \wedge (\neg P_{even}(i + 1) \vee \neg P_{odd}(i + 1))$$
$$\wedge (P_{even}(i) \wedge X(i) \rightarrow P_{odd}(i + 1)) \wedge (P_{odd}(i) \wedge X(i) \rightarrow P_{even}(i + 1))$$
$$\wedge (P_{even}(i) \wedge \neg X(i) \rightarrow P_{even}(i + 1)) \wedge (P_{odd}(i) \wedge \neg X(i) \rightarrow P_{odd}(i + 1))$$

The first capture result that started this field is due to Fagin [Fag74], who showed that existential second-order logic captures precisely the complexity class NP. His proof has a similar structure to Cook's original proof of NP-completeness of propositional formula satisfiability, but instead of propositional variables Fagin used existentially quantified second-order variables. A notable feature of Fagin's result is that it holds over all structures, whereas $AC^0$ vs. first-order logic only holds for arithmetic structures. A big open problem is whether there is a logic capturing on all (including unordered) structures the complexity class P of polynomial-time decidable languages.

## 3.1   Logics between First-Order and Existential Second-Order

In the data complexity setting, first-order logic captures $AC^0$ and existential second-order logic already captures NP. However, many interesting complexity classes lie between these two, in particular classes P (polynomial time) and NL (non-deterministic logspace). The two ways to capture these classes (in the presense of order) are either by extending first-order logic with additional predicates, or restricting second-order logic. In the first approach, a logic for NL is obtained by adding a transitive closure operator to first-order logic, and P is captured by first-order logic together with a least fixed point operator [Imm83, Imm82, Var82]. Here we will concentrate on the second approach, due to Grädel [Grä91].

**Definition 2.** Restricted second-order *formulae are of the form* $\exists P_1 \ldots P_k$ $\forall x_1 \ldots x_l \psi(\bar{P}, \bar{x}, \bar{a}, \bar{Y})$, *where $\psi$ is quantifier-free.*

Two important types of restricted second-order formulae are $SO\exists$-Horn and $SO\exists$-Krom:

- $SO\exists$-Horn: $\psi$ is a CNF with no more than one positive literal of the form $P_i(t)$ per clause.
- $SO\exists$-Krom: $\psi$ is a CNF with no more than two $P_i$ literals per clause.

In particular, the formula for Parity in the example above is both a second-order Horn and a second-order Krom formula.

**Theorem 1 ([Grä92]).** *Over structures with successor, $SO\exists$-Horn and $SO\exists$-Krom capture complexity classes* P *and* NL, *respectively.*


## 4   Bounded Arithmetic

Just like in complexity classes P and NP the computation length is bounded by a polynomial, in bounded arithmetic quantified variables are bounded by a term in the language (e.g., a polynomial in free variables of a formula). Here, instead of describing functions by formulae the goal is to prove their totality within a system; we will say that a system of arithmetic captures a function class if it proves totality of all and only functions in this class. In this setting, arithmetic reasoning with formulas having an unbounded existential quantifier captures primitive recursive functions in the same sense that reasoning with an appropriate bounded version of these formulae captures polynomial-time functions.


### 4.1   The Language and Translation from the Finite Model Theory Setting

There are two notions of hardness appearing in the setting of theories of arithmetic. One of them is a "descriptive" notion of arithmetic formulae *representing* (that is, describing) a property; another is a more recursion-theoretic notion in terms of the *provable* totality of functions (note that term "defining" is used in both contexts in different literature). In this section we show how results in descriptive complexity, translated into the framework of bounded arithmetic to become representability results, give us provability of properties.

**Definition 3.** *Define $\Sigma_0^B$ to be the class of bounded formulas with no second order quantifiers over $\mathcal{L}_2$, and $\Sigma_1^B$ as a closure of $\Sigma_0^B$ under bounded existential second-order quantification. $\Sigma_i^B$ is defined inductively in a natural way.*

It seems that the correspondence of logics in finite model theory framework with representability in bounded arithmetic is folklore. This translation works most naturally in the setting of second-order systems of arithmetic such as systems $V_i$. Here, by "second-order" systems we mean two-sorted, with one sort for numbers, and another for strings (viewed as sets of numbers). For a thorough treatment of this framework, please see [Coo03].

Let $\phi$ be a (possibly second-order) formula with free relational variables $R_1, \ldots, R_k$ over a vocabulary that contains arithmetic. The corresponding $\Sigma_i^B$ formula has $R_i$ as parameters, where non-monadic relational variables are represented using a pairing function, as well as an additional parameter $n$ denoting the size of the structure. It is easy to see that this formula holds on its free variables iff the corresponding structure is a model of the original formula in the finite model theory setting.

In particular, arithmetic versions of $SO\exists$-Horn and $SO\exists$-Krom are subsets of $\Sigma_1^B$, with no existential first-order quantifier and Horn (resp. Krom) restriction on the occurrences of quantified second-order variables. In this paper we will abuse the notation and say $SO\exists$-Horn and $SO\exists$-Krom meaning their translation into the language of arithmetic.

*Example 2.* The Parity formula defined in example 1 can be written as follows in the bounded arithmetic setting:

$$\text{PARITY}(X) \equiv \exists P_{even} \exists P_{odd} \forall i < |X|$$
$$P_{even}(0) \wedge \neg P_{odd}(0) \wedge P_{odd}(|X|) \wedge (\neg P_{even}(i+1) \vee \neg P_{odd}(i+1))$$
$$\wedge (P_{even}(i) \wedge X(i) \rightarrow P_{odd}(i+1)) \wedge (P_{odd}(i) \wedge X(i) \rightarrow P_{even}(i+1))$$
$$\wedge (P_{even}(i) \wedge \neg X(i) \rightarrow P_{even}(i+1)) \wedge (P_{odd}(i) \wedge \neg X(i) \rightarrow P_{odd}(i+1))$$

Here, second-order variables are implicitly bounded by the largest value of the indexing term ($i + 1 = |X|$ in this example). Note that here it is possible to reference $P_{odd}(|X|)$, and so we do to simplify the formula. Of course, a direct translation would not account for such details.

### 4.2 Systems of Bounded Arithmetic

Early study of weak systems of arithmetic concentrated on restricted fragments of Peano Arithmetic, e.g., $I\Delta_0$ in which induction is over bounded first-order formulae [Par71]. This system, $I\Delta_0$, was used by Ajtai [Ajt83] to obtain lower bounds for the Parity Principle, which implied lower bounds for the complexity class $\mathsf{AC}^0$. A different approach was used by Cook: in 1975 he presented a system $PV$ for polynomial-time reasoning [Coo75]. $PV$ is an equational system with a function for every polynomial-time computable function.

The major development in bounded arithmetic came in the 1985 PhD thesis of S. Buss [Bus86]. There, several (classes of) systems of bounded arithmetic were described, capturing major complexity classes such as P and EXP (viewed as classes of functions). The best known system is $S_2^1$, which is a first-order system capturing P. To capture higher complexity classes such as PSPACE and EXP, Buss extends his systems to second-order.

In second-order systems we consider here, the richer language of Buss's first-order systems is simulated using second-order objects. Second-order quantified variables are strings of bounded length; the notation $\exists Z \leq b$ corresponds to $\exists Z \, |Z| \leq b$. First-order objects or numbers are index variables: their values are bounded by a term in number variables and lengths of second-order variables. The translation between first and second-order system is given by the RSUV isomorphism due to [Raz93, Tak93]. Here, the isomorphism is between first-order systems (R and S) and second-order systems (U and V). In particular, it translates Buss's $S_2^1$ into $V_1$, a system that reasons with existential second-order definable predicates.

Let $\mathcal{L}_2$ be the language of Peano Arithmetic with added terms $|X|$ (length of $X$) and $X(t)$ (membership of $t$ in $X$), where $X$ is second-order. We look at the systems axiomatized by Peano axioms on the number variables, together with the axioms defining length of second order variables: L1: $X(y) \to y < |X|$ and L2: $y + 1 = |X| \to X(y)$. Additionally, there is a comprehension axiom

$$\exists Z \leq b \forall i < b(Z(i) \leftrightarrow \phi(i, \bar{a}, \bar{X})), \qquad \text{(Comprehension)}$$

where $\phi$ is a class of formulae such as $\Sigma_1^B$ or $SO\exists$-Horn. Such systems of arithmetic reason with objects of complexity allowed in the comprehension axiom. For example, reasoning in $V_0$ is limited to reasoning with $\Sigma_0^B$-definable objects.

**Definition 4.** *For an integer $i \geq 0$, define $V_i$ to be the system with comprehension over $\Sigma_i^B$ formulas (e.g., $V_1$ has comprehension over $\exists SO$ formulae). For a general class $\Phi$ of formulas, $V$-$\Phi$ is the system with comprehension over $\Phi$. In particular, $V$-Horn and $V$-Krom are the systems with comprehension over $SO\exists$-Horn and $SO\exists$-Krom, respectively.*

Note that even in the weakest of this class of systems, $V_0$ with its comprehension over formulae with no second-order quantifiers, it is possible to prove induction and minimization principles for the respective class of formulae from comprehension and length axioms.

An interested reader can find more information in [Kra95, Bus86, Coo03] and an upcoming book by Stephen Cook and Phuong Nguyen.

## 5   Defining Functions in the Bounded Arithmetic Setting

In the setting of bounded arithmetic, the "hardness" is usually taken to be the complexity of properties *provable* in this system. In particular, the correspondence with complexity-theoretic notion of hardness is via the provability of the function totality. That is, the strength of a system of arithmetic is associated with the computational complexity of functions that this system proves total.

For example, a version of Peano Arithmetic with one unbounded existential quantifier allowed in induction formulae $(I\Sigma_1)$ proves the totality of primitive recursive functions, and $V_0$ where all quantifiers are bounded does the same for $\mathsf{AC}^0$ functions (in the second-order setting).

**Definition 5.** *(Capture by a system of arithmetic) A system of arithmetic captures a function class if it proves totality of all and only functions in this class.*

Traditionally, functions are introduced by their recursion-theoretic characterization (see [Cob65] for the original such result or [Zam96]). For example, Cobham's characterization of $\mathsf{P}$ uses limited recursion on notation:

$$F(0,\bar{x},\bar{Y}) = G(\bar{x},\bar{Y}) \tag{1}$$
$$F(z+1,\bar{x},\bar{Y}) = cut(p(z,\bar{x},\bar{Y}), H(z,\bar{x},\bar{Y},F(z,\bar{x},\bar{Y}))). \tag{2}$$

Here, the function $cut(x,y)$ cuts out the rest of $H(..)$ beyond the bound $p(z,\bar{x},\bar{Y})$, where $p()$ is a polynomial (that is, a term in the language).

Since we are trying to relate the expressive power of the formulas in comprehension and complexity of functions, we introduce function symbols by setting their bitgraphs to be formulas from the comprehension scheme as follows.

**Definition 6.** *Let $\Phi$ be a logic capturing a complexity class $C$ in the descriptive setting. We define a corresponding function class $FC$ by defining functions $f$ and $F$ in $FC$ as follows:*

$$z = f(\bar{x},\bar{Y}) \leftrightarrow \phi(z,\bar{x},\bar{Y}) \qquad F(\bar{x},\bar{Y})(i) \leftrightarrow i < t \wedge \phi(i,\bar{x},\bar{Y})$$

*Here, $f$ and $F$ are number and string functions, respectively, and $\phi \in \Phi$. That is, define functions by formulae from $\Phi$ by stating that graphs of number functions and bitgraphs of string functions are representable by formulae from $\Phi$.*

In particular, $\mathsf{NL}$ functions are the ones definable by $SO\exists$-Krom formulae and bitgraphs of polynomial-time functions are described by $SO\exists$-Horn formulae.

With these definitions we obtain the following capture results.

- System $V_0$ captures complexity class $\mathsf{AC}^0$. [Zam96, Coo02]
- Systems $V$-Horn and $V$-Krom with comprehension over $SO\exists$-Horn and $SO\exists$-Krom, respectively, capture $\mathsf{P}$ and $\mathsf{NL}$. [CK01, CK03]
- System $V_1$ also captures $\mathsf{P}$. [Zam96]

Note that the system $V_1$, although likely stronger than $V$-Horn because it reasons with $\mathsf{NP}$-predicates, also captures $\mathsf{P}$. That is, $\Sigma_1^B$-theorems of $V_1$ and $V$-Horn are the same, but it is likely that $V$-Horn does not prove the comprehension axiom of $V_1$, which is a $\Sigma_2^B$ statement. This leads to the following question:

## 6  When Do Systems Based on Formulas Describing a Complexity Class Capture the Same Class?

What makes systems such as $V_0$, $V$-Horn and $V$-Krom "minimal" among systems capturing the corresponding classes? They operate only with objects from the class, and yet prove totality of all functions they can define. The informal answer is *provable closure under first-order operations.* If a system can prove its own closure under $AC^0$ reductions such as conjunction, disjunction and complementation, then, with some additional technicalities, this system can prove totality of all functions definable by objects in its comprehension axiom.

Let $C$ be a complexity class. Suppose that $\Phi_C$ is a class of (existential second-order) formulas that captures $C$ in the descriptive complexity setting. We define a theory of bounded arithmetic $V\text{-}\Phi_C$ to be $V_0$ together with comprehension over bounded $\Phi_C$. The following is an informal statement of the general result:

*Claim.* [Kol04, Kol05] Let $AC^0 \subseteq C \subseteq P$. Suppose that $\Phi_C$ is closed under first-order operations provably in $V\text{-}\Phi_C$. Also, suppose that for every $\phi(\bar{x}, \bar{Y}) \in \Phi_C$, if $V\text{-}\Phi_C \vdash \phi$ then there is a function $F$ on free variables of $\phi$ which is computable in $C$ and witnesses existential quantifiers of $\phi$. Then the class of provably total functions of $V\text{-}\Phi_C$ is the class of functions computable in $C$.

Two examples of such systems are $V$-Horn and $V$-Krom. In particular, $V$-Horn formalizes and proves correctness of Horn formula satisfiability algorithm. Although $V$-Horn is provably in the system equivalent to limited recursion-based systems of polynomial-time reasoning, it has an additional nice feature: it is finitely axiomatizable. The system $V$-Krom formalizes the non-trivial inductive counting algorithm of Immerman [Imm88, Sze88], used to prove the closure of NL (and thus $SO\exists$-Krom) under complementation. However, we don't know whether NP = coNP and thus whether the class of predicates in the comprehension axiom of $V_1$ is closed under complementation. Moreover, even if it is proven that NP = coNP, the proof has to be constructive enough to be formalizable within $V_1$ to allow us to apply the claim.

This brings up an interesting meta-question: when are the properties of a complexity class itself provable within this class? We were able to prove the basic properties of P and NL with reasoning no more complex then the class itself. However, for classes such as symmetric logspace, now proven to be equal to logspace [Rei04], it is not clear whether the proof of complementation (or the proof of equivalence with logspace) are formalizable using only reasoning within this class. This line of research is related to other meta-questions in complexity theory such as what proof techniques are needed to prove complexity separations. For example, natural proofs of Razborov and Rudich [RR97] address the question of NP vs. P/poly, motivated by the P vs. NP question.

# 7   Conclusions

In this paper we touch upon one example of a connection between two different notions of hardness: the hardness of describing a property versus the hardness of proving a property. Another line of research that should be mentioned here is proof complexity. In proof complexity, the object of study is proof systems such as resolution system and Frege system; there, the lengths of proofs is the main complexity measure. For many systems of bounded arithmetic there are proof systems that are their non-uniform counterparts (that is, the system of arithmetic proves soundness of the proof system and the proof systems proves the axioms of the system of arithmetic). The line of research exploring connections between finite model theory and proof complexity is pursued by Atserias. He uses his results in finite model theory to obtain resolution proof system lower bounds [Ats02], and, in his later work with Kolaitis and Vardi, uses constraint satisfaction problems as a generic basis for a class of proof system [AKV04].

Complexity in logic is a broad area of research, with many problems still unsolved. Little is known about connections among different settings and notions of hardness. Here we have given one such example and have mentioned a couple more; other connections among various notions of complexity in logic are waiting to be discovered.

*Acknowledgements.* The results on connections between finite model theory and bounded arithmetic mentioned here come from my joint work with Stephen Cook (which resulted in my PhD thesis). I am very grateful to him for suggesting many of the ideas, as well as the framework in which these connections became natural.

# References


[Ajt83]     Ajtai, M.: $\Sigma_1^1$-Formulae on finite structures. Annals of Pure and Applied Logic 24(1), 1–48 (1983)

[AKV04]   Atserias, A., Kolaitis, P.G., Vardi, M.Y.: Constraint propagation as a proof system. In: Wallace, M. (ed.) CP 2004. LNCS, vol. 3258, pp. 77–91. Springer, Heidelberg (2004)

[Ats02]    Atserias, A.: Fixed-point logics, descriptive complexity and random satisfiability. PhD thesis, UCSC (2002)

[Bus86]    Buss, S.: Bounded Arithmetic. Bibliopolis, Naples (1986)

[CK01]     Cook, S.A., Kolokolova, A.: A second-order system for polynomial-time reasoning based on Grädel's theorem. In: Proceedings of the Sixteens annual IEEE symposium on Logic in Computer Science, pp. 177–186 (2001)

[CK03]     Cook, S.A., Kolokolova, A.: A second-order system for polytime reasoning based on Grädel's theorem. Annals of Pure and Applied Logic 124, 193–231 (2003)

[Cob65]    Cobham, A.: The intrinsic computational difficulty of functions. In: Bar-Hillel, Y. (ed.) Logic, Methodology and Philosophy of Science, pp. 24–30. North-Holland, Amsterdam (1965)

[Coo75]    Cook, S.A.: Feasibly constructive proofs and the propositional calculus. In: Proceedings of the Seventh Annual ACM Symposium on Theory of Computing, pp. 83–97 (1975)

[Coo02]    Cook, S.A.: CSC 2429S: Proof Complexity and Bounded Arithmetic. Course notes (Spring 1998-2002), http://www.cs.toronto.edu/~sacook/csc2429h

[Coo03]    Cook, S.: Theories for complexity classes and their propositional translations. In: Krajicek, J. (ed.) Complexity of computations and proofs, pp. 175–227. Quaderni di Matematica (2003)

[Fag74]    Fagin, R.: Generalized first-order spectra and polynomial-time recognizable sets. Complexity of computation, SIAM-AMC proceedings 7, 43–73 (1974)

[Grä91]  Grädel, E.: The Expressive Power of Second Order Horn Logic. In: Jantzen, M., Choffrut, C. (eds.) STACS 1991. LNCS, vol. 480, pp. 466–477. Springer, Heidelberg (1991)

[Grä92]  Grädel, E.: Capturing Complexity Classes by Fragments of Second Order Logic. Theoretical Computer Science 101, 35–57 (1992)

[Imm82]  Immerman, N.: Relational queries computable in polytime. In: 14th ACM Symp.on Theory of Computing, pp. 147–152. Springer, Heidelberg (1982)

[Imm83]  Immerman, N.: Languages that capture complexity classes. In: 15th ACM STOC symposium, pp. 347–354 (1983)

[Imm88]  Immerman: Nondeterministic space is closed under complementation. In: SCT: Annual Conference on Structure in Complexity Theory (1988)

[Imm99]  Immerman, N.: Descriptive complexity. Springer, New York (1999)

[Kol04]  Kolokolova, A.: Systems of bounded arithmetic from descriptive complexity. PhD thesis, University of Toronto (October 2004)

[Kol05]  Kolokolova, A.: Closure properties of weak systems of bounded arithmetic. In: Ong, L. (ed.) CSL 2005. LNCS, vol. 3634, pp. 369–383. Springer, Heidelberg (2005)

[Kra95]  Krajíček, J.: Bounded Arithmetic, Propositional Logic, and Complexity Theory. Cambridge University Press, New York (1995)

[Par71]  Parikh, R.: Existence and feasibility of arithmetic. Journal of Symbolic Logic 36, 494–508 (1971)

[Raz93]  Razborov, A.: An equivalence between second-order bounded domain bounded arithmetic and first-order bounded arithmetic. In: Clote, P., Krajíček, J. (eds.) Arithmetic, proof theory and computational complexity, pp. 247–277. Clarendon Press, Oxford (1993)

[Rei04]  Reingold, O.: Undirected ST-Connectivity in Log-Space. Electronic Colloquium on Computational Complexity, ECCC Report TR04-094 (2004)

[RR97]  Razborov, A.A., Rudich, S.: Natural proofs. Journal of Computer and System Sciences 55, 24–35 (1997)

[Sze88]  Szelepcsényi, R.: The method of forced enumeration for nondeterministic automata. Acta Informatica 26, 279–284 (1988)

[Tak93]  Takeuti, G.: RSUV isomorphism. In: Clote, P., Krajíček, J. (eds.) Arithmetic, proof theory and computational complexity, pp. 364–386. Clarendon Press, Oxford (1993)

[Tra50]  Trahtenbrot, B.: The impossibility of an algorithm for the decision problem for finite domains. Doklady Academii Nauk SSSR, 70:569–572 (in Russian, 1950)

[Var82]  Vardi, M.Y.: The complexity of relational query language. In: 14th ACM Symp.on Theory of Computing, Springer, Heidelberg (1982)

[Zam96]  Zambella, D.: Notes on polynomially bounded arithmetic. The Journal of Symbolic Logic 61(3), 942–966 (1996)

# On the Computational Power of Enhanced Mobile Membranes

Shankara Narayanan Krishna[1] and Gabriel Ciobanu[2]

[1] IIT Bombay, Powai, Mumbai, India 400 076
`krishnas@cse.iitb.ac.in`
[2] A.I.Cuza University, 700506 Iaşi, Romania
`gabriel@info.uaic.ro`

**Abstract.** The enhanced mobile membranes is a variant of membrane systems which has been proposed for describing some biological mechanisms of the immune system. In this paper, we study the computational power of the enhanced mobile membranes. In particular, we focus on the power of mobility given by the operations *endo*, *exo*, *fendo* and *fexo*. The computational universality is obtained with 12 membranes, while systems with 8 membranes subsume $ET0L$, and those with 3 membranes are contained in $MAT$.

## 1 Introduction

Membrane systems (called also P systems) are introduced in [8] as a class of parallel computing models inspired from the way the living cells process chemical compounds, energy, and information. The definition of these computing models starts from the observation that any biological system is a complex hierarchical structure with a flow of materials and information which underlies their functioning. The membrane computing deals with the evolution of systems composed by objects, rules and membranes nested in other membranes. The basic elements of membrane computing are presented in [10]; for the state-of-the art of the domain, the reader may consult the bibliography from the web page `http://psystems.disco.unimib.it`.

Over the years, several variants of the basic model introduced in [8] have been studied. One such model is defined by the mobile membranes [5], introduced with the motivation to describe the movement of membranes by using simpler operations than the model with active membranes [9]. Similar operations on membranes have been considered in [2].

The expressive power of mobile membranes has been studied in [6,7]. This paper is devoted to a variant of systems with mobile membranes, namely systems with enhanced mobile membranes. These systems were introduced in [1] as an useful model for describing the immune system. However, the computational power of the operations used in this model has not been studied so far. We study the operations of this variant, giving importance to the ones governing the mobility of membranes. We show that using only mobility, 12 membranes

A. Beckmann, C. Dimitracopoulos, and B. Löwe (Eds.): CiE 2008, LNCS 5028, pp. 326–335, 2008.
© Springer-Verlag Berlin Heidelberg 2008

provide the computational universality. It should be noted that unlike the results obtained in [6,7], we have not used the context-free evolution of objects in any of the results (proofs) of this paper.

The operations governing the mobility of the enhanced mobile membranes are endocytosis (endo), exocytosis (exo), forced endocytosis (fendo) and forced exocytosis (fexo). The interplay between these four operations is quite powerful, and get universality with 12 membranes. However, the exact computational power of the four operations against a subset of the operations (using more than one operation) is not clear. However, we show that, in the case of 3 membranes, the combination of endo and exo is equivalent to the combination of fendo and fexo. For systems with more membranes, this is not clear and remains an interesting problem to explore.

We refer to [3] and [12] for the elements of formal language theory we use here. For an alphabet $V$, we denote by $V^*$ the set of all strings over $V$; $\lambda$ denotes the empty string. $V^*$ is a monoid with $\lambda$ as its unit element. The length of a string $x \in V^*$ is denoted by $|x|$, and $|x|_a$ denotes the number of occurrences of symbol $a$ in $x$. A multiset over an alphabet $V$ is represented by a string over $V$ (together with all its permutations), and each string precisely identifies a multiset; the Parikh vector associated with the string indicates the multiplicities of each element of $V$ in the corresponding multiset. We now briefly recall details of the following:

1. **Parikh Vector:** For $V = \{a_1, \ldots, a_n\}$, the Parikh mapping associated with $V$ is $\psi_V : V^* \to \mathbf{N}^n$ defined by $\psi_V(x) = (|x|_{a_1}, \ldots, |x|_{a_n})$, for all $x \in V^*$. For a language $L$, its Parikh set $\psi_V(L) = \{\psi_V(x) \mid x \in L\}$ is the set of all Parikh vectors of all words $x \in L$. For a family $FL$ of languages, we denote by $PsFL$ the family of Parikh sets of vectors associated with languages in $FL$.

2. **E0L systems:** An E0L system is a context-free pure grammar with parallel derivations : $G = (V, T, \omega, R)$ where $V$ is the alphabet, $T \subseteq V$ is the terminal alphabet, $\omega \in V^*$ is the axiom, and $R$ is a finite set of rules of the form $a \to v$ with $a \in V$ and $v \in V^*$ such that for each $a \in V$ there is at least one rule $a \to v$ in $R$. For $w_1, w_2 \in V^*$, we say that $w_1 \Rightarrow w_2$ if $w_1 = a_1 \ldots a_n$, $w_2 = v_1 v_2 \ldots v_n$ for $a_i \to v_i \in R$, $1 \le i \le n$. The generated language is $L(G) = \{x \in T^* \mid \omega \Rightarrow^* x\}$.

3. **ET0L systems:** An ET0L system is a construct $G = (V, T, \omega, R_1, \ldots R_n)$ such that each quadruple $(V, T, \omega, R_i)$ is an E0L system; each $R_i$ is called a table, $1 \le i \le n$. The generated language is defined as $L(G) = \{x \in T^* \mid \omega \Rightarrow_{R_{j_1}} w_1 \Rightarrow_{R_{j_2}} \ldots \Rightarrow_{R_{j_m}} w_m = x\}$, where $m \ge 0, 1 \le j_i \le n, 1 \le i \le m$. In the sequel, we make use of the the following normal form for ET0L systems: each language $L \in ET0L$ can be generated by an ET0L system $G = (V, T, \omega, R_1, R_2)$ having only two tables. Moreover, from the proof of Theorem V.1.3 in [11], we see that any derivation with respect to $G$ starts by several steps of $R_1$, then $R_2$ is used exactly once, and the process is iterated; the derivation ends by using $R_2$.

4. **Matrix grammars:** A context-free matrix grammar without appearance checking is a construct $G = (N, T, S, M)$ where $N, T$ are disjoint alphabets of non-terminals and terminals, $S \in N$ is the axiom, and $M$ is a finite set of matrices of the form $(A_1 \to x_1, \ldots, A_n \to x_n)$ of context-free rules. For a string $w$, a matrix $m : (r_1, \ldots, r_n)$ is executed by applying the productions $r_1, \ldots, r_n$ one after the another, following the order in which they appear in the matrix. We write $w \Rightarrow_m z$ if there is a matrix $m : (A_1 \to x_1, \ldots, A_n \to x_n)$ in $M$ and the strings $w_1, \ldots, w_{n+1}$ in $(N \cup T)^*$ such that $w = w_1, w_{n+1} = z$, and for each $i = 1, 2, \ldots, n$ we have $w_i = w_i' A_i w_i'', w_{i+1} = w_i' x_i w_i''$. The language generated by $G$ is $L(G) = \{x \in T^* \mid S \Rightarrow^* x\}$. The family of languages generated by context-free matrix grammars is denoted by $MAT$.

A matrix grammar with appearance checking has an additional component $F$, a set of occurrences of rules in $M$. For $w, z \in (N \cup T)^*$, we write $w \Rightarrow_m z$ if (i) there is a matrix in $M$ whose rules can be applied in order to obtain $z$ from $w$, or if (ii) the $j$th rule $r_j$ of $M$ is not applicable to $w_j$, $(w \Rightarrow_m w_j$ in $j$ steps), and in which case $r_j$ can be skipped obtaining $w_{j+1} = w_j$.

In [4], it has been shown that each recursively enumerable language can be generated by a matrix grammar in the *strong binary normal form*. Such a grammar is a construct $G = (N, T, S, M, F)$, where $N = N_1 \cup N_2 \cup \{S, \#\}$, with these three sets mutually disjoint, two distinguished symbols $B^{(1)}, B^{(2)} \in N_2$, and the matrices in $M$ of one of the following forms:

(1) $(S \to XA)$, with $X \in N_1, A \in N_2$,
(2) $(X \to Y, A \to x)$, with $X, Y \in N_1, A \in N_2, x \in (N_2 \cup T)^*$,
(3) $(X \to Y, B^{(j)} \to \#)$, with $X, Y \in N_1, j = 1, 2$,
(4) $(X \to \lambda, A \to x)$, with $X \in N_1, A \in N_2, x \in T^*$.

Moreover, there is only one matrix of type 1, and $F$ consists of all the rules $B^{(j)} \to \#$, $j = 1, 2$ appearing in matrices of type 3. $\#$ is a trap-symbol, once introduced it is never removed. Clearly, a matrix of type 4 is used only once, in the last step of a derivation. The corresponding family of generated languages is denoted by $MAT_{ac}^\lambda$.

We denote by $RE, E0L, ET0L, MAT, MAT_{ac}^\lambda, CS$ the families of languages generated by arbitrary, extended 0L, extended tabled 0L, matrix, matrix with appearance checking, and context-sensitive grammars, respectively. It is known that $PsE0L \subseteq PsET0L \subset PsCS \subset PsRE$; $PsMAT \subset PsCS \subset PsRE$.

## 2   Systems with Enhanced Mobile Membranes

We follow the notations used in [1] for P systems with enhanced mobile membranes. A *P system with enhanced mobile membranes* is a construct

$$\Pi = (V, H, \mu, w_1, \ldots, w_n, R, i),$$

where: $n \geq 1$ (the initial *degree* of the system); $V$ is an alphabet (its elements are called *objects*); $H$ is a finite set of *labels* for membranes; $\mu$ is a *membrane structure*, consisting of $n$ membranes, labelled (not necessarily in a one-to-one manner) with elements of $H$; $w_1, w_2, \ldots, w_n$ are strings over $V$, describing the *multisets of objects* placed in the $n$ regions of $\mu$, $i$ is the output membrane of the system, and $R$ is a finite set of *developmental rules* of the following forms:

(a) $[a \rightarrow v]_m$, for $m \in H, a \in V, v \in V^*$; object evolution rules.

(b) $[a]_h[\ ]_m \rightarrow [[w]_h]_m$, for $h, m \in H, a \in V, w \in V^*$;
   *endocytosis*: an elementary membrane labelled $h$ enters the adjacent membrane labelled $m$, under the control of object $a$; the labels $h$ and $m$ remain unchanged during this process, however, the object $a$ may be modified to $w$ during the operation; $m$ is not necessarily an elementary membrane.

(c) $[[a]_h]_m \rightarrow [w]_h[\ ]_m$, for $h, m \in H, a \in V, w \in V^*$;
   *exocytosis*: an elementary membrane labelled $h$ is sent out of a membrane labelled $m$, under the control of object $a$; the labels of the two membranes remain unchanged, but the object $a$ from membrane $h$ may be modified during this operation; membrane $m$ is not necessarily elementary.

(d) $[\ ]_h[a]_m \rightarrow [[\ ]_h w]_m$, for $h, m \in H, a \in V, w \in V^*$;
   *forced endocytosis*: an elementary membrane labelled $h$ enters the adjacent membrane labelled $m$, under the control of object $a$ of $m$; the labels $h$ and $m$ remain unchanged during this process, however, the object $a$ may be modified to $w$ during the operation; $m$ is not necessarily an elementary membrane.

(e) $[a[\ ]_h]_m \rightarrow [\ ]_h[w]_m$, for $h, m \in H, a \in V, w \in V^*$;
   *forced exocytosis*: an elementary membrane labelled $h$ is sent out of a membrane labelled $m$, under the control of object $a$ of $m$; the labels of the two membranes remain unchanged, but the object $a$ from membrane $m$ may be modified during this operation; membrane $m$ is not necessarily elementary.

(f) $[[a]_j[b]_h]_k \rightarrow [[w]_j[b]_h]_k$ for $h, j, k \in H, a, b \in V, w \in V^*$;
   *contextual evolution* : an object $a$ in membrane $m$ evolves into $w$ when membranes $m, h$ are adjacent to each other inside membrane $k$.

(g) $[a]_h \rightarrow [b]_h[c]_h$, for $h \in H, a, b, c \in V$;
   *division for elementary membranes*: in reaction with an object $a$, the membrane labelled $h$ is divided into two membranes labelled $h$, with the object $a$ replaced in the two new membranes by possibly new objects.

The rules are applied according to the following principles:

1. All rules are applied in parallel, non-deterministically choosing the rules, the membranes, and the objects, but in such a way that the parallelism is maximal; this means that in each step we apply a set of rules such that no further rule can be added to the set, no further membranes and objects can evolve at the same time.

2. The membrane $m$ from each type (a) – (g) of rules as above is said to be *passive*, while the membrane $h$ is said to be *active*. In any step of a computation, any object and any active membrane can be involved in at most one rule, but the passive membranes are not considered involved in the use of rules (hence they can be used by several rules at the same time as passive membranes); for instance, a rule $[a \rightarrow v]_m$, of type (a), is considered to involve only the object $a$, not also the membrane $m$.

3. The evolution of objects and membranes takes place in a bottom-up manner. After having a (maximal) set of rules chosen, they are applied starting from the innermost membranes, level by level, up to the skin membrane (all these sub-steps form a unique evolution step, called a *transition* step).

4. When a membrane is moved across another membrane, by rules (b)-(e), its whole contents (its objects) are moved; because of the bottom-up way of using the rules, the inner objects first evolve (if there are rules applicable for them), and then any membrane is moved with the contents as obtained after this inner evolution.

5. All objects and membranes which do not evolve at a given step (for a given choice of rules which is maximal) are passed unchanged to the next configuration of the system.

By using the rules in this way, we get transitions among the configurations of the system. A sequence of transitions is a computation, and a computation is successful if it halts (it reaches a configuration where no rule can be applied).

The multiplicity vector of the multiset from a special membrane called output membrane is considered as a result of the computation. Thus, the result of a halting computation consists of all the vectors describing the multiplicity of objects from the output membrane; a non-halting computation provides no output. The set of vectors of natural numbers produced in this way by a system $\Pi$ is denoted by $Ps(\Pi)$. A computation can produce several vectors, all of them considered in the set $Ps(\Pi)$. The family of all sets $Ps(\Pi)$ generated by systems of degree at most $n$ using rules $\alpha \subseteq \{exo, endo, fendo, fexo, cevol\}$, is denoted by $PsEM_n(\alpha)$. Here $endo$ and $exo$ represent endocytosis and exocytosis, $fendo$ and $fexo$ represent forced endocytosis and forced exocytosis, and $cevol$ represents contextual evolution.

<u>Note</u>: In this paper, we do not allow exo rules where a membrane exits the system.

## 3   Computational Power

**Theorem 1.** $PsEM_{12}(endo, exo, fendo, fexo) = PsRE.$

*Proof.* Consider a matrix grammar $G = (N, T, S, M, F)$ with appearance checking in the improved strong binary normal form ($N = N_1 \cup N_2 \cup \{S, \#\}$), having $n_1$ matrices $m_1, \ldots, m_{n_1}$ of types 2 and 4 and $n_2$ matrices of type 3. The initial matrix is $m_0 : (S \rightarrow XA)$. Let $B^{(1)}$ and $B^{(2)}$ be two objects in $N_2$ for which there are rules $B^{(j)} \rightarrow \#$ in matrices of $M$. The matrices of the form $(X \rightarrow Y, B^{(j)} \rightarrow \#)$ are labelled by $m'_i$, $1 \leq i \leq n_2$ with $i \in lab_j$, for $j \in \{1, 2\}$, such that $lab_1, lab_2$, and $lab_0 = \{1, 2, \ldots, n_1\}$ are mutually disjoint sets.

Construct a P system $\Pi = (V, \{1, \ldots, 12\}, \mu, w_1, \ldots, w_{12}, R, 7)$ with

$$V = N_1 \cup N_2 \cup T \cup \{X'_{0i}, A'_{0i} \mid X \in N_1, A \in N_2, 1 \leq i \leq n_1\} \cup \{Z, \#\}$$

$$\cup \{X_{ji}, A_{ji} \mid 0 \leq i, j \leq n_1\} \cup \{X_i^{(j)}, X_j \mid X \in N_1, j \in \{1, 2\}, 1 \leq i \leq n_2\},$$

$$\mu = [\ [\ ]_7 [\ ]_8 [\ ]_9 [\ ]_{10} [\ ]_{11} \ [\ [\ ]_3 [\ ]_5]_1 \ [\ [\ ]_4 [\ ]_6]_2 ]_{12},$$

$$w_7 = \{XA \mid m_0 : (S \rightarrow XA)\}, w_i = \emptyset, i \neq 7.$$

1. Simulation of a matrix $m_i : (X \rightarrow Y, A \rightarrow x), 1 \leq i \leq n_1$.

   1. $[X]_7[\ ]_8 \rightarrow [[X_{ii}]_7]_8$, $[[A]_7]_8 \rightarrow [A_{ii}]_7[\ ]_8$, (endo, exo),
   2. $[X_{ji}]_7[\ ]_9 \rightarrow [[X_{j-1i}]_7]_9$, $[[A_{ji}]_7]_9 \rightarrow [A_{j-1i}]_7[\ ]_9, j > 0$, (endo, exo),
   3. $[\ ]_{10}[X_{0i}]_7 \rightarrow [X'_{0i}[\ ]_{10}]_7$, $[\ ]_{11}[A_{0i}]_7 \rightarrow [A'_{0i}[\ ]_{11}]_7$, (fendo),
   4. $[\ ]_{10}[X_{ji}]_7 \rightarrow [\#[\ ]_{10}]_7$, $[\ ]_{11}[A_{ji}]_7 \rightarrow [\#[\ ]_{11}]_7, j > 0$, (fendo),
      $[[A_{0i}]_7]_9 \rightarrow [\#]_7[\ ]_9$, (exo),
   5. $[X'_{0i}[\ ]_{10}]_7 \rightarrow [\ ]_{10}[Y]_7$, $[A'_{0i}[\ ]_{11}]_7 \rightarrow [\ ]_{11}[x]_7$, (fexo)

By rule 1, membrane 7 enters membrane 8, replacing $X \in N_1$ with $X_{ii}$. A symbol $A \in N_2$ is replaced with $A_{jj}$, and membrane 7 comes out of membrane 8. The suffixes $i, j$ represent the matrices $m_i(m_j)$, $1 \leq i, j \leq n_1$ corresponding to which $X, A$ have a rule. Next, rule 2 is used until $X_{ii}$ and $A_{jj}$ become $X_{0i}$ and $A_{0j}$, respectively. If $i = j$, then we have $X_{0i}$ and $A_{0i}$ simultaneously in membrane 7. Then rule 3 is used, by which membranes 10 and 11 enter membrane 7 replacing $X_{0i}$ and $A_{0i}$ with $X'_{0i}$ and $A'_{0i}$, respectively. This is then followed by rule 5, when membranes 10 and 11 exit membrane 7 by fexo rules replacing $X'_{0i}$ and $A'_{0i}$ with $Y$ and $x$. If $i > j$, then we obtain $A_{0j}$ before $X_{0i}$. In this case, we have a configuration where membrane 7 is inside membrane 9 containing $A_{0j}$. Then rule 4 is used, replacing $A_{0j}$ with $\#$, and an infinite computation will be obtained (rule 13). If $j > i$, then we obtain $X_{0i}$ before $A_{0j}$. In this case, we reach a configuration with $X_{0i}A_{kj}, k > 0$ in membrane 7, and membrane 7 is in the skin membrane. Rule 2 cannot be used now, and the only possibility is to use rule 4, which leads to an infinite computation (due to rule 13). Thus, if $i = j$, then we can correctly simulate a type-2 matrix.

2. Simulation of a matrix $m'_i : (X \rightarrow Y, B^{(j)} \rightarrow \#), j \in \{1, 2\}, 1 \leq i \leq n_2$.

   6. $[X]_7[\ ]_j \rightarrow [[X_i^{(j)}]_7]_j$, (endo),
   7. $[\ ]_{j+2}[X_i^{(j)}]_7 \rightarrow [X_i^{(j)}[\ ]_{j+2}]_7$, $[\ ]_{j+4}[B^{(j)}]_7 \rightarrow [\#[\ ]_{j+4}]_7$, (fendo),
   8. $[X_i^{(j)}[\ ]_{j+2}]_7 \rightarrow [\ ]_{j+2}[Y_j]_7$, (fexo),
   9. $[Y_j[\ ]_{j+4}]_7 \rightarrow [\ ]_{j+4}[Y_j]_7$ (fexo),  10. $[[Y_j]_7]_j \rightarrow [Y]_7[\ ]_j$, (exo).

To simulate a matrix of type 3, we start with rule 6. Membrane 7 enters either membrane 1 or membrane 2, depending on whether it has the symbol $B^{(1)}$ or $B^{(2)}$ associated to it. Inside membrane $j$, rule 7 is used, by which membrane $j+2$ enters membrane 7, and membrane $j+4$ enters membrane 7 if the symbol $B^{(j)}$ is present. In this case, $B^{(j)}$ is replaced with $\#$. Otherwise, membrane $j+2$ comes out of membrane 7 replacing $X_i^{(j)}$ with $Y_j$. Membrane 7 exits membrane $j$, replacing $Y_j$ by $Y$, successfully simulating a matrix of type 3.

3. Halting and simulation of $m_i : (X \rightarrow \lambda, A \rightarrow x), 1 \leq i \leq n_1$. We begin with rules 1-5 as before and simulate the matrix $(X \rightarrow Z, A \rightarrow x)$ in place of $m_i$,

where $Z$ is a new symbol. The special symbol $Z$ is obtained in membrane 7 at the end.

> 11. $[\ ]_8[\,Z\,]_7 \to [\lambda[\ ]_8]_7$, (fendo),
>
> 12. $[A[\ ]_8]_7 \to [\ ]_8[\,\#\,]_7, A \in N_2$, (fexo),
>
> 13. $[\ ]_8[\,\#\,]_7 \to [\#[\ ]_8]_7, [\#[\ ]_8]_7 \to [\ ]_8[\,\#\,]_7$, (fendo, fexo).

Now, use rule 11 to erase the symbol $Z$ while membrane 8 enters membrane 7. This is followed by rule 12 if there are any more symbols $A \in N_2$ remaining, in which case they are replaced with the trap symbol $\#$. An infinite computation is obtained if the symbol $\#$ is present in membrane 7. It is clear that if the computation proceeds correctly, then membrane 7 contains a multiset of terminal symbols $x \in T^*$ such that $\psi_V(x) \in PsL(G)$.    □

**Theorem 2.** $PsEM_3(cevol) = PsRE$.

**Theorem 3.** $PsET0L \subseteq PsEM_8(endo, exo, fendo, fexo)$.

*Proof.* Let $G = (V, T, \omega, R_1, R_2)$ be an ET0L system in the normal form. We construct the P system $\Pi = (V', H, \mu, w_0, \ldots, w_7, R, 0)$ as follows:

$$V' = \{\dagger, g_i, g_i', g_i'', g_i''', j_i, h_i, k_i, j_i', j_i'', j_i''', h_i', h_i'', h_i''' \mid i = 1, 2\}$$
$$\cup \{a_i \mid a \in V, i = 1, 2\}, H = \{0, 1, \ldots, 7\},$$
$$\mu = [[\ ]_0[\ ]_1[\ ]_2\ [[\ ]_4[\ ]_5[\ ]_6]_3]_7,\ w_0 = g_1\omega, w_i = \emptyset, i > 0.$$

*Simulation of table $R_i$, $i \in \{1, 2\}$*

> 1. $[g_i]_0[\ ]_i \to [[g_i]_0]_i$, (endo),
>
> 2. $[[a]_0]_i \to [w_i]_0[\ ]_i$, if $a \to w \in R_i$, (exo),
>
> 3. $[g_i]_0[\ ]_3 \to [[g_i']_0]_3$, (endo),
>
> 4. $[\ ]_4[\,g_i']_0 \to [g_i''[\ ]_4]_0$, $[\ ]_5[\,a]_0 \to [\dagger[\ ]_5]_0$, (fendo),
>
> 5. $[g_i''[\ ]_4]_0 \to [\ ]_4[g_i''']_0$, (fexo), $[\dagger[\ ]_5]_0 \to [\dagger]_0[\ ]_5$ (fexo),
>
> 6. $[\ ]_5[\dagger]_0 \to [\dagger[\ ]_5]_0$, (fendo),
>
> 7. $[g_i''']_0[\ ]_5 \to [[g_i''']_0]_5$, (endo),
>
> 8. $[[a_i]_0]_5 \to [a]_0[\ ]_5$, (exo),
>
> 9. $[[g_i''']_0]_5 \to [j_i h_i k_i]_0[\ ]_5$, (exo),
>
> 10. $[\ ]_4[\,j_i]_0 \to [j_i'[\ ]_4]_0, [\ ]_5[\,h_i]_0 \to [h_i'[\ ]_5]_0,$
>     $[\ ]_6[\,k_i]_0 \to [\lambda[\ ]_6]_0$, (fendo),
>
> 11. $[j_i'[\ ]_4]_0 \to [\ ]_4[j_i'' h_i'']_0, [h_i'[\ ]_5]_0 \to [\ ]_5[\lambda]_0,$
>     $[a_i[\ ]_6]_0 \to [\ ]_6[\dagger]_0$, (fexo),
>
> 12. $[j_i''[\ ]_6]_0 \to [\ ]_6[j_i''']_0$, (fexo), $[\ ]_5[h_i'']_0 \to [h_i'''[\ ]_5]_0$, (fendo),
>
> 13. $[h_i'''[\ ]_5]_0 \to [\ ]_5[\lambda]_0$, (fexo),
>
> 14. $[[j_1''']_0]_3 \to [g_i]_0[\ ]_3, i = 1, 2, [[j_2''']_0]_3 \to [g_1]_0[\ ]_3$, (exo),
>     $[[j_2''']_0]_3 \to [\lambda]_0[\ ]_3$, (exo),
>
> 15. $[\dagger]_0[\ ]_1 \to [[\dagger]_0]_1$, (endo), $[[\dagger]_0]_1 \to [\dagger]_0[\ ]_1$, (exo).

In the initial configuration, the string $g_1\omega$ is in membrane 0, where $\omega$ is the axiom, and $g_1$ indicates that table 1 should be simulated first. The simulation begins with rule 1, with membrane 0 entering membrane 1. In membrane 1 the only applicable rule is the exo rule 2, by which the symbols $a \in V$ are replaced by $w_1$ corresponding to the rule $a \to w_1 \in R_1$. Rules 1 and 2 can be repeated until all the symbols $a \in V$ are replaced according to a rule in $R_1$. Finally, if all the symbols $a \in V$ have been replaced by rules of $R_1$, only symbols of $V_1$ and the symbol $g_1$ are in membrane 0. Rule 3 can be used anytime after this; symbol $g_1$ is replaced with $g_1'$, and membrane 0 enters membrane 3. No rules of $R_i$ can be simulated until membrane 0 comes out of membrane 3.

Inside membrane 3, rule 4 is used. Membrane 4 enters membrane 0 by an fendo rule replacing $g_1'$ with $g_1''$; in parallel, membrane 5 enters membrane 0 if any symbol $a \in V$ is present in membrane 0, i.e, if some symbol $a \in V$ has been left out from the simulation using a rule from $R_1$. The trap symbol $\dagger$ is introduced, and this triggers a never ending computation (rules 5 and 6). Next, membrane 4 comes out of membrane 0 replacing $g_1''$ with $g_1'''$. Membrane 0 now enters membrane 5 using rule 7. Rules 8 and 7 are applied as long as required until all the symbols of $V_1$ are replaced with the corresponding symbols of $V$. When all the symbols of $V_1$ are replaced with symbols of $V$, rule 9 can be used to replace $g_1'''$ with $j_1 h_1 k_1$. We now need to check if all the symbols of $V_1$ are indeed replaced with symbols of $V$. For this, rule 10 is used. The membranes 4,5 and 6 enter membrane 0 in parallel by fendo rules replacing $j_1, h_1$ and $k_1$ with $j_1', h_1'$ and $\lambda$, respectively. Next, by fexo rule 11, membranes 4 and 5 come out of membrane 0 replacing $j_1'$ with $j_1'' h_1''$ and $h_1'$ with $\lambda$, respectively. If any symbol of $a_1 \in V_1$ is present in membrane 0, then membrane 6 also comes out (fexo rule 11), replacing it with the trap symbol $\dagger$. In case there are no symbols of $V_1$ in membrane 0, membrane 6 stays inside membrane 0 until we obtain $j_1'' h_1''$. Using the fexo rule 12, membrane 6 comes out replacing $j_1''$ with $j_1'''$, while in parallel, membrane 5 enters membrane 0 using the fendo rule 12 replacing $h_1''$ with $h_1'''$ (Note that if we choose to use the fendo rule 4 instead of 12, the trap symbol $\dagger$ is introduced in membrane 0). Membrane 5 comes out of membrane 0 erasing $h_1'''$. Once membrane 0 becomes elementary, it comes out of membrane 3 replacing $j_1'''$ with $g_i$. If $j_1'''$ is replaced with $g_1$, then table 1 is simulated again, else, table 2 is simulated.

The computation can stop only after simulating table 2. If table 2 is simulated, we obtain $j_2'''$ at the end of the simulation. $j_2'''$ is replaced with either (i) $g_1$, in which case we simulate table 1 again, or (ii) $\lambda$, in which case membrane 0 remains idle inside the skin membrane, in the absence of the trap symbol $\dagger$. It is clear that $Ps(\Pi)$ contains only the vectors in $Ps(L(G))$.                □


**Corollary 1.** $PsE0L \subseteq PsEM_7(endo, exo, fendo, fexo)$.


We can interpret the multiset of objects present in the output membrane as a set of strings $x$ such that the multiplicity of symbols in $x$ is same as the multiplicity of objects in the output membrane. This way, the multiset of objects in the output membrane gives rise to a language. For a system $\Pi$, let $L(\Pi)$ represent this language (all strings computed by $\Pi$), and let $LEM_n(\alpha)$ represent the family of languages $L(\Pi)$ generated by systems having $\leq n$ membranes, using a set of operations $\alpha \subseteq \{endo, exo, fendo, fexo\}$. Then, we have the following:

**Lemma 1.** $LEM_8(endo, exo, fendo, fexo) - ET0L \neq \emptyset$.

*Proof.* $L = \{x \in \{a,b\}^* \mid |x|_b = 2^{|x|_a}\} \notin ET0L$ [11]. We construct $\Pi = (\{a, b, b', c, \eta, \eta', \eta'', \eta''', \dagger\}, \{0, \ldots, 7\}, [\; [\eta b]_1 \; [\; [\; ]_3 \; [\; ]_4 \;]_2 \; [\; [\; ]_6 [\; ]_7 \;]_5 \;]_0, R, 1)$ with rules as given below to generate $L$.

1. $[\eta]_1[\;]_2 \rightarrow [[\eta]_1]_2, [[b]_1]_2 \rightarrow [b'b']_1[\;]_2$, (endo, exo),
   $[[\eta]_1]_2 \rightarrow [\eta]_1[\;]_2$, (exo),
2. $[\eta]_1[\;]_2 \rightarrow [[\eta'ac]_1]_2$, (endo),
3. $[\;]_3[c]_1 \rightarrow [[\;]_3\lambda]_1, [\;]_4[\eta']_1 \rightarrow [[\;]_4\eta']_1$, (fendo),
4. $[b[\;]_3]_1 \rightarrow [\;]_3[\dagger]_1, [\eta'[\;]_4]_1 \rightarrow [\;]_4[\eta'']_1$, (fexo),
5. $[\eta''[\;]_3]_1 \rightarrow [\;]_3[\eta'']_1$, (fexo),
6. $[\eta'']_1[\;]_4 \rightarrow [[\eta'']_1]_4, [[b']_1]_4 \rightarrow [b]_1[\;]_4$, (endo, exo),
7. $[[\eta'']_1]_4 \rightarrow [\eta''']_1[\;]_4$, (exo),
8. $[[\eta''']_1]_2 \rightarrow [\eta''']_1[\;]_2$, (exo),
9. $[\eta''']_1[\;]_5 \rightarrow [[\eta''']_1]_5$, (endo),
10. $[\;]_6[b']_1 \rightarrow [\dagger[\;]_6]_1, [\;]_7[\eta''']_1 \rightarrow [\eta[\;]_7]_1$, (fendo),
11. $[\dagger[\;]_6]_1 \rightarrow [\;]_6[\dagger]_1, [\eta[\;]_7]_1 \rightarrow [\;]_7[\eta]_1$, (fexo),
12. $[[\eta]_1]_5 \rightarrow [\lambda]_1[\;]_5$, (exo),
13. $[\eta[\;]_5]_1 \rightarrow [\;]_5[\eta]_1$, (fexo),
14. $[\dagger]_1[\;]_2 \rightarrow [[\dagger]_1]_2$, (endo), $[[\dagger]_1]_2 \rightarrow [\dagger]_1[\;]_2$, (exo).

The system works as follows: Rule 1 is used to replace every $b$ with $b'b'$. Rule 2 can be used at any point replacing $\eta$ (guessing that all $b$'s have been replaced). The rules 3-5 check that every $b$ has been replaced with $b'b'$, and then all the $b'$ are replaced with $b$ by rules 6-8. Next, rules 9-11 check that all $b'$ are replaced with $b$. The computation can halt using rule 12, and can continue using rule 13. An infinite computation is obtained using rule 14 when (i) membrane 1 enters membrane 2 using rule 2 before replacing all $b$'s, or (ii) membrane 1 enters membrane 5 before replacing all the $b'$. It is easy to see that membrane 1 contains strings of $L$ at the end of a halting computation. $\square$

**Theorem 4.** $PsEM_3(endo, exo) = PsEM_3(fendo, fexo)$.

*Proof.* Consider a P system $\Pi$ with 3 membranes having $endo, exo$ rules. Let the initial configuration be $\mu = [[w_1]_1[w_2]_2]_3$. It is easy to construct a P system $\Pi'$ using only $fendo, fexo$ rules having initial configuration $\mu' = \mu$ as follows: For every endo rule $[a]_i[\;]_j \rightarrow [[w_a]_i]_j$ in $\Pi$, add the fendo rule $[\;]_j[a]_i \rightarrow [[\;]_j w_a]_i$, and for every exo rule $[[a]_i]_j \rightarrow [w_a]_i[\;]_j$, add an fexo rule $[[\;]_j a]_i \rightarrow [\;]_j[w_a]_i$ in $\Pi'$. Note that the computation starts in $\Pi$ using an endo rule only, so in $\Pi'$, we can start with the corresponding fendo rule. If the initial configuration of $\Pi$ was $\mu = [\,[\,w_2[\,w_1]_1]_2]_3$, then the first applicable rule is an exo rule. In this case, construct $\Pi' = \Pi_\alpha$ with initial configuration $\mu' = [\,[\,\alpha]_1[\,w_2]_2]_3$ where $\alpha$ is a string dependent on $w_1$ and the set of exo rules applicable to $w_1$ in $\Pi$. After this preprocessing step, the set of fendo, fexo rules are constructed similarly as above. $\qquad\square$

**Theorem 5.** $PsEM_3(endo, exo, fendo, fexo) \subseteq PsMAT$.

## 4   Conclusion

Living matter becomes a source of inspiration for defining new models of computation. The enhanced mobile membranes were introduced in [1] to describe biological mechanisms of the immune system. In this paper, we have studied the computational power of the membrane systems with enhanced mobile operations. The universality results uses 12 membranes and the four mobility operations endo, exo, fendo and fexo. An interesting problem to look at is whether systems using all the four rules can be simulated by systems using only a pair of rules from $\{endo, exo, fendo, fexo\}$, namely either pair $(endo, exo)$ or pair $(fendo, fexo)$. The optimality of the results presented in this paper is also an open problem.

## References


1. Aman, B., Ciobanu, G.: Describing the Immune System Using Enhanced Mobile Membranes. In: Proceedings FBTC, CONCUR 2007, pp. 1–14 (2007)
2. Cardelli, L.: Brane Calculi, Interactions of biological membranes. In: Danos, V., Schachter, V. (eds.) CMSB 2004. LNCS (LNBI), vol. 3082, pp. 257–278. Springer, Heidelberg (2005)
3. Dassow, J., Paun, G.: Regulated Rewriting in Formal Language Theory. Springer, Heidelberg (1989)
4. Freund, R., Păun, G.: On the number of non-terminals in graph-controlled, programmed, and matrix grammars. In: Margenstern, M., Rogozhin, Y. (eds.) MCU 2001. LNCS, vol. 2055, pp. 214–225. Springer, Heidelberg (2001)
5. Krishna, S.N., Paun, G.: P Systems with Mobile Membranes. Natural Computing 4(3), 255–274 (2005)
6. Krishna, S.N.: The Power of Mobility: Four Membranes Suffice. In: Cooper, S.B., Löwe, B., Torenvliet, L. (eds.) CiE 2005. LNCS, vol. 3526, pp. 242–251. Springer, Heidelberg (2005)
7. Krishna, S.N.: Upper and Lower Bounds for the Computational Power of P Systems with Mobile Membranes. In: Beckmann, A., Berger, U., Löwe, B., Tucker, J.V. (eds.) CiE 2006. LNCS, vol. 3988, pp. 526–535. Springer, Heidelberg (2006)
8. Paun, G.: Computing with membranes. Journal of Computer and System Sciences 61(1), 108–143 (2000)
9. Paun, G.: P Systems with Active Membranes: Attacking NP-Complete Problems. Journal of Automata, Languages and Combinatorics 6(1), 75–90 (2001)
10. Paun, G.: Membrane Computing. An Introduction. Springer, Heidelberg (2002)
11. Rozenberg, G., Salomaa, A. (eds.): The Mathematical Theory of L Systems. Academic Press (1980)
12. Salomaa, A.: Formal Languages. Academic Press (1973)

# Recursion in Higher Types and Resource Bounded Turing Machines


Lars Kristiansen

Department of Mathematics, University of Oslo
P.O. Box 1053, Blindern, NO-0316 Oslo, Norway
larsk@math.uio.no
http://www.iu.hio.no/~larskri



**Abstract.** We prove that neat and natural fragments of Gödel's T and Plotkin's PCF capture complexity classes defined by imposing resource bounds on Turing machines.


## 1 Introduction

Some years ago the author, and others, noted that interesting things tend to happen when successor-like functions and operations are removed from a standard model of computation, e.g. a function algebra, a rewriting system or a programming language. See e.g. Kristiansen [7,9], Kristiansen & Voda [11], Jones [5,6]. Our investigations of such successor-free models of computation have turned out to be fruitful and have spawned some surprising theorems, particularly when the models permit recursion in higher types. Gödel's system T and Plotkin's programming language PCF are prime examples of models of computation permitting recursion in higher type. We will refer to their successor-free variants as respectively $T^-$ and $PCF^-$.

This is a compact and technical paper where we show that neat and natural fragments of $T^-$ and $PCF^-$ capture complexity classes defined by imposing resource bounds on Turing machines. $PCF^-$ is studied for the first time in the present paper. The results on $T^-$ are published in joint papers with Paul Voda, but we reprove some of these results here in order to elucidate the relationship between $T^-$ and $PCF^-$, and in general, we have tried to tailor our definitions and proofs such that the nature of this relationship becomes apparent. Besides, the proof given here are more direct, and presumably more transparent, than those published elsewhere. Our results on $PCF^-$ are related to the results published in Jones [5].

We expect the reader to be familiar with the typed $\lambda$-calculus; Gödel's T and primitive recursion in finite types (see e.g. Avigad & Feferman [1]); PCF and the fundamentals of domain theory (see e.g. Streicher [16]); and Turing machines and the fundamentals of complexity theory (see e.g. Odifreddi [15]). Subjects related to $T^-$ are studied in Barra et al. [2], Kristiansen & Voda [12,13,14], Kristiansen & Barra [10], Kristiansen [8,9]. The author wants to thank Dag Normann for his help in preparing this paper and for numerous insightful lectures on PCF and domain theory.


A. Beckmann, C. Dimitracopoulos, and B. Löwe (Eds.): CiE 2008, LNCS 5028, pp. 336–348, 2008.

## 2   A Brief Discussion

Before we turn to the technical work, we will briefly discuss the relevance of the results proved in this paper.

Let TIME $2_i^{\text{LIN}}$ (SPACE $2_i^{\text{LIN}}$) be the set of problems decidable by a deterministic Turing machine working in time (space) $2_i^{c|x|}$ for some fixed $c \in \mathbb{N}$ (where $2_0^x = x$ and $2_{i+1}^x = 2^{2_i^x}$). It is trivial that TIME $2_i^{\text{LIN}} \subseteq$ SPACE $2_i^{\text{LIN}}$ and SPACE $2_i^{\text{LIN}} \subseteq$ TIME $2_{i+1}^{\text{LIN}}$, and thus, we have an *alternating space-time hierarchy*

$$\text{SPACE } 2_0^{\text{LIN}} \subseteq \text{TIME } 2_1^{\text{LIN}} \subseteq \text{SPACE } 2_1^{\text{LIN}} \subseteq \text{TIME } 2_2^{\text{LIN}} \subseteq \text{SPACE } 2_2^{\text{LIN}} \subseteq \text{TIME } 2_3^{\text{LIN}} \subseteq \dots .$$

The three classes at the bottom of the hierarchy are often called respectively LINSPACE, EXP, and EXPSPACE in the literature. It is well known, and quite obvious, that we have SPACE $2_i^{\text{LIN}} \subset$ SPACE $2_{i+1}^{\text{LIN}}$ and TIME $2_i^{\text{LIN}} \subset$ TIME $2_{i+1}^{\text{LIN}}$ for any $i \in \mathbb{N}$. Thus, we know that at least one of the two inclusions

$$\text{SPACE } 2_i^{\text{LIN}} \subseteq \text{TIME } 2_{i+1}^{\text{LIN}} \subseteq \text{SPACE } 2_{i+1}^{\text{LIN}}$$

is strict; similarly, we know that at least one of the inclusions

$$\text{TIME } 2_i^{\text{LIN}} \subseteq \text{SPACE } 2_i^{\text{LIN}} \subseteq \text{TIME } 2_{i+1}^{\text{LIN}}$$

is strict, and the general opinion is that they all are. Still, no one has ever been able to prove that any particular of the inclusions actually is strict.

The classes in the alternating space-time hierarchy are defined by imposing *explicit bounds* on a particular machine model, but the classes are *not uniformly defined* as some of the classes are defined by imposing space-bounds whereas others are defined by imposing time-bounds. In contrast, the classes in our $\text{T}^-$-hierarchy $\mathcal{G}_0 \subseteq \mathcal{G}_1 \subseteq \mathcal{G}_2 \subseteq \dots$ are *uniformly defined*. They are also defined *without referring to explicit bounds*. (Each class $\mathcal{G}_i$ is defined as the set of problems decidable in a natural fragment of $\text{T}^-$.) Thus, it is surprising and interesting that the $\text{T}^-$-hierarchy matches, level by level, the alternating space-time hierarchy. How can it be that such a uniformly defined hierarchy contains both space and time classes?

The match between the $\text{T}^-$-hierarchy and the alternating space-time hierarchy was proved for the first time in [13]. In this paper we introduce a $\text{PCF}^-$-hierarchy $\mathcal{P}_0 \subseteq \mathcal{P}_1 \subseteq \mathcal{P}_2 \subseteq \dots$. The definition of the $\text{PCF}^-$-hierarchy is very similar to definition of the $\text{T}^-$-hierarchy, and it is natural to as how these two hierarchies relate to each other. We will prove that $\mathcal{P}_{n+1} = \mathcal{G}_{2n+1}$ for $n \in \mathbb{N}$. Hence, only time classes occur in the $\text{PCF}^-$-hierarchy.


### $\text{T}^-$ and $\text{PCF}^-$

**Definition (Types).** We define the *types* recursively: $\iota$ is a type (primitive type); $\sigma \times \tau$ is a type if $\sigma$ and $\tau$ are types (product types); $\sigma \to \tau$ is a type if $\sigma$ and $\tau$ are types (arrow types). We define the *cardinality of the type $\sigma$ at base*

$b$, written $|\sigma|_b$, the *domain height of the type* $\sigma$ *at base* $b$, written $\lceil\sigma\rceil_b$, and the *degree of the type* $\sigma$, written $\mathrm{dg}(\sigma)$, by recursion on the structure of the type $\sigma$: $|\iota|_b = b$; $\lceil\iota\rceil_b = 1$; $\mathrm{dg}(\iota) = 0$; $|\rho \times \tau|_b = |\rho|_b \times |\tau|_b$; $\lceil\rho \times \tau\rceil_b = \lceil\rho\rceil_b + \lceil\tau\rceil_b$; $\mathrm{dg}(\rho \times \tau) = \max(\mathrm{dg}(\rho), \mathrm{dg}(\tau))$; $|\rho \to \tau|_b = |\tau|_b^{|\rho|_b}$; $\lceil\rho \to \tau\rceil_b = |\rho|_{b+1} \times \lceil\tau\rceil_b$; and $\mathrm{dg}(\rho \to \tau) = \max(\mathrm{dg}(\rho) + 1, \mathrm{dg}(\tau))$. The notation $\sigma_1, \sigma_2, \ldots, \sigma_n \to \tau$ is shorthand for $\sigma_1 \to (\sigma_2 \to (\ldots(\sigma_n \to \tau)\ldots))$. $\qquad\square$

The proof of the next lemma is straightforward.

**Lemma 1 (Growth Rates).** *Let* $2_0^x = x$ *and* $2_{i+1}^x = 2^{2_i^x}$. *(i) For any type* $\sigma$ *of degree* $n$, *there exists a polynomial* $p(x)$ *such that* $|\sigma|_x < 2_n^{p(x)}$. *(ii) For any* $n \in \mathbb{N}$ *and polynomial* $p(x)$, *there exists a type* $\sigma$ *of degree* $n$ *such that* $2_n^{p(x)} < |\sigma|_{\max(x,2)}$. *(iii) For any type* $\sigma$ *of degree* $n+1$ *there exists a type* $\tau$ *of degree* $n$ *such that* $\lceil\sigma\rceil_b < |\tau|_b$.

**Definition (Calculi).** We define the *terms* of *the typed* $\lambda$-*calculus*: We have an infinite supply of variables $x_0^\sigma, x_1^\sigma, x_2^\sigma, \ldots$ for each type $\sigma$. A variable of type $\sigma$ is a term of type $\sigma$; $\lambda x M$ is a term of type $\sigma \to \tau$ if $x$ is a variable of type $\sigma$ and $M$ is a term of type $\tau$ *($\lambda$-abstraction)*; $(MN)$ is a term of type $\tau$ if $M$ is a term of type $\sigma \to \tau$ and $N$ is a term of type $\sigma$ *(application)*; $\langle M, N\rangle$ is a term of type $\sigma \times \tau$ if $M$ is a term of type $\sigma$ and $N$ is a term of type $\tau$ *(pairing)*; $\mathbf{fst}M$ *($\mathbf{snd}M$)* is a term of type $\sigma$ *($\tau$)* if $M$ is a term of type $\sigma \times \tau$ *(projections)*. Next we define the *reduction rules* of the typed $\lambda$-calculus. We have the following $\beta$-conversions: $(\lambda x M)N \triangleright M[x := N]$ if $x \notin FV(N)$; $\mathbf{fst}\langle M, N\rangle \triangleright M$; $\mathbf{snd}\langle M, N\rangle \triangleright N$. Further, we have $\alpha$-conversion, $\eta$-conversion and all the other standard reduction rules, i.e., $(MN) \triangleright (MN')$ if $N \triangleright N'$; $(MN) \triangleright (M'N)$ if $M \triangleright M'$; $\ldots$ etcetera. The calculus $\mathrm{T}^-$ is the typed $\lambda$-calculus extended with (i) for each $i \in \mathbb{N}$ a constant $k_i$ of type $\iota$; (ii) *recursor terms* $\mathrm{R}_\sigma(G^\sigma, F^{\iota, \sigma \to \sigma}, N^\iota)$ of type $\sigma$, that is, $\mathrm{R}_\sigma(G, F, N)$ is a term of type $\sigma$ if $G$ and $F$ and $N$ are terms of, respectively, types $\sigma$ and $\iota, \sigma \to \sigma$ and $\iota$; (iii) reduction rules of the form

$$\mathrm{R}_\sigma(G, F, k_0) \ \triangleright \ G \text{ and } \mathrm{R}_\sigma(G, F, k_{n+1}) \ \triangleright \ F(k_n, \mathrm{R}_\sigma(G, F, k_n)) \ .$$

The calculus $\mathrm{PCF}^-$ is the calculus $\mathrm{T}^-$ extended with with a *fixed-point term* $(\mathrm{Y}_\sigma M)$ of type $\sigma$ for each term $M$ of type $\sigma \to \sigma$ and reduction rules of the form $\mathrm{Y}_\sigma M \triangleright M(\mathrm{Y}_\sigma M)$.

We will use the standard conventions in the literature: $M[x := N]$ denotes the term $M$ where every occurrence of the variable $x$ is replaced by the term $N$; the notations $M^\sigma$ and $M : \sigma$ are two alternative ways to signify that the term $M$ is of type $\sigma$; $M(M_1, M_2)$ denotes the term $((MM_1)M_2)$, etcetera. If we write $M^k$ where $M$ is a term and $k$ is a natural number, then $M^k$ denotes the term $M$ repeated $k$ times in a row , e.g., $M^3 N$ denotes the term $M(M(M(N)))$. We will use $\overset{\star}{\triangleright}$ to denote the transitive-reflexive closure of the reducibility relation $\triangleright$; and $=$ to denote the symmetric-transitive-reflexive closure of $\triangleright$; and $\equiv$ to denote syntactical equality between terms.

We define the *total terms* of type $\sigma$ recursively over the structure of $\sigma$: the term $M : \iota$ is total if $M \overset{\star}{\triangleright} k_n$ for some constant $k_n$; the term $M : \sigma \to \tau$ is total if for every total term $N:\sigma$ there exists a total term $P:\tau$ such that $(MN) \overset{*}{\rhd} P$; and the term $M:\sigma \times \tau$ is total if **fst** $M$ and **snd** $M$ are total terms. A term $M$ is *hereditary total* if any subterm of $M$ is total. (The reader should note that our notion of hereditary total is stronger than the standard notion of hereditary total.)

A term $M$ *is of rank* $n$ if $\mathrm{dg}(\sigma) \leq n$ for any recursor term $\mathrm{R}_\sigma(\ldots)$ and any fixed-point term $\mathrm{Y}_\sigma(\ldots)$ occurring in $M$. We use $\mathrm{Rk}(M)$ to denote the smallest rank of $M$. $\qquad\qquad\square$

To achieve a laconic presentation, we have defined $\mathrm{PCF}^-$ as a parsimonious extension of $\mathrm{T}^-$. Our main results would still hold if $\mathrm{PCF}^-$ were based on a standard version of PCF, that is, PCF without recursor terms, but with boolean types, predecessor terms, etcetera. Note that any $\mathrm{T}^-$-term is a hereditary total $\mathrm{PCF}^-$-term.

There are various different definitions of $\mathrm{T}^-$ in the literature. The reason for this is mathematical convenience, and all the definitions are essentially equivalent to the definition given above.

Every hereditary total $\mathrm{PCF}^-$-term $M : \sigma$ has a natural interpretation as a functional of type $\sigma$ over the finite domain $\{0, 1, \ldots, b-1\}$. *We will interpret terms as natural numbers.* The interpretation $\mathbf{val}_b(M)$ defined below evaluates the closed term $M$ to a natural number, moreover, if $M$ is of type $\sigma$ and $b > 1$, we have $\mathbf{val}_b(M) < |\sigma|_b$. There is a natural bijection between the finite functionals of type $\sigma$ over the domain $\{0, 1, \ldots, b-1\}$ and the set $\{0, \ldots, |\sigma|_b - 1\}$. Our interpretation is nothing but the composition of this bijection with the obvious interpretation of a hereditary total term as a finite functional.

**Definition (Interpretation).** If $a < |\sigma \to \tau|_b$, then $a$ can be viewed as a $|\sigma|_b$ digit number in base $|\tau|_b$, and hence, uniquely written in the form

$$v_0 + v_1|\tau|_b^1 + \ldots + v_k|\tau|_b^k$$

where $k = |\sigma|_b - 1$ and $v_j < |\tau|_b$ for $j = 0, \ldots, k$. The numbers $v_0, \ldots, v_k$ are the *digits* in $a$, and for any $i < |\sigma|_b$, we denote the $i$'th digit of $a$ by $a[i]_b$, that is, $a[i]_b = v_i$. A *valuation* $\mathcal{V}$ is a total function from the set of variables into $\mathbb{N}$. We define the valuation $\mathcal{V}_i^x$ by $\mathcal{V}_i^x(x) = i$ and $\mathcal{V}_i^x(y) = \mathcal{V}(y)$ for any variable $y$ different from $x$. We define the *value of the term $M$ at the base $b$ under valuation $\mathcal{V}$*, written $\mathbf{val}_b^{\mathcal{V}}(M)$, recursively over the structure of $M$:

1. Let $\mathbf{val}_b^{\mathcal{V}}(x^\sigma) = \mathcal{V}(x) \pmod{|\sigma|_b}$.
2. Let $\mathbf{val}_b^{\mathcal{V}}(\lambda x^\sigma M^\tau) = \sum_{i<|\sigma|_b} \mathbf{val}_b^{\mathcal{V}_i^x}(M) \times |\tau|_b^i$.
3. Let $\mathbf{val}_b^{\mathcal{V}}((MN)) = \mathbf{val}_b^{\mathcal{V}}(M)[\mathbf{val}_b^{\mathcal{V}}(N)]_b$.
4. Let $\mathbf{val}_b^{\mathcal{V}}((\langle M^\sigma, N^\tau \rangle)) = \mathbf{val}_b^{\mathcal{V}}(M) \times |\tau|_b + \mathbf{val}_b^{\mathcal{V}}(N)$.
5. Let $\mathbf{val}_b^{\mathcal{V}}(\mathbf{fst}\,M^{\sigma \times \tau}) = \mathbf{val}_b^{\mathcal{V}}(M)\,\mathrm{div}\,|\tau|_b$ (integer division).
6. Let $\mathbf{val}_b^{\mathcal{V}}(\mathbf{snd}\,M^{\sigma \times \tau}) = \mathbf{val}_b^{\mathcal{V}}(M) \pmod{|\tau|_b}$.
7. Let $\mathbf{val}_b^{\mathcal{V}}(k_n) = n \pmod{|\iota|_b}$
8. Let $\mathbf{val}_b^{\mathcal{V}}(\mathrm{R}_\sigma(G, F, N)) = f(\mathbf{val}_b^{\mathcal{V}}(N))$ where $f(k+1) = (\mathbf{val}_b^{\mathcal{V}}(F)[k]_b)[f(k)]_b$ and $f(0) = \mathbf{val}_b^{\mathcal{V}}(G)$.
9. Let $\mathbf{val}_b^{\mathcal{V}}(\mathrm{Y}_\sigma M) = f^{\lceil \sigma \rceil_b}(0)$ where $f(x) = \mathbf{val}_b^{\mathcal{V}}(M)[x]_b$.

For any closed term $M$, we have $\mathbf{val}_b^{\mathcal{V}_0}(M) = \mathbf{val}_b^{\mathcal{V}_1}(M)$ for any valuations $\mathcal{V}_0, \mathcal{V}_1$, and we will simply write $\mathbf{val}_b(M)$ when $M$ is closed. $\qquad\square$

Any reader that finds Clause 9 in the definition of $\mathbf{val}_b^{\mathcal{V}}$ surprising, should study the proof of the next lemma.

**Lemma 2 (Adequacy).** *(i) Let $M$ and $M'$ be hereditary total $PCF^-$-terms such that $M \triangleright M'$, and let $b$ be such that $b > m$ for every constant $k_m$ occurring in $M$. We have $\mathbf{val}_b^{\mathcal{V}}(M) = \mathbf{val}_b^{\mathcal{V}}(M')$. (ii) Let $M : \iota$ be a a closed hereditary total $PCF^-$-term, and let $b = \max\{m \mid k_m$ is a subterm of $M\}$. Then we have $\mathbf{val}_{b+1}(M) = n$ iff $M \overset{\star}{\triangleright} k_n$.*

*Proof.* Now, (ii) follows straightforwardly from (i) since a total term of type $\iota$ normalise to a unique constant $k_n$. We have to prove (i) for each reduction rule $M \triangleright N$. All the cases are straightforward apart from the case $Y_\sigma M \triangleright M(Y_\sigma M)$. In this nontrivial case we need the following claim.

(**Claim**) Let $M, M_1, \ldots, M_n$ be arbitrary $PCF^-$-terms of proper types, and assume that
$$(Y_\sigma M)(M_1, \ldots, M_n) \overset{\star}{\triangleright} k_m \ .$$
Then we have
$$(M^k(\Omega_\sigma))(M_1, \ldots, M_n) \overset{\star}{\triangleright} k_m$$
for any $k \geq \lceil \sigma \rceil_b$ where
- $b = \max\{k_\ell \mid k_\ell$ occurs in the terms $M, M_1, \ldots, M_n\}$
- $\Omega_\sigma$ is the totally undefined term defined by

$$\Omega_\iota \equiv Y_\iota(\lambda x^\iota.x) \qquad \Omega_{\sigma\to\tau} \equiv \lambda X^\sigma.\Omega_\tau \qquad \Omega_{\sigma\times\tau} \equiv \langle \Omega_\sigma, \Omega_\tau \rangle \ .$$

Let $PCF_b^-$ be $PCF^-$ where the set of constants is restricted to $\{k_0, \ldots, k_{b-1}\}$, and let $D_{\sigma_b}$ be the standard Scott domain interpreting the type $\sigma$ when the base type $\iota$ is interpreted by $D_{\iota_b} = \{\bot, 0, 1, \ldots, b-1\}$. The reader should note that domain height of the type $\sigma$ at base $b$, i.e. $\lceil \sigma \rceil_b$, is the greatest $m$ such that there exist $d_0, \ldots, d_m \in D_{\sigma_b}$ where $d_i \sqsubseteq d_{i+1}$ and $d_i \neq d_{i+1}$ for all $i < m$.

The claim suggests the following interpretation $[\![ \cdot ]\!]_{\mathcal{V}}$ of the $PCF_b^-$-terms: let $[\![ Y_\sigma M ]\!]_{\mathcal{V}} = [\![ M ]\!]_{\mathcal{V}}^k(\bot)$ where $k = \lceil \sigma \rceil_b$; let $[\![ k_\ell ]\!]_{\mathcal{V}} = \ell$ for any constant $k_\ell$; let $[\![ x ]\!]_{\mathcal{V}} = \mathcal{V}(x)$ for any variable $x$; let $[\![ (MN) ]\!]_{\mathcal{V}} = [\![ M ]\!]_{\mathcal{V}}([\![ N ]\!]_{\mathcal{V}})$; let $[\![ \lambda x^\sigma M^\tau ]\!]_{\mathcal{V}} = f$ where the function $f : D_{\sigma_b} \to D_{\tau_b}$ is given by $f(d) = [\![ M ]\!]_{\mathcal{V}_d^x}$; etcetera. (We omit the rather obvious interpretations of pairs, projections and recursor terms.) In this interpretation of $PCF_b^-$-terms, we have

- $[Y_\sigma M]_{\mathcal{V}} = [M]_{\mathcal{V}}^k(\bot)$ for any $k \geq \lceil \sigma \rceil_b$
- $[\![ M ]\!]_{\mathcal{V}}^j(\bot) \sqsubseteq [\![ M ]\!]_{\mathcal{V}}^j(d)$ for any $j \in \mathbb{N}, d \in D_{\sigma_b}$. $\qquad\qquad$ (†)

Furthermore, there exists a natural bijection $\|\cdot\|$ between the total elements of $D_{\sigma_b}$ and the set $\{n \mid n < |\sigma|_b\}$, and it is straightforward to prove that

$$\|[\![ M ]\!]_{\mathcal{V}}\| \ = \ \mathbf{val}_b^{\mathcal{V}}(M) \qquad\qquad (\ddagger)$$

for any $PCF_b^-$-term $M$ not containing fixed-point terms.

We are now ready to prove that $\mathbf{val}_b^{\mathcal{V}}(Y_\sigma M) = \mathbf{val}_b^{\mathcal{V}}(M(Y_\sigma M))$ where $Y_\sigma M$ is an arbitrary hereditary total $\mathrm{PCF}_b^-$-term. Let $f(x) = \mathbf{val}_b^{\mathcal{V}}(M)[x]_b$ and let $k = \lceil \sigma \rceil_b$. Hence, we have $\mathbf{val}_b^{\mathcal{V}}(Y_\sigma M) = f^k(0)$ by the definition of $\mathbf{val}_b^{\mathcal{V}}$. Pick a total $d_0 \in D_{\sigma_b}$ such that $\|d_0\| = 0$. By (†), we have $[\![Y_\sigma M]\!]_{\mathcal{V}} = [M]_{\mathcal{V}}^k(\bot) \sqsubseteq [M]_{\mathcal{V}}^k(d_0)$ and $[\![Y_\sigma M]\!]_{\mathcal{V}} = [M]_{\mathcal{V}}^{k+1}(\bot) \sqsubseteq [M]_{\mathcal{V}}^{k+1}(d_0)$, and thus, as $Y_\sigma M$ is total, we also have

$$[M]_{\mathcal{V}}^k(d_0) \;=\; [\![Y_\sigma M]\!]_{\mathcal{V}} \;=\; [M]_{\mathcal{V}}^{k+1}(d_0) \,. \tag{$*$}$$

Furthermore, by (‡) we have

$$\|[\![M]\!]_{\mathcal{V}}\| \;=\; \mathbf{val}_b^{\mathcal{V}}(M) \;=\; f \tag{$**$}$$

if $M$ contains no fixed-point terms. If $M$ contains fixed-point terms, we still have ($**$), now by an easy induction argument. Finally, by ($*$) and ($**$), we have

$$\mathbf{val}_b^{\mathcal{V}}(Y_\sigma M) \;=\; f^k(0) \;=\; \|[M]_{\mathcal{V}}^k(d_0)\| \;=\; \|[M]_{\mathcal{V}}^{k+1}(d_0)\| \;=$$
$$ff^k(0) \;=\; \mathbf{val}_b^{\mathcal{V}}(M(Y_\sigma M)) \,. \qquad \square$$

## Arithmetic in $\mathrm{T}^-$ and $\mathrm{PCF}^-$

**Lemma 3 (Conditionals).** *For any type $\sigma$ there exists a $\mathrm{T}^-$-term $\mathrm{Cond}_\sigma$ of type $\iota, \sigma, \sigma \to \sigma$ such that $\mathrm{Cond}_\sigma(k_0, M_1, M_2) = M_1$ and $\mathrm{Cond}_\sigma(k_n, M_1, M_2) = M_2$ when $n > 0$. Moreover, $\mathrm{Rk}(\mathrm{Cond}_\sigma) = 0$.*

A proof of Lemma 3 can be found in [13]. Proofs of the two next lemmas can also be found in [13], but we will repeat the core of the proofs here.

**Lemma 4 (Long Iterations in $\mathrm{T}^-$).** *For all types $\sigma$ and $\tau$ there exists a $\mathrm{T}^-$-term $\mathrm{It}_\tau^\sigma : (\iota, \tau \to \tau, \tau) \to \tau$ such that $\mathrm{It}_\tau^\sigma(k_b, F, G) = F^{|\sigma|_{b+1}}G$. Moreover, we have $\mathrm{Rk}(\mathrm{It}_\tau^\sigma) = \mathrm{dg}(\sigma) + \mathrm{dg}(\tau)$. (We will call $\mathrm{It}_\tau^\sigma$ an iterator.)*

*Proof.* We prove the lemma by induction on the structure of $\sigma$.

Assume $\sigma = \iota$. Let $\mathrm{It}_\tau^\iota \equiv \lambda n^\iota Y^{\tau \to \tau} X^\tau . R_\tau(Y(X), \lambda x^\iota . Y, n)$. Obviously, we have $\mathrm{Rk}(\mathrm{It}_\tau^\iota) = \mathrm{dg}(\iota) + \mathrm{dg}(\tau)$, and it is straightforward to prove by induction on $b$ that $\mathrm{It}_\tau^\sigma(k_b, F, G) = F^{b+1}(G)$. Thus, the lemma holds when $\sigma = \iota$ since $|\iota|_{b+1} = b + 1$.

Assume $\sigma = \sigma_1 \to \sigma_2$. Let $\mathrm{It}_\tau^\sigma \equiv \lambda x^\iota Y^{\tau \to \tau} X^\tau . (\mathrm{It}_{\tau \to \tau}^{\sigma_1}(x, \mathrm{It}_\tau^{\sigma_2}(x), Y)X)$. We have

$$\begin{aligned}
\mathrm{Rk}(\mathrm{It}_\tau^\sigma) &= \max(\mathrm{Rk}(\mathrm{It}_{\tau \to \tau}^{\sigma_1}), \mathrm{Rk}(\mathrm{It}_\tau^{\sigma_2})) && \text{def. of Rk}\\
&= \max(\mathrm{dg}(\sigma_1) + \mathrm{dg}(\tau \to \tau), \mathrm{dg}(\sigma_2) + \mathrm{dg}(\tau)) && \text{ind. hyp.}\\
&= \max(\mathrm{dg}(\sigma_1) + \mathrm{dg}(\tau) + 1, \mathrm{dg}(\sigma_2) + \mathrm{dg}(\tau)) && \text{def. of dg}\\
&= \max(\mathrm{dg}(\sigma_1) + 1, \mathrm{dg}(\sigma_2)) + \mathrm{dg}(\tau) &&\\
&= \mathrm{dg}(\sigma) + \mathrm{dg}(\tau) \,. && \text{def. of dg}
\end{aligned}$$

So, the iterator has the right rank. We will now prove that we indeed have $\mathrm{It}_\tau^\sigma(k_b, F, G) = F^{|\sigma|_{b+1}}G$. Let $A \equiv (\mathrm{It}_\tau^{\sigma_2}k_b)$. We prove by induction on $k$ that

$(A^k F)G = F^{|\sigma_2|^k_{b+1}}G$ (*). We have $(A^0 F) = F$, and hence $(A^0 F)G = F^{|\sigma_2|^0_{b+1}}G$. Moreover, $(A^{k+1}F)G = (A(A^k F))G = \text{It}^{\sigma_2}_\tau(k_b, A^k F, G) = (A^k F)^{|\sigma_2|_{b+1}}G = F^{|\sigma_2|^{k+1}_{b+1}}G$. The two last equalities hold respectively by induction hypothesis on $\sigma_2$ and by induction hypothesis on $k$. This proves (*). Furthermore, we have

$$\text{It}^\sigma_\tau(k_b, F, G) = \text{It}^{\sigma_1}_{\tau \to \tau}(k_b, (\text{It}^{\sigma_2}_\tau k_b), F)G =$$

$$(\text{It}^{\sigma_2}_\tau k_b)^{|\sigma_1|_{b+1}}F)G \overset{(*)}{=} F^{|\sigma_2|^{|\sigma_1|_{b+1}}_{b+1}}G = F^{|\sigma|_{b+1}}G.$$

The first equality holds by the definition of $\text{It}^\sigma_\tau$, the second by induction hypothesis on $\sigma_1$.

When $\sigma = \sigma_1 \times \sigma_2$, let $\text{It}^\sigma_\tau \equiv \lambda x^\iota Y^{\tau \to \tau} X^\tau . \text{It}^{\sigma_1}_\tau(x, \text{It}^{\sigma_2}_\tau(x, Y), X)$ and the lemma holds. We omit the details. $\qquad\square$

**Lemma 5 (Arithmetic in $\mathbf{T^-}$).** *For any type $\sigma$ there exists $T^-$-terms*

$$0_\sigma : \iota,\ \text{Suc}_\sigma : \iota, \sigma \to \sigma,\ \text{Pred}_\sigma : \iota, \sigma \to \sigma,\ \text{Le}_\sigma : \iota, \sigma, \sigma \to \iota\ \text{ and } \ \text{Eq}_\sigma : \iota, \sigma, \sigma \to \iota$$

*of rank $2\text{dg}(\sigma)\dot{-}2$ such that (i) $\mathbf{val}_{b+1}(0_\sigma) = 0$*

*(ii)* $\mathbf{val}_{b+1}(\text{Suc}_\sigma(k_b, M)) = \mathbf{val}_{b+1}(M) + 1 \pmod{|\sigma|_{b+1}}$
*(iii)* $\mathbf{val}_{b+1}(\text{Pred}_\sigma(k_b, M)) = \mathbf{val}_{b+1}(M) - 1 \pmod{|\sigma|_{b+1}}$
*(iv)* $\text{Le}_\sigma(k_b, M_1, M_2) = 0$ *iff* $\mathbf{val}_{b+1}(M_1) \leq \mathbf{val}_{b+1}(M_2)$
*(v)* $\text{Eq}_\sigma(k_b, M_1, M_2) = 0$ *iff* $\mathbf{val}_{b+1}(M_1) = \mathbf{val}_{b+1}(M_2)$

*for any closed T-terms $M$, $M_1$ and $M_2$.*

*Proof.* Let $0_\iota \equiv k_0$; $0_{\pi \times \tau} \equiv \langle 0_\pi, 0_\tau \rangle$; and $0_{\pi \to \tau} \equiv \lambda x^\pi 0_\tau$. Obviously, we have $\mathbf{val}_{b+1}(0_\sigma) = 0$ and $\text{Rk}(0_\sigma) = 0 \leq 2\text{dg}(\sigma)\dot{-}2$ for any $\sigma$.

We will define $\text{Suc}_\sigma$, $\text{Le}_\sigma$ and $\text{Eq}_\sigma$ in parallel recursively over the structure of $\sigma$. We omit the definition of $\text{Pred}_\sigma$ as this definition is very similar to the definition of $\text{Suc}_\sigma$.

Let $\sigma = \iota$. We omit the details for this case. See [13].
Let $\sigma = \pi \to \tau$. We define $F$ by

$$F \equiv \lambda b^\iota X^\sigma Y^\sigma z^{\iota \times \pi} . \text{Cond}_{\iota \times \pi}(\text{Eq}_\tau(b, X(\mathbf{snd}z), Y(\mathbf{snd}z)),$$

$$\langle \mathbf{fst}z, \text{Suc}_\pi(b, \mathbf{snd}z) \rangle, \text{Cond}_{\iota \times \pi}(\text{Le}_\tau(b, X(\mathbf{snd}z), Y(\mathbf{snd}z)),$$

$$\langle k_0, \text{Suc}_\pi(b, \mathbf{snd}z) \rangle, \langle k_1, \text{Suc}_\pi(b, \mathbf{snd}z) \rangle))) .$$

By the induction hypothesis, we have

$$F(k_b, M, N, \langle i, j \rangle) =$$

$$\begin{cases} \langle i, \text{Suc}_\pi(j) \rangle & \text{if } \mathbf{val}_{b+1}(M)[\mathbf{val}_{b+1}(j)]_{b+1} = \mathbf{val}_{b+1}(N)[\mathbf{val}_{b+1}(j)]_{b+1} \\ \langle 0, \text{Suc}_\pi(j) \rangle & \text{if } \mathbf{val}_{b+1}(M)[\mathbf{val}_{b+1}(j)]_{b+1} < \mathbf{val}_{b+1}(N)[\mathbf{val}_{b+1}(j)]_{b+1} \\ \langle 1, \text{Suc}_\pi(j) \rangle & \text{otherwise.} \end{cases}$$

Thus, we have

$$\mathbf{fst}(F(k_b, M, N)^{|\pi|_b}(\langle k_0, 0_\pi \rangle)) = \begin{cases} k_0 & \text{if } \mathbf{val}_{b+1}(M) \leq \mathbf{val}_{b+1}(N) \\ k_1 & \text{otherwise.} \end{cases}$$

and then (iii) holds when

$$\mathrm{Le}_\sigma \equiv \lambda bXY.\mathbf{fst}\,\mathrm{It}^\pi_{\iota\times\pi}(b, F(b, X, Y), \langle k_0, 0_\pi\rangle). \tag{\dag}$$

Now, let $\mathrm{Eq}_\sigma \equiv \lambda bXY.\mathrm{Cond}_\iota(\mathrm{Le}_\sigma(b, X, Y), \mathrm{Cond}_\iota(\mathrm{Le}_\sigma(b, Y, X), k_0, k_1), k_1)$ and (iv) holds. We will now argue that $\mathrm{Le}_\sigma$ and $\mathrm{Eq}_\sigma$ have the required rank. First, we note that $\mathrm{Rk}(F) \le \max(2\mathrm{dg}(\pi)\dot{-}2, 2\mathrm{dg}(\tau)\dot{-}2)$ (*). (Lemma 3 states that $\mathrm{Rk}(\mathrm{Cond}_{\iota\times\pi}) = 0$, and then (*) follows from the induction hypothesis.)

$$
\begin{aligned}
\mathrm{Rk}(\mathrm{Eq}_\sigma) &= \mathrm{Rk}(\mathrm{Le}_\sigma) && \text{def. of Rk, def. of } \mathrm{Eq}_\sigma \\
&= \max(\mathrm{Rk}(F), \mathrm{Rk}(\mathrm{It}^\pi_{\iota\times\pi})) && \text{def. of Rk, def. of } \mathrm{Le}_\sigma \\
&\le \max(\mathrm{Rk}(F), \mathrm{dg}(\pi) + \mathrm{dg}(\iota\times\pi)) && \text{Lemma 4} \\
&\le \max(2\mathrm{dg}(\pi)\dot{-}2, 2\mathrm{dg}(\tau)\dot{-}2, 2\mathrm{dg}(\pi)) && (*) \\
&= 2(\max(\mathrm{dg}(\pi)+1, \mathrm{dg}(\tau))\dot{-}1) \\
&= 2\mathrm{dg}(\sigma)\dot{-}2\,. && \text{def. of dg, } \sigma = \pi\to\tau
\end{aligned}
$$

Thus, $\mathrm{Eq}_\sigma$ and $\mathrm{Le}_\sigma$ have the required rank. Next we define $\mathrm{Suc}_\sigma$. Any number $a < |\sigma|_b$ can be uniquely written in the form $a = v_0|\tau|_b^0 + v_1|\tau|_b^1 + \cdots + v_k|\tau|_b^k$ where $k = |\pi|_b - 1$ and $v_j < |\tau|_b$ for $j = 1, \ldots, k$. There exists $i \le k$ such that

$$a + 1 \;=\; v_0'|\tau|_b^0 \;+\; \cdots \;+\; v_i'|\tau|_b^i \;+\; v_{i+1}|\tau|_b^{i+1} \;+\; \cdots \;+\; v_k|\tau|_b^k \quad (\mathrm{mod}\ |\sigma|_b)$$

where $v_j' = v_j + 1 \ (\mathrm{mod}\ |\tau|_b)$ for $j = 0, \ldots, i$. We call such an $i$ the *carry border* of the number $a$. Let

$$\mathrm{C}_\sigma \equiv \lambda bX.\mathbf{snd}\,\mathrm{It}^\pi_{\iota\times\pi}(b, G(b, X), \langle 0, 0_\pi\rangle) \tag{\ddag}$$

where

$$
\begin{aligned}
G \equiv \lambda b^\iota X^{\pi\to\tau} z^{\iota\times\pi}.\mathrm{Cond}_\iota\big(&\mathbf{fst}\,z, \mathrm{Cond}_{\iota\times\pi}(\mathrm{Eq}_\tau(b, \mathrm{Suc}_\tau(X(\mathbf{snd}\,z))), 0_\tau), \\
&\langle 0, \mathrm{Suc}_\pi(\mathbf{snd}\,z)\rangle), \langle 1, \mathrm{Suc}_\pi(\mathbf{snd}\,z)\rangle\big), \langle 1, \mathrm{Suc}_\pi(\mathbf{snd}\,z)\rangle\big)\,.
\end{aligned}
$$

Then $\mathbf{val}_{b+1}(\mathrm{C}_\sigma(k_b, M))$ equals the carry border of $\mathbf{val}_{b+1}(M)$ when $M : \sigma$ is a closed term. Let $\mathrm{Suc}_\sigma \equiv \lambda b^\iota X^\sigma i^\pi.\mathrm{Cond}_\tau(\mathrm{Le}_\pi(b, i, \mathrm{C}_\sigma(b, X)), \mathrm{Suc}_\tau(X(i)), X(i))$ and (i) holds. An argument similar to the one showing that the ranks of $\mathrm{Eq}_\sigma$ and $\mathrm{Le}_\sigma$ are bounded by $2\mathrm{dg}(\sigma)\dot{-}2$, will show that also the rank of $\mathrm{C}_\sigma$ is bounded by $2\mathrm{dg}(\sigma)\dot{-}2$. The rank of $\mathrm{Suc}_\sigma$ equals the rank of $\mathrm{C}_\sigma$.

Let $\sigma = \pi\times\tau$. Define $\mathrm{Suc}_\sigma$ such that

$$
\mathrm{Suc}_\sigma(b, \langle F, G\rangle) \;=\;
\begin{cases}
\langle \mathrm{Suc}_\pi(F), \mathrm{Suc}_\tau(G)\rangle & \text{if } \mathrm{Eq}_\tau(b, \mathrm{Suc}_\tau(G)) = 0_\tau \\
\langle F, \mathrm{Suc}_\tau(G)\rangle & \text{otherwise.}
\end{cases}
$$

Define $\mathrm{Le}_\sigma$ such that $\mathrm{Le}_\sigma(b, \langle F, G\rangle, \langle F', G'\rangle) = 0$ iff

$$
(\mathrm{Le}_\pi(b, F, F') = 0 \;\wedge\; \mathrm{Eq}_\pi(b, F, F') > 0) \;\vee\;
$$
$$
(\mathrm{Le}_\pi(b, G, G') = 0 \;\wedge\; \mathrm{Eq}_\pi(b, F, F') = 0)
$$

Define $\mathrm{Eq}_\sigma$ as above. It is easy to construct the required terms, and we skip the details. $\qquad\square$

**Lemma 6 (Arithmetic in PCF⁻).** *For any type $\sigma$ there exists PCF⁻-terms*

$$0_\sigma : \iota\,, \ \mathrm{Suc}_\sigma : \iota, \sigma \to \sigma\,, \ \mathrm{Pred}_\sigma : \iota, \sigma \to \sigma\,, \ \mathrm{Le}_\sigma : \iota, \sigma, \sigma \to \iota \ \text{ and } \ \mathrm{Eq}_\sigma : \iota, \sigma, \sigma \to \iota$$

*of rank $\mathrm{dg}(\sigma)$ such that (i), (ii), (iii), (iv) and (v) of Lemma 5 hold.*

*Proof.* This proof is nearly identical to the proof of Lemma 5. We define the terms $\mathrm{Suc}_\sigma$, $\mathrm{Pred}_\sigma$, $\mathrm{Le}_\sigma$ and $\mathrm{Eq}_\sigma$ as we do in the proof of Lemma 5, i.e., in parallel recursively over the structure of $\sigma$, but now we will use fixed-point terms in place of the iterators. This will reduce the ranks of the terms we are defining.

When $\sigma = \pi \to \tau$, the statement marked (†) in the proof of Lemma 5 defines the T⁻-term $\mathrm{Le}_\sigma$ by $\mathrm{Le}_\sigma \equiv \lambda bXY.\mathbf{fst}\,\mathrm{It}^\pi_{\iota\times\pi}(b, F(b, X, Y), \langle k_0, 0_\pi\rangle)$ where $F$ is a term such that

$$\mathbf{fst}(F(k_b, M, N)^{|\pi|_b}(\langle k_0, 0_\pi\rangle)) = \begin{cases} k_0 & \text{if } \mathbf{val}_{b+1}(M) \leq \mathbf{val}_{b+1}(N) \\ k_1 & \text{otherwise.} \end{cases} \qquad (*)$$

and $\mathrm{Rk}(F) = \max(\mathrm{Rk}(\mathrm{Eq}_\tau), \mathrm{Rk}(\mathrm{Suc}_\pi))$ (**).

Given a term $F$ with the this properties, we can define a PCF⁻-term $\mathrm{Le}_\sigma$ by $\mathrm{Le}_\sigma \equiv \lambda bXY.\mathbf{fst}((Y_{\pi\to\iota\times\pi}A)0_\pi)$ where

$$A \equiv \lambda U^{\pi\to\iota\times\pi}W^\pi.\mathrm{Cond}_{\iota\times\pi}(\ \mathrm{Eq}_\pi(b, \mathrm{Suc}_\pi(b, W), 0_\pi)\,,$$
$$\langle k_0, 0_\pi\rangle\,, \ F(b, X, Y)U(\mathrm{Suc}_\pi(b, W))\ )\,.$$

It follows from (*) and the induction hypotheses that clause (iv) of the lemma holds. Furthermore $\mathrm{Le}_\sigma$ has the required rank since

$$\begin{aligned}
\mathrm{Rk}(\mathrm{Le}_\sigma) &= \max(\mathrm{Rk}(\mathrm{Eq}_\tau), \mathrm{Rk}(\mathrm{Suc}_\pi), \mathrm{dg}(\pi \to \iota \times \pi)) &&(**), \text{ def. of } \mathrm{Le}_\sigma \\
&= \max(\mathrm{Rk}(\mathrm{Eq}_\tau), \mathrm{Rk}(\mathrm{Suc}_\pi), \mathrm{dg}(\pi) + 1) &&\text{def. of dg} \\
&= \max(\mathrm{dg}(\tau), \mathrm{dg}(\pi), \mathrm{dg}(\pi) + 1) &&\text{ind. hyp.} \\
&= \mathrm{dg}(\sigma)\,. &&\text{def. of dg}
\end{aligned}$$

Along this line, we can also find a PCF⁻-term $\mathrm{Suc}_\sigma$ satisfying the lemma by eliminating the iterator $\mathrm{It}^\pi_{\iota\times\pi}$ from the formula marked (‡) in the proof of Lemma 5. The definition of the term $\mathrm{Pred}_\sigma$ is very similar to the definition of $\mathrm{Suc}_\sigma$. Let $\mathrm{Eq}_\sigma \equiv \lambda bXY.\mathrm{Cond}_\iota(\mathrm{Le}_\sigma(b, X, Y), \mathrm{Cond}_\iota(\mathrm{Le}_\sigma(b, Y, X), k_0, k_1), k_1)$.  $\square$

## Complexity Classes and Fragments of T⁻ and PCF⁻

**Definition.** A *problem* is a subset of $\mathbb{N}$. A term $M : \iota \to \iota$ *decides* a problem $A$ when $M(k_x) \overset{*}{\rhd} k_0$ if $x \in A$, and $M(k_x) \overset{*}{\rhd} k_1$ if $x \notin A$. We define the sets of problems $\mathcal{G}_k$ and $\mathcal{P}_k$ by respectively (i) $A \in \mathcal{G}_k$ iff $A$ is decided by a T⁻-term $M$ where $k \leq \mathrm{Rk}(M)$ and (ii) $A \in \mathcal{P}_k$ iff $A$ is decided by a PCF⁻-term $M$ where $k \leq \mathrm{Rk}(M)$. A deterministic Turing machine $M$ *decides* a problem $A$ when $M$ on input $x \in \mathbb{N}$ halts in a distinguished accept state if $x \in A$, and in a distinguished reject state if $x \notin A$. The input $x \in \mathbb{N}$ should be represented in binary on the Turing machine's input tape. We will use $|x|$ to denote the length of the standard binary representation of the natural number $x$. For $i \in \mathbb{N}$, we define TIME $2_i^{\text{LIN}}$ (SPACE $2_i^{\text{LIN}}$) to be the set of problems decidable by a deterministic Turing machine working in time (space) $2_i^{c|x|}$ for some fixed $c \in \mathbb{N}$.        $\square$

**Theorem 1.** *We have* SPACE $2_n^{\text{LIN}} \subseteq \mathcal{G}_{2n}$ TIME $2_{n+1}^{\text{LIN}} \subseteq \mathcal{G}_{2n+1}$. *Furthermore, we have* TIME $2_{n+1}^{\text{LIN}} \subseteq \mathcal{P}_{n+1}$.

*Proof.* Let $\mathfrak{m}$ be a one-way 1-tape deterministic Turing machine solving some problem $A$ in time $t(|x|)$ and space $s(|x|)$. The input $x$ is a natural number, and we assume that $s(|x|) < 2_n^x$ for some fixed $n \in \mathbb{N}$. Let $\{a_0, \ldots, a_i\}$ and $\{q_0, \ldots, q_j\}$ be respectively $\mathfrak{m}$'s alphabet and $\mathfrak{m}$'s set of states. Let $q_0$ and $q_1$ be respectively the reject and accept state.

First we will construct a $\text{T}^-$-term $\text{T}_{\pi,\sigma}^{\mathfrak{m}}$ and a $\text{PCF}^-$-term $\text{PCF}_{\pi,\sigma}^{\mathfrak{m}}$ which simulate $\mathfrak{m}$. There exists a type $\sigma$ such that $s(|x|) < |\sigma|_x$ for $x > 1$. Fix such a $\sigma$. We can w.l.o.g. assume that the input $x$ is greater that 1, and thus, we can represent a configuration of $\mathfrak{m}$ by a closed term $\langle T^{\sigma \to \iota}, \langle H^\sigma, S^\iota \rangle \rangle$ of type $\xi = (\sigma \to \iota) \times (\sigma \times \iota)$ where $T$ represent the tape, $H$ the current position of the head, and $S$ the current state:

- $\mathbf{val}_x(T)[i]_x = j$ iff the $i$'th cell of the tape contains $a_j$
- $\mathbf{val}_x(H) = i$ iff the head scans the $i$'th cell of the tape
- $\mathbf{val}_x(S) = i$ iff $q_i$ is the current state.

Furthermore, we can use the functionals given by Lemma 5 and Lemma 6 to simulate the execution of $\mathfrak{m}$ on input $x$. The functionals $\text{Pred}_\sigma(k_x) : \sigma \to \sigma$ and $\text{Suc}_\sigma(k_x) : \sigma \to \sigma$ will move the head back and forth, and the functional

$$\text{Md} \equiv \lambda F^{\sigma \to \iota} X^\sigma V^\iota Y^\sigma . \text{Cond}_\iota (\text{Eq}_\sigma(k_x, X, Y), V, F(Y))$$

will modify the tape, e.g., $\text{Md}(T, H, k_{17})$ writes the symbol $a_{17}$ in the scanned cell. We construct terms $\text{Step}_\sigma : \iota, \xi \to \xi$ and $\text{Init}_\sigma : \iota \to \xi$ such that $\text{Init}_\sigma(k_x)$ represents the initial configuration of $\mathfrak{m}$ on input $x$, and $\text{Step}_\sigma(k_x, \text{Init}_\sigma(k_x))$ represents the configuration after one transition, $\text{Step}_\sigma(k_x, \text{Step}_\sigma(k_x, \text{Init}_\sigma(k_x)))$ represents the configuration after two transitions, and so on. We construct $\text{Step}_\sigma$ such that $\text{Step}_\sigma(k_x, C) = C$ when $C$ represents a halt configuration.

It is easy to see that these terms can be constructed such that $\text{Rk}(\text{Step}_\sigma) = \text{Rk}(\text{Init}_\sigma) = \text{Rk}(\text{Suc}_\sigma) = \text{Rk}(\text{Pred}_\sigma) = \text{Rk}(\text{Eq}_\sigma)$. Thus, by Lemma 5, $\text{Step}_\sigma$ and $\text{Init}_\sigma$ are of rank $2\text{dg}(\sigma) - 2$ if we are working in $\text{T}^-$, and by Lemma 6, $\text{Step}_\sigma$ and $\text{Init}_\sigma$ are of rank $\text{dg}(\sigma)$ if we are working in $\text{PCF}^-$.

Now, fix a type $\pi$ such that $t(|x|) < |\pi|_x$ for $x > 1$. Assume we are working in $\text{T}^-$. Let $\text{T}_{\pi,\sigma}^{\mathfrak{m}} \equiv \lambda y^\iota \text{It}_\xi^\pi(y, \text{Step}_\sigma(y), \text{Init}_\sigma(y))$ where the $\text{T}^-$-term $\text{It}_\xi^\sigma$ is given by Lemma 4. We have $\text{T}_{\pi,\sigma}^{\mathfrak{m}}(k_x) = \text{Step}_\sigma(k_x)^{|\pi|_x}(\text{Init}_\sigma(k_x))$, and thus, $\text{T}_{\pi,\sigma}^{\mathfrak{m}}$ simulates $\mathfrak{m}$ since $\mathfrak{m}$ runs in time $|\pi|_x$. Furthermore, since $\text{dg}(\xi) = \text{dg}(\sigma) + 1$, we have

$$\text{Rk}(\text{T}_{\pi,\sigma}^{\mathfrak{m}}) = \text{Rk}(\text{It}_\xi^\pi) = \text{dg}(\pi) + \text{dg}(\xi) = \text{dg}(\pi) + \text{dg}(\sigma) + 1 . \qquad (*)$$

Next, assume we are working in $\mathrm{PCF}^-$. Let $\mathrm{PCF}^{\mathfrak{m}}_{\pi,\sigma} \equiv \lambda x^{\iota}.((Y_{\pi\to\xi}P)0_{\pi})$ where

$$P \equiv \lambda X^{\pi\to\xi} Z^{\pi}.\mathrm{Cond}_{\xi}(\mathrm{Eq}_{\pi}(x^{\iota},\mathrm{Suc}_{\pi}(x^{\iota},Z),0_{\pi}),\mathrm{Init}_{\sigma}(x^{\iota}),\mathrm{Step}_{\sigma}(x^{\iota},XZ))) \ .$$

Then $\mathrm{PCF}^{\mathfrak{m}}_{\pi,\sigma}$ simulates $\mathfrak{m}$ and

$$\mathrm{Rk}(\mathrm{PCF}^{\mathfrak{m}}_{\pi,\sigma}) = \mathrm{dg}(\pi \to \xi) = \max(\mathrm{dg}(\pi)+1,\mathrm{dg}(\xi)) =$$
$$\max(\mathrm{dg}(\pi)+1,\mathrm{dg}(\sigma\to\iota)) = \max(\mathrm{dg}(\pi),\mathrm{dg}(\sigma))+1 \ . \quad (**)$$

Let $A \in \textsc{time}\ 2^{\textsc{lin}}_{n+1}$, and let $\mathfrak{m}$ be a Turing machine which decides $A$ in time, and thus also in space, $2^{k|x|}_{n+1}$ for some fixed number $k$. We have $2^{k|x|}_{n+1} \le 2^{k\log_2(x+2)}_{n+1} \le 2^{\log_2(x+2)^k}_{n+1} \le 2^{(x+2)^k}_{n}$, and hence, there exists a type $\sigma$ of degree $n$ such that $2^{k|x|}_{n+1} \le |\sigma|_x$ for $x > 1$. Let $M \equiv \lambda x^{\iota}.\mathbf{snd}\,\mathbf{snd}\ \mathrm{T}^{\mathfrak{m}}_{\sigma,\sigma}(x)$ and $M' \equiv \lambda x^{\iota}.\mathbf{snd}\,\mathbf{snd}\ \mathrm{PCF}^{\mathfrak{m}}_{\sigma,\sigma}(x)$. Both the $\mathrm{T}^-$-term $M$ and the $\mathrm{PCF}^-$-term $M'$ decide the problem $A$. We have $\mathrm{Rk}(M) = \mathrm{Rk}(\mathrm{T}^{\mathfrak{m}}_{\sigma,\sigma}) \overset{(*)}{=} \mathrm{dg}(\sigma)+\mathrm{dg}(\sigma)+1 = 2n+1$, and thus $A \in \mathcal{G}_{2n+1}$. Furthermore, $\mathrm{Rk}(M') = \mathrm{Rk}(\mathrm{PCF}^{\mathfrak{m}}_{\sigma,\sigma}) \overset{(**)}{=} \mathrm{dg}(\sigma)+1 = n+1$, and thus $A \in \mathcal{P}_{n+1}$. This proves the inclusions $\textsc{time}\ 2^{\textsc{lin}}_{n+1} \subseteq \mathcal{G}_{2n+1}$ and $\textsc{time}\ 2^{\textsc{lin}}_{n+1} \subseteq \mathcal{P}_{n+1}$ for all $n \in \mathbb{N}$.

Let $A \in \textsc{space}\ 2^{\textsc{lin}}_{n+1}$. Thus, there exists a Turing machine $\mathfrak{m}$ deciding $A$ in space $2^{k|x|}_{n+1}$ and time $2^{k'|x|}_{n+2}$ for some $k, k'$. We can find a type $\sigma$ of degree $n$ and a type $\rho$ of degree $n+1$ such that $2^{k|x|}_{n+1} \le |\sigma|_x$ and $2^{k'|x|}_{n+2} \le |\rho|_x$. Let $M \equiv \lambda x^{\iota}.\mathbf{snd}\,\mathbf{snd}\ \mathrm{T}^{\mathfrak{m}}_{\rho,\sigma}(x)$. The $\mathrm{T}^-$-term $M$ decides the problem $A$, and $\mathrm{Rk}(M) = 2n+2$. Hence, the inclusion $\textsc{space}\ 2^{\textsc{lin}}_{n} \subseteq \mathcal{G}_{2n}$ holds for all $n > 0$. The inclusion $\textsc{space}\ 2^{\textsc{lin}}_{0} \subseteq \mathcal{G}_{0}$ requires a tailored proof, and we omit the details. $\qquad\square$

**Theorem 2.** *We have $\mathcal{P}_{n+1} \subseteq \textsc{time}\ 2^{\textsc{lin}}_{n+1}$.*

*Proof.* We need the following claim.

> **(Claim)** Let $M:\sigma$ be a $\mathrm{PCF}^-$-term where $\mathrm{dg}(\sigma) \le n+1$ and $\mathrm{Rk}(M) \le n+1$, and let $\mathcal{V}$ be any valuation. The value $\mathbf{val}^{\mathcal{V}}_x(M)$ can be computed by a Turing machine running in time $2^{k|x|}_{n+1}$ for some $k \in \mathbb{N}$.

Assume $A \in \mathcal{P}_{n+1}$. The definition of $\mathcal{P}_{n+1}$ says that there exists a closed $\mathrm{PCF}^-$-term $M^{\iota\to\iota}$ of rank $n+1$ such that $Mk_x \overset{\star}{\rhd} k_0$ if $x \in A$, and $Mk_x \overset{\star}{\rhd} k_1$ if $x \notin A$. We can w.l.o.g. assume that $M$ is hereditary total. Let $f_A(x) = \mathbf{val}_{\max(x,m)+1}(Mk_x)$ where $m$ is the greatest $m$ such that $k_m$ occurs in $M$. By Lemma 2 (ii), $f_A(x) = 0$ if $x \in A$, and $f_A(x) = 1$ if $x \notin A$. By (Claim), $f_A(x)$ can be computed by a Turing machine running in time $2^{k|x|}_{n+1}$ for some $k \in \mathbb{N}$. Hence, we have $A \in \textsc{time}\ 2^{\textsc{lin}}_{n+1}$ and the inclusion $\mathcal{P}_{n+1} \subseteq \textsc{time}\ 2^{\textsc{lin}}_{n+1}$ holds.

Next we prove the claim. We will give an informal recursive algorithm for computing the number $\mathbf{val}^{\mathcal{V}}_x(M)$ where $M$ has the properties stated in the claim. The algorithm is meant to be carried out by pen and paper, and we will argue that the number of symbols written down during the execution is bounded by $2^{p(x)}_{n}$ for some polynomial $p$. It is to easy to see that the informal algorithm can be implemented by a Turing machine $\mathfrak{m}$ running in time $2_n^{p_0(x)}$ for some polynomial $p_0$, and hence, there exists $k \in \mathbb{N}$ such that $\mathfrak{m}$ runs in time $2_{n+1}^{k|x|}$.

We can w.l.o.g. assume that $\mathrm{dg}(\pi) = 0$ for any recursor term $\mathrm{R}_\pi(\ldots)$ occurring in $M$, and if $\mathrm{dg}(\xi) > n+1$ for some subterm $N : \xi$ of $M$, then $N$ is of the form $N \equiv \lambda X^\sigma . P^\sigma$ and occur in the context $\mathrm{Y}_\sigma(N)$ where $\mathrm{dg}(\sigma) = n+1$.

Let $y_1, \ldots, y_\ell$ be an enumeration of the (bound and unbound) variables occurring in $M$. The algorithm keeps track of the values assigned to the variables in a list $y_1/a_1, \ldots, y_\ell/a_\ell$ where the value $a_i$ is the value currently assign to the variable $y_i$. The number $\ell$ is fixed (rename variables to avoid name conflicts), and we have $a_i < |\sigma|_x$ if $a_i$ is assigned to a variable of type $\sigma$. Since we have $\mathrm{dg}(\tau) \leq n+1$ for any variable $y_i^\tau$ in the list, there exists a polynomial $p_0$ such that $2_{n+1}^{p_0(x)}$ bounds bounds every $a_i$ in the list, and thus each $a_i$ can be represented by a bit string of length $2_n^{p_0(x+2)}$. Thus, there exists a polynomial $p(x)$ such that the algorithm can assign values to variables, and retrieve values assigned to variables, in time $2_n^{p(x)}$.

The algorithm computes the value $\mathbf{val}_x^\mathcal{V}(M)$ recursively over the structure of $M$. (Note that the structure of $M$ does not depend on the input $x$.) *Case* $M \equiv k_m$: `return` $m \pmod{x+1}$. The number of symbols we will write down is obviously bounded by $2_n^{p(x)}$ for some polynomial $p$. *Case* $M \equiv y_i$: `return` $a_i$. The number $a_i$ is stored as a bit string in the assignment list, and the number of symbols we have to write down to retrieve this number is bounded by $2_n^{p(x)}$ for some polynomial $p$. *Case* $M^\sigma \equiv \lambda z^\pi N^\tau$:

```
sum:=0;
for i = 0,...,|π|_x − 1 do { assign i to z; sum:= sum × |τ|_x + val_x^V(N) };
return sum
```

We have $\mathrm{dg}(\pi) \leq n$ and $\mathrm{dg}(\tau) \leq n+1$ since $\mathrm{dg}(\sigma) \leq n+1$. Hence, there exist polynomials $p_0$ and $p_1$ such that the loop will be executed $2_n^{p_0(x)}$ times and any number assigned to `sum` will be bounded by $2_{n+1}^{p_1(x)}$. The number of bits required to represent such a number will be bounded by $2_n^{p_1(x+2)}$. Assume inductively that there exists a polynomial $p_2$ such that the number of symbols we have to write down to compute $\mathbf{val}_x^\mathcal{V}(N)$ is bounded by $2_n^{p_2(x)}$. It follows that there exists a polynomial $p$ such that the number of symbols we have to write down to compute $\mathbf{val}_x^\mathcal{V}(M)$ is bounded by $2_n^{p(x)}$. *Case* $M^\sigma \equiv \mathrm{Y}_\sigma(\lambda z^\sigma P^\sigma)$:

```
a:=0; for i = 0,...,⌈σ⌉_x do { assign a to z; a:=val_x^V(P) }; return a
```

The argument that this algorithm satisfies our efficiency requirements is similar to the argument for the case $M \equiv \lambda z N$, and we leave the details to the reader. Note that there exists a polynomial $p_0$ such that $\lceil \sigma \rceil_x < 2_n^{p_0(x)}$.

The remaining cases, i.e. $M \equiv (M_1 M_2)$, $M \equiv \mathbf{fst}\, M_1$, $M \equiv \mathbf{snd}\, M'$, $M \equiv \langle M_1, M_2 \rangle$ and $M \equiv \mathrm{R}_\pi(G, F, N)$, are fairly straightforward, and we omit the details.                                                                 $\square$

We have not proved the inclusions $\mathcal{G}_{2n} \subseteq \text{SPACE } 2_n^{\text{LIN}}$ and $\mathcal{G}_{2n+1} \subseteq \text{TIME } 2_{n+1}^{\text{LIN}}$, but both inclusions hold. Fairly detailed proofs based on an idea of Goerdt & Seidl [3,4], are published in Kristiansen & Voda [13]. Alternative, and essentially different, proofs can be found in Kristiansen & Voda [14].

These two inclusions and the theorems above yield the following corollary.

**Corollary 1.** *We have* $\text{TIME } 2_{n+1}^{\text{LIN}} = \mathcal{P}_{n+1} = \mathcal{G}_{2n+1}$ *and* $\text{SPACE } 2_n^{\text{LIN}} = \mathcal{G}_{2n}$.

# References


1. Avigad, J., Feferman, S.: Gödel's functional interpretation. In: Buss, S. (ed.) Handbook of Proof Theory. Elsevier (1998)
2. Barra, G., Kristiansen, L., Voda, P.: Nondeterminism without Turing machines. In: Cooper, S.B., Löwe, B., Sorbi, A. (eds.) CiE 2007. LNCS, vol. 4497. Springer, Heidelberg (2007), http://www.mat.unisi.it/~sorbi/Proceedings.pdf
3. Goerdt, A., Seidl, H.: Characterizing complexity classes by higher type primitive recursive definitions, part ii. In: Dassow, J., Kelemen, J. (eds.) Aspects and prospects of theoretical computer science, Smolenice. LNCS, vol. 464, pp. 148–158. Springer, Heidelberg (1990)
4. Goerdt, A.: Characterizing complexity classes by higher type primitive recursive definitions. Theoretical Computer Science 100, 45–66 (1992)
5. Jones, N.: The expressive power of higher-order types or, life without CONS. Journal of Functional Programming 11, 55–94 (2001)
6. Jones, N.: LOGSPACE and PTIME characterized by programming languages. Theoretical Computer Science 228, 151–174 (1999)
7. Kristiansen, L.: Neat function algebraic characterizations of LOGSPACE and LINSPACE. Computational Complexity 14, 72–88 (2005)
8. Kristiansen, L.: Complexity-theoretic hierarchies. In: Beckmann, A., Berger, U., Löwe, B., Tucker, J.V. (eds.) CiE 2006. LNCS, vol. 3988, pp. 279–288. Springer, Heidelberg (2006)
9. Kristiansen, L.: Complexity-theoretic hierarchies induced by fragments of Gödel's T. In: Theory of Computing Systems. Springer, Heidelberg (2007)
10. Kristiansen, L., Barra, G.: The small Grzegorczyk classes and the typed λ-calculus. In: Cooper, S.B., Löwe, B., Torenvliet, L. (eds.) CiE 2005. LNCS, vol. 3526, pp. 252–262. Springer, Heidelberg (2005)
11. Kristiansen, L., Voda, P.: Complexity classes and fragments of C. Information Processing Letters 88, 213–218 (2003)
12. Kristiansen, L., Voda, P.: The surprising power of restricted programs and Gödel's functionals. In: Baaz, M., Makowsky, J.A. (eds.) CSL 2003. LNCS, vol. 2803, pp. 345–358. Springer, Heidelberg (2003)
13. Kristiansen, L., Voda, P.: Programming languages capturing complexity classes. Nordic Journal of Computing 12, 1–27 (2005); Special issue for NWPT 2004
14. Kristiansen, L., Voda, P.: The trade-off theorem and fragments of Gödel's T. In: Cai, J.-Y., Cooper, S.B., Li, A. (eds.) TAMC 2006. LNCS, vol. 3959, pp. 654–674. Springer, Heidelberg (2006)
15. Odifreddi, P.: Classical recursion theory II. Studies in Logic and the Foundations of Mathematics, vol. 143. North-Holland Publishing Co., Amsterdam (1999)
16. Streicher, T.: Domain-theoretic foundations of functional programming. World Scientific Publishing Co. Pte. Ltd, Singapore (2006)

# Computability and Complexity in Self-assembly


James I. Lathrop, Jack H. Lutz*, Matthew J. Patitz, and Scott M. Summers**

Department of Computer Science
Iowa State University
Ames, IA 50011, U.S.A.
{jil,lutz,mpatitz,summers}@cs.iastate.edu



**Abstract.** This paper explores the impact of geometry on computability and complexity in Winfree's model of nanoscale self-assembly. We work in the two-dimensional tile assembly model, i.e., in the discrete Euclidean plane $\mathbb{Z} \times \mathbb{Z}$. Our first main theorem says that there is a roughly quadratic function $f$ such that a set $A \subseteq \mathbb{Z}^+$ is computably enumerable if and only if the set $X_A = \{(f(n), 0) | n \in A\}$ – a simple representation of $A$ as a set of points on the $x$-axis – self-assembles in Winfree's sense. In contrast, our second main theorem says that there are decidable sets $D \subseteq \mathbb{Z} \times \mathbb{Z}$ that do *not* self-assemble in Winfree's sense.

Our first main theorem is established by an explicit translation of an arbitrary Turing machine $M$ to a modular tile assembly system $\mathcal{T}_M$, together with a proof that $\mathcal{T}_M$ carries out concurrent simulations of $M$ on all positive integer inputs.

**Keywords:** computability, computational complexity, molecular computing, self-assembly.


## 1 Introduction

A major objective of nanotechnology is to engineer systems that, like many systems in nature, autonomously assemble themselves from molecular components. One promising approach to this objective, pioneered by Seeman in the 1980s [9], is *DNA tile self-assembly*. This approach exploits the information-processing and structural properties of DNA to construct nanoscale components (DNA tiles) that, with moderate control of physical conditions (temperature, concentration in solution, etc.), spontaneously assemble themselves into larger structures. These structures, which may be complex and aperiodic, are *programmed*, in the sense that they are determined by a designer's selection of the short, single-strand nucleotide sequences ("sticky ends") that appear on each side of each type of DNA tile to be used.


---

* Corresponding author. This author's research was supported in part by National Science Foundation Grants 0344187, 0652569 and 0728806 and by Spanish Government MEC Project TIN 2005-08832-C03-02

** This author's research was supported in part by NSF-IGERT Training Project in Computational Molecular Biology Grant number DGE-0504304.



A. Beckmann, C. Dimitracopoulos, and B. Löwe (Eds.): CiE 2008, LNCS 5028, pp. 349–358, 2008.
© Springer-Verlag Berlin Heidelberg 2008

In the late 1990s, Winfree [13] introduced a mathematical model of DNA tile assembly called the Tile Assembly Model. This model, which was soon refined by Rothemund and Winfree [7,8], is an extension of Wang tiling [11,12]. (see also [1,6,10].) The Tile Assembly Model, which is described in section 2 below, uses square tiles with various types and strengths of "glue" on their edges as abstractions of DNA tiles adsorbing to a growing structure on a very flat surface such as mica. Winfree [13] proved that the Tile Assembly Model is computationally universal, i.e., that any Turing machine can be encoded into a finite set of tile types whose self-assembly simulates that Turing machine.

The computational universality of the Tile Assembly Model implies that self-assembly can be algorithmically directed, and hence that a very rich set of structures can be formed by self-assembly. However, as we shall see, this computational universality does not seem to imply a simple characterization of the class of structures that can be formed by self-assembly. The difficulty is that self-assembly (like sensor networks, smart materials, and other topics of current research [2,3]) is a phenomenon in which the *spatial* aspect of computing plays a crucial role. Two processes, namely self-assembly and the computation that directs it, must take place in the same space and time.

In this paper, we focus on the self-assembly that takes place in the discrete Euclidean plane $\mathbb{Z}^2 = \mathbb{Z} \times \mathbb{Z}$, where $\mathbb{Z}$ is the set of integers. Roughly, a *tile assembly system* $\mathcal{T}$ (defined precisely in section 2) is a finite set $T$ of tile types, together with a finite seed assembly $\sigma$ consisting of one or more tiles of these types. Self-assembly is then the process in which tiles of the types in $T$ successively adsorb to $\sigma$ (more precisely, to the assembly which is thereby dynamically growing from $\sigma$) in any manner consistent with the rules governing the glue types on the edges of the given tile types. Note that self-assembly is, in general, a nondeterministic, asynchronous process.

We say that a set $X \subseteq \mathbb{Z}^2$ *self-assembles* in a tile assembly system $\mathcal{T}$ if every possible "run" of self-assembly in $\mathcal{T}$ results in the placement of black tiles on the set $X$ and only on the set $X$. (Some of the tile types in $\mathcal{T}$ are designated to be black. Non-black tiles may appear on some or all points in the complement $\mathbb{Z}^2 - X$.) This is the sense in which Winfree [13] has demonstrated the self-assembly of the standard discrete Sierpinski triangle. (In [5] this is called *weak self-assembly* to contrast it with a stricter notion in which, essentially, all tiles are required to be black.)

This paper presents two main theorems on the interplay between geometry and computation in tile self-assembly. To explain our first main theorem, define the function $f : \mathbb{Z}^+ \to \mathbb{Z}^+$ by

$$f(n) = \binom{n+1}{2} + (n+1)\lfloor \log n \rfloor + 6n - 2^{1+\lfloor \log(n) \rfloor} + 2.$$

Note that $f$ is a reasonably simple, strictly increasing, roughly quadratic function of $n$. For each set $A \subseteq \mathbb{Z}^+$, the set

$$X_A = \{(f(n), 0) | n \in A\}$$

is thus a straightforward representation of $A$ as a set of points on the positive $x$-axis.

Our first main theorem says that *every* computably enumerable set $A$ of positive integers (decidable or otherwise) self-assembles in the sense that there is a tile assembly system $\mathcal{T}_A$ in which the representation $X_A$ self-assembles. Conversely, the existence of such a tile assembly system implies the computable enumerability of $A$.

In contrast, our second main theorem says that there are *decidable* sets $D \subseteq \mathbb{Z}^2$ that do *not* self-assemble in any tile assembly system. In fact, we exhibit such a set $D$ for which the condition $(m, n) \in D$ is decidable in time polynomial in $|m| + |n|$.

Taken together, our two main theorems indicate that the interaction between geometry and computation in self-assembly is not at all simple. Further investigation of this interaction will improve our understanding of tile self-assembly and, more generally, spatial computation.

The proof of our first main theorem has two features that may be useful in future investigations. First, we give an explicit transformation (essentially a compiler, implemented in C++) of an arbitrary Turing machine $M$ to a tile assembly system $\mathcal{T}_M$ whose self-assembly carries out concurrent simulations of $M$ on (the binary representation of) all positive integer inputs. Second, we prove two lemmas – a pseudoseed lemma and a multiseed lemma – that enable us to reason about tile assembly systems in a modular fashion. This modularity, together with the local determinism method of Soloveichik and Winfree [10], enables us to prove the correctness of $\mathcal{T}_M$.

## 2   Preliminaries

We work in the 2-dimensional discrete Euclidean space $\mathbb{Z}^2$. We write $U_2$ for the set of all *unit vectors*, i.e., vectors of length 1, in $\mathbb{Z}^2$. We regard the 4 elements of $U_2$ as (names of the cardinal) *directions* in $\mathbb{Z}^2$.

We now give a brief and intuitive sketch of the Tile Assembly Model that is adequate for reading this paper. More formal details and discussion may be found in [7,8,13].

Intuitively, a tile type $t$ is a unit square that can be translated, but not rotated, so it has a well-defined "side $\boldsymbol{u}$" for each $\boldsymbol{u} \in U_2$. Each side $\boldsymbol{u}$ is covered with a "glue" of "color" $col_t(\boldsymbol{u})$ and "strength" $str_t(\boldsymbol{u})$ specified by its type $t$. If two tiles are placed with their centers at adjacent points $\boldsymbol{m}, \boldsymbol{m} + \boldsymbol{u} \in \mathbb{Z}^2$, where $\boldsymbol{u} \in U_2$, and if their abutting sides have glues that match in both color and strength, then they form a *bond* with this common strength. If the glues do not so match, then no bond is formed between these tiles. In this paper, all glues have strength 0, 1, or 2. When drawing a tile as a square, each side's glue strength is indicated by whether the side is a dotted line (0), a solid line (1), or a double line (2). Each side's "color" is indicated by an alphanumeric label.

Given a set $T$ of tile types and a "temperature" $\tau \in \mathbb{N}$, a $\tau$-$T$-*assembly* is a partial function $\alpha : \mathbb{Z}^2 \dashrightarrow T$ - intuitively, a placement of tiles at some locations in $\mathbb{Z}^2$ - that is *stable* in the sense that it cannot be "broken" into smaller assemblies without breaking bonds of total strength at least $\tau$. If $\alpha$ and $\alpha'$ are assemblies, then $\alpha$ is a *subassembly* of $\alpha'$, and we write $\alpha \sqsubseteq \alpha'$, if dom $\alpha \subseteq$ dom $\alpha'$ and $\alpha(\boldsymbol{m}) = \alpha'(\boldsymbol{m})$ for all $\boldsymbol{m} \in$ dom $\alpha$.

Self-assembly begins with a *seed assembly* $\sigma$ and proceeds asynchronously and nondeterministically, with tiles adsorbing one at a time to the existing assembly in any manner that preserves stability at all times. A *tile assembly system* (*TAS*) is an ordered triple $\mathcal{T} = (T, \sigma, \tau)$, where $T$ is a finite set of tile types, $\sigma$ is a seed assembly with finite domain, and $\tau \in \mathbb{N}$. In this paper we always have $\tau = 2$. A *generalized tile assembly system* (*GTAS*) is defined similarly, but without the finiteness requirements. We write $\mathcal{A}[\mathcal{T}]$ for the set of all assemblies that can arise (in finitely many steps or in the limit) from $\mathcal{T}$. An assembly $\alpha$ is *terminal*, and we write $\alpha \in \mathcal{A}_\square[\mathcal{T}]$, if no tile can be stably added to it. A GTAS $\mathcal{T}$ is *directed*, or *produces a unique assembly*, if it has exactly one terminal assembly.

A set $X \subseteq \mathbb{Z}^2$ *(weakly) self-assembles* if there exist a TAS $\mathcal{T} = (T, \sigma, \tau)$ and a set $B \subseteq T$ such that $\alpha^{-1}(B) = X$ holds for every terminal assembly $\alpha$. That is, there is a set $B$ of "black" tile types such that every terminal assembly has black tiles on $X$ and only on $X$.

An *assembly sequence* in a TAS $\mathcal{T} = (T, \sigma, \tau)$ is an infinite sequence $\boldsymbol{\alpha} = (\alpha_0, \alpha_1, \alpha_2, \ldots)$ of assemblies in which $\alpha_0 = \sigma$ and each $\alpha_{i+1}$ is obtained from $\alpha_i$ by the addition of a single tile. In general, even a directed TAS may have a very large (perhaps uncountably infinite) number of different assembly sequences leading to its terminal assembly. This seems to make it very difficult to prove that a TAS is directed. Fortunately, Soloveichik and Winfree [10] have recently defined a property, *local determinism*, of assembly sequences and proven the remarkable fact that, if a TAS $\mathcal{T}$ has *any* assembly sequence that is locally deterministic, then $\mathcal{T}$ is directed.


## 3    Pseudoseeds and Multiseeds

This section introduces two conceptual tools that enable us to reason about tile assembly systems in a modular fashion.

The idea of our first tool is intuitive. Suppose that our objective is to design a tile assembly system $\mathcal{T}$ in which a given set $X \subseteq \mathbb{Z}^2$ self-assembles. The set $X$ might have a subset $X^*$ for which it is natural to decompose the design into the following two stages.

(i) Design a TAS $\mathcal{T}_0 = (T_0, \sigma, \tau)$ in which an assembly $\sigma^*$ with domain $X^*$ self-assembles.
(ii) Extend $T_0$ to a tile set $T$ such that $X$ self-assembles in the TAS $\mathcal{T}^* = (T, \sigma^*, \tau)$

We would then like to conclude that $X$ self-assembles in the TAS $\mathcal{T} = (T, \sigma, \tau)$. This will not hold in general, but it does hold if (i) continues to hold with $T$ in place of $T_0$ and $\sigma^*$ is a pseudoseed in the following sense.

**Definition 1.** *Let $\mathcal{T} = (T, \sigma, \tau)$ be a GTAS. A* pseudoseed *of $\mathcal{T}$ is an assembly $\sigma^* \in \mathcal{A}[\mathcal{T}]$ with the property that, if we let $\mathcal{T}^* = (T, \sigma^*, \tau)$, then, for every assembly $\alpha \in \mathcal{A}[\mathcal{T}]$, there exists an assembly $\alpha' \in \mathcal{A}[\mathcal{T}^*]$ such that $\alpha \sqsubseteq \alpha'$.*

The following lemma says that the above definition has the desired effect.

**Lemma 1 (pseudoseed lemma).** *If $\sigma^*$ is a pseudoseed of a GTAS $\mathcal{T} = (T, \sigma, \tau)$ and $\mathcal{T}^* = (T, \sigma^*, \tau)$, then $\mathcal{A}_\square[\mathcal{T}] = \mathcal{A}_\square[\mathcal{T}^*]$.*

Note that the pseudoseed lemma entitles us to *reason as though* the self-assembly proceeds in stages, even though this may not actually occur. (E.g., the pseudoseed $\sigma^*$ may itself be infinite, in which case the self-assembly *of $\sigma^*$* and the self-assembly *from $\sigma^*$* must occur concurrently.)

Our second tool for modular reasoning is a bit more involved. Suppose that we have a tile set $T$ and list $\sigma_0, \sigma_1, \sigma_2, \ldots$ of seeds that, for each $i$, the TAS $\mathcal{T}_i = (T, \sigma_i, \tau)$ has a desired assembly $\alpha_i$ as its result. If the assemblies $\alpha_0, \alpha_1, \alpha_2, \ldots$ have disjoint domains, then it might be possible for *all* these assemblies to grow from a "multiseed" $\sigma^*$ that has $\sigma_0, \sigma_1, \sigma_2, \ldots$ embedded in it. We now define a sufficient condition for this.

**Definition 2.** *Let $T$ and $T'$ be sets of tile types with $T \subseteq T'$, and let $\boldsymbol{\sigma} = (\sigma_i \mid 0 \le i < k)$ be a sequence of $\tau$-$T$-assemblies, where $k \in \mathbb{Z}^+ \cup \{\infty\}$. A $\boldsymbol{\sigma}$-$T$-$T'$-multiseed is a $\tau$-$T'$ assembly $\sigma^*$ with the property that, if we write*

$$\mathcal{T}^* = (T', \sigma^*, \tau)$$

*and*

$$\mathcal{T}_i = (T, \sigma_i, \tau)$$

*for each $0 \le i < k$, then the following four conditions hold.*

1. *For each $i$, $\sigma_i \sqsubseteq \sigma^*$.*
2. *For each $i \ne j$, $\alpha \in \mathcal{A}[\mathcal{T}_i]$, $\alpha' \in \mathcal{A}[\mathcal{T}_j]$, $\boldsymbol{m} \in \operatorname{dom} \alpha$, and $\boldsymbol{m}' \in \operatorname{dom} \alpha'$, $\boldsymbol{m} - \boldsymbol{m}' \in U_2 \cup \{\boldsymbol{0}\} \Rightarrow \boldsymbol{m}, \boldsymbol{m}' \in \operatorname{dom} \sigma^*$. (Recall that $U_2$ is the set of unit vectors in $\mathcal{Z}^2$.)*
3. *For each $i$, and each $\alpha \in \mathcal{A}[\mathcal{T}_i]$, there exists $\alpha^* \in \mathcal{A}[\mathcal{T}^*]$ such that $\alpha \sqsubseteq \alpha^*$.*
4. *For each $\alpha^* \in \mathcal{A}[\mathcal{T}^*]$, there exists, for each $0 \le i < k$, $\alpha_i \in \mathcal{A}[\mathcal{T}_i]$ such that $\alpha^* \sqsubseteq \sigma^* + \sum_{0 \le i < k} \alpha_i$.*
5. *For each $\alpha \in \mathcal{A}[\mathcal{T}^*]$, $\alpha^{-1}(T' - T) \subseteq \operatorname{dom} \sigma^*$.*

Note: In condition 4 we are using the operation $+$ defined as follows. If $\alpha, \alpha' : \mathbb{Z}^2 \dashrightarrow T$ are *consistent*, in the sense that they agree on $\operatorname{dom} \alpha \cap \operatorname{dom} \alpha'$, then $\alpha + \alpha' : \mathbb{Z}^2 \dashrightarrow T$ is the unique partial function satisfying $\operatorname{dom}(\alpha + \alpha') = \operatorname{dom} \alpha \cup \operatorname{dom} \alpha'$, $\alpha \sqsubseteq \alpha + \alpha'$, and $\alpha' \sqsubseteq \alpha + \alpha'$. This is extended to summations $\sum_{0 \le i < k} \alpha_i$ in the obvious way. The assemblies being summed in condition 4 are consistent by conditions 1 and 2.

Intuitively, the four conditions in the above definition can be stated as follows.

1. The seeds $\sigma_i$ are embedded in $\sigma^*$.
2. Assemblies in $\mathcal{A}[\mathcal{T}_i]$ and assemblies in $\mathcal{A}[\mathcal{T}_j]$ do not interfere with each other.

3. $\sigma^*$ does not interfere with assemblies in $\mathcal{A}[\mathcal{T}_i]$.
4. $\sigma^*$ does not produce anything other than what its embedded seeds $\sigma_i$ produce.
5. Tile types in $T' - T$ cannot occur outside $\sigma^*$.

The following lemma says that the multiseed definition has the desired effect.

**Lemma 2 (multiseed lemma).** *Let $T \subseteq T'$ be sets of tile types, and let $\boldsymbol{\sigma} = (\sigma_i \mid 0 \leq i < k)$ be a sequence of $\tau$-$T$-assemblies, where $k \in \mathbb{Z}^+ \cup \{\infty\}$. If $\sigma^*$ is a $\boldsymbol{\sigma}$-$T$-$T'$-multiseed of $\mathcal{T}^*$ and $\mathcal{T}_i$ $(0 \leq i < k)$ are defined as in the multiseed definition, then*

$$\mathcal{A}_\square[\mathcal{T}^*] = \left\{ \sigma^* + \sum_{0 \leq i < k} \alpha_i \;\middle|\; \text{each } \alpha_i \in \mathcal{A}_\square[\mathcal{T}_i] \right\}.$$

## 4   Self-assembly of Computably Enumerable Sets

In [13], Winfree proved that the Tile Assembly Model is Turing universal in two dimensions. In this section, we prove a stronger result: for every TM $M$, there exists a directed TAS that simulates $M$ on (the binary representation of) *every* input $x \in \mathbb{N}$ in the two dimensional discrete Euclidean plane. We state our result precisely in the following theorem.

**Theorem 1 (first main theorem).** *If $f : \mathbb{Z}^+ \rightarrow \mathbb{Z}^+$ is defined as in section 1, then, for all $A \subseteq \mathbb{Z}^+$, $A$ is computably enumerable if and only if the set $X_A = \{(f(n), 0) \mid n \in A\}$ self-assembles.*

The "$\Leftarrow$" direction is easy. To prove the "$\Rightarrow$" direction, we exhibit, for any TM $M$, a directed TAS $\mathcal{T}_M = (T, \sigma, \tau)$ that correctly simulates $M$ on all inputs $x \in \mathbb{Z}^+$ in the two dimensional discrete Euclidean plane. We sketch our construction in the remainder of this section. Note that the full details of our construction can be found at the following url: `http://www.cs.iastate.edu/~lnsa`.

### 4.1   Overview of Construction

Intuitively, $\mathcal{T}_M$ self-assembles a "gradually thickening bar", immediately below the positive $x$-axis with upward growths emanating from well-defined intervals of points. For each $x \in \mathbb{Z}^+$, there is an upward growth that simulates $M$ on $x$. If $M$ halts on $x$, then (a portion of) the upward growth associated with the simulation of $M(x)$ eventually stops, and sends a signal down along the right side of the upward growth via a one-tile-wide-path of tiles to the point $(f(x), 0)$, where a black tile is placed. See Figure 1 for a finite, yet intuitive snapshot of this infinite process.

Our tile assembly system $\mathcal{T}_M$ is divided into three modules: the ray, the planter, and the TM module, which control the spacing between successive simulations, the initiation of upward growth, and the actual simulations of $M$ on each positive integer, respectively.

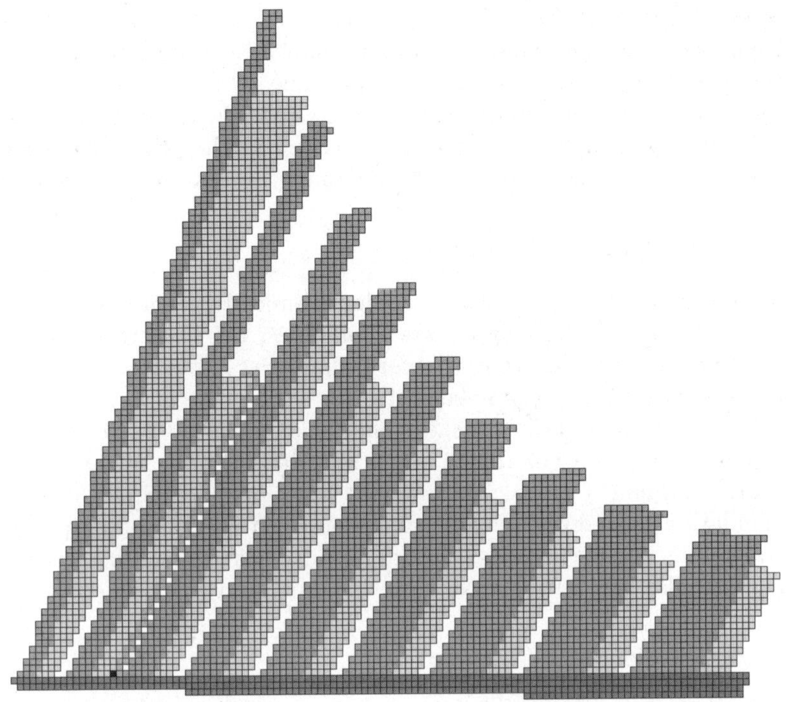

**Fig. 1.** Simulation of $M$ on every input $x \in \mathbb{N}$. Notice that $M(2)$ halts - indicated by the black tile along the $x$-axis. Note that this image should be viewed in color.

## 4.2   The Ray Module

The first module in our construction is the ray module (middle shade of gray squares in Figure 1). For any $3 \leq w \in \mathbb{Z}^+$, a ray of width $w$ is a fixed-width, periodic, binary counter that repeatedly counts from 0 to $2^w - 1$, such that each integer is counted once, and then immediately copied once before the value of the binary counter is incremented. Essentially, a ray of width $w$ is a discrete line of constant thickness $w$, having a kind of "slope" that depends on $w$ in the following way. In every other row (except for two special cases), the first tile to attach does so on top of the second-to-left most tile in the previous row. Thus, a ray of width $w$ will have a slope of $\frac{2^w}{2^{w-1}-1} = 2 + \frac{2}{2^{w-1}-1}$. This implies that the set of points occupied by properly spaced, consecutive rays of strictly increasing width, will not only be disjoint but the width of the gap in between such rays will increase without limit.

## 4.3   The Planter Module

The next module is the planter module because it "plants the seeds" from which the ray modules will ultimately grow (the darkest gray squares in Figure 1). At the core of the planter module is a log-width, horizontal binary counter that counts every positive integer, starting at 1, in order. A key feature of the binary counter embedded in the planter module is that, after each integer is counted, a number of columns, equal to the current value of the binary counter, plus the number of bits in the binary representation of this value, plus a few extra "dummy" spacing columns, self-assemble. This has the effect of spacing out successive ray modules according to the function $f$ (given in the introduction).

## 4.4    The Computation Module

The final module is, for any TM $M$, an algorithmically generated tile set that, in conjunction with the ray and planter modules, achieves the simulation of $M$ on (the binary representation of) every input $x \in \mathbb{Z}^+$. The simulation of $M$ on $\mathrm{bin}(x)$ proceeds vertically, immediately above the planter, while following the contour defined by the rightmost edge of the ray of width $x+2$ (Note that by our construction of the planter module, there is one, and only one ray of such width). As with other standard Turing machine constructions (see [8,10,13]), each row in our simulation represents a configuration of $M$. However, the frequency with which transitions occur is a novel feature of our construction, and is controlled by "color" signals that are received from the abutting ray module.

## 4.5    Sketch of Correctness Proof

Let $f : \mathbb{Z}^+ \to \mathbb{Z}^+$ be defined as in section 1, and stipulate that $f(0) = -1$. For each $n \in \mathbb{Z}^+$, let $\sigma_n$ be the portion of the planter lying in the rectangle

$$Q_n = \{f(n-1) + 2, \ldots, f(n) - 1\} \times \{-2, -1\},$$

and let $\rho_n$ be the ray of width $n + 2$, translated so that its base is the leftmost $2(n + 2)$ tiles of $\sigma_n$. Let $\sigma_n^* = \sigma_n + \rho_n$.

Let $\mathcal{T} = (T, \sigma, \tau)$ be our TAS, noting that $\sigma$ consists of a single tile at the origin. For each $n \in \mathbb{N}$, let $\mathcal{T}_n = (T, \sigma_n, \tau)$. We use local determinism to prove that each $\sigma_n^*$ is a pseudoseed of $\mathcal{T}_n$ and that each $\mathcal{T}_n^* = (T, \sigma_n^*, \tau)$ has the unique terminal assembly $\alpha_n = \sigma_n^* + \gamma_n$, where $\gamma_n$ is the assembly that simulates $M(n)$ as in section 4.4. It follows by the pseudoseed lemma that each $\alpha_n$ is the unique terminal assembly of $\mathcal{T}_n$. We then use local determinism to prove that our planter $\sigma^*$ has the following two properties.

(i)  $\sigma^*$ is a pseudoseed of $\mathcal{T}$.
(ii) $\sigma^*$ is a $\boldsymbol{\sigma}$-$T$-multiseed, where $\boldsymbol{\sigma} = (\sigma_0, \sigma_1, \ldots)$.

Let $\mathcal{T}^* = (T, \sigma^*, \tau)$. By (ii) and the multiseed lemma, we now have that

$$\alpha = \sigma^* + \sum_{n=0}^{\infty} \alpha_n$$

is the unique terminal assembly of $\mathcal{T}^*$. It follows by (i) and the pseudoseed lemma that $\alpha$ is the unique terminal assembly of $\mathcal{T}$.

## 5   A Decidable Set That Does Not Self-assemble

We now show that there are decidable sets $D \subseteq \mathbb{Z}^2$ that do not self-assemble in the Tile Assembly Model.

For each $r \in \mathbb{N}$, let

$$D_r = \{\, (m,n) \in \mathbb{Z}^2 \,\big|\, |m| + |n| = r \}.$$

This set is a "diamond" in $\mathbb{Z}^2$ with radius $r$ (i.e., a sphere of radius $r$ with respect to the "taxicab metric" on $\mathbb{Z} \times \mathbb{Z}$) and center at the origin. For each $A \subseteq \mathbb{N}$, let

$$D_A = \bigcup_{r \in A} D_r.$$

This set is the "system of concentric diamonds" centered at the origin with radii in $A$.

**Lemma 3.** *If $A \subseteq \mathbb{N}$ and $D_A$ self-assembles, then $A \in DTIME\left(2^{4n}\right)$.*

The proof of this lemma exploits the fact that a tile assembly system in which $D_A$ self-assembles must, for sufficiently large $r$, decide the condition $r \in A$ from *inside* the diamond $D_r$.

We now have the following result.

**Theorem 2 (second main theorem).** *There is a decidable set $D \subseteq \mathbb{Z}^2$ that does not self-assemble.*

*Proof.* By the time hierarchy theorem [4], there is a set $A \subseteq \mathbb{N}$ such that

$$A \in \text{DTIME}\left(2^{5n}\right) - \text{DTIME}\left(2^{4n}\right).$$

Let $D = D_A$. Then $D$ is decidable and, by Lemma 3, $D$ does not self-assemble.

## 6   Conclusion

Our first main theorem says that, for every computably enumerable set $A \subseteq \mathbb{Z}^+$, the representation $X_A = \{(f(n),0)|n \in A\}$ self-assembles. This representation of $A$ is somewhat sparse along the $x$-axis, because our $f$ grows quadratically. A linear function $f$ would give a more compact representation of $A$. We conjecture that our first main theorem does *not* hold for any linear function, but we do not know how to prove this.

Let $D$ be the set presented in the proof of our second main theorem. It is easy to see that the condition $(m,n) \in D$ is decidable in time polynomial in $|m| + |n|$, but $|m| + |n|$ is exponential in the length of the binary representation of $(m,n)$, so this only tells us that $D \in \text{E} = \text{DTIME}\left(2^{\text{linear}}\right)$. Is there a set $D \subseteq \mathbb{Z}^2$ such that $D \in \text{P}$, and $D$ does not self-assemble?

More generally, we hope that our results lead to further research illuminating the interplay between geometry and computation in self-assembly.

**Acknowledgment.** The authors wish to thank Dave Doty and Aaron Sterling for useful discussions. We also thank the reviewers for useful suggestions.

# References


1. Adleman, L.: Towards a mathematical theory of self-assembly, Tech. report, University of Southern California (2000)
2. Bachrach, J., Beal, J.: Building spatial computers, Tech. report, MIT CSAIL (2007)
3. Beal, J., Sussman, G.: Biologically-inspired robust spatial programming, Tech. report, MIT (2005)
4. Hartmanis, J., Stearns, R.E.: On the computational complexity of algorithms. Transactions of the American Mathematical Society 117, 285–306 (1965)
5. Lathrop, J.I., Lutz, J.H., Summers, S.M.: Strict self-assembly of discrete Sierpinski triangles. In: Proceedings of The Third Conference on Computability in Europe, Siena, Italy, June 18-23, 2007 (2007)
6. Reif, J.H.: Molecular assembly and computation: From theory to experimental demonstrations. In: Proceedings of the Twenty-Ninth International Colloquium on Automata, Languages and Programming, pp. 1–21 (2002)
7. Paul, W.K.: Rothemund, Theory and experiments in algorithmic self-assembly, Ph.D. thesis, University of Southern California (December 2001)
8. Rothemund, P.W.K., Winfree, E.: The program-size complexity of self-assembled squares (extended abstract). In: Proceedings of the Thirty-Second Annual ACM Symposium on Theory of Computing, pp. 459–468 (2000)
9. Seeman, N.C.: Nucleic-acid junctions and lattices. Journal of Theoretical Biology 99, 237–247 (1982)
10. Soloveichik, D., Winfree, E.: Complexity of self-assembled shapes. SIAM Journal on Computing 36, 1544–1569 (2007)
11. Wang, H.: Proving theorems by pattern recognition – II. The Bell System Technical Journal XL(1), 1–41 (1961)
12. Wang, H.: Dominoes and the AEA case of the decision problem. In: Proceedings of the Symposium on Mathematical Theory of Automata New York, 1962, Polytechnic Press of Polytechnic Inst. of Brooklyn, pp. 23–55 (1963)
13. Winfree, E.: Algorithmic self-assembly of DNA, Ph.D. thesis, California Institute of Technology (June 1998)

# Extraction in Coq: An Overview

Pierre Letouzey

Laboratoire PPS, Université Paris Diderot - Paris 7
Case 7014, F-75205 Paris Cedex 13, France
`letouzey@pps.jussieu.fr`

**Abstract.** The extraction mechanism of Coq allows one to transform
Coq proofs and functions into functional programs. We illustrate the
behavior of this tool by reviewing several variants of Coq definitions for
Euclidean division, as well as some more advanced examples. We then
continue with a more general description of this tool: key features, main
examples, strengths, limitations and perspectives.

## 1   Introduction

This article describes the current status of the extraction mechanism available
in the Coq proof assistant [7, 8]. The extraction mechanism of Coq is a
tool for automatic generation of programs out of Coq proofs and functions.
These extracted programs are expressed in functional languages such as Ocaml,
Haskell or Scheme, these three languages being the ones currently supported
by Coq extraction. The main motivation for this extraction mechanism is to
produce certified programs: each property proved in Coq will still be valid after
extraction.

Through a series of examples about Euclidean division, we will review several
alternatives that allow the user to express in Coq an algorithm that does not fit
naturally in this system. We will also see how these alternatives influence the
shape of the program obtained by extraction. We will then mention two advanced
situations that illustrate the fact that Coq's current extraction can handle any
Coq objects, even the ones defined via high-end features of Coq and without
direct counterpart in Ocaml or Haskell. We will summarize the key features of
Coq extraction, mention some significant Coq developments taking advantage of
the extraction, and conclude on the current strengths of this tool, its limitations
and future research perspectives.

## 2   Extraction in Practice : Div

In this section, we illustrate the use of Coq extraction on a small yet revealing
example: Euclidean division amongst natural numbers. For sake of simplicity, we
will use the unary `nat` datatype for representing these natural numbers: every
number is stored as either zero or the successor `S` of another number. Even if this
representation is inherently inefficient, the discussion that follows would be quite

A. Beckmann, C. Dimitracopoulos, and B. Löwe (Eds.): CiE 2008, LNCS 5028, pp. 359–369, 2008.

similar with more clever coding of numbers. Coq's Standard Library provides basic operations and relations on nat, such as +, *, <, ≤. In Coq, logical relations do not necessarily have corresponding boolean test functions, but here a result named le_lt_dec, noted ≤? afterwards, can be used as an effective comparison function for determining whether n≤m or m<n for any numbers n and m.

Each Coq snippet proposed below is taken verbatim from a valid session[1] with Coq 8.2, including unicode notations and other syntactic improvements.

## 2.1   A Division That Fulfills the Structural Constraint

One usual algorithm for division on natural numbers is to proceed by successive subtractions: div x y = 0 when x<y and div x y = S (div (x-y) y) otherwise. But this cannot be written directly in Coq. Due to the intimate relationship between proofs and programs in Coq, no Coq objects may be allowed to trigger infinite computations. A rather drastic constraint is hence required on recursive functions in order to ensure their termination: they should have at least one inductive parameter such that recursive calls are done on an immediate subterm of this parameter[2]. Here, our recursive call fails this criterion, since (x-y) is not an immediate subterm of x, and second parameter y has not changed. Even worse, trying this algorithm with y=0 leads to an infinite computation: Coq's rejection is here quite legitimate.

For defining nonetheless our division in Coq, a first solution is to try to live with this structural constraint, and adapt our algorithm accordingly. For instance:

```
Fixpoint div x y := match x with
  | 0 => 0
  | S x' =>
     let z := div x' y in
     if (S z)*y ≤? x then S z else z
end.
```

Knowing the quotient for the predecessor of x can indeed be used to infer the quotient for x. But proceeding this way leads to a costly test repeated x times. This is a common situation with Coq: intended algorithms can be adapted to be "structural", but this may result in an awkward and/or less efficient algorithm.

Command **Extraction** div can then be used to convert this division to Ocaml code[3]:

```
let rec div x y =
  match x with
    | 0 -> 0
    | S x' ->
        let z = div x' y in if le_lt_dec (mult (S z) y) x then S z else z
```

---

[1] See http://www.pps.jussieu.fr/~letouzey/download/examples_CiE2008.v

[2] For a more precise definition of this structural constraint, see Chap. 4 of Coq Reference Manual at http://coq.inria.fr.

[3] The complete list of extraction commands can be found in the Coq Reference Manual.

This first extracted `div` highlights the fact that on basic Coq functions, extraction is mainly performing a straightforward syntactic translation. But even on such a simple function, some proof elimination occurs during extraction. In fact, comparison `le_lt_dec a b` is not producing a mere boolean, but rather a proof-carrying boolean type {a≤b}+{b<a}, which is an inductive type internally named `sumbool`, with two constructors `left` and `right` both having a proof as parameter, here respectively a proof of a≤b and a proof of b<a. Extraction removes these proofs, hence obtaining an extracted `sumbool` datatype with two constant constructors, isomorphic to `bool`. In order to get precisely the extracted code shown above, one could then teach Coq to take advantage of this isomorphism, via : `Extract Inductive sumbool => bool [ true false ]`.

One should note that the proof elimination done during extraction is based on earlier declarations by the user (or by the library designer). Here, proof-carrying boolean {a≤b}+{b<a} is exactly isomorphic to logical disjunction a≤b ∨ b<a (instead of `left` and `right`, constructors are named `or_introl` and `or_intror`). Simply, the former is declared in the logical world named `Prop` and is pruned during extraction whereas the latter is declared in `Set`, the world of Coq programs, and simply loses at extraction the logical parameters of its constructors. Similarly, two existential types coexist in Coq: the logical one ∃x:A,P x and the informative one { x:A | P x }.

## 2.2   A Division with an Explicit Counter

Let's now try to implement a function closer to our intended division algorithm, instead of the ad-hoc structural version of the last section. A solution is to artificially add a new structurally decreasing parameter that will control the number of allowed recursive calls. Here for instance, if y≠0, it is clear that at most x successive subtractions can occur before the algorithm stops. A common presentation is to separate the function to iterate `div_F` from the actual counter-based recursive iterator `div_loop`. The main function `div` is then a simple call to `div_loop` with the right initial counter value.

```
Definition div_F div x y := if y ≤? x then S (div (x-y) y) else 0.
```

```
Fixpoint div_loop (n:nat) :=
  match n with
    | 0 => fun _ _ => 0
    | S n => div_F (div_loop n)
  end.
```

```
Definition div x y := div_loop x x y.
```

One more time, extraction is straightforward and mostly amounts to replacing Coq keywords with Ocaml ones. The counter, whose type is `nat`, is kept by the extraction, even though it is morally useless for the computation. At the same time, removing it and replacing `div_loop` by an unbounded loop would change the semantics of the program at least for y=0: with the above definition, `div 5 0` computes to 5, while a Ocaml version without counter would loop forever. As a consequence, the extraction cannot be expected to detect and remove automatically such a "useless" parameter.

Using such an explicit counter is often an interesting compromise: the written Coq code is not exactly what we intended in the first place, but is close to it, there is no complex internal Coq object as with the methods we will study in the next sections, computations can be done both in Coq and after extraction, and the additional cost induced by the presence of the counter is often modest. Here for instance the `x` value would have been computed anyway. Another example of this technique can be found in module `Numtheory` of the Standard Library, where a `gcd` function is defined on binary numbers thanks to a counter that can be the depth (i.e. the logarithm) of these binary numbers.

## 2.3   A Division by General Recursion, Historical Approach

We can in fact build a Coq `div` function that will produce exactly the intended algorithm after extraction. Before presenting the modern ways of writing such a function with two frameworks recently added to Coq, let us first mention the historical approach. For a long time, the only possibility has been to play with accessibility predicates and induction principles such as `well_founded_induction`[4]. In this case, recursive functions do satisfy the structural constraint of Coq, not via their regular arguments, but rather via an additional logical argument expressing that some quantity is accessible. Recursive calls can then be done on quantities that are "more easily accessible" than before. This extra logical parameter is then meant to disappear during extraction. In practice, non-trivial functions are impossible to write as a whole with this approach, due to the numerous logical details to provide. Such functions are hence built piece by piece using Coq interactive tactics, as for proofs. Reasoning a posteriori on the body of such functions is also next to impossible, so key properties of these functions are to be attached to their output, via post-conditions `{ a:A | P a }`. Pre-conditions can also be added to restrict functions on a certain domain: for instance, `div` will be defined only for $y\neq0$. Here comes the complete specification of our `div` and its implementation in a proof-like style:

```
Definition div : ∀x y, y≠0 → { z | z*y ≤ x < (S z)*y }.
Proof.
 induction x as [x Hrec] using (well_founded_induction lt_wf).
 intros y Hy.
 destruct (y ≤? x) as [Hyx|Hyx]. (* do we have y≤x or x<y ? *)
 (* first case: y≤x *)
 assert (Hxy : x-y < x) by omega.
 destruct (Hrec (x-y) Hxy y Hy) as [z Hz]. (* ie: let z = div (x-y) y *)
 exists (S z); simpl in *; omega. (* ie:  z+1 fits as (div x y) *)
 (* second case: x<y *)
 exists 0; omega.
Defined.
```

_______________

[4] See for instance Chap. 1 of [8] for more details on this topic.

We use `lt_wf`, which states that < is well-founded on natural numbers. When combined with `well_founded_induction`, this allows us to perform recursive calls at will on any strictly smaller numbers. Doing such a recursive call can be quite cumbersome: for calling `Hrec` on `x-y`, we need to have already built a proof `Hxy` stating that `x-y < x`. Without additional help such as comments, it is also very tedious to keep track on the algorithm used in such a proof. Fortunately, extraction can do it for us:

```
let rec div x y =
  if le_lt_dec y x then S (div (minus x y) y) else O
```

## 2.4  A Division by General Recursion with the Russell Framework

The function-as-proof paradigm of the last section can be used on a relatively large scale, see for instance `union` and the few other non-structural operations on well-balanced trees in early versions of module `FSetAVL` in the Standard Library. But such Coq functions are hardly readable and maintainable, consume lots of resources during their definitions and in practice almost always fail to compute in Coq.

Recent versions of Coq include Russell, a framework due to M. Sozeau [10] that greatly eases the design of general recursive and/or dependently-typed functions. With this framework, bodies of functions can be written without being bothered by proof parts or by structural constraints. Simply, such definitions are fully accepted by Coq only when some corresponding proof obligations have been proved later on. For instance:

```
Definition id (n:nat) := n.

Program Fixpoint div (x:nat)(y:nat|y≠0) { measure id x }
  : { z | z*y ≤ x < (S z)*y }
  := if y ≤? x then S (div (x-y) y) else 0.

Next Obligation. (* Measure decreases on recursive call : x-y < x *)
 unfold id; simpl; omega.
Qed.

Next Obligation. (* Post-condition enforcement : z*y ≤ x < (S z)*y *)
 destruct_call div; simpl in *; omega.
Qed.
```

After this definition and the proofs of corresponding obligations, a Coq object `div` is added to the environment, mixing the pure algorithm and the logical obligations. This object is similar to the dependently-typed `div` of the previous section, and its extraction produces the very same Ocaml code.

Russell framework can be seen as a sort of anti-extraction, in the spirit of C. Parent's earlier works [9]. Even if it is still considered as experimental, it is already quite usable. For instance, we have a version of `FSetAVL` where the aforementioned non-structural operations on well-balanced trees are written and proved using Russell.

## 2.5   A Division by General Recursion with the Function Framework

An alternative framework can also be used to define our `div` function: Function, due to J. Forest and alii [4]. It is similar to Russell to some extent: algorithms can be written in a natural way, while proof obligations may have to be solved afterwards. Here, as for Russell, these proof obligations are trivial:

```
Function div (x y:nat)(Hy:y≠0) { measure id x } : nat :=
 if y ≤? x then S (div (x-y) y Hy) else 0.
Proof.
 intros; unfold id; omega.
Defined.
```

Moreover, as for Russell, this framework builds complex internal Coq objects, and extraction of these objects produces back precisely the expected code. But unlike Russell, Function is not meant to manipulate dependent types: in particular the y≠0 pre-condition is possible here only since it is passed unmodified to the recursive call. On the contrary, Function focuses on the ease of reasoning upon functions defined with it, see for instance the `functional induction` tactics, allowing to prove separately properties of `div` that would have been post-conditions with Russell. Once again, the sensitive operations on well-balanced trees have be successfully tried and defined using Function.

## 3   Examples Beyond ML Type System

All our experiments on defining and extracting a division algorithm lead to legitimate Ocaml (or Haskell) code. But the type system of Coq allows us to build objects that have no counterparts in Ocaml nor Haskell. In this case, the type-checkers of these systems are locally bypassed by unsafe type casts (`Obj.magic` or `unsafeCoerce`). These unsafe type casts are now automatically inserted by the extraction in the produced code. We present now two of the various situations where such type casts are required.

### 3.1   Functions of Variable Arity

In Coq, a type may depend on an object such as a number. This allows us to write the type `nArrow` of n-ary functions (over `nat`), such that `nArrow 0 = nat` and `nArrow 1 = nat → nat` and so on.

```
Fixpoint nArrow n : Set := match n with
  | 0 => nat
  | S n => nat → nArrow n
end.
```

Furthermore, we can write a function `nSum` whose first parameter determines the number of subsequent parameters this function will accept (and sum together):

```
Fixpoint nSum n : nArrow (S n) :=
  match n return nArrow (S n) with
    | O => fun a => a
    | S m => fun a b => nSum m (a+b)
  end.
```

```
Eval compute in (nSum 2) 3 8 5.
```

The example (nSum 2) expects (S 2) = 3 arguments and computes here
3+8+5=16. Without much of a surprise nSum cannot be typecheked in ML,
so unsafe type casts are inserted during extraction:

```
let rec nSum n x =
  match n with
    | O -> Obj.magic x
    | S m -> Obj.magic (fun b -> nSum m (plus x b))
```

## 3.2   Existential Structures

Another situation is quite common in developments on algebra: records can
be used in Coq to define structures characterized by the existence of various
elements and operations upon a certain type, with possibly some constraints on
these elements and operations. For instance, let's define a structure of monoid,
and show that (nat,0,+) is indeed a monoid:

```
Record aMonoid : Type :=
  { dom : Type;
    zero : dom;
    op : dom → dom → dom;
    assoc : ∀x y z:dom, op x (op y z) = op (op x y) z;
    zerol : ∀x:dom, op zero x = x;
    zeror : ∀x:dom, op x zero = x }.
```

```
Definition natMonoid :=
 Build_aMonoid nat 0 plus plus_assoc plus_0_l plus_0_r.
```

Proofs concerning monoids can then be done in a generic way upon an abstract
object of type aMonoid, and be applied to concrete monoids such as natMonoid.
This kind of approach is heavily used in CoRN development at Nijmegen. For
the point of view of extraction, this aMonoid type hides from the outside the
type placed in its dom field. Such dependency is currently translated to unsafe
cast by extraction:

```
let natMonoid =
  { zero = (Obj.magic 0); op = (Obj.magic plus) }
```

In the future, it might be possible to exploit recent and/or planned extensions
of Haskell and Ocaml type-checkers to allow a nicer extraction of this example.
Considering Haskell Type Classes and/or Ocaml's objects might also help.

## 4   Key Features of Extraction

Let us summarize now the current status of Coq extraction. The theoretical extraction function described in [7] is still relevant and used as the core of the extraction system. This function collapses (but cannot completely remove) both logical parts (living in sort `Prop`) and types. A complete removal would induce dangerous changes in the evaluation of terms, and can even lead to errors or non-termination in some situations. Terms extracted by this theoretical function are untyped $\lambda$-terms with inductive constructions, they cannot be accepted by Coq in general, nor by ML-like languages. Two separate studies of correctness have been done for this theoretical phase.

The correctness of this theoretical phase is justified in several steps. First, we prove that the reduction of an extracted term is related to the reduction of the initial term in a bisimulation-alike manner (see [7] or Sect. 2.3 of [8]. Since this first approach is really syntactic and cannot cope for instance with the presence of axioms, we then started a semantical study based on realizability (see Sect. 2.4 of [8]). Finally, differences between theoretical reduction rules and the situation in real languages have been investigated, especially in the case of Haskell (see Sect. 2.6 of [8]).

Even if the actual implementation of the extraction mechanism is based on this theoretical study, it also integrates several additional features. First, the untyped $\lambda$-terms coming from the theoretical phase are translated to Ocaml, Haskell or Scheme syntax. In addition, several simplifications and optimizations are performed on extracted terms, in order to compensate the frequent awkward aspect of terms due to the incomplete pruning of logical parts. Indeed, complete removal of proof parts is often unsafe. Consider for instance a partial application of the `div` function of section 2.5, such as `div 0 0 : 0≠0→nat`. This partial application is quite legal in Coq, even if it does not produce much, being blocked by the need of an impossible proof of $0 \neq 0$. On the opposite, an extraction that would brutally remove all proof parts would produce `div 0 0 : nat` for this exemple, leading to an infinite computation. The answer of our theoretical model of extraction is to be very conservative and produce anonymous abstractions corresponding to all logical preconditions such as this `Hy:y≠0`. The presence of these anonymous abstractions permits a simple and safe translation of all terms, including partial applications. At the same time, dangerous partial applications are quite rare, so our actual implementation favors the removal of these anonymous abstractions, at least in head position of extracted functions, leading here to the expected `div` of type `nat→nat→nat`, whereas a special treatment is done for corresponding partial applications: any occurrences of `div 0 0` would become `fun _ -> div 0 0`, preventing the start of an infinite loop during execution.

Moreover, the extraction embeds an type-checker based on [5] whose purpose is to identify locations of ML type errors in extracted code. Unsafe type cast `Obj.magic` or `unsafeCoerce` are then automatically inserted at these locations. This type-checking is done accordingly to a notion of approximation of Coq types into ML ones (see Chap. 3 of [8]). In addition, Coq modules and functors are supported by the Ocaml extraction, while coinductive types can be extracted into Ocaml, Haskell or Scheme.

## 5   Some Significant Coq Developments Using Extraction

A list of user contributions related to extraction can be found at `http://coq.inria.fr/contribs/extraction-eng.html`. Let us highlight some of them, and also mention some developments not (yet?) in this list.

- **CoRN:** This development done in Nijmegen contains in particular a constructive proof of the fundamental theorem of algebra. But all attempts made in order to compute approximations of polynomial roots by extraction have been unsuccessful up to now [2]. This example illustrates how a large, stratified, mathematically-oriented development with a peculiar notion of logical/informative distinction can lead to a nightmare in term of extracted code efficiency and readability.
- **Tait:** This proof of strong normalization of simply typed $\lambda$-calculus produces after extraction a term interpretor [1]. This study with H. Schwichtenberg et alii has allowed us to compare several proof assistants and their respective extractions. In particular Minlog turned out to allow a finer control of what was eliminated or kept during extraction, while Coq `Prop/Set` distinction was rather rigid. At the same time, Coq features concerning proof management were quite helpful, and the extracted code was decent, even if not as nice as the one obtained via Minlog.
- **FSets:** Started with J.C. Filliâtre some years ago [3], this certification of Ocaml finite set and map libraries is now included in the Coq Standard Library. This example has allowed us to investigate a surprisingly wide range of questions, in particular concerning specifications and implementations via Coq modules, or concerning the best style for expressing delicate algorithms (tactics or Fixpoint or `Russell` or `Function`). It has been one of the first large example to benefit from extraction of modules and functors.
- **CompCert:** X. Leroy and alii have certified in Coq a compiler from C (with minor restrictions) to powerpc assembly [6]. While this development is quite impressive, its extraction is rather straightforward, since Coq functions have been written in a direct, structural way. The compiler obtained by extraction is performing quite well.
- **Fingertrees:** In [10], M. Sozeau experiments with his `Russell` framework. The fingertrees structure, relying heavily on dependent types, is a good test-case for both this framework and the extraction. In particular, the code obtained by extraction contains several unsafe type casts, its aspect could be improved but at least it can be executed and is reasonably efficient.

## 6   Conclusion and Future Works

Coq extraction is hence a rich framework allowing to obtain certified programs expressed in Ocaml, Haskell or Scheme out of Coq developments. Even if some details can still be improved, it is already quite mature, as suggested by the variety of examples mentioned above. This framework only seems to reach its limit when one tries to discover algorithm buried in large mathematical development such as CoRN, or when one seeks a fine control a la Minlog on the elimination performed by extraction. Most of the time, the `Prop/Set` distinction, which is a rather simple type-based elimination criterion, is quite efficient at producing reasonable extracted terms with little guidance by the user. Moreover, new tools such as `Russel` or `Function` now allow to easily define general recursive functions in Coq, hence allowing a wider audience to play with extraction of non-trivial Coq objects.

The correctness of this extraction framework currently rely on the theoretical studies made in [7, 8]. The next perspective is to obtain a mechanically-checked guarantee of this correctness. Work on this topic has already started with a student, S. Glondu. Starting from B. Barras CCI-in-Coq development, he has already defined a theoretical extraction in this framework and proved one of the main theorem of [7]. Another interesting approach currently under investigation is to use a Coq-encoded Mini-ML syntax as output of the current uncertified extraction, and have an additional mechanism try to build a proof of semantic preservation for each run of this extraction. Such extracted terms expressed in Mini-ML could then be fed to the certified ML compiler which is currently being built in the CompCert project of X. Leroy.

Some additional work can also be done concerning the typing of extracted code. For instance, thanks to advanced typing aspects of Haskell and/or Ocaml, examples such as the existential structure `aMonoid` may be typed some day without unsafe type casts. This would help getting some sensible program out of CoRN, which make extensive use of such structures. Manual experiments seem to show that Ocaml object-oriented features may help in this prospect. At the same time, some preliminary work has started in Coq in order to propose Haskell-like type classes, adding a support for these type classes to the Haskell extraction may help compensating the lack of module and functor extraction to Haskell.

# References


[1] Berger, U., Berghofer, S., Letouzey, P., Schwichtenberg, H.: Program extraction from normalization proofs. Studia Logica 82 (2005); Special issue
[2] Cruz-Filipe, L., Letouzey, P.: A Large-Scale Experiment in Executing Extracted Programs. In: 12th Symposium on the Integration of Symbolic Computation and Mechanized Reasoning, Calculemus 2005 (2005)
[3] Filliâtre, J.-C., Letouzey, P.: Functors for Proofs and Programs. In: Schmidt, D. (ed.) ESOP 2004. LNCS, vol. 2986. Springer, Heidelberg (2004)
[4] Pichardie, D., Barthe, G., Forest, J., Rusu, V.: Defining and reasoning about recursive functions: a practical tool for the coq proof assistant. In: Hagiya, M., Wadler, P. (eds.) FLOPS 2006. LNCS, vol. 3945. Springer, Heidelberg (2006)
[5] Lee, O., Yi, K.: Proofs about a folklore let-polymorphic type inference algorithm. ACM Transactions on Programming Languages and Systems 20(4), 707–723 (1998)

[6] Leroy, X.: Formal certification of a compiler back-end, or: programming a compiler with a proof assistant. In: 33rd symposium Principles of Programming Languages, pp. 42–54. ACM Press, New York (2006)

[7] Letouzey, P.: A New Extraction for Coq. In: Geuvers, H., Wiedijk, F. (eds.) TYPES 2002. LNCS, vol. 2646. Springer, Heidelberg (2003)

[8] Letouzey, P.: Programmation fonctionnelle certifiée – L'extraction de programmes dans l'assistant Coq. PhD thesis, Université Paris-Sud (July 2004), `http://www.pps.jussieu.fr/ letouzey/download/these_letouzey_English.ps.gz`

[9] Parent, C.: Synthèse de preuves de programmes dans le Calcul des Constructions Inductives. thèse d'université, École Normale Supérieure de Lyon (January 1995)

[10] Sozeau, M.: Program-ing Finger Trees in Coq. In: ICFP 2007, pp. 13–24. ACM Press (2007)

# Joining to High Degrees*

Jiang Liu and Guohua Wu

School of Physical and Mathematical Sciences
Nanyang Technological University, Singapore 639798

**Abstract.** Cholak, Groszek and Slaman proved in [1] that there is a nonzero computably enumerable (c.e.) degree cupping every low c.e. degree to a low c.e. degree. In the same paper, they pointed out that every nonzero c.e. degree can cup a $low_2$ c.e. degree to a $nonlow_2$ degree. In [2], Jockusch, Li and Yang improved the latter result by showing that every nonzero c.e. degree $c$ is cuppable to a high c.e. degree by a $low_2$ c.e. degree $b$. It is natural to ask in which subclass of $low_2$ c.e. degrees $b$ in [2] can be located. Wu proved [6] that $b$ can be cappable. We prove in this paper that $b$ in Jockusch, Li and Yang's result can be noncuppable, improving both Jockusch, Li and Yang, and Wu's results.

## 1  Introduction

Cholak, Groszek and Slaman proved in [1] the existence of almost deep degrees. That is, there is a nonzero computably enumerable (c.e.) degree cupping every low c.e. degree to a low c.e. degree. In the paper, they pointed out that every nonzero c.e. degree can cup a $low_2$ c.e. degree to a $nonlow_2$ c.e. degree. In [2], Jockusch, Li and Yang improved the latter result by showing that every nonzero c.e. degree $c$ is cuppable to a high c.e. degree by a $low_2$ c.e. degree $b$. It is natural to ask in which subclass of $low_2$ c.e. degrees $b$ in [2] can be located. The existence of almost deep degrees says that the class of low c.e. degrees cannot be a candidate of such a subclass. In [6], Wu proved that $b$ in [2] can be $low_2$ and cappable. It is well-known that the cappable degrees form a definable ideal of the computably enumerable degrees, which is denoted as $M$. We ask whether $b$ in [2] can be further restricted to a subideal of $M$ — the ideal consisting of all noncuppable degrees. In this paper, we give a positive answer to this question.

**Theorem 1.** *Given a nonzero c.e. degree $c$, there is a noncuppable $low_2$ degree $b$ such that $c \vee b$ is high.*

Note that if $c$ is noncuppale, the $c \vee b$ is a high noncuppable degree, whose existence was first proved by Harrington (see Miller [4]). We actually prove that following theorem, and Theorem 1 follows from it immediately, as every nonzero c.e. degree bounds a nonzero cappable degree.

**Theorem 2.** *If $c$ is a nonzero cappable degree, then there is a $low_2$ noncuppable degree $b$ such that $c \vee b$ is high.*

---

* Wu is partially supported by a research grant No. RG58/06 (M52110023.710079) from NTU.

A. Beckmann, C. Dimitracopoulos, and B. Löwe (Eds.): CiE 2008, LNCS 5028, pp. 370–378, 2008.

We note that in [2], Jockusch, Li and Yang considered whether $\mathbf{c} > \mathbf{0}$ is cappable or not. If $\mathbf{c}$ is cappable, then they proved that there is a $\mathrm{low}_2$ degree $\mathbf{b}$ such that $\mathbf{c} \vee \mathbf{b}$ is high. If $\mathbf{c}$ is noncappable, then as pointed out in [2], by a well-known fact of Ambos-Spies, Jockusch, Shore and Soare, $\mathbf{c}$ cups a low c.e. degree to $\mathbf{0}'$. Our Theorem 1 says that no matter what $\mathbf{c}$ is, if it is incomplete, then $\mathbf{c} \vee \mathbf{b}$ is always incomplete.

We present in this paper the basic idea of how to prove Theorem 2. Our notation and terminology are standard and generally follow Soare [5]. A number is referred to as big in the construction if it is the least natural number (in an effective way) bigger than any number mentioned so far.

## 2  Requirements and Strategies

Let $\mathbf{c} > \mathbf{0}$ be a given cappable degree and $C$ be a fixed c.e. set in $\mathbf{c}$. We will construct a $\mathrm{low}_2$ noncuppable set $B$, a c.e. set $F$ (an auxiliary set) and partial computable functionals $\Gamma$ and $\Delta_e$ for all $e \in \omega$ satisfying the following requirements:

$$\mathcal{N}_e\colon \Phi_e^{B,W_e} = K \oplus F \Longrightarrow \exists \Delta_e (K = \Delta_e^{W_e}),$$
$$\mathcal{P}_e\colon Tot^B(e) = \lim_{x \to \infty} \Gamma^{B,C}(e,x),$$

where $\{(\Phi_e, W_e) : e \in \omega\}$ is an effective enumeration of $\{(\Phi_i, W_j) : i, j \in \omega\}$, where $\Phi_i$ is a partial computable functional and $W_j$ is a c.e. set. $Tot^B = \{e : \Phi_e^B \text{ is total}\}$ is a $\Pi_2^B$-complete set.

By the $\mathcal{N}$ requirements, $B$ is noncuppable. By the $\mathcal{P}$ requirements and Shoenfield Limit Lemma,

$$Tot^B \leq_T (B \oplus C)' \leq_T \emptyset''.$$

Therefore, $B'' \equiv_T (B \oplus C)' \equiv_T \emptyset''$, $B$ is $\mathrm{low}_2$ and $B \oplus C$ is high.

### 2.1  An $\mathcal{N}$ Strategy

An $\mathcal{N}_e$-strategy, $\beta$ say, is devoted to the construction of a partial computable functional $\Delta_\beta$ such that if $\Phi_e^{B,W_e} = K \oplus F$, then $K = \Delta_\beta^{W_e}$. As usual, we have the length of agreement function between $\Phi_e^{B,W_e}$ and $K \oplus F$ as follows:

- $l(\beta,s) = \max\{x < s : \forall y < x[\Phi_e^{B,W_e}(y)[s] = (K \oplus F)(y)[s]]\}$,
- $m(\beta,s) = \max\{l(\beta,t) : t < s \text{ and } \mathrm{t} \text{ is a } \beta\text{-stage}\}$.

Say a stage $s$ is a $\beta$-expansionary if $s = 0$ or $s$ is a $\beta$-stage with $l(\beta,s) > m(\beta,s)$. We only define $\Delta_\beta$ at $\beta$-expansionary stages. That is, if $s$ is $\beta$-expansionary, and $\Delta_\beta^{W_e}(x)[s]$ is not defined with $2x + 1 < l(\beta,s)$, then define $\Delta_\beta^{W_e}(x) = K(x)[s]$ with use $s$, which is bigger than both $\varphi_e(2x)[s]$ and $\varphi_e(2x + 1)[s]$. After stage $s$, $\Delta_\beta^{W_e}(x)$ can be undefined only when $W_e$ changes below $\varphi_e(2x + 1)$. So, in case that $x$ enters $K$ later, we should force $W_e$ to change below $\varphi_e(2x + 1)$ so that we can redefine $\Delta_\beta^{W_e}(x)$ as $K(x)$ afterwards. If $W_e$ does not change below $\varphi_e(2x + 1)$, then $K \oplus F$ and $\Phi_e^{A,W_e}$ will differ at $2x + 1$, and $\mathcal{N}_e$ is satisfied vacuously.

$\beta$ has two possible outcomes: infinitary outcome $i$ and finitary outcome $f$.

Problem arises when numbers are enumerated into $B$ by the $\mathcal{P}$-strategies (to make $B \vee C$ high) below the infinitary outcome $i$ of $\beta$. Suppose that at stage $s$, $\Delta_\beta^{W_e}(x)[s]$ is defined and a number $n$ less than $\varphi_e(2x)[s]$ is enumerated into $B$, and such an enumeration can lift the use $\varphi_e(2x)$ up to a bigger number. Now suppose that $x$ enters $K$, before the next $\beta$-expansionary stage $s'$ say. Then

$$\Phi_e^{B,W_e}(2x)[s'] = 1 = (K \oplus F)(2x)[s'].$$

However, as $W_e$ has no changes below $\varphi_e(2x)[s]$, $\Delta_\beta^{W_e}(x)[s']$ is defined as $\Delta_\beta^{W_e}(x)[s]$, and is equal to 0, so $\Delta_\beta^{W_e}$ is not correct at $x$.

To avoid such a scenario, when we want to put a number, $n$ say, into $B$, at stage $s$, where $n$ is selected at stage $s_0 < s$, we want to ensure that no $\Delta_\beta^{W_e}(x)$ defined at stage $s$ is actually defined after stage $s_0$. Thus, if $\Delta_\beta^{W_e}(x)[s]$ has definition at stage $s$, then it is defined before stage $s_0$, and the enumeration of $n$ into $B$ will not change the computation $\Phi_e^{B,W_e}(2x)$, as $n$ is bigger than the corresponding use.

Now we explain the idea of ensuring that all $\Delta_\beta^{W_e}(x)$ having definition at stage $s$ are actually defined before stage $s_0$. When we choose $n$ at stage $s_0$, we also choose an auxiliary number, $a_\beta^n$ big, and we think that the next $\beta$-expansionary stage, $t$ say, should be a stage with $l(\beta, t) > m(\beta, t)$ and also $l(\beta, t) > 2a_\beta^n + 1$ (we extend the definition of $\Delta_\beta$ at this stage). In this way, the construction is delayed a little bit, but if there are infinitely many $\beta$-expansionary stages, then it makes no difference.

Assume that $\Delta_\beta^{W_e}(x)$ is defined at stage $t$, but before stage $s$ (remember that we want to put $n$ into $B$ at stage $s$), then we first put $a_\beta^n$ into $F$. There are two cases. The first case is that $W_e$ does not change below $\varphi_e(2a_\beta^n + 1)$, then $\Phi_e^{B,W_e}(2a_\beta^n + 1) = 0$ and $(K \oplus F)(2a_\beta^n + 1) = 1$, and $\beta$ is satisfied vacuously, and of course, there are no more $\beta$-expansionary stages. The other case is that a new $\beta$-expansionary stage appears, which means that $W_e$ does change below $\varphi_e(2a_\beta^n + 1)$. This $W_e$-change undefines $\Delta_\beta^{W_e}(x)$, which are defined after stage $t$. As no $\Delta_\beta^{W_e}(x)$ is defined between stages $s_0$ and $t$, at stage $s' > s$, the next expansionary stage, no $\Delta_\beta^{W_e}(x)$ is defined between stages $s_0$ and $s'$, and as pointed above, we can now put $n$ into $B$ as wanted. Definitely, this enumeration keeps the definition of $\Delta_\beta^{W_e}(x)$ correct.

We now consider a general situation. On the priority tree, when we want to put a number, a $\gamma$-use, into $B$ at a $\mathcal{P}$-strategy, $\alpha$ say, we need to make sure that $\alpha$ works consistently with all the $\mathcal{N}$-strategies with priority higher than $\alpha$. Without loss of generality, we assume that $\beta_0^\frown i \subseteq \beta_1^\frown i \subseteq \cdots \subseteq \beta_n^\frown i \subseteq \alpha$ are the $\mathcal{N}$-strategies with priority higher than $\alpha$ and outcome $i$. Then, when $\alpha$ defines a $\gamma$-use, at stage $s_0$, say, it also defines a sequence of auxiliary numbers $z_0, z_1, \cdots, z_n$. Now, for each $i \leq n$, say that a stage $s$ is $\beta_i$-expansionary only when the length of agreement between $\Phi_{\beta_i}^{B,W_{\beta_i}}$ and $K \oplus F$ is greater than $2z_i + 1$. When $\alpha$ wants to put $\gamma(e(\alpha), y)$ into $B$, $\alpha$ puts $z_n$ into $F$ first, and waits for the next $\beta_n$-expansionary stage. We create a link between $\alpha$ and $\beta_n$, and at the next $\beta_n$-expansionary stage, we go to $\alpha$ via this link, and do further actions. If there is no such a $\beta_n$ expansionary stage, then $\beta_n$ has finitary outcome, and we do not need to satisfy $\alpha$ at all. Otherwise, at the next $\beta_n$-expansionary stage, we cancel the link between $\alpha$ and $\beta_n$, put $z_{n-1}$ into $F$, and create a link between $\alpha$ and $\beta_{n-1}$. Again, we wait for the next $\beta_{n-1}$-expansionary stage, and so on. Such a process can be iterated at most $n + 1$ many times, each of which will have the corresponding $W_{\beta_i}$-changes on some small numbers, undefining those $\Delta_{\beta_i}^{W_{\beta_i}}$ defined after stage $s_0$. So if eventually we cancel the link between $\alpha$ and $\beta_0$, we have actually forced that all of the $\Delta_{\beta_i}^{W_{\beta_i}}$ defined after stage $s_0$ are undefined, and now, we can enumerate $\gamma(e(\alpha), y)$ into $B$, and this enumeration is consistent with all the $\mathcal{N}$-strategies with higher priority.

## 2.2  A $\mathcal{P}$ Strategy

The basic idea of a $\mathcal{P}_e$ strategy is to approximate $Tot^B(e)$ via $\Gamma^{B,C}(e, -)$ at the limit, where $\Gamma$ is a (global) partial computable functional defined in the whole construction. With this in mind, we need to ensure that $\Gamma^{B,C}$ is totally defined, and that for all $e \in \omega$,

$$\Phi_e^B \text{ is total iff } \lim_{x \to \infty} \Gamma^{B,C}(e, x) = 1.$$

We first consider a single $\mathcal{P}$ strategy. This is a modified version of the gap-cogap argument developed in [2] by Jockusch, Li and Yang.

Let $\alpha$ be a $\mathcal{P}_e$ strategy. Assume that $|\alpha| = e$. For convenience, we write $\Phi_\alpha$ for $\Phi_e$. When we can know from the context that $\alpha$ is a $\mathcal{P}_e$ strategy, we just write $\gamma(e, y)$ for $\gamma(e(\alpha), y)$. In needed for clarity, we will still write $\gamma(e(\alpha), y)$ from time to time.

$\alpha$ will do two jobs simultaneously. $\alpha$'s first job is to define $\Gamma^{B,C}(e, x)$ for almost $x$ to make sure that $\Gamma^{B,C}(e, x)$ has a limit and computes $Tot^B(e)$ correctly. That is, the following equality is guaranteed:

$$Tot^B(e) = \lim_{x \to \infty} \Gamma^{B,C}(e, x).$$

In the construction, whenever $\alpha$ defines $\Gamma^{B,C}(e, x)$ at stage $s$, the $\gamma$-use $\gamma(e, x)$ is defined as a big number. In particular, $\gamma^{B,C}(e, x) > s$. $\Gamma^{B,C}(e, x)$ is undefined automatically if some number less than or equal to $\gamma^{B,C}(e, x)$ is enumerated into $B$. As a consequence, $\gamma(e, x)$ may be lifted when $\Gamma^{B,C}(e, x)$ is redefined later. If $C$ has a change below $\gamma^{B,C}(e, x)$ (it can happen when we get a cogap permission as specified later), unless we explicitly set $\Gamma^{B,C}(e, x)$ to be undefined, $\Gamma^{B,C}(e, x)$ is redefined with same value and the same use (it is not necessary to lift $\gamma^{B,C}(e, x)$ to big number) to ensure that $\Gamma^{B,C}(e, x)$ is defined eventually. Such rules also apply to the following $\Theta_\alpha$-functionals defined later.

As in our construction, a $\mathcal{P}_e$ needs to be consistent with the $\mathcal{N}$-strategies with higher priority, when we define $\gamma(e, x)$, we also select several other big numbers, $z_j$, associated. We do so, because when we want to enumerate $\gamma(e, x)$ into $B$, we will put these numbers into $F$ one by one, as described in the $\mathcal{N}$-strategy, to force changes of $W$ to undefine $\Delta_\beta^W$ defined after stage $s$.

$\alpha$'s second job is to preserve $\Phi_e^B(x)$, to ensure that if $\Gamma^{B,C}(e, x)$ has limit 1, then $\Phi_e^B$ is total. At stage $s$, we define

- $l(\alpha, s) = \max\{x < s : \Phi_\alpha^B(y)[s] \downarrow \text{ for all } y < x\}$,
- $m(\alpha, s) = \max\{0, l(\alpha, t) : t < s \text{ and } t \text{ is an } \alpha\text{-stage}\}$.

Say that $s$ is an $\alpha$-expansionary stage if $s = 0$ or $m(\alpha, s) < l(\alpha, s)$.

If there are only finitely many $\alpha$-expansionary stages, then $\Phi_\alpha^B$ is obviously not total. Thus, $Tot^B(e) = 0$, and $\alpha$ will define $\Gamma^{B,C}(e, x) = 0$ for (almost) all $x \in \omega$, and eventually, we have that $\lim_{x \to \infty} \Gamma^{B,C}(e, x) = 0 = Tot^B(e)$.

Suppose that there are infinitely many $\alpha$-expansionary stages. Then we should ensure that $\Phi_\alpha^B$ is total, and $\Gamma^{B,C}(e, x)$ are defined to be 1 for (almost) all $x \in \omega$ such that $Tot^B(e) = \lim_{x \to \infty} \Gamma^{B,C}(e, x) = 1$. Here comes a direct conflict: to preserve a computation $\Phi_\alpha^B(x)$, we need to preserve $B$ on the use $\varphi_\alpha(x)$, and to change $\Gamma^{B,C}(e, x)$ from 0, defined by a strategy on the right, to 1, we may need to enumerate $\gamma(e, x)$ into $B$ to undefine $\Gamma^{B,C}(e, x)$ first. Fortunately, it is not a fatal conflict, as we can use the $C$-changes to undefine $\Gamma^{B,C}(e, x)$. With this in mind, to preserve a computation $\Phi_\alpha^B(x)$, we introduce the following substrategies, $\mathcal{S}_{\alpha,i}, i \in \omega$, of $\alpha$, to undefine $\Gamma^{B,C}(e, x)$, and to preserve $\Phi_\alpha^B(x)$, if needed.

For the sake of the consistency between defining $\Gamma^{B,C}$ and preserving $\Phi_\alpha^B(x)$, $x \in \omega$, $\alpha$ will construct an auxiliary c.e. set $E_\alpha$ and a partial functional $\Theta_\alpha^{E_\alpha,C}$, which attempts to satisfy the following requirements:

$$\mathcal{S}_{\alpha,i} : \ E_\alpha'(i) = \Theta_\alpha^{E_\alpha,C}(i).$$

An $\mathcal{S}_{\alpha,i}$-strategy works at $\alpha$-expansionary stages. It defines $\Theta_\alpha^{E_\alpha,C}(i)$ with $\Theta_\alpha^{E_\alpha,C}(i) = E_\alpha'(i)$, and if it fails, then it will ensure that $\Phi_\alpha^B$ is total and $\lim_{x \to \infty} \Gamma^{B,C}(e, x) = 1$, and $Tot^B(e) = \lim_{x \to \infty} \Gamma^{B,C}(e, x)$ satisfying $\mathcal{P}_e$. As $C$ is given as a set with cappable degree, not every $\mathcal{S}_{\alpha,i}$ can be satisfied, as otherwise, $E_\alpha$ would be a low set, and $C$ is low-cuppable, which is a contradiction. Therefore, there is a least $i$ such that $\mathcal{S}_{\alpha,i}$ is not satisfied.

Fix $i$. Say that $\mathcal{S}_{\alpha,i}$ is in the $x$-turn (or $x$-turn is in progress) if $\mathcal{S}_{\alpha,i}$ attempts to make $\Phi_\alpha^B(x)$ clear of the $\gamma(e, -)$ uses. Here we say that $\mathcal{S}_{\alpha,i}$ attempts to make $\Phi_\alpha^B(x)$ clear of the $\gamma(e, -)$ uses at stage $s$, we mean that $\mathcal{S}_{\alpha,i}$ sees $\Phi_\alpha^B(x)[s] \downarrow$, and $\mathcal{S}_{\alpha,i}$ wants to prevent the construction of $\Gamma$ from enumerating smaller numbers into $B$ to injure this computation. The basic idea of an $\mathcal{S}_{\alpha,i}$-strategy is to force a $C$-change to lift the $\gamma$-uses to bigger numbers or to undefine $\Theta_\alpha^{E_\alpha,C}(i)$. In the first case, the computation $\Phi_\alpha^B(x)[s]$ is clear of the $\gamma$-uses, and hence $\Phi_\alpha^B(x)$ converges. In the latter case, $\Theta_\alpha^{E_\alpha,C}(i)$ is undefined, and we can make $E_\alpha'(i) = \Theta_\alpha^{E_\alpha,C}(i)$, satisfying $\mathcal{S}_{\alpha,i}$.

$\mathcal{S}_{\alpha,i}$ works as follows:

1. Suppose that the $x$-turn is in progress.

   Wait for an $\alpha$-expansionary stage $s_1 > 0$ with $l(\alpha, s_1) > x$. If $\Theta_\alpha^{E_\alpha,C}(i)$ is undefined at stage $s_1$, then define $\Theta_\alpha^{E_\alpha,C}(i)[s_1]$ as 0 if $\Phi_i^{E_\alpha}(i)[s_1] \uparrow$, and 1 otherwise. In both cases, set use $\theta_\alpha(i)[s_1]$ big.

2. Wait for an $\alpha$-expansionary stage $s_2 > s_1$ with $\Phi_i^{E_\alpha}(i)[s_2] \downarrow$ and $\Theta_\alpha^{E_\alpha,C}(i)[s_2] \downarrow= 0$. We say that we are ready to open a gap at stage $s_2$, and we will put those numbers associated to $\gamma(e, x)$ into $F$ one by one, as described in the $\mathcal{N}$-strategy.

   *When $\gamma(e, x)$ is defined by a strategy $\alpha'$, for the sake of the consistency between the $\mathcal{P}_e$-strategies and the $\mathcal{N}$-strategies, we also select several numbers $z_0^\xi, z_1^\xi, \cdots, z_e^\xi$, where $\xi$ is a $\mathcal{S}_{\beta,j}$-strategy to the left of $\alpha'$, in case when $\gamma(e, x)$ is enumerated into $B$ by $\xi$, $\xi$ needs to enumerate the associated numbers $z_0^\xi, z_1^\xi, \cdots, z_e^\xi$ into $F$ to force* *the corresponding $W$ s to have needed changes. As there are only finitely many such $\xi$-strategies on the left of $\alpha'$, it is fine for us to select these auxiliary numbers. We point out here that $\alpha$, a $\mathcal{P}_e$ strategy wanting to put $\gamma(e, x)$ into $B$, can be the same as $\alpha'$, or on the left of $\alpha'$, as the $\mathcal{P}_e$ strategy on the true path will be responsible for the definition of $\Gamma^{B,C}(e, x)$ for almost all $x$, and hence, it needs to first undefine $\Gamma^{B,C}(e, x)$ defined by those strategies on the right.*

*When $\alpha$ wants to put $\gamma(e, x)$ into $B$, $\alpha$ will do as in the $\mathcal{N}$-strategies to put numbers $z_0^\alpha, z_1^\alpha, \cdots, z_e^\alpha$ into $F$ in reverse order, and creates and cancels the corresponding links. After the last link is cancelled, $\alpha$ puts $\gamma(e, x)$ into $B$. This delays the opening of a gap — we open a gap only when $\gamma(e, x)$ is put into $B$, because before $\gamma(e, x)$ is put into $B$, a $C$-change below $\gamma(e, x)$ can always undefine $\Gamma^{B,C}(e, x)$, and we can protect $\Phi_\alpha^B(x)$ successfully. Obviously, this delay does not affect the $\mathcal{S}_{\alpha,i}$-strategy. So at a stage $s_3 > s_2$ at which the last link between $\alpha$ and the highest $\mathcal{N}$-strategy above $\alpha$ is cancelled, we open a gap. This is a new, crucial feature of our construction.*

*It may happen that $\alpha'$ is on the left of $\alpha$, then we will not allow $\alpha$ to enumerate $\gamma(e, x)$ into $B$, as $\alpha$ has lower priority. If $\alpha$ is on the true path, then only finitely many $\mathcal{S}$-strategies on the left can be accessible during the whole construction, and as a consequence, there are only finitely many $x$ with $\Gamma^{B,C}(e, x)$ defined by these higher priority strategies. This will not affect the limit $\lim_{x \to \infty} \Gamma^{B,C}(e, x)$, which is in the control of $\alpha$.*

3. Let $s_3 > s_2$ be a stage at which the last link between $\alpha$ and the highest $\mathcal{N}$-strategy above $\alpha$ is cancelled. Open a gap as follows:

   - Set $r(\alpha, i) = 0$. Define $f_{\alpha,i}^x(y) = C_{s_3}(y)$ for those $y < \theta_\alpha(i)[s_2]$ with $f_{\alpha,i}^x(y)$ not defined yet.

     (*Here, $f_{\alpha,i}^x$ is an auxiliary partial computable function defined during the $x$-turn, threatening the incomputability of $C$ if the $x$-turn do not terminate.*)
   - Enumerate $\gamma(e, x)$ into $B$.

4. Wait for the least $\alpha$-expansionary stage $s_4 > s_3$. $\alpha$ closes the gap which is opened at stage $s_3$. There are two cases, depending on whether $C$ has a wanted change.

   - (Successful close) $C$ has a change below $\theta_\alpha(i)[s_2]+1$ between stages $s_2$ and $s_4$. Then redefine $\Theta_\alpha^{E_\alpha,C}(i) = 1$. Note that this $C$-change undefines all $\Theta_\alpha^{E_\alpha,C}(j)$, $j \geq i$. Declare that the gap is closed successfully and that $\mathcal{S}_{\alpha,i}$ is satisfied. ($\mathcal{S}_{\alpha,i}$ succeeds in defining $\Theta_\alpha^{E_\alpha,C}(i) = E_\alpha'(i)$ because the computation $\Phi_i^{E_\alpha}(i)[s_2]$ is preserved from now on and hence $\Phi_i^{E_\alpha}(i) \downarrow$.)
   - (Unsuccessful close) $C$ has no change below $\theta_\alpha(i)[s_2] + 1$ between stages $s_2$ and $s_4$. Then define $r(\alpha, i) = s_4$ and enumerate $\theta_\alpha(i)[s_2]$ into $E_\alpha$. This enumeration undefines all $\Theta_\alpha^{E_\alpha,C}(j)$, $j \geq i$. Declare that the gap is closed unsuccessfully. Go to (1) and simultaneously, wait for a $C$-change on a number in $\text{dom}(f_{\alpha,i}^x)$, till the stage when the next gap is open. If so, go to (5).

     (*Here, we set $r(\alpha, i) = s_4$ as a restraint to preserve the computation $\Phi_e^B(x)$ the same as $\Phi_e^B(x)[s_4]$ until this restraint is cancelled in (3), when a new gap is open, in which case, $\Phi_e^B(x)$ may be injured by the enumeration of the $\gamma(e, x)$, or when (5) is reached. If before we open a new gap, $C$ does have a change below $\theta_\alpha(i)[s_2] + 1$, as this $C$ change lifts the $\gamma$-uses to big numbers, $\mathcal{S}_{\alpha,i}$ succeeds in preserving the computation $\Phi_e^B(x)$ as $\Phi_e^B(x)[s_4]$.*)

In any case, find the least $y$ with $\Gamma^{B,C}(e, y)$ undefined, and define $\Gamma^{B,C}(e, y) = 1$ with use $\gamma^{B,C}(e, y)$ big.

5. Define $r(\alpha, i) = 0$. Declare that the computation $\Phi_e^B(x)$ is preserved by $\mathcal{S}_{\alpha,i}$. Terminate the $x$-turn (and hence stop defining $f_{\alpha,i}^x$, as it is not correct anymore), and start the $x + 1$-turn (to preserve $\Phi_e^B(x + 1)$). We call such a $C$-change a cogap permission, because it happens inside a cogap. The action performed at (5) is called a cogap permission action.

Say that $\mathcal{S}_{\alpha,i}$ requires attention at an $\alpha$-expansionary stage $s$ if one of the following holds:

1. $\mathcal{S}_{\alpha,i}$ is inside a gap, and it is ready to close a gap (at step 4). $\alpha$ will act by closing this gap, no matter whether it is a successful close or not.
2. $\mathcal{S}_{\alpha,i}$ is inside a cogap and $\Theta_\alpha^{E_\alpha,C}(i)[s] \uparrow$ (we will define $\Theta_\alpha^{E_\alpha,C}(i)$). $\alpha$ will act by defining $\Theta_\alpha^{E_\alpha,C}(i) = 1$ if $\Phi_i^{E_\alpha}(i)[s] \downarrow$, or 0 if $\Phi_i^{E_\alpha}(i)[s] \uparrow$.
3. $\mathcal{S}_{\alpha,i}$ is inside a cogap and is ready to open a gap (at step 2, and will put the associated numbers into $F$ one by one). $\alpha$ will act by enumerating the numbers $z_j^\alpha, j \le e$ into $F$ one by one, as described in the $\mathcal{N}$-strategy.
4. $\mathcal{S}_{\alpha,i}$ is at a stage at which a last link between $\alpha$ and an $\mathcal{N}$-strategy above $\alpha$ with the highest priority is cancelled (at step 3, and will open a new gap). $\alpha$ will act by enumerating $\gamma(e, x)$ into $B$ and open a gap.
5. $\mathcal{S}_{\alpha,i}$ is at a cogap and $C$ changes below $f_{\alpha,i}^x$ (step 5 is reached). We will start the $x + 1$-turn.

$\mathcal{S}_{\alpha,i}$ has three possible outcomes:

$s_i$: $\mathcal{S}_{\alpha,i}$ waits at step 1 or closes a gap successfully. In the latter case, $\mathcal{S}_{\alpha,i}$ succeeds in defining $\Theta_\alpha^{E_\alpha,C}(i) = E_\alpha'(i)$. $\mathcal{S}_{\alpha,i}$ is satisfied. We will consider the definition of $\Theta_\alpha^{E_\alpha,C}(i + 1)$, so we will not put outcome on the construction tree. In the former case, there are only finitely many $\alpha$-expansionary stages and $\alpha$ is satisfied, because $\Phi_\alpha^B$ is not total. In this case, we have an outcome $f$, under which $\Gamma^{B,C}(e, x)$ will be defined as 0 for almost all $x$. $\mathcal{S}_{\alpha,i}$ is again satisfied, because $\lim_{x \to \infty} \Gamma^{B,C}(e, x) = 0$, and equals to $Tot^B(e)$.

$g_i$: $\alpha$ opens infinitely many gaps and closes these gaps unsuccessfully, and for each $x \in \omega$, the $x$-turn stops by reaching step 5 (a wanted $C$-change appears). As a consequence, $\mathcal{S}_{\alpha,i}$ ensures that $\Phi_e^B(x)$ converges. So $\Phi_e^B$ is total. As we always define $\Gamma^{B,C}(e, x) = 1$ for $x \in \omega$ when we open a gap at step 3, $\lim_{x \to \infty} \Gamma^{B,C}(e, x) = 1$. Thus, $Tot^B(e) = \lim_{x \to \infty} \Gamma^{B,C}(e, x)$. Thus, $\mathcal{P}_e$ is satisfied;

$f_i$: After some sufficiently large stage, $\mathcal{S}_{\alpha,i}$ has an $x$-turn opening and closing (unsuccessfully) infinitely many gaps and never reaching step 5. Then $\Theta_\alpha^{E_\alpha,C}(i)$, and $\Phi_\alpha^B(x)$ diverge, but $f_{\alpha,i}^x$ is totally defined and computes $C$ correctly because $C$ does not change in a gap (otherwise we can undefine $\Theta_\alpha^{E_\alpha,C}(i)$ and preserve $\Phi_i^{E_\alpha}(i)$) nor in a cogap (otherwise, the $x$-turn is terminated and the $x + 1$-turn would be started). So $C$ is computable, contradicting our assumption on $C$. In the construction, as $f_i$ will not happen, we do not put this outcome on the priority tree.

In the construction, we do not put the $\mathcal{S}_{\alpha,i}$-substrategy on the priority tree. We just attach the outcome of $\mathcal{S}_{\alpha,i}$ to $\alpha$. As explained above, for each $i \in \omega$, $\mathcal{S}_{\alpha,i}$ has exactly one outcome $g_i$ listed on the priority tree. As a consequence, $\alpha$ has $\omega + 1$ many outcomes,

$$g_0 <_L g_1 <_L \cdots <_L g_i <_L \cdots <_L f,$$

where $f$ denotes the case that there only finitely many $\alpha$-expansionary stages and $g_i$ denotes the case that $\mathcal{S}_{\alpha,i}$ requires attention (and hence receive attentions) infinitely often.

Now, we consider how to make two (and more) $\mathcal{P}$-strategies consistent. Note that at step 5 of the $\mathcal{S}_{\alpha,i}$-strategy, $\alpha$ successfully ensures that the computation $\Phi_\alpha^B(x)$ is clear of the $\gamma(e, -)$ uses, and set $r(\alpha, i) = 0$ when step 5 is reached. It may happen that a computation $\Phi_\alpha^B(x)$ can still be changed by another $\mathcal{P}$-strategy (the enumeration of other $\gamma$-uses) below outcome $g_i$. We will explain how we can deal with such disturbance from lower priority strategies. In the construction, we will ensure that for any $x$, $\Phi_\alpha^B(x)$ can be changed by the strategies below outcome $g_i$ at most finitely often, and finally if we have a new $\alpha$-expansionary stage, at which $\Phi_\alpha^B(x)$ converges, then again, no strategy can change this computation and hence it is preserved forever.

Here is the point. Let $\alpha, \alpha'$ be two $\mathcal{P}$-strategies with $\alpha ^\frown g_i \subseteq \alpha'$. Suppose that $\alpha'$ wants to define $\Gamma^{B,C}(e(\alpha'), y)$ for some $y$, we first set $k^y$ as a big number, and wait for a stage, $s$ say, such that $\mathcal{S}_{\alpha,i}$ has completed the $k^y$-turn, and then define $\Gamma^{B,C}(e(\alpha'), y)$. That is, at stage $s$, all computations, $\Phi_\alpha^B(z)$, $z < k^y$, are clear of the $\gamma(e(\alpha), -)$-uses. Again, it delays the definition of $\Gamma^{B,C}(e(\alpha'), y)$ and we know that such an $s$ exists because $\alpha'$ guesses that $\alpha$ has outcome $g_i$, and hence $\Phi_\alpha^B$ is total. We will call $k^y$ an $\alpha$-bound at $y$. Obviously, $\gamma(e(\alpha'), y)$ is defined as a number bigger than $s$, and hence if it is enumerated into $B$, it will not change the computations $\Phi_\alpha^B(z)$, $z < k^y$. So for a fixed $x < k^y$, after stage $s$, $\Phi_\alpha^B(x)$ can be changed by $\alpha'$ at most finitely often by the enumeration of those $\gamma(e(\alpha'), y')$, $y' < y$, and if there are infinitely many $\alpha$-expansionary stages, then we will finally have a computation $\Phi_\alpha^B(x)$ not injured by $\alpha'$. Also note that by stage $s$, as the $x$-turn has already been completed, $\Phi_\alpha^B(x)$ is clear of the $\gamma(e(\alpha), -)$-uses, whenever $\alpha'$ enumerates a number into $B$, changing a computation $\Phi_\alpha^B(x)$, the corresponding $\gamma(e(\alpha), -)$-uses are also lifted up to even bigger numbers, and once $\Phi_\alpha^B(x)$ settles down, we can make sure keep these $\gamma(e(\alpha), -)$-uses the same. Therefore, $\alpha'$ is consistent with $\alpha$.

Now we consider how $\alpha'$ can work below $\alpha ^\frown g_i$, where $\alpha'$ knows that $\alpha$ will open infinitely many gaps and hence put infinitely numbers into $B$. As usual, when $\alpha$ sees a computation, $\alpha'$ first check whether all $\gamma(e(\alpha), z)$s (for $\Gamma^{B,C}(e(\alpha), z) = 0$) below the associated use of this computation have been enumerated into $B$. If yes, then $\alpha'$ knows that this computation will not be changed by the enumeration of $\alpha$, and will preserve this computation, as described above in the $\mathcal{N}$ and $\mathcal{P}$-strategies. In this case, we say that this computation is believable at $\alpha'$. Otherwise, because $\alpha'$ is below the outcome $g_i$ of $\alpha$, and $\alpha'$ knows that all these $\gamma(e(\alpha), -)$-uses will be put into $B$ sooner or later, $\alpha'$ will just wait until all these uses are bigger than the associated use — wait till a computation is believable at $\alpha'$. Again, it is a kind of delaying, and such a delay will not affect the whole construction at all.

Now, for a fixed $y$, we define $\gamma(e(\alpha'), y)$ only when $\Phi_{\alpha'}^B(y)$ is a believable computation at $\alpha'$. So $\alpha$'s enumeration will not affect the definition of $\Phi_{\alpha'}^B(y)$, and we can argue as usual that eventually, after all these $\gamma(e(\alpha'), y')$ become fixed, $y' < y$, settles down, $\gamma(e(\alpha'), y)$ can be changed only when the corresponding substrategy of $\alpha'$ at step 3, or by a $C$-change at step 5. As $C$ is assumed to be incomputable, step 5 is eventually reached, after which we will not change $\gamma(e(\alpha'), y)$ anymore.

This gives the basic ideas of strategies of satisfying $\mathcal{N}$ and $\mathcal{P}$ strategies, together with the idea of making these strategies work consistently. The full construction and the verification part will appear in [3], where more detail will be included.

# References


1. Cholak, P., Groszek, M., Slaman, T.: An almost deep degree. J. Symbolic Logic 66, 881–901 (2001)
2. Jockusch Jr., C.G., Li, A., Yang, Y.: A join theorem for the computably enumerable degrees. Trans. Amer. Math. Soc. 356, 2557–2568 (2004)
3. Liu, J., Wu, G.: Joining to high degrees via noncuppables (in preparation)
4. Miller, D.: High recursively enumerable degrees and the anti-cupping property. In: Lerman, Schmerl, Soare (eds.) Logic Year 1979-80: University of Connecticut. Lecture Notes in Mathematics, vol. 859, pp. 230–245 (1981)
5. Soare, R.I.: Recursively enumerable sets and degrees. Springer, Heidelberg (1987)
6. Wu, G.: Quasi-complements of the cappable degrees. Math. Log. Quart. 50, 189–201 (2004)

# Factoring Out Intuitionistic Theorems: Continuity Principles and the Uniform Continuity Theorem

Iris Loeb

Department of Mathematics and Statistics, University of Canterbury,
Private Bag 4800, Christchurch, New Zealand
I.Loeb@math.canterbury.ac.nz

**Abstract.** We prove the equivalence between some intuitionistic theorems and the conjunction of a continuity principle and a compactness principle over Bishop's Constructive Mathematics within the programme of Constructive Reverse Mathematics. To clarify our line of thought, we first point out the relation between quasi-equicontinuity, quasi-uniform convergence, and the continuity principle saying that the limit of a convergent sequence of continuous functions is again continuous. Finally, as a spin-off, we conclude that we have found a new, more economic proof of the statement that every convergent sequence of functions on a compact metric space converges uniformly.

**Keywords:** Constructive Mathematics, Reverse Mathematics, Intuitionistic Mathematics, Continuity Principle, Compactness Principle.

## 1 Introduction

Intuitionistic mathematics has two clearly separable traits. Firstly, that all functions from a complete, separable metric space are continuous (a property that is also true in the recursive interpretation) [12], and secondly that all continuous functions on a compact metric space are uniformly continuous (a property that is also classically valid).

This cognitive distinction can be found back in the results obtained within the programme of Constructive Reverse Mathematics [13]. Many theorems have been proved equivalent over Bishop's constructive mathematics (**BISH**, [7]) to either a continuity principle [12] or a compactness principle [3,4,5,6,8,10,14] but not much is known about the interplay between principles of these two classes.

In this paper we will study convergent sequences of functions and we will single out theorems that are equivalent to the *conjunction* of a continuity and a compactness principle over **BISH**[1]. This contributes not only to the programme of Constructive Reverse Mathematics, but also to our understanding of intuitionistic mathematics in itself: One of our results is a new intuitionistic proof of the

---

[1] However, we do not adopt Bishop's terminology: for our purposes it is important to make the distinction between continuity and uniform continuity.

A. Beckmann, C. Dimitracopoulos, and B. Löwe (Eds.): CiE 2008, LNCS 5028, pp. 379–388, 2008.
© Springer-Verlag Berlin Heidelberg 2008

statement that every pointwise convergent sequence of real-valued (continuous) functions on a compact metric space converges uniformly. We will compare the principles used in the new proof with the ones used in [15] and conclude that the new proof is more economic.

Before giving the main results (Sections 3 and 4), we will first take a closer look at the continuity principles and compactness principle that play an important role here (Section 2).

## 2   The Continuity and Compactness Principles

The continuity principles that we consider are the *Continuity Principle for Limits on Compact Spaces* (**CPL$^{cp}$**) and the *Continuity Principle for Compact Spaces* (**CONT$^{cp}$**).

> **CPL$^{cp}$:** Every pointwise convergent sequence of continuous functions from a compact metric space to a metric space has a continuous limit.

> **CONT$^{cp}$:** Every function from a compact metric space to a metric space is continuous.

To our knowledge, these principles have not been examined before, although the principles **CONT$^{c}$** (a continuity principle for complete spaces) and **CONT$^{cs}$** (a continuity principle for complete, separable spaces), which imply **CONT$^{cp}$**, were dealt with in [12]. Remark that it thus follows from that paper that **CONT$^{cp}$** and **CPL$^{cp}$** are true intuitionistically.

Furthermore, we will use the compactness principle *Uniform Continuity Theorem* (**UCT**, [8,10]),

> **UCT:** Every continuous function from a compact metric space to a metric space is uniformly continuous.

Because the principle **CPL$^{cp}$** has not been studied before, we will now show some important equivalents of it that are well-known to hold classically [1,2]. This also connects more clearly the results in this paper with the ones of [10], where we investigated (uniform) equicontinuity and (uniform) convergence.

Let $X$ and $Y$ be metric spaces. A sequence $(f_n)_{n \geq 0}$ of functions from $X$ to $Y$ is called **quasi-equicontinuous** if for each $\epsilon > 0$ and each $x \in X$ we can find $\delta > 0$ such that for each $y \in X$ there exists $n \in \mathbb{N}$ with the property that for all $m \geq n$:

> If $\rho(x, y) < \delta$, then $\rho(f_m(x), f_m(y)) < \epsilon$.

Clearly the notion of quasi-equicontinuity on a compact metric space $X$ is weaker than that of equicontinuity on $X$, as not all terms of a quasi-equicontinuous sequence need be continuous. Even if we require all terms of a quasi-equicontinuous sequence to be continuous, this will not be enough make the sequence equicontinuous, as we show in Theorem 3.

The next theorem shows a connection between converging sequences having a continuous limit and being quasi-equicontinuous. Note that we do not require that the terms of the sequence are continuous.

**Theorem 1.** *Let $X$ and $Y$ be metric spaces, and let $(f_n)_{n \geq 0}$ be a sequence of functions from $X$ to $Y$ that converges to a limit $f$. Then the following statements are equivalent.*

(i)  *The limit function $f$ is continuous on $X$.*
(ii)  *The sequence $(f_n)_{n \geq 1}$ is quasi-equicontinuous.*

*Proof.* Assume (i), let $\epsilon > 0$ and $x \in X$. Determine $\delta > 0$ such that if $\rho(x, y) < \delta$, then $\rho(f(x), f(y)) < \epsilon/3$. Consider $y \in X$ such that $\rho(x, y) < \delta$. We can find $n_0, n_1$ such that $\rho(f_m(x), f(x)) < \epsilon/3$ for all $m > n_0$, and $\rho(f_m(y), f(y)) < \epsilon/3$ for all $m > n_1$. Setting

$$N := \max\{n_0, n_1\},$$

we see that if $m > N$, then

$$\rho(f_m(x), f_m(y)) \leq \rho(f_m(x), f(x)) + \rho(f(x), f(y)) + \rho(f(y), f_m(y))$$
$$< \frac{\epsilon}{3} + \frac{\epsilon}{3} + \frac{\epsilon}{3} = \epsilon.$$

Thus (i) implies (ii).

For the reverse implication, assume (ii), and again let $\epsilon > 0$ and $x \in X$. Determine $\delta > 0$ such that for each $y \in X$ there exists $n$ such that if $\rho(x, y) < \delta$, then $\rho(f_m(x), f_m(y)) < \epsilon/3$ for all $m > n$. Consider such a point $y$ and the corresponding $n$. There exist $n_0, n_1$ such that $\rho(f_m(x), f(x)) < \epsilon/3$ for all $m > n_0$, and $\rho(f_m(y), f(y)) < \epsilon/3$ for all $m > n_1$. Setting

$$N := \max\{n, n_0, n_1\},$$

we see that if $m > N$, then

$$\rho(f(x), f(y)) \leq \rho(f(x), f_m(x)) + \rho(f_m(x), f_m(y)) + \rho(f_m(y), f(y))$$
$$< \frac{\epsilon}{3} + \frac{\epsilon}{3} + \frac{\epsilon}{3} = \epsilon.$$

Thus (ii) implies (i).

Let $X$ and $Y$ be metric spaces, and let $(f)_{n \geq 0}$ be a sequence of functions from $X$ to $Y$ converging pointwise to a limit $f$. We say that the sequence is **quasi-uniformly convergent** if for each $\epsilon > 0$ and each $x \in X$ we can find $\delta > 0$ and $n \in \mathbb{N}$ with the property that for each $y \in X$:

If $\rho(x, y) < \delta$, then $\rho(f_n(y), f(y)) < \epsilon$.

Usually the definition of quasi-uniformly convergence is formulated in the following way (see for example [11]), which we will refer to as **finite-cover quasi-uniform convergence**: Let $X$ and $Y$ be metric spaces, and let $(f)_{n \geq 0}$

be a sequence of functions from $X$ to $Y$ converging pointwise to a limit $f$. We say that the sequence is **finite-cover quasi-uniformly convergent** if for each $\epsilon > 0$, there exists a finite family $\{V_0, V_1, \ldots, V_k\}$ of open sets of $X$ and a finite family $\{n_0, n_1, \ldots, n_k\}$ of natural numbers[2], such that

$$X = V_0 \cup V_1 \cup \ldots \cup V_k, \tag{1}$$

and

$$V_i \subseteq \{x \in X : \rho(f_{n_i}(x), f(x)) < \epsilon\} \qquad (i = 0, 1, \ldots, k). \tag{2}$$

The next theorem shows that the notions of quasi-uniform convergence and finite-cover quasi-uniform convergence coincide on compact spaces in both classical and intuitionistic mathematics.

Let the **Principle of Choice for Compact Spaces** be the following statement:

> For every compact metric space $X$, if $P \subseteq X \times \mathbb{N}$, and for each $x \in X$ there exists $n \in \mathbb{N}$ such that $(x, n) \in P$, then there is a function $f : X \to \mathbb{N}$ such that $(x, f(x)) \in P$ for all $x \in X$.

Note that this principle follows from the Principle of Continuous Choice, and is hence intuitionistically valid. The Axiom of Choice ensures that it is also classically valid.

**Theorem 2.** *Let $X$ be compact metric space, $Y$ a metric space, and $(f_n)_{n \geq 0}$ a sequence of continuous functions from $X$ to $Y$ that converges to a limit $f$. Then the following statements are equivalent under the assumption of the Heine-Borel Covering Lemma and the Principle of Choice for Compact Spaces.*

(i)  *The sequence $(f_n)_{n \geq 0}$ converges quasi-uniformly.*
(ii) *The sequence $(f_n)_{n \geq 0}$ converges finite cover quasi-uniformly.*

*Proof.* Supposing that (i) holds, let $\epsilon > 0$. For every $x \in X$ determine (with the Principle of Choice for Compact Spaces) $m_x, n_x \in \mathbb{N}$ such that for all $y \in X$

$$\rho(x, y) < 2^{-m_x} \Rightarrow \rho(f_{n_x}(y), f(y)) < \epsilon.$$

Define for every $x \in X$

$$V_x := \{y \in X : \rho(x, y) < 2^{-m_x}\}.$$

Suppose that $y \in V_x$ for a certain $x \in X$. Then $\rho(x, y) < 2^{-m_x}$ and thus

$$y \in \{z \in X : \rho(f_{n_x}(z), f(z)) < \epsilon\}.$$

This means that if $(V_x)_{x \in X}$ has a finite subcover, then (2) holds. We get a finite subcover by the Heine-Borel Covering Lemma, because $(V_x)_{x \in X}$ is indeed a cover of $X$. Note that (1) holds by the definition of $V_x$. Hence (i) implies (ii).

---

[2] We remark that [11] requires additionally that these numbers be arbitrarily large. This requirement can be omitted [9].

To prove the converse, assume (ii), let $\epsilon > 0$ and $x \in X$. With $V_0, \ldots, V_k$ as in the definition of "finite-cover quasi-uniform convergence", pick $i$ such that $x \in V_i$. Because $V_i$ is open, we can find $\delta > 0$ such that

$$\{y \in X : \rho(x, y) < \delta\} \subseteq V_i.$$

Let $y \in X$ be such that $\rho(x, y) < \delta$. Then $y \in V_i$ and thus $\rho(f_{n_i}(y), f(y)) < \epsilon$.

In the rest of the paper we will only use quasi-uniform convergence. The next theorem connects the property of having a continuous limit to the property of being quasi-uniformly convergent. This also shows that for sequences of continuous functions, quasi-uniform convergence (resp., quasi-equicontinuity) is strictly weaker than uniform convergence (resp., equicontinuity), since classically there are convergent sequences of continuous functions on $[0, 1]$ with a continuous limit that are not uniformly convergent (resp., equicontinuous).

**Theorem 3.** *Let $X$ and $Y$ be metric space, and let $(f_n)_{n \geq 0}$ be a sequence of continuous functions from $X$ to $Y$ that converges to a limit $f$. Then the following statements are equivalent.*

(i) *The limit function $f$ is continuous.*
(ii) *The sequence $(f_n)_{n \geq 0}$ converges quasi-uniformly.*

*Proof.* Assuming (i), let $\epsilon > 0$ and $x \in X$. Determine $\delta_0 > 0$ such that for all $y \in X$ if $\rho(x, y) < \delta_0$ then $\rho(f(x), f(y)) < \epsilon/3$. Also determine $n$ such that $\rho(f_m(x), f(x)) < \epsilon/3$ for all $m \geq n$. Using the continuity of $f_n$, we can find $\delta_1 > 0$ such that for all $y \in X$ if $\rho(x, y) < \delta_1$, then $\rho(f_n(x), f_n(y)) < \epsilon/3$. Let

$$\delta := \min\{\delta_0, \delta_1\},$$

and consider $y \in X$ such that $\rho(x, y) < \delta$. We have

$$\rho(f_n(y), f(y)) \leq \rho(f_n(y), f_n(x)) + \rho(f_n(x), f(x)) + \rho(f(x), f(y))$$
$$< \frac{\epsilon}{3} + \frac{\epsilon}{3} + \frac{\epsilon}{3} = \epsilon.$$

Thus (i) implies (ii).

Now assume (ii), and again let $\epsilon > 0$ and $x \in X$. Determine $\delta_0 > 0$ and $n \in \mathbb{N}$ such that for each $y \in X$, if $\rho(x, y) < \delta_0$, then $\rho(f_n(y), f(y)) < \epsilon/3$. By the continuity of $f_n$, there exists $\delta_1 > 0$ such that for all $y \in X$, if $\rho(x, y) < \delta_1$, then $\rho(f_n(x), f_n(y)) < \epsilon/3$. Setting

$$\delta := \min\{\delta_0, \delta_1\},$$

consider $y \in X$ such that $\rho(x, y) < \delta$. We have

$$\rho(f(x), f(y)) \leq \rho(f(x), f_n(x)) + \rho(f_n(x), f_n(y)) + \rho(f_n(y), f(y))$$
$$< \frac{\epsilon}{3} + \frac{\epsilon}{3} + \frac{\epsilon}{3} = \epsilon.$$

This completes the proof that (ii) implies (i).

**Corollary 1.** *The following are equivalent over* ***BISH.***

(i) **CPL$^{\mathbf{cp}}$**.

(ii) *Every convergent sequence of continuous functions on a compact metric space is quasi-equicontinuous.*

(iii) *Every convergent sequence of continuous functions on a compact metric space is quasi-uniformly convergent.*

*Proof.* Immediate, by Theorems 1 and 3.

## 3    Equivalents of UCT $\wedge$ CPL$^{\mathbf{cp}}$

In this section and the next we present some equivalents of a **UCT** and a continuity principle. These theorems are a strengthening of results in [10]. There we proved the following equivalences:

**Theorem 4.** *The following statements are equivalent over* ***BISH****:*

(i) **UCT**.

(ii) *For each equicontinuous sequence $(f_n)_{n \geqslant 0}$ of real-valued continuous functions on $[0,1]$, if $\{f_i(x) : i \in \mathbb{N}\}$ is totally bounded for every $x \in [0,1]$, then $(f_n)_{n \geqslant 0}$ is uniformly equicontinuous.*

(iii) *Every equicontinuous, convergent sequence of real-valued continuous functions on $[0,1]$ has a uniformly continuous limit.*

(iv) *Every equicontinuous, convergent sequence of real-valued continuous functions on $[0,1]$ is uniformly convergent.*

This will be used to obtain equivalences to **UCT** $\wedge$ **CPL$^{\mathbf{cp}}$**. If we compare Theorem 4 with Theorem 5, we see illustrated how **CPL$^{\mathbf{cp}}$** in the latter replaces the requirement of equicontinuity in the former.

**Theorem 5.** *The following statements are equivalent over* ***BISH****:*

(i) **UCT** $\wedge$ **CPL$^{\mathbf{cp}}$**.

(ii) *For every compact metric space $X$: for each sequence $(f_n)_{n \geq 0}$ of continuous functions from $X$ to a metric space, if $\{f_i(x) : i \in \mathbb{N}\}$ is totally bounded for every $x \in X$, then $(f_n)_{n \geq 0}$ is uniformly equicontinuous.*

(iii) *Every convergent sequence of continuous functions from a compact metric space to a metric space has a uniformly continuous limit.*

(iv) *Every convergent sequence of continuous functions from a compact metric space to a metric space is uniformly convergent.*

Before giving the proof of Theorem 5, we generalise two results of [10].

**Lemma 1.** *Let $X, Y$ be two metric spaces. Then every uniformly equicontinuous, convergent sequence of functions from $X$ to $Y$ has a uniformly continuous limit.*

*Proof.* Let $X, Y$ be two metric spaces, and let $(f_n)_{n \geq 0}$ be a uniformly equicontinuous sequence of functions from $X$ to $Y$ that converges to a limit $f$. Given $\epsilon > 0$, determine $\delta$ such that for all $i \in \mathbb{N}$ and all $x, y \in X$, if $\rho(x, y) < \delta$, then $\rho(f_i(x), f_i(y)) < \frac{1}{3}\epsilon$. Let $x, y \in X$ such that $\rho(x, y) < \delta$. Determine $m_0, m_1$ such that $\rho(f_n(x), f(x)) < \frac{1}{3}\epsilon$ for all $n \geq m_0$, and $\rho(f_n(y), f(y)) < \frac{1}{3}\epsilon$ for all $n \geq m_1$. Define

$$p := \max\{m_0, m_1\}.$$

Then

$$\rho(f(x), f(y)) \leq \rho(f_p(x), f(x)) + \rho(f_p(y), f(y)) + \rho(f_p(x), f_p(y))$$
$$< \frac{1}{3}\epsilon + \frac{1}{3}\epsilon + \frac{1}{3}\epsilon = \epsilon.$$

**Lemma 2.** *Let $X$ be a totally bounded metric space, $Y$ a metric space, and $(f_n)_{n \geq 0}$ a convergent sequence of functions from $X$ to $Y$. If $(f_n)_{n \geq 0}$ is uniformly equicontinuous, then $(f_n)_{n \geq 0}$ converges uniformly.*

*Proof.* Let $X$ be a totally bounded metric space, $Y$ a metric space, and $(f_n)_{n \geq 0}$ a sequence of functions from $X$ to $Y$ that converges to a limit $f$. Suppose that $(f_n)_{n \geq 0}$ is uniformly equicontinuous. Then $f$ is uniformly continuous (by Lemma 1). Given $\epsilon > 0$, determine $\delta_0$ and $\delta_1$ such that for all $x, y \in X$ and each $i \in \mathbb{N}$, if $\rho(x, y) < \delta_0$, then $\rho(f_i(x), f_i(y)) < \frac{1}{3}\epsilon$ , and if $\rho(x, y) < \delta_1$, then $\rho(f(x), f(y)) < \frac{1}{3}\epsilon$. Determine $k$ such that $2^{-k} < \min\{\delta_0, \delta_1\}$. Let $A$ be a $2^{-k}$-approximation of $X$. For every $j \in A$ pick $n_j$ such that

$$\rho(f(j), f_m(j)) < \frac{1}{3}\epsilon$$

for every $m > n_j$. Define

$$N := \max\{n_j : j \in A\}.$$

Let $r \geq N$ and $x \in X$. Determine $p \in A$ such that $\rho(p, x) < 2^{-k}$. Then

$$\rho(f(x), f_r(x)) \leq \rho(f(p), f(x)) + \rho(f_r(p), f_r(x)) + \rho(f_r(p), f(p))$$
$$< \frac{1}{3}\epsilon + \frac{1}{3}\epsilon + \frac{1}{3}\epsilon = \epsilon$$

Now we give the proof of Theorem 5.

*Proof (of Theorem 5).* We will first show that (i) implies (ii). Assume (i); let $X$ be a compact metric space; let $Y$ be a metric space; let $(f_n)_{n \geq 0}$ be a sequence of continuous functions from $X$ to $Y$ such that $\{f_i(x) : i \in \mathbb{N}\}$ is totally bounded for every $x \in X$. Then

$$\{\rho(f_i(x), f_i(y)) : i \in \mathbb{N}\}$$

is totally bounded as well. Hence

$$\sup\{\rho(f_i(x), f_i(y)) : i \in \mathbb{N}\}$$

exists for every $x, y \in X$. Define a sequence $(g_n)_{n \geq 0} : X \times X \to \mathbb{R}$ by:

$$g_n(x, y) := \max_{i \leq n} \rho(f_i(x), f_i(y))$$

Then $g_n$ is continuous for every $n$. Moreover for every $x, y \in X$, the sequence $(g_n(x, y))_{n \geq 0}$ converges to

$$g(x, y) := \sup\{\rho(f_i(x), f_i(y)) : i \in \mathbb{N}\}.$$

Therefore, by $\mathbf{CPL^{cp}}$, $g$ is continuous. So $g$ is uniformly continuous by $\mathbf{UCT}$. Let $\epsilon > 0$. Determine $\delta$ such that for all $(x, y), (x', y') \in X \times X$, if $\rho((x, y), (x', y')) < \delta$, then $|g(x, y) - g(x', y')| < \epsilon$. Let $x, y \in X$ be such that $\rho(x, y) < \delta$, and let $i \in \mathbb{N}$. Then

$$\rho(f_i(x), f_i(y)) \leq \sup\{\rho(f_j(x), f_j(y)) : j \in \mathbb{N}\}$$
$$= g(x, y) = |g(x, y) - g(x, x)| < \epsilon$$

Because it is immediately clear that (iii) follows from (i), we continue with the proof of (iv) from (ii). So assume (ii). Let $X$ be a compact metric space, let $Y$ be a metric space, and let $(f_n)_{n \geq 0}$ be a convergent sequence (with a limit $f$) of continuous functions from $X$ to $Y$. Note that, because the sequence converges, $\{f_i(x) : i \in \mathbb{N}\}$ is totally bounded for every $x \in X$. Hence, by statement (ii), the sequence $(f_n)_{n \geq 0}$ is uniformly equicontinuous. So we conclude by Lemma 2 that $(f_n)_{n \geq 0}$ converges uniformly.

We now prove the reverse direction. It follows directly from Theorem 4 that the statements (ii), (iii), and (iv) imply $\mathbf{UCT}$. Remark that it is also immediately clear that each of the statements (iii) and (iv) implies $\mathbf{CPL^{cp}}$. To prove that (ii) implies $\mathbf{CPL^{cp}}$, let $X$ be a compact metric space, let $Y$ be a metric space, and let $(f_n)_{n \geq 0}$ be a sequence of continuous functions from $X$ to $Y$ converging to a limit $f$. Then $\{f_i(x) : i \in \mathbb{N}\}$ is totally bounded for every $x \in X$. Thus, by (ii), the sequence is uniformly equicontinuous. Therefore $f$ is continuous.

This completes the proof.

Note that the theorem

> *Every convergent quasi-equicontinuous sequence of continuous functions from a compact metric space to a metric space has a uniformly continuous limit.*

is equivalent to $\mathbf{UCT}$, but that neither of the following two statements is so equivalent:

> *For every compact metric space $X$: for each quasi-equicontinuous sequence $(f_n)_{n \geq 0}$ of continuous functions from $X$ to a metric space, if $\{f_i(x) : i \in \mathbb{N}\}$ is totally bounded for every $x \in X$, then $(f_n)_{n \geq 0}$ is uniformly equicontinuous.*

> *Every quasi-equicontinuous convergent sequence of continuous functions from a compact metric space to a metric space is uniformly convergent.*

# 4   Equivalents of UCT $\wedge$ CONT$^{\mathbf{cp}}$

The next theorem follows naturally from the results of the previous section.

**Theorem 6.** *The following statements are equivalent over* ***BISH****:*

(i) **UCT** $\wedge$ **CONT**$^{\mathbf{cp}}$.

(ii) *For every compact metric space $X$: for each sequence $(f_n)_{n\geq 0}$ of functions from $X$ to a metric space, if $\{f_i(x) : i \in \mathbb{N}\}$ is totally bounded for every $x \in X$, then $(f_n)_{n\geq 0}$ is uniformly equicontinuous.*

(iii) *Every convergent sequence of functions from a compact metric space to a metric space has a uniformly continuous limit.*

*Proof.* The fact that (i) implies the statements (ii), and (iii) is an immediate consequence of Theorem 5. That these statements imply **UCT** follows also from the results in the previous section. We will now show that they also imply **CONT**$^{\mathbf{cp}}$. Let $X$ be a compact metric space, $Y$ a metric space, and let $f$ be a function from $X$ to $Y$. Define a sequence $(f_n)_{n\geq 0}$ by $f_n = f$ for every $n \in \mathbb{N}$. Observe that this sequence converges. Then the conclusions that the sequence is uniformly equicontinuous or that $f$ is uniformly continuous, both imply that $f$ is continuous.

We have only a partial generalisation of Theorem 5 item (iv) in the sense of Theorem 6, which is an immediate consequence of the former theorem.

**Theorem 7. UCT $\wedge$ CONT$^{\mathbf{cp}}$** *implies over* ***BISH*** *that every convergent sequence of functions from a compact metric space to a metric space is uniformly convergent.*

Let us now come back to the original proof of the statement that every pointwise convergent sequence of real-valued functions on a compact metric space converges uniformly [15]. Without stating explicitly that it does so, this proof uses **CONT**$^{\mathbf{cp}}$ and the Full Fan Theorem.

Our results do not show whether we can weaken the continuity assumption without strengthening another hypothesis. However, if we assume that the functions of the sequence are continuous, then a weaker continuity principle suffices (Theorem 5, (iv)).

On the other hand, it is possible to economise on the strength of the compactness principle. Instead of using the Full Fan Theorem, we may assume just **UCT** (Theorem 7), which is believed to be weaker [10], even under the assumption of **CONT**$^{\mathbf{cp}}$. Note that we could not have concluded this directly from Theorem 4, as we do not have a proof that **CONT**$^{\mathbf{cp}}$ implies that all convergent sequences of functions on a compact space are equicontinuous.

**Acknowledgments.** The author thanks the Marsden Fund of the Royal Society of New Zealand for supporting her by a Postdoctoral Research Fellowship. She also thanks Douglas Bridges, Hannes Diener and Josef Berger for useful discussions, and the anonymous reviewers for useful comments.

# References


1. Arzelà, C.: Intorno alla continuità della somma d'infinità di funzioni continue. Rend. dell'Accad. di Bologna, 79–84 (1883-1884)
2. Bartle, R.G.: On Compactness in Functional Analysis. Transactions of the American Mathematical Society 79, 35–57 (1955)
3. Berger, J.: The fan theorem and uniform continuity. In: Cooper, S.B., Löwe, B., Torenvliet, L. (eds.) CiE 2005. LNCS, vol. 3526, pp. 18–22. Springer, Heidelberg (2005)
4. Berger, J.: The logical strength of the uniform continuity theorem. In: Beckmann, A., Berger, U., Löwe, B., Tucker, J.V. (eds.) CiE 2006. LNCS, vol. 3988, pp. 35–39. Springer, Heidelberg (2006)
5. Berger, J., Bridges, D.: A fan-theoretic equivalent of the antithesis of specker's theorem. Indag. Math. 18(2), 195–202 (2007)
6. Berger, J., Bridges, D.S.: A Bizarre Property Equivalent to the $Pi_1{}^0$-fan theorem. Logic Journal of the IGPL 14(6), 867–871 (2006)
7. Bishop, E., Bridges, D.: Constructive Analysis. Springer, Heidelberg (1985)
8. Bridges, D., Diener, H.: The pseudocompactness of $[0,1]$ is equivalent to the uniform continuity theorem. J. Symb. Log. 72(4), 1379–1384 (2007)
9. Charazišvili, A.B.: Strange Functions in Real Analysis. CRC Press, Boca Raton (2006)
10. Diener, H., Loeb, I.: Sequences of real-valued functions on $[0,1]$ in constructive reverse mathematics (submitted), http://www.math.canterbury.ac.nz/~i.loeb/SRF.pdf
11. Hazewinkel, M. (ed.): Encyclopaedia of Mathematics. Springer, Heidelberg (2002)
12. Ishihara, H.: Continuity properties in constructive mathematics. J. Symb. Log. 57(2), 557–565 (1992)
13. Ishihara, H.: Reverse Mathematics in Bishop's Constructive Mathematics. Philosophia Scientiae, Cahier Spécial 6, 43–59 (2004)
14. Ishihara, H., Schuster, P.: Compactness under constructive scrutiny. Math. Log. Q. 50(6), 540–550 (2004)
15. de Swart, H.: Elements of Intuitionistic Analysis: Rolle's theorem and complete, totally bounded, metric spaces. Zeitschrift für mathematische Logik und Grundlagen der Mathematik Bd. 22, 289–298 (1976)

# Interpreting Localized Computational Effects Using Operators of Higher Type

John Longley

Laboratory for Foundations of Computer Science
School of Informatics, University of Edinburgh
The King's Buildings, Mayfield Road
Edinburgh EH9 3JZ, UK

**Abstract.** We outline a general approach to providing intensional models for languages with computational effects, whereby the problem of interpreting a given effect reduces to that of finding an operator of higher type satisfying certain equations. Our treatment consolidates and generalizes an idea that is already implicit in the literature on game semantics. As an example, we work out our approach in detail for the case of fresh name generation, and discuss some particular models to which it applies.

## 1   Introduction

This paper explores a way in which computable operations of higher type can be useful in giving *denotational semantics* for programming languages. In broad terms, a denotational semantics for a language $\mathcal{L}$ consists of a mathematical model of the behaviour of programs in $\mathcal{L}$, given by assigning to each program $P$ a "meaning" $[\![P]\!]$ within some mathematical structure $\mathcal{M}$ which can be defined and studied independently of the syntax of $\mathcal{L}$. Not only does this result in a rigorous mathematical definition of the programming language, but if $\mathcal{M}$ itself enjoys good mathematical properties, these can be used to reason about programs of $\mathcal{L}$. Furthermore, a particularly well-behaved model $\mathcal{M}$ might even inspire the design of a new and better programming language. For information on the mathematical aspects of denotational semantics, we recommend [7].

Much of the foundational work in denotational semantics focused initially on purely functional languages such as Plotkin's PCF [26]. The essence of such languages is that the behaviour of a program can be adequately modelled by a mathematical *function*, so that the language is amenable to a denotational description in terms of some well-understood mathematical class of functions, such as Scott's partial continuous functionals of higher type. However, whilst such purely functional settings are simple and mathematically appealing, virtually all real-world programming languages abound in "impure" features that break the simple-minded functional paradigm, such as exceptions, state, continuations, fresh name generation, input/output and nondeterminism. It is therefore natural to seek appropriate mathematical theories for modelling such computational *effects* (as they are generically known).

A. Beckmann, C. Dimitracopoulos, and B. Löwe (Eds.): CiE 2008, LNCS 5028, pp. 389–402, 2008.
© Springer-Verlag Berlin Heidelberg 2008

To date, there have broadly been two approaches to the denotational semantics of computational effects. The first, and more widely established, was pioneered by Moggi [20] in his investigation of the use of *monads* to model effects of various kinds. Here, a term $M : \sigma$, possibly involving some computational effect, is modelled by an element not of the usual object $[\![\,\sigma\,]\!]$, but of some richer domain $T[\![\,\sigma\,]\!]$, where $T$ is a monadic functor chosen to match the effect in question. For instance, we may take $T(X) = X + E$ if the evaluation of $M$ may result in an exception drawn from the set $E$, or $T(X) = (S \times X)^S$ if the evaluation may have a side-effect on some state of type $S$, and so on. In this way, an essentially "functional" treatment of programs is maintained, at the cost of complicating the types of the functions involved: typically, a program of type $\sigma \to \tau$ will be modelled by a function of type $[\![\,\sigma\,]\!] \to T[\![\,\tau\,]\!]$ rather than $[\![\,\sigma\,]\!] \to [\![\,\tau\,]\!]$. The monadic approach has had a wide influence and some notable successes — in particular, it underpins the model of effects employed in the Haskell programming language [24]. More recently, a closely related approach has been developed by Plotkin and Power [27], emphasizing the primacy of Lawvere-style *algebraic theories* rather than monads — this promises, among other things, a more satisfactory account of how different computational effects may be combined in a principled way.

By contrast, the second main approach (advocated explicitly in [1]) seeks to model programs with effects not by functions of more complicated types, but by more fine-grained, *intensional* semantic objects than ordinary mathematical functions — typically algorithms, strategies or even programs of some kind. Such intensional notions of "computable operation" occasionally featured in the earlier literature on higher types (*e.g.* the non-extensional type structures HRO and ICF derived from Kleene's first and second models respectively [30]); in the computer science literature, an important early example was the *sequential algorithm* model of Berry and Curien [8]. A wealth of further interesting models were subsequently introduced by the literature on *game semantics* [3,4,5,12], where numerous full abstraction results were obtained for languages with control features, state, and non-determinism in various combinations (see [11] for a survey, and [19] for a useful taxonomy of models). Whilst these intensional models may appear unfamiliar at first, experience shows that many of them lead to beautiful mathematical structures, carry a persuasive intuition, and (in the author's view) provide good candidates for notions of higher type computability in the spirit of [16].

Although numerous examples of intensional models for languages with effects have now been collected, they have so far not conformed to much of a general pattern. Our purpose in the present article is to outline a somewhat uniform approach to the interpretation of computational effects in intensional models by means of operators of higher type. Particular instances of our approach may be discerned in the existing literature on game semantics, but as far as we know, the general idea has not hitherto been spelt out as a uniformly applicable method. Our presentation will, moreover, be at a level of generality which renders the ideas applicable to other intensional models besides game models.

The basic idea is as follows. Each kind of computational effect is typically associated with some characteristic syntactic operators: *e.g.* `raise` and `handle` in the case of exceptions; `read` and `write` in the case of store cells; `new` and `eq` in the case of the generation of fresh names (with equality testing), and so on. For the sake of discussion let us work with the example of fresh name generation, though the same basic strategy will clearly make sense for many other effects. In the context of a higher order (say call-by-value) language, we may naturally ascribe *types* to the relevant operators: *e.g.* in the spirit of the $\nu$-*calculus* of Pitts and Stark [25], we have operations `new : unit` $\to$ `name` and `eq : name` $*$ `name` $\to$ `bool`. However, rather than attempting to model what these operators actually do, let us choose to regard them simply as *variables* that may appear in a term, with the same formal status as ordinary program variables. Thus, no special technology is needed at this stage to model terms involving such operators. However, in an intensional model, the denotation of such a term $M$ may typically record information concerning when, and in what order, the characteristic operators are invoked, and how the results affect the subsequent computation. This means that (in good cases) the denotation of $M$ (or equivalently of its closure $\overline{M} = \lambda \texttt{new}, \texttt{eq}.\, M$) will in principle contain enough information to determine how $M$ *would* behave if genuine implementations of the appropriate operators were supplied.[1] Furthermore, in many cases, one can find within the model itself a higher order operator $\Phi$ which transforms the denotation of $\overline{M}$ to an element modelling the desired actual behaviour of $M$. One may informally think of $\Phi$ as modelling the behaviour of the program $\lambda F.\, F$ `New Eq`, where `New`, `Eq` are actual implementations of the relevant operators; note that such a $\Phi$ may exist even though `New`, `Eq` themselves have no standalone interpretation in the model.

This naturally raises the question: what properties must an operator $\Phi$ satisfy in order to give rise to a correct semantics for fresh name generation? Our "reference semantics" for freshness will presumably be derived from our operational understanding of `New` and `Eq`, but we would also like a denotational (*e.g.* an equational) condition on $\Phi$ within the model which is sufficient and (ideally) necessary for the soundness of our interpretation. An operator $\Phi$ satisfying this condition may then be dubbed a *freshness operator*.

Having arrived at this general definition, it is then natural to ask which particular intensional models possess a freshness operator. We may think of this property as capturing something interesting about the innate computational power of a model (somewhat akin, say, to the property of having a fixed point operator or a modulus of continuity operator of some sort — see *e.g.* [30]), as well as its potential usefulness in denotational semantics. Moreover, by formulating this notion uniformly for a class of models, we facilitate the task of comparing and classifying models, thus contributing to the author's project of mapping out the landscape of computability notions [16,17].

---

[1] In the case of store cells with `read` and `write` operations, this is very much how the interpretation *e.g.* in [3] works; this is perhaps the clearest manifestation in the existing literature of our basic idea.

Typically, our approach will work at its best for uses of computational effects that are *localized* to some block of code $M$ (*cf.* [15]). A program that makes global use of some effect will be modelled as the denotation of an open term with free variables for the characteristic operations; the operation of localizing this effect then corresponds to abstracting over these variables and applying the appropriate operator $\Phi$. Whilst this in principle allows us to interpret both complete "closed" programs and "open" fragments thereof, a common situation will be that our interpretation is fully abstract for closed programs, but very far from this for open ones.[2] A natural methodology is therefore to focus initially on the well-behaved situation for closed programs, and then to consider how the benefits of our interpretation might be extended to open programs. We will return to this issue in Section 4.

In the present article, we make a modest start on demonstrating the viability of our programme, focusing in particular on name generation. This seems an interesting example to consider for two reasons. Firstly, it cannot be straightforwardly modelled by monads on familiar categories of domains. This led Moggi originally to suggest using a monad on a *functor category* [21], an approach which has subsequently proved rather difficult to combine with other language features [28]. (In a different guise, functor categories are also an important ingredient in the Plotkin-Power approach to name generation — see [27,29].) Secondly, virtually all other efforts to model name generation have, in some way, made essential use of another idea: that of a set of names acted on by a permutation group in order to make them "indistinguishable" [2,13,28,31]. These approaches have achieved significant success, *e.g.* in terms of full abstraction results; however, we believe it is also interesting to explore how much can be achieved without resorting to the machinery of either functor categories or permutation actions. In particular, our approach shows that many models of computation that have proved to be of interest for other reasons (*e.g.* game models) already have what it takes to model name generation without any specialized additional technology.

The rest of the paper is structured as follows. In Section 2 we sketch a general framework (based on Moggi's notion of a $\lambda_c$-*model*) within which the general idea works out smoothly. In Section 3, as a concrete example, we consider the case of fresh name generation in some detail, including the definition of freshness operators and some technical results validating this definition. In Section 4 we survey some particular examples of models in which freshness operators are available, and in Section 5 we mention some avenues for further investigation.

The ideas in this paper arose rather naturally in the course of an attempt to design a programming language based around the structure available in a certain game model. For an account of this work in progress, see [18].

I am grateful to the CiE organizers for the invitation to present this material, to the reviewers for their helpful comments, and to Ian Stark and Nicholas Wolverson for valuable discussions. The research was supported by EPSRC Grant GR/T08791: "A programming language based on game semantics".

---

[2] In the case of store cells, this is related to the problem of "bad variables" — see [3].

## 2   The General Framework

Although we will not be using monads themselves to model effects, we are indebted to the monadic tradition for a general notion of (intensional) model that is suited to our purposes. Because of the special role played by *values* (*i.e.* fully evaluated expressions) in programming languages with effects, it is convenient to frame our ideas in a call-by-value setting, and here a very suitable notion of model is provided by Moggi's work on *computational $\lambda$-calculus* [20]. The definition is most compactly presented in categorical terms. Formally, a $\lambda_c$-*model* is a category $\mathcal{C}$ with finite products, equipped with a strong monad $(T, \eta, \mu, t)$, such that for any $A, B \in \mathcal{C}$ the exponential $TB^A$ exists (we henceforth denote $TB^A$ by $A \Rightarrow B$).[3] For convenience, we assume our $\lambda_c$-models come equipped with objects $1, 2$ representing unit and boolean types. We abbreviate $f : 1 \to A$ to $f \in A$.

The intuition is that whereas an object $A$ may serve for modelling *values* of some type, the corresponding object $TA$ will model more general expressions of this type whose evaluation may involve some effect. (For a detailed explanation of why strong monads are an appropriate choice of structure here, we refer the reader to [22].) In Moggi's work, one considers a range of different monads to capture different computational effects. By contrast, here we will be concerned almost exclusively with *lifting* monads representing potentially non-terminating computations — the interest for us lies in varying the base category $\mathcal{C}$ to capture different "levels of intensionality". However, it is worth remarking that computability models that are *too* finely intensional (such as those based on Kleene's models $K_1$ and $K_2$) fail even to be $\lambda_c$-models, and it is not clear whether our approach can be made to work at all in these settings.

Rather than writing lengthy categorical expressions, we use shall use a familiar lambda-calculus notation for denoting morphisms of $\mathcal{C}$ as in [14]; the precise intention in any given instance will be clear from the types involved. We shall supplement this with some meta-notation borrowed from [20]: we write $[e]$ for the inclusion of $e : A$ into $TA$ via $\eta_A$, and let $x = e$ in $e'$ for the "Kleisli composition" of $e$ and $e'$.[4] The essential point about the latter is that it captures the call-by-value discipline of forcing the evaluation of $e$ whether or not $x$ appears in $e'$. We write let $\boldsymbol{a} = e^r$ in $e'$ to abbreviate let $a_1 = e$ in $\cdots$ let $a_r = e$ in $e'$.

We also introduce some syntactic machinery intended to embody the general notion of a "programming language interpretable in $\mathcal{C}$". We represent the syntax of such a language as a category *à la* Lawvere, with composition corresponding to syntactic substitution. Formally, an *object language* $\mathcal{L}$ consists of:

---

[3] The "mono requirement" mentioned in [20] is not needed for our purposes.

[4] More precisely, if $e$ is a meta-expression of type $TA$ involving metavariables $y_i : C_i$, and $e'$ a meta-expression of type $TB$ involving metavariables $y_i : C_i$ and $x : A$, then let $x = e$ in $e'$ denotes the morphism

$$\Pi C_i \xrightarrow{\langle id, e \rangle} \Pi C_i \times TA \xrightarrow{t_{\Pi C_i, A}} T(\Pi C_i \times A) \xrightarrow{Te'} T(TB) \xrightarrow{\mu_B} TB$$

For a more formal treatment, see [22].

- a collection of *types* $\sigma, \tau$, equipped with binary operations $*, \rightarrow$ and including for convenience the types `unit` and `bool`;
- a category with finite products whose objects are finite tuples of types $(\sigma_1, \ldots, \sigma_n)$ (the product of two objects being their concatenation as tuples).

Morphisms $(\sigma_1, \ldots, \sigma_n) \rightarrow (\tau)$ should be thought of as equivalence classes of terms-in-context $x_1 : \sigma_1, \ldots, x_n : \sigma_n \vdash M : \tau$ modulo renaming of variables. We shall generally use terms-in-context to denote morphisms of $\mathcal{L}$ and blur the distinction between the two notions. We shall also require that there are constant terms $\langle \rangle :$ `unit` and `tt`, `ff` : `bool`, and for each $\sigma, \tau$ there exist pairing and application terms $x : \sigma, y : \tau \vdash \langle x, y \rangle : \sigma * \tau$ and $f : \sigma \rightarrow \tau, x : \sigma \vdash f \bullet x : \tau$. (We warn the reader against confusion between the object language syntax and the meta-notation conventions introduced above.)

A *programming language* will be an object language $\mathcal{L}$ endowed with an operational semantics which includes a notion of "symbolic evaluation" for open terms as well as the usual notion of evaluation for closed terms. Formally, we shall require:

- for each $\Gamma, \tau$, a reflexive-transitive *evaluation* relation $\twoheadrightarrow$ (more properly written $\Gamma \vdash - \twoheadrightarrow - : \tau$) on terms $\Gamma \vdash M : \tau$;
- for each $\Gamma, \sigma$, a set of terms of $\mathcal{L}$ in context $\Gamma, - : \sigma$ designated as *evaluation contexts* $E[-]$ (where '$-$' is a distinguished free variable).

In typical cases, these will satisfy further properties, *e.g.*:

- if $M \twoheadrightarrow N$ and $E[-]$ is an evaluation context then $E[M] \twoheadrightarrow E[N]$;
- for any $\Gamma, \tau, \Gamma'$, the term $\Gamma, -, \Gamma' : \tau \vdash - : \tau$ is an evaluation context;
- if $E[-], E'[-]$ are appropriately typed evaluation contexts then so is their evident composition $E'[E[-]]$;
- for any $M$ and $x$, the terms $- \bullet M, x \bullet -, \langle -, M \rangle$ and $\langle x, - \rangle$ (in any suitable context $\Gamma, - : \tau$) are evaluation contexts.

A language satisfying these properties will be called *standard*. Surprisingly, however, none of these properties will be required for our main results.

An *interpretation of types of $\mathcal{L}$ in a $\lambda_c$-model $\mathcal{C}$* is a mapping $[\![ - ]\!]$ from types of $\mathcal{L}$ to objects of $\mathcal{C}$ satisfying the expected properties: $[\![ \text{unit} ]\!] = 1$, $[\![ \text{bool} ]\!] = 2$, $[\![ \text{nat} ]\!] = N$, $[\![ \sigma * \tau ]\!] = [\![ \sigma ]\!] \times [\![ \tau ]\!]$, $[\![ \sigma \rightarrow \tau ]\!] = [\![ \sigma ]\!] \Rightarrow [\![ \tau ]\!]$. By convention, if $\Gamma = x_1 : \sigma_1, \ldots, x_r : \sigma_r$ we write $[\![ T\Gamma ]\!]$ for the object $T[\![ \sigma_1 ]\!] \times \cdots \times T[\![ \sigma_r ]\!]$. Note the appearance of $T$ here, which contrasts with the modelling of contexts *e.g.* in [20]. This reflects the fact that the role of variables here is purely to allow us to talk about the compositional structure of terms rather than to model any kind of object-level variable binding.

Relative to such a mapping, an *interpretation* $[\![ - ]\!]$ *of $\mathcal{L}$* maps each term $\Gamma \vdash M : \tau$ of $\mathcal{L}$ to a morphism $[\![ M ]\!]_\Gamma : [\![ T\Gamma ]\!] \rightarrow T[\![ \tau ]\!]$, in such a way that

- variables, constants, application and pairing receive the expected (left-strict) interpretation, and the evident weakening and contraction properties hold;

- $\llbracket - \rrbracket$ is *compositional*: that is, if $\Delta_i \vdash N_i : \sigma_i$ for each $i$ (with the $\Delta_i$ disjoint) and $\Gamma = x_1 : \sigma_1, \ldots, x_r : \sigma_r \vdash M : \tau$, then

$$\llbracket M[\boldsymbol{N}/\boldsymbol{x}] \rrbracket_{\boldsymbol{\Delta}} \;=\; \llbracket M \rrbracket_\Gamma \circ (\llbracket N_1 \rrbracket_{\Delta_1} \times \ldots \times \llbracket N_r \rrbracket_{\Delta_r})$$

- if $\Gamma, - : \sigma \vdash E[-] : \tau$ is an evaluation context then $\llbracket E[-] \rrbracket_{\Gamma,-:\sigma}$ is strict in '$-$', that is:

$$\llbracket E[-] \rrbracket(\boldsymbol{x}, y) \;=\; \mathsf{let}\ z = y\ \mathsf{in} \llbracket E \rrbracket(\boldsymbol{x}, [z])$$

Clearly, a global interpretation gives rise to a functor $\mathcal{L} \to \mathcal{C}$ which preserves finite products. Note, however, that products in $\mathcal{L}$ do *not* correspond to object-level product types, the difference being that between $TA \times TB$ and $T(A \times B)$.

An interpretation $\llbracket - \rrbracket$ is called *sound* if $\Gamma \vdash M \twoheadrightarrow N$ implies $\llbracket M \rrbracket_\Gamma = \llbracket N \rrbracket_\Gamma$. We also say $\llbracket - \rrbracket$ is *sound for* a class $\mathcal{T}$ of $\mathcal{L}$-terms if this property holds whenever $M \in \mathcal{T}$.

## 3   Fresh Name Generation

As our main example, we now work out our approach in some detail for the case of fresh name generation in the spirit of the $\nu$-calculus. We first present our operational understanding of name generation by describing how any programming language $\mathcal{L}$ may be extended to a language $\mathcal{L}^+$ with (localized) name generators. We then investigate the conditions under which an interpretation for $\mathcal{L}$ may be extended to one of $\mathcal{L}^+$.

First, let us assume for convenience that our original object language $\mathcal{L}$ comes already equipped with an infinite supply of *name types*, ranged over by $\nu$. (These types need play no active role in $\mathcal{L}$ beyond what is implied by the fact of being a finite-product category.) We may then freely extend $\mathcal{L}$ (as a finite-product category) to a language $\mathcal{L}^+$ by means of the following term formation rule:

$$\frac{\Gamma,\ new : \mathtt{unit} \to \nu,\ eq : \nu * \nu \to \mathtt{bool},\ a_1 : \nu, \ldots, a_r : \nu\ \vdash\ M : \tau}{\Gamma\ \vdash\ \mathsf{gen}_\nu, new, eq, (a_1, \ldots, a_r)\ \mathsf{in}\ M\ : \tau}\ \nu \notin \Gamma, \tau$$

where $\nu$ ranges over name types and $new, eq, \boldsymbol{a}$ may be any variables.[5] If $\mathcal{G} \subseteq \mathcal{T}$ is any set of types, we also obtain a restricted language $\mathcal{L}_{\mathcal{G}}^+$ by requiring that $\tau \in \mathcal{G}$ in the above rule. This construct allows us to introduce a localized name generator with characteristic operations represented by *new* and *eq*, whose generated names are prevented by the type system from being confused with names arising from other generators. (The side-condition also ensures that the generated names cannot leak out of their scope.) This kind of potentially nested block structure sets the pattern for our general approach to localized effects; in the present context it also models the situation *e.g.* in Standard ML, where a

---

[5] We regard $\nu, new, eq, \boldsymbol{a}$ as being *bound* by this construction, and should also require in the above rule that they do not occur bound within $M$ itself. Technically, we view $\alpha$-equivalent expressions as defining the same term in $\mathcal{L}^+$, so that for the purpose of substitution and evaluation we may freely apply $\alpha$-conversion as necessary.

local datatype declaration implicitly introduces a localized name generator for the corresponding **ref** type. The $a_i$ play the role of names that have already been created using the generator in question, and which may feature in $M$.

We let $E^*[-]$ range over $\mathcal{L}^+$-substitution instances of evaluation contexts in $\mathcal{L}$, and let $G[-]$ range over *gen-contexts*: that is, compositions of zero or more contexts of the form **gen** $new, eq, (\boldsymbol{a})$ **in** $-$. As evaluation contexts of $\mathcal{L}^+$ or $\mathcal{L}_G^+$, we take all contexts of the form $G[E^*[-]]$ (note that these are *not* closed under composition), and let $F[-]$ range over these. As an operational semantics, we let $\twoheadrightarrow^+$ be the evaluation relation generated by the following:

- if $M \twoheadrightarrow N$ in $\mathcal{L}$ and $\boldsymbol{P}$ is a list of $\mathcal{L}^+$-terms, then $G[M[\boldsymbol{P}/\boldsymbol{x}]] \twoheadrightarrow^+ G[N[\boldsymbol{P}/\boldsymbol{x}]]$;
- $\mathbf{gen}_\nu \, new, eq, (\boldsymbol{a})$ **in** $F[new \bullet \langle\rangle] \twoheadrightarrow^+ \mathbf{gen}_\nu \, new, eq, (\boldsymbol{a}, b)$ **in** $F[a']$ ($a'$ fresh);
- $\mathbf{gen}_\nu \, new, eq, (\boldsymbol{a})$ **in** $F[eq \bullet \langle a_i, a_i \rangle] \twoheadrightarrow^+ \mathbf{gen}_\nu \, new, eq, (\boldsymbol{a})$ **in** $F[\mathbf{tt}]$;
- $\mathbf{gen}_\nu \, new, eq, (\boldsymbol{a})$ **in** $F[eq \bullet \langle a_i, a_j \rangle] \twoheadrightarrow^+ \mathbf{gen}_\nu \, new, eq, (\boldsymbol{a})$ **in** $F[\mathbf{ff}]$ ($i \neq j$);
- $G[E^*[\mathbf{gen}_\nu \, new, eq, (\boldsymbol{a})$ **in** $M]] \twoheadrightarrow^+ G[\mathbf{gen}_\nu \, new, eq, (\boldsymbol{a})$ **in** $E^*[M]]$;
- $G[\mathbf{gen}_\nu \, new, eq, (\boldsymbol{a})$ **in** $M] \twoheadrightarrow^+ G[M]$ if $new, eq, \boldsymbol{a}$ do not appear in $M$.

We will take the definition of $\twoheadrightarrow^+$ as our reference semantics for fresh name generation, and ask when an interpretation in a $\lambda_c$-model accords with this. The last of the above rules is a "garbage collection rule" — the definition of $\twoheadrightarrow^+$ admits some non-determinism regarding exactly when this rule is to be applied, but this does not matter for our purposes, since Propositions 1 and 2 below will apply *a fortiori* to any reduction strategy included in $\twoheadrightarrow^+$. Note also that if $\mathcal{L}$ is standard, the following are always evaluation contexts:

$$new \bullet - \qquad eq \bullet \langle -, M \rangle \qquad eq \bullet \langle a_i, - \rangle$$

and this enables us to make progress with the evaluation of programs such as

$$\mathbf{gen}_\nu \, new, eq, () \text{ in } eq \bullet \langle new \bullet \langle\rangle, (\mathbf{fn} \text{ x} => new \bullet \langle\rangle) \bullet \langle\rangle\rangle$$

Next, suppose we are given an interpretation $[\![-]\!]$ of $\mathcal{L}$ in $\mathcal{C}$, and assume for simplicity that $[\![\nu]\!]$ is the same object $A_{name}$ for all name types $\nu$. Let $A_{new} = 1 \Rightarrow A_{name}$ and $A_{eq} = A_{name} \times A_{name} \Rightarrow 2$. Suppose moreover that for each $\tau \in \mathcal{G}$ we are given some operator $\Phi_\tau \in (A_{new} \Rightarrow A_{eq} \Rightarrow [\![\tau]\!]) \Rightarrow [\![\tau]\!]$, and let $\Phi$ denote the indexed family $\{\Phi_\tau \mid \tau \in \mathcal{G}\}$. Relative to $\Phi$, we may define an interpretation $[\![-]\!]^\Phi$ for terms of $\mathcal{L}_G^+$ as follows:

- $[\![M]\!]_\Gamma^\Phi = [\![M]\!]_\Gamma$ if $\Gamma \vdash M : \tau$ in $\mathcal{L}$;
- $[\![\mathbf{gen}_\nu \, new, eq, (a_1, \ldots, a_r) \text{ in } M \,]\!]_\Gamma^\Phi$ is taken to be

$$\lambda\boldsymbol{x} : [\![T\Gamma]\!]. \, \Phi_\tau \, (\lambda new, eq. \, \mathsf{let} \, \boldsymbol{a} = (new\, \langle\rangle)^r \, \mathsf{in}$$
$$[\![M]\!]_{\Gamma'}^\Phi \, (\boldsymbol{x}, [new], [eq], [a_1], \cdots, [a_r]))$$

where $\Gamma' = \Gamma, new : \mathbf{unit} \to \nu, eq : \nu * \nu \to \mathbf{bool}, a_1 : \nu, \ldots, a_r : \nu$;
- $[\![-]\!]^\Phi$ extends to arbitrary terms of $\mathcal{L}_G^+$ via compositionality.

Under what conditions is this interpretation a reasonable one? We propose the following semantic definition; note that only *equations* between higher type operators are involved. For readability, we take a few small liberties in our meta-notation, *e.g.* writing $f \, new \, eq \, \langle b, \boldsymbol{a} \rangle$ in place of let $f' = f$ in $f' \, new \, eq \, \langle b, \boldsymbol{a} \rangle$.

**Definition 1.** *A freshness operator for a type $\tau$ is an operator $\Phi_\tau \in (A_{new} \Rightarrow A_{eq} \Rightarrow [\![\,\tau\,]\!]) \Rightarrow [\![\,\tau\,]\!]$ satisfying the following equations.*

1. *If $A_g = T[\![\,\tau\,]\!]$ then*

$$\lambda g : A_g.\ \Phi_\tau(\lambda\mathsf{new}, \mathsf{eq}.\ \mathsf{let}\ a = (\mathsf{new}\,\langle\rangle)^r\ \mathsf{in}\ g)\ \ =\ \ \lambda g : A_g.\ g$$

2. *For all $r$ and all $i \leq r$, if $A_f = T(A_{new} \Rightarrow A_{eq} \Rightarrow (2 \times A^r_{name}) \Rightarrow [\![\,\tau\,]\!])$ then*

$$\lambda f : A_f.\ \Phi_\tau(\lambda\mathsf{new}, \mathsf{eq}.\ \mathsf{let}\ a = (\mathsf{new}\,\langle\rangle)^r\ \mathsf{in}\ \mathsf{let}\ b = \mathsf{eq}\,(a_i, a_i)\ \mathsf{in}\ f\ \mathsf{new}\ \mathsf{eq}\ \langle b, a\rangle)$$
$$=\ \ \lambda f : A_f.\ \Phi_\tau(\lambda\mathsf{new}, \mathsf{eq}.\ \mathsf{let}\ a = (\mathsf{new}\,\langle\rangle)^r\ \mathsf{in}\ f\ \mathsf{new}\ \mathsf{eq}\ \langle \mathsf{tt}, a\rangle)$$

3. *For all $r$ and all $i, j \leq r$ with $i \neq j$, if $A_f$ is as above then*

$$\lambda f : A_f.\ \Phi_\tau(\lambda\mathsf{new}, \mathsf{eq}.\ \mathsf{let}\ a = (\mathsf{new}\,\langle\rangle)^r\ \mathsf{in}\ \mathsf{let}\ b = \mathsf{eq}\,(a_i, a_j)\ \mathsf{in}\ f\ \mathsf{new}\ \mathsf{eq}\ \langle b, a\rangle)$$
$$=\ \ \lambda f : A_f.\ \Phi_\tau(\lambda\mathsf{new}, \mathsf{eq}.\ \mathsf{let}\ a = (\mathsf{new}\,\langle\rangle)^r\ \mathsf{in}\ f\ \mathsf{new}\ \mathsf{eq}\ \langle \mathsf{ff}, a\rangle)$$

Although a certain amount can be achieved even with a single freshness operator, more can be done with a family of such operators that fit well together. A family $\Phi = \{\Phi_\tau \mid \tau \in \mathcal{G}\}$ of freshness operators is called *coherent* if for any $\sigma, \tau \in \mathcal{G}$, writing $A_c = T[\![\,\sigma \to \tau\,]\!]$ and $A_p = T(A_{new} \Rightarrow A_{eq} \Rightarrow [\![\,\sigma\,]\!])$ we have

$$\lambda c : A_c.\ \lambda p : A_p.\ \Phi_\tau(\lambda\mathsf{new}, \mathsf{eq}.\ \mathsf{let}\ x = p\ \mathsf{new}\ \mathsf{eq}\ \mathsf{in}\ c\,x)$$
$$=\ \ \lambda c : A_c.\ \lambda p : A_p.\ \mathsf{let}\ x = \Phi_\sigma p\ \mathsf{in}\ c\,x$$

(In particular, the $\Phi_\tau$ form a natural transformation $(A_{new} \Rightarrow A_{eq} \Rightarrow -) \to -$ considered as functors on the relevant portion of the Kleisli category $\mathcal{C}_T$.)

We now state a series of results which collectively validate the above definitions, and also confirm the appropriateness of our framework as a whole. (We omit the rather straightforward proofs.) We henceforth assume a fixed sound interpretation $[\![-]\!]$ of $\mathcal{L}$ in $\mathcal{C}$. By a *primary term* we mean an $\mathcal{L}^+$-term of the form $G[M]$ where $M \in \mathcal{L}$.

**Proposition 1.** *(i) If $\Phi_\tau$ is a freshness operator for $\tau$, then $[\![-]\!]^\Phi$ as defined above constitutes an interpretation of $\mathcal{L}^+_{\{\tau\}}$ which is sound for primary terms of type $\tau$.*

*(ii) Moreover, if $\Phi$ is a coherent family of freshness operators for $\mathcal{G}$, then $[\![-]\!]^\Phi$ is a sound interpretation of $\mathcal{L}^+_{\mathcal{G}}$.*

As an immediate consequence, we have the following:

**Proposition 2.** *(i) Suppose $\Phi$ is a freshness operator for $\tau$, $M$ is a closed primary term of type $\tau$, and $N$ a closed $\mathcal{L}$-term of type $\tau$. Then $\emptyset \vdash M \twoheadrightarrow^+ N : \tau$, implies $[\![\,M\,]\!]^\Phi = [\![\,N\,]\!]$.*

*(ii) Suppose $\Phi$ is a coherent family of freshness operators for $\mathcal{G}$, $M$ is an arbitrary closed $\mathcal{L}^+_{\mathcal{G}}$-term of type $\tau \in \mathcal{G}$, and $N$ a closed $\mathcal{L}$-term of type $\sigma$. Then $\emptyset \vdash E[M] \twoheadrightarrow^+ N$ implies $[\![\,E[M]\,]\!]^\Phi = [\![\,N\,]\!]$.*

The moral of Proposition 2 is roughly as follows. Typically, we will be interested in languages $\mathcal{L}$ with a designated class of syntactic values $V$, and a class of *ground types* $\gamma$ with the property that a value $V$ of ground type cannot contain any variables $new, eq, a_i$ with types as above. Thinking of $N$ in Proposition 2 as ranging over values, the proposition implies (in typical cases) that to obtain a sound interpretation for programs containing only **gen** expressions of ground type, the existence of the corresponding freshness operators is sufficient; however, to give an interpretation programs involving non-ground type localizations which correctly accounts for ground type observations on them, coherence is required. This phenomenon is not special to the case of name generation, but appears to be typical of our approach.

The converse half of computational adequacy requires stronger hypotheses, such as a "syntactic continuity" property, though we will not go into the details here. For our present purposes, a more interesting kind of converse is the following, which can be seen as validating our definition of freshness operator:

**Proposition 3.** *(i) Suppose $\Phi_\tau$ is an operator such that for every programming language $\mathcal{L}'$ with a sound interpretation $[\![-]\!]'$ in $\mathcal{C}$ (agreeing on types with $[\![-]\!]$), the interpretation $[\![-]\!]'^{\Phi_\tau}$ of $\mathcal{L}'{}^{\,|}_{\{\tau\}}$ is sound for primary terms of type $\tau$. Then $\Phi$ is a freshness operator for $[\![\tau]\!]$.*

*(ii) Suppose $\Phi$ is a family of operators over $\mathcal{G}$ such that for every programming language $\mathcal{L}'$ with a sound interpretation $[\![-]\!]'$ in $\mathcal{C}$, $[\![-]\!]'^{\Phi}$ is a sound interpretation of $\mathcal{L}'^{+}_{G}$. Then $\Phi$ is a coherent family of freshness operators.*

Note also that in the above setting, the freshness operators are themselves syntactically definable in $\mathcal{L}^+$, since $\Phi_\tau = [\![\,\mathbf{gen}_\nu\, new, eq, ()\,\mathbf{in}\,(- \bullet new \bullet eq)\,]\!]^{\Phi}$. In fact, it seems reasonable to suppose that for *any* natural interpretation of $\mathcal{L}^+$, the operator defined in this way will be a freshness operator. Thus, if we are seeking to interpret a language with name generation in an intensional model, little or no useful generality appears to be lost by assuming the existence of freshness operators.

## 4   Models

We now briefly review what we know concerning particular $\lambda_c$-models that possess freshness operators. Firstly, any of the known game models that suffice for modelling local store of integer type (see *e.g.* [3,5,6]) will also yield a model for fresh name generation according to our scheme, for the simple reason that (taking names to be just integers) a freshness operator may be readily implemented using a local integer store cell. (Note that the game models in question may be transformed into suitable $\lambda_c$-models by means of a standard construction [4].)

Some idea of the landscape may be gained by considering a few particular game models that are relatively simple to construct. We content ourselves here with a bare sketch of the relevant points, referring to the literature for further details. Let $\mathcal{G}$ denote the basic game model introduced by Lamarche (see [10]): here, a game $G$ consists of sets $O_G, P_G$ of *opponent* and *player* moves respectively, together with a non-empty prefix-closed set $L_G$ of *legal positions* of the form $o_1 p_1 \ldots o_n p_n$ ($n \geq 0$) or $o_1 p_1 \ldots o_n$ ($n \geq 1$), where $o_i \in O_G$, $p_i \in P_G$. Such games (with suitable morphisms) form a symmetric monoidal closed category, on which one may consider several different linear exponentials '!' embodying different notions of "reusability". From any of these exponentials we may obtain a category $\mathcal{G}_!$ with the same objects as $\mathcal{G}$, in which morphisms $G \to H$ are simply morphisms $!G \to H$ of $\mathcal{G}$. This gives a cartesian-closed category with a lifting monad, and hence a suitable $\lambda_c$-model.

Some exponentials of particular interest are the following (*cf.* [19]):

- The "Lamarche exponential" $!_1$. Here moves in $!_1 G$ are certain finite subtrees of $L_G$, and a play in $!_1 G$ consists of an "exploration" of $L_G$ in which one new position $s \in L_G$ is added to the subtree at each stage. From the corresponding category $\mathcal{G}_{!_1}$ one recovers essentially the world of sequential algorithms [10]. However, this does *not* yield a model for either ground type store or freshness, essentially because repetitions of earlier moves are not accounted for in $!_1 G$.
- The (more powerful) "Hyland exponential" $!_2$ of [12]. Here $!_2 G$ essentially consists of $\omega$ copies of $G$ side by side, with the stipulation that one cannot play a move in the $(i+1)$th copy unless one has already played in the $i$th copy. The category $\mathcal{G}_{!_2}$ gives a good model for ground type store and more besides [32], and in particular has a coherent family of freshness operators.
- A still more powerful exponential $!_3$ may be defined, where plays in $!_3 G$ explore trees of *justified sequences* of moves in $G$. This essentially coincides with the exponential given in [5], except that we do not impose a visibility condition on our plays. Again, the corresponding model supports groundtype store and freshness operators.

We are also aware of one model which is *not* a game model, and which supports freshness operators but not local store. This provides an encouraging sign that our general approach is applicable beyond the class of models that motivated it. The model in question is based on a "resource-sensitive" model for linear logic, in which *multisets* are used to keep track of the number of times some argument is invoked in a computation, but without imposing a temporal order on these invocations as the game models do. Specifically, we have in mind the category **MRel**, whose objects are sets and whose morphisms $f : S \to T$ are relations $f \subseteq \mathcal{M}_f(S) \times T$, where $\mathcal{M}_f(S)$ is the set of finite multisets over $S$. (An explicit description of this category and its cartesian closed structure is given in [9].) We may also endow **MRel** with a (rather crude) lifting monad which simply adds to each set a new token $*$ signalling "definedness", and again apply the construction of [4] to obtain a $\lambda_c$-model. Within this model, it is possible to "probe" an operation $p : A_{new} \Rightarrow A_{eq} \Rightarrow X$ in order to discover what it does when all invocations of $A_{new}$ yield different answers. Using this idea, we obtain a coherent family of freshness operators within the model.

We now return to the question of full abstraction mentioned in the Introduction. According to the setup of Section 3, if the interpretation $[\![-]\!]$ of $\mathcal{L}$ in $\mathcal{C}$ satisfies full abstraction and definability, then so will the resulting interpretation

$[\![-]\!]^{\Phi}$ of $\mathcal{L}^{+}$. However, this relies on the fact that, in $\mathcal{L}^{+}$, the characteristic operators *new* and *eq* are just ordinary variables, whereas in more realistic languages they will be hard-wired in as language primitives, as in the original $\nu$-calculus [25]. In the latter case, we can still get a semantic interpretation

$$M : \sigma \ \mapsto \ [\![ M ]\!] \in A_{new} \Rightarrow A_{eq} \Rightarrow [\![ \sigma ]\!]$$

by treating *new* and *eq* as free variables, though this will (in game models, for instance) not even validate such simple observational equivalences as

```
let (x,y)=(new(),new()) in M  ≃  let (y,x)=(new(),new()) in M
```

To do better than this, an alternative (and still compositional) interpretation

$$M : \sigma \ \mapsto \ [\![ M ]\!]^{\dagger} \in (A_{new} \Rightarrow A_{eq} \Rightarrow T[\![ \sigma ]\!] \Rightarrow 1) \Rightarrow 1$$

may cheaply be defined from $[\![-]\!]$ as follows:

$$[\![ M ]\!]^{\dagger} \ = \ \lambda P. \ \Phi \, (\lambda \mathsf{new}, \mathsf{eq}. \ (P \, \mathsf{new} \, \mathsf{eq})([\![ M ]\!] \, \mathsf{new} \, \mathsf{eq}))$$

In typical models, $[\![-]\!]^{\dagger}$ will validate simple equivalences such as the one above, at least at low types. Whether a fully abstract semantics for (extensions of) the $\nu$-calculus can be given along these lines is an interesting outstanding question.

## 5   Further Work and Conclusions

The next step in our programme is to carry out a similar analysis for other computational effects. We have informally verified that a similar story can be told for exceptions and for (ground or higher type) local store. In the case of exceptions, two options are available. The first is to consider fourth-order exception operators of types such as

$$(A_{raise} \Rightarrow A_{handle} \Rightarrow [\![ \tau ]\!]) \Rightarrow [\![ \tau ]\!]$$

where $A_{raise} = 1 \Rightarrow 1$ and $A_{handle} = (1 \Rightarrow [\![ \sigma ]\!]) \Rightarrow (1 \Rightarrow [\![ \sigma ]\!]) \Rightarrow [\![ \sigma ]\!]$. This affords a very general treatment of exceptions allowing us to model complex dynamic scoping phenomena; however, the relevant exception operators are only available in relatively powerful game models such as $\mathcal{G}_{!_3}$. Another option is to restrict attention to a somewhat more disciplined class of uses of exceptions, in which an independent *raise* operator is eschewed and instead the first argument to *handle* explicitly incorporates the relevant invocations of *raise*: consider *e.g.* $A'_{handle} = ((1 \Rightarrow 1) \Rightarrow [\![ \sigma ]\!]) \Rightarrow (1 \Rightarrow [\![ \sigma ]\!]) \Rightarrow [\![ \sigma ]\!]$. This conveniently accounts for those uses of exceptions that can be reasonably modelled *e.g.* in $\mathcal{G}_{!_1}$, and also gives us the rare pleasure of finding a use for a fifth-order operator!

An important difference arises when one considers local store. Whilst a sensible notion "store cell operator" may be formulated, it turns out that non-trivial coherent families of such operators never exist. The issue here is that programs of non-ground type involving local store can define functions with *persistent* internal state, which may behave differently each time they are called, and one cannot hope to model such a thing in a $\lambda_c$-model. However, it turns out that one can do better on this front by working in a *linear* rather than an intuitionistic framework, as is often done in game semantics to account for stateful behaviour (see *e.g.* [32]).

We also expect a similar treatment of continuations to be possible using suitable higher order operators. However, our approach seems to have very little to offer in the case of non-determinism or input/output, since there is no evident specification for the relevant operators in these cases.

Once some further instances of our approach have been worked out, a detailed comparison of the merits and demerits of the monadic (or algebraic theory) and intensional approaches will be possible. There is, however, one suggestion we would like to make at this stage. In the monadic approach, each effect featuring in a complex language typically requires a separate increment to the model construction. Moreover, if *local* state is to be treated, maybe functor categories must also be added to the mix; for name generation, perhaps nominal sets are required too. By contrast, in the intensional approach, one may be able to account for all these effects using operators that are naturally to hand in a single model — indeed, it seems that there *are* fairly simple models (such as $\mathcal{G}_{!_3}$, or more accurately its linear counterpart) that support virtually all the effect operators one might hope for. If this is so, we regard it as an important point in favour of intensional semantics.

Finally, we note that there are some tantalizing resemblances between our approach and the mechanism employed in Haskell for localization of effects using the `runST` operator [15,23]. It would be interesting to explore this connection.


# References


1. Abramsky, S.: Semantics of interaction: an introduction to game semantics. In: Pitts, A.M., Dybjer, P. (eds.) Semantics and Logics of Computation, CUP, pp. 1–31 (1997)
2. Abramsky, S., Ghica, D., Murawski, A., Ong, C.-H., Stark, I.: Nominal games and full abstraction for the nu-Calculus. In: Proc. 19th LICS, pp. 150–159. IEEE Press, Los Alamitos (2004)
3. Abramsky, S., Honda, K., McCusker, G.: A fully abstract game semantics for general references. In: Proc. 13th LICS, pp. 334–344. IEEE Press, Los Alamitos (1998)
4. Abramsky, S., McCusker, G.: Call-by-value games. In: Nielsen, M. (ed.) CSL 1997. LNCS, vol. 1414, pp. 1–17. Springer, Heidelberg (1998)
5. Abramsky, S., McCusker, G.: Game semantics. In: Proc. 1997 Marktoberdorf Summer School, pp. 1–56. Springer, Heidelberg (1999)
6. Abramsky, S., McCusker, G.: Full Abstraction for Idealized Algol with passive expressions. Theor. Comp. Sci. 227, 3–42 (1999)
7. Amadio, R.M., Curien, P.-L.: Domains and Lambda-Calculi, CUP (1998)
8. Berry, G., Curien, P.-L.: Sequential algorithms on concrete data structures. Theor. Comp. Sci. 20, 265–321 (1982)

9. Bucciarelli, A., Ehrhard, T., Manzonetto, G.: Not enough points is enough. In: Duparc, J., Henzinger, T.A. (eds.) CSL 2007. LNCS, vol. 4646, pp. 298–312. Springer, Heidelberg (2007)

10. Curien, P.-L.: On the symmetry of sequentiality. In: Main, M.G., Melton, A.C., Mislove, M.W., Schmidt, D., Brookes, S.D. (eds.) MFPS 1993. LNCS, vol. 802, pp. 29–71. Springer, Heidelberg (1994)

11. Curien, P.-L.: Notes on game semantics. From the author's web page (2006)

12. Hyland, J.M.E.: Game semantics. In: Pitts, A.M., Dybjer, P. (eds.) Semantics and Logics of Computation, CUP, pp. 131–194 (1997)

13. Laird, J.: A game semantics of local names and good variables. In: Walukiewicz, I. (ed.) FOSSACS 2004. LNCS, vol. 2987, Springer, Heidelberg (2004)

14. Lambek, J., Scott, P.: Introduction to Higher-Order Categorical Logic, CUP (1986)

15. Launchbury, J., Peyton Jones, S.: State in Haskell. Lisp Symb. Comp. 8, 193–341 (1995)

16. Longley, J.R.: Notions of computability at higher types I. In: Longley, J.R. (ed.) Proc. Logic Colloquium 2000. Lecture Notes in Logic, ASL, vol. 200, pp. 32–142 (2005)

17. Longley, J.R.: On the ubiquity of certain total type structures. Math. Struct. in Comp. Science 17, 841–953 (2007)

18. Longley, J.R., Wolverson, N.: Eriskay: a programming language based on game semantics. GaLoP III, Budapest (to be presented) (2008)

19. Melliès, P.-A.: Comparing hierarchies of types in models of linear logic. Inf. Comp. 189(2), 202–234 (2004)

20. Moggi, E.: Computational lambda-calculus and monads. In: LFCS report ECS-LFCS-88-66, University of Edinburgh (1988); A shorter version appeared. In: Proc. 4th LICS, pp.14-23. IEEE Press (1989)

21. Moggi, E.: An abstract view of programming languages. LFCS report ECS-LFCS-90-113, University of Edinburgh (1989)

22. Moggi, E.: Notions of computation and monads. Inf. and Comp. 93, 55–92 (1991)

23. Moggi, E., Sabry, A.: Monadic encapsulation of effects: a revised approach. J. Funct. Prog. 11(6), 591–627 (2001)

24. Peyton Jones, S., Wadler, P.: Imperative functional programming. In: Proc. 20th POPL, ACM Press, New York (1993)

25. Pitts, A., Stark, I.: Observable Properties of Higher Order Functions that Dynamically Create Local Names, or: What's *new*? In: Borzyszkowski, A.M., Sokolowski, S. (eds.) MFCS 1993. LNCS, vol. 711, pp. 122–141. Springer, Heidelberg (1993)

26. Plotkin, G.D.: LCF considered as a programming language. Theor. Comp. Sci. 5, 223–255 (1977)

27. Plotkin, G.D., Power, A.J.: Algebraic operations and generic effects. Appl. Categ. Struct. 11(1), 69–94 (2003)

28. Shinwell, M.R., Pitts, A.M.: On a monadic semantics for freshness. Theor. Comp. Sci. 342, 28–55 (2005)

29. Stark, I.: Free-algebra models for the $\pi$-calculus. Theor. Comp. Sci. 390(2-3), 248–270 (2008)

30. Troelstra, A.S. (ed.): Metamathematical Investigation of Intuitionistic Arithmetic and Analysis. LNM, vol. 344. Springer, Heidelberg (1973)

31. Tzevelekos, N.: Full abstraction for nominal general references. In: Proc. 22nd LICS, pp. 399–410. IEEE Press, Los Alamitos (2007)

32. Wolverson, N.: Game semantics for object-oriented languages. PhD thesis, University of Edinburgh (submitted, 2007)

# Uniform Algebraic Reducibilities between Parameterized Numeric Graph Invariants

J.A. Makowsky[*]

Department of Computer Science
Technion–Israel Institute of Technology
Haifa, Israel
`janos@cs.technion.ac.il`

**Abstract.** We report about our ongoing study of inter-reducibilities of parameterized numeric graph invariants and graph polynomials. The purpose of this work is to systematize recent emerging work on graph polynomials and various partition functions with respect to their combinatorial expressiveness and computational complexity.

## 1 Graph Polynomials and Partition Functions

Motivated by problems in statistical mechanics, computational chemistry, computational biology and quantum physics, the recent years have produced many papers dealing with graph polynomials and various combinatorial counting and partition functions with their specific idiosyncrasies and intricacies, [ABS04a], [ABS04b], [BCL$^+$06], [BR99], [BG05], [CG95], [Cou], [DG00] [FLS07], [JVW90], [LM04], [Sok05], [Sze07]. The interrelations between the various new graph polynomials and functions introduced in these papers are not well understood. They tend to be computationally extremely complex and rather well beyond the complexity levels described as potentially feasible ($\sharp P$ rather than $NP$).

Although there is an abundance of literature about specific graph polynomials, cf. [Mak07], no systematic study of their complexities and inter-reducibilities has been presented so far.

We have created a WEB-page

$$\texttt{http://www.cs.technion.ac.il/}\sim\texttt{janos/RESEARCH/gp-homepage.html}$$

which documents our efforts in unifying research in graph polynomials and parametrized numeric graph invariants. It also contains links to our collaborators and to researchers interested in related topics.

## 2 Complexity and Reducibilities

In this talk we concentrate on algebraic and complexity theoretic aspects. The main goal is to uncover common features for wide classes of parameterized nu-

[*] Funded by ISF-Grant "Model Theoretic Interpretations of Counting Functions" 2007-2010 and the Grant for Promotion of Research by the Technion–Israel Institute of Technology.

A. Beckmann, C. Dimitracopoulos, and B. Löwe (Eds.): CiE 2008, LNCS 5028, pp. 403–406, 2008.

meric graph invariants in the spirit of the recently developed theory for graph homomorphisms and partition functions.

Various notions of reducibilities can be applied in these situations. However, the reducibilities by polynomial time Turing machines do not capture the inherent combinatorial content of these functions, especially when these functions are describing infinite families in a uniform way.

In the complexity analysis of graph polynomials the point-to-point reductions of evaluations were defined as non-uniform, although in the proofs a certain uniformity appeared naturally. In previous work we introduced several notions of uniform reducibilities between parameterized numeric graph invariants and used these to study interpretability and complexity of graph polynomials. In particular, our aim was

(i) To study in great detail these uniform reductions and explore them on selected examples from the existing literature;

(ii) To measure the combinatorial content of a graph polynomial. Traditionally, a graph polynomial subsumes another if the first can be obtained from the second by a substitution instance and multiplication by a pre-factor. This notions seems to be too weak. However, polynomial time Turing reducibility gives a notion which is to strong for this purpose. Combining combinatorial and algebraic techniques we plan to establish the "right" notions to capture combinatorial content.

## 3   The Difficult Point Conjecture

We were capable of recasting classical results on the Tutte polynomial in a general framework, and show that, from a complexity point of view, other graph polynomials share the same behavior. This has led us to formulate a rather general conjecture, which postulates similar behavior for a large class of parameterized numeric graph invariants, the Difficult Point Conjecture. Roughly speaking, the Difficult Point Conjecture states that, for the class of graph polynomials definable in Monadic Second Order Logic, either all its evaluations are uniformly computable in polynomial time, or, if there is an evaluation which is hard to compute, then, up to a small exception set, all its evaluation are equally hard to compute. Furthermore, there is a uniform algebraic reduction, which relates all the hard evaluations to each other. The exception set is small in the sense that the set of evaluation points for which the polynomial is easy to evaluate has strictly lower dimension than the space of evaluation points. More details can be found in [Mak07].

The class of graph polynomials definable in Monadic Second Order Logic (MSOL) comprises all of the graph polynomials we could find in the literature with two exceptions: The weighted graph polynomial introduced by S. Noble and D. Welsh [NW99], and the graph polynomial counting the number of harmonious colorings. Harmonious graph colorings were introduced in [HK83]. All the partition functions (vertex coloring models) studied in [FLS07] are evaluations of graph polynomials definable in Monadic Second Order Logic, however, the edge coloring models studied in [Sze07] need not be definable in this way. The conjecture is formulated for the class of MSOL-definable polynomials for convenience only. If it holds for MSOL, it is likely to hold also for larger classes of graph polynomials.

The difficulty in proving this conjecture consists in finding a *unique* reason for many graph polynomials to behave in this way. Partial progress in proving the Difficult Point Conjecture was recently achieved in the following papers: In [BD07] the complexity of the cover polynomial of [CG95] is completely analyzed in the spirit of [JVW90]. The same is accomplished in [BH07] for the interlace polynomial, and in [BDM08] for the multivariate Tutte polynomials. All these results rely on, sometimes intricate, variations of the original ideas used in [JVW90]. But it is not clear how to extend this to the general case of MSOL-definable polynomials.

**Acknowledgment.** I would like to thank my co-authors I. Averbouch, M. Bläser, B. Courcelle, H. Dell, B. Godlin, T. Kotek and C. Hoffmann for their contributions to this talk.

# References


[ABS04a]  Arratia, R., Bollobás, B., Sorkin, G.B.: The interlace polynomial of a graph. Journal of Combinatorial Theory Series B 92, 199–233 (2004)

[ABS04b]  Arratia, R., Bollobás, B., Sorkin, G.B.: A two-variable interlace polynomial. Combinatorica 24.4, 567–584 (2004)

[BCL+06]  Borgs, C., Chayes, J., Lovász, L., Sós, V.T., Vesztergombi, K.: Counting graph homomorphisms. In: Klazar, M., Kratochvil, J., Loebl, M., Matousek, J., Thomas, R., Valtr, P. (eds.) Topics in Discret mathematics, pp. 315–371. Springer, Heidelberg (2006)

[BD07]  Bläser, M., Dell, H.: Complexity of the cover polynomial. In: Arge, L., Cachin, C., Jurdziński, T., Tarlecki, A. (eds.) ICALP 2007. LNCS, vol. 4596, pp. 801–812. Springer, Heidelberg (2007)

[BDM08]  Bläser, M., Dell, H., Makowsky, J.A.: Complexity of the Bollobás-Riordan polynomia, exceptional points and uniform reductions. In: CSR 2008 (accepted for presentation) (2008)

[BG05]  Bulatov, A., Grohe, M.: The complexity of partition functions. Theoretical Computer Science 348, 148–186 (2005)

[BH07]  Bläser, M., Hoffmann, C.: On the complexity of the interlace polynomial. arXive 0707.4565 (2007)

[BR99]  Bollobás, B., Riordan, O.: A Tutte polynomial for coloured graphs. Combinatorics, Probability and Computing 8, 45–94 (1999)

[CG95]  Chung, F.R.K., Graham, R.L.: On the cover polynomial of a digraph. Journal of Combinatorial Theory Ser. B 65(2), 273–290 (1995)

[Cou]  Courcelle, B.: A multivariate interlace polynomial (preprint, December 2006)

[DG00]  Dyer, M., Greenhill, C.: The complexity of counting graph homomorphisms. Random Structures and Algorithms 17(3-4), 260–289 (2000)

[FLS07]  Freedman, M., Lovász, L., Schrijver, A.: Reflection positivity, rank connectivity, and homomorphisms of graphs. Journal of AMS 20, 37–51 (2007)

[HK83]    Hopcroft, J.E., Krishnamoorthy, M.S.: On the harmonious coloring of graphs. SIAM J. Algebraic Discrete Methods 4, 306–311 (1983)

[JVW90]   Jaeger, F., Vertigan, D.L., Welsh, D.J.A.: On the computational complexity of the Jones and Tutte polynomials. Math. Proc. Camb. Phil. Soc. 108, 35–53 (1990)

[LM04]    Lotz, M., Makowsky, J.A.: On the algebraic complexity of some families of coloured Tutte polynomials. Advances in Applied Mathematics 32(1-2), 327–349 (2004)

[Mak07]   Makowsky, J.A.: From a zoo to a zoology: Towards a general theory of graph polynomials. In: Theory of Computing Systems (on-line first) (2007)

[NW99]    Noble, S.D., Welsh, D.J.A.: A weighted graph polynomial from chromatic invariants of knots. Ann. Inst. Fourier, Grenoble 49, 1057–1087 (1999)

[Sok05]   Sokal, A.: The multivariate Tutte polynomial (alias Potts model) for graphs and matroids. In: Survey in Combinatorics, 2005. London Mathematical Society Lecture Notes, vol. 327, pp. 173–226 (2005)

[Sze07]   Szegedy, B.: Edge coloring models and reflection positivity. arXiv: math.CO/0505035 (2007)

# Updatable Timed Automata with Additive and Diagonal Constraints

Lakshmi Manasa, Shankara Narayanan Krishna, and Kumar Nagaraj

Department of Computer Science and Engineering
Indian Institute of Technology, Bombay,
Powai, Mumbai, India 400 076
{manasa,krishnas,kumar}@cse.iitb.ac.in

**Abstract.** Timed automata [1] are a well established theoretical framework used for modeling real time systems. In this paper, we introduce a class of timed automata with updates of the form $x :\sim d - y$. We show that these updates preserve the decidability of emptiness of classical timed automata with diagonal constraints for $\sim \in \{=, <, \leq\}$, while emptiness is undecidable for $\sim \in \{>, \geq, \neq\}$. When used along with additive constraints, these updates give decidable emptiness with 2 or lesser number of clocks, and become undecidable for 3 clocks. This provides a partial solution to a long standing open problem [5].

## 1  Introduction

Timed automata, introduced in [1] are a well established model for the verification of real time systems. Since their introduction, several variants of timed automata have been studied. Some of them are : [4], wherein new operations on clock updates were considered, [5] where new kinds of guards were introduced, [7] where a set of clocks could be freezed, and [6], where silent transitions were studied. A very interesting result in the theory of timed automata is that the emptiness problem is decidable. This has paved the way for using timed automata in verification and many tools viz., UPPAAL [8], KRONOS [9] and HyTech [2] were built using this.

In this paper, we consider a variant of timed automata having new update operations. The updates we consider are of the form $x :\sim d - y$, $\sim \in \{<, \leq, =, >, \geq, \neq\}$. It has been shown [4] that updating a clock based on another clock's value $(x :\sim y, x :\sim y + c)$ is very powerful and often leads to undecidability of emptiness. We show here that updates of the kind $x :\sim d - y \sim \in \{=, \leq, <\}$ preserve the decidability, and thus behave very differently from the known set of updates. However, when these kinds of updates are used along with additive constraints, emptiness remains decidable only when the number of clocks used are two or less. In fact, our proof uses only deterministic updates $x := d - y$. This also gives a partial solution to the problem of deciding emptiness for the class of timed automata with additive constraints and 3 clocks [5].

It is well known that the expressive power of classical timed automata [1] does not change when location invariants $(x \sim c)$ are used. It can be shown that using constraints $x - y \sim c$ as location invariants and guards, emptiness remains

A. Beckmann, C. Dimitracopoulos, and B. Löwe (Eds.): CiE 2008, LNCS 5028, pp. 407–416, 2008.

decidable for the class of timed automata using $x :< c$ as updates ([4] has proved this without using location invariants; however, adapting the proof to the case of using $x - y \sim c$ as location invariants is not hard). However, if we allow additive constraints $x + y = c$ as location invariants and guards, then emptiness becomes undecidable when used along with updates $x :< c$, in the presence of 3 or more clocks.

## 2   Prerequisites

For any set $S$, $S^*$ ($S^\omega$) denotes the set of all finite (infinite) strings over $S$. $S^\infty = S^* \cup S^\omega$. We consider as time domain $\mathbf{T}$ the set $\mathbf{Q}^+$ or $\mathbf{R}^+$ of non-negative rationals or reals, and $\Sigma$ a finite set of actions. A time sequence over $\mathbf{T}$ is a finite (infinite) non-decreasing sequence $\tau = (t_i)_{i \geq 1}$ ; for simplicity $t_0$ is taken to be zero always. A timed word over $\Sigma$ is defined as $\rho = (\sigma, \tau)$, where $\sigma = (\sigma_i)_{i \geq 1}$ is a finite (infinite) sequence of symbols in $\Sigma$ and $\tau = (t_i)_{i \geq 1}$ is a finite (infinite) sequence in $\mathbf{T}^\infty$. A *timed language* $L$ is a set of timed words.

We consider a finite set of variables $X$ called *clocks*. A clock valuation over $X$ is a map $\nu : X \to \mathbf{T}$ mapping each clock $x \in X$ to a time value. $\nu(x)$ represents the value assigned to the clock $x$ by $\nu$ and $frac(\nu(x))$ represents the fractional part of the value $\nu(x)$. For $t \in \mathbf{T}$, the valuation $\nu + t$ is defined as $(\nu + t)(x) = \nu(x) + t, \forall x \in X$. The set of all clock valuations over $X$ is denoted by $\mathbf{T}^X$.

For the set of clocks $X$, the set of constraints (guards) over $X$, denoted by $C_-(X)$ is given as follows: $\varphi ::= x \sim c | x - y \sim c | \varphi \wedge \varphi | \varphi \vee \varphi$ where $x, y \in X, c \in \mathbf{Q}^+$ and $\sim \in \{<, \leq, >, \geq, =, \neq\}$. Constraints $x - y \sim c$ are called *diagonal constraints*. Clock constraints are interpreted over clock valuations. The relation $\nu \models \varphi$ (valuation $\nu$ satisfies constraint $\varphi$) is defined as follows: $\nu \models x \sim c$ if $\nu(x) \sim c$, $\nu \models x - y \sim c$ if $\nu(x) - \nu(y) \sim c$ and so on.

Clock constraints allow us to test the values of clocks. To change the value of a clock $x$ we use a *simple update* of the form $up_x :: x := 0$. $U_0(X)$ denotes the set of updates $up \in U_0(X)$ defined as $up = \bigwedge_{x \in X} up_x$ where $up_x$ are *simple updates*. These updates are functions from $\mathbf{T}^X$ to $\mathcal{P}(\mathbf{T}^X)$.

Let $\nu$ be a valuation and let $up_z$ be a simple update over clock $z$. A valuation $\nu'$ is in $up_z(\nu)$ if $\nu'(y) = \nu(y), y \neq z$ and $\nu'(z) = 0$. For two valuations $\nu, \nu' \in \mathbf{T}^X$, $\nu' \in up(\nu)$ if for every $x \in X$, $\nu'$ coincides with $\nu'' \in up_x(\nu)$ over the value of $x$.

### 2.1   Timed Automata

A *timed automaton* [1] is a tuple $\mathcal{A} = (L, L_0, \Sigma, X, E, F)$ where $L$ is a finite set of locations; $L_0 \subseteq L$ is a set of initial locations; $\Sigma$ is a finite set of symbols; $X$ is a finite set of clocks; $E \subseteq L \times L \times \Sigma \times C_-(X) \times U_0(X)$ is the set of transitions and $F \subseteq L$ is a set of final locations. $C_-(X)$ and $U_0(X)$ are the set of clock constraints and clock updates as described above. An edge $e = (l, l', a, \varphi, \phi)$ represents a transition from $l$ to $l'$ on symbol $a$, with the valuation $\nu \in \mathbf{T}^X$ satisfying the guard $\varphi$, and then $\phi$ gives the updates of certain clocks.

A path is a finite (infinite) sequence of consecutive transitions. The path is said to be accepting if it starts in an initial location ($l_0 \in L_0$) and ends in a final location (or repeats a final location infinitely often). A run through a path from a valuation $\nu_0'$ (with $\nu_0'(x) = 0$ for all $x$) is a sequence $(l_0, \nu_0') \xrightarrow{t_1} (l_0, \nu_1) \xrightarrow{(\sigma_1, \varphi_1, \phi_1)} (l_1, \nu_1') \xrightarrow{t_2} (l_1, \nu_2) \xrightarrow{(\sigma_2, \varphi_2, \phi_2)} (l_2, \nu_2') \cdots (l_n, \nu_n')$. Note that $\nu_i = \nu_{i-1}' + (t_i - t_{i-1})$, $\nu_i \models \varphi_i$, and that $\nu_i' \in up(\nu_i)$, $i \geq 1$. A timed word $\rho$ is accepted by $\mathcal{A}$ iff there exists an accepting run (through an accepting path) over $\mathcal{A}$, the word corresponding to which is $\rho$. The timed language $L(\mathcal{A})$ accepted by $\mathcal{A}$ is defined as the set of all timed words accepted by $\mathcal{A}$.

## 2.2   Region Automata

Given a set $X$ of clocks, let $\mathcal{R}$ be a finite partitioning of $\mathbf{T}^X$. Each partition contains a set (possibly infinite) of clock valuations. Given $\alpha \in \mathcal{R}$, the successors of $\alpha$ represented by $Succ(\alpha)$ are defined as

$$\alpha' \in Succ(\alpha) \text{ if } \exists \nu \in \alpha, \exists t \in \mathbf{T} \text{ such that } \nu + t \in \alpha'$$

The finite partition $\mathcal{R}$ is said to be a *set of regions* iff

$$\alpha' \in Succ(\alpha) \Longleftrightarrow \forall \nu \in \alpha, \exists t \in \mathbf{T} \text{ such that } \nu + t \in \alpha'.$$

A set of regions is consistent with time elapse if two valuations which are equivalent (within the same partition) stay equivalent with time elapse. A region $\alpha \in \mathcal{R}$ is said to satisfy a clock constraint $\varphi \in C_-(X)$ denoted as $\alpha \models \varphi$, if $\forall \nu \in \alpha, \nu \models \varphi$. A clock update $\phi \in U_0(X)$ maps a region $\alpha$ to a set of regions $\phi(\alpha) = \{\alpha' \mid \alpha' \cap \phi(\nu) \neq \emptyset \text{ for some } \nu \in \alpha\}$.

A set of regions $\mathcal{R}$ is said to be *compatible* with a finite set of clock constraints $C_-(X)$ iff $\forall \; \varphi \in C_-(X)$ and $\forall \alpha \in \mathcal{R}$ either $\alpha \models \varphi$ or $\alpha \models \neg\varphi$. A set of regions $\mathcal{R}$ is said to be *compatible* with a finite set of clock updates $U_0(X)$ iff $\alpha' \in \phi(\alpha) \Rightarrow \forall \nu \in \alpha, \exists \nu' \in \alpha'$ such that $\nu' \in \phi(\nu)$.

Given a timed automaton $\mathcal{A}$, and a set of regions $\mathcal{R}$ compatible with $C_-(X)$ and $U_0(X)$, the *region automaton* $\mathcal{R}(\mathcal{A}) = (Q, Q_0, \Sigma, E', F')$ is defined as follows: $Q = L \times \mathcal{R}$ the set of locations; $Q_0 = L_0 \times \mathcal{R} \subseteq Q$ the set of initial locations; $F' = F \times \mathcal{R} \subseteq Q$ the set of final locations; $E' \subseteq (Q \times \Sigma \times Q)$ is the set of edges. $(l, \alpha) \xrightarrow{a} (l', \alpha')$ if $\exists \alpha'' \in \mathcal{R}$ and a transition $(l, l', a, \varphi, \phi) \in E$ such that (a) $\alpha'' \in Succ(\alpha)$, (b) $\alpha'' \models \varphi$ and (c) $\alpha' \in \phi(\alpha'')$.

The region automaton is an abstraction of the timed automaton accepting $\text{Untime}(L(\mathcal{A}))$ [1]. As a consequence, the following theorem was proved [1].

**Theorem 1.** *Let $\mathcal{A}$ be a timed automaton. Then the problem of checking emptiness of $L(\mathcal{A})$ is decidable.*

We say that a class of timed automata is *decidable* if there exists a *decider* for the emptiness problem.

## 3   Diagonal Constraints

In this section, we consider a class of timed automata called *updatable timed automata with diagonal constraints*. This class of automata has $C_-(X)$ as the set of clock constraints. The set of updates of this class of automata denoted by $U_{\leq}(X)$ are defined as $up = \bigwedge_{x \in X} up_x$ where $up_x$ are *simple updates* as follows: $up_x ::= x := 0 \mid x :\sim d - y,$ where $x, y \in X, d \in \mathbf{Q}^+$ and $\sim \in \{=, <, \leq\}$. Here, $d \leq c^{max}$ where $c^{max} = max\{c \mid x \sim c \in C_-(X) \text{ or } x - y \sim c \in C_-(X)\}$. Accordingly, we have $\nu' \in up_z(\nu)$ if $\nu'(y) = \nu(y), y \neq z$ and $\nu'(z) = 0$ if $up_z ::= z := 0$ and $\nu'(z) \sim d - \nu(y) \wedge \nu'(z) \geq 0$ if $up_z ::= z :\sim d - y$.

For every clock $x \in X$, define a constant $c_x^{max} = c^{max} + d_x^{max}$ where $d_x^{max} = max\{c \mid x - y \sim c \in C_-(X) \wedge y \in X\}$, and a set of intervals $\mathcal{I}_x$ as

$$\mathcal{I}_x = \{[c] | 0 \leq c \leq c_x^{max}\} \cup \{(c, c+1) | 0 \leq c < c_x^{max}\} \cup \{(c_x^{max}, \infty)\}.$$

For every pair of clocks $x, y \in X$, define $D_{xy} = max\{c \mid x - y \sim c \in C_-(X)\}$, and the set of intervals $\mathcal{J}_{xy}$ as

$$\mathcal{J}_{xy} = \{(-\infty, -d_{yx})\} \cup \{[d]\} \cup \{(d', d'+1)\} \cup \{(d_{xy}, +\infty)\} \text{ where } -d_{yx} \leq d \leq d_{xy} \text{ and } -d_{yx} \leq d' < d_{xy}.$$

Let $\alpha$ be a tuple $((I_x)_{x \in X}, (J_{xy})_{x,y \in X}, \prec)$ where

1. $I_x \in \mathcal{I}_x, J_{xy} \subset \mathcal{J}_{xy}$, and
2. $\prec$ is a total preorder on $X_0 = \{x \in X \mid I_x \text{ is of the form } (c, c+1)\}$.

$\alpha$ defines the following subset of $\mathbf{T}^X$ : valuations $\nu \in \mathbf{T}^X$ such that $\nu(x) \in I_x$ for all $x \in X$, $\nu(x) - \nu(y) \in J_{xy}$ for all $x, y \in X$, and $frac(\nu(x)) \leq frac(\nu(y))$ if $x \prec y$ for all $x, y \in X_0$. The set of all such tuples $\alpha$ partitions $\mathbf{T}^X$ and is represented by $\mathcal{R}^-$.

*Remark 1.* When using updates $x :\sim d - y$, the value $\nu'(x)$ must be non-negative, as mentioned above. This can be ensured by adding a guard $y \leq d$? while making the transition.

The following are easy to observe.

**Lemma 1.** *Set $\mathcal{R}^-$ forms a set of regions.*

*Proof.* Let $\alpha = ((I_x)_{x \in X}, (J_{xy})_{x,y \in X}, \prec) \in \mathcal{R}^-$. If for all $x, I_x = (c_x^{max}, \infty)$, then $Succ(\alpha) = \{\alpha\}$ as time elapse would not change $J_{xy}$. If there is atleast one $I_x$ such that $I_x \neq (c_x^{max}, \infty)$, then there exists a region $\alpha' \neq \alpha$ such that $\alpha' \in Succ(\alpha)$. We define the closest successor of $\alpha$ to be $\alpha'$ such that $\forall \nu \in \alpha, \forall t \in \mathbf{T}$, if $\nu + t \notin \alpha$, then $\exists t' \leq t$ such that $\nu + t' \in \alpha'$.

Let $Z = \{x \in X \mid I_x = [c]\}$. In this case, $I'_x = (c, c+1)$ for $x \in Z$ and $c < c_x^{max}$; $I'_x = (c_x^{max}, \infty)$ for $x \in Z$ and $c = c_x^{max}$ while $I'_x = I_x$ for $x \notin Z$. $x \prec' y$ if $x \prec y$ or if $I_x = [c], c < c_x^{max}$ and $I_y = (d, d+1)$. $J'_{xy} = J_{xy}$ for all $x, y$.

In case $Z = \emptyset$, then pick the clock(s) $x \in X_0$ having the maximal fractional part. For these, $I'_x = [c+1]$ if $I_x = (c, c+1), c < c_x^{max}$, and for the rest of clocks $y \in X, I'_y = I_y$. Here, $\prec'$ is the restriction of $\prec$ to clocks $x$ such that $I'_x = (d, d+1)$. Again, $J'_{xy} = J_{xy}$ for all $x, y$.

It is easy to see that for all $\nu \in \alpha$, there exists a $t \in \mathbf{T}$ such that $\nu + t \in \alpha'$.   $\square$

**Lemma 2.** *The set of regions $\mathcal{R}^-$ is compatible with the set of clock constraints $C_-(X)$.*

*Proof.* It is easy to see that with the choice of regions $\mathcal{R}^-$, any $\alpha \in \mathcal{R}^-$ satisfies either $\varphi$ or $\neg\varphi$. $\qquad\qquad\qquad\qquad\qquad\qquad\qquad\qquad\qquad\qquad\qquad\square$

**Lemma 3.** *The set of regions $\mathcal{R}^-$ is compatible with any simple update $z := 0$ or $z :\sim d - y$, $\sim \in \{=, <, \leq\}$.*

*Proof.* Let $\alpha = ((I_x)_{x \in X}, (J_{xy})_{x,y \in X}, \prec)$ be a region in $\mathcal{R}^-$ where $\prec$ is a total preorder on $X_0 = \{x \mid I_x$ is of the form $(c, c+1)\}$. Let $up_z$ be a simple update over clock $z \in X$. Let $\alpha' = ((I'_x)_{x \in X}, (J'_{xy})_{x,y \in X}, \prec')$. Then, $\alpha' \in up_z(\alpha)$ if $I'_x = I_x \forall x \in X \setminus \{z\}$, $J'_{xy} = J_{xy} \forall x, y \in X \setminus \{z\}$, and

1. $\phi = z := 0$
   - $I'_z = [0]$, $X'_0 = X_0 \setminus \{z\}$, $\prec' = \prec \cap (X'_0 \times X'_0)$,
   - $J'_{xz} = \begin{cases} [c] & \text{if } I_x = [c] & \text{and } c \leq d_{xz}, \\ (c, c+1) & \text{if } I_x = (c, c+1) \text{ and } c < d_{xz}, \\ (d_{xz}, \infty) & \text{otherwise} \end{cases}$
   - $J'_{zx} = \begin{cases} -[c] & \text{if } I_x = [c] & \text{and } c \leq d_{xz}, \\ (-c-1, -c) & \text{if } I_x = (c, c+1) \text{ and } c < d_{xz}, \\ (-\infty, -d_{xz}) & \text{otherwise} \end{cases}$
2. $\phi = z := d - y$
   Since $d - y \geq 0 \wedge d \leq c^{max}$, $I_y \neq (c_y^{max}, \infty)$, $I'_z \neq (c_z^{max}, \infty)$.
   - If $I_y = [c] \wedge 0 \leq c \leq d \leq c^{max}$ then $I'_z = [d - c]$
   - If $I_y = (c, c+1) \wedge 0 \leq c < d \leq c^{max}$ then $I'_z = (d - c - 1, d - c)$
   If $I'_z = [d - c]$ and let $e = d - c$ then
   - $X'_0 = X_0 \setminus \{z\}$, $\prec' = \prec \cap (X'_0 \times X'_0)$
   - $J'_{xz} = $
     $\begin{cases} (-\infty, -d_{zx}) & \text{if } I_x = [c'] \text{ or } (c', c'+1) \text{ and } c' - e < -d_{zx}, \\ [c' - e] & \text{if } I_x = [c'] \text{ and } -d_{zx} \leq c' - e \leq d_{xz}, \\ (c' - e, c' - e + 1) & \text{if } I_x = (c', c'+1) \text{ and } -d_{zx} \leq c' - e < d_{xz}, \\ (d_{xz}, \infty) & \text{otherwise} \end{cases}$
   - $J'_{zx}$ can be calculated similarly.
   If $I'_z = (d - c - 1, d - c)$ then
   - $X'_0 = X_0 \cup \{z\}$, $\prec'$ is same as $\prec$ except that if $frac(v(y)) = 0.5$ then $y \prec' z \wedge z \prec' y$, if $frac(v(y)) < 0.5$ then $y \prec' z$. Otherwise, $z \prec' y$.
   - $J'_{xz}$ and $J'_{zx}$ calculated similar to the case $I'_z = [d - c]$.
3. $\phi = z :< d - y$
   Note that as $d - y > 0 \wedge d \leq c^{max}$, $I_y \neq (c_y^{max}, \infty)$ and also $I'_z \neq (c_z^{max}, \infty)$.
   - If $I_y = [e] \wedge 0 \leq e \leq d \leq c^{max}$ then $I'_z = [e'] \vee I'_z = (e', e'+1)$ such that $e' < d - e$. If $I'_z = (e', e'+1)$, then $X'_0 = X_0 \cup \{z\}$, and $\prec'$ is any total preorder that coincides with $\prec$ on $X'_0 \setminus \{z\}$. Otherwise, $X'_0 = X_0 \setminus \{z\}$ and $\prec' = \prec \cap (X'_0 \times X'_0)$.
   - If $I_y = (e, e+1) \wedge 0 \leq e < d \leq c^{max}$ then $I'_z = [e']$ such that $e' < d - e$. $X'_0 = X_0 \setminus \{z\}$ and $\prec' = \prec \cap (X'_0 \times X'_0)$, or $I'_z = (e'', e''+1)$ such that $e'' \leq d - e - 1$. If $I'_z \cap (d - I_y) = \emptyset$, then $\prec'$ is any total preorder that coincides with $\prec$ on $X'_0 \setminus \{z\}$. If $I'_z \cap (d - I_y) \neq \emptyset$, $\prec'$ coincides with $\prec$ on $X'_0 \setminus \{z\}$ and $z \prec' y$, $y \not\prec' z$.
   $J'_{xz}$ and $J'_{zx}$ can be calculated in a similar manner to the above case.
4. $\phi = z :\leq d - y$. This is similar to the above case.

It is easy to see that for any $\nu \in \alpha$, and any $\alpha' \in up_z(\alpha)$, there exists $\nu' \in \alpha' \cap up_z(\nu)$.    $\square$

**Lemma 4.** *The set of regions $\mathcal{R}^-$ is compatible with updates $U_\leq(X)$.*

*Proof.* Let $\alpha = ((I_x)_{x \in X}, (J_{xy})_{x,y \in X}, \prec)$ and $\alpha_z = ((I_x^z)_{x \in X}, (J_{xy}^z)_{x,y \in X}, \prec_z)$ be two regions of $\mathcal{R}^-$ and $up_z$ be a simple update such that $\alpha_z \in up_z(\alpha)$. Let $\nu$ be a valuation in $\alpha$. By Lemma 3, $\mathcal{R}^-$ is compatible with $up_z$. Thus, for any valuation $\nu \in \alpha$, there exists some valuation $\nu_z \in up_z(\nu) \cap \alpha_z$. We need to show that for $up \in U_\leq(X)$, and region $\alpha$, we can find $\alpha' = ((I_x')_{x \in X}, (J_{xy}')_{x,y \in X}, \prec') \in up(\alpha)$ such that for any $\nu \in \alpha$, there exists $\nu' \in \alpha' \cap up(\nu)$.

Given $\nu \in \alpha$, define a valuation $\nu'$ as : (i) $\nu'(y) = \nu_y(y)$ for any clock $y$; (ii) for a pair $(y, z)$ of clocks, calculate $\nu'(y) - \nu'(z)$ as in Lemma 3, and (iii) $\prec'$ can be calculated from $\prec_z, z \in X$ as follows: Let $X' = \{x' \mid x \in X\}$ be a copy of clocks in $X$. Define $\prec_x'$ from $\prec_x$ by replacing $x$ with $x'$. $\prec_x'$ is a preorder on $X \cup X'$. Then $\prec'$ is obtained by taking the union of all $\prec_x', x \in X$; restricting it to $X'$; and then replacing $X'$ with $X$. Then $\nu' \in \alpha' \cap up(\nu)$.

Thus, from $\alpha_z \in up_z(\alpha), z \in X$, we can obtain $\alpha' \in up(\alpha)$ such that for any $\nu \in \alpha$, there exists $\nu' \in up(\nu) \cap \alpha'$.    $\square$

**Theorem 2.** *The class of updatable timed automata is decidable.*

*Proof.* Lemmas 1, 2, 3, 4 indicate that $\mathcal{R}^-$ forms a set of regions and is compatible with $C_-(X)$ and $U_\leq(X)$ and a region automaton can be constructed as explained in Section 2.1 and hence according to Theorem 1 this class is decidable.    $\square$

*Remark 2.* Note that if we consider updates $U_\nleq(X)$ of the form $up = \bigwedge_{x \in X} up_x$ with $up_x ::= x := 0 \mid x :\sim d - y$, where $x, y \in X, d \in \mathbf{Q}^+$ and $\sim \in \{\neq, >, \geq\}$, $d \leq c^{max}$, then emptiness is no longer decidable (The Undecidability result of section 4 in [4] can be modified [10]). Recall that allowing updates of the kind $x :< y$ or $x :> y$ or $x :\sim y + c$, $c \in \mathbf{Q}^+$, with constraints from $C_-(X)$, emptiness is not decidable [4]. It is interesting to note here that the updates $x :< d - y$ and $x :> d - y$ behave differently.

## 4    Additive Constraints

In this section, we consider *updatable timed automata with additive constraints*. The constraints $C_+(X)$ of the form $\varphi ::= x \sim c \mid x + y \sim c \mid \varphi \wedge \varphi \mid \varphi \vee \varphi$ where $x, y \in X, c \in \mathbf{Q}^+$ and $\sim \in \{<, \leq, >, \geq, =, \neq\}$. Constraints $x + y \sim c$ are called *additive constraints*. Timed automata with constraints $C_+(X)$ and updates $U_0(X)$ have been studied in [5]. It has been shown that with 2 or lesser number of clocks, emptiness is decidable, while 4 clocks gives undecidability. The case of 3 clocks has been open since.

In this section, we give a partial answer to this open problem by looking at the class of timed automata with constraints $C_+(X)$ and updates $\mathcal{U}_=(X)$ consisting of updates $up = \bigwedge_{x \in X} up_x$ where $up_x ::= x := 0 \mid x := d - y$, $x, y \in X, d \in \mathbf{Q}^+$. Here, $d \leq c^{max}$ where $c^{max} = max\{c \mid x \sim c \in C_+(X)$ or $x + y \sim c \in C_+(X)\}$.

We show that emptiness is undecidable if 3 or more clocks are used, while it is decidable with 2 or lesser number of clocks.

For every clock $x \in X$, define $c_x^{max} = c^{max}$. $\mathcal{I}_x$ is defined as done before. For every pair of clocks $x, y \in X$, define $i_{xy} = i_{yx} = max\{c \mid x + y \sim c \in \mathcal{C}_+(X)\}$ and the set of intervals $\mathcal{K}_{xy}$ as

$$\mathcal{K}_{xy} = \{[d] \mid 0 \le d \le i_{xy}\} \cup \{(d, d+1) \mid 0 \le d < i_{xy}\} \cup \{(i_{xy}, +\infty)\}$$

Let $\alpha$ be a tuple $((I_x)_{x \in X}, (K_{xy})_{x,y \in X}, \prec)$ where $I_x \in \mathcal{I}_x, K_{xy} \in \mathcal{K}_{xy}$ and $\prec$ is a total preorder on $X_0 = \{x \in X \mid I_x \text{ is of the form } (c, c+1)\}$. The region associated with $\alpha$ is defined as the set of all valuations $\nu$ such that $\nu(x) \in I_x$ for all $x \in X$, $\nu(x) + \nu(y) \in K_{xy}$ and $frac(\nu(x)) \le frac(\nu(y))$ if $x \prec y$, for all $x, y \in X_0$. Let $\mathcal{R}^+$ represent the set of all tuples $\alpha$. $\mathcal{R}^+$ forms a finite partition of $\mathbf{T}^X$.

**Theorem 3.** *The class of timed automata with 2 clocks, updates $U_=(X)$ and clock constraints $C_+(X)$ are decidable.*

*Proof.* The proof of this theorem follows by the extension of the classical region construction as given above. It is easy to see that the partition $\mathcal{R}^+$ is a set of regions (consistent with time in the sense that from two equivalent valuations, the same set is reached as time progresses), consistent with the set of constraints (2 equivalent valuations satisfy the same constraints), and updates (updates from 2 equivalent valuations yield the same region). □

**Theorem 4.** *The class of timed automata with 3 clocks, updates $U_=(X)$ and clock constraints $C_+(X)$ is undecidable.*

*Proof.* The proof lies in the simulation of a deterministic two counter machine. Such a machine consists of a finite sequence of labeled instructions which handle two counters $i$ and $j$ and end at a special instruction labeled *Halt*. The instructions are as follows:

1. $l_k : \quad x = x + 1; \ goto \ l_{k+1};$
2. $l_k : \quad if \ x = 0 \ goto \ l_{k+1} \ else \ x = x - 1; \ goto \ l_{k+2}.$

Without loss of generality, assume that the instructions are labeled $l_0, \ldots, l_n$ where $l_n = Halt$ and that in the initial configuration, both counters have value zero. Further, assume that the first instruction increments the counter $i$. The behaviour of the machine is described by a possibly infinite sequence of configurations $< l_0, 0, 0 >, < l_1, i_1, j_1 >, \cdots < l_k, i_k, j_k > \ldots$ where $i_k$ and $j_k$ are the respective counter values and $l_k$ is the label of the $k$th instruction. The halting problem of such a machine has been shown to be undecidable [11].

A timed automaton $\mathcal{A}_M$ with 3 clocks can be built which simulates a two counter machine $M$ and reaches a final state iff $M$ halts. Thus the halting problem of two counter machine is reduced to the problem of emptiness of $L(\mathcal{A}_M)$. The alphabet of $\mathcal{A}_M$ is $\{*\}$, it has 3 clocks $\{x_1, x_2, x_3\}$ and the set of locations of $\mathcal{A}_M$ is $Q = \{l'_i, l''_i \mid 0 \le i \le n\} \cup \{l_i^k \mid 1 \le k \le 7, 0 \le i < n\}$. The set of final states is $F = \{l'_n, l''_n\}$. There are 5 basic modules : (i)an initialization module, (ii) module to increment counter $i$, (iii) module to check for zero and decrement counter $i$, (iv)module to increment counter $j$ and (v)module to check for zero and decrement counter $j$. Initially, all clocks have value 0. At the end of the initialization module[1], $x_1 = x_2 = 0$ while $x_3 = 1$. The control shifts to the location $l'_0$, which represents the label of the first instruction. The values of the counters $i$ and $j$ are encoded in values of the clocks. The normal form for the clock values is as follows: $x_1 = \left(1 - \frac{1}{2^i}\right) + \left(1 - \frac{1}{2^j}\right), x_2 = \left(1 - \frac{1}{2^j}\right), x_3 = \frac{1}{2^j}$ or 0. Each module simulating instruction $l_i, i \geq 0$ has a unique initial location labeled by either $l'_i$ or $l''_i$, and has two last locations labeled by $l'_{i+1}, l''_{i+1}$, where $l_{i+1}$ is the next instruction to be simulated. At the beginning of each module, the clocks are assumed to be in normal form, and in the last location of each module, clocks will be in normal form. At location $l''_{i+1}$ in each module, the value of $x_3$ will be necessarily zero, whereas at $l'_{i+1}$ it will be $\frac{1}{2^j}$ where $j$ represents the value of counter $j$ at that point of time.

*Incrementing counter $i$*: The module in Fig. 1 simulates the instruction $l_k : i = i + 1;\ goto\ l_{k+1}$. It can be seen that if the clock values are $x_1 = \left(1 - \frac{1}{2^i}\right) + \left(1 - \frac{1}{2^j}\right), x_2 = \left(1 - \frac{1}{2^j}\right)$ and $x_3 = \frac{1}{2^j}$ while entering this module, then at the end, we will have the value of $x_1$ to be $\left(1 - \frac{1}{2^{i+1}}\right) + \left(1 - \frac{1}{2^j}\right)$, $x_2$ remains unchanged, and $x_3$ will retain its earlier value at $l'_{k+1}$ and is reset to zero at $l''_{k+1}$.

*Incrementing counter $j$*: Fig. 2 depicts the module for incrementing counter $j$. At the initial state of this module, $x_3$ must be zero. It is clear that after the transition, $x_2 = \left(1 - \frac{1}{2^{j+1}}\right)$, $x_1 = \left(1 - \frac{1}{2^i}\right) + \left(1 - \frac{1}{2^{j+1}}\right)$ and $x_3 = \frac{1}{2^{j+1}}$ or $x_3 = 0$ depending on the location.

*Decrementing counter $i$*: The module in figure 3 decrements counter $i$. The module begins in location $l''_k$. The control reaches any of $l'_{k+2}$ or $l''_{k+2}$ only if value of counter $i$ is 0. If counter $i$ is non-zero, then the location $l^2_k$ is reached. It can be seen that at $l'_{k+1}$, the value of $x_2$ will be unchanged, $x_3$ will be $1 - x_2$ and $x_1$ would be $\left(1 - \frac{1}{2^{i-1}}\right) + \left(1 - \frac{1}{2^j}\right)$. Similar is the case of $l''_{k+1}$ ($x_3 = 0$ here).

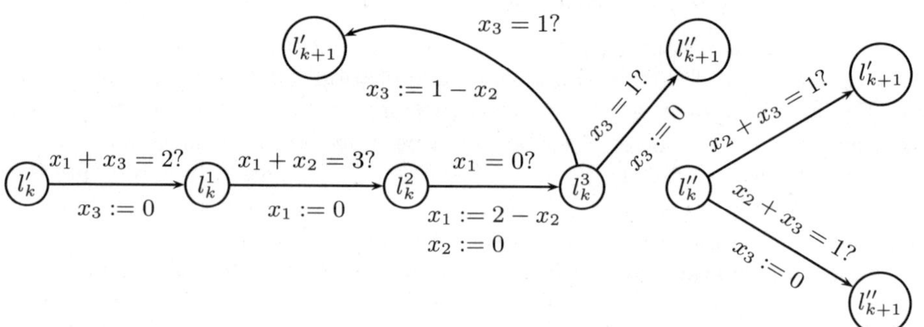


Fig. 1. Increment counter $i$          Fig. 2. Increment $j$

[1] The initialization module can be easily constructed by having a timed automaton with two locations with the constraint $x_3 = 1?$ on the transition between them. Clocks $x_1$ and $x_2$ are reset on transition.

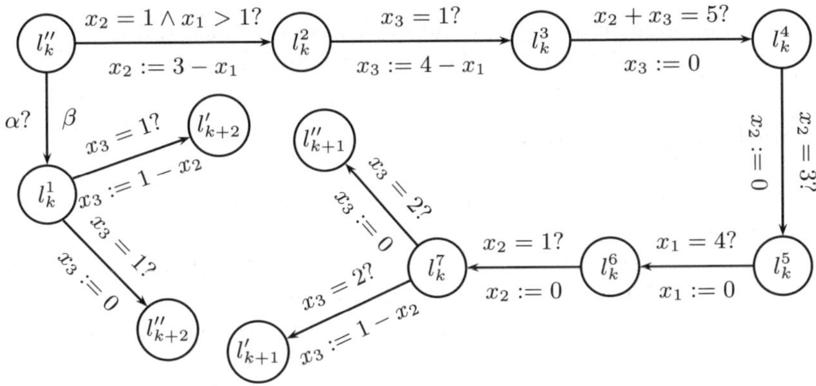

**Fig. 3.** Dec. counter $i$: $\alpha = (x_1 = 1 \wedge x_2 = 1)$ and $\beta = (x_1 := 0, x_2 := 0)$

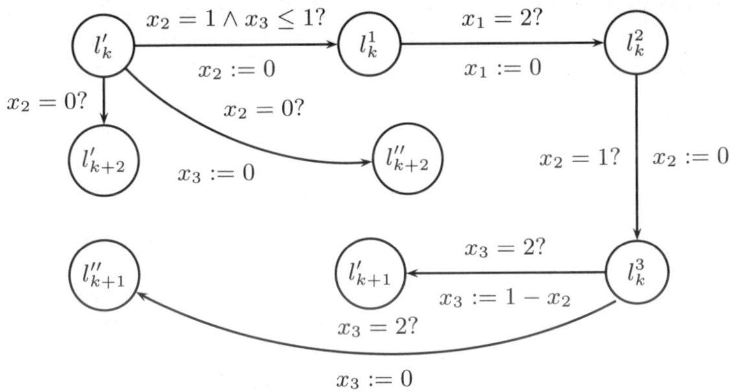

**Fig. 4.** Module to decrement counter $j$

_Decrementing counter $j$_: The module to decrement counter $j$ is shown in Fig. 4. The value of $x_3$ is $\frac{1}{2^j}$ at the start of the module.

The simulation of the 2-counter machine is done as follows: The automaton $\mathcal{A}_\mathcal{M}$ starts in location $q_0$ with all clocks initialized to zero. When $x_3 = 1$, the control goes to $l_0'$, resetting $x_1, x_2$ to zero. Each instruction is implemented by the corresponding module. $\mathcal{A}_\mathcal{M}$ is built according to the sequence of instructions of the 2-counter machine. The different modules are linked by choosing the appropriate end state of a module to be the initial state of the next module. Modules can be linked this way until the _Halt_ instruction is encountered. The set of final states is $\{l_n', l_n''\}$. The language accepted by $\mathcal{A}_\mathcal{M}$ is empty iff $\mathcal{M}$ does not halt, which concludes the proof. $\qquad\square$

_Remark 3._ It should be noted that Theorems 3,4 are in agreement with [3] wherein it has been shown that reachability is decidable for dynamical systems with piecewise-constant derivatives in the case of 2 dimensions, while it is not for higher dimensions. Even though our systems allow much simpler guards and have a constant rate of evolution, they differ from [3] due to updates.

## 5   Location Invariants

It is well known that adding location invariants of the form $x \sim c$ does not increase the expressive power of classical timed automata. It is easy to see that the same holds good even when we allow as location invariants the elements of $C_-(X)$ or $C_+(X)$ (invariants can be removed and the same condition appended to all outgoing transitions [10]). [4] has considered a class of timed automata with $U_1(X)$ with constraints from $C_-(X)$ and proved that this class is decidable. This result will hold even when we allow elements of $C_-(X)$ as constraints and location invariants (region automtaton can include loc. invariants [10]). However, note that for $U_1(X)$ along with location invariants and constraints from $C_+(X)$, the decidability will hold good only when the number of clocks is 2 or less. The undecidability result given in Theorem 4 can easily be changed to prove this (replacing update $x := d - y$ with $x :< d + 1$ followed by the loc.invariant $x + y = d$ [10]).

## References


1. Alur, R., Dill, D.L.: A Theory of Timed Automata. Theoretical Computer Science 126(2) (1994)
2. Alur, R., Henzinger, T.A., Ho, P.-H.: Automatic Symbolic Verification of Embedded Systems. IEEE Transactions on Software Engineering 22, 181–201 (1996)
3. Asarin, E., Maler, O., Pnueli, A.: Reachability analysis of dynamical sytems having piecewise-constant derivatives. Theoretical Computer Science 138, 35–65 (1995)
4. Bouyer, P., Duford, C., Fleury, E., Petit, A.: Updatable Timed Automata. Theoretical Computer Science 321(2-3), 291–345 (2004)
5. Bérard, B., Duford, C.: Timed automata and additive clock constraints. Information Processing Letters 75(1-2), 1–7 (2000)
6. Bérard, B., Gastin, P., Petit, A.: On the power of non observable actions in timed automata. In: Puech, C., Reischuk, R. (eds.) STACS 1996. LNCS, vol. 1046, pp. 257–268. Springer, Heidelberg (1996)
7. Demichelis, F., Zielonka, W.: Controlled Timed Automata. In: Sangiorgi, D., de Simone, R. (eds.) CONCUR 1998. LNCS, vol. 1466, pp. 455–469. Springer, Heidelberg (1998)
8. Behrmann, G., David, A., Larsen, K.G., Mller, O., Pettersson, P., Yi, W.: Uppaal - Present and Future. In: Proceedings of the 40th IEEE Conference on Decision and Control, Orlando, Florida, USA, December 4-7, 2001 (2001)
9. Daws, C., Olivero, A., Tripakis, S., Yovine, S.: The Tool KRONOS, Hybrid Systems, pp. 208–219 (1995)
10. Lakshmi Manasa, G., Krishna, S.N., Nagaraj, K.: Updatable Timed automata, Technical Report, http://www.cse.iitb.ac.in/krishnas/uta.pdf
11. Minsky, M.L.: Computation: finite and infinite machines. Prentice-Hall, USA (1967)

# First-Order Model Checking Problems
# Parameterized by the Model*

Barnaby Martin

Department of Computer Science, University of Durham,
Science Labs, South Road, Durham DH1 3LE, U.K.

**Abstract.** We study the complexity of the model checking problem, for
fixed models $A$, over certain fragments $\mathcal{L}$ of first-order logic, obtained by
restricting which of the quantifiers and boolean connectives we permit.
These are sometimes known as the expression complexities of $\mathcal{L}$. We obtain various full and partial complexity classification theorems for these
logics $\mathcal{L}$ as each ranges over models $A$, in the spirit of the dichotomy
conjecture for the Constraint Satisfaction Problem – which itself may be
seen as the model checking problem for existential conjunctive positive
first-order logic.

## 1   Introduction

The *model checking problem* over a logic $\mathcal{L}$ takes as input a model $A$ and a
sentence $\varphi$ of $\mathcal{L}$, and asks whether $A \models \varphi$. The problem can also be parameterised, either by the sentence $\varphi$, in which case the input is simply $A$, or by the
model $A$, in which case the input is simply $\varphi$. Vardi has studied the complexity
of this problem, principly for logics which subsume first-order logic (**FO**), in
[16], and their bounded variable fragments [17]. He describes the complexity of
the unrestricted problem as the *combined complexity*, and the complexity of the
parameterisation by the sentence (respectively, model) as the *data complexity*
(respectively, *expression complexity*). For the majority of his logics, when there
is no variable bound, the expression and combined complexities are comparable,
and are one exponential higher than the data complexity.

In this paper, we will be interested in fragments $\mathcal{L}$ of **FO** obtained by restricting which of the quantifiers and boolean connectives we permit. For these $\mathcal{L}$, we
then study the complexities of the parameterisation of the model checking problem by $A$, that is the expression complexities for certain $A$. When $\mathcal{L}$ is the *existential conjunctive positive* fragment of **FO**, $\{\exists, \wedge\}$-**FO**, the model checking problem
is equivalent to the much-studied *constraint satisfaction problem* (CSP). The parameterisation of this problem by $A$ is equivalent to what is sometimes described
as the *non-uniform* constraint satisfaction problem, CSP($A$) [10]. It has been
conjectured [7] that the class of CSPs exhibits *dichotomy* – that is, CSP($A$) is
always either in P or is NP-complete, depending on the model $A$. This is tan-
tamount to the condition that the expression complexity for $\{\wedge, \exists\}$-**FO** on $A$ is

---

* This research was supported by EPSRC grant EP/C54384X/1.

A. Beckmann, C. Dimitracopoulos, and B. Löwe (Eds.): CiE 2008, LNCS 5028, pp. 417–427, 2008.

always either in P or is NP-complete. While in general this conjecture remains open, it has been proved for certain classes of models $A$. Of particular interest to us is Hell and Nešetřil's dichotomy for undirected graphs $H$: in [8] it is proved that $\mathrm{CSP}(H)$ is in P, if $H$ has a self-loop or is bipartite, and is NP-complete, if $H$ is any other undirected graph. When $\mathcal{L}$ is the *(quantified) conjunctive positive* fragment of **FO**, $\{\exists, \forall, \wedge\}$-**FO**, the model checking problem is equivalent to the well-studied *quantified constraint satisfaction problem* (QCSP). No overarching polychotomy has been conjectured for the non-uniform $\mathrm{QCSP}(A)$, although the only known attainable complexities are P, NP-complete and Pspace-complete (see trichotomies in [1,13] and the results of [3,4]).

It is known, for each model $A$, that there is a digraph $H_A$ such that $\mathrm{CSP}(A)$ and $\mathrm{CSP}(H_A)$ are polynomial time equivalent [7]. Thus, in the case of the CSP, digraphs comprise a microcosm of all models. (For isomorphism problems, undirected graphs provide such a microcosm; however, for the CSP – a homomorphism problem – it is not clear this is the case.) A similar microcosm holds for the QCSP. Owing to space considerations, convenience and this observation, we work primarily with digraphs, as opposed to relational models in general. Where easily applicable, we indicate how our method may be extended to account for general relational models. We also work in the more general framework of logics without equality, since the situation in which we have equality may be simulated by the addition of a special binary relation to the model, whose digraph contains exactly every self-loop.

We will be interested in fragments of **FO** derived from restricting which of the logical symbols $\Gamma_0 := \{\neg, \exists, \forall, \wedge, \vee\}$ we permit, subject to there being at least one quantifier and one binary connective. For $\Gamma \subseteq \Gamma_0$, let $\Gamma$-**FO** be that fragment of **FO** obtained by allowing only the logical symbols of $\Gamma$. Owing to de Morgan duality, it follows that there is only one fragment we need consider with negation: $\Gamma_0$-**FO** itself. The fragments we consider may be classified as follows.

| Class I | Class II | Class III | Class IV | Class V | Class VI |
|---|---|---|---|---|---|
|  | (CSP) |  |  | (QCSP) |  |
| $\{\neg, \exists, \forall, \wedge, \vee\}$ | $\{\exists, \vee\}$ | $\{\exists, \wedge\}$ | $\{\exists, \wedge, \vee\}$ | $\{\exists, \forall, \wedge\}$ | $\{\forall, \exists, \wedge, \vee\}$ |
|  | $\{\forall, \wedge\}$ | $\{\forall, \vee\}$ | $\{\forall, \wedge, \vee\}$ | $\{\exists, \forall, \vee\}$ |  |

For each of our $\Gamma \subseteq \Gamma_0$, and for some model $A$, we define the *model checking problem* $\Gamma$-$\mathrm{MC}(A)$ to have as input a sentence $\varphi$ of $\Gamma$-**FO**, and as yes-instances those sentences such that $A \models \varphi$.

The model checking problems have been more-or-less studied for the Classes I, III and V. As Class III corresponds to the CSP, so Class V corresponds to the QCSP. A full complexity classification for these classes resists despite considerable efforts. On the other hand, we will see that the complexity classification is nearly trivial for the Classes I, II and IV. This leaves Class V as a candidate for non-trivial results. Indeed the consideration of $\{\forall, \exists, \wedge, \vee\}$-$\mathrm{MC}(A)$, in light of the study of $\mathrm{CSP}(A)$ and $\mathrm{QCSP}(A)$, is the main justification for this paper. The Classes II–IV are studied in the manuscript [12], and the results on those classes here presented may be found with more detailed proofs in that paper.

The paper is organised as follows. After the preliminaries, we consider in Section 3 the Class I fragment $\{\neg, \exists, \forall, \wedge, \vee\}$-**FO**, reproving the folklore dichotomy its model checking problem begets. A brief Section 4 considers the low complexity 'trivial' fragments of Class II. Sections 5 and 7, also very brief, exist only to place Class III (CSP) and Class V (QCSP) in the context of our work. Section 6 deals with the logics of Class IV, giving a full, albeit not very interesting, complexity classification for the model checking problems on these fragments. Finally, Section 8 addresses the Class VI fragment $\{\forall, \exists, \wedge, \vee\}$-**FO**. For this fragment, we derive some general hardness results before fully classifying its model checking problem on the boolean digraphs.

## 2    Preliminaries

In this paper, we consider only finite, non-empty relational models. A *model* (or *structure*) $A$ consists of a finite, non-empty set $|A|$ – the universe or domain of $A$ – together with some sets $R_i^A \subseteq |A|^{r_i}$ interpreting each $r_i$-ary relation $R_i$ of the underlying signature. We generally use $x, y, z$ to refer to elements of $A$, and $u, v$ to refer to variables that range over those elements. Where $x \in |A|$, we term the model $A$ as $x$-valid if every non-empty relation $R_i^A \subseteq |A|^{r_i}$ contains the tuple $(x^{r_i}) := (x, \ldots, x)$. A model $A$ is *trivial* if every relation $R_i^A$ is either empty ($R_i^A := \emptyset$) or contains all tuples ($R_i^A := |A|^{r_i}$); otherwise $A$ is *non-trivial*. For a model $A$, we define its complement $\overline{A}$ to be the model having the same universe as $A$, but whose relations are the (set-theoretic) complements of the relations of $A$. That is, for each $R_i^A$, the relation $R_i^{\overline{A}}$ is defined by $\mathbf{x} \in R_i^{\overline{A}}$ iff $\mathbf{x} \notin R_i^A$. A *digraph* $H$ is any model having a single, binary (edge) relation $E$; we refer to elements of $|H|$ as *vertices*. An *undirected graph* is a digraph whose edge relation is symmetric. A digraph $H$ is an *equivalence relation* if $E$ is reflexive, symmetric and transitive; an equivalence relation is *non-trivial* if it contains more than one equivalence class. A *self-loop* in a digraph $H$ is an edge $(x, x)$, where $x \in |H|$ (from the presence of such an edge, it follows that $H$ is $x$-valid). $H$ is *antireflexive* if it contains no self-loops and is *reflexive* if it contains all self-loops. The *n-clique* $K_n$ is the complete antireflexive digraph on $n$ vertices (i.e. $(x, y) \in E^{K_n}$ iff $x \neq y$). The complement digraph $\overline{K}_n$ is exactly the disjoint union of $n$ self-loops (equivalently, the *digraph of equality* on $n$ vertices). The digraph $K_n^r$ is the complete reflexive digraph on $n$ vertices (i.e. $(x, y) \in E^{K_n^r}$ for all $x, y$). $K_n^r$ is a trivial equivalence relation on $n$ vertices. Let $\mathbf{t} := (t_1, \ldots, t_m)$ be a tuple of positive integers; define $U(\mathbf{t})$ to be the digraph given by the disjoint union $K_{t_1}^r \uplus \cdots \uplus K_{t_m}^r$. The digraphs $U(\mathbf{t})$ are exactly the equivalence relations. If $(1^n)$ is an $n$-tuple of ones, then note that $U(1^n)$ is the digraph of equality $\overline{K}_n$. A vertex $x$ in a digraph $H$ is *isolated* if, for all $y \in |H|$, neither $(x, y)$ nor $(y, x)$ is in $E^H$. Dually, a vertex $x$ is *dominating* if, for all $y \in |H|$, both $(x, y)$ and $(y, x)$ is in $E^H$ (note that $H$ has an isolated vertex iff $\overline{H}$ has a dominating vertex).

For a digraph $H$, let sym-clos($H$) and tran-clos($H$) be the symmetric and transitive closures of $H$, respectively. Let doub($H$) be the subdigraph induced by the double edges of $H$; that is, $(x, y) \in E^{\text{doub}(H)}$ iff $(x, y) \in E^H$ and $(y, x) \in E^H$ (whereas $(x, y) \in E^{\text{sym-clos}(H)}$ iff $(x, y) \in E^H$ or $(y, x) \in E^H$).

Since we may convert any sentence $\varphi$ of $\Gamma$-**FO** to an equivalent one in prenex normal form in polynomial time, we will henceforth assume all sentences in such form. We will also refer to the atoms of $\varphi$, when we more properly mean the atoms in the quantifier-free part of $\varphi$.

## 3   Class I : $\{\neg, \exists, \forall, \wedge, \vee\}$-FO

The following result is probably folklore. (In [16], it is claimed that the **Pspace**-hard cases are proved in [2], but we are unable to find their proof.)

**Proposition 1.** *In full generality, the class of problems $\{\neg, \exists, \forall, \wedge, \vee\}$-MC$(A)$ exhibits dichotomy: if $A$ is trivial, then the problem is in* **P***, otherwise it is* **Pspace**-*complete.*

In general, **Pspace** membership follows by a simple evaluation procedure inward through the quantifiers (see [16]). If $A$ is trivial, then the existential and universal quantifiers are semantically equivalent – indeed, each atom of the input $\varphi$ may be evaluated to boolean true or false, independently of the quantifiers. What remains is a boolean sentence evaluation problem, known to be in **P** (see [11]).[1]

We now set out ot demonstrate why $\{\neg, \exists, \forall, \wedge, \vee\}$-MC$(A)$ is **Pspace**-complete when $A$ is non-trivial, proving this in the case of digraphs. Since we will wish to use some intermediate results in the later section on $\{\exists, \forall, \wedge, \vee\}$-**FO**, we will put off the use of negation until absolutely necessary. This will enable us to illustrate an important duality, although it will make the proof a little longer than necessary. We begin with the following observation.

**Lemma 1.** *For $n \geq 2$, $\{\exists, \forall, \wedge, \vee\}$-MC$(K_n)$ is* **Pspace**-*complete.*

*Proof.* For $n \geq 3$, $\{\exists, \forall, \wedge, \vee\}$-MC$(K_n)$ contains the **Pspace**-complete problem $\{\exists, \forall, \wedge\}$-MC$(K_n)$, a.k.a. QCSP$(K_n)$, (see [1]) as a special instance.

For $n = 2$, we use a reduction from the problem $\{\exists, \forall, \wedge\}$-MC$(B_{NAE})$, where $B_{NAE}$ is the boolean structure with a single ternary relation $NAE := \{0, 1\}^3 \setminus \{(0, 0, 0), (1, 1, 1)\}$. This problem is equivalent to the *quantified not-all-equal 3-satisfiability* problem, well-known to be **Pspace**-complete (see [15]). Let $\varphi$ be an input for $\{\exists, \forall, \wedge\}$-MC$(B_{NAE})$. Let $\varphi'$ be built from $\varphi$ by substituting all instances of $NAE(v, v', v'')$ by $E(v, v') \vee E(v', v'') \vee E(v, v'')$. It is easy to see that $B_{NAE} \models \varphi$ iff $K_2 \models \varphi'$, and the result follows.

---

[1] In fact, this problem is solvable in non-deterministic log space **NLogspace** (but is not thought to be complete for the class under logspace reductions). It is not hard to see that all our tractability results for the Classes I, II, IV and VI actually place the respective problems in **NLogspace** and not just in **P**. However, it is known for Classes III and V (respectively relating to CSP and QCSP) that the tractable problems display greater richness: there being examples that are complete under logspace reduction for each of the classes **P** and **NLogspace**.

Consider a (prenex) sentence $\psi_0$ of $\{\exists, \forall, \wedge, \vee\}$-**FO**, recalling that all atoms of $\psi_0$ are positive. By de Morgan's laws, it is clear that $\psi_0$ is logically equivalent to the sentence $\neg\psi_1$ where $\psi_1$ is derived from $\psi_0$ by I.) swapping all instances of $\exists$ and $\forall$, II.) swapping all instances of $\vee$ and $\wedge$ and III.) negating all atoms (in the quantifier-free part). Let $\psi_2$ be derived from $\psi_0$ in a similar manner, but without the execution of part III (negating the atoms). Let $A$ be some structure, then

$$(*)\ A \models \psi_0 \ \Leftrightarrow\ A \models \neg\psi_1 \ \Leftrightarrow\ A \not\models \psi_1 \ \Leftrightarrow\ \overline{A} \not\models \psi_2.$$

Since $\psi_2$ remains in the logic $\{\exists, \forall, \wedge, \vee\}$-**FO**, we are now in a position to derive the following.

**Lemma 2.** *Let $H$ be a digraph s.t. $\{\exists, \forall, \wedge, \vee\}$-MC$(H)$ is in* P *(resp., is* Pspace*-complete), then $\{\exists, \forall, \wedge, \vee\}$-MC$(\overline{H})$ is in* P *(resp., is* Pspace*-complete).*

*Proof.* The result holds since both P and Pspace are closed under complementation (see [15]). We reduce the complement of the problem $\{\exists, \forall, \wedge, \vee\}$-MC$(H)$ to $\{\exists, \forall, \wedge, \vee\}$-MC$(\overline{H})$ by the mapping $\psi_0 \mapsto \psi_2$. The result follows from $(*)$.

We refer to $(*)$ above as the *principle of duality*, and will make frequent use of it in this work, for various of our fragments of **FO**. Another interpretation of $(*)$ is that the problems $\{\exists, \forall, \wedge, \vee\}$-MC$(H)$ and $\{\exists, \forall, \wedge, \vee\}$-MC$(\overline{H})$ are polynomial time Turing equivalent (generally reductions used in this paper are polynomial time many-to-one; for more on the difference between many-to-one and Turing reductions, see [15]). As has already been observed, if $(1^n)$ is a $n$-tuple of ones, then $\overline{K}_n$ is the digraph $U(1^n)$. One can easily imagine that the logic $\{\neg, \exists, \forall, \wedge, \vee\}$-**FO** – not having equality – could not distinguish between $\overline{K}_n = U(1^n)$ and any other $U(\mathbf{t})$ in which $\mathbf{t}$ is a $n$-tuple of positive integers. This is indeed the case: when $\mathbf{t}$ is a $n$-tuple of positive integers, then $U(1^n)$ and $U(\mathbf{t})$ agree on all sentences of $\{\neg, \exists, \forall, \wedge, \vee\}$-**FO** (a proof of this appears in [14]). Since $\{\exists, \forall, \wedge, \vee\}$-**FO** is a fragment of $\{\neg, \exists, \forall, \wedge, \vee\}$-**FO**, we may derive the following corollary from the previous two lemmas.

**Corollary 1.** *For $n \geq 2$ and $\mathbf{t}$ a $n$-tuple of positive integers, both $\{\exists, \forall, \wedge, \vee\}$-MC$(\overline{K}_n)$ and $\{\exists, \forall, \wedge, \vee\}$-MC$(U(\mathbf{t}))$ are* Pspace*-complete.*

Now let $H$ be a digraph. For propositions $P$ and $Q$, let $P \leftrightarrow Q$ be an abbreviation for $(P \wedge Q) \vee (\neg P \wedge \neg Q)$. Consider the following relation $\sim$ on the vertices of $H$.

$$x \sim y\ :=\ \forall z\ (E(x, z) \leftrightarrow E(y, z)) \wedge (E(z, x) \leftrightarrow E(z, y))$$

It is straightforward to verify that $\sim$ is an equivalence relation on the vertices of $H$, s.t. $\sim$ is non-trivial if $H$ is non-trivial.

*Proof (of Proposition 1).* Let $H$ be a non-trivial digraph. We know that $\sim$ induces a non-trivial equivalence relation on $H$ whose underlying digraph is $U(\mathbf{t}_\sim)$ s.t. $\{\exists, \forall, \wedge, \vee\}$-MC$(U(\mathbf{t}_\sim))$ is Pspace-complete. The result follows by reducing the latter to $\{\neg, \exists, \forall, \wedge, \vee\}$-MC$(H)$ by systematically replacing instances of $E(u, v)$ in an input $\varphi$ in the former with instances of $u \sim v$ in the latter (taking care that the introduced universally quantified variable – that replaces $z$ in the definition of $\sim$ – is new for each $(u, v)$).

The situation for models $A$ that are not necessarily digraphs is very similar. The equivalence relation $\sim$ is defined over the conjunction of all positions of all relations, and $A$'s non-triviality remains sufficient to guarantee more than one equivalence class.

## 4   Class II: Low Complexity Fragments

We consider the problem $\{\exists, \vee\}$-MC$(H)$, for some digraph $H$ of size $n$. An input for this problem will be of the form:

$$\varphi := \exists \mathbf{v}\; E(v_1, v_1') \vee \ldots \vee E(v_m, v_m')$$

where $v_1, v_1', \ldots, v_m, v_m'$ are the not necessarily distinct variables that comprise $\mathbf{v}$. Now, $H \models \varphi$ iff $H$ has some edge of the form $(v_i, v_i')$ for some $i$. This is equivalent to $H$ having any edge (unless, for all $i$, $v_i = v_i'$, in which case $H$ must have some self-loop). In any case, the condition is easily checked, and one may easily infer that $\{\exists, \vee\}$-MC$(H)$ is solvable in polynomial time for all digraphs $H$. One may argue similarly for the problems $\{\vee, \wedge\}$-MC$(H)$; or appeal to the principle of duality. We sum up as follows (note that the result for general relational models may be argued in essentially the same manner).

**Proposition 2.** *For all digraphs $H$, the problems $\{\exists, \vee\}$-MC$(H)$ and $\{\forall, \wedge\}$-MC$(H)$ are in* P.

## 5   Class III: CSP and Its Dual

From the principle of duality, we may deduce that the complement of a problem $\{\exists, \wedge\}$-MC$(H)$ is polynomial time equivalent to $\{\forall, \vee\}$-MC$(\overline{H})$. It follows that the former class has a dichotomy (between P and NP-complete) iff the latter class has a dichotomy (between P and co-NP-complete). It may be imagined that either classification is difficult to resolve.

## 6   Class IV

A digraph *homomorphism* from $G$ to $H$ is a function $h : |G| \to |H|$ that preserves the edge relation, i.e. if $(x, y) \in E^G$ then $(h(x), h(y)) \in E^H$. If there exist homomorphisms both from $G$ to $H$ and from $H$ to $G$, then we say that $G$ and $H$ are *homomorphically equivalent*. It is well-known (see, e.g., [9]) that there exists a minimal digraph (say, w.r.t. size) in the equivalence class induced by homomorphic equivalence, and this is necessarily an induced subdigraph of all members of that equivalence class (such a digraph is termed a *core*). An undirected graph is *bipartite* iff its core is either $K_1$ or $K_2$. We make the following observation.

**Lemma 3.** *The following are equivalent.*

(*i*)  *The structures $G$ and $H$ are homomorphically equivalent.*
(*ii*)  *The problems $\{\exists, \wedge\}$-MC($G$) and $\{\exists, \wedge\}$-MC($H$) coincide.*
(*iii*)  *The problems $\{\exists, \wedge, \vee\}$-MC($G$) and $\{\exists, \wedge, \vee\}$-MC($G$) coincide.*

*Proof.* The equivalence of (*i*) and (*ii*) is well-known [7]. Since (*iii*) $\Rightarrow$ (*ii*), it remains only to prove (*i*) $\Rightarrow$ (*iii*). We prove directly that the existence of a homomorphism $h : G \to H$ implies $\{\exists, \wedge, \vee\}$-MC($G$) $\subseteq$ $\{\exists, \wedge, \vee\}$-MC($H$), by appealing to the monotonicity of (the quantifier-free part of) any sentence of $\{\exists, \wedge, \vee\}$-**FO**. Indeed any witnesses $x_1, \ldots, x_m \in |G|$ to the (existentially quantified) variables of an input $\varphi$ may be matched by witnesses $h(x_1), \ldots, h(x_m) \in |H|$ for the same $\varphi$. The same applies for a homomorphism from $H$ to $G$, and the result follows.

**Lemma 4.** *Let $H$ be an antireflexive digraph whose edge relation is non-empty. Then $\{\exists, \wedge, \vee\}$-MC($H$) is* NP-*complete.*

*Proof.* Membership of NP is elementary (guess a satisfying assignment and verify); we prove hardness. We may assume w.l.o.g. that $H$ is undirected (symmetric), since otherwise we may define the symmetric closure of the edge relation via $E(u, v) \vee E(v, u)$. More formally, $\{\exists, \wedge, \vee\}$-MC(sym-clos($H$)) reduces to $\{\exists, \wedge, \vee\}$-MC($H$) under the reduction which substitutes instances $E^{\text{sym-clos}(H)}(u, v)$ in the former by $E^H(u, v) \vee E^H(v, u)$ in the latter.

Let $H'$ be the core of $H$. Note that the NP-hardness of $\{\exists, \wedge\}$-MC($H'$) (a.k.a. CSP($H'$)) immediately implies the NP-hardness of both $\{\exists, \wedge\}$-MC($H$) and $\{\exists, \wedge, \vee\}$-MC($H$). Since $H$ is antireflexive and undirected, its core $H'$ is either $K_1$ or $K_2$ or some non-bipartite $H'$.

The core $H'$ can not be $K_1$, since then the edge relation of $H$ would have been empty. If the core $H'$ is non-bipartite, then, by Hell and Nešetřil's theorem [8], the problem $\{\exists, \wedge\}$-MC($H'$) is NP-complete, hence NP-hardness of both $\{\exists, \wedge\}$-MC($H'$) and $\{\exists, \wedge, \vee\}$-MC($H$) follows.

It remains for us to consider the case where the core $H'$ is $K_2$. By the previous lemma, it suffices for us to prove that $\{\exists, \wedge, \vee\}$-MC($K_2$) is NP-hard. We proceed in a similar manner as in the proof of Lemma 1. We define the ternary not-all-equal $NAE$ relation on $K_2$ in $\{\exists, \wedge, \vee\}$-**FO**, whereupon we may appeal to the NP-hardness of *not-all-equal 3-satisfiability* (whose inputs may readily be expressed in $\{\exists, \wedge, \vee\}$-**FO**). We give $NAE(v, v', v'') := E(v, v') \vee E(v', v'') \vee E(v, v'')$.

If $H$ contains no edges, then $\{\exists, \wedge, \vee\}$-MC($H$) contains only no-instances. If $H$ contains a self-loop $(x, x)$, then $\{\exists, \wedge, \vee\}$-MC($H$) contains only yes-instances (evaluate all variables to $x$). We are now in a position to the derive the following, whose statement for the dual problem $\{\forall, \wedge, \vee\}$-MC($H$) follows from the principle of duality.

**Proposition 3.** *In full generality, the class of problems $\{\exists, \wedge, \vee\}$-MC($H$) exhibits dichotomy: if $H$ contains no edges or a self-loop, then the problem is in* P *(and is trivial), otherwise it is* NP-*complete. Similarly, the class of problems $\{\forall, \wedge, \vee\}$-MC($H$) exhibits dichotomy: if $\overline{H}$ contains no edges or a self-loop, then the problem is in* P, *otherwise it is* co-NP-*complete.*

The following, for general relational models, may be obtained as a corollary [12].

**Corollary 2.** *If $A$ is $x$-valid, for some $x$, then $\{\exists, \wedge, \vee\}$-MC($A$) is in* P*; otherwise it is* NP*-complete.*

## 7   Class V: QCSP and Its Dual

From the principle of duality, we may deduce that the complement of a problem $\{\exists, \forall, \wedge\}$-MC($H$) is polynomial time equivalent to $\{\exists, \forall, \vee\}$-MC($\overline{H}$). The former problem is an instance of the QCSP. It follows from the discussion in the introduction that $\{\exists, \forall, \vee\}$-MC($H$) may attain each of the complexities P, co-NP-complete and Pspace-complete. The classification for each of these fragments may be imagined difficult.

## 8   Class VI: $\{\exists, \forall, \wedge, \vee\}$-FO

We commence with some basic hardness results.

**Lemma 5.** *If $H$ is a digraph with an isolated vertex, then $\{\exists, \forall, \wedge, \vee\}$-MC($H$) is polynomial time Turing equivalent to the problem $\{\exists, \wedge, \vee\}$-MC($H$) and is in* NP. *If $H$ is a digraph with a dominating vertex (equivalently, $\overline{H}$ has an isolated vertex), then $\{\exists, \forall, \wedge, \vee\}$-MC($H$) is polynomial time Turing equivalent to the problem $\{\forall, \wedge, \vee\}$-MC($H$) and is in* co-NP.

*Proof.* We prove the first statement, whereupon the second follows from the principle of duality. Let $x \in |H|$ be an isolated vertex, and let $\varphi$ be a (prenex) sentence of $\{\exists, \forall, \wedge, \vee\}$-**FO** that we aim to evaluate on $H$. Consider the (positive) atoms of $\varphi$ that involve variables that are universally quantified. These atoms are always made false when the universally quantified variables are set to $x$. It is easily seen that $\varphi$ is equivalent on $H$ to the sentence $\varphi'$ in which the atoms that involve universally quantified variables are set to false, and the universally quantified variables removed. Clearly, $\varphi'$ is either a sentence of $\{\exists, \wedge, \vee\}$-**FO** or the boolean false, and the result follows. (Only because $\varphi'$ may be the boolean false is this a Turing, and not many-to-one, reduction.)

**Lemma 6.** *Let $H$ be a digraph such that any of $\{\exists, \forall, \wedge, \vee\}$-MC($sym\text{-}clos(H)$), $\{\exists, \forall, \wedge, \vee\}$-MC($tran\text{-}clos(H)$) or $\{\exists, \forall, \wedge, \vee\}$-MC($doub(H)$) is* Pspace*-complete. Then $\{\exists, \forall, \wedge, \vee\}$-MC($H$) is* Pspace*-complete.*

*Proof.* We reduce any of the first three to the last. The case of sym-clos($H$) may be done as in the first part of the proof of Lemma 4, and the case of doub($H$) is very similar (but with $E(x, y) \wedge E(y, x)$ instead of $E(x, y) \vee E(y, x)$). For tran-clos($H$), the method is similar but involves the substitution of $E^{\text{tran-clos}(H)}(x, y)$ by a formula that asserts the existence of a path in $H$, from $x$ to $y$, of any of the lengths 1 to $n - 1$ (where $||H|| = n$).

A digraph $H$ is *strongly connected* if, for every $x, y \in |H|$, there is a directed path from $x$ to $y$ (each $x$ may or may not possess a self-loop). If $H$ is not strongly connected then it may be partitioned into strongly connected components in the obvious manner.

**Proposition 4.** *If $H$ is not strongly connected and every strongly connected component of $H$ either contains more than one vertex or is a self-loop, then $\{\exists, \forall, \wedge, \vee\}$-MC$(H)$ is* Pspace-*complete.*

*Proof.* It is easily seen that doub(tran-clos($H$)) is a non-trivial equivalence relation. The result follows from the previous lemma and Corollary 1.

## 8.1   Boolean Digraphs

We now give the complexities of $\{\exists, \forall, \wedge, \vee\}$-MC$(H)$ for the boolean digraphs $H$, whose isomorphism classes are drawn in Figure 1. The classification is not too difficult, though the following may be a little surprising.

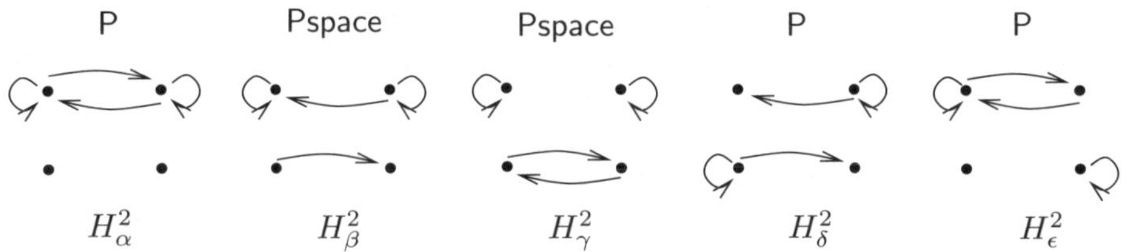


**Fig. 1.** The boolean digraphs: $H_\alpha^2$, $H_\beta^2$, $H_\gamma^2$, $H_\delta^2$, $H_\epsilon^2$, together with their complements. Above is the complexity of the problem $\{\exists, \forall, \wedge, \vee\}$-MC$(H)$ as $H$ ranges thereover. For Pspace read Pspace-complete.

**Lemma 7.** *Let $H_\delta^2$ be the boolean digraph with edge set $\{(x, x), (x, y)\}$. The problems $\{\exists, \forall, \wedge, \vee\}$-MC$(H_\delta^2)$ and $\{\exists, \forall, \wedge, \vee\}$-MC$(\overline{H}_\delta^2)$ are both in* P.

*Proof.* We prove the first statement, whereupon the second follows from the principle of duality. Let $\varphi$ be an input sentence for $\{\exists, \forall, \wedge, \vee\}$-MC$(H_\delta^2)$. We claim that $\varphi$ is a yes-instance iff the quantifier-free $\varphi'$ is a yes-instance, where $\varphi'$ is obtained from $\varphi$ by instantiating all universal variables as $y$ and all existential variables as $x$. The evaluation of $\varphi'$ on $H_\delta^2$ is nothing other than a boolean sentence value problem, known to be in P. That $\varphi$ and $\varphi'$ are equivalent on $H_\delta^2$ follows from $\varphi$'s being positive, since every forward-neighbour of $y$ is also a forward-neighbour of $x$ and every backward-neighbour of $y$ is also a backward-neighbour of $x$ (of course, $y$ has no forward-neighbours). Let us dwell on this briefly. Essentially, we may assume that all universal variables of $\varphi$, in turn, are set to $y$, since any existential witness to $y$ is also a witness to $x$. Thereafter, we may assume that all remaining (existential) variables are set to $x$, because $x$ acts as a witness to everything that $y$ does.

**Proposition 5.** *On the class of boolean digraphs* $\{\exists, \forall, \wedge, \vee\}$*-MC$(H)$ exhibits dichotomy. When $H$ is $H_\alpha^2$, $\overline{H}_\alpha^2$, $H_\delta^2$, $\overline{H}_\delta^2$, $H_\epsilon^2$, $\overline{H}_\epsilon^2$ the problem is in* P, *otherwise it is* Pspace-*complete.*

*Proof.* For $H_\alpha^2$ and $\overline{H}_\alpha^2$ the problem is trivial (as it is also for the loopless vertex $K_1$ and the self-loop $\overline{K}_1 = K_1^r$). For $H_\delta^2$, $\overline{H}_\delta^2$ the result follows from the previous lemma. For $H_\epsilon^2$ the result follows from Lemma 5 and Proposition 3, whereupon the result for $\overline{H}_\epsilon^2$ follows from the principle of duality.

The cases $H_\gamma^2$ and $\overline{H}_\gamma^2$ have been proved in Lemma 1 and Corollary 1, whereupon, since $H_\gamma^2 = \text{sym-clos}(H_\beta^2)$ the cases $H_\beta^2$, $\overline{H}_\beta^2$ follow from Lemma 6 and the principle of duality.

## 9    Conclusions

In this paper we have studied the model checking problem, parameterised by the model, for various fragments of **FO** derived by restricting which of the quantifiers and boolean connectives we permit. It is possibly the case that only our results for Class IV ($\{\exists, \wedge, \vee\}$-**FO** and its dual) and Class VI ($\{\exists, \forall, \wedge, \vee\}$-**FO**) are genuinely new (although, as stated, we are unaware of a published proof of the result for Class I, that appears as our Proposition 1).

It has been pointed out ([5]) that the method of Galois connections, so useful in the study of the CSP and QCSP, may be used to obtain a proof of a restricted version of Corollary 2, relating to the logic $\{\exists, \wedge, \vee, =\}$-**FO**. This is equivalent to considering the problems $\{\exists, \wedge, \vee\}$-MC$(A)$ in which $A$ must contain a binary relation that is the digraph of equality. This proof is simple, once the algebraic machinery has been introduced, and has the advantage of dispensing with the need for the highly non-trivial theorem of Hell and Nešetřil. Note that the dichotomy occurs under the same conditions for $\{\exists, \wedge, \vee\}$-MC$(A)$ and $\{\exists, \wedge, \vee, =\}$-MC$(A)$, because equality induces a digraph containing exactly self-loops. While there is an applicable Galois connection for the equality-free $\{\exists, \wedge, \vee\}$-**FO**, it is not clear how to use it to derive Corollary 2 in full.

A simple dichotomy may be observed for the class $\{\exists, \forall, \wedge, \vee, =\}$-MC$(A)$, equivalently $\{\exists, \forall, \wedge, \vee\}$-MC$(A)$ in which $A$ must contain a binary relation that is the digraph of equality. It follows from Proposition 4 that $\{\exists, \forall, \wedge, \vee, =\}$-MC$(A)$ is Pspace-complete if $\|A\| \geq 2$; whereupon it may be argued as with the Class I logic $\{\neg, \exists, \forall, \wedge, \vee\}$-**FO** that the problem is in P if $\|A\| = 1$. Again, there is an applicable Galois connection for $\{\exists, \forall, \wedge, \vee\}$-**FO**, but we have been unable to use it for a classification result.

It seems likely that the dichotomy of Proposition 5 may be extended beyond boolean digraphs to all boolean models – similar dichotomy results are known here for the CSP and QCSP (see [6]).

For digraphs $H$ of size $\leq 3$, each of the complexities P, NP-complete, co-NP-complete and Pspace-complete may be attained by the problem $\{\exists, \forall, \wedge, \vee\}$-MC$(H)$. The first and last are exemplified in Proposition 5; for NP-complete consider $H_3 := K_1 \uplus K_2$ and for co-NP-complete consider $\overline{H}_3$ – that these are of the claimed complexities follows from Lemma 5 and Proposition 3, and the principle of duality. In ongoing work, we believe that we have completely classified the complexity of $\{\exists, \forall, \wedge, \vee\}$-MC$(H)$, for $\|H\| \leq 3$, into each of these four complexity classes. The separating criteria appear to be non-trivial – for example, there appear to be $H$ without isolated vertices such that $\{\exists, \forall, \wedge, \vee\}$-MC$(H)$ is NP-complete. We conclude with the following conjecture, stated for general relational models.

*Conjecture 1.* In full generality, the class of problems $\{\exists, \forall, \wedge, \vee\}$-MC$(A)$ exhibits tetrachotomy. Each such problem is either in P, is NP-complete, is co-NP-complete or is Pspace-complete.

# References


1. Borner, F., Krokhin, A., Bulatov, A., Jeavons, P.: Quantified constraints and surjective polymorphisms. Tech. Rep. PRG-RR-02-11, Oxford (2002)
2. Chandra, A., Merlin, P.: Optimal implementation of conjunctive queries in relational databases. In: STOC (1979)
3. Chen, H.: Quantified constraint satisfaction and 2-semilattice polymorphisms. In: Principles and Practice of Constraint Programming (2004)
4. Chen, H.: The complexity of quantified constraint satisfaction: Collapsibility, sink algebras, and the three-element case. CoRR abs/cs/0607106. SIAM J. Comp. (to appear, 2006)
5. Chen, H.: Private communication (2007)
6. Creignou, N., Khanna, S., Sudan, M.: Complexity classifications of Boolean Constraint Satisfaction Problems. SIAM Monographs (2001)
7. Feder, T., Vardi, M.Y.: The computational structure of monotone monadic SNP and constraint satisfaction: a study through datalog and group theory. SIAM J. Comp. 28 (1999)
8. Hell, P., Nešetřil, J.: On the complexity of H-coloring. J. Combin. Theory Ser. B 48 (1990)
9. Hell, P., Nešetřil, J.: Graphs and Homomorphisms. OUP (2004)
10. Kolaitis, P., Vardi, M.: Conjunctive-query containment and constraint satisfaction. In: PODS (1998)
11. Lynch, N.: Log space recognition and translation of parenthesis languages. Journal of the ACM 24, 583–590 (1977)
12. Martin, B.: Dichotomies and duality in first-order model checking problems. CoRR abs/cs/0609022 (2006)
13. Martin, B., Madelaine, F.R.: Towards a trichotomy for quantified h-coloring. In: Beckmann, A., Berger, U., Löwe, B., Tucker, J.V. (eds.) CiE 2006. LNCS, vol. 3988, pp. 342–352. Springer, Heidelberg (2006)
14. Martin, B.D.: Logic, Computation and Constraint Satisfaction. PhD thesis, University of Leicester (2005)
15. Papadimitriou, C.: Computational Complexity. Addison-Wesley, Reading (1994)
16. Vardi, M.: Complexity of relational query languages. In: STOC (1982)
17. Vardi, M.: On the complexity of bounded-variable queries. In: PODS (1995)

# Domain Theory and the Causal Structure of Space-Time


Keye Martin[1] and Prakash Panangaden[2]

[1] Naval Research Laboratory, Center for High Assurance Computer Systems,
Washington D.C. 20375, USA
kmartin@itd.nrl.navy.mil
[2] School of Computer Science, McGill University, Montreal, Quebec, Canada
prakash@cs.mcgill.ca



**Abstract.** We prove that a globally hyperbolic spacetime with its causality relation is a bicontinuous poset whose interval topology is the manifold topology. From this one can show that from only a countable dense set of events and the causality relation, it is possible to reconstruct a globally hyperbolic spacetime in a purely order theoretic manner. The ultimate reason for this is that globally hyperbolic spacetimes belong to a category that is equivalent to a special category of domains called *interval domains*. We obtain a mathematical setting in which one can study causality independently of geometry and differentiable structure, and which also suggests that spacetime emerges from something discrete.


Domains [AJ94, GKK$^+$03] are special types of posets that have played an important role in theoretical computer science since the late 1960s when they were discovered by Dana Scott [Sco70] for the purpose of providing a semantics for the lambda calculus. They are partially ordered sets that carry intrinsic (order theoretic) notions of completeness and approximation. The basic intuition is that the order relation captures the idea of approximation qualitatively. There is an abstract notion of finite piece of information – or of finite approximation – which plays a key role in the analysis of computation.

These posets have a number of topologies defined on them: the Scott topology, the Alexandrov topology, the interval topology and the Lawson topology, all of which play a role. The Scott topology is particularly important in that continuity with respect to this topology captures some of the information processing aspects of computability. In particular, a Scott continuous function has the following property: a finite piece of information about the output requires only a finite piece of information about the input. While this does not completely reduce Turing computability to topology, it captures a very crucial information-processing aspect of computable functions.

General relativity is Einstein's theory of gravity in which gravity is understood not in terms of mysterious "universal" forces but rather as part of the geometry of spacetime. Einstein's general relativity is profoundly beautiful, and beautifully profound, from both the physical and mathematical viewpoints. It teaches us clear lessons about the universe in which we live that are easily explainable. For


A. Beckmann, C. Dimitracopoulos, and B. Löwe (Eds.): CiE 2008, LNCS 5028, pp. 428–430, 2008.
© Springer-Verlag Berlin Heidelberg 2008

example, it offers a wonderful explanation of gravity: if an apple falls from a tree, the path it takes is not determined by the Newtonian ideal of an "invisible force" but instead by the curvature of the space in which the apple resides – gravity is the curvature of spacetime. In addition, the presence of matter in spacetime causes it to "bend" and Einstein even gives us an equation that relates the curvature of spacetime to the matter present within it.

The study of spacetime structure from an abstract viewpoint – not from the viewpoint of solving differential equations – was initiated by Penrose [Pen65] in a dramatic paper in which he showed a fundamental inconsistency of gravity. It was known since Chandrasekhar [Cha31] that, since everything attracts everything else, a gravitating mass of sufficient size will eventually collapse. What Penrose showed was that any such collapse eventually leads to a singularity where the mathematical description of spacetime as a continuum breaks down. This leads to the need to reformulate gravity, it is hoped that the elusive quantum theory of gravity will resolve this problem.

Since the first singularity theorems [Pen65, HE73], causality has played a key role in understanding spacetime structure. The analysis of causal structure relies heavily on techniques of differential topology [Pen72]. For the past decade Sorkin and others [Sor91] have pursued a program for quantization of gravity based on causal structure. In this approach the causal relation is regarded as the fundamental ingredient and the topology and geometry are secondary.

In a paper that appeared in 2006 [MP06], we prove that the causality relation is much more than a relation – it turns a globally hyperbolic spacetime into what is known as a *bicontinuous poset*. The order on a bicontinuous poset allows one to define an intrinsic topology called *the interval topology*[1]. On a globally hyperbolic spacetime, the interval topology is the manifold topology. Theorems that reconstruct the spacetime topology have been known [Pen72] and Malament [Mal77] has shown that the class of time-like curves determines the causal structure. We establish these results as well though in a purely order theoretic fashion: there is no need to know what "smooth curve" means.

Our more abstract stance also teaches us something *new*: the fact that a globally hyperbolic spacetime is bicontinuous implies that it can be reconstructed in a purely order-theoretic manner, beginning from only a countable dense set of events and the causality relation. The ultimate reason for this is that the category of globally hyperbolic posets, which contains the globally hyperbolic spacetimes, is *equivalent* to a very special category of posets called *interval domains*.

From a certain viewpoint, then, the fact that the category of globally hyperbolic posets is equivalent to the category of interval domains is surprising, since globally hyperbolic spacetimes are usually not order theoretically complete. This equivalence also explains why spacetime can be reconstructed order theoretically from a countable dense set: each $\omega$-continuous domain is the ideal completion of a countable abstract basis, i.e., the interval domains associated to globally hyperbolic spacetimes are the systematic 'limits' of discrete sets.

---

[1] Other people use this term for a different topology: what we call the interval topology has been called the biScott topology.

Measurements were introduced by Martin [Mar00] as a way of incorporating quantitative information into domain theory. More recently, we have shown how the geometry of spacetime can also be viewed as being given by a measurement. There are some surprising connections between Lorentz invariance and measurements that will be described in a longer article in preparation.

# References

[AJ94]      Abramsky, S., Jung, A.: Domain theory. In: Maibaum, T.S.E., Abramsky, S., Gabbay, D.M. (eds.) Handbook of Logic in Computer Science, vol. III. Oxford University Press, Oxford (1994)

[Cha31]     Chandrasekhar, S.: The maximum mass of ideal white dwarfs. Astrophysical Journal 74, 81–82 (1931)

[GKK$^+$03]  Gierz, G., Hoffman, K.H., Keimel, K., Lawson, J.D., Mislove, M., Scott, D.S.: Continuous lattices and domains. Encyclopedia of Mathematics and its Applications, vol. 93. Cambridge University Press, Cambridge (2003)

[HE73]      Hawking, S.W., Ellis, G.F.R.: The large scale structure of space-time. Cambridge Monographs on Mathematical Physics. Cambridge University Press, Cambridge (1973)

[Mal77]     Malement, D.: The class of continuous timelike curves determines the topology of spacetime. J. Math. Phys. 18(7), 1399–1404 (1977)

[Mar00]     Martin, K.: The measurement process in domain theory. In: Welzl, E., Montanari, U., Rolim, J.D.P. (eds.) ICALP 2000. LNCS, vol. 1853, pp. 116–126. Springer, Heidelberg (2000)

[MP06]      Martin, K., Panangaden, P.: A domain of spacetime intervals in general relativity. Communications in Mathematical Physics 267(3), 563–586 (2006)

[Pen65]     Penrose, R.: Gravitational collapse and space-time singularities. Phys. Rev. Lett. 14, 57–59 (1965)

[Pen72]     Penrose, R.: Techniques of differential topology in relativity. Society for Industrial and Applied Mathematics (1972)

[Sco70]     Scott, D.: Outline of a mathematical theory of computation. Technical Monograph PRG-2, Oxford University Computing Laboratory (1970)

[Sor91]     Sorkin, R.: Spacetime and causal sets. In: D'Olivo, J., et al. (eds.) Relativity and Gravitation: Classical and Quantum. World Scientific, Singapore (1991)

# Recursion on Nested Datatypes in Dependent Type Theory

Ralph Matthes

Institut de Recherche en Informatique de Toulouse (IRIT)
C. N. R. S. et Université Paul Sabatier (Toulouse III)
118 route de Narbonne, F-31062 Toulouse Cedex 9

**Abstract.** Nested datatypes are families of datatypes that are indexed over all types and where the datatype constructors relate *different* members of the family. This may be used to represent variable binding or to maintain certain invariants through typing.

In dependent type theory, a major concern is the termination of all expressible programs, so that types that depend on object terms can still be type-checked mechanically. Therefore, we study iteration and recursion schemes that have this termination guarantee throughout. This is not based on syntactic criteria (recursive calls with "smaller" arguments) but just on types ("type-based termination"). An important concern are reasoning principles that are compatible with the ambient type theory, in our case induction principles.

In previous work, the author has proposed an abstract description of nested datatypes together with a mapping operation (like map for lists) and an iterator on the term side and an induction principle on the logical side that could all be implemented within the Coq system (with impredicative Set that is just needed for the justification, not for the definition and the examples). For verification purposes, it is important to have naturality theorems for the obtained iterative functions. Although intensional type theory does not provide naturality in general, criteria for naturality could be established that are met in case studies on "bushes" and representations of lambda terms (also with explicit flattening).

The new contribution is an extension of this abstract description to full primitive recursion and its illustration by way of examples that have been carried out in Coq. Unlike the iterative system, we do not yet have a justification within Coq.

## 1 Introduction

Nested datatypes [1] are families of datatypes that are indexed over all types and where the datatype constructors relate different members of the family (i. e., at least one datatype constructor constructs a family member from data of a type that refers to a different member of the family). Let $\kappa_0$ stand for the universe of (mono-)types that will be interpreted as sets of computationally relevant objects. Then, let $\kappa_1$ be the kind of type transformations, hence $\kappa_1 := \kappa_0 \rightarrow \kappa_0$. A typical example would be *List* of kind $\kappa_1$, where *List A* is the type of finite

A. Beckmann, C. Dimitracopoulos, and B. Löwe (Eds.): CiE 2008, LNCS 5028, pp. 431–446, 2008.

lists with elements from type $A$. But *List* is not a nested datatype since the recursive equation for *List*, i. e., $List\,A = 1 + A \times List\,A$, does not relate lists with different indices. A simple example of a nested datatype where an invariant is guaranteed through its definition are the powerlists [2], with recursive equation $PList\,A = A + PList(A \times A)$, where the type $PList\,A$ represents trees of $2^n$ elements of $A$ with some $n \geq 0$ (that is not fixed) since, throughout this article, we will only consider the least solutions to these equations. The basic example where variable binding is represented through a nested datatype is a higher-order de Bruijn representation of untyped lambda calculus, following ideas of [3,4,5]. The lambda terms with free variables taken from $A$ are given by $Lam\,A$, with recursive equation

$$Lam\,A = A + Lam\,A \times Lam\,A + Lam(option\,A) \ .$$

The first summand gives the variables, the second represents application of lambda terms and the interesting third summand stands for lambda abstraction: An element of $Lam(option\,A)$ (where $option\,A$ is the type that has exactly one more element than $A$) is seen as an element of $Lam\,A$ through lambda abstraction of that designated extra variable that need not occur freely in the body of the abstraction.

In dependent type theory, a major concern is the termination of all expressible programs. This may be seen as a heritage of polymorphic lambda calculus (system $F^\omega$) that, by the way, is able to express nested datatypes and many algorithms on them [6]. But termination is also of practical concern with dependent types, namely that type-checking should be decidable: If types depend on object terms, object terms have to be evaluated in order to verify types, as expressed in the convertibility rule. Note, however, that this only concerns evaluation within the definitional equality (i. e., convertibility), henceforth denoted by $\simeq$. Except from the above intuitive recursive equations, $=$ will denote propositional equality throughout: this is the equality type that requires proof and that satisfies the Leibniz principle, i. e., that validity of propositions is not affected by replacing terms by equal (w. r. t. $=$) terms.

Here, we study iteration and recursion schemes that have this termination guarantee throughout. Termination is not based on syntactic criteria such as strict positivity and that all recursive calls are done with "smaller" arguments, but just on types (called "type-based termination" in [7]). The article with Abel and Uustalu [6] presents a variety of iteration principles on nested datatypes in this spirit, all within the framework of system $F^\omega$. However, no reasoning principles, in particular no induction principles, were studied there. Newer work by the author [8] integrates rank-2 Mendler iteration into the Calculus of Inductive Constructions [9,10,11] that underlies the Coq theorem prover [12] and also justifies an induction principle for them. This is embodied in the system *LNMIt*, the "logic for natural Mendler-style iteration", defined in Section 3.1.

The articles [6,8] only concern plain iteration. While an extension of primitive Mendler-style recursion [13] to nested datatypes has been described earlier [14], we will present here an extension *LNMRec* of system *LNMIt* by an enriched

Mendler-style recursor where the step term additionally has access to a map term for the unknown type transformation $X$ that occurs there. By way of examples, its merits will be studied. An overview of extensions to *LNMRec* of results established for *LNMIt* in [8] is given. However, the main theorem of [8] is not carried over to the present setting, i. e., we do not yet have a justification within the Calculus of Inductive Constructions. Nevertheless, all the concepts and results have been formalised in the Coq system, using module functors with parameters of a module type that specifies our extension of *LNMRec*. The Coq code is available [15] and is based on [16].

The next section describes two examples of truly nested datatypes. The first with "bushes", treated in Section 2.1, motivates primitive recursion instead of plain iteration and the second about lambda calculus with explicit flattening, treated in Section 2.2, motivates the access to a map term in the defining clauses of an iterative function. Section 3.1 completes the precise definition of *LNMIt* from [8], while Section 3.2 defines the new system *LNMRec* and shows theorems about it. Section 4 describes when and how to define iterative functions with access to a map term in *LNMIt* and establishes a precise relation with the alternative within *LNMRec*.

## 2   Motivating Examples

A nested datatype will be called "truly nested" (non-linear [17]) if the intuitive recursive equation for the inductive family has at least one summand with a nested call to the family name, i. e., the family name appears somewhere inside the type argument of a family name occurrence of that summand. Our two examples will be the bushes [1] and the lambda terms with explicit flattening [18], described as follows:

$$Bush\,A = 1 + A \times Bush(Bush\,A)\ ,$$
$$LamE\,A = A + LamE\,A \times LamE\,A + LamE(option\,A) + LamE(LamE\,A)\ .$$

The last summand in both examples qualifies them as truly nested datatypes; it is even the same nested call pattern. Truly nested datatypes cannot be directly represented in the current version of the Calculus of Inductive Constructions (CIC), as it is implemented in Coq, while the examples of *PList* and *Lam*, mentioned in the first paragraph of the introduction, are now (since version 8.1) fully supported with recursion and induction principles. In these cases, our proposal is more generic but offers less comfort since it has neither advanced pattern matching nor guardedness checking. *PList* and *Lam* are strictly positive, but *Bush* and *LamE* are not even considered to be positive [19] (see [14] for a notion based on polarity that covers these examples). Since there was no system that combined the termination guarantee for recursion schemes on truly nested datatypes with a logic to reason about the defined functions, it seems only natural that examples like *Bush* and *LamE* did not receive more attention. They are studied in detail in [8]. Here, they are recapitulated and developed so as to motivate our new extension *LNMRec* of *LNMIt* that will be defined in Section 3.2.

## 2.1  Bushes

In order to fit the above intuitive definition of *Bush* into the setting of Mendler-style recursion, the notion of rank-2 functor is needed. Let $\kappa_2 := \kappa_1 \to \kappa_1$. Any constructor $F$ of kind $\kappa_2$ qualifies as rank-2 functor for the moment, and $\mu F : \kappa_1$ denotes the generated nested datatype.[1] For bushes, set

$$BushF := \lambda X^{\kappa_1} \lambda A^{\kappa_0}.\, 1 + A \times X(X\,A)$$

and $Bush := \mu\,BushF$. In general, there is just one datatype constructor for $\mu F$, namely $in : F(\mu F) \subseteq \mu F$, with the abbreviation $X \subseteq Y := \forall A^{\kappa_0}.\, X A \to Y A$ for any $X, Y : \kappa_1$. For bushes, more clarity comes from two derived datatype constructors

$$bnil : \forall A^{\kappa_0}.\, Bush\,A \ ,$$
$$bcons : \forall A^{\kappa_0}.\, A \to Bush(Bush\,A) \to Bush\,A \ ,$$

defined by $bnil := \lambda A^{\kappa_0}.\, in\,A\,(inl\,tt)$ (with $tt$ the inhabitant of 1 and left injection $inl$) and $bcons := \lambda A^{\kappa_0} \lambda a^A \lambda b^{Bush(Bush\,A)}.\, in\,A\,(inr(a,b))$ (with right injection $inr$ and pairing notation $(\cdot,\cdot)$).

Our first example of an iterative function on bushes is the function $BtL :$ $Bush \subseteq List$ ($BtL$ is a shorthand for $BushToList$) that gives the list of all elements in the bush and that obeys to the following specification:

$$BtL\,A\,(bnil\,A) \quad \simeq [] \ ,$$
$$BtL\,A\,(bcons\,A\,a\,b) \simeq a :: flat\_map_{Bush\,A,A}\,(BtL\,A)(BtL\,(Bush\,A)\,b) \ .$$

Here, we denoted by $[]$ the empty list and by $a :: \ell$ the cons operation on lists, and $flat\_map_{B,A}\,f\,\ell$ is the concatenation of all the $A$-lists $f\,b'$ for the elements $b'$ of the $B$-list $\ell$. See below why $BtL$ is to be called an iterative function.

With the *length* function for lists, we get a function that calculates the size of bushes: $size_i := \lambda A \lambda t^{Bush\,A}.\, length(BtL_A\,t)$. Note that we write the type parameter to $BtL$ just as an index, which we will do frequently in the sequel for type-indexed functions—if we do not omit it altogether, e. g., for $size_i$. The definition of $size_i$ is not iterative[2], but an easy induction on $BtL_{Bush\,A}\,b$ reveals

$$size_i(bcons_A\,a\,b) = S\big(fold\_right_{nat,Bush\,A}\,(\lambda x \lambda s.\,size_i\,x + s)\,0\,(BtL_{Bush\,A}\,b)\big) \ ,$$

with $S$ the successor function on the type $nat$ of natural numbers and

$$fold\_right : \forall A^{\kappa_0} \forall B^{\kappa_0}.\,(B \to A \to A) \to A \to List\,B \to A$$

with $fold\_right_{A,B}\,f\,a\,[] \simeq a$ and

$$fold\_right_{A,B}\,f\,a\,(b :: \ell) \simeq f\,b\,(fold\_right_{A,B}\,f\,a\,\ell) \ .$$

Since we used induction on bushes above, the recursive equation only holds for propositional equality and not for the definitional equality $\simeq$. But we might desire just that, i. e., we might want a recursive version $size_r$ of $size_i$ such that

$$size_r(bcons_A\,a\,b) \simeq S\big(fold\_right_{nat,Bush\,A}\,(\lambda x \lambda s.\,size_r\,x + s)\,0\,(BtL_{Bush\,A}\,b)\big) \ ,$$

---

[1] Strictly speaking, this includes *List* since nesting is not required.

[2] The index in the name $size_i$ stands for *indirect*, not for iterative.

but this is no longer within the realm of iteration in Mendler's style, as we will argue right now. Mendler iteration of rank 2 [6] can be described as follows: There is a constant

$$MIt : \forall G^{\kappa_1}.\,(\forall X^{\kappa_1}.\,X \subseteq G \to FX \subseteq G) \to \mu F \subseteq G$$

and the iteration rule

$$MIt\,G\,s\,A\,(in\,A\,t) \simeq s\,(\mu F)\,(MIt\,G\,s)\,A\,t\ .$$

In a properly typed left-hand side, $t$ is of type $F(\mu F)A$ and $s$ of type

$$\forall X^{\kappa_1}.\,X \subseteq G \to FX \subseteq G\ .$$

The term $s$ is called the *step term* of the iteration since it provides the inductive step that extends the function from the type transformation $X$ that is to be viewed as approximation to $\mu F$ (although this is not expressed here!), to a function from $FX$ to $G$.

Given a step term $s$, one gets $s\,G\,(\lambda A^{\kappa_0}\lambda x^{GA}.\,x) : FG \subseteq G$ – an $F$-algebra. Conversely, given an $F$-algebra $s_0 : FG \subseteq G$, one can construct a step term $s$ if there is a term $M : \forall X^{\kappa_1}\forall G^{\kappa_1}.\,X \subseteq G \to FX \subseteq FG$:

$$s := \lambda X^{\kappa_1}\lambda it^{X \subseteq G}\lambda A^{\kappa_0}\lambda t^{FXA}.\,s_0\,A\,(M\,X\,G\,it\,A\,t).$$

However, a typical feature of truly nested datatypes is that there is no such (closed) term $M$ [6, Lemma 5.3] (but see the notion of relativized basic monotonicity in Section 4). Moreover, the traditional approach with $F$-algebras does not display the operational behaviour as much as Mendler's style does.

The function $BtL$ is an instance of this iteration scheme with

$$BtL := MIt\,List\,\big(\lambda X^{\kappa_1}\lambda it^{X \subseteq List}\lambda A^{\kappa_0}\lambda t^{BushF\,X\,A}.\,\text{match}\,t\,\text{with}\,inl\,\_ \mapsto [\,]$$
$$|\,inr(a^A, b^{X(X\,A)}) \mapsto a :: flat\_map_{X\,A,\,A}\,(it\,A)(it\,(X\,A)\,b)\big)\ .$$

Note that when the term $t$ of type $BushF\,X\,A$ is matched with $inr(a, b)$, the variable $b$ is of type $X(X\,A)$.[3] This is the essence of Mendler's style: the recursive calls come in the form of uses of $it$ that does not have type $Bush \subseteq List$ but just $X \subseteq List$, and the type arguments of the datatype constructors are replaced by variants that only mention $X$ instead of $Bush$. So, the definitions have to be uniform in that type transformation variable $X$, but this is already sufficient to guarantee termination (for the rank-1 case of inductive *types*, this has been discovered in [20] by syntactic means and, independently, by the author with a semantic construction [21]).

We conclude that $BtL$ is an iterative function in the sense of Mendler but also in a more general sense since Mendler iteration can be simulated by impredicative encodings in system $F^\omega$. In a less technical sense, $BtL$ is iterative as opposed

---

[3] The pattern matching could easily be replaced by case analysis on sums and projections for products.

to primitive recursive since the recursive argument $b$ of *bcons* is only used as an argument of *BtL* itself. The recursive equation for $size_r$, however, uses $b$ as an argument not of $size_r$, but the previously defined *BtL*, whose result is then fed element-wise into $size_r$. It seems very unlikely that there is a direct definition of $size_r$ by help of *MIt*: If, through pattern matching, $b$ is only available with type $X(X\,A)$, the function $BtL_{Bush\,A}$ just cannot be applied to it. Neither could $BtL_{XA}$. The way out is provided already by Mendler for inductive types [13] and has been generalized to nested datatypes in [14]: Express in the step term in addition that $X$ is an approximation of $\mu F$, in the sense of an "injection" $j : X \subseteq \mu F$ that is available in the body of the definition. So, we assume a constant (the minus sign indicates that it is a preliminary version)

$$MRec^- : \forall G^{\kappa_1}. (\forall X^{\kappa_1}. X \subseteq \mu F \to X \subseteq G \to FX \subseteq G) \to \mu F \subseteq G$$

and the recursion rule

$$MRec^-\, G\, s\, A\, (in\, A\, t) \simeq s\, \mu F\, (\lambda A^{\kappa_0} \lambda x^{\mu FA}.\, x)\, (MRec^-\, G\, s)\, A\, t\ .$$

(In a more traditional formulation instead of Mendler's style, one would require a recursive $F$-algebra of type $F(\lambda A^{\kappa_0}.\, \mu F A \times G A) \subseteq G$ instead of a step term $s$, see [22] for the case of inductive *types*. Again, Mendler's style does not necessitate any consideration of monotonicity of $F$.) $MRec^-$ allows to program $size_r$ with $G := \lambda A.\, nat$ as

$$MRec^-\, G\, \Big(\lambda X^{\kappa_1} \lambda j^{X \subseteq Bush} \lambda rec^{X \subseteq G} \lambda A^{\kappa_0} \lambda t^{BushF\ X\ A}.\, \text{match}\, t\, \text{with}\, inl\, \_ \mapsto 0$$

$$|\ inr(a^A, b^{X(X\ A)}) \mapsto S\big(fold\_right_{nat,XA}\, (\lambda x \lambda s.\, rec_A\, x + s)\, 0\, (BtL_{XA}\, (j_{XA}\, b)))\big)\Big).$$

The injection $j$ is an artifact of Mendler's method to enforce termination. In the recursion rule, the term $j_{XA}\, b$ is instantiated by $(\lambda A^{\kappa_0} \lambda x^{Bush\ A}.\, x)\, (Bush\, A)\, b \simeq b$. Therefore, only $b$ appears in the displayed right-hand side of the recursive equation.[4] Viewed from a different angle that already accepts to use Mendler's style, the variable $j$ is there only for type-checking purposes: It allows to get a well-typed step term although it will not be visible afterwards. The advantage of this artifact is that no modification of the ambient type system is needed.

Certainly, we would now like to prove that $\forall A^{\kappa_0} \forall t^{Bush\ A}.\, size_i\, t = size_r\, t$, but this will require our new system *LNMRec*, to be defined in Section 3.2.

Looking back at this little example, we may say that the original function $size_i$ is not itself an instance of Mendler iteration. But there is a recursive equation that can even be made to hold definitionally (with respect to $\simeq$), in form of the instance $size_r$ of the primitive recursor $MRec^-$. But the essential question is how to prove that both functions agree.

---

[4] In an example that is only mentioned at the end of Section 3.2, the author encountered a natural situation where $j$ is not directly applied to a recursive argument of a datatype constructor, showing again the flexibility of type-based termination.

## 2.2   Untyped Lambda Calculus with Explicit Flattening

The untyped lambda terms with explicit flattening are obtained as $LamE :=$ $\mu\, LamEF$ with

$$LamEF := \lambda X^{\kappa_1}\lambda A^{\kappa_0}.\, A + XA \times XA + X(\text{option } A) + X\,(X\,A)\ .$$

Since the Calculus of Inductive Constructions allows a direct representation of $Lam$ (described in the first paragraph of the introduction), we go without $LNMIt$ for $Lam$ and just assume that we already have an implementation of renaming $lam : \forall A^{\kappa_0}\forall B^{\kappa_0}.\,(A \to B) \to Lam\,A \to Lam\,B$ and parallel substitution

$$subst : \forall A^{\kappa_0}\forall B^{\kappa_0}.\,(A \to Lam\,B) \to Lam\,A \to Lam\,B\ ,$$

where for a *substitution rule* $f : A \to Lam\,B$, the term $subst_{A,B}\,f\,t : Lam\,B$ is the result of substituting every variable $a : A$ in the term representation $t : Lam\,A$ by the term $f\,a : Lam\,B$. From now on we will no longer decorate variables $A, B$ with their kind $\kappa_0$.

The interesting new datatype constructor for $LamE$,

$$flat : \forall A.\, LamE(LamE\,A) \to LamE\,A\ ,$$

is obtained by composing $in$ with the right injection $inr$ (we assume that $+$ in the definition of $LamEF$ associates to the left). Its interpretation is an explicit (not executed) form of an integration of the lambda terms that constitute its free variable occurrences into the term itself. This is the monad multiplication form of explicit substitution, as opposed to parallel explicit substitution that would have the type of $subst$, with $Lam$ replaced by $LamE$. That other approach would need a quantifier in the rank-2 functor, but also an embedded function space which is not covered by $LNMIt/LNMRec$ in their current form due to extensionality problems. Moreover, the arising $LamE$ would not be a truly nested datatype.

We will review the definition of the iterative function $eval : LamE \subseteq Lam$ that evaluates all the explicit flattenings and thus yields the representation of a usual lambda term. As for bushes, we might want to enforce recursive equations with respect to $\simeq$ that are originally only provable. We do this first with an important property of $subst$ and then with naturality (having been proven for Mendler iteration). This will again lead out of the realm of Mendler iteration and motivate a second and conceptually new extension in an orthogonal direction.

Define $eval$ by Mendler iteration as

$$eval := MIt\big(\lambda X^{\kappa_1}\lambda it^{X \subseteq Lam}\lambda A\lambda t^{FXA}.\,\text{match}\,t\,\text{with}\,\dots$$
$$\mid inr\,e^{X(XA)} \mapsto subst_{XA,A}\,it_A\,(it_{XA}\,e)\big)\ ,$$

where the routine part has been omitted. This function, introduced in [8], evidently satisfies

$$eval_A(flat_A\,e) \simeq subst_{Lam\,A,A}\,eval_A\,(eval_{LamE\,A}\,e)\ .$$

In view of the equation (see [8])

$$\forall A\forall B\forall f^{A \to Lam\,B}\forall t^{Lam\,A}.\,subst_{Lam\,B,B}\,(\lambda x^{Lam\,B}.\,x)\,(lam\,f\,t) = subst_{A,B}\,f\,t,$$

the right-hand side is propositionally equal to

$$subst_{Lam\ A,A}\ (\lambda x^{Lam\ A}.\,x)\ (lam\ eval_A\ (eval_{LamE\ A}\ e))\ .$$

We can easily enforce the latter term to be definitionally equal to the outcome on $flat_A\ e$ for a variant $eval1$ of $eval$, fitting better with what will come later:

$$eval1 := MIt\big(\lambda X^{\kappa_1}\lambda it^{X\subseteq Lam}\lambda A\lambda t^{FXA}.\,\mathrm{match}\,t\,\mathrm{with}\,\ldots$$
$$\mid inr\ e^{X(XA)}\mapsto subst\ (\lambda x.\,x)\ (lam\ it_A\ (it_{XA}\ e))\ .$$

It is an easy exercise to prove in $LNMIt$ that $eval1$ and $eval$ agree propositionally on all arguments.

A natural question for polymorphic functions $j$ of type $X\subseteq Y$ is whether they behave—propositionally—as a natural transformation from $(X,mX)$ to $(Y,mY)$, given map functions[5] $mX:mon\ X$ and $mY:mon\ Y$, where, for any type transformation $X:\kappa_1$, we define

$$mon\ X := \forall A\forall B.\,(A\to B)\to XA\to XB\ .$$

Here, the pair $(X,mX)$ is seen as a functor although no functor laws are required. The proposition that defines $j$ to be such a natural transformation is

$$j\in\mathcal{N}(mX,mY) := \forall A\forall B\forall f^{A\to B}\forall t^{XA}.\,j_B\,(mX\ A\ B\ f\ t) = mY\ A\ B\ f\,(j_A\ t)\ ,$$

and $LNMIt$ allows to prove $eval\in\mathcal{N}(lamE,lam)$, with $lamE$ the canonical renaming operation for $LamE$, to be provided by $LNMIt$ (see Section 3.1). Since [23], naturality is seen as free in pure functional programming because naturality with respect to parametric equality is an instance of the parametricity theorem for types of the form $X\subseteq Y$, but in intensional type theory such as our $LNMIt$, naturality with respect to propositional equality has to be proven on a case by case basis. Since naturality of $j$ only depends propositionally on the values of $j_A\ t$, we also have $eval1\in\mathcal{N}(lamE,lam)$.

It is time to address the second extension of $MIt$ that we consider desirable from the point of view of computational behaviour of algorithms on nested datatypes. It does not seem to have been considered previously but is somehow implicit in the author's [8, section 2.3]. It does not even make sense for inductive types but is confined to inductive families.

An instance of naturality of $eval1$ is (for $e:LamE(LamE\ A)$)

$$eval1_{Lam\ A}(lamE\ eval1_A\ e) = lam\ eval1_A\ (eval1_{LamE\ A}\ e)\ .$$

In the same spirit as for $size_r$, we might now desire to have the left-hand side as subterm of the definitional outcome of evaluation of $flat_A\ e$ instead of the right-hand side, but $subst\ (\lambda x.\,x)\ (it_{Lam\ A}(lamE\ it_A\ e))$ does not type-check in place of the right branch of the definition of $eval1$. As noted in [8], we would need to replace $lamE$ by some term $m:mon\ X$, hence a map term for $X$, but

<hr>

[5] The name map function comes from the function $map$ on lists that is of type $mon\ List$ and that we prefer to call $list$.

this did not seem available. In fact, it is not available in general (see Section 4 for a discussion under which circumstances it can be constructed). Our proposal is now simply to add such an $m$ to the bound variables that are accessible in the body of the step term. We thus require a constant

$$MIt^+ : \forall G^{\kappa_1}. (\forall X^{\kappa_1}. mon\, X \to X \subseteq G \to FX \subseteq G) \to \mu F \subseteq G$$

and the extended iteration rule

$$MIt^+ \, G\, s\, A\, (in\, A\, t) \simeq s\, \mu F\, map_{\mu F}\, (MIt^+ \, G\, s)\, A\, t \ ,$$

with a canonical map term $map_{\mu F} : mon\, \mu F$ that is anyhow provided by $LNMIt$ and which is $lamE$ in our example. Now, define $eval1' := MIt^+\, s_{eval1'}$ with

$$s_{eval1'} := \lambda X^{\kappa_1} \lambda m^{mon\, X} \lambda it^{X \subseteq Lam} \lambda A \lambda t^{FXA}. \, match\, t\, with\, \dots$$
$$|\, inr\, e^{X(XA)} \mapsto subst\, (\lambda x.\, x)\, (it_{Lam}\, A(m\, it_A\, e)) \ ,$$

which enjoys the desired operational behaviour

$$eval1'_A(flat_A\, e) \simeq subst\, (\lambda x.\, x)\, (eval1'_{Lam\, A}(lamE\, eval1'_A\, e)) \ .$$

The termination of such a function can be proven by sized nested datatypes [24], but there do not yet exist induction principles for reasoning on programs with sized nested datatypes. A *logic* for Mendler-style recursion that covers our examples will now be described.

Before that, let us note that this second example showed us a situation where a recursive equation with the map term of the nested datatype on the right-hand side may require to abstract over an arbitrary term of type $mon\, X$ in the step term of an appropriately extended notion of Mendler iteration.

## 3   Logic for Natural Mendler-Style Recursion of Rank 2

First, we recall $LNMIt$ from [8], then we describe its modifications in order to obtain its extension $LNMRec$.

### 3.1   $LNMIt$

In $LNMIt$, for a nested datatype $\mu F$, we require that $F : \kappa_2$ preserves *extensional functors*. In the Calculus of Inductive Constructions, we may form for $X : \kappa_1$ the dependently-typed record $\mathcal{E}X$ that contains a map term $m : mon\, X$, a proof $e$ of extensionality of $m$, defined by

$$ext\, m := \forall A \forall B \forall f^{A \to B} \forall g^{A \to B}. (\forall a^A. \, fa = ga) \to \forall r^{XA}. \, m\, A\, B\, f\, r = m\, A\, B\, g\, r \ ,$$

and proofs $f_1, f_2$ of the first and second functor laws for $(X, m)$, defined by the propositions

$$fct_1\, m := \forall A \forall x^{XA}. \, m\, A\, A\, (\lambda y.y)\, x = x \ ,$$
$$fct_2\, m := \forall A \forall B \forall C \, \forall f^{A \to B} \, \forall g^{B \to C} \, \forall x^{XA}. \, m\, A\, C\, (g \circ f)\, x = m\, B\, C\, g\, (m\, A\, B\, f\, x).$$

Parameters:
$$F \quad\quad : \kappa_2$$
$$Fp\mathcal{E} \quad : \forall X^{\kappa_1}.\,\mathcal{E}X \to \mathcal{E}(FX)$$
Constants:
$$\mu F \quad\quad : \kappa_1$$
$$map_{\mu F} : mon(\mu F)$$
$$In \quad\quad : \forall X^{\kappa_1}\,\forall ef^{\mathcal{E}X}\,\forall j^{X\subseteq\mu F}.\,j \in \mathcal{N}(m\,ef,\,map_{\mu F}) \to FX \subseteq \mu F$$
$$MIt \quad\quad : \forall G^{\kappa_1}.\,(\forall X^{\kappa_1}.\,X \subseteq G \to FX \subseteq G) \to \mu F \subseteq G$$
$$\mu FInd \;\; : \forall P : \forall A.\,\mu F A \to Prop.\,\Big(\forall X^{\kappa_1}\forall ef^{\mathcal{E}X}\forall j^{X\subseteq\mu F}\forall n^{j\in\mathcal{N}(m\,ef,\,map_{\mu F})}.$$
$$\big(\forall A\forall x^{XA}.\,P_A(j_A\,x)\big) \to \forall A\forall t^{FXA}.\,P_A(In\,ef\,j\,n\,t)\Big)$$
$$\to \forall A\forall r^{\mu FA}.\,P_A\,r$$
Rules:
$$map_{\mu F}\,f\,(In\,ef\,j\,n\,t) \simeq In\,ef\,j\,n\,(m(Fp\mathcal{E}\,ef)\,f\,t)$$
$$MIt\,s\,(In\,ef\,j\,n\,t) \simeq s\,\big(\lambda A.\,(MIt\,s)_A \circ j_A\big)\,t$$
$$\lambda A\lambda x^{\mu FA}.\,(MIt\,s)_A\,x \simeq MIt\,s$$

**Fig. 1.** Specification of *LNMIt*

Given a record $ef$ of type $\mathcal{E}X$, Coq's notation for its field $m$ is $m\,ef$, and likewise for the other fields. We adopt this notation instead of the more common $ef.m$. Preservation of extensional[6] functors for $F$ is required in the form of a term of type $\forall X^{\kappa_1}.\,\mathcal{E}\,X \to \mathcal{E}(FX)$, and the full definition of *LNMIt* is given as the extension of the predicative Calculus of Inductive Constructions (= pCIC) [12] by the constants and rules in Figure 1, adopted from [8].[7] In *LNMIt*, one can show the following theorem [8, Theorem 3] about canonical elements: There are terms $ef_{\mu F} : \mathcal{E}\mu F$ and $InCan : F(\mu F) \subseteq \mu F$ (the *canonical* datatype constructor that constructs canonical elements) such that the following convertibilities hold:

$$m\,ef_{\mu F} \simeq map_{\mu F}\ ,$$
$$map_{\mu F}\,f\,(InCan\,t) \simeq InCan(m\,(Fp\mathcal{E}\,ef_{\mu F})\,f\,t)\ ,$$
$$MIt\,s\,(InCan\,t) \simeq s\,(MIt\,s)\,t\ .$$

Some explanations are in order: The datatype constructor $In$ is way more complicated than our previous $in$, but we get back $in$ in the form of $InCan$ that only constructs the "canonical elements" of the nested datatype $\mu F$. The map term $map_{\mu F}$ for $\mu F$, which does renaming in the example of $LamE$, is an integral part of the system definition since it occurs in the type of $In$. The Mendler iterator $MIt$ has not been touched at all; there is just a more general iteration rule that also covers non-canonical elements, but for the canonical elements, we get the same behaviour, i.e., the same equation with respect to $\simeq$. The

---

[6] While the functor laws are certainly an important ingredient of program verification, the extensionality requirement is more an artifact of our intensional type theory that does not have extensionality of functions in general.

[7] If the rules were formulated with $=$ instead of $\simeq$, *LNMIt* would just be a signature within the pCIC and only a logic without any termination guarantees for the rules.

crucial part is the induction principle $\mu FInd$, where *Prop* denotes the universe of propositions (all our propositional equalities and their universal quantifications belong to it). Without access to the argument $n$ that assumes naturality of $j$ as a transformation from $(X, m\,ef)$ to $(\mu F, map_{\mu F})$, one would not be able to prove naturality of *MIt s*, i.e., of iteratively defined functions on the nested datatype $\mu F$. The author is not aware of ways how to avoid non-canonical elements and nevertheless have an induction principle that allows to establish naturality of *MIt s* [8, Theorem 1].

*BushF* and *LamEF* are easily seen to fulfill the requirement of *LNMIt* to preserve extensional functors (using [8, Lemma 1 and Lemma 2]).

## 3.2   *LNMRec*

Let *LNMRec* be the modification of *LNMIt*, where *MIt* and its two rules are replaced by *MRec* and its proper two rules: the recursor, the rule of primitive recursion and the extensionality rule

$$MRec : \forall G^{\kappa_1}.\,(\forall X^{\kappa_1}.\,mon\,X \to X \subseteq \mu F \to X \subseteq G \to FX \subseteq G) \to \mu F \subseteq G\ ,$$

$$MRec\,s\,(In\,ef\,j\,n\,t) \simeq s\,(m\,ef)\,j\,\big(\lambda A.\,(MRec\,s)_A \circ j_A\big)\,t\ ,$$

$$\lambda A\lambda x^{\mu F\,A}.\,(MRec\,s)_A\,x \simeq MRec\,s\ .$$

The first general consequence of the definition of *LNMRec* is the analogue of [8, Theorem 3] with the primitive recursion rule (generalizing both the rules of $MRec^-$ and $MIt^+$; recall $in := InCan$)

$$MRec\,G\,s\,(in\,A\,t) \simeq s\,\mu F\,map_{\mu F}\,(\lambda A^{\kappa_0}\lambda x^{\mu F A}.\,x)\,(MRec\,G\,s)\,A\,t\ ,$$

where we are more explicit about all the type parameters.

Evidently, the examples of Section 2 can be formulated in *LNMRec* by not using the argument of type $mon\,X$ in the case of bushes and by not using the injection of type $X \subseteq \mu F$ in the case of evaluation. By using neither monotonicity nor injection, we get back *MIt* and hence may view *LNMRec* as an extension of *LNMIt*.

We continue with two new results that are analogues of results in [8].

**Theorem 1 (Naturality of** *MRec s***).** *Assume* $G\ :\ \kappa_1$, $mG\ :\ mon\,G$, $s\ :$ $\forall X^{\kappa_1}.\,mon\,X \to X \subseteq \mu F \to X \subseteq G \to FX \subseteq G$ *and that the following holds:*

$$\forall X^{\kappa_1}\forall ef^{\mathcal{E}X}\forall j^{X\subseteq\mu F}\forall rec^{X\subseteq G}.\,j \in \mathcal{N}(m\,ef,\,map_{\mu F}) \to$$
$$rec \in \mathcal{N}(m\,ef,\,mG) \to s\,(m\,ef)\,j\,rec \in \mathcal{N}(m(Fp\mathcal{E}\,ef),\,mG)\ .$$

*Then* $MRec\,s \in \mathcal{N}(map_{\mu F},\,mG)$, *hence* $MRec\,s$ *is a natural transformation for the respective map terms.*

*Proof.* By the induction principle $\mu FInd$, as for [8, Theorem 1].

We abbreviate $bush := map_{\mu BushF}$. By using that $BtL \in \mathcal{N}(bush,\,list)$ [8], one can immediately use Theorem 1 to show that $size_r(bush\,f\,t) = size_r\,t$ because this is a naturality statement.

442     R. Matthes

Trivially, the condition on $s$ is simpler for $MIt^+ s$, namely

$$(*) \quad \forall X^{\kappa_1} \forall ef^{\mathcal{E}X} \forall it^{X \subseteq G}. \, it \in \mathcal{N}(m \, ef, mG) \to s \, (m \, ef) \, it \in \mathcal{N}(m(Fp\mathcal{E} \, ef), mG) \ .$$

This condition is fulfilled for $s_{eval1'}$ in place of $s$, hence $eval1' \in \mathcal{N}(lamE, lam)$ follows.

**Theorem 2 (Uniqueness of $MRec \, s$).** *Assume $G : \kappa_1$, $s$ of the type as in the preceding theorem and $h : \mu F \subseteq G$ (the candidate for being $MRec \, s$). Assume further the following extensionality property of $s$:*

$$\forall X^{\kappa_1} \forall ef^{\mathcal{E}X} \forall j^{X \subseteq \mu F} \forall f, g : X \subseteq G. \, (\forall A \forall x^{XA}. \, f_A \, x = g_A \, x) \to$$
$$\forall A \forall t^{FXA}. \, s \, (m \, ef) \, j \, f \, t = s \, (m \, ef) \, j \, g \, t \ .$$

*Assume finally that $h$ satisfies the equation for $MRec \, s$:*

$$\forall X^{\kappa_1} \forall ef^{\mathcal{E}X} \forall j^{X \subseteq \mu F} \forall n^{j \in \mathcal{N}(m \, ef, \, map_{\mu F})} \forall A \forall t^{FXA}.$$
$$h_A(In \, ef \, j \, n \, t) = s \, (m \, ef) \, j \, (\lambda A. \, h_A \circ j_A) \, t \ .$$

*Then, $\forall A \forall r^{\mu F \, A}. \, h_A \, r = MRec \, s \, r$.*

*Proof.* By the induction principle $\mu FInd$, as for [8, Theorem 2].

The final question of Section 2.1 is precisely of the form of the conclusion of Theorem 2 (taking into account that $MRec^-$ is just an instance of $MRec$), and its conditions can be shown to be fulfilled, hence $\forall A \forall t^{Bush \, A}. \, size_i \, t = size_r \, t$ follows.[8] By using that $eval1 \in \mathcal{N}(lamE, lam)$, one can also apply Theorem 2 in order to show $\forall A \forall t^{LamE \, A}. \, eval1_A \, t = eval1'_A \, t$. Alternatively, one can combine [8, Theorem 2] and $eval1' \in \mathcal{N}(lamE, lam)$ for that result. For details, see [15], as before.

The author has carried out other case studies with $LNMRec$. For example, in order to show injectivity of the datatype constructors of $LamE$ for application and lambda abstraction, one seems to need to go beyond $LNMIt$, but with $MRec \, s$ in the particular form where the body of $s$ neither uses $m$ nor $rec$, just the injection $j$. This might be termed Mendler-style *inversion*. There is also a natural example (more natural than the examples in Section 2) – namely a nicer representation of substitution than in [25] – that uses all the ingredients: the map term $m$, the injection $j$ and the recursive call $rec$. However, this last example needs the ideas of [25] as well and is only an instance of $LNMRec$ when $Set$ is made impredicative.


# 4    Access to Map Terms within $MIt$?

Since there is not yet a justification of $LNMRec$, one might want to have the extra liberty of $MIt^+$ even within $LNMIt$. That is, can one have access to the

---

[8] Note that, as a function that is only defined by help of $MIt$ but not as an instance of $MIt$, the function $size_i$ does not have a characterization that could prove this equation.

map term $m : mon\,X$ in the body of $s$, despite doing a definition with $MIt$? There is no answer in the general situation of $LNMIt$, but under the following conditions that are met in the example of evaluation in Section 2.2.

Assume that $F$ does not only preserve extensional functors, but is also monotone in the following sense (introduced as such in [8] and called relativized basic monotonicity of rank 2): There is a closed term

$$Fmon2br : \forall X^{\kappa_1} \forall Y^{\kappa_1}.\, mon\,Y \to X \subseteq Y \to FX \subseteq FY \ .$$

Assume an extensional functor $ef_G : \mathcal{E}G$ and set $mG := m\,ef_G$. Assume a step term for $MIt^+\,G$, hence $s : \forall X^{\kappa_1}.\,mon\,X \to X \subseteq G \to FX \subseteq G$.

Define $MMIt\,s := MIt\,G\,s' : \mu F \subseteq G$ with

$$s' := \lambda X^{\kappa_1} \lambda it^{X \subseteq G} \lambda A \lambda t^{FXA}.\,s\,G\,mG\,(\lambda A \lambda x^{GA}.\,x)\,(Fmon2br_{X,G}\,mG\,it\,A\,t) \ .$$

In continuation of the example in Section 2.2, we can define

$$eval1'' := MMIt\,s_{eval1'} : LamE \subseteq Lam \ ,$$

since $lam$ is extensional and satisfies the functor laws and $LamEF$ is monotone in the above sense. With this data, one could see that

$$eval1''_A(flat_A\,e) \simeq subst\,(\lambda x.\,x)\,\big(lam\,(\lambda x.\,x)(lam\,eval1''_A\,(eval1''\,e))\big) \ .$$

This is not too convincing, as far as $\simeq$ is concerned. By the first functor law for $lam$, one immediately gets a propositionally equal right-hand side that corresponds to the recursive equation for $eval1$. Recall that $eval1'$ has been derived from $eval1$ with the idea of taking profit from naturality, which then justified that they always produce propositionally equal values. If we want to have a general insight about the relation between $MMIt\,s$ and $MIt^+\,s$, we need to ensure naturality. So, assume further condition $(*)$ on $s$, given after Theorem 1. Moreover, we impose that $Fmon2br$ preserves naturality in the following sense: it fulfills

$$\forall X^{\kappa_1} \forall ef^{\mathcal{E}X} \forall it^{X \subseteq G}.\,it \in \mathcal{N}(m\,ef,\,mG) \to$$
$$Fmon2br\,mG\,it \in \mathcal{N}(m(Fp\mathcal{E}\,ef),\,m(Fp\mathcal{E}\,ef_G)) \ .$$

Then, the naturality theorem for $MIt$ [8, Theorem 1] proves naturality of $MMIt\,s$, i.e., $MMIt\,s \in \mathcal{N}(map_{\mu F},\,mG)$.

We are heading towards a proof of $\forall A \forall t^{\mu FA}.\,MMIt\,s\,t = MIt^+\,s\,t$ by Theorem 2. This imposes on us to require the respective extensionality assumption (for an $s$ that does not use the injection facility):

$$\forall X^{\kappa_1} \forall ef^{\mathcal{E}X} \forall f, g : X \subseteq G.\,(\forall A \forall x^{XA}.\,f_A\,x = g_A\,x) \to$$
$$\forall A \forall t^{FXA}.\,s\,(m\,ef)\,f\,t = s\,(m\,ef)\,g\,t \ .$$

The final condition is that

$$\forall X^{\kappa_1} \forall ef^{\mathcal{E}X} \forall it^{X \subseteq G}.\,it \in \mathcal{N}(m\,ef,\,mG) \to$$
$$\forall A \forall t^{FXA}.\,s\,(m\,ef)\,it\,t = s\,mG\,(\lambda A \lambda x^{GA}.\,x)\,(Fmon2br\,mG\,it\,t) \ ,$$

where the right-hand side is just the body of the definition of $s'$.

**Theorem 3.** *Under all the assumptions of the present section, we have that* $\forall A \forall t^{\mu FA} . MMIt\, s\, t = MIt^+\, s\, t$.

*Proof.* By Theorem 2.

With this long list of conditions, it is reassuring to verify that Theorem 3 is sufficient to prove that *eval1″* and *eval1′* yield propositionally equal values, and this is the case. It can be shown that, in presence of the other conditions, $(*)$ already follows from its special case $s\, m_G\, (\lambda A \lambda x^{GA}.x) \in \mathcal{N}(m(Fp\mathcal{E}\, ef_G), m_G)$.

While the extensions of *MIt* towards *MRec* were dictated by algorithmic concerns, more precisely, by behaviour with respect to definitional equality $\simeq$, this last section contributes more to "type-directed programming": If an argument $m$ of type *mon X* is additionally available in the body of the step term, one may try to use it, just being guided by the wish to produce a term of the right type. We already know that *MIt* can only define terminating functions. So, *MMIt s* will be some terminating function of the right type. If all the requirements of Theorem 3 are met, we even know that *MMIt s* calculates values that we can understand better through their description by $MIt^+ s$. Of course, if there were already a justification of *LNMRec*, we would prefer to use $MIt^+ s$, but this is not yet achieved.

As a final remark, the idea to try to define *MMIt s* arose when studying the article [26] by Johann and Ghani since their type of interpreter transformers `InterpT` only quantifies over all `Functor y`, where the type class mechanism of Haskell is used to express the existence of a map term for the type transformation `y`. However, in that article, this map term is never used. Moreover, instead of monotonicity in the sense of *Fmon2br*, an unrelativized version is used that cannot treat truly nested datatypes such as *Bush* and *LamE*. Finally, the target constructor $G$ was restricted to be a right Kan extension (in order to have the expressive power of generalized refined iteration [6]), and only an algebra—a term of type $FG \subseteq G$—was constructed and not the step term for Mendler-style iteration.

## 5   Conclusion

Already for the sake of perspicuous behaviour of iterative functions on nested datatypes, one is led to consider extensions of Mendler-style iteration towards Mendler-style primitive recursion with access to a map term for the nested datatype. The logical system *LNMIt* of earlier work by the author is extended to system *LNMRec*, without changing the induction principle that is the crucial element for verification. *LNMIt* could be defined within the Calculus of Inductive Constructions (CIC) with impredicative universe $\kappa_0 := Set$ and propositional proof irrelevance, i. e., with $\forall P : Prop \forall p_1^P \forall p_2^P . p_1 = p_2,$[9] and this definition does respect definitional equality $\simeq$. For *LNMRec*, there does not yet exist a justification like that, not even any justification, despite several attempts by the author

---

[9] Proof irrelevance is needed for proofs of naturality statements and in order to have injectivity of the first projection out of a strong sum.

to extend the construction of [8]. It is the built-in naturality that can not yet be treated. We may call *LMRec* the version of *LNMRec* where the references to naturality are deleted. The ideas of the first half of [8, Section 2.3] are sufficient to define *LMRec* in the CIC with impredicative *Set*. But naturality is important for verification purposes, and thus *LMRec* is not satisfying as a logical system. To conclude, a justification of the logic of *natural* Mendler-style recursion of rank 2 is strongly needed. It would be an instance of the more general problem of adding an impredicative version of simultaneous inductive-recursive definitions [27] to the CIC.

*Acknowledgement.* I am grateful to the referees whose valuable feedback helped to improve the presentation and will influence future work even more.

# References


1. Bird, R., Meertens, L.: Nested datatypes. In: Jeuring, J. (ed.) MPC 1998. LNCS, vol. 1422, pp. 52–67. Springer, Heidelberg (1998)
2. Bird, R., Gibbons, J., Jones, G.: Program optimisation, naturally. In: Davies, J., Roscoe, B., Woodcock, J. (eds.) Millenial Perspectives in Computer Science, Proceedings (2000)
3. Bellegarde, F., Hook, J.: Substitution: A formal methods case study using monads and transformations. Science of Computer Programming 23, 287–311 (1994)
4. Bird, R.S., Paterson, R.: De Bruijn notation as a nested datatype. Journal of Functional Programming 9(1), 77–91 (1999)
5. Altenkirch, T., Reus, B.: Monadic presentations of lambda terms using generalized inductive types. In: Flum, J., Rodríguez-Artalejo, M. (eds.) CSL 1999. LNCS, vol. 1683, pp. 453–468. Springer, Heidelberg (1999)
6. Abel, A., Matthes, R., Uustalu, T.: Iteration and coiteration schemes for higher-order and nested datatypes. Theoretical Computer Science 333(1–2), 3–66 (2005)
7. Barthe, G., Frade, M.J., Giménez, E., Pinto, L., Uustalu, T.: Type-based termination of recursive definitions. Mathematical Structures in Computer Science 14, 97–141 (2004)
8. Matthes, R.: An induction principle for nested datatypes in intensional type theory. Journal of Functional Programming (to appear, 2008)
9. Coquand, T., Paulin, C.: Inductively defined types. In: Martin-Löf, P., Mints, G. (eds.) COLOG 1988. LNCS, vol. 417, pp. 50–66. Springer, Heidelberg (1990)
10. Paulin-Mohring, C.: Inductive definitions in the system Coq – rules and properties. In: Bezem, M., Groote, J.F. (eds.) TLCA 1993. LNCS, vol. 664, pp. 328–345. Springer, Heidelberg (1993)
11. Paulin-Mohring, C.: Définitions Inductives en Théorie des Types d'Ordre Supérieur. Habilitation thesis, Université Claude Bernard Lyon I (1996)
12. Coq Development Team: The Coq Proof Assistant Reference Manual Version 8.1. Project LogiCal, INRIA (2006), `coq.inria.fr`
13. Mendler, N.P.: Recursive types and type constraints in second-order lambda calculus. In: Proceedings of the Second Annual IEEE Symposium on Logic in Computer Science, Ithaca, N.Y, pp. 30–36. IEEE Computer Society Press, Los Alamitos (1987)

14. Abel, A., Matthes, R.: Fixed points of type constructors and primitive recursion. In: Marcinkowski, J., Tarlecki, A. (eds.) CSL 2004. LNCS, vol. 3210, pp. 190–204. Springer, Heidelberg (2004)
15. Matthes, R.: Coq development for Recursion on nested datatypes in dependent type theory (January 2008),
`http://www.irit.fr/~Ralph.Matthes/Coq/MendlerRecursion/`
16. Matthes, R.: Coq development for An induction principle for nested datatypes in intensional type theory (January 2008),
`http://www.irit.fr/~Ralph.Matthes/Coq/InductionNested/`
17. Bird, R., Paterson, R.: Generalised folds for nested datatypes. Formal Aspects of Computing 11(2), 200–222 (1999)
18. Abel, A., Matthes, R. (Co-)iteration for higher-order nested datatypes. In: Geuvers, H., Wiedijk, F. (eds.) TYPES 2002. LNCS, vol. 2646, pp. 1–20. Springer, Heidelberg (2003)
19. Pfenning, F., Paulin-Mohring, C.: Inductively defined types in the calculus of constructions. In: Main, M.G., Melton, A., Mislove, M.W., Schmidt, D.A. (eds.) MFPS 1989. LNCS, vol. 442, pp. 209–228. Springer, Heidelberg (1990)
20. Uustalu, T., Vene, V.: A cube of proof systems for the intuitionistic predicate $\mu$-, $\nu$-logic. In: Haveraaen, M., Owe, O. (eds.) Selected Papers of the 8th Nordic Workshop on Programming Theory (NWPT 1996), May 1997. Research Reports, Department of Informatics, University of Oslo, vol. 248, pp. 237–246 (1997)
21. Matthes, R.: Naive reduktionsfreie Normalisierung (translated to English: naive reduction-free normalization). Slides of talk on December 19, 1996, given at the Bern Munich meeting on proof theory and computer science in Munich, available at the author's homepage (December 1996)
22. Geuvers, H.: Inductive and coinductive types with iteration and recursion. In: Nordström, B., Pettersson, K., Plotkin, G. (eds.) Proceedings of the Workshop on Types for Proofs and Programs, Båstad, Sweden, pp. 193–217 (1992),
`http://www.cs.chalmers.se/Cs/Research/Logic/Types/proc92.ps`
23. Wadler, P.: Theorems for free! In: Proceedings of the fourth international conference on functional programming languages and computer architecture, Imperial College, London, England, September 1989, pp. 347–359. ACM Press, New York (1989)
24. Abel, A.: A Polymorphic Lambda-Calculus with Sized Higher-Order Types. Doktorarbeit (PhD thesis), LMU München (2006)
25. Matthes, R.: Nested datatypes with generalized Mendler iteration: map fusion and the example of the representation of untyped lambda calculus with explicit flattening. In: Paulin-Mohring, C. (ed.) Mathematics of Program Construction, Proceedings. LNCS, Springer, Heidelberg (2008) (Accepted for publication)
26. Johann, P., Ghani, N.: Initial algebra semantics is enough! In: Ronchi Della Rocca, S. (ed.) TLCA 2007. LNCS, vol. 4583, pp. 207–222. Springer, Heidelberg (2007)
27. Dybjer, P.: A general formulation of simultaneous inductive-recursive definitions in type theory. The Journal of Symbolic Logic 65(2), 525–549 (2000)

# Perfect Local Computability
# and Computable Simulations


Russell Miller[1] and Dustin Mulcahey[2]

[1] Queens College of CUNY
65-30 Kissena Blvd., Flushing, NY 11367 USA
and
The CUNY Graduate Center
Russell.Miller@qc.cuny.edu
http://qcpages.qc.cuny.edu/~rmiller
[2] The CUNY Graduate Center
365 Fifth Avenue, New York NY 10016 USA
dustinmulcahey@gmail.com
https://wfs.gc.cuny.edu/DMulcahey/www/



**Abstract.** We study perfectly locally computable structures, which are
(possibly uncountable) structures $\mathcal{S}$ that have highly effective presentations of their local properties. We show that every such $\mathcal{S}$ can be simulated, in a strong sense and even over arbitrary finite parameter sets, by
a computable structure. We also study the category theory of a perfect
cover of $\mathcal{S}$, examining its connections to the category of all finitely generated substructures of $\mathcal{S}$.

**Keywords:** Category theory, computability, local computability, perfect
local computability, simulation.


## 1   Introduction

Locally computable structures were introduced in [3], as a method for effective
presentation of uncountable structures. One considers a structure $\mathcal{S}$ locally rather
than globally, giving presentations of all finitely generated substructures of $\mathcal{S}$,
up to isomorphism, along with a description of the ways in which these "pieces"
of $\mathcal{S}$ fit together to form the entire structure. We review these definitions below.
The entire package, including both the pieces and the ways they fit together,
comprise a *cover* of $\mathcal{S}$, which is said to be *uniformly computable* if it can be
given in a sufficiently effective manner.

The notion of a *perfect cover* was also defined in [3], and it is stated there
(and proven in [4]) that a countable structure has a perfect cover iff it is computably presentable. This suggests that for uncountable structures, perfect local
computability (i.e. having a perfect cover) is a reasonable analogue for computable presentability. In this paper we further explore perfectly locally computable structures, showing that all such structures have *computable simulations*,
or computably presentable elementary substructures realizing the same types.


A. Beckmann, C. Dimitracopoulos, and B. Löwe (Eds.): CiE 2008, LNCS 5028, pp. 447–456, 2008.

Such substructures may be used in natural ways to simulate operations in the larger structure – which in general is not countable, hence not itself computably presentable.

A cover bears a natural resemblance to a category, and indeed is closely related to the category $\mathbf{FGSub}(\mathcal{S})$ of all finitely generated substructures of $\mathcal{S}$ under inclusion maps. Of course, $\mathbf{FGSub}(\mathcal{S})$ will be uncountable if $\mathcal{S}$ is, whereas uniformly computable covers are always countable. Moreover, Definition 2.2 does not require that the morphisms among objects (i.e. the embeddings of one finitely generated substructure into another) be closed under composition. However, for the specific case of a perfect cover, we may assume closure under composition, and therefore we find ourselves working in a (computable) category. We examine the category-theoretic properties of such covers in Section 4.

Our computability-theoretic terminology coincides with that of [5], the standard reference for the subject, and the definitions in local computability from [3] are used here without alteration. For category theory, we recommend [2].

## 2    Local Computability Definitions

Let $T$ be a $\forall$-axiomatizable theory in a signature of size $n < \omega$. (The theory of fields is a good example to keep in mind, skolemized to have function symbols for negation and inversion.) We first consider simple covers of a model $\mathcal{S}$ of $T$. These describe only the finitely generated substructures of $\mathcal{S}$, with no attention paid to any relations between those substructures.

**Definition 2.1.** A *simple cover of* $\mathcal{S}$ is a (finite or countable) collection $\mathfrak{A}$ of finitely generated models $\mathcal{A}_0, \mathcal{A}_1, \ldots$ of $T$, such that:

- every finitely generated substructure of $\mathcal{S}$ is isomorphic to some $\mathcal{A}_i \in \mathfrak{A}$; and
- every $\mathcal{A}_i \in \mathfrak{A}$ embeds isomorphically into $\mathcal{S}$.

This simple cover is *uniformly computable* if every $\mathcal{A}_i \in \mathfrak{A}$ is a computable structure and the sequence $\langle (\mathcal{A}_i, \bar{a}_i) \rangle_{i \in \omega}$ can be given uniformly: there must exist a single computable function which, on input $i$, outputs a tuple of elements $\langle e_1, \ldots, e_n, \langle a_0, \ldots, a_{k_i} \rangle \rangle \in \omega^n \times \mathcal{A}_i^{<\omega}$ such that $\{a_0, \ldots, a_{k_i}\}$ generates $\mathcal{A}_i$ and $\varphi_{e_j}$ computes the $j$-th function, constant, or relation in $\mathcal{A}_i$.

**Definition 2.2.** A *cover of* $\mathcal{S}$ consists of a simple cover $\mathfrak{A} = \{\mathcal{A}_0, \mathcal{A}_1, \ldots\}$ of $\mathcal{S}$, along with sets $I_{ij}^{\mathfrak{A}}$ (for all $\mathcal{A}_i, \mathcal{A}_j \in \mathfrak{A}$) of injective homomorphisms $f : \mathcal{A}_i \hookrightarrow \mathcal{A}_j$, such that:

- for all finitely generated substructures $\mathcal{B} \subseteq \mathcal{C}$ of $\mathcal{S}$, there exist $i, j \in \omega$ and $f \in I_{ij}^{\mathfrak{A}}$ and isomorphisms $\beta : \mathcal{A}_i \twoheadrightarrow \mathcal{B}$ and $\gamma : \mathcal{A}_j \twoheadrightarrow \mathcal{C}$ with $\beta = \gamma \circ f$; and
- for every $i$ and $j$ and every $f \in I_{ij}^{\mathfrak{A}}$, there exist substructures $\mathcal{B} \subseteq \mathcal{C}$ of $\mathcal{S}$ and isomorphisms $\beta : \mathcal{A}_i \twoheadrightarrow \mathcal{B}$ and $\gamma : \mathcal{A}_j \twoheadrightarrow \mathcal{C}$ with $\beta = \gamma \circ f$.

This cover is *uniformly computable* if $\mathfrak{A}$ is a uniformly computable simple cover of $\mathcal{S}$ and there exists a c.e. set $W$ such that for all $i, j \in \omega$,

$$I_{ij}^{\mathfrak{A}} = \{\varphi_e \restriction \mathcal{A}_i : \langle i, j, e \rangle \in W\}.$$

A structure $\mathcal{B}$ is *locally computable* if it has a uniformly computable cover.

We will be concerned with a particularly nice class of locally computable structures, so nice that we call them *perfectly locally computable*. Here we have much stronger connections between the substructures of $\mathcal{S}$ and the structures $\mathcal{A}_i$ in our effective description.

**Definition 2.3.** Let $\mathfrak{A}$ be a uniformly computable cover for a structure $\mathcal{S}$. A set $M$ is a *correspondence system* for $\mathfrak{A}$ and $\mathcal{S}$ if it satisfies the following five rules:

1. Each element of $M$ is an embedding of some $\mathcal{A}_i \in \mathfrak{A}$ into $\mathcal{S}$; and
2. For every $\mathcal{A}_i \in \mathfrak{A}$, there exists a $\beta \in M$ with domain $\mathcal{A}_i$; and
3. For every finitely generated substructure $\mathcal{B}$ of $\mathcal{S}$, there exists a $\beta \in M$ with image $\mathcal{B}$; and
4. For every $\mathcal{A}_i \in \mathfrak{A}$, every $\beta \in M$ with domain $\mathcal{A}_i$, every $\mathcal{A}_j \in \mathfrak{A}$, and every $f \in I_{ij}^{\mathfrak{A}}$, there exists a $\gamma \in M$ with domain $\mathcal{A}_j$ such that $\beta = \gamma \circ f$ (and hence $\beta(\mathcal{A}_i) \subseteq \gamma(\mathcal{A}_j)$); and
5. For every $\mathcal{A}_i \in \mathfrak{A}$, every $\beta \in M$ with domain $\mathcal{A}_i$, and every finitely generated $\mathcal{B} \subseteq \mathcal{S}$ such that $\beta(\mathcal{A}_i) \subseteq \mathcal{B}$, there exists an $\mathcal{A}_j \in \mathfrak{A}$, a $\gamma \in M$ with domain $\mathcal{A}_j$ and image $\mathcal{B}$, and an $f \in I_{ij}^{\mathfrak{A}}$ such that $\beta = \gamma \circ f$.

A correspondence system is *perfect* if it also satisfies:

6. For every finitely generated $\mathcal{B} \subseteq \mathcal{S}$, if $\beta : \mathcal{A}_i \twoheadrightarrow \mathcal{B}$ and $\gamma : \mathcal{A}_j \twoheadrightarrow \mathcal{B}$ both lie in $M$, then $\gamma^{-1} \circ \beta \in I_{ij}^{\mathfrak{A}}$.

If a perfect correspondence system exists, then its elements are called *perfect matches* between their domains and their images. The uniformly computable cover $\mathfrak{A}$ is then called a *perfect cover* for $\mathcal{S}$, and $\mathcal{S}$ itself is said to be *perfectly locally computable*.

For example, it is proven in [3] that every algebraically closed field is perfectly locally computable. The field of real numbers, on the other hand, is not perfectly locally computable, and the ordered field of real numbers is not even locally computable.

Other uncountable examples are known. In the language of linear orders, Cantor space is perfectly locally computable. Likewise, if Cantor space is viewed as the top level of the tree $2^{<\omega+1}$ in the language of partial orders – with a additional unary predicate $C$ identifying that top level, and also with an immediate-predecessor function on the elements of all lengths strictly between $0$ and $\omega$, if one wishes – then this entire tree, under the additional predicate and function symbol, is perfectly locally computable. However, if one adds the linear order on the top level of that tree, then the new structure loses its perfect local computability, even though both of the structures merged together had it; in fact the new structure is barely even locally computable.

Another theorem from [3] and [4] shows that perfect local computability can be viewed as a natural extension of the notion of computability – or more precisely, computable presentability – for countable structures.

**Theorem 2.4.** *Let $S$ be any countable structure. Then $S$ is computably presentable iff $S$ is perfectly locally computable.*     $\square$

In the following, we consider perfectly locally computable structures, whether countable or uncountable, and show that in a certain sense, we can dispense with their covers and indeed with the structures themselves, and replace them with structures that actually are countable and computably presented. The notion of a *computable simulation* will be the precise version of this substitution.

## 3   Computable Simulations

**Definition 3.1.** Let $S$ be any structure. A *simulation of $S$* is an elementary substructure $B \preceq S$ such that $B$ and $S$ realize exactly the same finitary types. We often refer to any $A$ isomorphic to such a $B$ as a simulation of $S$, even if $A$ is not itself a substructure of $S$. Hence a *computable simulation of $S$* is a computable structure isomorphic to a simulation of $S$.

**Lemma 1.** *Let $S$ be locally computable, with a correspondence system $N$ over a uniformly computable cover $\mathfrak{A}$. Then $S$ has a countable substructure $B$ with its own correspondence system $M \subseteq N$ over $\mathfrak{A}$. If $N$ was a perfect correspondence system for $S$, then $M$ is perfect for $B$ as well.*

*Proof.* $B$ will be a countable union of countable substructures $B_s$ of $S$. To start, we fix for each $i \in \omega$ one map $\alpha_i \in N$ with domain $A_i$, Let $M_0 = \{\alpha_i : i \in \omega\}$, and let $B_0$ be the substructure of $S$ generated by the union of all the images of these $\alpha_i$. The conditions for a perfect cover are $\forall \exists$ conditions, so now we will be able to keep $B$ countable as we close $M$ under those conditions, using the analogous conditions in the correspondence system $N$.

Assume we have defined a countable $B_s$ and $M_s$. First, for every $i$, every $\alpha \in M_s$ with domain $A_i$, and every $f \in I_{ij}^{\mathfrak{A}}$ (for any $j$), there exists some $\gamma \in N$ with domain $A_j$ such that $f$ lifts via $\alpha$ and $\gamma$ to the inclusion $\alpha(A_i) \subseteq \gamma(A_j)$. Form $M_s' \supseteq M_s$ by adjoining one such $\gamma$ to $M_s$ for each such $i$, $\alpha$ and $f$. Also, let $B_s'$ be generated by the union of the images of the maps in $M_s'$. Clearly both $B_s'$ and $M_s'$ remain countable.

Next, for every $i$, every $\alpha \in M_s$ with domain $A_i$, and every finitely generated $C \subseteq B_s$ with $\alpha(A_i) \subseteq C$, there exists some $j$, some $f \in I_{ij}^{\mathfrak{A}}$, and some $\gamma \in N$ with domain $A_j$ such that $f$ lifts to the inclusion $\alpha(A_i) \subseteq C$ via $\alpha$ and $\gamma$. Adjoin to $M_s'$ one such $\gamma$ for each such $i$, $\alpha$, and $C$, to form $M_s''$.

Finally, for every finitely generated substructure $C \subseteq B_s$, there exists a $\gamma \in N$ with image $C$ (since $C \subseteq S$). Form $M_{s+1}$ by adjoining to $M_s''$ one such $\gamma$ for each such $C$. Since $B_s$ was countable, it has only countably many finitely generated substructures, and so $M_{s+1}$ is still countable.

It is clear that the union $\mathcal{B} = \cup_s \mathcal{B}_s$ is a countable substructure of $\mathcal{S}$, with cover $\mathfrak{A}$, and that $M = \cup_s M_s$ is a correspondence system for this $\mathcal{B}$ over $\mathfrak{A}$. Our $\mathcal{B}_0$ already satisfied item (2) of Definition 2.3, and our ensuing adjoinments satisfied (4), (5), and (3), in that order, without ever violating (1). (Of course $\mathfrak{A}$ is still uniformly computable as well; that definition has nothing to do with the structure covered by $\mathfrak{A}$.)

It remains to see that this $M$ is perfect for $\mathcal{B}$ whenever $N$ is perfect for $\mathcal{S}$. But this is easy: if $\alpha$ and $\gamma$ lie in $M$ and have the same image in $\mathcal{B}$, then they lie in $N$ and have the same image in $\mathcal{S}$. Since $N$ is perfect, $\gamma^{-1} \circ \alpha$ must then lie in the appropriate $I_{ij}^{\mathfrak{A}}$. $\qquad\square$

In this situation, $\mathcal{B}$ will be an elementary substructure of $\mathcal{S}$. The next lemma extends this observation. (If $\mathcal{B}$ is as in Lemma 1, and $P$ is empty, then in the proof of Lemma 2 we may show that at every step $\psi_s$ is just inclusion.)

**Lemma 2.** *Let $\mathcal{B}$ and $\mathcal{S}$ be two structures, each with a correspondence system over the same uniformly computable cover. Assume that $\mathcal{B}$ is countable. Then $\mathcal{B}$ is a simulation of $\mathcal{S}$. Indeed, for any countable set $P \subseteq \mathcal{S}$ of parameters, we can elementarily embed $\mathcal{B}$ into $\mathcal{S}$ so that its image contains $P$ and realizes the same finitary types as $\mathcal{S}$ over every finite $P_0 \subseteq P$.*

*Proof.* Let $\mathfrak{A}$ be a common uniformly computable cover of $\mathcal{S}$ and $\mathcal{B}$, with correspondence systems $M$ for $\mathcal{B}$ and $N$ for $\mathcal{S}$. Our embedding is built step by step, so we start by enumerating the domain of $\mathcal{B}$ as $\{b_0, b_1, \ldots\}$, and $P$ as $\{p_0, p_1, \ldots\}$. Fix an $\alpha \in M$ whose image is the substructure $\mathcal{B}_0 \subseteq \mathcal{B}$ generated by $b_0$, and a $\gamma \in N$ with the same domain as $\alpha$, and define $\psi_0$ to be $\gamma \circ \alpha^{-1}$, with $\mathcal{B}_0 = \mathrm{dom}(\psi_0) \subseteq \mathcal{B}$ and $\mathcal{C}_0 = \mathrm{range}(\psi_0) \subseteq \mathcal{S}$.

At stage $t+1 = 2s+1$, we extend $\psi_t$ so that its range contains $p_s$. By induction $\psi_t = \gamma \circ \alpha^{-1}$ for some $\gamma \in N$ and $\alpha \in M$ with common domain $\mathcal{A}_i$ in $\mathfrak{A}$. Let $\mathcal{C}_{t+1}$ be the substructure of $\mathcal{S}$ generated by $\mathcal{C}_t$ and $p_s$. By induction $\mathcal{C}_t$ is finitely generated, so there is a $\delta \in N$ with some domain $\mathcal{A}_j \in \mathfrak{A}$, and an $f \in I_{ij}^{\mathfrak{A}}$, such that $f$ lifts via $\gamma$ and $\delta$ to the inclusion $\mathcal{C}_t \subseteq \mathcal{C}_{t+1}$. In turn there is a $\beta \in M$ with domain $\mathcal{A}_j$ such that $f$ lifts via $\alpha$ and $\beta$ to the inclusion $\mathcal{B}_t \subseteq \beta(\mathcal{A}_j)$. Set $\mathcal{B}_{t+1} = \mathrm{range}(\beta)$ and $\psi_{t+1} = \delta \circ \beta^{-1}$.

At stage $t+1 = 2s+2$, we extend the embedding $\psi_t$ from its current domain $\mathcal{B}_t$ to the structure $\mathcal{B}_{t+1}$ generated by $\mathcal{B}_t$ and $b_s$. By induction $\mathcal{B}_t$ is finitely generated, and $\psi_t = \gamma \circ \alpha^{-1}$ for some $\gamma \in N$ and $\alpha \in M$ with common domain $\mathcal{A}_i$ in $\mathfrak{A}$. So there is a $\beta \in M$ with some domain $\mathcal{A}_j \in \mathfrak{A}$, and an $f \in I_{ij}^{\mathfrak{A}}$, such that $f$ lifts via $\alpha$ and $\beta$ to the inclusion $\mathcal{B}_t \subseteq \mathcal{B}_{t+1}$. In turn there is a $\delta \in M$ with domain $\mathcal{A}_j$ such that $f$ lifts via $\gamma$ and $\delta$ to the inclusion $\mathcal{C}_t \subseteq \delta(\mathcal{A}_j)$. Set $\mathcal{C}_{t+1} = \mathrm{range}(\delta)$ and $\psi_{t+1} = \delta \circ \beta^{-1}$.

Now we define $\psi = \cup_t \psi_t$. Clearly $\psi$ has domain $\mathcal{B}$ and range $\subseteq \mathcal{S}$ containing $P$, and $\psi$ must be an embedding. To see that it is elementary, suppose that $\exists x \theta(\psi(b_0), \ldots, \psi(b_s), x)$ is an existential formula true in $\mathcal{S}$. Now $\psi_{2s+2} = \gamma \circ \alpha^{-1}$ for some $\alpha \in M$ and $\gamma \in N$ with common domain $\mathcal{A}_i \in \mathfrak{A}$. Since $N$ is a correspondence system, there is a $\delta \in N$ (with some domain $\mathcal{A}_j$) and an $a \in \mathcal{A}_j$ and an $f \in I_{ij}^{\mathfrak{A}}$, such that $\theta(f(\alpha^{-1}(b_0)), \ldots, f(\alpha^{-1}(b_s)), a)$ holds in $\mathcal{A}_j$. But since $M$ is also a correspondence system, there is a $\beta \in M$ with the same domain $\mathcal{A}_j$ such that $f$ lifts to the inclusion $\mathcal{B}_s \subseteq \beta(\mathcal{A}_j)$ via $\alpha$ and $\beta$. Therefore $\theta(b_0, \ldots, b_s, \beta(a))$ holds in $\mathcal{B}$. The dual argument shows that existential formulas true in $\mathcal{B}$ also hold in $\mathcal{S}$. But since existential formulas with parameters go back and forth between $\mathcal{B}$ and $\mathcal{S}$ this way, $\psi$ must be an elementary embedding.

Finally, given any $n$-type $\Gamma$ over any finite parameter set $P_0 \subseteq P$, such that $\Gamma$ is realized in $\mathcal{S}$ by a tuple $(d_1, \ldots, d_n)$, we start with the substructure $\mathcal{P}_0 \subseteq \mathcal{S}$ generated by $P_0$. Since $P_0 \subseteq \mathrm{range}(\psi)$, we have a $t$ for which $P_0 \subseteq \mathrm{range}(\psi_t)$. Let $\psi_t = \gamma \circ \alpha^{-1}$ with $\alpha \in M$ and $\gamma \in N$. There must be a $\delta \in N$ with some domain $\mathcal{A}_j \in \mathfrak{A}$ and an $f \in I_{ij}^{\mathfrak{A}}$ such that $f$ lifts via $\gamma$ and $\delta$ to the inclusion of $\mathcal{P}_0$ into the substructure generated by $\mathcal{P}_0$ and $d_0, \ldots, d_n$. But now there is also some $\beta \in M$ with domain $\mathcal{A}_j$ such that $f$ lifts via $\alpha$ and $\beta$ to the inclusion $\mathcal{B}_t \subseteq \beta(\mathcal{A}_j)$, and we set $b_i = \beta(\delta^{-1}(d_i))$ and $c_i = \psi(b_i)$ for each $i$. Then $(c_1, \ldots, c_n)$ is an $n$-tuple within the image of $\psi$ which realizes the type $\Gamma$ over $P_0$, by standard arguments using $M$ and $N$. $\qquad\square$

When we have a parameter set $P$ as in Lemma 2, we refer to the image of $\mathcal{B}$ as a *simulation of $\mathcal{S}$ over $P$*. We might also refer to $\mathcal{B}$ itself the same way, but only when the embedding $\psi : \mathcal{B} \hookrightarrow \mathcal{S}$ is clear, because we need to know which elements $\psi^{-1}(p) \in \mathcal{B}$ correspond to the elements of $P$ in this simulation. Later we will discuss the extent to which $\mathcal{B}_P$ can be said to be uniform in $P$.

**Corollary 3.2.** *Two countable structures with correspondence systems over the same uniformly computable cover are isomorphic.*

*Proof.* Since $\mathcal{S}$ is countable, we simply set $P = \mathcal{S}$ and apply Lemma 2, whose proof may now be regarded as a back-and-forth construction of an isomorphism from $\mathcal{B}$ onto $\mathcal{S}$. $\qquad\square$

We are now ready for the main result of this section.

**Theorem 3.3.** *Every perfectly locally computable structure $\mathcal{S}$ has a computable simulation $\mathcal{A}$, which can be embedded into $\mathcal{S}$ so as to simulate $\mathcal{S}$ over arbitrary countable parameter sets. Specifically, there is a set of elementary embeddings $\psi_p : \mathcal{A} \hookrightarrow \mathcal{S}$, one for each function $p : \omega \to \mathcal{S}$ which enumerates a countable parameter set $Q_p = \mathrm{range}(p) \subseteq \mathcal{S}$, such that:*

- $Q_p \subseteq \psi_p(\mathcal{A})$; *and*
- $\psi_p(\mathcal{A})$ *is a simulation of $\mathcal{S}$ over $Q_p$; and*
- *if $p$ and $p'$ are two such functions and $p{\restriction}n = p'{\restriction}n$, then for all $k < n$,*

$$\psi_p^{-1}(p(k)) = \psi_{p'}^{-1}(p'(k)).$$

*As a partial converse, every structure which has a computable simulation $\mathcal{A}$ with embeddings $\psi_p$ satisfying these conditions has a uniformly computable cover with a correspondence system.*

To make this last claim an actual converse, we would need to show that the correspondence system for $\mathcal{S}$ is perfect. Whether this is true remains open. We also note that it would be equivalent to give the same statement only for finite parameter enumerations $p$, since the last condition would allow a simulation over a countable parameter set $P$ to be built by taking successive nested finite enumerations $p_m \subseteq p_{m+1}$ with $P = \cup_m \mathrm{range}(p_m)$, and setting $\psi = \lim_m \psi_{p_m}$.

*Proof.* When we assume that $\mathcal{S}$ is perfectly locally computable, the existence of a computable simulation of $\mathcal{S}$ follows from Lemma 1, which also ensures that $\mathcal{S}$ and its computable simulation $\mathcal{A}$ both have perfect correspondence systems over the same uniformly computable cover. Therefore Lemma 2 shows that $\mathcal{A}$ can be elementarily embedded into $\mathcal{S}$ so as to simulate $\mathcal{S}$ over any parameter set $Q_p$ enumerated by a function $p : \omega \to \mathcal{S}$. Moreover, an examination of the proof of Lemma 2 shows that the embedding chooses the $a \in \mathcal{A}$ with $\psi_p(a) = p(k)$ using only the common cover $\mathfrak{A}$, its correspondence systems for $\mathcal{A}$ and $\mathcal{S}$, and the elements $p(0), p(1), \ldots, p(k)$ in $\mathcal{S}$. This proves the claim about parameter enumerations $p$ and $p'$ which agree up to $n$.

The proof of the partial converse is longer and more technical, and we relegate it to Appendix A. $\qquad\qquad\square$

We can think of $\mathcal{B}_P$ as being built uniformly in the parameter set $P$ if the elements of $P$ are named as elements in different $\mathcal{A}_i$ in the cover $\mathfrak{A}$ of $\mathcal{S}$. That is, suppose that we are given a computable enumeration $\langle (i_k, a_k, f_k) \rangle_{k \in \omega}$ for which there exist maps $\beta_k \in N$ with $a_k \in \mathcal{A}_{i_k} = \mathrm{dom}(\beta_k)$ such that

- each $f_k \in \mathcal{I}^{\mathfrak{A}}_{i_k, i_{k+1}}$; and
- $\beta_{k+1} \circ f_k = \beta_k$; and
- $\{\beta_k(a_k) : k \in \omega\} = P$.

Then we could build a computable copy of the simulation $\mathcal{B}_P$ of $\mathcal{S}$ over $P$, uniformly in the perfect cover of $\mathcal{S}$ and the enumeration $\langle (i_k, a_k, f_k) \rangle_{k \in \omega}$, and enumerate the image of $P$ in $\mathcal{B}_P$. More generally, if the enumeration $\langle (i_k, a_k, f_k) \rangle_{k \in \omega}$ has Turing degree $\boldsymbol{d}$, then with a $\boldsymbol{d}$-oracle we can build a copy of $\mathcal{B}_P$ in which the image of $P$ will be computably enumerable in $\boldsymbol{d}$. It is awkward to think of the set $P$ itself as having Turing degree $\boldsymbol{d}$, because an infinite set $P$ will have distinct enumerations with distinct Turing degrees, but within the cover $\mathfrak{A}$ of $\mathcal{S}$, we can view $P$ as being computably enumerable in $\boldsymbol{d}$, as well as in the degrees of other enumerations. Of course, $P$ itself, viewed as a subset of $\mathcal{S}$, does not admit effective enumeration in any obvious way.

## 4   Category Theory

In general, a cover of a structure $\mathcal{S}$ need not be a category. The cover has objects (the structures $\mathcal{A}_i$) and morphisms among them, but the definition does not require that the composition of two morphisms be a morphism, nor that the identity morphisms be included. Adding the identity morphisms would not be a problem, but it can happen that when one closes the sets of morphisms under composition, the resulting category is no longer a cover of $\mathcal{S}$, since the new morphisms may not correspond to any inclusion maps within $\mathcal{S}$. However, for perfectly locally computable structures, this difficulty vanishes.

**Lemma 3.** *Let $\mathcal{S}$ be a perfectly locally computable structure. Then there is a perfect cover $\mathfrak{A}$ of $\mathcal{S}$, called the* derived cover *of $\mathcal{S}$, with the properties that every $I_{ii}^{\mathfrak{A}}$ contains the identity map on $\mathcal{A}_i$ and that for all $i, j, k \in \omega$ and all $f \in I_{ij}^{\mathfrak{A}}$ and $g \in I_{jk}^{\mathfrak{A}}$, we have $g \circ f \in I_{ik}^{\mathfrak{A}}$. Thus $\mathfrak{A}$ may be viewed as a computable category, with objects $\{\mathcal{A}_i : i \in \omega\}$ and with maps in $I_{ij}^{\mathfrak{A}}$ as the morphisms from $\mathcal{A}_i$ to $\mathcal{A}_j$.*

To be careful, we should speak of *the* derived cover only when some perfect cover has already been specified.

*Proof.* $\mathfrak{A}$ is just the closure of the original perfect cover of $\mathcal{S}$ under the required properties. It is clear from Definition 2.3 that this $\mathfrak{A}$ is still a perfect cover of $\mathcal{S}$, and the uniform computability of the cover $\mathcal{A}$ follows from that of the original cover, since the underlying simple cover has not changed and the new sets $I_{ij}^{\mathfrak{A}}$ are derived from the old ones by existential conditions. $\qquad\square$

Category theorists have long known the category $\mathbf{FGSub}(\mathcal{S})$ of all finitely generated substructures of a structure $\mathcal{S}$, where the morphisms are just the inclusion maps within $\mathcal{S}$. Of course, if $\mathcal{S}$ is uncountable, then so is $\mathbf{FGSub}(\mathcal{S})$, so this presentation is not of immediate use for computability theorists. However, we do have a connection.

**Proposition 4.1.** *If $\mathcal{S}$ is perfectly locally computable, then there exists a faithful functor $R$ mapping $\mathbf{FGSub}(\mathcal{S})$ into the derived cover $\mathfrak{A}$ of $\mathcal{S}$. Moreover, there exists a natural isomorphism $\beta : (I_{\mathfrak{A}} \circ R) \to I_{\mathbf{FGSub}(\mathcal{S})}$ where the $I_{-}$ denote the appropriate inclusions into the category of all $L$-structures under embeddings.*

*Proof.* Let $\mathfrak{A}$ be a perfect cover of $\mathcal{S}$, with correspondence system $M$. For each $\mathcal{B} \in \mathbf{FGSub}(\mathcal{S})$, fix $R(\mathcal{B})$ to be any $\mathcal{A}_i \in \mathfrak{A}$ which is the domain of some $\beta \in M$ whose image is $\mathcal{B}$. $\mathcal{A}_i$ and $\beta$ need not be unique, and it is startling that we can make such an arbitrary choice and have both a functor and a natural isomorphism, but the perfection of the cover allows it to work. Let $N \subseteq M$ contain exactly the maps $\beta$ chosen in this process. Thus each $\mathcal{B} \in \mathbf{FGSub}(\mathcal{S})$ is the image of a unique $\beta_{\mathcal{B}} \in N$, whose domain is $R(\mathcal{B})$.

Now every morphism $\mathcal{B} \hookrightarrow \mathcal{C}$ in $\mathbf{FGSub}(\mathcal{S})$ is an inclusion $\mathcal{B} \subseteq \mathcal{C}$ within $\mathcal{S}$, and with $\beta_{\mathcal{B}} \in N$ as above, we have some $f \in I_{ij}^{\mathfrak{A}}$ for some $j$, and some $\gamma \in M$ with domain $\mathcal{A}_j$ and image $\mathcal{C}$, such that $\gamma \circ f = \beta_{\mathcal{B}}$. Now we also have some $\beta_{\mathcal{C}} \in M$ mapping $R(\mathcal{C})$ onto $\mathcal{C}$, say with $R(\mathcal{C}) = \mathcal{A}_k$. Since $M$ is perfect, the map $\beta_{\mathcal{C}}^{-1} \circ \gamma$ must lie in $I_{jk}^{\mathfrak{A}}$. We then define $R$ to take the inclusion morphism $\mathcal{B} \subseteq \mathcal{C}$ to the morphism $R(\mathcal{B} \subseteq \mathcal{C}) = \beta_{\mathcal{C}}^{-1} \circ \gamma \circ f$, which lies in $I_{ik}^{\mathfrak{A}}$, since $\mathfrak{A}$ is a derived cover. Again, an arbitrary choice of $f$ and $\gamma$ may have been involved here. However, for any $x \in R(\mathcal{B}) = \mathcal{A}_i$, we have

$$R(\mathcal{B} \subseteq \mathcal{C})(x) = \beta_{\mathcal{C}}^{-1} \circ \gamma \circ f(x) = \beta_{\mathcal{C}}^{-1} \circ \beta_{\mathcal{B}}(x),$$

so that in fact the definition of $R(\mathcal{B} \subseteq \mathcal{C})$ was independent of this choice. Moreover, this will show that $R$ respects composition of morphisms, as functors must. Suppose that $\mathcal{B} \subseteq \mathcal{C} \subseteq \mathcal{D}$ are objects of **FGSub**$(\mathcal{S})$. Then

$$R(\mathcal{C} \subseteq \mathcal{D}) \circ R(\mathcal{B} \subseteq \mathcal{C}) = (\beta_{\mathcal{D}}^{-1} \circ \beta_{\mathcal{C}}) \circ (\beta_{\mathcal{C}}^{-1} \circ \beta_{\mathcal{B}}) = \beta_{\mathcal{D}}^{-1} \circ \beta_{\mathcal{B}} = R(\mathcal{B} \subseteq \mathcal{D}).$$

Thus $R$ really is a functor.

The naturality of $\beta$ is evident from its construction. $\qquad\qquad\square$

**Corollary 4.2.** $colim(I_{\mathfrak{A}} \circ R) \simeq \mathcal{S}$

*Proof.* Since $\beta$ is a natural isomorphism, we have $colim(I_{\mathbf{FGSub}(\mathcal{S})}) \simeq colim(I_{\mathfrak{A}} \circ R)$. The former is easily seen to be $\mathcal{S}$. $\qquad\qquad\square$

The functor $R$ need not be onto. It is possible, for instance, for two distinct $\mathcal{A}_i$ and $\mathcal{A}_j$ in $\mathfrak{A}$ to map onto the same $\mathcal{B}$ via maps $\gamma, \delta \in M$, and also possible that these are the only maps in $M$ with domains $\mathcal{A}_i$ or $\mathcal{A}_j$. In such a case, one of $\mathcal{A}_i$ or $\mathcal{A}_j$ will not lie in the image of $R$. Of course, since $M$ is perfect, $\delta^{-1} \circ \gamma \in I_{ij}^{\mathfrak{A}}$ in this case, and this map is an isomorphism, as is $\gamma^{-1} \circ \delta \in I_{ji}^{\mathfrak{A}}$.

This can be done in a bit more canonically using $\beta$ as above:

**Lemma 4.** *The functor $R : \mathbf{FGSub}(\mathcal{S}) \to \mathfrak{A}$ defined above is essentially surjective.*

*Proof.* Fix any $\mathcal{A}_i \in \mathfrak{A}$. Then there is some $\alpha \in M$ with domain $\mathcal{A}_i$ and image $\mathcal{B} = \alpha(\mathcal{A}_i)$, and some $\beta_{\mathcal{B}} \in N \subseteq M$ with domain $\mathcal{A}_j = R(\mathcal{B})$ for some $j$. By perfection of $M$, $\beta_{\mathcal{B}}^{-1} \circ \alpha \in I_{ij}^{\mathfrak{A}}$, and this embedding maps $\mathcal{A}_i$ onto $R(\mathcal{B})$. $\qquad\square$

# References


1. Borceux, F.: Handbook of Categorical Algebra, 3 volumes. Cambridge University Press, Cambridge (1994-1995)
2. Mac Lane, S.: Categories for the Working Mathematician. 2nd edn. Springer, New York (1997)
3. Miller, R.G.: Locally computable structures. In: Cooper, B., Löwe, B., Sorbi, A. (eds.) CiE 2007. LNCS, vol. 4497, pp. 575–584. Springer, Heidelberg (2007), `qcpages.qc.cuny.edu/~rmiller/research`
4. Miller, R.G.: Local computability and uncountable structures (to appear)
5. Soare, R.I.: Recursively Enumerable Sets and Degrees. Springer, New York (1987)


# Appendix A

Here we complete the proof of Theorem 3.3 by proving the partial converse there stated. Suppose that $\mathcal{A}$ is a computable simulation of $\mathcal{S}$ and that there is a set of elementary embeddings $\psi_p : \mathcal{A} \hookrightarrow \mathcal{S}$, one for each function $p : \omega \to \mathcal{S}$ which enumerates a countable parameter set $Q_p = \text{range}(p) \subseteq \mathcal{S}$, such that:

- $Q_p \subseteq \psi_p(\mathcal{A})$; and

$-$ $\psi_p(A)$ is a simulation of $\mathcal{S}$ over $Q_p$; and
$-$ if $p$ and $p'$ are two such functions and $p\!\restriction\! n = p'\!\restriction\! n$, then for all $k < n$,

$$\psi_p^{-1}(p(k)) = \psi_{p'}^{-1}(p'(k)).$$

We have to show that the existence of such an $A$ implies that $\mathcal{S}$ has a uniformly computable cover with a correspondence system. Now $A$ has a perfect cover $\mathfrak{A}$, by Theorem 2.4. Let $M$ be a perfect correspondence system for $A$ and $\mathfrak{A}$. The correspondence system $N$ will consist of all maps of the form $\psi_p \circ \alpha$, for all finite $p : n \to \mathcal{S}$ and all $\alpha \in M$ such that range$(\alpha)$ is generated by $\{\psi_p^{-1}(p(i)) : i < n\}$. (Here we think of a finite function $p : n \to \mathcal{S}$ as a function from $\omega$ into $\mathcal{S}$ by repeating its image over and over: $p(k + nm) = p(k)$ for all $k$ and $m$.)

Now each finitely generated $\mathcal{C} \subseteq \mathcal{S}$ with generators enumerated by $p$ lies within the image of $\psi_p$, and the finitely generated substructure $\psi_p^{-1}(\mathcal{C}) \subseteq A$ must be the image of some $A_i \in \mathfrak{A}$ under some $\alpha \in M$, since $M$ is a perfect cover of $A$. Hence $\mathcal{C} = (\psi_p \circ \alpha)(A_i)$ is the image of some map in $N$. Likewise, each $A_i \in \mathfrak{A}$ is the domain of some $\alpha \in M$, hence also of some map in $N$. Moreover, each $f \in I_{ij}^{\mathfrak{A}}$, for any $i$ and $j$, lifts to an inclusion map within $A$, and then lifts further to an inclusion map within $\mathcal{S}$, via any $\psi_p$ we like. Conversely, any inclusion $\mathcal{C}' \subseteq \mathcal{C}$ of finitely-generated substructures of $\mathcal{S}$ is the lift (via $\psi_p$, where $p$ enumerates first the generators of $\mathcal{C}'$, and then the generators of $\mathcal{C}$) of an inclusion in $A$, which in turn is the lift of some $f$ in some $I_{ij}^{\mathfrak{A}}$ via some $\alpha, \beta \in M$. If $p'$ is the restriction of $p$ to the generators of $\mathcal{C}'$, then the inclusion $\mathcal{C}' \subseteq \mathcal{C}$ is the lift of $f$ via $(\psi_{p'} \circ \alpha)$ and $(\psi_p \circ \beta)$, which both lie in $N$. Thus $\mathfrak{A}$ is a uniformly computable cover of $\mathcal{S}$.

The preceding remarks also proved the first three parts in Definition 2.3. For part 4, fix any $f \in I_{ij}^{\mathfrak{A}}$ for any $i$ and $j$, along with any $\beta \in N$ with domain $A_i$. Then $\beta = \psi_p \circ \alpha$ for some $\alpha \in M$ and some $p : n \to \mathcal{S}$ for which $\{\psi_p^{-1}(p(k) : k < n\}$ generates range$(\alpha)$. Since $M$ is a correspondence system, there is a $\gamma \in M$ with domain $A_j$ such that $\alpha = \gamma \circ f$. But now there is a finite $q$ such that $q\!\restriction\! n = p$ and $q(n + k) = \psi_p(a_k)$, where $a_0, \ldots, a_m$ generate $\gamma(A_j)$ within $A$. So $(\psi_q \circ \gamma) \in N$, and

$$(\psi_q \circ \gamma \circ f) = (\psi_q \circ \alpha) = (\psi_p \circ \alpha),$$

with the last equality following because $p\!\restriction\! n = q\!\restriction\! n$ and range$(\alpha)$ is generated by the elements $\psi_p^{-1}(p(k)) = \psi_q^{-1}(q(k))$ for $k < n$. This proves part 4.

For part 5 of Definition 2.3, fix any $\beta \in N$ with domain $A_i$ and any finitely generated $\mathcal{C} \subseteq \mathcal{S}$ with $\beta(A_i) \subseteq \mathcal{C}$. Now $\beta = \psi_p \circ \alpha$ for some $\alpha \in M$ and some finite $p : n \to \mathcal{S}$, with the elements $\psi_p^{-1}(p(k))$ generating range$(\alpha)$. Let $q(k) = p(k)$ for $k < n$, and let $q(n), \ldots, q(n + m - 1)$ enumerate the generators of $\mathcal{C}$ in $\mathcal{S}$. By assumption, $\psi_q$ is an elementary embedding of $A$ into $\mathcal{S}$ whose image contains range$(q)$. Let $\mathcal{D} = \langle \psi_q^{-1}(q(k)) : k < m \rangle \subseteq A$. Since $\psi_q^{-1}(q(k)) = \psi_p^{-1}(p(k))$ for all $k < n$, we know that $\alpha(A_i) \subseteq \mathcal{D}$, and since $M$ is a correspondence system, there is some $\beta \in M$ and some $j$ and $f \in I_{ij}^{\mathfrak{A}}$ with $\mathcal{D} = $ range$(\beta)$ and $\beta \circ f = \alpha$. But then

$$(\psi_q \circ \beta \circ f) = (\psi_q \circ \alpha) = (\psi_p \circ \alpha),$$

proving part 5 of Definition 2.3, since $(\psi_q \circ \beta) \in N$. Thus $\mathfrak{A}$ is a uniformly computable cover of $\mathcal{S}$ with correspondence system $N$. $\qquad\square$

# Complete Determinacy and Subsystems of Second Order Arithmetic


Takako Nemoto[*]

Mathematical Institute, Tohoku University, Sendai 980-8578, Japan
sa4m20@math.tohoku.ac.jp, nmt0731@yahoo.co.jp



**Abstract.** This paper investigates the determinacy and the complete determinacy of infinite games, following reverse mathematics program whose purpose is to find the set comprehension axioms that are necessary and sufficient for these statements in the frame of second order arithmetic. In some sense, this research clarifies how complex oracles we need to obtain the algorithms which give a winning strategies and which determine the winning positions for the players. It will be shown that, depending on the complexity of the rules of games, the complexity of the oracles changes drastically and that determinacy and complete determinacy statements are not always equivalent.

**keywords:** infinite game, determinacy, complete determinacy, Wadge class, reverse mathematics, second order arithmetic.


## 1   Introduction

We consider the following type of game: Two players, say player I and player II, alternatively choose an element of $X$ to form an infinite sequence $f$. Player I wins iff a given formula $\varphi(f)$ of $f$ holds. Player II wins iff I does not win.

A determinacy statement asserts that one of the players has a winning strategy in such games. This kind of statements has been of great interest in many fields of mathematics and informatics, e.g., infinitary combinatorics, set theory and automata theory.

It is quite natural to ask the complexity of a winning strategy and of an algorithm to determine at which position a player wins (has a winning strategy) in such a game. Because the determinacy of complex games, i.e., the games whose rules $\varphi(f)$ are complex (e.g., analytic), cannot be proved even in ZFC, such a kind of research must be on games with lower complexity.

Even the determinacy for such games (e.g., $\Sigma_1^0$, $\Sigma_2^0$ and $\Sigma_3^0$) is quite interesting (e.g., from the viewpoint of admissible set theory, [11]) and has strong utility (e.g., in automata theory and second order arithmetic). As applications, we can list [7], which solves the problem whether the language recognized by an automaton on infinite trees is empty, the proof in [9] of the implication of $\Sigma_1^1$ choice from ATR$_0$, and [4], which compares systems between ACA$_0$ and $\Pi_1^1$-CA$_0$.

---


[*] Partly supported by the Grant-in-Aid for JSPS Fellows, The Ministry of Education, Culture, Sports, Science and Technology, Japan.



A. Beckmann, C. Dimitracopoulos, and B. Löwe (Eds.): CiE 2008, LNCS 5028, pp. 457–466, 2008.

The research on the strength of such weak determinacy from the viewpoint of *reverse mathematics* (cf. [8]) is on going: In [3], [5], [6], [9] and [10], necessary and sufficient set comprehension axioms for the determinacy statements are investigated in the framework of second order arithmetic. These results can be seen as answers to the question: How complex does a winning strategy become for games with rules of a certain complexity?

For applications, it is also needed to know how complex the algorithm becomes which recognizes the winning positions. To answer this question in a similar way, we can use complete determinacy statements, which guarantee the existence of the winning set. Here, we call $W$ the *winning set for player I* if

$$(s \in W \to \text{player I wins at a position } s) \land (s \notin W \to \text{player II wins at } s).$$

This paper will measure the strength of complete determinacy statements from the viewpoint of reverse mathematics.

It is known that the strength changes depending also on the set $X$ of moves that the players can choose at each turn. In the previous works, $X$ is either $\mathbb{N}$ or $2 = \{0, 1\}$, i.e., they treat the games in the Baire space $\mathbb{N}^{\mathbb{N}}$ and in the Cantor space $2^{\mathbb{N}}$. Accordingly, here we treat complete determinacy in these two spaces.

We also use a finer hierarchy (consisting of $(\Sigma^0_1)_2$, $\mathsf{Bisep}(\Delta^0_1, \Sigma^0_1)$, $\mathsf{Sep}(\Sigma^0_1, \Sigma^0_2)$, etc.) of formulas than arithmetical hierarchy $\Sigma^0_n$'s, since, as shown in the previous works, the determinacy statement becomes drastically stronger even if the rules of the games become stronger only slightly (for details, see Subsection 2.2).

The results of the previous works and of this paper are summarized as follows, where $\mathsf{Det}$ (resp. $\mathsf{Det}^*$) stands for determinacy in the Baire (resp. the Cantor) space, where $\mathsf{comp.Det}$ (resp. $\mathsf{comp.Det}^*$) stands for complete determinacy in the Baire (resp. the Cantor) space and where the statements are equivalent to the system on the same row over $\mathsf{RCA}_0$. Our contribution is the establishment of the new two columns for $\mathsf{comp.Det}$ and $\mathsf{comp.Det}^*$ and this shows that complete determinacy statement is sometimes strictly stronger than the corresponding determinacy statement: we see that $\Sigma^0_1\text{-}\mathsf{comp.Det}^*$, $\Sigma^0_2\text{-}\mathsf{comp.Det}^*$ and $\Sigma^0_1\text{-}\mathsf{comp.Det}$ are strictly stronger than $\Sigma^0_1\text{-}\mathsf{Det}^*$, $\Sigma^0_2\text{-}\mathsf{Det}^*$ and $\Sigma^0_1\text{-}\mathsf{Det}$, respectively.

| system | -Det* | -comp.Det* | -Det | -comp.Det |
|---|---|---|---|---|
| $\mathsf{WKL}_0$ | $\Delta^0_1$<br>$\Sigma^0_1$<br>$\mathsf{Bisep}(\Delta^0_1, \Sigma^0_1)$ | $\Delta^0_1$ | | |
| $\mathsf{ACA}_0$ | $(\Sigma^0_1)_2$ | $\Sigma^0_1$<br>$\mathsf{Bisep}(\Delta^0_1, \Sigma^0_1)$<br>$(\Sigma^0_1)_2$ | | |
| $\mathsf{ATR}_0$ | $\Delta^0_2$<br>$\Sigma^0_2$<br>$\mathsf{Bisep}(\Delta^0_1, \Sigma^0_2)$ | $\Delta^0_2$ | $\Delta^0_1$<br>$\Sigma^0_1$ | $\Delta^0_1$ |
| $\Pi^1_1\text{-}\mathsf{CA}_0$ | $\mathsf{Bisep}(\Sigma^0_1, \Sigma^0_2)$<br>$\mathsf{Sep}(\Sigma^0_1, \Sigma^0_2)$ | $\Sigma^0_2$<br>$\mathsf{Bisep}(\Delta^0_1, \Sigma^0_2)$<br>$\mathsf{Bisep}(\Sigma^0_1, \Sigma^0_2)$<br>$\mathsf{Sep}(\Sigma^0_1, \Sigma^0_2)$ | $\mathsf{Bisep}(\Delta^0_1, \Sigma^0_1)$<br>$(\Sigma^0_1)_2$ | $\Sigma^0_1$<br>$\mathsf{Bisep}(\Delta^0_1, \Sigma^0_1)$<br>$(\Sigma^0_1)_2$ |

# 2   Preliminaries

## 2.1   Subsystems of Second Order Arithmetic

The readers are assumed to be familiar with the basic definitions and results in second order arithmetic. The language $L_2$ consists of the language of first order arithmetic plus set variables, set quantifiers and $\in$. Small letters, $x, y, z, ...$, are number variables. Capital letters, $X, Y, Z, ...$, are set variables. Classes $\Sigma_n^0$, $\Pi_n^0$ and $\Delta_n^0$ are defined as usual (the superscripts are omitted in first order arithmetic), where formulas may contain parameters. Formulas without set quantifiers are called *arithmetical*. A formula is $\Sigma_1^1$ (resp. $\Pi_1^1$) if it is of the form $\exists X\varphi(X)$ (resp. $\forall X\varphi(X)$), where $\varphi(X)$ is arithmetical. For a class $\Gamma$ of formulas, we loosely say that a formula is $\Gamma$ if it is equivalent, within the theory in which we are working, to a $\Gamma$ formula.

For a class $\Gamma$ of formulas, $\Gamma$ *comprehension* is the following schema

$$\exists X\forall n(n \in X \leftrightarrow \varphi(n)),$$

where $\varphi(x)$ is any $\Gamma$ formula in which $X$ does not occur freely.

For the classes (e.g., $\Delta_1^0$) which are defined within the theory in which we are working, some modifications are needed. E.g., $\Delta_1^0$ comprehension is defined as follows:

$$\forall x(\varphi(x) \leftrightarrow \neg\psi(x)) \rightarrow \exists X\forall x(x \in X \leftrightarrow \varphi(x)),$$

where $\varphi(x)$ and $\psi(x)$ are $\Sigma_1^0$.

This paper employs $\mathsf{RCA}_0$ as a base theory. Within $\mathsf{RCA}_0$, all recursive functions are definable. The smallest $\omega$-model (i.e., the model whose first order part is standard) is the class **REC** of recursive sets. The first order part of $\mathsf{RCA}_0$ is $\mathrm{I}\Sigma_1$, discrete ordered semi-ring axioms for $(\mathbb{N}, +, \cdot, 0, 1, <)$ plus $\Sigma_1$ induction.

**Definition 1** ($\mathsf{RCA}_0$). $\mathsf{RCA}_0$ *is the* $L_2$-*theory consisting of discrete ordered semi-ring axioms for* $(\mathbb{N}, +, \cdot, 0, 1, <)$ *plus* $\Delta_1^0$ *comprehension and* $\Sigma_1^0$ *induction.*

For any class $\Gamma$ of $L_2$ formulas, $\neg\Gamma$ is the dual class of $\Gamma$, namely, $\varphi$ is $\neg\Gamma$ iff $\neg\varphi$ is $\Gamma$. $\Gamma$ and $\neg\Gamma$ comprehensions are equivalent over $\mathsf{RCA}_0$.

A finite sequence of natural numbers can be coded by a natural number in $\mathsf{RCA}_0$. Basic operations, such as *evaluation* $s(n)$, *length* $|s|$, *segmentation* $s[n]$, $f[n]$ and *concatenation* $s * t$, $s * f$, on finite and infinite sequences of natural numbers are available in $\mathsf{RCA}_0$ in a suitable sense. By convention, if $|s| < n$, $s[n] = s$. Most notations in this paper are fairly standard (see [8, II.2]). We also use the following abbreviations: $2^{\leq n} = \{s \in 2^{<\mathbb{N}} : |s| \leq n\}$, $2^{<n} = \{s \in 2^{<\mathbb{N}} : |s| < n\}$, and $0^n$ is the sequence of length $n$ whose all components are 0.

A set $T \subseteq 2^{<\mathbb{N}}$ is *tree* if, for any $s \in 2^{<\mathbb{N}}$, $s \in T$ implies $t \in T$ for any $t \subseteq s$. $T$ is *finite* if the lengths of sequences in $T$ are bounded by some $n$.

Now we introduce and overview the supersystems of $\mathsf{RCA}_0$ that are widely used in the researches of second order arithmetic. For the details, see [8].

**Definition 2** ($\mathsf{WKL}_0$). *Weak König's lemma asserts that every infinite tree* $T \subseteq 2^{<\mathbb{N}}$ *has an infinite path i.e.,* $f : \mathbb{N} \to \{0, 1\}$ *such that* $f[n] \in T$ *for all* $n$.
  $\mathsf{WKL}_0$ *is the system* $\mathsf{RCA}_0$ + *weak König's lemma.*

Clearly $\mathsf{WKL_0}$ includes $\mathsf{RCA_0}$. Although the first order part of $\mathsf{WKL_0}$ is the same as that of $\mathsf{RCA_0}$, this inclusion is strict. A path of a recursive infinite tree is not always recursive, but it can be *low* in terms of recursion theory (cf. [8, VIII.2]). Therefore, if the determinacy of a class of games is equivalent to weak König's lemma over $\mathsf{RCA_0}$, then we can find a winning strategy, which is low, but no recursive winning strategy in general. Similarly, if the complete determinacy of the class is equivalent to it, then we can have a low oracle that allows an algorithm to determine the winning positions.

**Definition 3** ($\mathsf{ACA_0}$). $\mathsf{ACA_0}$ *is the system* $\mathsf{RCA_0} + \Sigma^0_1$ *comprehension.*

In this system we can treat number quantifiers freely. Actually, $\mathsf{ACA_0}$ proves arithmetical comprehension. The first order part of $\mathsf{ACA_0}$ is *Peano arithmetic*. The smallest $\omega$-model of $\mathsf{ACA_0}$ is the class **ARITH** of *arithmetical sets*. Thus, the equivalence to $\Sigma^0_1$ comprehension means that we can find an arithmetical winning strategy and an arithmetical oracle to the algorithm.

**Definition 4** ($\mathsf{ATR_0}$). *For a class $\Gamma$ of formulas, $\Gamma$ transfinite recursion consists of all axioms of the following form:*

$$\mathrm{WO}(W, \prec) \to \exists X \forall x \forall n (n \in X_x \leftrightarrow \theta(n, \{m : \exists y \prec x \theta(m, X_y)\})),$$

*where* $\mathrm{WO}(W, \prec)$ *asserts that $\prec$ is a linear ordering on $W$ without $\prec$-descending chains, where $\theta$ is any $\Gamma$ formula and where $X_x = \{n : (x, n) \in X\}$.*
    $\mathsf{ATR_0}$ *is the system* $\mathsf{RCA_0}$ + *arithmetical transfinite recursion.*

The first order part of $\mathsf{ATR_0}$ is Feferman's system $\mathsf{IR}$ of predicative analysis.

**Definition 5** ($\Pi^1_1$-$\mathsf{CA_0}$). $\Pi^1_1$-$\mathsf{CA_0}$ *is the system* $\mathsf{RCA_0} + \Pi^1_1$ *comprehension.*

The equivalence to $\Pi^1_1$ comprehension means that we can find an *analytic* or *coanalytic* winning strategy and an analytic or coanalytic oracle to the algorithm.
    It is known $\mathsf{RCA_0} \subsetneq \mathsf{WKL_0} \subsetneq \mathsf{ACA_0} \subsetneq \mathsf{ATR_0} \subsetneq \Pi^1_1$-$\mathsf{CA_0}$.

## 2.2 Finer Hierarchy of Formulas

As mentioned in Introduction, we need a finer hierarchy of formulas than the ordinary arithmetical hierarchy. The author [6] introduced the new classes of formulas, imitating the description [1] of *Wadge classes* of *Polish spaces*.

**Definition 6** $((\Sigma^0_n)_2)$. *The class $(\Sigma^0_n)_2$ consists of formulas of the form $\xi \wedge \neg \eta$ where both $\xi$ and $\eta$ are $\Sigma^0_n$.*

**Definition 7** ($\mathsf{Bisep}, \mathsf{Sep}$). *Let $\Gamma$ and $\Gamma'$ be classes of formulas with a distinct function variable $f$.*
    $\mathsf{Bisep}(\Gamma, \Gamma')$ *consists of formulas of the form* $(\xi_0(f) \wedge \neg \eta_0(f)) \vee (\xi_1(f) \wedge \eta_1(f))$, *where $\eta_i(f)$'s are $\Gamma'$ and where $\xi_0(f)$ and $\xi_1(f)$ are disjoint $\Gamma$, i.e., (within the theory) there is no $f$ with $\xi_0(f) \wedge \xi_1(f)$. $\mathsf{Sep}(\Gamma, \Gamma')$ is the class of formulas of the form $(\xi(f) \wedge \neg \eta_0(f)) \vee (\neg \xi(f) \wedge \eta_1(f))$, where $\xi(f)$ is $\Gamma$ and where $\eta_i(f)$'s are $\Gamma'$.*

By [1], we have the following proper hierarchy of formulas. (Here we omit the dual classes, e.g., $\Pi_n^0$.) No intermediate class (corresponding to a Wadge class) exists, except for "..." areas:

$$\Delta_1^0 \subsetneq \Sigma_1^0 \subsetneq \mathsf{Bisep}(\Delta_1^0, \Sigma_1^0) \subsetneq (\Sigma_1^0)_2 \subsetneq \cdots$$
$$\subsetneq \Delta_2^0 \subsetneq \Sigma_2^0 \subsetneq \mathsf{Bisep}(\Delta_1^0, \Sigma_2^0) \subsetneq \mathsf{Bisep}(\Sigma_1^0, \Sigma_2^0) \subsetneq \cdots \subsetneq \mathsf{Sep}(\Sigma_1^0, \Sigma_2^0) \subsetneq \cdots$$
$$\subsetneq \mathsf{Bisep}(\Delta_2^0, \Sigma_2^0) \subsetneq (\Sigma_2^0)_2 \subsetneq \cdots$$

## 2.3   Games in Second Order Arithmetic

Let $X \subseteq \mathbb{N}$. For a given formula $\varphi(f)$ with a distinct functional variable $f \in X^{\mathbb{N}}$, the game $\varphi$ is as follows: Player I and player II alternatively choose an element of $X$ to form an infinite sequence $f \in X^{\mathbb{N}}$. Player I wins iff $\varphi(f)$ holds. Player II wins iff I does not win. We regard a class of formulas with a distinct functional variable as a class of games. $s \in X^{<\mathbb{N}}$ is called a *position*. In this paper, we assume that player I is male and II is female.

A *strategy* for player I (resp. II) is a function assigning an element of $X$ to each even-(resp. odd-)length sequence from $X$. A strategy for a player is a *winning strategy* if the player wins as long as he or she plays following it.

For a given class $\Gamma$ of games in $X^{\mathbb{N}}$, $\Gamma$ determinacy in $X^{\mathbb{N}}$ is the following statement: For every $\Gamma$ game $\varphi$ in $X^{\mathbb{N}}$, exactly one of the players has a winning strategy. Strictly, this is a schema rather than one axiom. See [8].

For a game $\varphi(f)$ in $X^{\mathbb{N}}$, *player I (resp. II) wins $\varphi(f)$ at $s \in X^{<\mathbb{N}}$ if*

(1) $|s|$ is even and I (resp. II) has a winning strategy in the game $\varphi(s * f)$, or
(2) $|s|$ is odd and II (resp. I) has a winning strategy in the game $\varphi(s * f)$.

Note that in case (2) players are exchanged.

If $s = \langle \rangle$, we simply say that player I (resp. II) wins $\varphi(f)$.

For a class $\Gamma$ of games, $\Gamma$ and $\Gamma'$ determinacies in $X^{\mathbb{N}}$ are equivalent over $\mathsf{RCA}_0$: Player I (resp. II) has a winning strategy in $\neg\varphi(f)$ if player II (resp. I) has a winning strategy in $\varphi(f')$, where $f'$ is defined by $f'(m) = f(m+1)$.

**Definition 8 (complete determinacy).** *Let $X \subseteq \mathbb{N}$. $W \subseteq X^{<\mathbb{N}}$ is the winning set for player I in $\varphi(f)$ if, for any $s \in X^{<\mathbb{N}}$, $s \in W$ implies that player I wins $\varphi(f)$ at $s$ and $s \notin W$ implies that player II wins $\varphi(f)$ at $s$.*

*For a class $\Gamma$ of formulas, $\Gamma$ complete determinacy in $X^{\mathbb{N}}$ consists of all axioms of the form*

$$\exists W (W \text{ is the winning set for player I in } \varphi(f)),$$

*where $\varphi(f)$ is a game in $X^{\mathbb{N}}$ which belongs to $\Gamma$ and $W$ is not free in $\varphi(f)$.*

Clearly $\Gamma$ complete determinacy in $X^{\mathbb{N}}$ implies $\Gamma$ determinacy in $X^{\mathbb{N}}$. $\Gamma$ and $\neg\Gamma$ complete determinacies in $X^{\mathbb{N}}$ are equivalent over $\mathsf{RCA}_0$.

$\Gamma$-Det (resp. $\Gamma$-Det$^*$) stands for $\Gamma$ determinacy in the Baire space $\mathbb{N}^{\mathbb{N}}$ (resp. in the Cantor space $2^{\mathbb{N}}$), and $\Gamma$-comp.Det (resp. $\Gamma$-comp.Det$^*$) stands for $\Gamma$ complete determinacy in the Baire space (resp. in the Cantor space).

## 3  WKL$_0$ and Complete Determinacy

In this section, we prove that $\Delta_1^0$-comp.Det$^*$, $\Delta_1^0$-Det$^*$, $\Sigma_1^0$-Det$^*$ and weak König's lemma are pairwise equivalent over RCA$_0$.

**Fact 1.** *([5, Theorems 3.5 and 3.6], [6, Corollary 3.8])* Bisep($\Delta_1^0, \Sigma_1^0$)-Det$^*$, $\Sigma_1^0$-Det$^*$, $\Delta_1^0$-Det$^*$ *and Weak König's lemma are pairwise equivalent over* RCA$_0$.

**Lemma 1 (Normal form theorem 1).** *([8, Theorem II.2.7]) For any $\Sigma_1^0$ formula $\varphi(X)$, there is a $\Pi_0^0$ formula $\theta$ with* RCA$_0 \vdash \forall X(\varphi(X) \leftrightarrow \exists n\theta(X[n]))$.

**Theorem 1.** WKL$_0 \vdash \Delta_1^0$-comp.Det$^*$.

*Proof.* Let $\varphi(f)$ be a $\Delta_1^0$ game in the Cantor space. By Lemma 1, we can find $\Pi_0^0$ formulas $\theta$ and $\theta'$ with $\forall f \in 2^{\mathbb{N}}(\varphi(f) \leftrightarrow \exists n\theta(f[n]))$ and $\forall f \in 2^{\mathbb{N}}(\neg\varphi(f) \leftrightarrow \exists n\theta'(f[n]))$. $\Delta_1^0$ comprehension yields $T = \{s \in 2^{<\mathbb{N}} : \forall t \subseteq s\neg(\theta(t) \vee \theta'(t))\}$. Intuitively, $T$ is the set of those positions at which none of the players has yet won. Since there is no $f \in 2^{\mathbb{N}}$ with $\varphi(f) \wedge \neg\varphi(f)$, $T$ has no infinite path.

WKL$_0$ proves the finiteness of $T$. Let $n$ be the maximal length of sequences in $T$. Note that any $s \in 2^{<\mathbb{N}}$ of length $n+1$ satisfies exactly one of $\exists t \subseteq s\theta(t)$ and $\exists t \subseteq s\theta'(t)$. For $s \in 2^{<\mathbb{N}}$, we can determine whether I wins at $s$ as follows:

- If $|s| > n$, player I wins at $s$ iff there exists $t \subseteq s$ with $\theta(t)$.
- If $|s| \le n$ and $|s|$ is even (resp. odd), player I wins $\varphi(f)$ at $s$ iff I wins $\varphi(f)$ at either $s * \langle 0 \rangle$ or $s * \langle 1 \rangle$ (resp. both $s * \langle 0 \rangle$ and $s * \langle 1 \rangle$).

Precisely, $\Sigma_1^0$ induction on $k$ yields $u_k : (2^{\le n+1} - 2^{<n+1-k}) \to \{0,1\}$ such that

$$u_k(s) = \begin{cases} 1 & \text{if } |s| = n+1 \text{ and } \exists t \subseteq s\theta(t), \\ 0 & \text{if } |s| = n+1 \text{ and } \exists t \subseteq s\theta'(t), \\ \max\{u_k(s * \langle 0 \rangle), u_k(s * \langle 1 \rangle)\} & \text{if } |s| \le n \text{ and } |s| \text{ is even, and} \\ \min\{u_k(s * \langle 0 \rangle), u_k(s * \langle 1 \rangle)\} & \text{if } |s| \le n \text{ and } |s| \text{ is odd.} \end{cases}$$

$\Delta_1^0$ comprehension yields $u : 2^{<\mathbb{N}} \to \{0,1\}$ with $\forall s \in 2^{<\mathbb{N}}(u(s) = u_{n+1}(s[n+1]))$ and $W = \{s : u(s) = 1\}$. We show that if $s \in W$ then player I wins $\varphi(f)$ at $s$ and otherwise II wins $\varphi(f)$ at $s$. Assume $s \in W$. We may assume that $|s|$ is even, since if $|s|$ is odd, $u(s) = 1$ implies $u(s * \langle 0 \rangle) = u(s * \langle 1 \rangle) = 1$, and since the claims for $s * \langle 0 \rangle$ and for $s * \langle 1 \rangle$ imply that for $s$.

For any even-length $|s|$, player I can win $\varphi(s * f)$ by playing as follows: At any even-length $t \in 2^{<\mathbb{N}}$, he chooses 0 if $u(s * t * \langle 0 \rangle) = 1$ and he chooses 1 otherwise. If he plays following this strategy, the resulting sequence $f$ satisfies that $g((s * f)[m]) = 1$ for all $m$, and so there exists $m \le n$ with $\theta((s * f)[m])$. Similarly, player II wins $\varphi(f)$ at $s$ if $s \notin W$. $\qquad \square$

*Remark 1 (for the readers who are familiar with* RCA$_0^*$). In the proof, $u_k$ is bounded by some iterated power of $n$, and so only $\Pi_0^0$ induction is required if the language has exp. Therefore, on the basis of RCA$_0^*$, $\Delta_1^0$-comp.Det$^*$ is equivalent to WKL$_0^*$. For the details of the equivalence over RCA$_0^*$, see [6].

**Corollary 1.** *We can add $\Delta_1^0$-comp.Det$^*$ to the list of Fact 1.*

# 4   $\mathbf{ACA_0}$ and Complete Determinacy

In this section we prove that $\Sigma_1^0$-comp.Det$^*$, $(\Sigma_1^0)_2$-Det$^*$ and $\Sigma_1^0$ comprehension (i.e., $\mathsf{ACA_0}$) are equivalent over $\mathsf{RCA_0}$.

$\mathsf{ACA_0}$ can be characterized game-theoretically as follows.

**Fact 2.** *([5, Theorems 3.7 and 3.8]) $(\Sigma_1^0)_2$-Det$^*$ and $\Sigma_1^0$ comprehension are pairwise equivalent over $\mathsf{RCA_0}$.*

For any $\Sigma_1^0$ game $\varphi(f)$ in the Cantor space, the statement "player I wins $\varphi(f)$ at $s$" can be written in a $\Sigma_1^0$ formula in $\mathsf{WKL_0}$ (cf. [5, Corollary 3.4]). Hence $\mathsf{ACA_0}$ yields $X = \{s \in 2^{\mathbb{N}} : \text{I wins } \varphi(f) \text{ at } s\}$. By $\Sigma_1^0$-Det$^*$, which is proved in $\mathsf{ACA_0}$, $s \notin X$ implies that II wins $\varphi(f)$ at $s$, and so $\mathsf{ACA_0}$ proves $\Sigma_1^0$-comp.Det$^*$. The next theorem is the converse.

**Theorem 2.** $\mathsf{RCA_0} \vdash \Sigma_1^0$-comp.Det$^* \to \Sigma_1^0$ *comprehension.*

*Proof.* Let $\psi(n)$ be a $\Sigma_1^0$ formula of the form $\exists m \theta(n, m)$, where $\theta$ is a $\Pi_0^0$ formula. Consider the following game $\varphi(f)$ in the Cantor space:

- player I chooses $n \in \mathbb{N}$ by playing his first 1 at his $(n+1)$-th turn.
- I wins iff there is $m$ such that $\theta(n, m)$ holds.

To illustrate, the game goes as follows:

$$\begin{array}{c} n \text{ times} \\ \overbrace{\qquad\qquad} \end{array}$$
$$\begin{array}{ll} \text{player I} & 0, ..., 0, 1 \,\cdots \\ \text{player II} & *, ..., * \; * \,\cdots \end{array}$$

Clearly $\varphi(f)$ is $\Sigma_1^0$. $\Sigma_1^0$-comp.Det$^*$ yields the winning set $W$ for player I in $\varphi(f)$. We can check that, for any $n$, player I wins at $0^{2n} * \langle 1 \rangle$ iff $\psi(n)$. $\Delta_1^0$ comprehension yields $X = \{n : 0^{2n} * \langle 1 \rangle \in W\}$. It is easy to see $X = \{n : \psi(n)\}$.     $\square$

$\Gamma$ determinacy does not always imply $\Gamma$ complete determinacy, as this theorem with Fact 2 separates $\Sigma_1^0$-comp.Det$^*$ and $\Sigma_1^0$-Det$^*$. Moreover the equivalence between $\Gamma$ and $\Gamma'$ determinacies does not necessarily imply that between $\Gamma$ and $\Gamma'$ complete determinacies. Indeed $\Delta_1^0$-comp.Det$^*$ and $\Sigma_1^0$-comp.Det$^*$ are not equivalent, while $\Delta_1^0$-Det$^*$ and $\Sigma_1^0$-Det$^*$ are.

The completion of $(\Sigma_1^0)_2$-Det$^*$ does not lift up the comprehension axiom:

**Theorem 3.** $\mathsf{ACA_0}$ *proves* $(\Sigma_1^0)_2$-comp.Det$^*$.

*Proof.* Let $\varphi(f)$ be a $(\Sigma_1^0)_2$ game $\exists n \theta(f[n]) \wedge \eta(f)$, where $\theta(x)$ is $\Pi_0^0$ and where $\eta(f)$ is $\Pi_1^0$. $V = \{s : \exists t \subseteq s \theta(t) \wedge (\text{I wins } \eta(f) \text{ at } s)\}$ is provided by $\Pi_1^0$-comp.Det$^*$, proved in $\mathsf{ACA_0}$. Define a new $\Sigma_1^0$ game $\varphi'(f)$ by $\exists n(f[n] \in V)$. Similarly to the proof of [5, Theorem 3.7] (where we use $\Pi_1^0$ choice, which is provable in $\mathsf{WKL_0}$), we see that $W = \{s : \text{player I wins } \varphi'(f) \text{ at } s\}$, provided by $\Sigma_1^0$-comp.Det$^*$, is the winning set for player I in $\varphi(f)$.     $\square$

By $\Sigma_1^0 \subsetneq \mathsf{Bisep}(\Delta_1^0, \Sigma_1^0) \subsetneq (\Sigma_1^0)_2$, $\mathsf{ACA_0}$ proves also $\mathsf{Bisep}(\Delta_1^0, \Sigma_1^0)$-comp.Det.

**Corollary 2.** *We can add* $(\Sigma_1^0)_2$-comp.Det$^*$, $\mathsf{Bisep}(\Delta_1^0, \Sigma_1^0)$-comp.Det$^*$ *and* $\Sigma_1^0$-comp.Det$^*$ *to the list of Fact 2.*

# 5    $\Pi_1^1$-CA$_0$ and Complete Determinacy

Before turning to $\Pi_1^1$-CA$_0$, we briefly mention complete determinacy for ATR$_0$.

**Fact 3.** *The arithmetical transfinite recursion is equivalent to $\Sigma_2^0$-Det*, $\Delta_2^0$-Det*, $\Sigma_1^0$-Det, $\Delta_1^0$-Det and $\Delta_1^0$-comp.Det over RCA$_0$:*
    *See [8] for $\Sigma_1^0$-Det, $\Delta_1^0$-Det, [5] for $\Sigma_2^0$-Det*, $\Delta_2^0$-Det*, [4] for $\Delta_1^0$-comp.Det.*

**Theorem 4.** ATR$_0 \vdash \Delta_2^0$-comp.Det*.

*Proof.* We can prove this in an almost same way as the proof of [9, Theorem 6.1], by replacing $\Pi_1^1$ transfinite recursion with arithmetical transfinite recursion. Note that, although the original theorem states the equivalence between $\Pi_1^1$ transfinite recursion and $\Delta_2^0$-Det, the proof actually shows that between $\Pi_1^1$ transfinite recursion and $\Delta_2^0$-comp.Det.    □

**Corollary 3.** *We can add $\Delta_2^0$-comp.Det* to the list of Fact 3.*

We turn to $\Pi_1^1$-CA$_0$. The following is a game-theoretical characterization.

**Fact 4.** *([8, VI. 5] and [6])* Sep$(\Sigma_1^0, \Sigma_2^0)$-Det*, Bisep$(\Sigma_1^0, \Sigma_2^0)$-Det*, $(\Sigma_1^0)_2$-Det, Bisep$(\Delta_1^0, \Sigma_1^0)$-Det *and $\Pi_1^1$ comprehension are pairwise equivalent over RCA$_0$.*

Over ACA$_0$, $\Sigma_1^1$ formulas have normal forms.

**Lemma 2 (Normal form theorem 2).** *([8, Lemma V.1.4]) For any $\Sigma_1^1$ $\varphi(X)$, there is a $\Pi_0^0$ formula $\theta$ with ACA$_0 \vdash \forall X(\varphi(X) \leftrightarrow \exists f \in \mathbb{N}^{\mathbb{N}} \forall n \theta(X[n], f[n]))$.*

The next theorem separates the strengths of $\Sigma_1^0$-comp.Det and of $\Delta_1^0$-comp.Det, similarly to the case of the Cantor space.

**Theorem 5.** RCA$_0 \vdash \Sigma_1^0$-comp.Det $\to \Pi_1^1$ *comprehension.*

*Proof.* Assume $\Sigma_1^0$-comp.Det. We prove $\Sigma_1^1$ comprehension, which is equivalent to $\Pi_1^1$ comprehension. Let $\psi(n)$ be $\exists f \forall m \theta(n, f[m])$, where $\theta$ is $\Pi_0^0$. We prove the existence of $X = \{n : \psi(n)\}$. Set a $\Pi_1^0$ game $\varphi(f)$ in the Baire space by $\forall m \theta(f(0), \widetilde{f}[m])$, where $\widetilde{f}$ is a sequence in $\mathbb{N}^{\mathbb{N}}$ defined by $\widetilde{f}(k) = f(2k+2)$ for all $k$. To illustrate, the game goes as follows:

$$\begin{array}{llllll} \text{player I} & n & \widetilde{f}(0) & \widetilde{f}(1) & \widetilde{f}(2) & \cdots \\ \text{player II} & * & * & * & \cdots \end{array}$$

Intuitively, when player I chooses $n$, he wins iff he constructs an infinite sequence witnessing for $\psi(n)$. $\Sigma_1^0$-comp.Det yields the winning set $W$ for player I in $\varphi(f)$. Then $\Delta_1^0$ comprehension yields $X = \{n : \langle n \rangle \in W\}$. For $n$ with $\psi(n)$, player II cannot win at $\langle n \rangle$ if he constructs $\widetilde{f}$ with $\forall m \theta(n, \widetilde{f}[m])$. Hence $n \in X$ if $\psi(n)$. Conversely, player I cannot win $\varphi(f)$ at $\langle n \rangle$ with $\neg\psi(n)$, since no $\widetilde{f}$ satisfies $\forall m \theta(n, \widetilde{f}[m])$. Thus $n \in X$ iff $\psi(n)$.    □

Now we modify the last theorem to get the Cantor space version. For that purpose, some definitions and notations are needed. $g \in 2^{\mathbb{N}}$ is *regular* if, for any $n$, there exists $m > n$ with $g(m) = 1$. Note that the statement "$f \in 2^{\mathbb{N}}$ is regular" is $\Pi_2^0$. For a regular sequence $g \in 2^{\mathbb{N}}$, $\bar{g}$ is the unique sequence in $\mathbb{N}^{\mathbb{N}}$ such that

$$g = \langle \underbrace{0, ..., 0}_{\bar{g}(0) \text{ times}}, 1, \underbrace{0, ..., 0}_{\bar{g}(1) \text{ times}}, 1, \underbrace{0, ..., 0}_{\bar{g}(2) \text{ times}}, 1, ... \rangle.$$

**Theorem 6.** $\mathsf{RCA}_0 \vdash \Sigma_2^0\text{-comp.Det}^* \to \Pi_1^1$ *comprehension.*

*Proof.* This can be proved similarly to the last theorem. Assume $\Sigma_2^0\text{-comp.Det}^*$. Let $\psi(n)$ be $\exists f \forall m \theta(n, f[m])$, where $\theta$ is $\Pi_0^0$.

Define a $\Pi_2^0$ game $\varphi(f)$ by $(f \text{ is regular}) \wedge \forall n \theta(\bar{f}(0), \langle \bar{f}(1), ..., \bar{f}(m+1) \rangle)$. Intuitively, I chooses $n$ by playing his first 1 at his $(n+1)$-th turn, and tries to construct the sequence $\langle \underbrace{0, ..., 0}_{g(0) \text{ times}}, 1, \underbrace{0, ..., 0}_{g(1) \text{ times}}, 1... \rangle$, instead of a sequence $g \in \mathbb{N}^{\mathbb{N}}$

with $\forall m \theta(n, g[m])$.

$\Sigma_2^0$ complete determinacy yields the winning set $W \subseteq 2^{<\mathbb{N}}$ for player I in $\varphi(f)$. $\Delta_1^0$ comprehension yields $X = \{n : 0^{2n} * \langle 1 \rangle \in W\}$. As in the proof of the last theorem, we can also check $X = \{n : \psi(n)\}$.                    $\square$

*Remark 2.* For a given $\Pi_2^0$ game $\varphi(f)$ in the Cantor space, the statement "I has a winning strategy in $\varphi(f)$" is $\Sigma_1^1$ over $\mathsf{ATR}_0$ (cf. [6, Remark 4.4]), and so $\Pi_1^1\text{-}\mathsf{CA}_0$ yields $W = \{s \in \mathbb{N}^{\mathbb{N}} : \text{I wins } \varphi(f) \text{ at } s\}$. By $\Sigma_2^0\text{-Det}$, proved in $\Pi_1^1\text{-}\mathsf{CA}_0$, $W$ is a winning set for player I in $\varphi(f)$. Hence $\Sigma_2^0\text{-comp.Det}^*$ is implied by $\Pi_1^1\text{-}\mathsf{CA}_0$.

Theorem 3 shows that the completion of $(\Sigma_1^0)_2\text{-Det}^*$ does not strengthen the system. Similarly, neither does that of $(\Sigma_1^0)_2\text{-Det}$ nor of $\mathsf{Sep}(\Sigma_1^0, \Sigma_2^0)\text{-Det}^*$.

**Theorem 7.** $\Pi_1^1\text{-}\mathsf{CA}_0$ *proves 1.* $(\Sigma_1^0)_2\text{-comp.Det}$ *and 2.* $\mathsf{Sep}(\Sigma_1^0, \Sigma_2^0)\text{-comp.Det}^*$.

*Proof.* 1. It can be proved in an almost same way to Theorem 3. Note that, for a $\Pi_1^0$ game $\varphi(f)$ in the Baire space, the assertion "player I has a winning strategy in $\varphi(f)$" is $\Sigma_1^1$ over $\mathsf{ATR}_0$ and $\mathsf{ATR}_0$ proves $\Sigma_1^1$ choice. (see [8, Lemma VI.5.2]).

2. Let $\varphi(f) \equiv (\exists n \theta(f[n]) \wedge \neg \eta_0(f)) \vee (\neg \exists n \theta(f[n]) \wedge \eta_1(f))$, where $\theta(x)$ is $\Pi_0^0$ and where $\eta_i(f)$'s are $\Sigma_2^0$. $V = \{s \in 2^{<\mathbb{N}} : \exists t \subseteq s \theta(t) \wedge (\text{I wins } \neg \eta_0(f) \text{ at } s)\}$ and $V' = \{s \in 2^{<\mathbb{N}} : \text{I wins } (\neg \exists n \theta(f[n]) \wedge \eta_1(f)) \vee \exists n f[n] \in V \text{ at } s\}$ are provided by $\Pi_1^1\text{-}\mathsf{CA}_0$. We can prove that player I wins $\varphi(f)$ at $s$ iff $s \in V'$.                    $\square$

By $\mathsf{Bisep}(\Delta_1^0, \Sigma_1^0) \subseteq (\Sigma_1^0)_2$ and 1 above, $\Pi_1^1\text{-}\mathsf{CA}_0$ proves also $\mathsf{Bisep}(\Delta_1^0, \Sigma_1^0)\text{-comp.Det}$, and by $\Sigma_2^0 \subseteq \mathsf{Bisep}(\Delta_1^0, \Sigma_1^0) \subseteq \mathsf{Bisep}(\Sigma_1^0, \Sigma_2^0) \subseteq \mathsf{Sep}(\Sigma_1^0, \Sigma_2^0)$, 2 and Theorem 6 show the equivalence among $\Sigma_2^0\text{-comp.Det}^*$, $\mathsf{Bisep}(\Delta_1^0, \Sigma_2^0)\text{-comp.Det}^*$, $\mathsf{Bisep}(\Sigma_1^0, \Sigma_2^0)\text{-comp.Det}^*$, $\mathsf{Sep}(\Sigma_1^0, \Sigma_2^0)\text{-comp.Det}^*$ and $\Pi_1^1$ comprehension.

**Corollary 4.** *We can add the following to the list of Fact 4:* $\mathsf{Sep}(\Sigma_1^0, \Sigma_2^0)\text{-comp.Det}^*$; $\mathsf{Bisep}(\Sigma_1^0, \Sigma_2^0)\text{-Det}^*$ $\Sigma_2^0\text{-comp.Det}^*$, $(\Sigma_1^0)_2\text{-comp.Det}$, $\mathsf{Bisep}(\Delta_1^0, \Sigma_1^0)\text{-comp.Det}$ *and* $\Sigma_1^0\text{-comp.Det}$.

# 6  Stronger Complete Determinacy Statements

In the previous works, stronger determinacy statements also have been investigated. Here we make brief comments on some stronger complete determinacy statements.

As mentioned in the proof of Theorem 4, [9, Theorem 6.1] essentially shows that $\Pi_1^1$-TR$_0$, the system RCA$_0 + \Pi_1^1$ transfinite recursion, proves $\Delta_2^0$-comp.Det.

$\Pi_1^1$-TR$_0$ also proves Bisep($\Delta_2^0, \Sigma_2^0$)-comp.Det$^*$. This is essentially shown in the proof of [6, Theorem 6.5], which proves Bisep($\Delta_2^0, \Sigma_2^0$)-Det$^*$ in $\Pi_1^1$-TR$_0$.

Together with the results of [6, Theorem 6.7 and 6.8] and [9], Bisep($\Delta_2^0, \Sigma_2^0$)-Det$^*$, Bisep($\Delta_2^0, \Sigma_2^0$)-comp.Det$^*$, $\Delta_2^0$-Det, $\Delta_2^0$-comp.Det and $\Pi_1^1$ transfinite recursion are pairwise equivalent over RCA$_0$.

*Acknowledgments.* The author would like to express her gratitude to Dr. Antonio Montalbán, who gave her a chance to consider some problems in this paper by his admirable talk at Kyoto University on August 2006. She is also greatly indebted to Prof. Kazuyuki Tanaka and to Mr. SATO Kentaro for their beneficent comments on the draft of this paper.

# References


1. Louveau, A.: Some results in the Wadge hierarchy of Borel sets, Cabal Seminar 79-81. Lecture Note in Mathematics, vol. 1019, pp. 28–55. Springer, Heidelberg (1983)
2. Medsalem, M.O., Tanaka, K.: $\Delta_3^0$-determinacy, comprehension and induction. J. Symbolic Logic 72, 452–462 (2007)
3. Medsalem, M.O., Tanaka, K.: Weak determinacy and iterations of inductive definitions. In: Proc. Computable Prospects of Infinity. World Scientific, Singapore (to appear)
4. Montalbán, A.: Indecomposable linear orderings and Theories of Hyperarithmetic Analysis. J. Math. Log. 6, 89–120 (2006)
5. Nemoto, T., MedSalem, M.O., Tanaka, K.: Infinite games in the Cantor space and subsystems of second order arithmetic. Math. Log. Q. 53, 226–236 (2007)
6. Nemoto, T.: Determinacy of Wadge classes and subsystems of second order arithmetic. Mathematical logic quarterly,
   http://www.math.tohoku.ac.jp/~sa4m20/wadge.pdf
7. Nießner, F.: Nondeterministic tree automata. In: Grädel, E., Thomas, W., Wilke, T. (eds.) Automata Logics, and Infinite games, pp. 152–155. Springer, Heidelberg (2002)
8. Simpson, S.G.: Subsystems of Second Order Arithmetic. Springer, Heidelberg (1999)
9. Tanaka, K.: Weak axioms of determinacy and subsystems of analysis I: $\Delta_2^0$-games Z. Math. Logik Grundlag. Math. 36, 481–491 (1990)
10. Tanaka, K.: Weak axioms of determinacy and subsystems of analysis II: $\Sigma_2^0$-games. Ann. Pure Appl. Logic 52, 181–193 (1991)
11. Welch, P.: Weak systems of determinacy and arithmetical quasi-inductive definitions, http://www.maths.bris.ac.uk/~mapdw/det7.pdf

# Internal Density Theorems for Hierarchies of Continuous Functionals

Dag Normann

Department of Mathematics, The University of Oslo, P.O. Box 1053,
Blindern N-0316 Oslo, Norway
dnormann@math.uio.no

**Abstract.** One standard way of constructing a hierarchy of total, continuous functionals over a fixed set of base types is to use a suitable cartesian closed category of domains where we may construct the corresponding hierarchy of partial continuous functionals, and then extract the hereditarily total ones. One important theorem, when available, is the *Density Theorem*: Each finitary domain object can be extended to a total one.

We will see how we in the context of limit spaces, may formulate and prove versions of the density theorems and avoid domain theory.

**Keywords:** Continuous functional, Limit space, Probabilistic selection, Density theorem.

## 1   Introduction

The countable and continuous functionals were independently defined by Kleene [4] and Kreisel [5]. Kleene defined the countable functionals as a substructure of the full hierarchy of total functionals over $\mathbb{N}$ of pure type, while Kreisel defined the continuous functionals as functionals operating in a continuous way on objects of lower types, using filters of formal neighborhoods to represent these objects. Each countable functional $\Phi$ is linked to a set of *associates* $\alpha$, functions from $\mathbb{N}$ to $\mathbb{N}$ coding how $\Phi$ acts on countable functionals $\phi$ one type down via defining a continuous function on the set of associates for such functionals.

Kleene and Kreisel seemed to agree that they had defined essentially the same typed structure of functionals, and this is now a well established fact. In the recent literature, the functionals are known as the *Kleene-Kreisel continuous functionals*, we will just call them *continuous* here.

Both authors proved what is later known as *the density theorem*, Kleene showed that uniformly in the type $n$ we can decide in a primitive recursive way the set of finite sequences of natural numbers that can be extended to an associate for a functional of type $n$, and, when possible, uniformly construct one such associate in a primitive recursive way. Kreisel showed that all formal neighborhoods are elements of total filters. The proofs have many similarities, and the consequence is that there is an effectively enumerable dense subset of the set of continuous functionals for each type. Kreisel used this density theorem in his

A. Beckmann, C. Dimitracopoulos, and B. Löwe (Eds.): CiE 2008, LNCS 5028, pp. 467–475, 2008.

description of the constructive content of a statement in analysis, and as a consequence, in the proof of the *Kreisel Representation Theorem*. The density theorem turned out to be a very important tool in analyzing the finer aspects of Kleene's [3] $S1 - S9$-based notion of computations relative to continuous functionals.

In the approximately 50 years that has passed since Kleene's and Kreisel's constructions, the study of the continuous functionals and corresponding hierarchies of partial functionals reflects all important aspects of CiE, from theorems that are of interest mainly due to their mathematical elegance to modeling real life computations (e.g. using Haskell). This will be discussed in more depth at CiE2008,

This wide range of approaches makes the continuous functionals an interesting structure from many points of view. In computer science we may see it as the quotient structure of the hereditarily total objects in a, for other purposes, interesting hierarchy of partial functionals (e.g. continuous or sequential), while other characterizations are based on direct constructions of mathematical interest. One of these is the approach via Kuratowski Limit Spaces [6] as suggested by Scarpellini [9]. It has later turned out that this characterization of the continuous functionals is rather useful. It can be used to show that the Kleene and Kreisel approaches are indeed equivalent. It can be used to show that the domain theoretic definition of the continuous functionals over the reals, generalizing Kreisel's construction, and the TTE-construction, generalizing Kleene's construction, are equivalent. See Normann [7] and Schröder [10] for the two characterizations.

In this note, our starting point is that the continuous functionals are of interest on their own, and that it thus is of interest to have a smooth construction from which we in an effective way may develop an infrastructure without introducing domain theory, fragments of category theory or other forms of external machinery. The motivation is a general view on issues of computability in data-types. If a data-type can be viewed as a definable class in the universe $HF$ of *hereditarily finite sets*, then the concept of computability is absolute. For other data-types, the strength of the chosen notion of computability may depend on how we represent the data. If we can avoid hiding a lot of strength in the representation of data, and still obtain the computability of some object or operator, we have obtained a better result.

The core of the paper is Section 2, where we give alternative proofs of effective density for the hierarchies of continuous functionals over $\mathbb{N}$ and over $\mathbb{R}$. There are not many proofs in Section 2, but one important aspect of our approach is that it actually simplifies the matter to the extent that the proofs in Section 2 safely can be left for the reader. Another aspect is that we actually get slightly better results than when using traditional methods, in particular when $\mathbb{R}$ is the base type.

If we consider $\mathbb{N}$ and $\mathbb{R}$ as two base types, we cannot expect to have a density theorem in the traditional domain-theoretical sense, because the total objects in the domain interpretation of $\mathbb{R} \to \mathbb{N}$ is not dense. However, the set of constant functions forms an effectively enumerable dense subset, so seen from within the structure, effective density holds. One may even say that this simple example visualizes the difference between the Kreisel density theorem and the Kleene density theorem, the modern variant of Kreisel's theorem does not hold, while the modern version of Kleene's theorem does hold. (This is extremely unfair to Kreisel, who would probably not have considered formal neighborhoods not extendable to total objects if he found it interesting to consider this set of functions at all.)

In Section 3 we will state and indicate the proof of a general result visualizing this kind of limitations of Domain Theory.

## 2   The Core

### 2.1   Limit Spaces

**Definition 1.** ( Kuratowski [6]) A *limit space* is a set $X$ together with a relation

$$x = \lim_{n \to \infty} x_n$$

between elements of $X$ and sequences from $X$ satisfying

i)  If $x_n = x$ for all $n$, then $x = \lim_{n \to \infty} x_n$.
ii)  If $x = \lim_{n \to \infty} x_n$ then $x = \lim_{n \to \infty} x_{f(n)}$ for all increasing $f : \mathbb{N} \to \mathbb{N}$.
iii)  If $\neg(x = \lim_{n \to \infty} x_n)$, there is an increasing function $f : \mathbb{N} \to \mathbb{N}$ such that for no increasing $g : \mathbb{N} \to \mathbb{N}$ we have that

$$x = \lim_{n \to \infty} x_{f(g(n))}.$$

i) says that constant sequences converge, ii) says that a subsequence of a convergent sequence is converging, and iii) says that if $x$ is not a limit point of a sequence, there is a subsequence that does not have $x$ as a cluster point.

The axioms allow for more than one limit of a sequence, but this will not be relevant to us.

**Definition 2.** Let $X$ and $Y$ be limit spaces (suppressing the notation for the limit structure).

a)  $f : X \to Y$ is *continuous* if

$$x = \lim_{n \to \infty} x_n \Rightarrow f(x) = \lim_{n \to \infty} f(x_n)$$

for all $x$ and $\{x_n\}_{n \in \mathbb{N}}$.
Let $Z = X \to Y$ be the set of continuous functions from $X$ to $Y$.
b)  If $f \in Z$ and $\{f_n\}_{n \in \mathbb{N}}$ is a sequence from $Z$, we let

$$f = \lim_{n \to \infty} f_n$$

if for all $x$ and $\{x_n\}_{n \in \mathbb{N}}$,

$$x = \lim_{n \to \infty} x_n \Rightarrow f(x) = \lim_{n \to \infty} f_n(x_n).$$

The condition that a sequence of functions converges along any convergent sequence of inputs is a generalization of the property of being pointwise convergent and equicontinuous for functions between metric spaces. It is not hard to see that this construction organizes the space $Z = X \to Y$ into a limit space.

Any limit space will induce a topology: A set $O$ is open if all sequences converging to an $x \in O$ will be almost contained in $O$. Conversely, every topological space induce a limit structure, the set of convergences in terms of the topology. A *topological limit space* is a limit space that is the limit space of its topology. The corresponding topological spaces are called *sequential*. All limit spaces considered in this paper will be topological.

**Definition 3.**  a) Let $\iota$ be the base type.

If $\tau$ and $\delta$ are types, we let $\sigma = (\tau \to \delta)$ be a new type.

b) By recursion on the type $\sigma$, we define the limit spaces $Ct_\mathbb{N}(\sigma)$ and $Ct_\mathbb{R}(\sigma)$ by

- $Ct_\mathbb{N}(\iota) = \mathbb{N}$ with the limit structure induced from the discrete topology.
- $Ct_\mathbb{R}(\iota) = \mathbb{R}$ with the limit structure of the Euclidian topology.
- If $\sigma = (\tau \to \delta)$, we let

$$Ct_\mathbb{P}(\sigma) = Ct_\mathbb{P}(\tau) \to Ct_\mathbb{P}(\delta)$$

as constructed in Definition 2, where $\mathbb{P} = \mathbb{N}$ or $\mathbb{P} = \mathbb{R}$

## 2.2  The Classical Continuous Functionals

For the typed hierarchy over $\mathbb{N}$, it is quite easy to construct a dense set leading to effective approximations. What is sort of new here is that we only need the limit space structure to establish the basic properties:

**Definition 4.** By recursion on the type $\sigma$ over $\iota$ we define the *n 'th approximation* $(F)_n$ to $F$ as follows:

If $\sigma = \iota$ (and $F = m$) we let

$$(m)_n = \begin{cases} m \text{ if } m \leq n \\ n \text{ otherwise} \end{cases}$$

If $\sigma = \tau \to \delta$ and $f \in Ct_\mathbb{N}(\tau)$ we let

$$(F)_n(f) = (F((f)_n))_n.$$

The following facts are easy to prove by non-simultaneous induction on $\sigma$. c) is proved by a simple use of the construction of typed limit spaces.

**Theorem 1.** *For each type $\sigma$ over $\iota$ we have*

a) *For each $n$, the set of $(F)_n$, when $F$ varies over $Ct_\mathbb{N}(\sigma)$, is finite.*
b) *If $F \in Ct_\mathbb{N}(\sigma)$ and $n \leq m$, then $((F)_m)_n = (F)_n$.*
c) *If $F = \lim_{n \to \infty} F_n$, then $F = \lim_{n \to \infty}(F_n)_n$.*

This theorem shows that for the hierarchy over $\mathbb{N}$, we will have effectively enumerable dense subsets of every space $Ct_{\mathbb{N}}(\sigma)$, and we can effectively in any object select a sequence from this set converging to the object.

From here on it is possible to interpret Gödel's $T$, typed $\mu$-recursion and even the full Kleene schemes [3] within the framework of limit spaces, but this will be to follow a side track.

## 2.3   The Hierarchy over the Reals

There is no way we in a continuous way can select a sequence converging to $x \in \mathbb{R}$ from a fixed countable subset of $\mathbb{R}$, simply because any such continuous selection will be constant for each index. In this section we will construct a set of "finitary" functionals in the hierarchy over the reals. The construction is not new, it is copied from Normann [8]. What is new is again that all essential properties can be proved in the context of limit spaces, and there is no need for external structures like domains or $TTE$-codings.

**Definition 5.** By recursion on the type $\sigma$ we define a finitary type-structure $\{X_n(\sigma)\}_{\sigma \text{ type}}$ for each $n \in \mathbb{N}$ as follows:

$$X_n(\iota) = \{\tfrac{k}{n!} \mid -(n+1)! \leq k \leq (n+1)!\}.$$
$$X_n(\tau \to \delta) = \{h \mid h : X_n(\tau) \to X_n(\delta)\}.$$

We will now, by simultaneous recursion for each $n$, define an embedding $\nu_{n,\sigma} : X_n(\sigma) \to Ct_{\mathbb{R}}(\sigma)$ and for each $a \in Ct_{\mathbb{R}}(\sigma)$ a probability distribution $\mu_{n,\sigma}(a)$ on $X_n(\sigma)$.

## Definition 6

1. Let $\sigma = \iota$. We let $\nu_{n,\iota}$ be the inclusion map.

   If $x \notin \langle -(n+1), n+1 \rangle$, we let $\mu_{n,\iota}(\tfrac{k}{n!}) = 1$ if $\tfrac{k}{n!}$ is the object in $X_n(\iota)$ closest to $x$, otherwise we let $\mu_{n,\iota}(\tfrac{k}{n!}) = 0$.

   If $x \in \langle -(n+1), n+1 \rangle$ there are unique $k \in \mathbb{Z}$ and $y \in [0,1)$ such that $x = \tfrac{k}{n!} + \tfrac{1-y}{n!}$.

   Let $\mu_{n,\iota}(\tfrac{k}{n!}) = y$, $\mu_{n,\iota}(\tfrac{k+1}{n!}) = 1 - y$ and $\mu_{n,\iota}(\tfrac{l}{n!}) = 0$ for all $l \neq k, k+1$.

2. Let $\sigma = \tau \to \delta$

   Let $h \in X_n(\tau \to \delta)$, $a \in Ct_{\mathbb{R}}(\tau)$.
   Let

   $$\nu_{n,\sigma}(h)(a) = \sum_{c \in X_n(\tau)} \mu_{n,\tau}(a)(c) \cdot \nu_{n,\delta}(h(c)).$$

   Now let $f \in Ct_R(\tau \to \delta)$ and $h \in X_n(\tau \to \delta)$.
   Let

   $$\mu_{n,\sigma}(f)(h) = \prod_{c \in X_n(\tau)} \mu_{n,\delta}(f(\nu_{n,\tau}(c)))(h(c)).$$

We observe that at type $\iota$ we are constructing explicit probability distributions, while at higher types we are using finite products of probability distributions, which again will be probability distributions. Thus $\mu_{n,\sigma}(a)$ is a probability distribution for all $n$, $\sigma$ and $a \in Ct_{\mathbb{R}}(\sigma)$. We also observe that each $Ct_{\mathbb{R}}(\sigma)$ will be a topological vector space (in the sequential topology). This observation is used in:

**Lemma 1.**   a) *For each type $\sigma$ and $n \in \mathbb{N}$, $\nu_{n,\sigma}$ maps $X_n(\sigma)$ into $Ct_{\mathbb{R}}(\sigma)$ and for each $F \in X_n(\sigma)$, the map*

$$G \mapsto \mu_{n,\sigma}(G)(F)$$

*is continuous as a map from $Ct_{\mathbb{R}}(\sigma)$ to $[0,1]$.*
*b) If $F \in X_n(\sigma)$, then*

$$\mu_{n,\sigma}(\nu_{n,\sigma}(F))(F) = 1.$$

The proof of a) is trivial, and the proof of b) is tedious but by direct calculation. Both arguments are by induction.

**Lemma 2.** *Let $\sigma$ be a type, and for each $n$, let $Y_n \subset Ct_{\mathbb{R}}(\sigma)$ be finite. Assume that for all $n \in \mathbb{N}$, $\mu_n$ is a probability distribution on $Y_n$ and that for all*

$$\{F_n\}_{n \in \mathbb{N}} \in \prod_{n \in \mathbb{N}} Y_n$$

*we have that*

$$F = \lim_{n \to \infty} F_n.$$

*Then*

$$F = \lim_{n \to \infty} \sum_{G \in Y_n} \mu_n(G) \cdot G.$$

*Proof*
We use induction on $\sigma$. For $\sigma = \iota$, this is simple, so let $\sigma = \tau \to \delta$.
Let $a = \lim_{n \to \infty} a_n$ in $Ct_R(\tau)$. We must show that

$$F(a) = \lim_{n \to \infty} \sum_{G \in Y_n} \mu_n(G)G(a_n).$$

This follows from the induction hypothesis by letting $\mu_n$ induce a suitable probability distribution on $Z_n = \{G(a_n) \,|\, G \in Y_n\}$.

**Theorem 2.** *Let $\sigma$ be a type and let $F, F_n \in Ct_{\mathbb{R}}(\sigma)$ be such that $F = \lim_{n \to \infty} F_n$. For each $n$, let $G_n \in X_n(\sigma)$ be such that*

$$\mu_{n,\sigma}(F_n)(G_n) > 0.$$

*Then*

$$F = \lim_{n \to \infty} \nu_{n,\sigma}(G_n).$$

*Proof*

We use induction on $\sigma$. For $\sigma = \iota$, the theorem is trivial, so assume that $\sigma = \tau \to \delta$.

Let $a = \lim_{n\to\infty} a_n$ in $Ct_{\mathbb{R}}(\tau)$.

We must show that $F(a) = \lim_{n\to\infty} \nu_{n,\sigma}(G_n)(a_n)$. We have that

$$\nu_{n,\sigma}(G_n)(a_n) = \sum_{b \in X_n(\tau)} \mu_{n,\tau}(a_n)(b) \cdot \nu_{n,\delta}(G_n(b)).$$

For every $\{b_n\}_{n\in\mathbb{N}} \in \prod_{n\in\mathbb{N}} X_n(\tau)$ and every $n$ we have that

$$(*) \quad \mu_{n,\delta}(F_n(\nu_{n,\tau}(b_n)))(G_n(b_n)) > 0$$

since $\mu_{n,\sigma}(F_n)(G_n) > 0$.

Now, let us restrict ourselves to those $\{b_n\}_{n\in\mathbb{N}}$ such that for all $n$,

$$\mu_{n,\tau}(a_n)(b_n) > 0.$$

Then, by the induction hypothesis for $\tau$, $a = \lim_{n\to\infty} \nu_{n,\tau}(b_n)$, so

$$F(a) = \lim_{n\to\infty} F_n(\nu_{n,\tau}(b_n)).$$

Then, by $(*)$ and the induction hypothesis for $\delta$ ,

$$F(a) = \lim_{n\to\infty} \nu_{n,\delta}(G_n(b_n)).$$

Now, by Lemma 2, the induction step follows.

**Corollary 1.**

$$\{\nu_{n,\sigma}(G) \mid n \in \mathbb{N} \wedge G \in X_n(\sigma)\}$$

*is dense in $Ct_{\mathbb{R}}(\sigma)$.*

The corollary is not new. The density theorem was proved in [7], and the corollary as formulated was essentially proved in DeJaeger [2], using the filter space structure on $Ct_{\mathbb{R}}(\sigma)$.


## 3   Moving on

It is natural to restrict investigations of computability aspects to the natural numbers and structures based on the natural numbers. After all, genuine data are discrete. There are however good reasons for extending these investigations to base data-types beyond $\mathbb{N}$ and even $\mathbb{R}$, e.g to $L^p$-spaces, fractals and so on. Of course, the $TTE$-program of Weihrauch and others, see [11], provides us with one way of investigating computability over structures appearing in analysis. Effective domain representations is another useful approach. We do however also find it useful to see how far we can get towards constructing an effective infrastructure on such spaces without introducing superstructures and imposing external notions of computability on the given structures, see the introduction for a discussion. One way to create a useful part of an infrastructure will be to isolate a dense subset that in some way is effectively dense. We have shown how to do it directly for two important special cases in Section 2, but though the methods are new, the results are not so, they have been proved by other methods, e.g. domain theory. It will be of interest to extend the formation of effectively dense subsets to type structures with an alternative set of base types, and then we believe that the methods developed in this paper will be useful.

The standard method for proving results like this using domain theory is to show that the total elements are effectively dense in the underlying domain. If we look at the hybrid type structure using both $\mathbb{N}$ and $\mathbb{R}$ as base types, this will not work because $\mathbb{N}$ is strictly disconnected and $\mathbb{R}$ is path connected. This turns out to be a rather general phenomenon.

A *Polish space* is a separable topological space that has a complete metric. If we extend our pool of base types with sets of separable Banach spaces, everything done in Section 2.3 extends smoothly. The main challenge is to provide the Banach spaces themselves with probabilistic selection from a prefixed countable dense subset, but to do so can be seen as an advanced exercise in a course on metric spaces. Every Polish space can be topologically embedded into a separable Banach space, actually into a subspace of $l^\infty$. However, our machinery fails when dealing with base spaces with nontrivial connectedness components but without some kind of convexity structure.

It is unclear to what extent we may construct useful countable dense subsets of function spaces using the approaches discussed in this paper, but we will make two observations showing that in the case of domain theory, we cannot expect much beyond what has already been obtained. We will use the dense domain representation of metric spaces as constructed in Blanck [1] in the definition and theorem below.

**Definition 7.** Let $X$ and $Y$ be two topological spaces. We say that $Y$ is *compactly saturated* over $X$ if, whenever $C \subseteq D$ are compact subsets of $X$ and $f : C \to Y$ is continuous, then $f$ can be extended to a continuous $g : D \to Y$.

One consequence of this property will be that if $X$ contains one non-trivial path, then $Y$ is path connected. If $X$ contains one nontrivial copy of a disc, then all loops in $Y$ can be "filled" and so forth.

If $X$ and $Y$ are metric spaces such that the total objects are dense in the domain representation of $X \to Y$, then any local property of connectedness in $X$ will result in a corresponding global one in $Y$:

**Theorem 3.** *Let $X$ and $Y$ be complete separable metric spaces and assume that the set of total objects in the domain representation of $X \to Y$ is dense. Then*

a) *Either $X$ is strongly disconnected or $Y$ is connected.*

b) *$Y$ is compactly saturated over $X$.*

Part a) is trivial, but the proof of b) is quite long and we will only sketch the idea behind the proof. Even the sketch cannot be read without the knowledge of some domain theory.

The significance of this result is that we cannot rely on domain theory if we want to find effectively enumerable dense subsets of function spaces, a more detailed understanding of how the spaces in question relate will be necessary.

*Idea of proof of b)*

Let $X$, $Y$, $C$, $D$ and $f$ be given as in the statements of Definition 7 and Theorem 3.

Let $\hat{f} = \bigsqcup_{n \in \mathbb{N}} \hat{f}_n$ be a domain representation of $f$, where $\{\hat{f}_n\}_{n \in \mathbb{N}}$ is an increasing sequence of finitary approximations.

Using the density property we will construct a sequence $\{g_n\}_{n \in \mathbb{N}}$ of total extensions of the finitary objects $\hat{g}_n$ where:

$\hat{f}_n \sqsubseteq \hat{g}_n$

$\hat{g}_{n+1}$ is constructed from the part of $g_n$ needed to determine the modulus of continuity of $g_n$ on $D$ with a precision of at least $2^{-n}$, but adjusted "as little as possible" in order to make $\hat{g}_{n+1}$ extend $\hat{f}_{n+1}$.

The details of the construction are quite lengthy, and so is the proof that they work.

# References


1. Blanck, J.: Domain representability of metric spaces. Annals of Pure and Applied Logic 83, 225–247 (1997)
2. DeJaeger, F.: Calculabilité sur les réels. Thesis. Paris VII (2003)
3. Kleene, S.C.: Recursive functionals and quantifiers of finite types I. Trans. Amer. Math. Soc. 91, 1–52 (1959)
4. Kleene, S.C.: Countable functionals. In: Heyting, A. (ed.) Constructivity in Mathematics, pp. 81–100. North-Holland, Amsterdam (1959)
5. Kreisel, G.: Interpretation of analysis by means of functionals of finite type. In: Heyting, A. (ed.) Constructivity in Mathematics, pp. 101–128. North-Holland, Amsterdam (1959)
6. Kuratowski, C.: Topologie, Warsawa, vol. I (1952)
7. Normann, D.: The continuous functionals of finite types over the reals. In: Keimel, K., Zhang, G.Q., Liu, Y., Chen, Y. (eds.) Domains and processes, pp. 103–124. Kluwer Academic Publishers, Dordrecht (2001)
8. Normann, D.: Applications of the Kleene-Kreisel Density Theorem to Theoretical Computer Science. In: Cooper, S.B., Löwe, B., Sorbi, A. (eds.) New Computational Paradigms: Changing Conceptions of what is Computable, pp. 119–138. Springer, Heidelberg (2008)
9. Scarpellini, B.: A model for Bar recursion of Higher Types. Comp. Math. 23, 123–153 (1971)
10. Schröder, M.: Admissible representations of limit spaces. In: Blanck, J., Brattka, V., Hertling, P., Weihrauch, K. (eds.) Computability and complexity in Analysis. Informatik Berichte, vol. 237, pp. 369–388 (2000)
11. Weihrauch, K.: Computable analysis. Texts in Theoretical Computer Science. Springer, Heidelberg (2000)

# Two-by-Two Substitution Systems and the Undecidability of the Domino Problem

Nicolas Ollinger

Laboratoire d'informatique fondamentale de Marseille (LIF)
Aix-Marseille Université, CNRS,
39 rue Joliot-Curie, 13 013 Marseille, France
`Nicolas.Ollinger@lif.univ-mrs.fr`

**Abstract.** Thanks to a careful study of elementary properties of two-by-two substitution systems, we give a complete self-contained elementary construction of an aperiodic tile set and sketch how to use this tile set to elementary prove the undecidability of the classical Domino Problem.

## Introduction

The *Domino Problem* is the following simple problem: given a finite set of tiles, copies of the unit square with colored edges, decide if it is possible to tile the whole euclidian plane using as many copies of each tile as you need ensuring that tiles colors match along edges, without scaling or rotating the tiles. This problem was first described by Wang to study a particular syntactical restricting of the Entscheidungsproblem (for a proof of the logical problem without using the Domino Problem and an explanation of the field, see [6]). The Domino Problem turns out to be undecidable, as it was proved in 1964 by Berger [1, 2], a student of Wang. The undecidability of the Domino Problem as since been used outside its original realm, providing a valuable tool to prove undecidability results, see for example the results of Kari [7] on cellular automata.

The historical proof, as found in [1], is very technical. One technical difficulty of the proof is the involvement of aperiodic tile sets: set of tiles that only admit aperiodic tilings. Several authors worked both to ameliorate the proof of the Domino Problem and to construct simpler aperiodic tile sets. The most quoted proof is certainly the one from Robinson [12], the first proof involving substitutions is the one of Mozes [10]. For an historical survey of the field, the reader might consult [5] and [11].

Recently, several authors have been independently interested into providing new proofs of the undecidability of the Domino Problem: Durand, Levin, and Shen [3] revisited the classical Robinson proof technic introducing new tools based on substitutions; Kari [8] proposed a completely new proof technic reducing the Domino Problem to the immortality problem rather than on the halting problem; Durand, Romashchenko, and Shen [4] proposed another new proof technic based on Kleene's fixed point theorem.

In the present paper, we give a complete self-contained elementary construction of an aperiodic tile set by combining tools from [3, 10, 12] with a careful

A. Beckmann, C. Dimitracopoulos, and B. Löwe (Eds.): CiE 2008, LNCS 5028, pp. 476–485, 2008.

study of two-by-two substitution systems. Our claim is that this proof both explains where the tile set comes from and why it works. Moreover, the number of tiles (104) is a reasonable compromise between classically big tile sets (more than 16000 tiles in [12, 3]) and very involved small tile sets (see [5] for details on the competition). Furthermore, the tile set as a nice property to be easily extendable to code any substitution, leading to a new and shorter proof of the undecidability of the Domino Problem sketched in 4.

## 1  Two-by-Two Substitution Systems

A *pattern* $\mathcal{P}$ is a subset of the discrete plane $\mathbb{Z}^2$. The *translation* $\mathcal{P} + u$ of a pattern $\mathcal{P}$ by a vector $u \in \mathbb{Z}^2$ is the pattern $\{z + u \mid z \in \mathcal{P}\}$. Let $\boxplus^i$ denote for all $i \in \mathbb{N}$ the pattern $\left\{x \in \mathbb{Z} \mid 0 \leqslant x < 2^i\right\}^2$: the square of size $2^i$ with south-west corner at $\binom{0}{0}$. The two-by-two square $\boxplus^1$ is abbreviated as $\boxplus$. The two-by-two *scaling* $\square(\mathcal{P})$ of a pattern $\mathcal{P}$ is the pattern $\{2z + c \mid z \in \mathcal{P}, c \in \boxplus\}$.

Let $\Sigma$ denote a finite set, or *alphabet*. A *coloring* $\mathcal{C} : \mathcal{P} \to \Sigma$ is a covering of a pattern $\mathcal{P}$, the support of $\mathcal{C}$ denoted as $\mathrm{Sup}(\mathcal{C})$, by letters of $\Sigma$. A *subcoloring* $\mathcal{C}'$ of a coloring $\mathcal{C}$ is a restriction of this coloring, formally $\mathcal{C}' = \mathcal{C}_{|\mathrm{Sup}(\mathcal{C}')}$. The *translation* $u \cdot \mathcal{C}$ of a coloring $\mathcal{C}$ by a vector $u \in \mathbb{Z}^2$ is the coloring with support $\mathrm{Sup}(\mathcal{C}) + u$ satisfying $u \cdot \mathcal{C}(z + u) = \mathcal{C}(z)$ for all $z \in \mathrm{Sup}(\mathcal{C})$. A coloring $\mathcal{C}$ *occurs* in a coloring $\mathcal{C}'$, denoted as $\mathcal{C} \prec \mathcal{C}'$, if a translation of $\mathcal{C}$ is a subcoloring of $\mathcal{C}'$.

A coloring $\mathcal{C}$ is *periodic*, with period vector $p \in \mathbb{Z}^2$, if, for all $z \in \mathrm{Sup}(\mathcal{C})$, if $z + p \in \mathrm{Sup}(\mathcal{C})$ then $\mathcal{C}(z+p) = \mathcal{C}(z)$. An *aperiodic* coloring is a coloring admiting no non-trivial period (*i.e.* other than the trivial period 0). A set of coloring is *aperiodic* if it is not empty and all its colorings are aperiodic.

Let $X$ be the set of colorings with support $\mathbb{Z}^2$. Endow $X$ with the product topology of the discrete topology on $\Sigma$. This topology is compatible with the metric $d$ defined for all colorings $\mathcal{C}, \mathcal{C}' \in X$ by $d(\mathcal{C}, \mathcal{C}') = 2^{-\min\{|z|,\, \mathcal{C}(z) \neq \mathcal{C}'(z)\}}$. Such topology is compact and perfect. A subset of $X$ both topologically closed and closed by translations is a *subshift* from symbolic dynamics [9].

A two-by-two *substitution system* is a pair $(\Sigma, s)$ where $\Sigma$ is a finite alphabet and $s : \Sigma \to \Sigma^{\boxplus}$ is called the *substitution rule*. The local rule $s$ is extended to a *global rule* $S : \Sigma^{\mathcal{P}} \to \Sigma^{\square(\mathcal{P})}$ mapping colorings into colorings by:

$$\forall z \in \mathcal{P}, \forall c \in \boxplus, \quad S(\mathcal{C})(2z + c) = s(\mathcal{C}(z))(c) \quad .$$

The restriction of the global rule to $X$ is a continuous map. The global rule weakly commutes with translations: $S(u \cdot \mathcal{C}) = 2u \cdot S(\mathcal{C})$ for all vector $u \in \mathbb{Z}^2$ and all coloring $\mathcal{C}$. The *i-level* image of a letter $a \in \Sigma$ by $s$ is the coloring $S^i(a)$ with support $\boxplus^i$.

*Example 1.* Figure 1 depicts a variation on the classical chair two-by-two substitution. This substitution will reappear later in this paper.                    $\square$

The literature provides several different methods to extend substitutions to colorings of the whole plane. A classical one is to consider the set $X_s$ of colorings

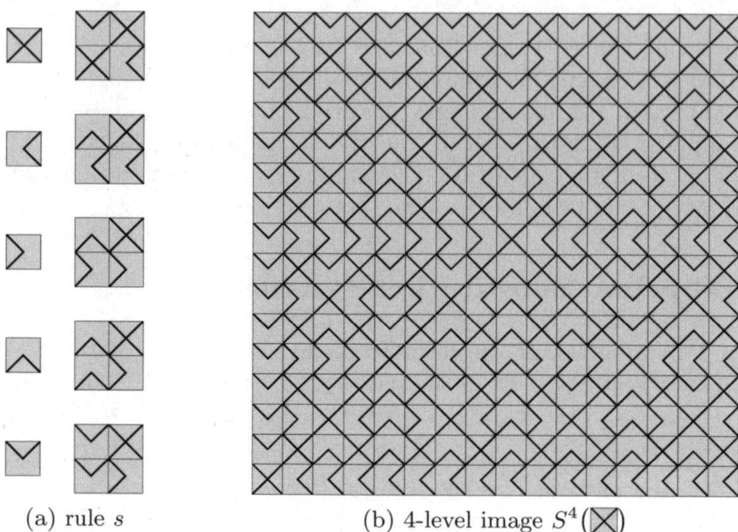

(a) rule $s$        (b) 4-level image $S^4(\boxtimes)$

**Fig. 1.** A sample two-by-two substitution with 5 letters

such that each of their finite subcolorings $\mathcal{C}$ occurs at some level $i$, that is there exists $a \in \Sigma$ such that $\mathcal{C} \prec S^i(a)$.

In this paper[1], we prefer to take a more dynamical point of view by considering the limit set $\Lambda_S$, the intersection $\bigcap_{n \in \mathbb{N}} \Lambda_S^n$ of a decreasing sequence of nonempty subshifts. Let $\Lambda_S^0 = X$ and $\Lambda_S^{n+1} = \{u \cdot S(\mathcal{C}) \mid \mathcal{C} \in \Lambda_s^n, u \in \boxplus\}$, for all $n \in \mathbb{N}$. As $S$ weakly commutes with translations, $\Lambda_S^n$ is precisely the closure by translation of the compact set $S^n(X)$. Notice that, depending on $s$, the sets $X_s$ and $\Lambda_S$ might be different.

*Example 2.* Consider the constant substitution on two letters $(\{a, b\}, \lambda x.\lambda z.x)$. The set $X_s$ contains 2 elements, but $\Lambda_S$ contains infinitely many elements (the closure by translation of 16 elementary elements). $\qquad\square$

An *history* for a coloring $\mathcal{C} \in X$ is a sequence $(\mathcal{C}_i, u_i) \in (X \times \boxplus)^{\mathbb{N}}$ such that $\mathcal{C}_0 = \mathcal{C}$ and $\mathcal{C}_i = u_i \cdot S(\mathcal{C}_{i+1})$, for all $i \in \mathbb{N}$.

**Proposition 1.** *The set $\Lambda_S$ is precisely the set of colorings admitting histories.*

*Proof.* Consider an history $(\mathcal{C}_i, u_i) \in (X \times \boxplus)^{\mathbb{N}}$. As $\mathcal{C}_0$ is a translation of $S^i(\mathcal{C}_i)$ for all $i \in \mathbb{N}$, straightforwardly $\mathcal{C}_0 \in \Lambda_S$. Conversely, let $\mathcal{C}$ be a coloring in $\Lambda_S$. By construction, one can find $(\mathcal{C}_i, u_i) \in (X \times \boxplus)^{\mathbb{N}}$ such that $\mathcal{C}_0 = \mathcal{C}$ and $\left(\sum_{j=0}^{i-1} 2^j u_j\right) \cdot S^i(\mathcal{C}_i) = \mathcal{C}$. By compacity of $X$, one can extract from the iterated images by $S$ of subcolorings of the family $(\mathcal{C}_i)$ a new family of colorings $(\mathcal{C}'_i)$ such that $\mathcal{C}'_i = u_i \cdot S(\mathcal{C}'_{i+1})$, for all $i \in \mathbb{N}$. $\qquad\blacksquare$

---

[1] In fact, even when claiming to construct tilings coding $X_s$, all the constructions we know about really code $\Lambda_S$ and can do it for all $s$, not only $s$ such that $X_s = \Lambda_S$.

Informally, to code elements of $\Lambda_S$ using only local rules, one can code a whole history and ensure that the validity of the history is locally checkable. One way to achieve that is to ensure that part of the *story* at each position is locally available. A story simply explains the value of at a particular position according to an history: it is the sequence of substitution rules applied to obtain the value at that particular position.

More formally, the *story* at position $z \in \mathbb{Z}^2$ for an history $(\mathcal{C}_i, u_i) \in (X \times \boxplus)^{\mathbb{N}}$ is the sequence $(a_i, v_i) \in (\Sigma \times \boxplus)^{\mathbb{N}}$ such that, for all $i \in \mathbb{N}$, $a_i = s(a_{i+1})(v_i)$ and $a_i = \mathcal{C}_i(z_i)$ where $z_i \in \mathbb{Z}^2$ is the only position such that $z$ is an element of the pattern $\mathcal{P}_i = \boxplus^i - \sum_{j=0}^{i-1} 2^j u_j - 2^i z_i$. Notice that the subcoloring of $\mathcal{C}$ of support $\mathcal{P}_i$ is a translation of $S^i(a_i)$.

Every story is computable from the story of any among two of its four neighbors, most of the time from any of them. Let $\left(a_i, \binom{x_i}{y_i}\right) \in (\Sigma \times \boxplus)^{\mathbb{N}}$ be a story at position $z \in \mathbb{Z}^2$, the story $(b_i, u_i) \in (\Sigma \times \boxplus)^{\mathbb{N}}$ at position $z + \binom{1}{0}$ can be constructed as follows. Let $k \in \mathbb{N}$ be the smallest $k$ such that $x_k = 0$. Let $u_k = \binom{1}{y_k}$. For all $i < k$, let $u_i = \binom{0}{y_i}$. For all $i > k$, let $u_i = \binom{x_i}{y_i}$ and $b_i = a_i$. For all $i \leqslant k$ let $b_i$ be such that $b_i = s(b_{i+1})(u_i)$. This procedure will always produce the story for position $z + \binom{1}{0}$, but when $k$ is not defined. Stories at positions $z + \left\{ -\binom{1}{0}, \binom{0}{1}, -\binom{0}{1} \right\}$ can be defined symmetrically.

**Proposition 2.** *Every history can be reconstructed from 1,2, or 4 of its stories.*

*Proof.* By construction, an history is completely defined by the set of all its stories. Consider any story $(a_i, \binom{x_i}{y_i})$ of a given history. Four different cases may occur depending on both sequences $(x_i)$ and $(y_i)$. If none of them is ultimately constant, the history can be reconstructed by reconstructing the stories of each position of the plane. If exactly one of them is ultimately constant, only half a plane of stories can be reconstructed and two different stories, with different ultimate constants, are needed. If both sequences are ultimately constant, only a quarter of the plane of stories can be reconstructed and four different stories, with different ultimate constants, are needed. ∎

A substitution is *aperiodic* if its limit set $\Lambda_S$ is aperiodic. A substitution is *unambiguous* if, for every coloring $\mathcal{C}$ from its limit set $\Lambda_S$, there exists a unique coloring $\mathcal{C}'$ and a unique translation $u \in \boxplus$ satisfying $\mathcal{C} = u \cdot S(\mathcal{C}')$. Notice that the injectivity of the local rule is not sufficient to enforce unambiguity. Every unambiguous substitution admits a unique history.

**Proposition 3.** *Every unambiguous substitution is aperiodic.*

*Proof.* Let $s$ be an unambiguous substitution. Let us assume that $s$ is not aperiodic. Let $p$ be the smallest non-trivial period of a coloring in the limit set of $s$ for the maximum norm. Let $\mathcal{C} \in \Lambda_S$ be $p$-periodic. Let $u$ and $\mathcal{C}'$ satisfy $\mathcal{C} = u \cdot S(\mathcal{C}')$. By construction, $(p + u) \cdot S(\mathcal{C}')$ is equal to $\mathcal{C}$. As $s$ is unambiguous, $p$ has to be a multiple of 2, $p = 2p'$, so that $(p + u) \cdot S(\mathcal{C}') = u \cdot S(p' \cdot \mathcal{C}')$ and $\mathcal{C}' = p' \cdot \mathcal{C}'$. Thus $\mathcal{C}' \in \Lambda_S$ is $p'$-periodic and $p'$ is smaller than $p$, contradicting our hypothesis. ∎

*Example 3.* The substitution on Figure 1 is unambiguous thus aperiodic.    □

A syntactical way to enforce unambiguity of a substitution $s$, as used in [3], is to ensure that it is injective and that one of the four projectors $s_i : a \mapsto s(a)(i)$ has an image disjoined from the images of the other three.

## 2    Tilings

A *domino relation* $\mathcal{R} \subseteq Y \times Y$ over a finite set $Y$ satisfies the domino property :

$$\forall a, b, c, d \in Y_?^4 \quad a\mathcal{R}c \wedge a\mathcal{R}d \wedge b\mathcal{R}d \rightarrow b\mathcal{R}c \tag{1}$$

The color set associated to a domino relation $\mathcal{R}$ is the set of equivalence classes of the equivalence relation $\sim_\mathcal{R}$ defined on $Y^2$ for all $a, b, c, d \in Y$ by $(a, c) \sim_\mathcal{R} (b, d)$ if $a\mathcal{R}c \wedge a\mathcal{R}d \wedge b\mathcal{R}d$. The right color of an element $a \in Y$ is the color $|a\rangle$ such that there exists $b$ satisfying $(a, b) \sim_\mathcal{R} |a\rangle$. Symmetrically, the left color of an element $b \in Y$ is the color $\langle b|$ satisfying $(a, b) \sim_\mathcal{R} \langle b|$. Straightforwardly, for all $a, b \in Y$, $a\mathcal{R}b$ if and only $|a\rangle = \langle b|$.

Tilings correspond to the extension of subshifts of finite type (SFT) [9] to $\mathbb{Z}^2$: colorings of the plane satisfying a finite set of local constraints. Several definitions of tiling constraints are possible. In this paper, we focus on so called Wang tiles, but instead of using the classical definition by colors, we use domino relations to simplify the discussions. A *tile set* $\tau$ is a triple $(T, \mathcal{H}, \mathcal{V})$ where $T$ is a finite alphabet of tiles, $\mathcal{H}$ and $\mathcal{V}$ are domino relations over $T$. A tile set is *degenerated* if two tiles $a, b \in T$ define the same quadruple of colors $(|a\rangle_\mathcal{V}, |a\rangle_\mathcal{H}, \langle a|_\mathcal{V}, \langle a|_\mathcal{H})$ and $(|b\rangle_\mathcal{V}, |b\rangle_\mathcal{H}, \langle b|_\mathcal{V}, \langle b|_\mathcal{H})$. A pair of tiles $(a, b) \in T^2$ *matches horizontally* if $a\mathcal{H}b$, *matches vertically* if $a\mathcal{V}b$. A *tiling* $\mathcal{T} \in X$ is a coloring satisfying the tiling constraints: for all $z \in \mathbb{Z}^2$, the pair $\left(\mathcal{T}(z), \mathcal{T}\left(z + \binom{1}{0}\right)\right)$ matches horizontally and the pair $\left(\mathcal{T}(z), \mathcal{T}\left(z + \binom{0}{1}\right)\right)$ matches vertically. The set of tilings $X_\tau$ of a tile set $\tau$ is a subshift (of finite type). A tile set is *aperiodic* if its set of tilings is aperiodic.

*Example 4.* The tile set $\tau_0 = \left(\boxplus, \left\{(u, v)\big| |v - u| = \binom{1}{0}\right\}, \left\{(u, v)\big| |v - u| = \binom{0}{1}\right\}\right)$ admits 4 tilings: the limit set of the simple substitution $\lambda x.\lambda z.z$.    □

A tile set $(T', \mathcal{H}', \mathcal{V}')$ *codes* a tile set $(T, \mathcal{H}, \mathcal{V})$, according to a *coding rule* $t : T \rightarrow T'^\boxplus$ if $t$ is injective and $X_{\tau'} = \{u \cdot t(\mathcal{C}) | \mathcal{C} \in X_\tau, u \in \boxplus\}$.

*Example 5.* A simple coding scheme is depicted on figure 2. Given a tile set $(T, \mathcal{H}, \mathcal{V})$, the coding tile set is constructed as a layered tile set : a synchronized product of several layers, upper layers being constrained by lower layers. Layer 1 is the tile set from example 4. Layer 2 is constrained according to layer 1 : on top of $\binom{0}{0}$ stack elements of $T$; on top of $\binom{1}{0}$ stack elements of $\mathcal{H}/ \sim_\mathcal{H}$; on top of $\binom{0}{1}$ stack elements of $\mathcal{V}/ \sim_\mathcal{V}$; on top of $\binom{1}{1}$ is an empty element. The matching rules for layer 2 are simple : a tile in $T$ is required to match horizontal and vertical colors of its four neighbor tiles thus propagating its colors to the next $\binom{0}{0}$ tile. The depicted coding rule $t$ maps every tile $a$ as follows. On layer 1, $t(a)$ is the

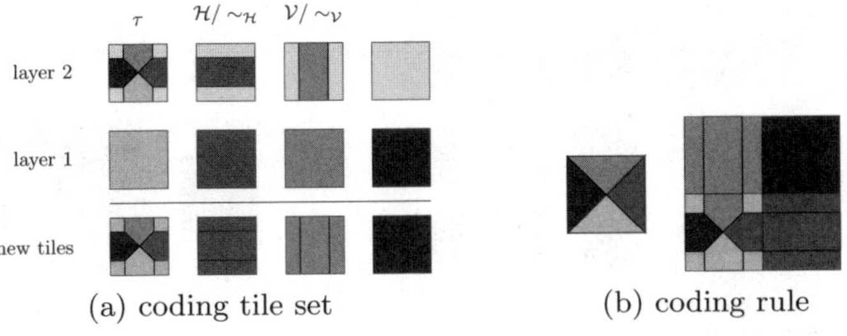



**Fig. 2.** A sample coding scheme

identity map. On layer 2, $t(a)$ puts $a$ on top of $\binom{0}{0}$, $|a\rangle_{\mathcal{H}}$ on top of $\binom{1}{0}$ and $|a\rangle_{\mathcal{V}}$ on top of $\binom{0}{1}$. $\qquad\qquad\Box$

A tile set $(T, \mathcal{H}, \mathcal{V})$ *codes* a substitution $s : T \to T^{\boxplus}$ if it codes itself according to the coding rule $s$.

**Proposition 4.** *A tile set both admitting a tiling and coding an unambiguous substitution is aperiodic.*

*Proof.* Let $(T, \mathcal{H}, \mathcal{V})$ be a tile set admitting a tiling and coding an unambiguous substitution $s : T \to T^{\boxplus}$. By construction, every tiling is the translated image of a tiling by $s$, thus $X_\tau \subseteq \Lambda_S$. As $X_\tau$ is not empty and $\Lambda_S$ is aperiodic, $X_\tau$ is aperiodic. $\qquad\blacksquare$

## 3   An Aperiodic Tile Set of 104 Tiles

To apply proposition 4, we construct a fixed point[2] of a coding scheme in the spirit of the one described in example 5 (this particular coding scheme will not help: the new tile set is always strictly larger than the original one).

One can refine the scheme, as depicted on Figure 3: as the coding scheme generates a layered tile set, one might forget about layer 1 on top of $\binom{0}{0}$ and code it all around on wires in $\binom{1}{0}$, $\binom{0}{1}$, and $\binom{1}{1}$. On top of $\binom{1}{1}$, 4 different corners can occur corresponding to the 4 different corners in $\tau_0$. On top of $\binom{0}{1}$, 4 possible pairs of wires propagates vertically crossing the $\mathcal{H}$-colors that propagates horizontally. The case of $\binom{1}{0}$ is symmetrical. The new matching rule on layer 2 still requires to match colors but also wires between $\binom{1}{1}$ and its neighbors.

*A priori*, this new coding scheme cannot be applied to every tile set. In a coded tiling, each tile is coded both by its layer 2 component on top of $\binom{0}{0}$ and its layer 1 component by a square wire around neighbors tiles. The only constraints between layer 1 and layer 2 values are checked on $\binom{0}{1}$ and $\binom{1}{0}$ restricting possible $\mathcal{H}$-colors and $\mathcal{V}$-colors for a given layer 1 tile. The coding scheme can only be

---

[2] The way we use fixed points with tilings in this paper, whereas sharing similarities with the approach in [4], is less sophisticated.

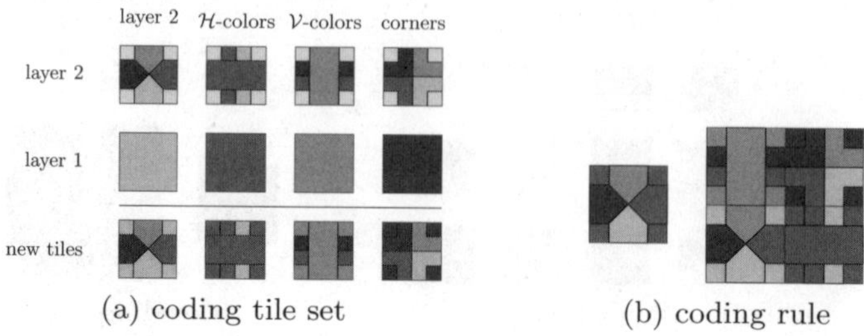

(a) coding tile set                     (b) coding rule

**Fig. 3.** A second coding scheme

applied to tile sets closed with respect to this property: if a layer 2 tile $a_2$ exists and its four $\mathcal{H}$-colors and $\mathcal{V}$-colors are compatible with a given layer 1 tile $a_1$, the tile $(a_1, a_2)$ occurs in the tile set.

The new coding scheme cannot be iterated: the coded tile set does not satisfy the closure hypothesis. This can be corrected by adding one bit of information on the wire pair edges of each tile to indicate on which side of the edge one can find the nearest corner: on top of $\binom{1}{1}$ corners are always inside; on top of $\binom{1}{0}$, the nearest corners are outside for both vertical edges; on top of $\binom{0}{1}$, the nearest corners are outside for both horizontal edges. The matching rule has to enforce that both sides of a wire pair edge agree on the direction. With these new bits of information, the coding scheme preserves the closure property and can be iterated. Moreover, it is not strictly increasing and admits a fixed point: a tile set $\tau$ of 104 tiles, the coding substitution of which is depicted on figure 4. The bits on the edges are considered as a third layer, an inside edge being represented

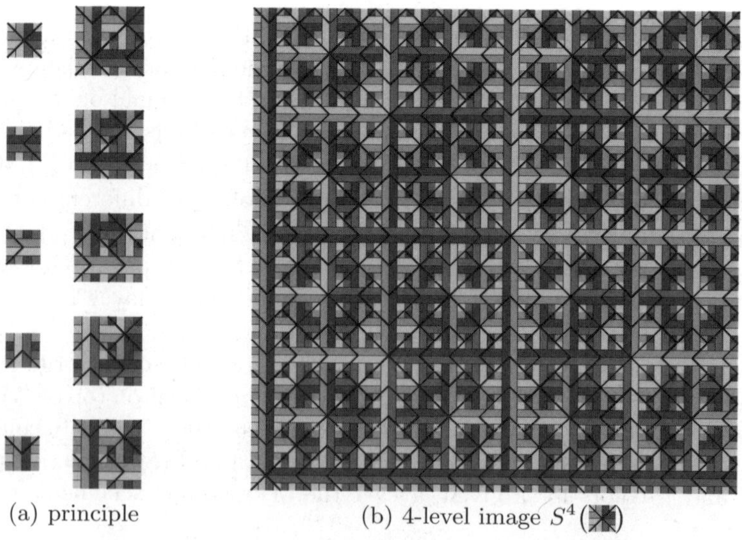

(a) principle                     (b) 4-level image $S^4(\text{⬛})$

**Fig. 4.** A 104 tiles aperiodic tile set $\tau$ coding an unambiguous substitution

by a V shape pointing the center of the tile: exactly five tiles occur on layer 3, the letters of figure 1.

Before proving the aperiodicity of $\tau$, let us first describe it more precisely. The tile set has three layers. Layer 1 is $\tau_0$. Layer 2 consists of X, H and V tiles (see below) transmitting pairs of wires from edge to edge, each wire being colored by an element of $\tau_0$ so that on edges, only pairs satisfying $\mathcal{H}_{\tau_0}$ and $\mathcal{V}_{\tau_0}$ are valid. The matching rule on layer 2 is to match the colors of the facing wires. Layer 3 consists of the five letters from figure 1. The matching rule on layer 3 is to agree on direction (exactly one V shape along each edge pointing in one of the two directions). Only the following synchronizations between layers occur inside $\tau$:

**8 X tiles.** layer 2 ⊞ with layer 1 $\binom{0}{0}$ or $\binom{1}{1}$ and layer 3 equal to ⨉.

**48 H tiles.** layer 2 ▤ with layer 1 $\binom{0}{0}$ or $\binom{1}{0}$ and layer 3 equal to ◁ or ▷.

**48 V tiles.** layer 2 ▥ with layer 1 $\binom{0}{0}$ or $\binom{0}{1}$ and layer 3 equal to △ or ▽.

There is only 48 tiles of type H and V (instead of 64) because the propagating direction of their layer 3 has to satisfy the wire color constraint: if one of the orthogonal wires is colored $\binom{1}{1}$, the direction has to point inside the $\binom{1}{1}$ boundary.

The associated substitution rule $s$ transforms a tile $a$ as follows. In position $\binom{0}{0}$, layer 1 is $\binom{0}{0}$ and layers 2 and 3 are the same as for $a$. In position $\binom{1}{1}$, there is an X tile with layer 1 $\binom{0}{0}$ and bottom-left wire color equal to the layer 1 of $a$. In position $\binom{1}{0}$, there is an H tile with layer 1 $\binom{1}{0}$: it propagates the wire colors of both its neighbors and points in the same direction as the right edge of the layer 3 of $a$. In position $\binom{0}{1}$, there is a V tile with layer 1 $\binom{0}{1}$: it propagates the wire colors of both its neighbors and points in the same direction as the top edge of the layer 3 of $a$. Notice that the image of a tile is always defined as the wire color constraint is always satisfied.

**Theorem 1.** *The tile set $\tau$ is aperiodic.*

*Proof.* The tile set $\tau$ admits at least one tiling. Consider the substitution $s$: the four inside edges of each image of a tile by the substitution satisfy the matching rules. Moreover, if two tiles $a$ and $b$ match either horizontally or vertically, $s(a)$ and $s(b)$ will still match in the same direction. Therefore, $\Lambda_S \cap X_\tau \neq \emptyset$.

The substitution $s$ is unambiguous: it is clearly injective and all the $s_i$ projectors have disjoined images.

The tile set $\tau$ codes $s$. Let $\mathcal{T}$ be a tiling of $\tau$. As layer 1 admits only 4 tilings, there exists $u \in \boxplus$ such that, for all $v \in \boxplus$, $u \cdot \mathcal{T}(v)$ has layer 1 $v$. Thus, $u \cdot \mathcal{T}$ has an X tile in $\binom{1}{1}$, an H tile in $\binom{1}{0}$, and a V tile in $\binom{0}{1}$. Thanks to the wire color constraint, $u \cdot \mathcal{T}_{|\boxplus}$ has to be the image of a tile by $s$. Repeating the argument on the $2\mathbb{Z}^2$ translations of $\mathcal{T}$, one conclude that $\mathcal{T}$ is the translated image of coloring by $s$. Moreover, this coloring is a valid tiling: as the information is propagated on layer 2 and 3 by H and V tiles, if two tile images $s(a)$ and $s(b)$ match, then $a$ and $b$ match. Therefore, as the image of a tiling by $s$ is also a tiling, $X_\tau = \{u \cdot s(\mathcal{C}) | \mathcal{C} \in X_\tau, u \in \boxplus\}$.

As $\tau$ admits a tiling and codes an unambiguous substitution, it is aperiodic. ∎

The tile set $\tau$ somehow uses its layers 1 and 2 to draw infinitely many infinite grids to code an history of the substitution of figure 1: the set of layer 3 of tilings is the limit set of the chair substitution. When comparing $\tau$ to aperiodic tile sets of the literature, the author found the following facts. The PhD dissertation of Berger [1] contains an aperiodic 104 tile set not found in the AMS memoir [2]. Berger's tile set also has three layers, the first two being isomorphic to the one used here... but the third layer is different leading to a more delicate proof and a different set of tilings! Let now $\tau'$ be the modified tile set where the colors $\binom{0}{0}$, $\binom{1}{0}$ and $\binom{0}{1}$ are merged in one unique color. The tile set $\tau'$ admits more tilings than just the repainted tilings of $\tau$, it has 56 tiles: it is the tile set of Robinson from [12], the aperiodicity proof of which is rather technical.

## 4   Enforcing Any Substitution

The tile set $\tau$ might be slightly modified to enforce the limit set of any substitution system $s'$: the idea is to use $\binom{1}{1}$ squares of all sizes to propagate the history of an element of $\Lambda_{S'}$. The transformed tile set $\tau(s')$ is constructed from $\tau$ by replacing $\boxplus$ with $\boxplus \times \Sigma$ on layer 1 and on the wires of layer 2. The matching rule is extended so that letters have to be equal on layer 1 on the four edges between $\binom{0}{0}$ and $\binom{1}{0}$, $\binom{0}{0}$ and $\binom{0}{1}$, $\binom{1}{0}$ and $\binom{1}{1}$, and $\binom{0}{1}$ and $\binom{1}{1}$. Then, both X tile and V tile wire colors are constrained. For any X tile, let $a$ be the letter on the $\binom{1}{1}$ wire and $b$ be the letter on layer 1, the constraint is $b = s'(a)(u)$ where $u$ depends on the position of the $\binom{1}{1}$ wire ($\binom{0}{0}$ for top-right, $\binom{1}{0}$ for bottom-right, $\binom{0}{1}$ for top-left, $\binom{1}{1}$ for bottom-left). For any V tile with two $\binom{1}{1}$ wire, let $a$ be the letter on the vertical $\binom{1}{1}$ wire and $b$ the letter on the horizontal one, the constraint is $b = s'(a)(u)$ where $u = \binom{x}{y}$ depends on the positions of both wires ($x = 0$ for right, $x = 1$ for left, $y = 0$ for bottom, $y = 1$ for top). Let $\pi$ map every tile of $\tau(s')$ to $s'(a)(u)$ where $a$ and $u$ are the letter and the value of $\boxplus$ on layer 1.

**Theorem 2.** *Let $s'$ be any substitution system. The tile set $\tau(s')$ enforces $s'$:* $\pi\left(X_{\tau(s')}\right) = \Lambda_{S'}.$

*Sketch of the proof.* Every tiling of $\tau(s')$ codes an history of $S'$ and every history of $S'$ can be encoded into a tiling of $\tau(s')$.

**Corollary 1 (Berger, 1964 [1]).** *The Domino Problem is undecidable.*

*Sketch of the proof.* Consider the following 6 letters substitution $\mathfrak{s}$:

$$T \mapsto \begin{matrix} t & t \\ t & t \end{matrix} \qquad V \mapsto \begin{matrix} v & v \\ v & v \end{matrix} \qquad H \mapsto \begin{matrix} h & h \\ h & h \end{matrix} \qquad t \mapsto \begin{matrix} T & V \\ H & T \end{matrix} \qquad v \mapsto \begin{matrix} T & V \\ H & V \end{matrix} \qquad h \mapsto \begin{matrix} T & V \\ H & H \end{matrix}$$

Consider the letter $T$ as a place for a tile ⊠, the letter $H$ as a horizontal color transmission path ▭, and the letter $V$ as a vertical color transmission path ▮▮. Every coloring $\mathcal{C}$ of the limit set $\Lambda_{\mathfrak{S}}$ containing letters $H, V, T$ has the following property: for all $i \in \mathbb{N}$, a $\boxplus^i$ square of $T$ letters, bordered horizontally by $V$ letters and vertically by $H$ letters, eventually spaced by $H$ and $V$ letters providing color transmission, occurs in $\mathcal{C}$. To prove the undecidability of the Domino Problem, recursively construct for every Turing machine a tile set with two layers: on layer 1 put $\tau(\mathfrak{s})$ with the additional constraint that it has only $H$, $V$, $T$ letters on layer 1 ; on layer 2 put tiles simulating Turing machine computations on $T$ tiles, so that the bottom left corner of each square (a $T$ connected to a $V$ on the left and the $H$ on the bottom) contains the initialization of the Turing computation. This tile set tiles the plane if and only if the machine does not halt starting from the empty string.

# References


[1] Berger, R.: The Undecidability of the Domino Problem, Ph.D. thesis, Harvard University (July 1964)

[2] Berger, R.: The undecidability of the domino problem. Memoirs American Mathematical Society 66 (1966)

[3] Durand, B., Levin, L., Shen, A.: Local rules and global order, or aperiodic tilings. Math. Intelligencer 27(1), 64–68 (2005)

[4] Durand, B., Romashchenko, A., Shen, A.: Fixed point and aperiodic tilings (preprint, 2007)

[5] Grünbaum, B., Shephard, G.C.: Tilings and patterns. A Series of Books in the Mathematical Sciences. W. H. Freeman and Company, New York (1989) (an introduction)

[6] Kahr, A.S., Moore, E.F., Wang, H.: Entscheidungsproblem reduced to the $\forall\exists\forall$ case. Proc. Natl. Acad. Science 48(3), 365–377 (1962)

[7] Kari, J.: The nilpotency problem of one-dimensional cellular automata. SIAM J. Comput. 21(3), 571–586 (1992)

[8] Kari, J.: The tiling problem revisited (extended abstract). In: Durand-Lose, J., Margenstern, M. (eds.) MCU 2007. LNCS, vol. 4664, pp. 72–79. Springer, Heidelberg (2007)

[9] Lind, D., Marcus, B.: An introduction to symbolic dynamics and coding. Cambridge University Press, Cambridge (1995)

[10] Mozes, S.: Tilings, substitution systems and dynamical systems generated by them. J. Analyse Math. 53, 139–186 (1989)

[11] Radin, C.: Miles of tiles. Student Mathematical Library, vol. 1. American Mathematical Society, Providence (1999)

[12] Robinson, R.M.: Undecidability and nonperiodicity for tilings of the plane. Inventiones Mathematicae 12, 177–209 (1971)

# The Relative Consistency of the Axiom of Choice — Mechanized Using Isabelle/ZF


Lawrence C. Paulson

Computer Laboratory, University of Cambridge, England
`LP15@cam.ac.uk`


Gödel [3] published a monograph in 1940 proving a highly significant theorem, namely that the axiom of choice (AC) and the generalized continuum hypothesis (GCH) are consistent with respect to the other axioms of set theory. This theorem addresses the first of Hilbert's famous list of unsolved problems in mathematics. I have mechanized this work [8] using Isabelle/ZF [5,6]. Obviously, the theorem's significance makes it a tempting challenge; the proof also has numerous interesting features. It is not a single formal assertion, as most theorems are. Gödel [3, p. 33] states it as follows, using $\Sigma$ to denote the axioms for set theory:

> What we shall prove is that, if a contradiction from the axiom of choice and the generalized continuum hypothesis were derived in $\Sigma$, it could be transformed into a contradiction obtained from the axioms of $\Sigma$ alone.

Gödel presents no other statement of this theorem. Neither does he introduce a theory of syntax suitable for reasoning about transformations on proofs, surely because he considers it to be unnecessary.

Gödel's work consists of several different results which, taken collectively, express the relative consistency of the axiom the choice. The concluding inference takes place at the meta-level and is not formalized. Standard proofs use meta-level reasoning extensively. Gödel writes [3, p. 34],

> However, the only purpose of these general metamathematical considerations is to show how the proofs for theorems of a certain kind can be accomplished by a general method. And, since applications to only a finite number of instances are necessary ..., the general metamathematical considerations could be left out entirely, if one took the trouble to carry out the proofs separately for any instance.

I decided to take the trouble, with the help of a mechanical theorem prover.

In brief, the proof goes as follows. We define a class model, called **L**, for the axioms of set theory.[1] **L** can be seen as containing just the sets that must exist because they can be defined by formulas. Since **L** is a proper class and not a set, we need to be careful about the notion of satisfaction. We cannot

---

[1] A class in ZF is simply a first-order formula in one variable. We typically endow classes with set notation, e.g. writing $a \in \mathbf{L}$ rather than $\mathbf{L}(a)$, but they exist only in the metalanguage.


A. Beckmann, C. Dimitracopoulos, and B. Löwe (Eds.): CiE 2008, LNCS 5028, pp. 486–490, 2008.
© Springer-Verlag Berlin Heidelberg 2008

talk within ZF about a formula being satisfied by a class model. Instead we transform the formula, restricting each quantifier to range over $\mathbf{L}$ instead of ranging over all sets. For example, we transform $\forall x\, \phi(x)$ into $\forall x\, [x \in \mathbf{L} \to \phi^{\mathbf{L}}(x)]$, where $\phi^{\mathbf{L}}(x)$ is the result of recursively transforming $\phi$. This transformation is called *relativization*. If the relativized formula is a theorem, then we say that the original formula is true in $\mathbf{L}$. We must prove that $\mathbf{L}$ satisfies (in that sense) all the axioms of set theory, and we must further prove that $\mathbf{L}$ satisfies the axiom of choice. Although we continue to work in first-order logic and ZF, relativizing all quantifiers to $\mathbf{L}$ has the effect of augmenting our axiom system with the axiom of choice.

- Provided we work entirely with formulas relativized to $\mathbf{L}$, we can prove all the consequences of the axioms of set theory including the axiom of choice.
- Because relativization is merely a syntactic transformation within first-order logic, every proof in $\mathbf{L}$ is also a proof in the original set theory, which lacks the axiom of choice.
- The relativization of **false** is **false**.

Thus, if we prove **false** using the axiom of choice, then we have also found a contradiction in the original set theory. This is a strong form of relative consistency. Gödel specifically notes that a contradiction in basic set theory "could actually be constructed" [3, p. 87] from a contradiction in $\mathbf{L}$. We merely have to express this contradiction using formulas relativized to $\mathbf{L}$. However, to show that a proof exists using relativized formulas seems to require a small amount of proof theory [8].

The main steps of the proof are as follows:

1. Define the class $\mathbf{L}$.
2. Prove that $\mathbf{L}$ satisfies the axioms of set theory. For ZF, the main difficulty is the axiom scheme of comprehension, also known as separation.
3. Prove that $\mathbf{L}$ satisfies the assertion "every set belongs to $\mathbf{L}$," which is traditionally written $\mathbf{V} = \mathbf{L}$.
4. Prove that $\mathbf{V} = \mathbf{L}$ implies AC.

Set-theoretic notation complicates the formalization. We are accustomed to writing unions, intersections, etc, with variable binding as in $\bigcup_{x \in A} B(x)$. But formally, the language of set theory consists of first-order logic plus the membership relation and equality. It has no terms other than individual variables. Before we can relativize an expression $E(x)$, we must translate it into a pure formula $\phi(x, y)$ such that $\phi(x, y) \leftrightarrow y = E(x)$. We must even translate the complicated expressions generated by Isabelle/ZF as it processes recursive definitions of sets and functions. In mathematical textbooks, relativization is done implicitly: all you have to do is put the superscript $\mathbf{L}$ on a term or formula. For example, the claim that $\mathbf{L}$ satisfies $\mathbf{V} = \mathbf{L}$ is trivially expressed by $(\mathbf{V} = \mathbf{L})^{\mathbf{L}}$. In the Isabelle/ZF proof, I have had to write out each relativized expression explicitly for each concept used in the construction of $\mathbf{L}$, in order to express $(\mathbf{V} = \mathbf{L})^{\mathbf{L}}$.

Proving that $\mathbf{L}$ satisfies $\mathbf{V} = \mathbf{L}$ is a key part of the proof, and despite first appearances, it is not trivial. It amounts to saying that the construction of $\mathbf{L}$ is idempotent. In other words, if starting in **L** we repeat the construction of **L**, then it will yield the whole of **L** and not some subclass of it. The underlying concept is called *absoluteness*, which expresses that a given notion or expression is the same in every transitive model of set theory.[2] Most constructions are absolute. For example, $A \subseteq B$ can only mean that each element of $A$ also belongs to $B$. The empty set, obviously, can only be a set containing no elements. If $A$ and $B$ are sets then their union can only be the set containing precisely the elements of those sets. Wellorderings and ordinals are absolute. Powersets however are not absolute, for there could be many subsets even of the natural numbers that cannot be shown to exist; they could exist in some models and not in others.

Skolem's paradox [4, p. 141] is a striking illustration that cardinality is not absolute. Set theorists naturally assume that models exist of the ZF axioms, from which it follows by the downward Löwenheim-Skolem theorem that there exists a countable model $M$ of ZF. The "paradox" is that this countable model "thinks" that it contains arbitrarily large cardinals. More precisely, if $\alpha$ is an uncountable cardinal according to $M$, then obviously $\alpha$ must be really be countable because $\alpha \subseteq M$. The point is that none of the bijections between $\alpha$ and $\omega$ belong to $M$; although the property of being a bijection is absolute, the property of being the set of all functions from $\alpha$ to $\omega$ is not. Neither is the property of being a cardinal.

Papers on formal verification often describe the work as "straightforward but tedious." The idempotence proof meets this description in the extreme. It has been necessary to relativize all the concepts of set theory, from the empty set to ordinals, recursive functions, etc. Then I had to prove that each of these concepts was absolute. In essence, this amounts to examining each definition to ensure that it uses only absolute constructions. Powersets are not absolute, but they appear surprisingly often, and then an alternative definition must be found and proved equivalent to the original one. The treatment of recursion was particularly difficult. I had to prove much of the foundations of recursion again from first principles. Having done this, we cannot merely note that all functions defined by recursion are absolute, as textbooks do. We must take each recursive definition, take it apart piece by piece, prove absoluteness for the pieces and feed those results into a theorem that will yield absoluteness for that particular function. I have done all of this with respect to an arbitrary transitive class model **M**, and later instantiated the proofs to **L**.

Further tedium arises from the need to internalize the notion of formula. A recursive datatype of formulas is defined, since it is needed to define **L**. Most of the relativized formulas mentioned in the previous paragraph have to be translated a second time into this datatype of formulas. Fortunately, some of the translations are done automatically.

Once the idempotence proof is done, we are justified in assuming **V** = **L**. I have separately proved that **V** = **L** implies the axiom of choice. This proof is straightforward both in concept and in execution. By transfinite induction, each level of the construction of **L** is well-ordered. The wellordering comes in an obvious way from the countability of the set of formulas. Gödel went on to prove

---

[2] **M** is *transitive* if $x \in$ **M** implies $x \subseteq$ **M**.

that $\mathbf{V} = \mathbf{L}$ implies the generalized continuum hypothesis. Although I omitted this step, it can probably be done with an acceptable amount of effort.

My formalization has two limitations. First, I am not able to prove that $\mathbf{L}$ satisfies the axiom scheme of comprehension. Although Isabelle/ZF handles schematic proofs easily, the proof of comprehension for the formula $\phi$ requires an instance of the reflection theorem for $\phi$. Each instance of reflection [7] involves recursion over the structure of $\phi$. Each instance of comprehension therefore has a different proof and must be proved separately. At the meta-level, of course, all of these proofs are instances of one algorithm, and they are generated by nearly identical proof scripts. Reasoning at the meta-level, we can see that all instances of the reflection theorem are available and that they imply all instances of the axiom of comprehension. But these meta-level inferences cannot be formalized in my framework. The inability to prove comprehension once and for all added further tedium to the project: in the absoluteness proofs, I had to keep track of each instance of comprehension that I used. Then, in order to instantiate these proofs to $\mathbf{L}$, I had to prove that each of those instances held in $\mathbf{L}$. There are about 35 such instances.

My formalization has another limitation. The proof that $\mathbf{L}$ satisfies $\mathbf{V} = \mathbf{L}$ cannot be combined with the proof that $\mathbf{V} = \mathbf{L}$ implies the axiom of choice in order to conclude that $\mathbf{L}$ satisfies the axiom of choice. The reason is that the two instances of $\mathbf{V} = \mathbf{L}$ are formalized very differently: one is relativized and the other is not. These problems arise because my work builds on the existing Isabelle/ZF formalization of set theory, comprising some 20 000 lines of proof scripts, rather than creating an new mechanized proof system specifically for Gödel's proof. Using Isabelle/ZF allows much of the work to be undertaken in the style of textbook proofs, and it enjoys the property that every proof involving relativized formulas (including one of **false**) is also a proof in ZF.

We could remedy both limitations by tackling Gödel's proof in a quite different way, working entirely in the metatheory. Unfortunately, experience shows that a formalized metatheory is convenient only for proving metatheorical results and not for proving, e.g., specific theorems of set theory. The formalization of the theorem statement would have to be done with care: the obvious $\mathrm{Con(ZF)} \rightarrow \mathrm{Con(ZF} + (\mathbf{V}{=}\mathbf{L}))$ sacrifices a crucial aspect of Gödel's result, namely that a contradiction in $\mathrm{ZF} + (\mathbf{V}{=}\mathbf{L})$ can be *effectively transformed* into a contradiction in ZF. Thus it appears necessary to introduce both proof theory and a model of computation, imposing two unenlightening technical layers onto Gödel's construction. I leave such issues as a challenge for the theorem-proving community.

A few other researchers have undertaken mechanized proof in set theory. Quaife [9] has generated proofs of hundreds of elementary results using the Otter resolution theorem prover, starting with a machine-oriented formalization of Bernays-Gödel set theory. BG set theory differs from ZF in that it replaces the axiom scheme of comprehension by a finite set of primitives that can be used to express particular comprehensions; these primitives are difficult to use, but they allow the axiom system to be finite. Building on Quaife's work, Belinfante has

implemented a Mathematica program for translating comprehensions into the
BG formalism; he submits these to Otter and thereby has proved facts about
the ordinals [2], for example. The Mizar system is based on Tarski-Grothendieck
set theory. It is designed for formalizing mathematics [1] and not merely for formalizing set theory; however, much set theory has been formalized using Mizar,
for example some elementary facts concerning large cardinal axioms [10].

In many respects, my formalization follows traditional ones. My development
is largely based on Kunen [4]. My use of native set theory (as embodied in
Isabelle/ZF) is very much in the spirit of those proofs, although it leads to the
difficulties mentioned above. A byproduct of the work is a general theory of
absoluteness for arbitrary class models of ZF. It could be used for other formal
investigations of inner models.

*Acknowledgements.* Krzysztof Grąbczewski devoted much effort to an earlier,
unsuccessful, attempt to formalize Gödel's proof. Isabelle is funded by the U.K.'s
Engineering and Physical Sciences Research Council, grants GR/M75440, GR/R
01156/01, GR/S57198/01, etc. Isabelle is developed in large part by Prof. Tobias
Nipkow's group at the Technical University of Munich.

# References


1. Bancerek, G., Rudnicki, P.: A compendium of continuous lattices in mizar.
   Journal of Automated Reasoning 29(3-4), 189–224 (2002)
2. Belinfante, J.G.F.: On computer-assisted proofs in ordinal number theory.
   Journal of Automated Reasoning 22(3), 341–378 (1999)
3. Gödel, K.: The consistency of the axiom of choice and of the generalized
   continuum hypothesis with the axioms of set theory. In: Feferman, S., et al. (eds.)
   Kurt Gödel: Collected Works, vol. II, pp. 33–101. Oxford University Press,
   Oxford (1990); First published in 1940 by Princeton University Press
4. Kunen, K.: Set Theory: An Introduction to Independence Proofs. North-Holland,
   Amsterdam (1980)
5. Paulson, L.C.: Set theory for verification: I. From foundations to functions.
   Journal of Automated Reasoning 11(3), 353–389 (1993)
6. Paulson, L.C.: Set theory for verification: II. Induction and recursion. Journal of
   Automated Reasoning 15(2), 167–215 (1995)
7. Paulson, L.C.: The reflection theorem: A study in meta-theoretic reasoning. In:
   Voronkov, A. (ed.) CADE 2002. LNCS (LNAI), vol. 2392, pp. 377–391. Springer,
   Heidelberg (2002)
8. Paulson, L.C.: The relative consistency of the axiom of choice — mechanized
   using Isabelle/ZF. LMS Journal of Computation and Mathematics 6, 198–248
   (2003), http://www.lms.ac.uk/jcm/6/lms2003-001/
9. Quaife, A.: Automated deduction in von Neumann-Bernays-Gödel set theory.
   Journal of Automated Reasoning 8(1), 91–147 (1992)
10. Urban, J.: Basic facts about inaccessible and measurable cardinals. Journal of
    Formalized Mathematics 12 (2000),
    http://mizar.uwb.edu.pl/JFM/Vol12/card_fil.html

# Upper Semilattices in Many-One Degrees

Sergei Podzorov[*]

Sobolev Institute of Mathematics
Akad. Koptyug pr., 4, 630090, Novosibirsk, Russian Federation
`podz@math.nsc.ru`

**Abstract.** The paper gives an overview over recent results of the author on various upper semilattices of many-one degrees. The local isomorphism type (i.e. the collection of isomorphism types of all principal ideals) of $m$-degrees belonging to any fixed class of arithmetical hierarchy is completely described. The description of the semilattices of simple, hypersimple and $\Delta_2^0$ $m$-degrees up to isomorphism is also given.

**Keywords:** Distributive Upper Semilattice, Many-One Degree, Lachlan Semilattice, Arithmetical Hierarchy, Computably Enumerable Set, Simple Set, Hypersimple Set, Immune Set, Hyperimmune Set.

## 1  Distributive Upper Semilattices

The notion of upper semilattice is well known. Let $\mathcal{L} = \langle L, \leqslant^{\mathcal{L}} \rangle$ be a partially ordered set. It is called an *upper semilattice* if for any $a, b \in L$ there exists $\sup\{a, b\}$ in $\mathcal{L}$. Similarly $\mathcal{L}$ is a *lower semilattice* if $\inf\{a, b\}$ exists for any $a, b \in L$. Partially ordered sets with both properties are called *lattices*.

It is usual to denote $\sup\{a, b\}$ as an union $a \cup b$ and $\inf\{a, b\}$ as an intersection $a \cap b$. So an upper semilattice $\mathcal{L}$ can be considered as a structure in the language with one binary operation and one can write $\mathcal{L} = \langle L, \leqslant^{\mathcal{L}}, \cup^{\mathcal{L}} \rangle$. Similarly if $\mathcal{L}$ is a lattice than $\mathcal{L} = \langle L, \leqslant^{\mathcal{L}}, \cup^{\mathcal{L}}, \cap^{\mathcal{L}} \rangle$.

A lattice $\mathcal{L}$ is called *distributive* if the distributivity identity $x \cap^{\mathcal{L}} (y \cup^{\mathcal{L}} z) = (x \cap^{\mathcal{L}} y) \cup^{\mathcal{L}} (x \cap^{\mathcal{L}} z)$ holds for any $x, y, z \in L$. It is well known [5] that the next conditions are equivalent:

1. $\mathcal{L}$ is distributive;
2. identity $x \cup^{\mathcal{L}} (y \cap^{\mathcal{L}} z) = (x \cup^{\mathcal{L}} y) \cap^{\mathcal{L}} (x \cup^{\mathcal{L}} z)$ holds for any $x, y, z \in L$;
3. $\mathcal{L}$ does not contain a sublattice isomorphic with either the pentagon or the diamond.

Upper semilattice is called *distributive* if for any $x, y \in L$ and $z \leqslant^{\mathcal{L}} x \cup^{\mathcal{L}} y$ there exist $z_x, z_y \in L$ such that $z_x \leqslant^{\mathcal{L}} x$, $z_y \leqslant^{\mathcal{L}} y$ and $z = z_x \cup^{\mathcal{L}} z_y$.

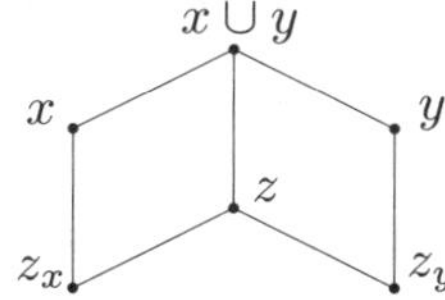


[*] Supported by RFBR grant 08-01-00336.

A. Beckmann, C. Dimitracopoulos, and B. Löwe (Eds.): CiE 2008, LNCS 5028, pp. 491–497, 2008.

One can easily prove [5, 4] that any lattice $\mathcal{L}$ is distributive as a lattice if and only if it distributive as an upper semilattice (for distributive lattices $z_x = z \cap^{\mathcal{L}} x$ and $z_y = z \cap^{\mathcal{L}} y$).

We are not going to consider lower semilattices. So by *semilattice* we often mean upper semilattice.

For any semilattice $\mathcal{L}$ the next four conditions are equivalent [5, 4]:

1. $\mathcal{L}$ is distributive;
2. any finite $F \subseteq L$ can be extended to a distributive finite subsemilattice $L_F \subseteq L$ (i.e. $F \subseteq L_F \subseteq L$, $L_F$ is finite and the union operation on $L_F$ is the same as in $\mathcal{L}$);
3. $\mathcal{L}$ is isomorphic to direct limit of some family of finite distributive semilattices (with corresponding homomorphisms);
4. the lattice $J(\mathcal{L})$ of ideals of $\mathcal{L}$ (w.r.t. inclusion) is distributive.

A semilattice is said to be *bounded* if it has the least element $\perp_{\mathcal{L}}$ and the greatest element $\top_{\mathcal{L}}$.

For any preordered set $\mathcal{A} = \langle A, \leqslant^{\mathcal{A}} \rangle$ one can define an associated poset $\widetilde{\mathcal{A}} = \langle \widetilde{A}, \leqslant^{\widetilde{A}} \rangle$ in a natural way. Elements of $\widetilde{A}$ arc classes of the equivalence relation $x \equiv^{\mathcal{A}} y \Leftrightarrow x \leqslant^{\mathcal{A}} y \,\&\, y \leqslant^{\mathcal{A}} x$. By $[x]_{\mathcal{A}}$ (or simply $[x]$ if $\mathcal{A}$ is clear from the context) we denote the class containing $x$. We also write $[x]_{\mathcal{A}} \leqslant^{\mathcal{A}} [y]_{\mathcal{A}}$ instead of $[x]_{\mathcal{A}} \leqslant^{\widetilde{A}} [y]_{\mathcal{A}}$, omitting tilde in the superscript.

A preordered set $\mathcal{P} = \langle P, \leqslant^{\mathcal{P}} \rangle$ is called a *presemilattice* (*prelattice*) if its associated poset $\widetilde{\mathcal{P}}$ is a semilattice (lattice). A presemilattice (prelattice) is said to be *distributive* if its associated semilattice (lattice) possesses the property of distributivity.

**Proposition 1.** *A partially ordered set* $\mathcal{L} = \langle L, \leqslant^{\mathcal{L}} \rangle$ *with (at most) countable nonempty universe* $L$ *is a bounded distributive semilattice if and only if there exists a sequence* $\{\mathcal{D}_i = \langle D_i, \leqslant_i \rangle\}_{i \in \mathbb{N}}$ *such that*

1. $D_0 \subseteq D_1 \subseteq \ldots$ *are finite subsets of* $\mathbb{N}$, $\{0, 1\} \subseteq D_0$ *and* $\bigcup_{i \in \mathbb{N}} D_i = \mathbb{N}$;
2. *for any* $i \in \mathbb{N}$ *the structure* $\mathcal{D}_i$ *is a distributive prelattice with the least element* $[0]_{\mathcal{D}_i}$ *and the greatest element* $[1]_{\mathcal{D}_i}$;
3. *for any* $i \in \mathbb{N}$ *and* $x, y \in D_i$ $x \leqslant_i y$ *implies* $x \leqslant_{i+1} y$; *the natural mapping from* $\widetilde{D}_i$ *to* $\widetilde{D}_{i+1}$ *given by the rule* $[x]_{\mathcal{D}_i} \mapsto [x]_{\mathcal{D}_{i+1}}$ *preserves the union operation;*
4. *the structure* $\langle \mathbb{N}, \leqslant_\omega \rangle$ *where* $x \leqslant_\omega y \Leftrightarrow (\exists i \in \mathbb{N})(x, y \in D_i \,\&\, x \leqslant_i y)$ *is a presemilattice and its associated semilattice is isomorphic to* $\mathcal{L}$.

*Proof.* Sufficiency is obvious because $\mathcal{L}$ is isomorphic to a direct limit of a family $\{\widetilde{\mathcal{D}}_i : i \in \mathbb{N}\}$ (with natural homomorphisms). For necessity fix the sequence $L_0 \subseteq L_1 \subseteq \ldots$ of finite sets such that $\bigcup_{i \in \mathbb{N}} L_i = L$, $\perp_{\mathcal{L}}, \top_{\mathcal{L}} \in L_0$ and for every $i \in \mathbb{N}$ $L_i$ is distributive semilattice with the order and the union operation from $\mathcal{L}$. Such a sequence could be found because $L$ is at most countable and every finite subset of $L$ could be extended to finite distributive subsemilattice. Every $L_i$ is a lattice because it is a finite semilattice with the least element.

Fix a surjective mapping $f$ from $\mathbb{N}$ onto $L$, such that $f(0) = \bot_{\mathcal{L}}$, $f(1) = \top_{\mathcal{L}}$ and $(f(x) = f(y)) \Rightarrow (x = y) \vee (f(x) = f(0))$. It only remains to define $D_i = f^{-1}(L_i \setminus \{\bot_{\mathcal{L}}\}) \cup \{n \leqslant i : f(n) = \bot_{\mathcal{L}}\}$ and $x \leqslant_i y \Leftrightarrow f(x) \leqslant^{\mathcal{L}} f(y)$ for any $x, y \in D_i$. $\qquad\square$

## 2  Many-One Reducibility and Degrees

Let $A, B \subseteq \mathbb{N}$. Say that $A$ is *m-reducible* to $B$ ($A \leqslant_m B$) if either $A$ is a computable set or there exists a total computable function $f$ such that $x \in A \Leftrightarrow f(x) \in B$ for all $x \in \mathbb{N}$.

This notion slightly differs from the classical notion of many-one reducibility [12] but the difference is inessential. The only cases when $A \leqslant_m B$ is in accordance to this definition but not in classical terms is when $A$ is computable and $B$ is equal $\varnothing$ or $\mathbb{N}$.

The $m$-reducibility relation is a preorder on $\mathcal{P}(\mathbb{N})$; elements of the associated order are called $m$-degrees. Many-one degrees with $m$-reducibility relation form a semilattice [12] which we denote by $\mathcal{L}_m = \langle L_m, \leqslant_m \rangle$. This semilattice is distributive [12, 4] and contains the least element which consists of all computable sets. There is a natural nontrivial automorphism of $\mathcal{L}_m$ defined by the rule $\deg_m(A) \mapsto \deg_m(\overline{A})$ where $\overline{A} = \mathbb{N} \setminus A$ is a complement of $A$.

Classes of arithmetical hierarchy are closed downward with respect to $m$-reducibility (i.e. if $B$ is in $\Sigma_n^0$, $\Pi_n^0$ or $\Delta_n^0$ for some $n \in \mathbb{N}$ and $A \leqslant_m B$ then $A$ belongs to the same class). So for each $\mathcal{C} \in \{\Sigma_n^0, \Pi_n^0, \Delta_n^0 : n \in \mathbb{N}\}$ one can easily define poset $\mathcal{L}_m^{\mathcal{C}}$ consisting of all $m$-degrees which are subsets of $\mathcal{C}$. Every such poset is an ideal in $\mathcal{L}_m$ so it is a distributive semilattice itself. For $\mathcal{C} = \Sigma_n^0$ or $\Pi_n^0$ this semilattice is bounded. It is also true that $\mathcal{L}_m^{\Sigma_n^0} \cong \mathcal{L}_m^{\Pi_n^0}$ due to natural automorphism of $\mathcal{L}_m$ mentioned above. The semilattice $\mathcal{L}_m^{\Sigma_1^0}$ consisting of all computably enumerable $m$-degrees is also denoted by $\mathcal{L}_m^{ce}$.

Say that $m$-degree is *immune* (*hyperimmune, hyperhyperimmune*) if it contains computable or immune (hyperimmune, hyperhyperimmune) set.

**Proposition 2.** *Immune (hyperimmune, hyperhyperimmune) m-degrees form an ideal in $\mathcal{L}_m$.*

*Proof.* If $A$ and $B$ are immune (hyperimmune, hyperhyperimmune) then the set $A \oplus B = \{2x : x \in A\} \cup \{2x + 1 : x \in B\}$ is also immune (hyperimmune, hyperhyperimmune). So if $m$-degrees **a** and **b** are immune (hyperimmune, hyperhyperimmune) then their union in $\mathcal{L}_m$ has the same property. It only remains to prove that these properties of degrees are closed downward with respect to $m$-reducibility.

Let $B$ is immune (hyperimmune, hyperhyperimmune) and $A \leqslant_m B$. If $A$ is computable then $\deg_m(A)$ is immune (hyperimmune, hyperhyperimmune) by definition. Suppose that $f$ is a total computable function such that $x \in A \Leftrightarrow f(x) \in B$ for any $x \in \mathbb{N}$. If $\mathrm{Range}(f) \cap B$ is finite then $A$ is computable and have the required property by definition. Suppose that $\mathrm{Range}(f) \cap B$ is infinite and fix a total computable function $g$ such that $\mathrm{Range}(f) = \mathrm{Range}(g)$ and $g$ is injective. It is easy to check that the set $g^{-1}(B)$ is immune (hyperimmune, hyperhyperimmune) and $m$-equivalent to $A$.                                         $\square$

A computably enumerable $m$-degree is said to be *simple* (*hypersimple, hyperhypersimple*) if it contains a computable or a simple (hypersimple, hyperhypersimple) set.

**Corollary 1.** *Simple (hypersimple, hyperhypersimple) $m$-degrees form an ideal in $\mathcal{L}_m^{ce}$.*

*Proof.* Immediately follows from proposition 2 and the automorphism of $\mathcal{L}_m$ mentioned above.                                         $\square$

Corollary 1 shows that we can introduce the distributive semilattices of simple, hypersimple and hyperhypersimple $m$-degrees. Denote them by $\mathcal{L}_m^s$, $\mathcal{L}_m^{hs}$ and $\mathcal{L}_m^{hhs}$ respectively.

# 3   Arithmetical Presentations of Upper Semilattices

Let $\mathcal{L}$ be a nonempty and at most countable semilattice. By a *presentation* of $\mathcal{L}$ we mean an arbitrary preorder $\leqslant_{\mathcal{L}}$ on $\mathbb{N}$ such that $\langle \mathbb{N}, \leqslant_{\mathcal{L}} \rangle$ is a presemilattice and the semilattice associated with it is isomorphic to $\mathcal{L}$.

For $n \in \mathbb{N}$ a presentation $\leqslant_{\mathcal{L}}$ is said to be $\Sigma_n^0$-*presentation* if the binary relation $\leqslant_{\mathcal{L}}$ belongs to the class $\Sigma_n^0$ of arithmetical hierarchy and there exists a total computable function $u$ which presents the union operation (i.e. $[u(x,y)] = [x] \cup [y]$ in the semilattice associated with $\langle \mathbb{N}, \leqslant_{\mathcal{L}} \rangle$ for any $x, y \in \mathbb{N}$).

Let us define a Lachlan presentation. A presentation $\leqslant_{\mathcal{L}}$ of a semilattice $\mathcal{L}$ is called $n$-*Lachlan* if there exists a sequence $\{\mathcal{D}_i = \langle D_i, \leqslant_i \rangle\}_{i \in \mathbb{N}}$ such that

1.  $D_0 \subseteq D_1 \subseteq \ldots$ are finite subsets of $\mathbb{N}$, $\{0,1\} \subseteq D_0$ and $\bigcup_{i \in \mathbb{N}} D_i = \mathbb{N}$;
2.  for any $i \in \mathbb{N}$ the structure $\mathcal{D}_i$ is a distributive prelattice with the least element $[0]_{\mathcal{D}_i}$ and the greatest element $[1]_{\mathcal{D}_i}$;
3.  for any $i \in \mathbb{N}$ and $x, y \in D_i$ $x \leqslant_i y$ implies $x \leqslant_{i+1} y$; the natural mapping from $\widetilde{D}_i$ to $\widetilde{D}_{i+1}$ given by the rule $[x]_{\mathcal{D}_i} \mapsto [x]_{\mathcal{D}_{i+1}}$ preserves the union operation;
4.  the sequence $\{D_i\}_{i \in \mathbb{N}}$ is strongly computable (i.e. one can compute a canonical index of $D_i$ uniformly on $i$);
5.  the ternary relation $x, y \in D_i$ & $x \leqslant_i y$ belongs to the class $\Pi_{n+2}^0$ of the arithmetical hierarchy;
6.  there exist functions $u_i, v_i : D_i^2 \to D_i$ which are uniformly computable in $i$ such that $u_i$ and $v_i$ present the union and the intersection operations on $\widetilde{D}_i$ respectively;
7.  $x \leqslant_{\mathcal{L}} y \Leftrightarrow (\exists i \in \mathbb{N})(x, y \in D_i \ \& \ x \leqslant_i y)$ for any $x, y \in \mathbb{N}$.

It is clear that every $n$-Lachlan presentation is a $\Sigma_{n+3}^0$-presentation. Proposition 1 implies that any semilattice with $n$-Lachlan presentation is distributive and bounded. The converse is not true but it can be proved that any distributive bounded semilattice which has a $\Sigma^0_{n+3}$-presentation also has $n$-Lachlan presentation. Moreover it can be proved that if $\leqslant^1_{\mathcal{L}}$ is a $\Sigma^0_{n+3}$-presentation of a distributive bounded semilattice $\mathcal{L}$ then there exist an $n$-Lachlan presentation $\leqslant^2_{\mathcal{L}}$ of $\mathcal{L}$ and a total computable function $f$ such that the mapping $[x] \mapsto [f(x)]$ is an isomorphism between the semilattices associated with $\langle \mathbb{N}, \leqslant^1_{\mathcal{L}} \rangle$ and $\langle \mathbb{N}, \leqslant^2_{\mathcal{L}} \rangle$ respectively [9].

Presentations which are 0-Lachlan are also called *Lachlan presentations*. Upper semilattice is called *Lachlan semilattice* if it has a Lachlan presentation. So results from [9] give us the next statement: an upper semilattice is Lachlan if and only if it is a bounded distributive semilattice with a $\Sigma^0_3$-presentation.

## 4    Principal Ideals in Many-One Degrees

Since $\mathcal{L}_m$ is distributive and has a minimal element every principal ideal in $\mathcal{L}_m$ is a distributive bounded semilattice. The converse is also true: Ershov showed [3, 4] that every at most countable bounded distributive semilattice is isomorphic to some principal ideal in $\mathcal{L}_m$.

Previously the same author investigated principal ideals generated by hyper-hypersimple $m$-degrees. It was proved [2] that $\mathcal{L}_m^{hhs}$ contains infinitely many minimal elements (above the least element) and atomless element, i.e. element with no minimal elements below (and also that this semilattice is not a lattice and have undecidable first-order theory).

In 1972 Lachlan [6] gave a full description of principal ideals in the computably enumerable $m$-degrees. He proved that an upper semilattice is isomorphic to a principal ideal in $\mathcal{L}_m^{ce}$ if and only if it is a Lachlan one (Lachlan semilattices were introduced in this paper). Due to results of [9] described in previous section it could be said that the principal ideals in $\mathcal{L}_m^{ce}$ are exactly the distributive bounded semilattices with $\Sigma^0_3$-presentations.

In [8] the author showed that the same is true for simple and hypersimple $m$-degrees. See further results on $\mathcal{L}_m^s$ and $\mathcal{L}_m^{hs}$ in the next section.

In 2007 generalization of Lachlan result to arithmetical $m$-degrees was obtained. The next theorem was proved by the author.

**Theorem 1 (see [11]).** *For any $n \in \mathbb{N}$ and an at most countable upper semilattice $\mathcal{L}$ the next four conditions are equivalent:*

1. *$\mathcal{L}$ is distributive bounded semilattice with a $\Sigma^0_{n+3}$-presentation;*

2. *$\mathcal{L}$ is isomorphic to a principal ideal in $\mathcal{L}_m^{\Sigma^0_{n+1}}$ generated by a coimmune or computable set from $\Sigma^0_{n+1}$;*

3. *$\mathcal{L}$ is isomorphic to a principal ideal in $\mathcal{L}_m^{\Pi^0_{n+1}}$ generated by an immune or computable set from $\Pi^0_{n+1}$;*

4. *$\mathcal{L}$ is isomorphic to a principal ideal in $\mathcal{L}_m^{\Delta^0_{n+2}}$ generated by an arbitrary set from $\Delta^0_{n+2}$.*

## 5    Universal Lachlan Semilattice and Ideals in $\mathcal{L}_m$

It is known that the greatest element of $\mathcal{L}_m^{ce}$ (formed by creative sets) is join-irreducible (i.e. is not the union of two elements which are strictly below it). So the structure $\mathcal{L}_m^{ce-} = \mathcal{L}_m^{ce} \setminus \{\top_{\mathcal{L}_m^{ce}}\}$ is distributive upper semilattice as well as $\mathcal{L}_m^{ce}$. It also true that $\top_{\mathcal{L}_m^{ce}}$ is not a minimal cover in $\mathcal{L}_m^{ce}$ that is for every $a < \top_{\mathcal{L}_m^{ce}}$ there exists $b \in \mathcal{L}_m^{ce}$ such that $a < b < \top_{\mathcal{L}_m^{ce}}$. So the semilattice $\mathcal{L}_m^{ce-}$ has no greatest element. The proof of both facts could be found in [4].

In 1978 Denisov [1] established that the natural $\Sigma_{n+3}^0$-presentation of $\mathcal{L}_m^{ce}$ (this natural presentation is $\leqslant_\pi$ defined by the formula $x \leqslant_\pi y \Leftrightarrow W_x \leqslant_m W_y$) possesses the special property of universality. Briefly this property means that for any Lachlan semilattice $\mathcal{L}$ any effective embedding of an ideal in $\mathcal{L}$ into $\mathcal{L}_m^{ce-}$ could be extended to an effective embedding of the whole semilattice $\mathcal{L}$ into $\mathcal{L}_m^{ce-}$ (see [1] for details). Denisov proved (see [1] and also [7, p. 534]) that this property defines $\mathcal{L}_m^{ce}$ up to isomorphism (more precisely any bounded distributive semilattice with universal $\Sigma_{n+3}^0$-presentation is isomorphic to $\mathcal{L}_m^{ce}$). Presently semilattices isomorphic to $\mathcal{L}_m^{ce}$ are called *universal Lachlan semilattices*.

Some of the semilattices introduced in section 2 coincide with the universal Lachlan semilattice with the greatest element excluded. They are listed in the next theorem proved by the author.

**Theorem 2 (see [10]).** *Upper semilattices* $\mathcal{L}_m^{hs}$, $\mathcal{L}_m^{s}$, $\mathcal{L}_m^{ce-}$ *and* $\mathcal{L}_m^{\Delta_2^0}$ *are isomorphic.*

The result on isomorphism between $\mathcal{L}_m^{ce-}$ and $\mathcal{L}_m^{\Delta_2^0}$ was announced by Denisov in 1978 (see [1] and also [7, p. 735]) but the proof has never been published before [10].

Notice that these semilattices are not only isomorphic to each other but also form the chain of ideals: $\mathcal{L}_m^{hs} \lhd \mathcal{L}_m^{s} \lhd \mathcal{L}_m^{ce-} \lhd \mathcal{L}_m^{\Delta_2^0}$. For the semilattice $\mathcal{L}_m^{hhs}$ of hyperhypersimple $m$-degrees problems solved for the semilattice of hypersimple $m$-degrees are still open. It is not known whether $\mathcal{L}_m^{hhs}$ is isomorphic to $\mathcal{L}_m^{ce-}$ and even not known whether any Lachlan semilattice could be embedded in $\mathcal{L}_m^{hhs}$ as an ideal.

## References


[1] Denisov, S.D.: Structure of the upper semilattice of recursively enumerable $m$-degrees and related questions. I. Algebra and Logic 17, 418–443 (1978)

[2] Ershov, Y.L.: Hyperhypersimple $m$-degrees. Algebra i Logika 8, 523–552 [in Russian] (1969)

[3] Ershov, Y.L.: The upper semilattice of numerations of a finite set. Algebra and Logic 14, 159–175 (1975)

[4] Ershov, Y.L.: Numbering theory. Nauka, Moskow [in Russian] (1977)

[5] Grätzer, G.: Generall lattice theory. Birkhäuser Verlag, Basel (1998)

[6] Lachlan, A.: Recursively enumerable many-one degrees. Algebra and Logic 11, 186–202 (1972)

[7] Odifreddi, P.: Classical recursion theory, vol. II. Elsevier, Amsterdam (1999)

[8] Podzorov, S.Y.: On the local structure of Rogers semilattices of $\Sigma^0_n$-computable numberings. Algebra and Logic 44, 82–94 (2005)

[9] Podzorov, S.Y.: Numbered distributive semilattices. Siberian Adv. in Math 17, 171–185 (2007)

[10] Podzorov, S.Y.: The universal Lachlan semilattice without the greatest element. Algebra and Logic 46, 163–187 (2007)

[11] Podzorov, S.Y.: Arithmetical $m$-degrees. Siberian Math. J. (submitted), http://www.nsu.ru/education/podzorov/Arithm.pdf

[12] Rogers, H.: Theory of recursive functions and effective computability. McGraw-Hill Book Company, New York (1967)

# Union of Reducibility Candidates for Orthogonal Constructor Rewriting

Colin Riba

Projet Everest
INRIA Sophia Antipolis[*]
Colin.Riba@sophia.inria.fr

**Abstract.** We revisit Girard's reducibility candidates by proposing a general of the notion of neutral terms. They are the terms which do not interact with some contexts called elimination contexts. We apply this framework to constructor rewriting, and show that for orthogonal constructor rewriting, Girard's reducibility candidates are stable by union.

## 1   Introduction

The most flexible termination proof methods for various extensions of typed $\lambda$-calculi use type interpretations [2, 5, 6, 7, 16]. Among them we distinguish three families: Girard's reducibility candidates [10], Tait's saturated sets [20], and interpretations based on biorthogonality [11, 16]. An interesting way to compare different type interpretations is to study their stability by union. This is even a necessary property in some cases [2, 7, 21].

This paper concerns the extension of the simply-typed $\lambda$-calculus with constructor rewriting. We are not interested in termination criteria by themselves, but by the investigation of the closure properties of types interpretations that allow to *formulate* different termination criteria. We focus on Girard's reducibility candidates and their stability by union.

We give a generalization of the notion of neutral terms that allows to define Girard's reducibility candidates in a generic way. Neutral terms are the terms that do not interact with some contexts called elimination contexts. Terms which are not neutral are *observable* since they interact with some elimination contexts. We call them *values*. We instantiate this framework with constructor rewriting. In order to get interesting values, we use elimination contexts with *destructors* to eliminate the constructors.

Next, we study the question of stability by union. By instantiating the condition of [18], we show that reducibility candidates are stable by union for orthogonal constructor rewriting. The proof uses a result of [13] on the existence of external redexes for orthogonal Context-sensitive Conditional Expression Systems (CCERSs).

The paper is organized as follows. We present our notations in Sect. 2.

---

[*] 2004 route des Lucioles, BP 93, 06902 Sophia Antipolis Cedex, France.

A. Beckmann, C. Dimitracopoulos, and B. Löwe (Eds.): CiE 2008, LNCS 5028, pp. 498–510, 2008.
© Springer-Verlag Berlin Heidelberg 2008

Section 3 presents a general definition of reducibility families and type interpretations. We apply it by briefly discussing Tait's saturated sets for the pure $\lambda$-calculus and one possible extension to deal with rewriting.

Section 4 is devoted to Girard's reducibility candidates, of which we suggest a generalization in Sect. 4.1. We instantiate it in Sect. 4.2 to the framework of $\lambda$-calculus plus constructor rewriting.

We then discuss stability by union in Sect. 5. We first briefly present the key problems and known results. Section 5.1 recalls a necessary and sufficient condition for the stability by union of Girard's candidates. In Sect. 5.2, we show that this condition is met with *orthogonal* constructor rewriting.

We assume familiarity with typed $\lambda$-calculus [4], reducibility [9, 14] and rewriting [22]. Concerning CCERSs, we refer to [12]. The paper (except Sect. 5.2) is based on parts of the Phd thesis of the author [17] (in French).

## 2   Simply Typed $\lambda$-Calculus with Constructor Rewriting

Given a set $A$, $\vec{a}$ denotes a finite sequence of elements of $A$ of length $|\vec{a}|$.

*Terms and types.* A *signature* $\Sigma$ is a family of sets $(\Sigma_n)_{n \in \mathbb{N}}$ such that $\Sigma_n$ contains algebraic symbols of arity $n$. We consider $\lambda$-terms with uncurried symbols $f$ in a signature $\Sigma$ and variables $x \in \mathcal{X}$:

$$t, u \in \Lambda(\Sigma) \quad ::= \quad x \quad | \quad \lambda x.t \quad | \quad t\,u \quad | \quad f(t_1, \ldots, t_n)\,,$$

where $f \in \Sigma_n$. Let $\Lambda$ be the set of *pure $\lambda$-terms* $\Lambda(\emptyset)$. A *substitution* is a function $\sigma : \mathcal{X} \to \Lambda(\Sigma)$ of finite domain. The capture avoiding application of $\sigma$ to the term $t$ is written $t\sigma$ or $t[\sigma(x_1)/x_1, \ldots, \sigma(x_n)/x_n]$ if $\mathcal{D}om(\sigma) = \{x_1, \ldots, x_n\}$.

Given *base types* $B \in \mathcal{B}$, simple types are defined as usual:

$$T, U \in \mathcal{T}(\mathcal{B}) \quad ::= \quad B \quad | \quad U \Rightarrow T\,.$$

*Typing contexts* are functions $\Gamma$ of finite domain from variables to types, written $x_1 : T_1, \ldots, x_n : T_n$. Given a *type assignment* $\tau : \Pi_{n \in \mathbb{N}}.\Sigma_n \to \mathcal{T}(\mathcal{B})^{n+1}$, the *typing relation* $\Gamma \vdash_\tau t : T$ is inductively defined by the following rules:

$$(\text{Ax}) \quad \frac{}{\Gamma, x : T \vdash_\tau x : T}$$

$$(\text{Symb}) \quad \frac{\Gamma \vdash_\tau t_1 : T_1 \quad \ldots \quad \Gamma \vdash_\tau t_n : T_n}{\Gamma \vdash_\tau f(t_1, \ldots, t_n) : T} \quad \tau(f) = (T_1, \ldots, T_n, T)$$

$$(\Rightarrow\text{I}) \quad \frac{\Gamma, x : U \vdash_\tau t : T}{\Gamma \vdash_\tau \lambda x.t : U \Rightarrow T} \qquad (\Rightarrow\text{E}) \quad \frac{\Gamma \vdash_\tau t : U \Rightarrow T \quad \Gamma \vdash_\tau u : U}{\Gamma \vdash_\tau t\,u : T}$$

*Constructor rewriting.* Assume given a set $\mathcal{C} \subseteq \Sigma$ of *constructor symbols* $c$ of type $(\vec{T}, B)$ with $B \in \mathcal{B}$. For normalization, $B$ must occur only at positive positions in $\vec{T}$ [15]. As we are not interested in strong normalization *conditions*, we do not care of this restriction here.

A *constructor rewrite system* (or *rewrite system*) on $\mathcal{C}$ is a set $\mathcal{R}$ of rewrite rules $\mathtt{f}(\vec{l}) \mapsto_{\mathcal{R}} r$ such that $r \in \Lambda(\Sigma)$, $FV(r) \subseteq FV(\mathtt{f}(\vec{l}))$, $\mathtt{f} \in \Sigma \backslash \mathcal{C}$ (defined symbols are not constructors) and $\vec{l}$ are terms of the grammar

$$p \quad ::= \quad x \quad | \quad c(p_1, \ldots, p_n) \,,$$

where $c \in \mathcal{C}$ (hence, $\vec{l}$ are patterns).

A rewrite system $\mathcal{R}$ is *typed* if for each rewrite rule $\mathtt{f}(\vec{l}) \mapsto_{\mathcal{R}} r$ with $\tau(\mathtt{f}) = (\vec{T}, T)$, there exists a (necessarily unique) context $\Gamma$ with $\mathcal{D}om(\Gamma) = FV(\mathtt{f}(\vec{l}))$ such that

$$\Gamma \vdash_\tau \mathtt{f}(\vec{l}) : T \qquad \text{and} \qquad \Gamma \vdash_\tau r : T \,.$$

**Example 2.1.** We consider the type $\mathtt{Nat}$ of Peano's numbers, with constructors $0 : \mathtt{Nat}$ and $S : (\mathtt{Nat}, \mathtt{Nat})$. The following system, defining addition, is a constructor rewrite system:

$$\mathtt{plus}(x, 0) \quad \mapsto \quad x \qquad\qquad \mathtt{plus}(x, S(y)) \quad \mapsto \quad \mathtt{plus}(S(x), y) \,.$$

*Reductions.* A rewrite relation is a binary relation $\to_R \subseteq (\Lambda(\Sigma) \setminus \mathcal{X}) \times \Lambda(\Sigma)$ which is stable by contexts and substitutions. We let $(t)_R =_{\text{def}} \{v \mid t \to_R v\}$ and say that a term $t$ is $R$-*reducible* (or *reducible*) if $(t)_R \neq \emptyset$. We define the *product extension* of $\to_R$ as $(t_1, \ldots, t_n) \to_R (u_1, \ldots, u_n)$ when there is $k \in \{1, \ldots, n\}$ such that $t_k \to_R u_k$ and $t_i = u_i$ for all $i \neq k$. We denote by $\mathcal{SN}_R$ the set of *strongly normalizing* terms for $\to_R$, which is the smallest set of terms such that

$$\forall t. \quad (\forall u. \quad t \to_R u \implies u \in \mathcal{SN}_R) \implies t \in \mathcal{SN}_R \,.$$

Given a constructor rewrite system $\mathcal{R}$, we let $\to_{\beta\mathcal{R}}$ be the smallest rewrite relation on $\Lambda(\Sigma)$ containing $\mapsto_{\mathcal{R}}$ and $\beta$-reduction: $(\lambda x.t)u \mapsto_\beta t[u/x]$.

## 3   Reducibility Families

In this section, we present a general notion of reducibility family and of type interpretation. We then briefly take a look at their instantiation to deal with the pure $\lambda$-calculus and with the combination of $\lambda$-calculus with rewriting.

**Definition 3.1.** *Let* $\to_R$ *be a rewrite relation on* $\Lambda(\Sigma)$.

  (i) *The* function space *is the function* $\_ \Rightarrow \_ : \mathcal{P}(\Lambda(\Sigma))^2 \to \mathcal{P}(\Lambda(\Sigma))$ *defined as*

$$A \Rightarrow B \quad =_{def} \quad \{t \mid \forall u. \quad u \in A \implies tu \in B\} \,.$$

  (ii) *A reducibility family for* $\to_R$ *is a set of sets* $\mathcal{R}ed \subseteq \{A \mid \mathcal{X} \subseteq A \subseteq \mathcal{SN}_R\}$ *which is closed by intersections and by the function space.*

  (iii) *A type interpretation in* $\mathcal{R}ed$ *is a map* $\llbracket \_ \rrbracket : \mathcal{T}(\mathcal{B}) \to \mathcal{R}ed$ *such that for all* $T, U \in \mathcal{T}(\mathcal{B})$ *we have* $\llbracket U \Rightarrow T \rrbracket = \llbracket U \rrbracket \Rightarrow \llbracket T \rrbracket$.

  (iv) *A type interpretation* $\llbracket \_ \rrbracket$ *is* adequate *if for all* $\Gamma$, $t$, $T$ *and* $\sigma$ *we have*

$$(\Gamma \vdash_\tau t : T \quad \wedge \quad \sigma \models_{\llbracket \_ \rrbracket} \Gamma) \implies t\sigma \in \llbracket T \rrbracket \,,$$

  *where* $\sigma \models_{\llbracket \_ \rrbracket} \Gamma$ *iff* $\sigma(x) \in \llbracket \Gamma(x) \rrbracket$ *for all* $x \in \mathcal{D}om(\Gamma)$.

*Pure $\lambda$-calculus.* Let $\mathfrak{Red}$ be a reducibility family and $[\![\_]\!] : \mathcal{T}(\mathcal{B}) \to \mathfrak{Red}$ be a type interpretation. Let us see, in the case of the pure $\lambda$-calculus, some sufficient conditions to ensure that $[\![\_]\!]$ is adequate. As usual, we reason by induction on $\Gamma \vdash t : T$ and by cases on the last applied typing rule.

We only have to check the rules (Ax), ($\Rightarrow$ E) and ($\Rightarrow$ I). The rule (Ax) is trivial while ($\Rightarrow$ E) is dealt with by definition of the function space $\_ \Rightarrow \_$. Concerning the rule ($\Rightarrow$ I), it is sufficient that for all $A \in \mathfrak{Red}$,

$$\forall t, u \in \Lambda. \quad (t[u/x] \in A \quad \wedge \quad u \in \mathcal{SN}_\beta) \quad \Longrightarrow \quad (\lambda x.t)u \in A . \qquad (1)$$

As for $\mathcal{X} \subseteq A \subseteq \mathcal{SN}_\beta$, condition (1) has to be preserved by $\_ \Rightarrow \_ : \mathfrak{Red}^2 \to \mathfrak{Red}$. We can conveniently formulate this by using *elimination contexts* [1]. For the pure $\lambda$-calculus, they are defined by the grammar

$$E[\ ] \in \mathcal{E}_\Rightarrow \quad ::= \quad [\ ] \quad | \quad E[\ ]t .$$

Then, we get the following clauses: for all $E[\ ] \in \mathcal{E}$, all $x \in \mathcal{X}$ and all $t, u \in \Lambda$,

$$E[\ ] \in \mathcal{SN}_\beta \quad \Longrightarrow \quad E[x] \in A , \qquad (2)$$
$$(E[t[u/x]] \in A \quad \wedge \quad u \in \mathcal{SN}_\beta) \quad \Longrightarrow \quad E[(\lambda x.t)u] \in A . \qquad (3)$$

The sets $A \subseteq \mathcal{SN}_\beta$ satisfying (2) and (3) are *Tait's saturated sets* [20]. The set of saturated sets, denoted by $\mathcal{SAT}_\beta$, forms a reducibility family.

*Remark 3.2.* Note that properties (2) and (3) use *call-by-name* evaluation contexts to prove strong normalization of the *full* reduction $\to_\beta$.

Showing that $\mathcal{SAT}_\beta$ is not empty amounts to showing that $\mathcal{SN}_\beta \in \mathcal{SAT}_\beta$. We must check properties (2) and (3) with $A = \mathcal{SN}_\beta$, which in this case are consequences of two important facts. First, a reduction step from a term of the form $E[(\lambda x.t)u]$ (resp. $E[x]$) occurs either in the elimination context $E[\ ]$ or in the term $(\lambda x.t)u$, but involves no interaction between them:

$$\forall v. \quad E[x] \to_\beta v \quad \Longrightarrow \quad (v = E'[x] \text{ with } E[\ ] \to_\beta E'[\ ]) \qquad (4)$$
$$\forall v. \ E[(\lambda x.t)u] \to_\beta v \quad \Longrightarrow \quad (v = E'[s] \text{ with } (E[\ ], (\lambda x.t)u) \to_\beta (E'[\ ], s)) \qquad (5)$$

Second, property (3) follows from (5) and the fact that $(\lambda x.t)u \in \mathcal{SN}_\beta$ as soon as $t[u/x] \in \mathcal{SN}_\beta$ and $u \in \mathcal{SN}_\beta$. This property holds in turn thanks to the *Weak Standardization Lemma*, which was used in [3] for extensions of the Calculus of Constructions. It is obvious for the pure $\lambda$-calculus.

**Lemma 3.3 (Weak Standardization).** *A reduct of a $\beta$-redex $(\lambda x.t)u$ is either* $t[u/x]$ *or a $\beta$-redex $(\lambda x.t')u'$ with $(t, u) \to_\beta (t', u')$.*

*$\lambda$-calculus with rewriting.* To deal with rewriting, we must consider the rule (SYMB). For constructors, we have to use specific interpretations of base types (eg. using inductive types, as in [6]). We concentrate on symbols $\mathbf{f} \in \Sigma \setminus \mathcal{C}$. Given $\mathbf{f}$ of type $(\vec{T}, T)$ and $\vec{t} \in [\![\vec{T}]\!]$, we have to make sure that $\mathbf{f}(\vec{t}) \in [\![T]\!]$. Sufficient

conditions for this are given by *termination criteria*, a subject that we do not treat in this paper (see for instance [6, 1, 2, 5, 7]).

Here, we are interested in the exploration of reducibility families that allow to *formulate* termination criteria. As for the $\lambda$-calculus, we can use a non-interaction property similar to (5):

$$\forall v. \quad E[\mathbf{f}(\vec{t})] \to_{\beta\mathcal{R}} v \quad \Longrightarrow \quad \left(v = E'[s] \text{ with } (E[\,], \mathbf{f}(\vec{t})) \to_{\beta\mathcal{R}} (E'[\,], s)\right) \ . \quad (6)$$

But rewrite systems do not satisfy in general the weak standardization lemma. Therefore, in order to get $\mathbf{f}(\vec{t}) \in \mathcal{SN}_{\beta\mathcal{R}}$, we need $v \in \mathcal{SN}_{\beta\mathcal{R}}$ *for all* $v$ such that $\mathbf{f}(\vec{t}) \to_{\beta\mathcal{R}} v$. This is subsumed by the clause

$$\left(\forall v. \quad E[\mathbf{f}(\vec{t})] \to_{\beta\mathcal{R}} v \quad \Longrightarrow \quad v \in A\right) \quad \Longrightarrow \quad E[\mathbf{f}(\vec{t})] \in A \ . \quad (7)$$

In this case, we also need saturated sets to be stable by reduction: if $t \in A$ and $t \to_{\beta\mathcal{R}} u$ then $u \in A$.

## 4    Neutral Terms and Reducibility Candidates

We now turn to Girard's reducibility candidates [10]. They form a reducibility family in which properties (3) and (7) can be formulated in a uniform and elegant way. This is due to *neutral terms*, that enjoy non-interaction properties such as (4), (5) and (6).

We first give a general formulation, and then apply it to constructor rewriting.

### 4.1    A General Formulation

We give a generalization of the original notion of neutral terms that allows to define reducibility candidates in a generic way. The key idea is that neutral terms are the terms that do not interact with *elimination contexts*. In the whole section, we assume given a rewrite relation $\to_R$.

Elimination contexts will be defined as a special case of *evaluation contexts*.

**Definition 4.1 (Evaluation Contexts).** *Let $[\,] \in \mathcal{X}$ be a distinguished variable. A set of evaluation contexts for $\to_R$ is a set $\mathcal{E}$ of terms $E[\,]$ which is*

(i) *stable by reduction: if $E[\,] \in \mathcal{E}$ and $E[\,] \to_R t$ then $t = F[\,] \in \mathcal{E}$;*
(ii) *stable by composition: if $E[\,] \in \mathcal{E}$ and $F[\,] \in \mathcal{E}$ then $E[F[\,]] \in \mathcal{E}$, where $E[t] =_{def} (E[\,])[t/[\,]]$.*

We now assume given a set $\mathcal{E}$ of evaluation contexts for $\to_R$.

**Definition 4.2 (Neutral Terms).** *A term $t$ is neutral for $\to_R$ in $\mathcal{E}$ if for all $E[\,] \in \mathcal{E}$,*

$$\forall v. \quad E[t] \to_R v \quad \Longrightarrow \quad (v = E'[t'] \quad with \quad (E[\,], t) \to_R (E'[\,], t')) \ .$$

*We denote by $\mathcal{N}_{R\mathcal{E}}$ the set of neutral terms for $\to_R$ in $\mathcal{E}$.*

The terms that are not neutral interact with evaluation contexts. They are therefore observable, and we think of them as being values.

**Definition 4.3 (Values).** *A value for $\to_R$ in $\mathcal{E}$ is a term which is not neutral. We denote by $\mathcal{V}_{R\mathcal{E}}$ the set of values for $\to_R$ in $\mathcal{E}$.*

**Example 4.4.** For the pure $\lambda$-calculus, taking $\mathcal{E}_\Rightarrow$ as evaluation contexts, the values are exactly the terms of the form $\lambda x.t$. Hence values are determined by the shape of evaluation contexts. Thanks to Weak Standardization (Lem. 3.3), we can use call-by-name evaluation contexts to prove the strong normalization of the full $\beta$-reduction (see also Rem. 3.2).

To build reducibility candidates, we are interested in neutral terms and evaluation contexts that enjoy some properties. This leads to the notion of *elimination contexts*.

**Definition 4.5 (Elimination Contexts).** *Let $\mathcal{E}$ be a set of evaluation contexts for $\to_R$. Then $\mathcal{E}$ is a set of* elimination contexts *for $\to_R$ if*

   *(i)*  *all variables are neutral:* $\mathcal{X} \subseteq \mathcal{N}_{R\mathcal{E}}$,
  *(ii)*  *if* $t \in \mathcal{N}_{R\mathcal{E}}$ *and* $E[\ ] \in \mathcal{E}$ *then* $E[t] \in \mathcal{N}_{R\mathcal{E}}$.

**Example 4.6.** For the pure $\lambda$-calculus, $\mathcal{E}_\Rightarrow$ is a set of elimination contexts.

We now define reducibility candidates in the usual way: our generalization regards neutral terms and their definition using elimination contexts. Assume that $\mathcal{E}$ is a set of elimination contexts for $\to_R$.

**Definition 4.7 (Reducibility Candidates).** *The set $\mathcal{CR}_{R\mathcal{E}}$ of reducibility candidates for $\to_R$ in $\mathcal{E}$ is the set of all $C \subseteq \mathcal{SN}_R$ such that*

$(\mathcal{CR}0)$  *if* $t \in C$ *and* $t \to_R u$ *then* $u \in C$,
$(\mathcal{CR}1)$  *if* $t \in \mathcal{N}_{R\mathcal{E}}$ *and* $\forall u.\ t \to_R u \implies u \in C$ *then* $t \in C$.

Note that $\mathcal{CR}_{R\mathcal{E}}$ is a complete lattice for $\subseteq$ whose top element is $\mathcal{SN}_R$ and whose greatest lower bounds are intersections. In order to verify that $\mathcal{X}$ is contained in any candidate, it is interesting to look at the least reducibility candidate. This is $\mathcal{HN}_{R\mathcal{E}}$, the set of *hereditary neutral terms*, defined as the smallest set of terms such that

$$\forall t \in \mathcal{N}_{R\mathcal{E}}.\ \ (\forall u.\ \ t \to_R u \ \implies\ u \in \mathcal{HN}_{R\mathcal{E}}) \ \implies\ t \in \mathcal{HN}_{R\mathcal{E}}\ .$$

Since variables are neutral terms in normal form (recall that by assumption $\to_R \subseteq (\Lambda(\Sigma) \setminus \mathcal{X}) \times \Lambda(\Sigma)$), we have $\mathcal{X} \subseteq \mathcal{HN}_{R\mathcal{E}} \subseteq C$ for every $C \in \mathcal{CR}_{R\mathcal{E}}$.

The non-interaction between neutral terms and elimination contexts has the following simple but fundamental consequence.

**Lemma 4.8.** *Let $t \in \mathcal{N}_{R\mathcal{E}}$ and $E[\ ] \in \mathcal{E} \cap \mathcal{SN}_R$. Then, for all $C \in \mathcal{CR}_{R\mathcal{E}}$,*

$$(\forall u.\ \ t \to_R u \ \implies\ E[u] \in C) \ \implies\ E[t] \in C\ .$$

*Proof.* First, since $t \in \mathcal{N}_{R\mathcal{E}}$ and $E[\ ] \in \mathcal{E}$, we have $E[t] \in \mathcal{N}_{R\mathcal{E}}$ by Def. 4.5.(ii). Hence, we only have to show that $(E[t])_R \subseteq C$.

We reason by induction on $E[\ ] \in \mathcal{SN}_R$. Let $v$ such that $E[t] \to_R v$. Since $t$ is neutral, we have $v = E'[t']$ with $(E[\ ], t) \to_R (E'[\ ], t')$, and there are two cases.

**Case of** $E[\ ] \to_R E'[\ ]$. We have $E'[\ ] \in \mathcal{E}$ by Def. 4.1.(i) and $E'[\ ] \in \mathcal{SN}_R$ since $E[\ ] \in \mathcal{SN}_R$. For all $u \in (t)_R$, since $E[u] \to_R E'[u]$ and $E[u] \in C$, we have $E'[u] \in C$ by $(\mathcal{CR}0)$. Hence, we can apply the induction hypothesis on $E'[\ ]$ and we conclude that $E'[t] \in C$.

**Case of** $t \to_R t'$. In this case, we have $E[t'] \in C$ by assumption. $\qquad\qquad\square$

We now give a sufficient condition for $\mathcal{CR}_{R\mathcal{E}}$ to be a reducibility family (in the sense of Def. 3.1) when $\mathcal{E}$ contains $[\ ]t$ for all $t \in \mathcal{SN}_R$. The key point is to show that $\_ \Rightarrow \_$ is a function from $\mathcal{CR}^2_{R\mathcal{E}}$ to $\mathcal{CR}_{R\mathcal{E}}$. We rely on Lem. 4.8.

**Lemma 4.9.** *If $\mathcal{E}$ contains $[\ ]t$ for all $t \in \mathcal{SN}_R$ and moreover $[\ ]t \in \mathcal{SN}_R$ for all $t \in \mathcal{SN}_R$, then $\mathcal{CR}_{R\mathcal{E}}$ is a reducibility family.*

*Proof.* It remains to show that $\_ \Rightarrow \_ : \mathcal{CR}^2_{R\mathcal{E}} \to \mathcal{CR}_{R\mathcal{E}}$. Let $A, B \in \mathcal{CR}_{R\mathcal{E}}$. First, we have $A \Rightarrow B \subseteq \mathcal{SN}_R$: for all $t \in A \Rightarrow B$, since $\mathcal{X} \subseteq A$ we have $tx \in B \subseteq \mathcal{SN}_R$, hence $t \in \mathcal{SN}_R$. Let us now check the clauses $(\mathcal{CR}0)$ and $(\mathcal{CR}1)$.

$(\mathcal{CR}0)$ Let $t \in A \Rightarrow B$ and $u \in (t)_R$. For all $a \in A$, we have $ta \in B$, hence $ua \in B$ by $(\mathcal{CR}0)$ applied to $B$. It follows that $u \in A \Rightarrow B$.

$(\mathcal{CR}1)$ Let $t \in \mathcal{N}_{R\mathcal{E}}$ such that $(t)_R \subseteq A \Rightarrow B$ and let $a \in A$. For all $u \in (t)_R$, we have $ua \in B$. Since $[\ ]a \in \mathcal{SN}_R$, it follows from Lem. 4.8 that $ta \in B$. We conclude that $t \in A \Rightarrow B$. $\qquad\square$

**Example 4.10.** For the pure $\lambda$-calculus, $\mathcal{CR}_{\beta\mathcal{E}_\Rightarrow}$ is the usual set of reducibility candidates. In particular, each $C \in \mathcal{CR}_{\beta\mathcal{E}_\Rightarrow}$ satisfies property (3).

## 4.2   Application to Constructor Rewriting

Let $\mathcal{R}$ be a constructor rewrite system on $\mathcal{C}$. If we use elimination contexts of the form $\mathcal{E}_\Rightarrow$, then the values are the terms of the form $\lambda x.t$.

However, we would like to build values from constructors. This is particularly useful with inductive types [6]. According to Def. 4.3, we have to make them observable. To this end, we introduce appropriate *destructors* in elimination contexts. To each $c \in \mathcal{C}$ of type $(\vec{T}, B)$ with $|\vec{T}| > 0$ and each $i \in \{1, \ldots, |\vec{T}|\}$, we associate a new unary *destructor symbol* $d_{c,i}$ defined by the rewrite rule

$$d_{c,i}(c(x_1, \ldots, x_n)) \quad \mapsto_{\mathcal{D}} \quad x_i \ .$$

Let $\mho$ be a new nullary symbol. For the elimination of a nullary constructor $c$, we use a new unary destructor $d_c$ defined by the rewrite rule

$$d_c(c) \quad \mapsto_{\mathcal{D}} \quad \mho \ .$$

**Lemma 4.11.** *Let $\mathcal{E}_{\Rightarrow C}$ be the set of terms defined by the grammar*

$$\mathsf{E}[\,] \in \mathcal{E}_{\Rightarrow C} \quad ::= \quad [\,] \quad | \quad \mathsf{E}[\,]\,\mathsf{t} \quad | \quad \mathsf{d}(\mathsf{E}[\,])\,,$$

*where $\mathsf{d}$ is a destructor of a constructor of $C$. Then,*

- *(i) $\mathcal{E}_{\Rightarrow C}$ is a set of evaluation contexts for $\to_{\beta\mathcal{RD}}$.*
- *(ii) The values in $\mathcal{E}_{\Rightarrow C}$ for $\to_{\beta\mathcal{RD}}$ are exactly the terms of the form*
    - *$\lambda\mathsf{x}.\mathsf{t}$; or*
    - *$\mathsf{c}(\vec{\mathsf{t}})$ with $\mathsf{c} \in C$.*
- *(iii) $\mathcal{E}_{\Rightarrow C}$ is a set of elimination contexts for $\to_{\beta\mathcal{RD}}$.*
- *(iv) Each $\mathsf{C} \in \mathbb{CR}_{\beta\mathcal{RD}\mathcal{E}_{\Rightarrow C}}$ satisfies properties (3) and (7).*

*Proof*

- (i) It is clear that $\mathcal{E}_{\Rightarrow C}$ is a set of evaluation contexts for $\to_{\beta\mathcal{RD}}$.
- (ii) It is clear that the terms of (ii) are values. We check that if $\mathsf{t}$ is a value, then it is in one of these forms. Assume now that $\mathsf{E}[\,] \in \mathcal{E}_{\Rightarrow C}$ is a minimal context that interacts with $\mathsf{t}$. Note that the top symbol of $\mathsf{E}[\,]$ is either an application or a destructor $\mathsf{d}$. We consider these two cases:
    **The top symbol of $\mathsf{E}[\,]$ is an application.** In this case, $\mathsf{E}[\,]$ is of the form $\mathsf{F}[\,]\,\mathsf{u}$. By minimality, $\mathsf{F}[\,] = [\,]$ and $\mathsf{t}$ is an abstraction.
    **The top symbol of $\mathsf{E}[\,]$ is a destructor.** In this case, $\mathsf{E}[\,]$ is of the form $\mathsf{d}(\mathsf{F}[\,])$. By minimality, $\mathsf{F}[\,] = [\,]$ and $\mathsf{t}$ is a constructor.
- (iii) The fact that $\mathcal{E}_{\Rightarrow C}$ is a set elimination contexts is a direct consequence of the shape of values (ii).
- (iv) From (ii) we know that terms of the form $(\lambda\mathsf{x}.\mathsf{t})\mathsf{u}$ and $\mathsf{f}(\vec{\mathsf{t}})$ with $\mathsf{f} \in \Sigma \setminus C$ are neutral. Properties (3) and (7) then follow from Lem. 4.8. $\quad\square$

**Example 4.12.** Consider the system presented at example 2.1. Its values are the terms of the form

$$\lambda\mathsf{x}.\mathsf{t} \qquad\qquad \mathsf{S}(\mathsf{t}) \qquad\qquad 0\,.$$

Indeed, we have

$$(\lambda\mathsf{x}.\mathsf{t})\,\mathsf{u} \to_\beta \mathsf{t}[\mathsf{u}/\mathsf{x}] \qquad\qquad \mathsf{d}_{\mathsf{s},1}(\mathsf{S}(\mathsf{t})) \to_{\mathcal{D}} \mathsf{t} \qquad\qquad \mathsf{d}_0(0) \to_{\mathcal{D}} \mathfrak{V}\,.$$

## 5   Stability by Union

A reducibility family $\mathfrak{Red}$ is *stable by union* if

$$\forall \mathcal{R}.\quad \mathcal{R} \subseteq \mathfrak{Red} \quad\Longrightarrow\quad \bigcup \mathcal{R} \in \mathfrak{Red}\,.$$

The main question on stability by union is the following: given a rewrite relation $\to_\mathsf{R}$, does there exists a reducibility family $\mathfrak{Red}$ for $\to_\mathsf{R}$ which is stable by union *and* leads to an adequate type interpretation?

For the pure $\lambda$-calculus, it is well-known that the answer is positive: Tait's saturated sets (presented in Sect. 3) are stable by union and lead to an adequate type interpretation. This has been exploited for instance in [2, 21].

The question becomes more difficult with rewriting. We have seen in Sect. 3 that rewrite systems do not satisfy in general the weak standardization lemma (Lem. 3.3), and that we need a reducibility family satisfying a clause like (7). But this is precisely what makes stability by union difficult. Assume given $\mathcal{R} \subseteq \mathcal{R}ed$ such that for all $v \in (E[f(\vec{t})])_{\beta\mathcal{R}}$, we have $v \in \bigcup \mathcal{R}$. Then, unless we find some $A \in \mathcal{R}$ such that $(E[f(\vec{t})])_{\beta\mathcal{R}} \subseteq A$, there is no reason to have $E[f(\vec{t})] \in \bigcup \mathcal{R}$.

Besides, using intersection and union types, we have shown in [19] that there are confluent typed rewrite systems for which no reducibility family that is stable by union leads to an adequate type interpretation.

However, we can in some cases obtain a reducibility family which is stable by union. In [18], we have given a necessary and sufficient condition for reducibility candidates to be stable by union; and in [19], we have given a necessary and sufficient condition for the closure by union of biorthogonals [11, 16, 8] to be reducibility candidates. The second condition is strictly more general than the first one.

We now recall the condition established in [18] for the stability by union of reducibility candidates, and then show that it is met with *orthogonal* constructor rewriting.

### 5.1    Reducibility Candidates

The study of stability by union of reducibility candidates of [18] carries over to our generalization of neutral terms and elimination contexts. The key observation is a characterization of the membership of a term to a candidate using a *weak observational preorder*.

**Definition 5.1.** *Let* $t \precsim_{\mathcal{N}} u$ *if and only if* $t, u \in \mathcal{SN}_R$ *and*

$$\forall v \in \mathcal{V}_{R\mathcal{E}}. \quad t \to_R^* v \quad \Longrightarrow \quad u \to_R^* v \,.$$

Every candidate $C \in \mathcal{CR}_{R\mathcal{E}}$ is a non-empty subset of $\mathcal{SN}_R$ which is downward-closed wrt. $\precsim_{\mathcal{N}}$. Reducibility candidates are stable by union exactly when the converse is also true.

**Theorem 5.2 ([18]).** *The following are equivalent:*

*(i) $\mathcal{CR}_{R\mathcal{E}}$ is stable by union;*

*(ii) $\mathcal{CR}_{R\mathcal{E}}$ is the set of all non-empty $C \subseteq \mathcal{SN}_R$ which are downward-closed wrt. the preorder $\precsim_{\mathcal{N}}$;*

*(iii) every strongly normalizable neutral term* $t$ *which is reducible has a reduct $u$ such that $t \precsim_{\mathcal{N}} u$. Such $u$ is a strong principal reduct[1] of $t$.*

*Proof.* For each $t \in \mathcal{SN}_R$, we let $\mathcal{CR}(t)$ be the smallest reducibility candidate containing $t$. Note that $\mathcal{CR}(t) = \{u \mid u \precsim_{\mathcal{N}} t\}$.

---

[1] Called "principal reduct" in [18].

**(i)** $\Longrightarrow$ **(ii).** Let $C \subseteq \mathcal{SN}_R$ be a non-empty set downward-closed wrt. $\lesssim_{\mathcal{N}}$. Since $\mathcal{CR}(t)$ is downward closed wrt. $\lesssim_{\mathcal{N}}$ for all $t \in \mathcal{SN}_R$, $C = \bigcup\{\mathcal{CR}(t) \mid t \in C\}$. Hence $C \in \mathcal{CR}$ because $\mathcal{CR}$ is stable by union.

**(ii)** $\Longrightarrow$ **(iii).** Let $t \in \mathcal{N} \cap \mathcal{SN}_R$ be reducible. For all $u \in (t)_R$, the set $\mathcal{CR}(u)$ is non-empty and downward-closed wrt. $\lesssim$. Therefore, the set $C$ of all $v$ such that $v \in \mathcal{CR}(u)$ for some $u \in (t)_R$ is non-empty and downward-closed wrt. $\lesssim_{\mathcal{N}}$. It follows that $C \in \mathcal{CR}$ and that $t \in C$ since $(t)_R \subseteq C$. Hence there is $u \in (t)_R$ such that $t \lesssim_{\mathcal{N}} u$.

**(iii)** $\Longrightarrow$ **(i).** Let $\emptyset \neq \mathcal{C} \subseteq \mathcal{CR}$. In order to show $\bigcup \mathcal{C} \in \mathcal{CR}$, the key-point is to show that if $t \in \mathcal{N}$ is such that $(t)_R \subseteq \bigcup \mathcal{C}$ then $t \in \bigcup \mathcal{C}$. If $(t)_R = \emptyset$ then $t \in C$ for all $C \in \mathcal{C}$ and we are done. Otherwise, we have $t \in \mathcal{SN}_R$ since $(t)_R \subseteq \bigcup \mathcal{C} \subseteq \mathcal{SN}_R$. Let $u$ be a strong principal reduct of $t$. There is $C \in \mathcal{C}$ such that $u \in C$, and since $t \lesssim_{\mathcal{N}} u$ and $C$ is downward-closed wrt. $\lesssim_{\mathcal{N}}$, we have $t \in C$, hence $t \in \bigcup \mathcal{C}$. $\qquad\square$

**Example 5.3 ([18, 21]).** For the pure $\lambda$-calculus, thanks to the weak standardization lemma (Lem. 3.3), $\mathcal{CR}_{\beta\mathcal{E}_{\Rightarrow}}$ is stable by union.

## 5.2   Application to Orthogonal Constructor Rewriting

Recall that a rewrite system $\mathcal{R}$ is *orthogonal* if it is left-linear and has no critical pairs [22]. For instance, the system of Ex. 2.1 is orthogonal.

In this section, we show that if $\mathcal{R}$ is an orthogonal constructor rewrite system, then $\mathcal{CR}_{\beta\mathcal{RDE}_{\Rightarrow c}}$ is stable by union. According to Thm. 5.2, this amounts to showing that every strongly normalizing reducible neutral term has a strong principal reduct. We prove it by using a general theorem on external redexes [13] (see also [12]). This result applies to the framework of orthogonal CCERSs, of which our higher-order rewrite systems $\beta \cup \mathcal{R} \cup \mathcal{D}$ are an instance.

To prove our result, we need some machinery to deal with the term structure. We use the following standard notations: a position in a term $t$ is a finite word on $\mathbb{N} \setminus \{0\}$, $\phi \cdot \psi$ denotes the concatenations of the words $\phi$ and $\psi$. Moreover, $t|_\phi$ is the subterm of $t$ at position $\phi$, and $t[u]_\phi$ is the term $t$ in which the subterm of $t$ at position $\phi$ is textually replaced by $u$.

The notion of descendant, which is standard in rewriting theory, is used to trace terms during reduction. We do not detail the definition which is quite technical and can be found in standard textbooks [22], but we provide an example.

**Example 5.4.** Consider the derivation

$$P: \quad (\lambda x.\mathtt{plus}(x, 0))\; 0\; \mathtt{plus}(y, S(z)) \quad \rightarrow_\beta \quad \mathtt{plus}(0, 0)\; \mathtt{plus}(y, S(z))$$

- The $\beta$-redex $(\lambda x.\mathtt{plus}(x, 0))\; 0$ has no descendant along $P$.
- The first occurrence of $0$ in $\mathtt{plus}(0, 0)$ is the only descendant along $P$ of the argument $0$ of the $\beta$-redex $(\lambda x.\mathtt{plus}(x, 0))\; 0$.
- $\mathtt{plus}(0, 0)$ is the only descendant of $\mathtt{plus}(x, 0)$ along $P$.
- $\mathtt{plus}(y, S(z))$ is the only descendant of $\mathtt{plus}(y, S(z))$ along $P$.

We use the notions of *redex-arguments* and of *external redexes* of [13, 12].

**Definition 5.5.** *A term* $u$ *is a* redex-argument *of* $t$ *if either*

   *(i)* $t$ *has a subterm of the form* $(\lambda x.t_1)t_2$ *and* $u$ *is a subterm of* $t_1$ *or* $t_2$, *or*
   *(ii)* $t$ *has a subterm of the form* $l\sigma$, *where* $l \mapsto_{\mathcal{R}} r$ *is a rewrite rule and* $\sigma$ *a substitution, and there is a variable* $x$ *of* $l$ *such that* $u$ *is a subterm of* $\sigma(x)$.

The key notion is that of *external redex*.

**Definition 5.6.** *A redex at position* $\phi$ *in a term* $t$ *is* external *if for any derivation* $P : t \to_{\beta\mathcal{RD}}^* v$, *no descendant of* $\phi$ *along* $P$ *appears inside redex-arguments.*

Hence every descendant of an external redex is external. Note that every external redex is outermost, but the converse is false.

Using external redexes, we have the following weak standardization lemma. We apply it in Lem. 5.8 to show that every neutral term which has an external redex has a strong principal reduct.

**Lemma 5.7 (Weak Standardization).** *Assume that* $\mathcal{R}$ *is orthogonal. Let* $t \to_{\beta\mathcal{RD}} u$ *by contracting an external redex of* $t$. *If* $t \to_{\beta\mathcal{RD}} v$ *by contracting a different redex, then there is* $w$ *such that* $u \to_{\beta\mathcal{RD}}^* w$ *and* $v \to_{\beta\mathcal{RD}} w$ *by contracting a descendant of* $t \to_{\beta\mathcal{RD}} u$ *(which is therefore external in* $v$).

*Proof.* Let $\phi$ and $\psi$ be the respective positions of the redexes contracted in $t \to_{\beta\mathcal{RD}} u$ and $t \to_{\beta\mathcal{RD}} v$. We show that there exists $w$ such that the redex contracted in $v \to_{\beta\mathcal{RD}} w$ is a descendant of $\phi$. It is an external redex of $v$ because $\phi$ is external in $t$.

Note that since $\phi$ is external in $t$, for every $\phi_1$, $\phi_2$ such that $\phi = \phi_1 \cdot \phi_2$, $\phi_2$ is external in $t|_{\phi_1}$: if a descendant $\phi_2'$ of $\phi_2$ appears in a redex argument of a reduct $u$ of $t|_{\phi_1}$, then $\phi_1 \cdot \phi_2'$, which is a descendant of $\phi$, appears in a redex argument of $t[u]_{\phi_1}$, which is a reduct of $t$.

We now reason by induction on $t$.

$t \in \mathcal{X}$. Not possible.

$t = \lambda x.t_1$. In this case, $u = \lambda x.u_1$ and $v = \lambda x.v_1$. Moreover we have $\phi = 1 \cdot \phi_1$, $\psi = 1 \cdot \psi_1$, $t_1 \to_{\beta\mathcal{RD}} u_1$ by contracting $\phi_1$ and $t_1 \to_{\beta\mathcal{RD}} v_1$ by contracting $\psi_1$. Since $\phi_1$ is external in $t_1$, by induction hypothesis there is $w_1$ such that $u_1 \to_{\beta\mathcal{RD}}^* w_1$ and $v_1 \to_{\beta\mathcal{RD}} w_1$ by contracting a descendant of $\phi_1$. It follows that $u \to_{\beta\mathcal{RD}}^* \lambda x.w_1$ and $v \to_{\beta\mathcal{RD}} \lambda x.w_1$ by contracting a descendant of $\phi$.

$t = t_1 t_2$ **with** $t_1$ **not an abstraction.** There are $i, j \in \{1, 2\}$ such that $\phi = i \cdot \phi_1$ and $\psi = j \cdot \psi_1$. Moreover, $u = u_1 u_2$ and $v = v_1 v_2$ with $t_i \to_{\beta\mathcal{RD}} u_i$, $t_j \to_{\beta\mathcal{RD}} v_j$, $u_j = t_j$ and $v_i = t_i$. There are two cases.

If $i = j$, then we reason as in the case of $t = \lambda x.t_1$.

**Otherwise** $i \neq j$. Let $w = w_1 w_2$ with $w_i = u_i$ and $w_j = v_j$. We have $u \to_{\beta\mathcal{RD}} w$ and $v \to_{\beta\mathcal{RD}} w$ by contracting a descendant of $\phi$.

$t = (\lambda x.t_1)t_2$. Since $\phi$ is external in $t$, $\phi$ is the root $\beta$-redex of $t$. Therefore $u = t_1[t_2/x]$ and $v = (\lambda x.t_1')t_2'$ with $(t_1, t_2) \to_{\beta\mathcal{RD}} (t_1', t_2')$. We are done by taking $w =_{\text{def}} t_1'[t_2'/x]$ since $(\lambda x.t_1')t_2'$ is a descendant of $(\lambda x.t_1)t_2$ and $t_1[t_2/x] \to_{\beta\mathcal{RD}}^* t_1'[t_2'/x]$.

$t = f(\vec{t})$ **and $t$ is not a $\mathcal{RD}$-redex.** We reason as in the case of $t_1\, t_2$ with $t_1$ not an abstraction.

$t = f(\vec{t})$ **and $t$ is a $\mathcal{RD}$-redex.** Since $\phi$ is external in $t$, it is the roof $\mathcal{RD}$-redex of $t$. Hence there is a rule $l \mapsto_{\mathcal{RD}} r$ and a substitution $\sigma$ such that $t = l\sigma$ and $u = r\sigma$. Because $l$ is linear, there is a substitution $\sigma'$ such that $\sigma \to_{\beta\mathcal{RD}} \sigma'$ and $v = l\sigma'$. We are done by taking $w =_{\text{def}} r\sigma'$ since $l\sigma'$ is a descendant of $l\sigma$ and $r\sigma \to^*_{\beta\mathcal{RD}} r\sigma'$. $\qquad\square$

**Lemma 5.8.** *Assume that $\mathcal{R}$ is orthogonal and let $t \in \mathcal{N}_{\beta\mathcal{RDE}\Rightarrow c}$. If $t \to^*_{\beta\mathcal{RD}} v$ with $v \in \mathcal{V}_{\beta\mathcal{RDE}\Rightarrow c}$, then for all $u$ such that $t \to_{\beta\mathcal{RD}} u$ by contracting an external redex of $t$, we have $u \to^*_{\beta\mathcal{RD}} v$.*

*Proof.* Let $u$ such that $t \to_{\beta\mathcal{RD}} u$ by contracting an external redex of $t$ and let $v \in \mathcal{V}_{\beta\mathcal{RDE}\Rightarrow c}$ such that $t \to^*_{\beta\mathcal{RD}} v$. Since $t$ is neutral, we have $t \to^+_{\beta\mathcal{RD}} v$. We reason by induction on the length of the derivation.

**Base Case.** Since $t$ is neutral, it follows from Lem. 4.11.(ii) that $t \to_{\beta\mathcal{RD}} v$ by contracting a top redex of $t$. Since every external redex is outermost, it follows that every external redex of $t$ is a root redex of $t$. By orthogonality, this is *the* redex contracted in $t \to_{\beta\mathcal{RD}} v$, and we have $u = v$.

**Induction Case.** Assume that $t \to_{\beta\mathcal{RD}} s \to^*_{\beta\mathcal{RD}} v$. If $s$ is a value, then reasoning as in the base case we have $s = u$, hence $u \to^*_{\beta\mathcal{RD}} v$. Otherwise $s$ is neutral. If $s$ is obtained from $t$ by contracting the same redex as in $t \to_{\beta\mathcal{RD}} u$, then $s = u$ and $u \to^*_{\beta\mathcal{RD}} v$. Otherwise, by weak standardization (Lem. 5.7), there is $w$ such that $u \to^*_{\beta\mathcal{RD}} w$ and $s \to_{\beta\mathcal{RD}} w$ by contracting an external redex of $s$. We then have $w \to^*_{\beta\mathcal{RD}} v$ by induction hypothesis, hence $u \to^*_{\beta\mathcal{RD}} v$. $\qquad\square$

If $t$ is neutral and $t \to_{\beta\mathcal{RD}} u$ by contracting an external redex of $t$, then we have $t \precsim_{\mathcal{N}} u$. Hence $u$ is a strong principal reduct of $t$. Then, by Thm. 5.2 $\mathcal{CR}_{\beta\mathcal{RDE}\Rightarrow c}$ is stable by union if all neutral terms have external redexes. It remains to show this last property, which follows from the next theorem, proved in [13]. We can apply it because the CCERS $\beta \cup \mathcal{R} \cup \mathcal{D}$ is orthogonal as soon as the rewrite system $\mathcal{R}$ is orthogonal.

**Theorem 5.9 ([13, 12]).** *If $\mathcal{R}$ is orthogonal then every reducible term has an external redex.*

**Theorem 5.10.** *If $\mathcal{R}$ is orthogonal then $\mathcal{CR}_{\beta\mathcal{RDE}\Rightarrow c}$ is stable by union.*

# References

[1] Abel, A.: Termination Checking with Types. RAIRO – Theoretical Informatics and Applications 38(4), 277–319 (2004); Special Issue (FICS 2003)

[2] Abel, A.: Semi-Continuous Sized Types and Termination. In: Ésik, Z. (ed.) CSL 2006. LNCS, vol. 4207, pp. 72–88. Springer, Heidelberg (2006)

[3] Altenkirch, T.: Constructions, Inductive Types and Strong Normalization. PhD thesis, University of Edinburgh (1993)

[4] Barendregt, H.P.: Lambda Calculi with Types. In: Abramsky, S., Gabbay, D.M., Maibaum, T.S.E. (eds.) Handbook of Logic in Computer Science, vol. 2, Oxford University Press, Oxford (1992)

[5] Barthe, G., Grégoire, B., Pastawski, F.: Type-Based Termination of Recursives Definitions in the Calculus of Inductive Constructions. In: Proceedings of LPAR 2006, pp. 257–271 (2006)

[6] Blanqui, F., Jouannaud, J.-P., Okada, M.: Inductive-Data-Types Systems. Theoretical Computer Science 271, 41–68 (2002)

[7] Blanqui, F., Riba, C.: Combining Typing and Size Constraints for Checking the Termination of Higher-Order Conditional Rewrite Systems. In: Hermann, M., Voronkov, A. (eds.) LPAR 2006. LNCS (LNAI), vol. 4246, pp. 105–119. Springer, Heidelberg (2006)

[8] Danos, V., Krivine, J.-L.: Disjunctive Tautologies as Synchronisation Schemes. In: Clote, P.G., Schwichtenberg, H. (eds.) CSL 2000. LNCS, vol. 1862, pp. 292–301. Springer, Heidelberg (2000)

[9] Gallier, J.H.: On Girard's Candidats de Reducibilité. In: Odifredi, P. (ed.) Logic and Computer Science. Academic Press, London (1989)

[10] Girard, J.-Y.: Interprétation Fonctionnelle et Élimination des Coupures de l'Arithmétique d'Ordre Supérieur. PhD thesis, Université Paris 7 (1972)

[11] Girard, J.-Y.: Linear Logic. Theoretical Computer Science 50, 1–102 (1987)

[12] Glauert, J., Kesner, D., Khasidashvili, Z.: Expression Reduction Systems and Extensions: An Overview. In: Middeldorp, A., van Oostrom, V., van Raamsdonk, F., de Vrijer, R. (eds.) Processes, Terms and Cycles: Steps on the Road to Infinity. LNCS, vol. 3838, pp. 496–553. Springer, Heidelberg (2005)

[13] Khasidashvili, Z., Ogawa, M., van Oostrom, V.: Perpetuality ans Uniform Normalization in Orthogonal Rewrite Systems. Information and Computation 164(1), 118–152 (2001)

[14] Krivine, J.-L.: Lambda-Calcul, Types et Modèles. Masson (1990)

[15] Mendler, N.P.: Recursive Types and Type Constraints in Second Order Lambda-Calculus. In: Proceedings of LiCS 1987, pp. 30–36. IEEE Computer Society, Los Alamitos (1987)

[16] Parigot, M.: Proofs of Strong Normalization for Second Order Classical Natural Deduction. Journal of Symbolic Logic 62(4), 1461–1479 (1997)

[17] Riba, C.: Definitions par Réécriture dans le λ-Calcul: Confluence, Réductibilité et Typage. PhD thesis, INPL (2007)

[18] Riba, C.: On the Stability by Union of Reducibility Candidates. In: Seidl, H. (ed.) FOSSACS 2007. LNCS, vol. 4423, pp. 317–331. Springer, Heidelberg (2007)

[19] Riba, C.: Strong Normalization as Safe Interaction. In: Proceedings of LiCS 2007, pp. 13–22. IEEE Computer Society, Los Alamitos (2007)

[20] Tait, W.W.: A Realizability Interpretation of the Theory of Species. In: Parikh, R. (ed.) Logic Colloquium. LNCS, vol. 453, pp. 240–251 (1975)

[21] Tatsuta, M.: Simple Saturated Sets for Disjunction and Second-Order Existential Quantification. In: Della Rocca, S.R. (ed.) TLCA 2007. LNCS, vol. 4583, pp. 366–380. Springer, Heidelberg (2007)

[22] Terese: Term Rewriting Systems. In: Bezem, M., Klop, J.W., de Vrijer, R.C. (eds.) Cambridge Tracts in Theoretical Computer Science. Cambridge University Press, Cambridge (2003)

# The Quantum Complexity of Markov Chain Monte Carlo*

Peter C. Richter

Laboratoire de Recherche en Informatique
Université de Paris-Sud XI
91405 Orsay, France
richterp@lri.fr

**Abstract.** Markov chain Monte Carlo (MCMC) is the widely-used classical method of random sampling from a probability distribution $\pi$ by simulating a Markov chain which "mixes" to $\pi$ at equilibrium. Despite the success quantum walks have been shown to have in speeding up random walk algorithms for search problems ("hitting") and simulated annealing, it remains to prove a general speedup theorem for MCMC sampling algorithms. We review the progress toward this end, in particular using decoherent quantum walks.

## 1 Introduction

Many important problems in computer science are solved most efficiently or elegantly by simulating a random walk on a Markov chain. Local search algorithms for graph connectivity and constraint satisfaction have complexity given by the *hitting time* of a Markov chain – the expected time for a random walk to reach a "marked" state. Markov chain Monte Carlo algorithms for approximating the permanent of a nonnegative matrix and the volume of a convex body have complexity tightly related to the *mixing time* of a Markov chain – the time at which a random walk reaches an almost $\pi$-distributed state, where $\pi$ is the Markov chain's stationary (equilibrium) distribution.

Quantum computing offers the possibility of implementing algorithms based on so-called *quantum walks*. Since such an algorithm must be more efficient than a classical randomized algorithm to be interesting, a central question is whether quantum walks have "hitting" and "mixing" times which are generically faster than their classical counterparts. In the case of hitting time, this question was answered in the affirmative by Szegedy [2], building on the work of Ambainis [3] who proved such a speedup on the Johnson graph and used it to show that the element distinctness problem has a quantum algorithm which is faster than any classical randomized algorithm. Despite much study of the mixing time case [4,5,6], it remains unknown whether a generic quantum speedup of the mixing

* This material is based upon work supported by the National Science Foundation under Grant No. 0523866 and is adapted in part from the author's PhD thesis at Rutgers University [1].

A. Beckmann, C. Dimitracopoulos, and B. Löwe (Eds.): CiE 2008, LNCS 5028, pp. 511–522, 2008.

time is achievable. Such a discovery would imply quantum speedups for a number of MCMC approximate sampling and counting algorithms. We survey the partial progress that has been achieved thus far.

## 2  Classical Markov Chains

The theory of Markov chains and random walks has a large literature. Here we present just enough background to discuss the *mixing time* and *hitting time* parameters relevant to algorithmic applications. See [7] for more information.

*Basic definitions.* Let $\mathcal{S}$ be a countable set of *states*. Unless otherwise stated, we take $N := |\mathcal{S}| < \infty$. An $\mathcal{S} \times \mathcal{S}$ matrix $P$ is a *transition probability matrix* if it is column-stochastic; i.e., if each of its columns is a distribution (probability vector), or equivalently, if it preserves $l_1$ norm of nonnegative vectors. Let $q$ be a distribution on $\mathcal{S}$. The sequence of random variables $(X_t)_{t \in [0..\infty)}$ distributed according to $P^t q$ is a *discrete-time Markov chain*. More frequently, we use the term *Markov chain* to refer to any valid transition probability matrix $P$, without specifying $q$. We can think of any Markov chain as a *random walk* on a graph or digraph. For example, the *simple random walk* on a $d$-regular graph $G = (V, E)$ is the Markov chain $P(y, x) = \frac{1}{d}$ if $(x, y) \in E$, $P(y, x) = 0$ if $(x, y) \notin E$.

A *stationary distribution* $\pi$ for a Markov chain $P$ is a probability vector satisfying $P\pi = \pi$. The Perron-Frobenius Theorem for nonnegative matrices guarantees the existence of a stationary distribution for any finite Markov chain. Moreover, uniqueness and strict positivity of the stationary distribution are guaranteed if $P$ is *irreducible*; i.e., if the digraph $G$ underlying $P$ is strongly connected. For example, the stationary distribution of the simple random walk on a regular graph is the uniform vector $u = [\frac{1}{N}, \ldots, \frac{1}{N}]^{\dagger}$.

*Reversible Markov chains.* Most MCMC applications involve so-called "reversible" Markov chains. Let $P$ be a Markov chain and $\pi > 0$ be a distribution satisfying the *detailed balance* condition $P(y, x)\pi(x) = P(x, y)\pi(y)$; or equivalently, suppose that the matrix

$$M(P) := \sqrt{R}^{-1} P \sqrt{R} \qquad \text{where } R := diag(\pi) \tag{1}$$

is symmetric. Then $P$ is *reversible* and $\pi$ is its stationary distribution. For example, a Markov chain $P$ that is *symmetric* (i.e., $P$ equals its transpose) is reversible with uniform stationary stationary distribution and satisfies $M(P) = P$.

A particularly famous example of a reversible Markov chain is the *Metropolis process*. Suppose we wish to sample from an arbitrary distribution $\pi > 0$ on a set $\mathcal{S}$ and are able to construct a single irreducible Markov chain $P$ on $\mathcal{S}$. When $\mathcal{S}$ is a set of combinatorial objects (e.g., microstates in statistical physics or data structures in computer science), one typically obtains $P$ by a local update rule (e.g., change a single spin or flip a single edge), called the *Glauber dynamics* in statistical physics. We obtain the Metropolis process $P'$ using $P$ as follows: at each state $x \in \mathcal{S}$, select an adjacent state $y$ with probability $P(y, x)$ and move to $y$ with probability $\min\{1, \frac{P(x,y)}{P(y,x)} \cdot \frac{\pi(y)}{\pi(x)}\}$; otherwise, stay put. Applications of the Metropolis process include approximating the permanent of a nonnegative matrix and the volume of a convex body.

*Mixing and hitting times.* How quickly, if at all, does an irreducible Markov chain converge to its stationary distribution $\pi$? Let us define the matrix $P^\infty :=$ $[\pi\pi\cdots\pi]$. Convergence $P^t \to P^\infty$ is guaranteed provided that $P$ is *aperiodic*; i.e., the lengths of all closed walks in the graph $G$ underlying $P$ must be mutually coprime. If $G$ is undirected, $P$ is aperiodic precisely when $G$ is non-bipartite. A Markov chain which is both irreducible and aperiodic is *ergodic*. The asymptotic behavior of an ergodic Markov chain is summarized by the *fundamental theorem of Markov chains*: If $P$ is an ergodic Markov chain, then it has a unique stationary distribution $\pi$, and $P^t \to P^\infty$ as $t \to \infty$.

The speed at which an ergodic Markov chain $P$ converges from an initial distribution $e_x$ concentrated at $x \in \mathcal{S}$ to its stationary distribution $\pi$ is captured by the *mixing time*

$$\tau_x := \min\{t : ||P^t e_x - \pi||_1 \leq 1/e\} \tag{2}$$

where $|| \cdot ||_1$ is the $l_1$ vector norm. The maximum mixing time over all initial states (or equivalently, distributions) is

$$\tau := \max_x \tau_x = \min\{t : ||P^t - P^\infty||_1 \leq 1/e\} \tag{3}$$

where $|| \cdot ||_1$ is the $l_1$ matrix norm. The parameter $1/e$ is somewhat arbitrary, in that we can change it to any $\epsilon > 0$ provided we multiply the mixing time by $O(\log 1/\epsilon)$. The mixing time of a reversible Markov chain with (real) eigenvalues $\{\lambda_k\}_{k=1}^N$ is closely related to its *spectral gap* $\delta := 1 - \lambda$, where $\lambda := \max\{\lambda_2, |\lambda_N|\}$ [8,9,10]:

**Theorem 1.** *Let $P$ be a reversible, ergodic Markov chain with stationary distribution $\pi$ and spectral gap $\delta$. Its mixing time satisfies (a) $\tau_x \leq \delta^{-1} (\log 1/\pi(x) + \log 2e)$, and (b) $\tau \geq \frac{1}{2}|\lambda|\delta^{-1}$.*

Suppose that some states $M \subseteq \mathcal{S}$ of an ergodic Markov chain $P$ are "marked," and we stop the random walk upon reaching a marked state. The *hitting time* is the expected time for this to occur. One can show [2]:

**Theorem 2.** *Let $P$ be a symmetric, ergodic Markov chain with uniform stationary distribution $u$ and spectral gap $\delta$. Let states $M \subseteq \mathcal{S}$ be "marked," where $|M| = \varepsilon N$. Then the hitting time $h_u$ is at most $1/(\delta\varepsilon)$.*

A somewhat different search method is simulated annealing; one can show that it too scales like $1/\delta$ [11].

## 3   Quantum Walk Constructions

Both discrete-time [12,3,2] and continuous-time [13,14,15] quantum walks have found application in quantum algorithms.

*Discrete-time quantum walks.* A *discrete-time quantum walk* is a unitary process $\{U^t|\psi\rangle\}$, or more frequently the unitary operator $U$ generating this process. Of particular relevance in algorithmic applications are those $U$ whose coefficients are derived naturally and efficiently from a corresponding classical Markov chain $P$ – e.g., with the same locality structure as $U$ [16]. This turns out to be possible only if we are willing to let $U$ operate on the Hilbert space generated by the state transitions (directed edges) of $P$ rather than the states themselves [17,18].[1] We describe the standard recipe for "quantizing" a Markov chain this way. For additional background on quantum walks, see the surveys of Ambainis [19], Kempe [20], and Kendon [21].

The *Grover walk* was invented by Watrous [12] as a quantization of the simple random walk on a regular graph to show that quantum logspace machines can simulate their classical counterparts. It has since found application in numerous quantum algorithms. The idea of the walk is to apply the Grover diffusion operator locally as a substitute for the classical state transitions. The Grover walk on a $d$-regular graph is given by

$$W := SC \tag{4}$$

where

$$S : |x\rangle|y\rangle \mapsto |y\rangle|x\rangle \tag{5}$$

is a "swap" of two registers (each a state in $\mathcal{S}$) and where

$$C := \sum_x |x\rangle\langle x| \otimes (2|u_x\rangle\langle u_x| - I) \tag{6}$$

is the "coin flip," a Grover reflection[2] about the vector:

$$|u_x\rangle := \frac{1}{\sqrt{d}} \sum_{y:(x,y)\in E} |y\rangle \tag{7}$$

The uniform superposition

$$|u\rangle := \frac{1}{\sqrt{dN}} \sum_{(x,y)\in E} |x\rangle|y\rangle \tag{8}$$

is a fixed point of $W$.

Szegedy [2] generalized the Grover walk to arbitrary Markov chains. The idea is to replace each local reflection about $|u_x\rangle$ by one about:

$$|p_x\rangle := \sum_{y\in\mathcal{S}} \sqrt{P(y,x)}|y\rangle \tag{9}$$

---

[1] For example, there is no unitary matrix with tridiagonal nonzero structure, so there is no discrete-time quantum walk $U : \mathcal{H}_\mathbb{Z} \to \mathcal{H}_\mathbb{Z}$ on the line $\mathbb{Z}$.

[2] Other quantum coin flips are possible, for example the Hadamard and DFT operators [20].

When $P$ is the simple random walk on a regular graph, this reduces to the Grover walk. If $P$ is reversible with stationary distribution $\pi$, then the eigenvector

$$|\tilde{\pi}\rangle := \sum_x \sqrt{\pi(x)}|x\rangle|p_x\rangle = \sum_x \sqrt{\pi(x)}|p_x\rangle|x\rangle \tag{10}$$

of $W_P$ is a fixed point.

*Continuous-time quantum walks.* The *continuous-time quantum walk* $\{e^{-iHt}|\psi\rangle\}$ generated by a time-independent Hamiltonian $H$ merits special focus beyond that of its discrete-time counterpart. This is because the precise relationship between discrete-time and continuous-time quantum walks, unlike that of their classical counterparts, is not well understood.

Obtaining a natural quantization of the simple random walk on a graph is an easier proposition in continuous time than in discrete time. Farhi and Gutmann [13,14] noted that the Laplacian matrix of a graph $G = (V, E)$ is symmetric and can therefore be used as a natural *walk Hamiltonian* on $G$. Aharonov and Ta-Shma [15] observed that another matrix

$$H_P := I - M(P) \tag{11}$$

can be used as a natural walk Hamiltonian for a reversible Markov chain, where $M(P)$ is the matrix (1). Their walk reduces to the Farhi/Gutmann walk when $P$ is the simple random walk on a regular graph. It is easy to see that the ground state (minimum-eigenvalue eigenvector) of the Aharonov/Ta-Shma walk Hamiltonian is:

$$|\pi\rangle := \sum_x \sqrt{\pi(x)}|x\rangle \tag{12}$$

*Quantum walk speedup theorems.* A key property of discrete- and continuous-time quantum walks useful for proving general quantum speedup theorems is the size of the *phase gap* $\Delta(P)$, the smallest nontrivial polar angle in the spectrum of $W_P$ (or $e^{iH_P}$, in continuous time). In either case, we have:

$$\Delta(P) = \Omega(\sqrt{\delta}) \tag{13}$$

Szegedy [2] used a stronger form of this property to show that a natural notion of "quantum hitting time" of a symmetric Markov chain is at most the square root of its classical counterpart:

**Theorem 3.** *Let $P$ be a symmetric, ergodic Markov chain and let $h_u$ be its hitting from the uniform distribution $u$ to the marked set $M$. Then the quantum hitting time is $O(\sqrt{h_u})$.*

A subtlety is that this is only a *decision* algorithm, not a *search* algorithm – i.e., the algorithm decides with high probability whether or not $M$ is empty, but does not necessarily output an element of $M$ with high probability. Magniez et al. [22] proved a weaker upper bound for a search algorithm that holds for all reversible Markov chains:

**Theorem 4.** *Let $P$ be a reversible, ergodic Markov chain with stationary distribution $\pi$ and spectral gap $\delta$. Let $M$, the set of marked states, satisfy $\varepsilon := \sum_{x \in M} \pi(x) > 0$. Then there is a quantum algorithm finding a marked state with constant success probability with cost $O(1/\sqrt{\delta \varepsilon})$.*

A similar $\sqrt{\delta^{-1}}$ speedup theorem was shown recently for simulated annealing [11]. What we would like to know is: can such a speedup be proven for the mixing time?

## 4    Sampling Via Quantum Walks

The ability of quantum walks to speed up mixing processes was first demonstrated by Nayak et al. [4,5] and Aharonov et al. [6] on cycles and by Moore and Russell [23] on hypercubes.

*Mixing time revisited.* Recall that the mixing time $\tau_x$ (2) of a classical random walk satisfies the inequality

$$\delta^{-1} \leq \tau_x \leq \delta^{-1} \log 1/\pi(x) \tag{14}$$

where $\delta$ is the spectral gap of $P$. Although this inequality is tight with respect to $\delta$, it is somewhat unsatisfactory in the following sense. Let $P$ be the simple random walk on a regular graph $G$ on $N$ vertices. It can be shown [24] that the diameter $d$ of $G$ satisfies:

$$d(G) = O(\sqrt{\delta^{-1}} \log N) \tag{15}$$

Clearly, $d(G)$ is a lower bound on any notion of the "quantum mixing time" from a worst-case start vertex $s \in V$. But (14) shows that the classical random walk on $P$ does *significantly* worse than this. For example, consider the simple case in which $P$ is the simple random walk on a line: its spectral gap is $\delta = \Theta(\frac{1}{N^2})$, so the random walk on $P$ samples from (close to) its uniform distribution only after $\Theta(N^2)$ walk steps.

Fundamentally, the random walk's mixing performance bottleneck is its Gaussian width. After $t$ steps, most of the probability mass is supported on only the middle $\Theta(\sqrt{t})$ vertices of the line. What we would like is an algorithm that spreads mass nearly uniformly across $\Theta(t)$ vertices after $t$ steps. More generally, we would like an algorithm that "mixes" in time:

$$T = O(\sqrt{\delta^{-1}} \log N) \tag{16}$$

A promising avenue that has received attention in recent years is replacing $P$ by a quantum walk.

*Quantum walks on cycles and hypercubes.* Figure 1 compares the behavior of random and quantum walks on the one-dimensional infinite line. In continuous

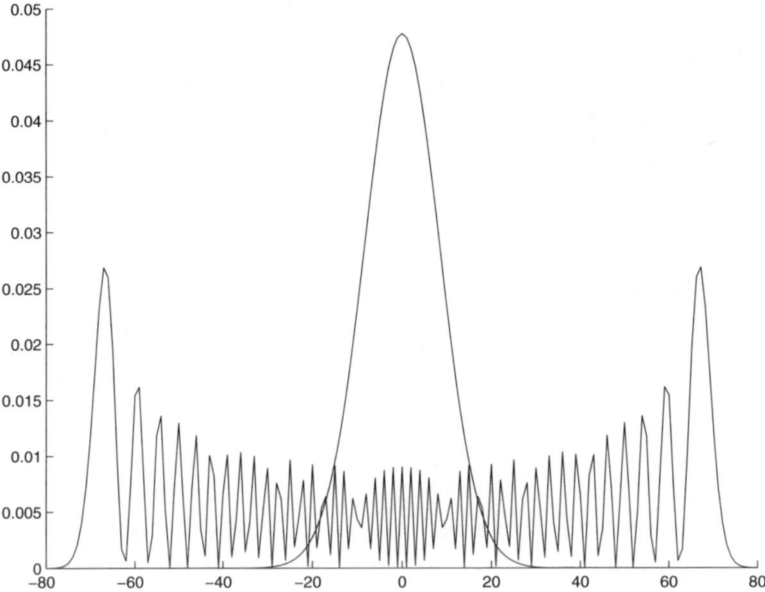

**Fig. 1.** The behavior of random vs. quantum walks in one dimension

time, a distribution $p_t$ initally concentrated at the origin propagates by the simple random walk $P$ according to

$$p_t(x) = e^{-(I-P)t}p_0 = e^{-2t}I_x(2t) \approx \frac{1}{\sqrt{4\pi t}}\exp(-x^2/4t) \tag{17}$$

where $I_x$ is the modified Bessel function of order $x$ and the latter expression is a Gaussian of width $\sqrt{2t}$. Using $P$ as the Hamiltonian for a continuous-time quantum walk, a wavefunction $|\psi_t\rangle$ initially concentrated at the origin spreads as

$$\langle x|\psi_t\rangle = e^{-iPt}(-i)^{|x|}J_{|x|}(2t) \tag{18}$$

where $J_{|x|}$ is a Bessel function of order $|x|$. For $|x| \gg 1$ the quantity $|J_{|x|}(t)|$ is (a) exponentially small in $|x|$ for $t < (1 - \epsilon) \cdot |x|$ and (b) of order $|x|^{-1/2}$ for $t > (1 + \epsilon) \cdot |x|$ (Childs [25]). The two "peaks" $t \in [(1 - \epsilon) \cdot |x|, (1 + \epsilon) \cdot |x|]$ of $|\langle x|\psi_t\rangle|^2$ in Figure 1 are relatively thin – indeed, the total amplitude of $|\psi_t\rangle$ is spread almost entirely and nearly uniformly over the interval between the two peaks. Nayak et al. [4,5] showed that the discrete-time quantum walk on the line with the so-called "Hadamard" coin flip also puts amplitude $\Omega(1/\sqrt{t})$ on the vertices within an interval of length $\Theta(t)$ around the origin at time $t$.

Nayak et al. [4,5] and Aharonov et al. [6] were the first to investigate analytically whether quantum walks might speed up classical mixing processes. Nayak et al. [4,5] showed the following:

**Theorem 5.** *There is a notion of "quantum mixing time" which is $O(n)$ for the Hadamard walk on the cycle $\mathbb{Z}_n$.*

Aharonov et al. [6] showed that under a different notion of mixing time, the bound is $O(n \log n)$. The classical mixing time is $\Theta(n^2)$. Moore and Russell [23] showed:

**Theorem 6.** *The "quantum mixing times" of the Grover and continuous-time quantum walks on the hypercube $\mathbb{Z}_2^d$ are $O(d)$.*

The classical mixing time is $O(d \log d)$.

*Obstacles for unitary walks.* Let $U$ be a discrete-time quantum walk operator, or let $U = e^{iH}$ where $H$ is a continuous-time quantum walk Hamiltonian, and define the $N \times N$ stochastic matrix

$$P_t(y, x) := |\langle y|U^t|x\rangle|^2 \tag{19}$$

induced by starting the quantum walk from a classical state, running it for time $t$, and measuring the walk register in the classical basis.[3] A key observation of Aharonov et al. [6] is that while $P_t$ does not converge to a limit (due to the underlying unitary dynamics), its time average does: if $U$ is a unitary operator and $\omega_T$ is the uniform distribution on $[0..T]$, then the following limit matrix exists.

$$\Pi := \lim_{T \to \infty} \mathrm{E}_{t \leftarrow \omega_T}[P_t] \tag{20}$$

However, it is typically the case that the quantum walk generated by a Markov chain $P$ with stationary distribution $\pi$ satisfies

$$\Pi \neq [\pi \pi \cdots \pi] = P^\infty \tag{21}$$

even though in the classical case we have $P^t \to P^\infty$. Moreover, the columns of $\Pi$ are not identical, so the limiting distribution of the time-averaged quantum walk depends on the initial state. Even when $P$ is symmetric (so that $\pi$ is the uniform distribution $u$), we rarely have an "ergodic theorem" (time-space average) $\Pi = [uu \cdots u]$ [6,26]. Consider for example the pathological case of the quantum walk on the complete graph $K_N$, for which $\Pi$ has diagonal elements $\approx 1 - \frac{1}{N}$ and off-diagonal elements $\approx \frac{1}{N}$.

## 5    Decoherent Quantum Walks

Decoherence was first identified as useful to quantum walks in numerical experiments performed by Kendon and Tregenna [27]. This discovery was supported by analytical estimates of Fedichkin et al. [28,29,30] and was proven in [31].

---

[3] If the walk is discrete-time and requires a second register, we consider its effect on a single register.

*Decoherence can be useful.* Non-unitary interaction with the surrounding environment, or *decoherence*, is a key physical obstacle to building a quantum computer. But in small amounts, decoherence was identified by Kendon and Tregenna [27] as a way to improve spreading and mixing properties of quantum walks. Under a Bernoulli decoherence model (i.e., wavefunction collapse occurs at each walk step with fixed probability $p$), the Hadamard walk on the $n$-cycle is observed to "mix" faster and more smoothly. On the other hand, high rates of decoherence in quantum walks have been shown to degrade mixing properties substantially by the quantum Zeno effect (Alagic and Russell [32]). Fedichkin et al. [28,29,30] gave analytical estimates suggesting $O(n)$ and $O(d \log d)$ scaling of decoherent quantum walks on the cycle $\mathbb{Z}_n$ and hypercube $\mathbb{Z}_2^d$, respectively. They conjectured the optimally-mixing quantum walk to be decoherent rather than unitary.

For an excellent survey of these and other aspects of decoherent quantum walks, see Kendon [21].

*General properties of decoherent walks.* It was proven in [31,33] that a decoherent quantum walk derived from a symmetric underlying Markov chain "mixes" to the correct (uniform) stationary distribution, and the "mixing time" is largely insensitive to the precise form of the decoherence model.

Let $\langle H_P, \omega_T \rangle$ denote the continuous-time quantum walk with Hamiltonian $H$ under decoherence model $\omega_T$, a $T$-parametrized family of probability mass (or density) functions on $[0, \infty)$ characterizing the (random) time at which a total measurement of the walk is performed in the classical basis. One can show [31,33]:

**Theorem 7.** *Let $P$ be a symmetric, irreducible Markov chain and $\omega_T$ be any "smooth" decoherence model.*[4] *For $T$ sufficiently large (but fixed), the $T'$-repeated quantum walk $\langle H_P, \omega_T \rangle$ "mixes" as $T' \to \infty$.*

Thus, decoherent quantum walks (which are non-unitary) circumvent the barrier to uniform-mixing identified by Aharonov et al. [6]. Moreover, the "mixing time" is robust [31]:

**Theorem 8.** *Up to logarithmic factors in $N$ and $T'$, the $T'$-repeated quantum walk $\langle H_P, \omega_T \rangle$ has the same "mixing time" under any "smooth" decoherence model.*

*Decoherent walks on lattices.* Let us turn our attention now to the special case in which $P$ is the simple random walk on the $d$-dimensional periodic lattice (torus) $\mathbb{Z}_n^d$. In this case, one can prove a quantum speedup for the mixing time [31]. Recall that the classical mixing time is $O(n^2 d \log d)$, which stays under the target upper bound (16) only when $\mathbb{Z}_n^d$ is quite high-dimensional: in particular, when $d$ is roughly of order $n^2$ or larger. That the low-dimensional case is a bottleneck is unsurprising considering the discussion immediately preceding (16).

---

[4] More precisely, $\omega_T$ is a family of distributions satisfying $E_{t \leftarrow \omega_T}[e^{i\theta t}] \to 0$ as $T \to \infty$ for any $\theta \neq 0$.

In [31] it is proven that the "quantum mixing time" of the decoherent quantum walk on the torus is $T \cdot T' = O(nd \log d)$:

**Theorem 9.** *Let $P$ be the simple random walk on $\mathbb{Z}_n^d$, the d-dimensional torus with $n \geq 2$ vertices per side. The $O(\log d)$-repeated continuous-time quantum walk $\langle H_P, \omega_{nd/2} \rangle$ mixes.*

This beats the target upper bound (16) and confirms the $O(n)$ and $O(d \log d)$ scaling estimates of Fedichkin et al. [28,29,30]. Previously, mixing speedups were known only for the unitary quantum walks of Nayak et al. [4,5] and Aharonov et al. [6] on the cycle and of Moore and Russell [23] on the hypercube – see Table 1. Thus, this result shows that introducing a small amount of decoherence to a quantum walk can simultaneously force convergence to the uniform distribution while preserving a quantum mixing speedup, an advantageous combination for algorithmic applications.

**Table 1.** Known mixing upper bounds for random vs. quantum walks

| Graph | Spectral estimate (16) | Random walk | Quantum walk |
|---|---|---|---|
| $\mathbb{Z}_n$ | $O(n \log n)$ | $O(n^2)$ | $O(n)$ [4,5,6] |
| $\mathbb{Z}_n^d$ | $O(d\sqrt{d}n \log n)$ | $O(n^2 d \log d)$ | $O(nd \log d)$ [31] |
| $\mathbb{Z}_2^d$ | $O(d\sqrt{d})$ | $O(d \log d)$ | $O(d)$ [23] |

A final remark: The continuous-time quantum walk can be simulated efficiently by a quantum circuit – in particular, without destroying the quantum speedup over the classical mixing time. Nevertheless, one would like to know whether the Grover walk mixes just as well as the continuous-time walk. There is some positive evidence for the torus $\mathbb{Z}_n^2$ [34,35,36] and a proof for the hypercube $\mathbb{Z}_2^n$ [23].

## 6   Outlook

Characterizing the quantum speedup achievable for the mixing time remains an active area of research. Is a generic quantum mixing speedup of $\sqrt{\delta^{-1}}$ provable? If so, is decoherence is a necessary ingredient? If not, which Markov chains admit no such speedup? Can a quantum speedup be demonstrated for mixing of a reversible (but asymmetric) Markov chain to its non-uniform stationary distribution?

*Acknowledgements.* Thanks to Mario Szegedy, Viv Kendon, and Todd Brun for their feedback on [31,33].

## References

1. Richter, P.: Quantum walks and ground state problems. PhD thesis, Rutgers University (2007)
2. Szegedy, M.: Quantum speed-up of Markov chain based algorithms. In: Proc. IEEE FOCS, pp. 32–41 (2004)

3. Ambainis, A.: Quantum walk algorithm for element distinctness. SIAM J. Comput. 37(1), 210–239 (2007)
4. Nayak, A., Vishwanath, A.: Quantum walk on the line. DIMACS TR 2000-43. quant-ph/0010117
5. Ambainis, A., Bach, E., Nayak, A., Vishwanath, A., Watrous, J.: One-dimensional quantum walks. In: Proc. ACM STOC, pp. 37–49 (2001)
6. Aharonov, D., Ambainis, A., Kempe, J., Vazirani, U.: Quantum walks on graphs. In: Proc. ACM STOC, pp. 50–59 (2001)
7. Lovász, L.: Random walks on graphs: a survey. In: Combinatorics: Paul Erdos is Eighty, Bolyai Society (1993)
8. Aldous, D.: Some inequalities for reversible Markov chains. J. Lond. Math. Soc. 25(2), 564–576 (1982)
9. Diaconis, P., Strook, D.: Geometric bounds for eigenvalues of Markov chains. Ann. Appl. Probab. 1, 36–61 (1991)
10. Sinclair, A.: Improved bounds for mixing rates of Markov chains and multicommodity flow. Combin. Probab. Comput. 1, 351–370 (1992)
11. Somma, R., Boixo, S., Barnum, H.: Quantum simulated annealing. arXiv:0712.1008 [quant-ph] (2007)
12. Watrous, J.: Quantum simulations of classical random walks and undirected graph connectivity. J. Comput. System Sci. 62(2), 376–391 (2001)
13. Farhi, E., Gutmann, S.: Quantum computation and decision trees. Phys. Rev. A 58, 915 (1998)
14. Childs, A., Farhi, E., Gutmann, S.: An example of the difference between quantum and classical random walks. Quantum Inf. Process. 1, 35 (2002)
15. Aharonov, D., Ta-Shma, A.: Adiabatic quantum state generation. SIAM J. Comput. 37(1), 47–82 (2007)
16. Aharonov, Y., Davidovich, L., Zagury, N.: Quantum random walks. Phys. Rev. A 48, 1687 (1993)
17. Meyer, D.: From quantum cellular automata to quantum lattice gases. J. Statist. Phys. 85, 551–574 (1996)
18. Meyer, D.: On the absence of homogeneous scalar unitary cellular automata. Phys. Lett. A 223(5), 337–340 (1996)
19. Ambainis, A.: Quantum walks and their algorithmic applications. Internat. J. Quantum Inf. 1, 507–518 (2003)
20. Kempe, J.: Quantum random walks – an introductory overview. Contemporary Physics 44(4), 307–327 (2003)
21. Kendon, V.: Decoherence in quantum walks – a review. quant-ph/0606016
22. Magniez, F., Nayak, A., Roland, J., Santha, M.: Search via quantum walk. In: Proc. ACM STOC, pp. 575–584 (2007)
23. Moore, C., Russell, A.: Quantum walks on the hypercube. In: Rolim, J.D.P., Vadhan, S.P. (eds.) RANDOM 2002. LNCS, vol. 2483, pp. 164–178. Springer, Heidelberg (2002)
24. Spielman, D.: Lecture notes for Spectral Graph Theory and its Applications (Yale University) (Fall 2004)
25. Childs, A.: Quantum information processing in continuous time. PhD thesis, Massachusetts Institute of Technology (2004)
26. Gerhardt, H., Watrous, J.: Continuous-time quantum walks on the symmetric group. In: Arora, S., Jansen, K., Rolim, J.D.P., Sahai, A. (eds.) RANDOM 2003 and APPROX 2003. LNCS, vol. 2764, pp. 290–301. Springer, Heidelberg (2003)
27. Kendon, V., Tregenna, B.: Decoherence can be useful in quantum walks. Phys. Rev. A 67, 042315 (2003)

28. Fedichkin, L., Solenov, D., Tamon, C.: Mixing and decoherence in continuous-time quantum walks on cycles. Quantum Inf. Comput. 6, 263–276 (2006)
29. Solenov, D., Fedichkin, L.: Continuous-time quantum walks on a cycle graph. Phys. Rev. A 73, 012313 (2006)
30. Solenov, D., Fedichkin, L.: Non-unitary quantum walks on hyper-cycles. Phys. Rev. A 73, 012308 (2006)
31. Richter, P.: Quantum speedup of classical mixing processes. Phys. Rev. A 76, 042306 (2007)
32. Alagic, G., Russell, A.: Decoherence in quantum walks on the hypercube. Phys. Rev. A 72, 062304 (2005)
33. Richter, P.: Almost uniform sampling via quantum walks. New J. Phys. 9(72) (2007)
34. Inui, N., Konishi, Y., Konno, N.: Localization of two-dimensional quantum walks. Phys. Rev. A 69, 052323 (2004)
35. Tregenna, B., Flanagan, W., Maile, R., Kendon, V.: Controlling discrete quantum walks: coins and initial states. New J. Phys. 5(83) (2003)
36. Mackay, T., Bartlett, S., Stephenson, L., Sanders, B.: Quantum walks in higher dimensions. J. Phys. A 35, 2745 (2002)

# Topological Dynamics of 2D Cellular Automata

Mathieu Sablik[1] and Guillaume Theyssier[2]

[1] UMPA, (UMR 5669 — CNRS, ENS Lyon), 46,
allée d'Italie 69364 Lyon cedex 07, France
and
LATP, (UMR 6632 — CNRS, Université de Provence), CMI, Université de Provence,
Technopôle Château-Gombert, 39, rue F. Joliot Curie, 13453 Marseille Cedex 13,
France
`mathieu.sablik@umpa.ens-lyon.fr`, `sablik@cmi.univ-mrs.fr`
[2] LAMA, (UMR 5127 — CNRS, Université de Savoie), Campus Scientifique, 73376
Le Bourget-du-lac cedex, France
`guillaume.theyssier@univ-savoie.fr`

**Abstract.** Topological dynamics of cellular automata (CA), inherited from classical dynamical systems theory, has been essentially studied in dimension 1. This paper focuses on 2D CA and aims at showing that the situation is different and more complex. The main results are the existence of non sensitive CA without equicontinuous points, the non-recursivity of sensitivity constants and the existence of CA having only non-recursive equicontinuous points. They all show a difference between the 1D and the 2D case. Thanks to these new constructions, we also extend undecidability results concerning topological classification previously obtained in the 1D case.

## 1   Introduction

Cellular automata were introduced by J. von Neumann as a simple formal model of cellular growth and replication. They consist in a discrete lattice of finite-state machines, called *cells*, which evolve uniformly and synchronously according to a local rule depending only on a finite number of neighboring cells. A snapshot of the states of the cells at some time of the evolution is called a *configuration*, and a cellular automaton can be view as a global action on the set of configurations.

Despite the apparent simplicity of their definition, cellular automata can have very complex behaviours. One way to try to understand this complexity is to endow the space of configurations with a topology and consider cellular automata as classical dynamical systems. With such a point of view, one can use well-tried tools from dynamical system theory like the notion of sensitivity to initial condition or the notion of equicontinuous point.

This approach has been followed essentially in the case of one-dimensional cellular automata. P. Kůrka has shown in [1] that 1D cellular automata are partitioned into two classes:

- $\mathcal{E}_q$, the set of cellular automata with equicontinuous points,
- $\mathcal{S}$, the set of sensitive cellular automata.

A. Beckmann, C. Dimitracopoulos, and B. Löwe (Eds.): CiE 2008, LNCS 5028, pp. 523–532, 2008.
© Springer-Verlag Berlin Heidelberg 2008

We stress that this partition result is false in general for classical (continuous) dynamical systems. Thus, it is natural to ask whether this result holds for the model of CA in any dimension, or if it is a "miracle" or an "anomaly" of the one-dimensional case due to the strong constraints on information propagation in this particular setting. One of the main contributions of this paper is to show that this is an anomaly of the 1D case (section 3): there exist a class $\mathcal{N}$ of 2D CA which are neither in $\mathcal{E}_q$ nor in $\mathcal{S}$.

Each of the sets $\mathcal{E}_q$ and $\mathcal{S}$ has an extremal sub-class: equicontinous and expansive cellular automata (respectively). This allows to classify cellular automata in four classes according to the degree of sensitivity to initial conditions. The dynamical properties involved in this classification have been intensively studied in the literature for 1D cellular automata (see for instance [1,2,3,4]). Moreover, in [5], the undecidability of this classification is proven, except for the expansivity class whose decidability remains an open problem.

In this paper, we focus on 2D CA and we are particularly interested in differences from the 1D case. As said above, we will prove in section 3 that there is a fundamental difference with respect to the topological dynamics classification, but we will also adopt a computational complexity point of view and show that some properties or parameters which are computable in 1D are non recursive in 2D (proposition 5 and 8 of section 4). To our knowledge, only few dimension-sensitive undecidability results are known for CA ([6,7]). However, we believe that such subtle differences are of great importance in a field where the common belief is that everything interesting is undecidable.

Moreover, we establish in section 4 several complexity lower bounds on the classes defined above and extend the undecidability result of [5] to dimension 2. Notably, we show that each of the class $\mathcal{E}_q$, $\mathcal{S}$ and $\mathcal{N}$ is neither recursively enumerable, nor co-recursively enumerable. This gives new examples of "natural" properties of CA that are harder than the classical problems like reversibility, surjectivity or nilpotency (which are all r.e. or co-r.e.).

## 2    Definitions

Let $\mathcal{A}$ be a finite set and $\mathbb{M} = \mathbb{Z}$ (for the one-dimensional case) or $\mathbb{Z}^2$ (for the two-dimensional case). We consider $\mathcal{A}^{\mathbb{M}}$, the *configuration space* of $\mathbb{M}$-indexed sequences in $\mathcal{A}$. If $\mathcal{A}$ is endowed with the discrete topology, $\mathcal{A}^{\mathbb{M}}$ is compact, perfect and totally disconnected in the product topology. Moreover one can define a metric on $\mathcal{A}^{\mathbb{M}}$ compatible with this topology:

$$\forall x, y \in \mathcal{A}^{\mathbb{M}}, \quad d_C(x, y) = 2^{-\min\{\|i\|_\infty : x_i \neq y_i \ i \in \mathbb{M}\}}.$$

Let $\mathbb{U} \subset \mathbb{M}$. For $x \in \mathcal{A}^{\mathbb{M}}$, denote $x_{\mathbb{U}} \in \mathcal{A}^{\mathbb{U}}$ the restriction of $x$ to $\mathbb{U}$. Let $\mathbb{U} \subset \mathbb{M}$ be a finite subset, $\Sigma$ is a *subshift of finite type of order* $\mathbb{U}$ if there exists $\mathcal{F} \subset \mathcal{A}^{\mathbb{U}}$ such that $x \in \Sigma \iff x_{m+\mathbb{U}} \in \mathcal{F} \quad \forall m \in \mathbb{M}$. In other word, $\Sigma$ can be viewed as a tiling where the allowed patterns are in $\mathcal{F}$.

In the sequel, we will consider *tile sets* and ask whether they can tile the plane or not. In our formalism, a tile set is a subshift of finite type: a set of states (the tiles) given together with a set of allowed patterns (the tiling constraints). We will restrict to $2 \times 1$ and $1 \times 2$ patterns (dominos) since it is sufficient to have the undecidability results of Berger [8].

A *cellular automaton* (CA) is a pair $(\mathcal{A}^{\mathbb{M}}, F)$ where $F : \mathcal{A}^{\mathbb{M}} \to \mathcal{A}^{\mathbb{M}}$ is defined by $F(x)_m = f((x_{m+u})_{u \in \mathbb{U}})$ for all $x \in \mathcal{A}^{\mathbb{M}}$ and $m \in \mathbb{M}$ where $\mathbb{U} \subset \mathbb{Z}$ is a finite set named *neighborhood* and $f : \mathcal{A}^{\mathbb{U}} \to \mathcal{A}$ is a *local rule*. The radius of $F$ is $r(F) = \max\{\|u\|_\infty : u \in \mathbb{U}\}$. By Hedlund's theorem [9], it is equivalent to say that $F$ is a continuous function which commutes with the shift (i.e. $\sigma^m \circ F = F \circ \sigma^m$ for all $m \in \mathbb{M}$).

We recall here general definitions of topological dynamics used all along the article. Let $(X, d)$ be a metric space and $F : X \to X$ be a continuous function.

- $x \in X$ is an *equicontinuous point* if for all $\varepsilon > 0$, there exists $\delta > 0$, such that for all $y \in X$, if $d(x, y) < \delta$ then $d(F^n(x), F^n(y)) < \varepsilon$ for all $n \in \mathbb{N}$.
- $(X, F)$ is *sensitive* if there exists $\varepsilon > 0$ such that for all $\delta > 0$ and $x \in X$, there exists $y \in X$ and $n \in \mathbb{N}$ such that $d(x, y) < \delta$ and $d(F^n(x), F^n(y)) > \varepsilon$.

## 3 Non Sensitive CA without Any Equicontinuous Point

In this section, we will construct a 2D CA which has no equicontinuous point and is not sensitive to initial conditions. This is in contrast with dimension 1 where any non-sensitive CA must have equicontinuous points as shown in [1].

The CA (denoted by $F$ in the following) is made of two components:

- an *obstacle component* (almost static) for which only finite type conditions are checked and corrections are made locally ;
- a *particle component* whose overall behaviour is to move left and to bypass obstacles.

Formally, $F$ has a Moore's neighborhood of radius 2 (25 neighbors) and a state set $\mathcal{A}$ with 12 elements : $\mathcal{A} = \{U, D, 0, 1, \downarrow, \uparrow, \leftarrow, \rightarrow, \swarrow, \searrow, \nwarrow, \nearrow\}$ where the subset $\mathcal{A}_F = \{1, \downarrow, \uparrow, \leftarrow, \rightarrow, \swarrow, \searrow, \nwarrow, \nearrow\}$ corresponds to the obstacle component and $\{U, D, 0\}$ to the particle component.

Let $\Sigma_F$ be the subshift of finite type of $\mathcal{A}^{\mathbb{Z}^2}$ defined by the set of allowed patterns constituted by all the $3 \times 3$ patterns appearing in the following set of finite configurations:

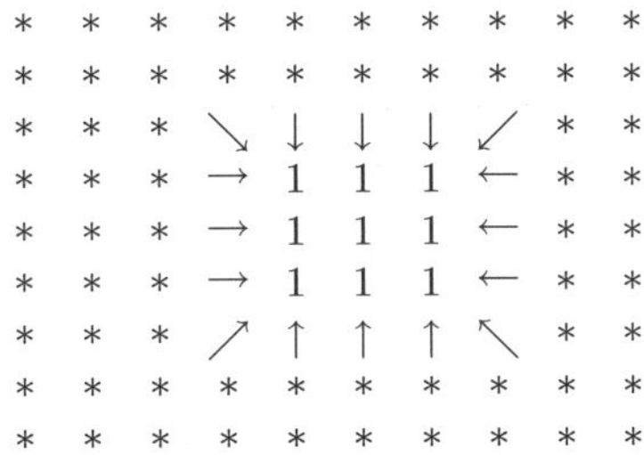

where $*$ stand for any state in $\mathcal{A} \setminus \mathcal{A}_F$.

In the sequel, a configuration $x$ is said to be *finite* if the set $\{z : x(z) \neq 0\}$ is finite. Moreover, in such a configuration, we call *obstacle* a maximal 4-connected region of states from $\mathcal{A}_F$.

The following lemma (the proof is straightforward) states that finite configurations from $\Sigma_F$ consist of rectangle obstacles inside a free $\mathcal{A} \setminus \mathcal{A}_F$ background. Moreover, obstacles are spaced enough to ensure that any position "sees" at most one obstacle in its $3 \times 3$ neighborhood.

**Lemma 1.** *Let $x \in \Sigma_F$ be a finite configuration. For any $z \in \mathbb{Z}^2$ we have the following:*

- *either $x(z) \in \mathcal{A}_F$ and $z$ belongs to a rectangular obstacle;*
- *or $x(z) \notin \mathcal{A}_F$ and the set of positions $\{z' : x(z') \in \mathcal{A}_F$ and $\|z' - z\|_\infty \leq 1\}$ is empty or belongs to the same obstacle.*

The local transition function of $F$ can be sketched as follows:

- states from $\mathcal{A}_F$ are turned into 0's if finite type conditions defining $\Sigma_F$ are violated locally and left unchanged in any other case ;
- states $U$ and $D$ behave like a left-moving particle when $U$ is just above $D$ in a background of 0's, and they separate to bypass obstacles, $U$ going over and $D$ going under, until they meet at the opposite position and recompose a left-moving particle (see figure 1).

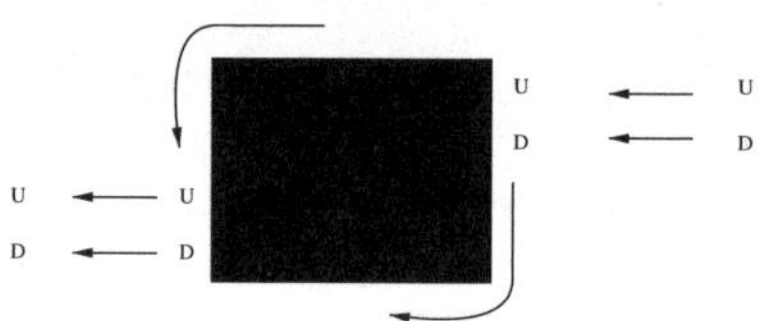


**Fig. 1.** A particle separating into two parts ($U$ and $D$) to bypass an obstacle (the black region)

A precise definition of the local transition function of $F$ is the following:

1. if the neighborhood ($5 \times 5$ cells) forms a pattern forbidden in $\Sigma_F$, then turn into state 0 ;
2. else, apply (if possible) one of the transition rules depending only on the $3 \times 3$ neighborhood detailed in figure 2
3. in any other case, turn into state 0.

The possibility to form arbitrarily large obstacles prevents $F$ from being sensitive to initial conditions.

**Proposition 1.** *$F$ is not sensitive to initial conditions.*

*Proof.* Let $\varepsilon > 0$. Let $c_\varepsilon$ be the configuration everywhere equal to 0 except in the square region of side $2\lceil -\log \varepsilon \rceil$ around the centre where there is an obstacle.

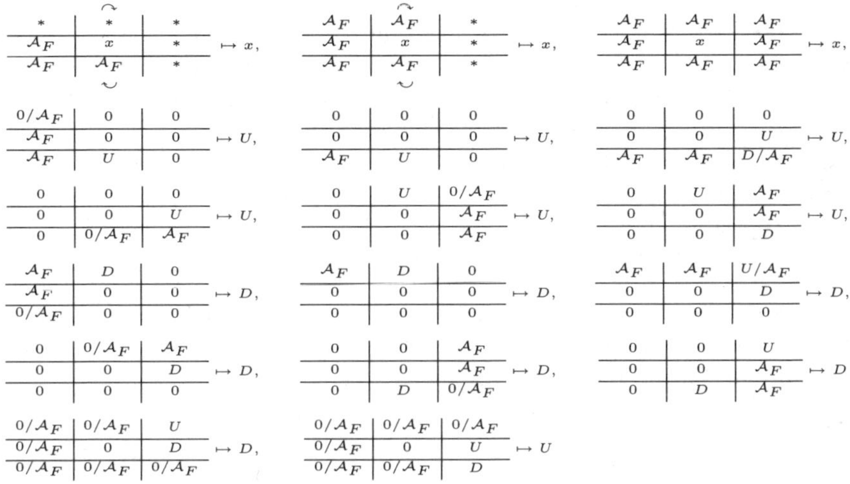

**Fig. 2.** Transition rule of $F$ where $x$ stands for any state in $\mathcal{A}_F$, '$*$' means any state in $\mathcal{A} \setminus \mathcal{A}_F$ (2 occurrences of $*$ are independent), and curved arrows mean that the transition is the same for any rotation of the neighborhood pattern

$\forall y \in \mathcal{A}^{\mathbb{Z}^2}$, if $d(y, c_\varepsilon) \le \varepsilon/4$ then $\forall t \ge 0$, $d\big(F^t(c_\varepsilon), F^t(y)\big) \le \varepsilon$ since a well-formed obstacle (precisely, a partial configuration that would form a valid obstacle when completed by 0 everywhere) is inalterable for $F$ provided it is surrounded by states in $\mathcal{A} \setminus \mathcal{A}_F$ (see the 3 first transition rules of case 2 in the definition of the local rule): this is guarantied for $y$ by the condition $d(y, c_\varepsilon) \le \varepsilon/4$. $\qquad\square$

The next lemma shows that $\Sigma_F$ attracts any finite configuration under the action of $F$.

**Lemma 2.** *For any finite configuration $x$, there exists $t_0$ such that $\forall t \ge t_0$ :* $F^t(x) \in \Sigma_F$.

The following lemma establishes the key property of the dynamics of $F$: particles can reach any free position inside a finite field of obstacles from arbitrarily far away from the field.

**Lemma 3.** *Let $x \in \Sigma_F \cap \big(\{0\} \cup \mathcal{A}_F\big)^{\mathbb{Z}^2}$ be a finite configuration. For any $z_0 \in \mathbb{Z}^2$ such that $x(z_0) = 0$ there exists a path $(z_n)$ such that:*

1. $\|z_n\|_\infty \to \infty$
2. $\exists n_0, \forall n \ge n_0$, *if $x_n$ is the configuration obtained from $x$ by adding a particle at position $z_n$ (precisely, $x_n(z_n) = U$ and $x_n\big(z_n + (0, -1)\big) = D$) then* $\big(F^n(x_n)\big)(z_0) \in \{U, D\}$.

*Proof.* First, since $x \in \Sigma_F$ and $x(z_0) = 0$, then either $x\big(z_0 + (0, 1)\big) = 0$ or $x\big(z_0 + (0, -1)\big) = 0$. We will consider only the first case since the proof for the second one is similar. Let $(z_n)$ be the path starting from $z_0$ defined as follows:

- If $x\big(z_n + (1,0)\big) = 0$ and $x\big(z_n + (1,-1)\big) = 0$ then $z_{n+1} = z_n + (1,0)$.
- Else, position $z_n + (1,0)$ and/or position $z_n + (1,-1)$ belongs to an obstacle $P$. Let $a$, $b$ and $c$ be the positions of the upper-left, upper-right and lower-right outside corners of $P$ and let $p$ be its half perimeter. Then define $z_{n+1}, \ldots, z_{n+p+1}$ to be the sequence of positions made of:
    - a (possibly empty) vertical segment from $z_n$ to $a$,
    - the segment $[a; b]$,
    - a (possibly empty) vertical segment from $b$ to $z_{n+p+1}$ where $z_{n+p+1}$ is the point on $[b; c]$ such that $z_n a + b z_{n+p+1} = bc$.

We claim that the path $(z_n)$ constructed above has the properties of the lemma. Indeed, one can check that for each case of the inductive construction of a point $z_m$ from a point $z_n$ we have:

- $\|z_m\|_\infty > \|z_n\|_\infty$,
- $\big(F^{m-n}(x_m)\big)(z_n) = U$ and $\big(F^{m-n}(x_m)\big)(z_n + (0,-1)) = D$ (straightforward from the definition of $F$).                                          $\square$

**Proposition 2.** *$F$ has no equicontinuous points.*

*Proof.* Assume $F$ has an equicontinuous point, precisely a point $x$ which verifies $\forall \varepsilon > 0, \exists \delta : \forall y, d(x,y) \le \delta \Rightarrow \forall t, d\big(F^t(x), F^t(y)\big) \le \varepsilon$.

Suppose that there is $z_0$ such that $x(z_0) = 0$ and let $\varepsilon = 2^{-\|z_0\|_\infty - 1}$. We will show that the hypothesis of $x$ being an equicontinuous point is violated for this particular choice of $\varepsilon$. Consider any $\delta > 0$ and let $y$ be the configuration everywhere equal to 0 except in the central region of radius $-\log\lceil \delta \rceil$ where it is identical to $x$. Since $y$ is finite, there exists $t_0$ such that $y_+ = F^{t_0}(y) \in \Sigma_F$ (by lemma 2). Moreover, the proof of lemma 2 guaranties that for any positive integer $t$, $\big(F^t(y_+)\big)(z_0) = x(z_0) = 0$. So we can apply lemma 3 on $y_+$ and position $z_0$ to get the existence of a path $(z_n)$ allowing particles placed arbitrarily far away from $z_0$ to reach the position $z_0$ after a certain time. For any sufficiently large $n$, we can construct a configuration $y'$ obtained from $y$ by adding a particle at position $z_n$. By the property of $(z_n)$, we have: $\big(F^n(y)\big)(z_0) \neq \big(F^n(y')\big)(z_0)$ and therefore $d\big(F^n(y), F^n(y')\big) > \varepsilon$. Since, if $n > -\log\lceil \delta \rceil$, both $y$ and $y'$ are in the ball of centre $x$ and radius $\delta$, we have the desired contradiction.

Assume now that $\forall z, x(z) \in \mathcal{A}_F$. There must exist some $z_0$ such that $x(z_0) \neq 1$ (since the uniform configuration everywhere equal to 1 is not an equicontinuous point). It follows from the definition of $\Sigma_F$ that $z_0$ belongs to a forbidden pattern for $\Sigma_F$. Therefore $\big(F(x)\big)(z_0) = 0$ and we are brought back to the previous case of this proof.                                          $\square$

## 4  Undecidability of Topological Classification Revisited

We will use simulations of Turing machines by tile sets in the classical way (originally suggested by Wang [10]): the tiling represents the space-time diagram of the computation and the transition rule of the Turing machine are converted into tiling constraints. Without loss of generality, we only consider Turing machines working on a semi-infinite tape with a single final state. The $i^{th}$ machine of this kind in a standard enumeration is denoted by $\mathcal{M}_i$. In the sequel we use the following notations. First, to each $\mathcal{M}_i$ we associate a tile set $T_i$ whose constraints ensure the simulation of $\mathcal{M}_i$ as mentioned above; Second, when constructing a CA $G$, we denote by $\Sigma_G$ the subshift of its admissible obstacles, which plays the same role as $\Sigma_F$ for $F$ with some differences detailed below.

In [5], the authors give a recursive construction which produce either a 1D sensitive CA or a 1D CA with equicontinuous points according to whether a Turing machine halts on the empty input. By noticing that a 1D CA is sensitive (resp. has equicontinuous points) in the 1D topology if and only if it is sensitive (resp. has equicontinuous points) in the 2D topology when viewed as a 2D CA (neighbors are aligned, e.g. horizontally), we get the following proposition.

**Proposition 3.** *There is a recursive function* $\Phi_1 : \mathbb{N} \to CA$ *such that* $\Phi_1(i) \in \mathcal{E}_q$ *if* $\mathcal{M}_i$ *halts on the empty input and* $\Phi_1(i) \in \mathcal{S}$ *otherwise.*

However, this is not enough to establish the overall undecidability of the topological classification of 2D CA. The main concern of this section is to complete proposition 3 in order to prove a stronger and more complete undecidability result summarized in the following theorem.

**Theorem 1.** *Each of the class* $\mathcal{E}_q$, $\mathcal{S}$ *and* $\mathcal{N}$ *is neither r.e. nor co-r.e. Moreover any pair of them is recursively inseparable.*

The proof of this theorem rely on different variants of the construction of the automaton $F$ above. Each time, the construction scheme is the same, and the desired property is obtained by adding various contents inside obstacles and slightly changing the rules of destruction of obstacles according to that content.

The next proposition can be established by using a mechanism to bound or not the size of admissible obstacles according to whether a Turing machine halts or not. The idea is to force the tiling representation of a computation on a blank tape in each obstacle (using the lower left corner) and to forbid the final state. The proof mechanism used for $F$ can be applied if there is no bound on admissible obstacles. Otherwise, we get a sensitive CA.

**Proposition 4.** *There is a recursive function* $\Phi_2 : \mathbb{N} \to CA$ *such that* $\Phi_2(i) \in \mathcal{S}$ *if* $\mathcal{M}_i$ *halts on the empty input and* $\Phi_2(i) \in \mathcal{N}$ *otherwise.*

Before going on with the different constructions needed to prove theorem 1, let us stress the dynamical consequence of the construction of proposition 4. It is well-known that for any 1D sensitive CA of radius $r$, $2^{-2r}$ is always the maximal admissible sensitivity constant (see for instance [1]). Thanks to the above construction it is easy to construct CA with tiny sensitivity constants as shown by the following proposition.

**Proposition 5.** *The (maximal admissible) sensitivity constant of sensitive 2D CA cannot be recursively (lower-)bounded in the number of states and the neighborhood size.*

*Proof.* It is straightforward to check that for each CA $\Phi_2(i)$ where $\mathcal{M}_i$ halts after $n$ steps on the empty input, the maximal admissible obstacle is of height $O(n)$ and of width at least $O(\log(n))$. The proposition follows since the sensitivity constant of any CA $\Phi_2(i) \in \mathcal{S}$ is precisely $2^{-l/2+1}$ where $l$ is the minimum between the largest height and the largest width of admissible obstacles.     $\square$

Back to the path towards theorem 1, the following proposition uses the same ideas as proposition 4 but it exchanges the role of halting and non-halting computations.

**Proposition 6.** *There is a recursive function $\Phi_3 : \mathbb{N} \to CA$ such that $\Phi_3(i) \in \mathcal{N}$ if $\mathcal{M}_i$ halts on the empty input and $\Phi_3(i) \in \mathcal{S}$ otherwise.*

The properties of the CA $F$ and the other constructions above rely on the fact that obstacles able to stop or deviate particles cannot be fit together to form larger obstacles. Thus, $F$ and other CA have no equicontinuous point. In the following, we will use a new kind of obstacles: they are protected from particles by a boundary as the classical obstacles of $F$, but they are made only of successive boundaries like onion skins. With this new construction it is not difficult to build an equicontinuous point provided there are arbitrarily large valid obstacles. The next proposition use this idea to reduce existence of equicontinuous point to a tiling problem.

**Proposition 7.** *There is a recursive function $\Phi_4$ which associate with any tile set $T$ a CA $\Phi_4(T)$ which is in class $\mathcal{E}_q$ if $T$ tiles the plane and in class $\mathcal{N}$ otherwise.*

*Proof (sketch).* Given a tile set $T$, the CA $\Phi_4(T)$ is identical to $F$, except that it has a second kind of obstacles, called $T$-obstacles. $T$-obstacles are square patterns of states from the set $E = T \times X$ with $X = \{\downarrow, \uparrow, \leftarrow, \rightarrow, \swarrow, \searrow, \nwarrow, \nearrow, -\}$ and where the $T$ component is a valid tiling and the $X$ component is made from the set of $2 \times 2$ patterns appearing in the following finite configuration:

$$
\begin{array}{ccccccc}
 & & \downarrow & \downarrow & & & \\
 & \searrow & \downarrow & \downarrow & \swarrow & & \\
\rightarrow & \rightarrow & \searrow & \downarrow & \swarrow & \leftarrow & \leftarrow \\
\rightarrow & \rightarrow & \rightarrow & - & \leftarrow & \leftarrow & \leftarrow \\
 & \rightarrow & \nearrow & \uparrow & \nwarrow & \leftarrow & \\
 & \nearrow & \uparrow & \uparrow & \uparrow & \nwarrow & \\
 & \uparrow & \uparrow & \uparrow & \uparrow & \uparrow & \\
 & & \uparrow & \uparrow & & &
\end{array}
$$

The $X$ component is used to give everywhere in $T$-obstacles a local notion of *inside* and *outside* as depicted by figure 3 (up to $\pi/2$ rotations): roughly speaking, arrows point to the inside region. Other constraints concerning $T$-obstacles are checked locally:

- any pair of obstacles must be at least 2 cells away from each other;
- their shape must be a square and this is ensured by requiring that any cell of a $T$-obstacle can have $\{0, U, D\}$ neighbors only in its outside region;

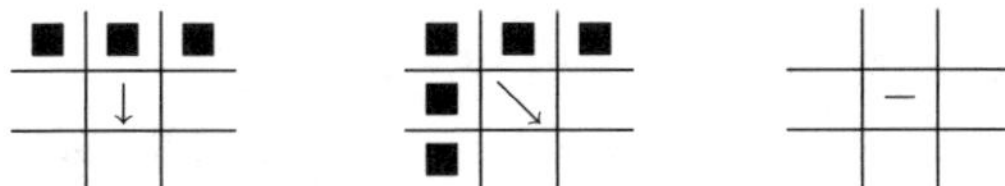

**Fig. 3.** Inside (white) and outside (black) positions for states of $X$

- the behaviour of particles with $T$-obstacles is the same as with classical obstacles (they can not cross them and states $U$ and $D$ separate to bypass them).

The overall dynamics of $\Phi_4(T)$ is similar to that of $F$ with the following exception, which is the key point of the construction: destruction of non-valid $T$-obstacles is done progressively to preserve as much as possible valid zones inside non-valid obstacles. More precisely any cell in a state from $E$ remains unchanged unless one of the following conditions is verified in which case it turns into state 0:

- if there is an error in the inside neighborhood;
- if there is a position in the outside neighborhood such that the pattern formed by that position together with the cell itself is forbidden;
- if there is a state from $E$ in the $5 \times 5$ neighborhood which is not connected to the cell by states from $E$.

One can check that proposition 1 is still true. Moreover proposition 2 is true if and only if their is a bound on the size of valid $T$-obstacle. Indeed, lemma 2 and 3 are always true and the only point which can be eventually false with the CA $\Phi_4(T)$ is the last case in the proof of proposition 2. Precisely, if a valid configuration $x$ such that $\forall z, x(z) \in E$ can be constructed, then it is an equicontinuous point. Otherwise, if such a $x$ is not valid, then it contains an error somewhere and the proof scheme of proposition 2 can be applied to $\Phi_4(T)$. The proposition follows since such a valid $x$ can be constructed if and only if $T$ can tile the plane.     □

The previous propositions give a set of reductions from Turing machines or tile sets to CA and one can easily check that the main theorem follows using Berger's theorem [8] and classical results of the set of halting Turing machines.

To finish this section, we will discuss another difference between 1D and 2D concerning the complexity of equicontinuous points. Let us first recall that equicontinuous point in 1D CA can be generated by finite words often called "blocking" words. Precisely, for any $F$ with equicontinuous points, there exists a finite word $u$ such that $^{\infty}u^{\infty}$ is an equicontinuous point for $F$ (proof in [1]). The previous construction can be used with the tile set of Myers [11] which can produce only non-recursive tilings of the plane. Therefore the situation is more complex in 2D, and we have the following proposition.

**Proposition 8.** *There exists a 2D CA having equicontinuous points, but only nonrecursive ones.*

## 5  Open Problems

It is well-known that equicontinuous CA are exactly ultimately periodic CA (if they are also bijective, they are periodic). The proof techniques developed by Kari in [6] allow to prove that there is no recursive lower-bound on the pre-period and period of 2D equicontinuous CA. An interesting open question in the continuation of this paper is to determine whether periods of 1D equicontinuous CA (bijective or not) can be recursively bounded or not. The only known result in 1D is that pre-periods are not recursively bounded (this is essentially the nilpotency problem).

It is interesting to notice that for 1D CA, classes $\mathcal{S}$ and $\mathcal{E}_q$ are easily definable in first-order arithmetic. This is due to the characterisation by blocking words mentioned above: the existential quantification over configurations can be replaced by a quantification over finite words in the definition of $\mathcal{E}_q$. Proposition 8 shows that first-order definability of $\mathcal{S}$, $\mathcal{E}_q$ or $\mathcal{N}$ for 2D CA is more challenging. We believe they are but at a higher level in the arithmetical hierarchy.

## References


1. Kůrka, P.: Languages, equicontinuity and attractors in cellular automata. Ergodic Theory and Dynamical Systems 17, 417–433 (1997)
2. Blanchard, F., Maass, A.: Dynamical properties of expansive one-sided cellular automata. Israel J. Math. 99 (1997)
3. Blanchard, F., Tisseur, P.: Some properties of cellular automata with equicontinuity points. Ann. Inst. Henri Poincaré, Probabilités et statistiques 36, 569–582 (2000)
4. Fagnani, F., Margara, L.: Expansivity, permutivity, and chaos for cellular automata. Theory of Computing Systems 31(6), 663–677 (1998)
5. Durand, B., Formenti, E., Varouchas, G.: On undecidability of equicontinuity classification for cellular automata. In: Morvan, M., Rémila, É. (eds.) DMCS 2003. Volume AB of DMTCS Proceedings, pp. 117–128 (2003)
6. Kari, J.: Reversibility and Surjectivity Problems of Cellular Automata. Journal of Computer and System Sciences 48(1), 149–182 (1994)
7. Bernardi, V., Durand, B., Formenti, E., Kari, J.: A new dimension sensitive property for cellular automata. In: Fiala, J., Koubek, V., Kratochvíl, J. (eds.) MFCS 2004. LNCS, vol. 3153, pp. 416–426. Springer, Heidelberg (2004)
8. Berger, R.: The undecidability of the domino problem. Mem. Amer. Math Soc. 66 (1966)
9. Hedlund, G.A.: Endomorphisms and Automorphisms of the Shift Dynamical Systems. Mathematical Systems Theory 3(4), 320–375 (1969)
10. Wang, H.: Proving theorems by pattern recognition ii. Bell System Tech. Journal 40(2) (1961)
11. Myers, D.: Nonrecursive tilings of the plane. ii. The Journal of Symbolic Logic 39(2), 286–294 (1966)

# Complexity of Aperiodicity for Topological Properties of Regular $\omega$-Languages

Victor L. Selivanov[1,*] and Klaus W. Wagner[2]

[1] A.P. Ershov Institute of Informatics Systems
Siberian Division Russian Academy of Sciences
vseliv@nspu.ru
[2] Institut für Informatik
Julius-Maximilians-Universität Würzburg
wagner@informatik.uni-wuerzburg.de

**Abstract.** We study the complexity of aperiodicity restricted to topological properties of regular $\omega$-languages (i.e. properties closed under the Wadge equivalence on the Cantor space of $\omega$-words) restricted to aperiodic sets. In particular, we show the PSPACE-completeness of such problems for several usual deterministic and non-deterministic automata representations of regular $\omega$-languages.

**Keywords:** Regular aperiodic $\omega$-language, Wadge reducibility, aperiodic automaton, deterministic Muller automaton, nondeterministic Büchi automaton, monadic second-order formula.

## 1  Introduction

The study of complexity questions for different types of automata on finite and infinite words is an important research topic in automata theory closely related to the synthesis and verification of finite-state systems. Whereas there are many complexity results on finite automata accepting finite words, there are only a few results on finite $\omega$-automata. As a prominent example of the latter let us mention the recent result that the problem of deciding membership in the class of regular aperiodic $\omega$-languages is PSPACE-complete when the $\omega$-languages are given by nondeterministic Büchi automata [DG08].

In this paper we study the complexity of aperiodicity restricted to topological properties. By *topological property* we mean here a property closed under the Wadge equivalence on the Cantor space $A^\omega$ of $\omega$-words. For $L, K \subseteq A^\omega$, $L$ is said to be *Wadge reducible* to $K$ (in symbols $L \leq_{\mathrm{W}} K$), if $L = g^{-1}(K)$ for some continuous function $g : A^\omega \to A^\omega$. By $\equiv_{\mathrm{W}}$ we denote the induced equivalence relation which gives rise to the corresponding quotient partial ordering. Following a well established jargon, we call this ordering the structure of Wadge degrees. The operation $L \oplus K = \{a \cdot \xi, b \cdot \eta \mid b \in A \smallsetminus \{a\}, \xi \in L, \eta \in K\}$, where $a$ is a

---

$^{\star}$ Supported by DFG Mercator program, by DFG-RFBR Grant 06-01-04002 and by RFBR grant 07-01-00543a.

A. Beckmann, C. Dimitracopoulos, and B. Löwe (Eds.): CiE 2008, LNCS 5028, pp. 533–543, 2008.

fixed letter in $A$, induces the operation of least upper bound in the structures of Wadge degrees.

To explain our results, let us recall some facts from [Wag79] where the Wadge degrees of regular $\omega$-languages (over any alphabet $A$ with at least two symbols) were determined, in particular the following results were established:

1) The structure $(\mathcal{R}; \leq_W)$ of regular $\omega$-languages under the Wadge reducibility is almost well-ordered with order type $\omega^\omega$, i.e. there are regular $\omega$-languages $A_\alpha \in \mathcal{R}$, for each ordinal $\alpha < \omega^\omega$, such that $A_\alpha <_W A_\alpha \oplus \overline{A}_\alpha <_W A_\beta$ for $\alpha < \beta < \omega^\omega$ and any regular set is Wadge-equivalent to one of the sets $A_\alpha, \overline{A}_\alpha, A_\alpha \oplus \overline{A}_\alpha (\alpha < \omega^\omega)$.

2) Any class $\mathcal{R}_\alpha = \{C \in \mathcal{R} \mid C \leq_W A_\alpha\}$ is decidable.

Let $\mathcal{A}$ be the class of regular aperiodic (or star-free) $\omega$-languages. This class is certainly the most important and popular proper subclass of $\mathcal{R}$ that has many interesting characterizations and plays a noticeable role in the literature on specification and verification. We are interested in the complexity of the topological properties restricted to $\mathcal{A}$, in particular of the classes $\mathcal{A}_\alpha = \mathcal{R}_\alpha \cap \mathcal{A}$. The classes $\mathcal{A}_\alpha$ were studied in detail in [Se07] where it was in particular observed that the sets $A_\alpha$ in 1) may be assumed aperiodic, hence the structure of the Wadge degrees of sets in $\mathcal{A}$ coincides with the structure of those in $\mathcal{R}$.

From a result in [DG08] it follows that the class $\mathcal{A}$ is decidable in PSPACE w.r.t. the deterministic Muller representation. Since the classes $\mathcal{R}_\alpha$ are decidable in polynomial time P w.r.t. the Muller representation [KPB95, WY95], each class $\mathcal{A}_\alpha$ is also decidable in PSPACE . Is this upper complexity bound optimal or can one find a clever more efficient algorithm for deciding $\mathcal{A}_\alpha$? We consider questions of this type and determine the complexity of many similar natural problems (for different known representations of regular $\omega$-languages and for different topological properties). For some cases we were able to obtain only partial results. One of our main results is the following

**Theorem 1.** *Let $C$ be a nonempty Boolean combination of classes from $\{\mathcal{A}_\alpha \mid 1 \leq \alpha < \omega^\omega\} \cup \{\text{co-}\mathcal{A}_\alpha \mid 1 \leq \alpha < \omega^\omega\}$ where $\overline{C} =_{\text{def}} \mathcal{A} \setminus C$. Then the problem of deciding, given a deterministic Muller (Rabin, Mostowski, Streett), or nondeterministic Büchi (Muller, Rabin, Mostowski) automaton $\mathcal{M}$, whether $L_\omega(\mathcal{M}) \in C$ is PSPACE-complete.*

Since $\mathcal{R}_0 = \mathcal{A}_0 = \{\emptyset\}$, from the well known facts it follows that $\mathcal{A}_0$ is decidable in P for any kind of automata representation. This is the reason for the absence of the ordinal 0 in the formulation of Theorem 1. This result and its proof below informally mean that the aperiodicity property is in a sense independent from the topological property (such an independence was already clear from the results in [Se07]). Surprisingly, the complexity estimation in Theorem 1 is the same for several popular automata representations of regular $\omega$-languages.

As is well known, along with the different kind of automata, the regular $\omega$-languages are characterized by several other popular formalisms including monadic second-order logic or $\omega$-regular expressions. We provide also some information on the complexity of classes from the last theorem for the corresponding representations of regular $\omega$-languages. E.g., we prove the following result where $L_\phi$ denotes the set of $\omega$-words satisfying a sentence $\phi$. For $k \geq 1$ define the tower-of-exponents function exp by $\exp(0, m) =_{\text{def}} m$ and $\exp(k + 1, m) =_{\text{def}} 2^{\exp(k,m)}$.

**Theorem 2.** *Let $\mathcal{C}$ be a Boolean combination of classes from $\{\mathcal{A}_\alpha \mid 1 < \alpha < \omega^\omega\} \cup \{\text{co-}\mathcal{A}_\alpha \mid 1 < \alpha < \omega^\omega\}$ where $\overline{\mathcal{C}} =_{\text{def}} \mathcal{A} \setminus \mathcal{C}$. Then the problem of deciding, given a weak monadic second-order formula $\phi$, whether $L_\phi \in \mathcal{C}$ is* $\text{DSPACE}(\exp(\mathcal{O}(n), 1))$-*complete.*

After recalling in the next section notation, notions and facts we rely upon we consider subsequently the complexity problems with respect to the automata representations, the representation by monadic second-order formulas, and the representations by $\omega$-regular expressions.

## 2 Preliminaries

We assume the reader to be acquainted with automata on finite and infinite words and standard notation about languages. *Regular languages* are languages $L(\mathcal{M})$ recognized by deterministic finite automata (dfa) $\mathcal{M}$ or, equivalently, by nondeterministic finite automata (nfa). *Aperiodic regular languages* are languages recognized by the so called aperiodic (or counter-free) dfa's (for details see e.g. [Th90, Th97, PP04] and Appendix). Moreover, for any dfa $\mathcal{M} = (Q, A, \delta, s_0, F)$, $L(\mathcal{M})$ is aperiodic iff $\mathcal{M}$ is aperiodic, i.e. $\delta(q, u^m) = q$ implies $\delta(q, u) = q$ for all states $q \in Q$, words $u \in A^*$ and $m \geq 1$.

Unlike automata on finite words, for automata on $\omega$-words the acceptance conditions were defined in different ways by different authors, and it is not clear which of these conditions are more natural than the others. As a result, there are several notions of automata accepting $\omega$-words: deterministic (or nondeterministic) Büchi (Muller, Rabin, Mostowski, Streett) automata. Regular $\omega$-languages coincide with the $\omega$-languages $L_\omega(\mathcal{M})$ recognized by deterministic Muller (Rabin, Mostowski, Streett) automata or nondeterministic Büchi (Muller, Rabin, Mostowski, Streett) automata $\mathcal{M}$. Regular aperiodic $\omega$-languages are the $\omega$-languages recognized by aperiodic deterministic Muller (equivalently, Mostowki) automata.

Relate to any alphabet $A = \{a, \dots\}$ the signature $\sigma_A = \{\leq, \mathbf{Q_a}, \dots, \}$ where $\leq$ is a binary relation symbol and $\mathbf{Q_a}$ (for each $a \in A$) is a unary relation symbol. A word $u = u_0 \dots u_n \in A^+$ may be considered as a structure $\mathbf{u} = (\{0, \dots, n\}; \leq, Q_a, \dots)$ of signature $\sigma_A$ where $\leq$ has its usual meaning and $Q_a(a \in A)$ are unary predicates on $\{0, \dots, n\}$ defined by $Q_a(i) \leftrightarrow u(i) = a$. For a sentence $\phi$ of $\sigma_A$, let $L_\phi = \{u \in A^* \mid \mathbf{u} \models \phi\}$ be the language defined by $\sigma_A$ (the empty word can also be included in consideration in a natural way). In automata theory people are mostly interested in definability by first-order sentences and by monadic second-order sentences (which, along with the syntax of first-order logic, may include unary predicate variables and the possibility to quantify over such variables).

Monadic second-order (resp., first-order) sentences define exactly the regular languages (resp., the aperiodic regular languages).

The logical formalism is adapted to the context of $\omega$-words in the obvious way. Relate to any $\xi \in A^\omega$ the structure $\tilde{\xi} = (\omega; \leq, Q_a, \ldots)$ of the signature $\sigma_A = \{\leq, \mathbf{Q_a}, \ldots, \}$ where $\leq$ is interpreted as the usual order on $\omega$ and $Q_a(i) \leftrightarrow \xi(i) = a$ for all $i < \omega$ and $a \in A$. For any sentence $\phi$ of $\sigma_A$, let $L_\phi = \{\xi \in A^\omega \mid \tilde{\xi} \models \phi\}$ be the $\omega$-language defined by $\phi$. As in the case in finite words, we will consider formulas of first-order and monadic second-order logic. Monadic second-order (resp., first-order) sentences of signature $\sigma_A$ define exactly the regular $\omega$-languages (resp., the aperiodic regular $\omega$-languages).

For $\mathcal{O} \subseteq \{\cup, \cap, ^-, \cdot, ^2, ^*\}$ *regular $\mathcal{O}$-expressions* are variable-free terms of the signature $\{\emptyset, \varepsilon\} \cup \{a \mid a \in A\} \cup \mathcal{O}$ where $\emptyset$, $\varepsilon$ and $a$ for each $a \in A$ are constant symbols, $\cup$, $\cap$, and $\cdot$ are binary function symbols, and $^-$, $^2$, and $^*$ are unary function symbols. Regular expressions are interpreted in the class $P(A^*)$ of all languages over $A$ according to the usual definitions of language operations, where we define $L^2 =_{\mathrm{def}} L \cdot L$. The constant symbols are interpreted respectively as $\emptyset$, $\{\varepsilon\}$, and $\{a\}, \ldots$. Thus, any regular $\mathcal{O}$-expression defines a language over $A$. Regular $\mathcal{O}$-expression where $\{\cup, \cdot, ^*\} \subseteq \mathcal{O} \subseteq \{\cup, \cap, ^-, \cdot, ^*\}$ define exactly the regular languages. Regular $\omega$-languages coincide with the finite union of sets $U \cdot V^\omega$ where $U, V$ are given by regular $\mathcal{O}$-expressions, $\{\cup, \cdot, ^*\} \subseteq \mathcal{O} \subseteq \{\cup, \cap, ^-, \cdot, ^*\}$. The regular $\{\cup, \cap, ^-, \cdot\}$-expressions define exactly the regular aperiodic languages. Regular aperiodic $\omega$-languages coincide with the finite union of sets $U \cdot V^\omega$ where $U, V$ are given by regular $\{\cup, \cap, ^-, \cdot\}$-expressions. More details on the material mentioned above may be found e.g. in [Th90, Th97, Sta97, PP04].

Next we introduce names for the above-mentioned representations of regular $\omega$-languages. We will also formulate the decision problems to be studied in the subsequent sections. For automata representations, B, M, R, P, and S stand for Büchi automata, Muller automata, Rabin automata, Mostowski (parity) automata, and Strett automata, resp., and D and N stand for deterministic and nondeterministic, resp. In this way, for example, NB is the name for representation by nondeterministic Büchi automata. For $\mathcal{O} \subseteq \{\cup, \cap, ^-, \cdot, ^2, ^*\}$, the name for the representation by regular $\mathcal{O}$-expressions is simply $\mathcal{O}$. The name for the representation by monadic second-order formulas is MSO.

Let $\mathcal{C}$ be a class of $\omega$-languages., and let $D$ be one of the formalisms to represent regular $\omega$-languages. We consider the

**Problem:** $(\mathcal{C})_D$

*Given:* A $D$-representation $H$.
*Question:* Does $H$ represent a set in $\mathcal{C}$?

Sometimes one wants to consider such a problem not for all $\omega$-languages but only for such from a subclass $\mathcal{B}$. For example, one could be interested to decide aperiodicity only for open $\omega$-languages. This could be less complicated than deciding aperiodicity for all $\omega$-languages. This leads to the following type of problems.

**Problem:** $(\mathcal{C}|\mathcal{B})_D$

*Given:* A $D$-representation $H$ which represents a set in $\mathcal{B}$.
*Question:* Does $H$ represent a set in $\mathcal{C}$?

In particular, we consider the problems $(\mathcal{R}_\alpha)_D$, $(\mathcal{A})_D$, $(\mathcal{A}_\alpha)_D$, $(\text{co-}\mathcal{A}_\alpha)_D$, $(\mathcal{A}|\mathcal{R}_\alpha)_D$, $(\mathcal{A}|\text{co-}\mathcal{R}_\alpha)_D$ and the boolean combinations thereof. The following relationships are obvious:

**Proposition 1.** *Let $D$ be any formalism for the representation of $\omega$-regular languages, and let $\alpha < \omega^\omega$. There holds*
1. $(\mathcal{A}|\mathcal{R}_\alpha)_D \leq_{\mathrm{m}}^{\mathrm{P}} (\mathcal{A})_D$ *and* $(\mathcal{A}|\mathcal{R}_\alpha)_D \leq_{\mathrm{m}}^{\mathrm{P}} (\mathcal{A}_\alpha)_D$.
2. $(\mathcal{A}_\alpha)_D = (\mathcal{A})_D \cap (\mathcal{R}_\alpha)_D$.

Because of the duality of the deterministic Rabin acceptance and the deterministic Streett acceptance we have

**Proposition 2.** *If $\mathcal{C}$ is a class of $\omega$-languages then $(\mathcal{C})_{\mathrm{DS}} \equiv_{\mathrm{m}}^{\mathrm{P}} (\text{co-}\mathcal{C})_{\mathrm{DR}}$.*

All the introduced classes of $\omega$-automata (besides deterministic Büchi automata) are equivalent in the sense that they recognize the same $\omega$-languages. Moreover, the well known proofs of these equivalences are effective, i.e. from a given automaton of some type one can compute an equivalent automaton of any other type. When one is interested in complexity considerations (as we are here), the computational resources needed for finding the equivalent automaton and its size become important. For the purposes of this paper, the computability in (deterministic) polynomial time would be sufficient.

We say that a type $D$ of $\omega$-automata is polynomial time reducible to a type $D'$ of $\omega$-automata (for short $D \leq^{\mathrm{P}} D'$) if there exists a polynomial time computable function $f$ such that, for every automaton $\mathcal{M}$ of type $D$, the result $f(\mathcal{M})$ is an automaton of type $D'$ which accepts the same $\omega$-language as $\mathcal{M}$. The following relationship to decision problems is obvious:

**Proposition 3.** *Let $D$ and $D'$ be two types of $\omega$-automata, and let $P$ be a problem on $\omega$-languages. Then $D \leq^{\mathrm{P}} D'$ implies $(P)_D \leq_{\mathrm{m}}^{\mathrm{P}} (P)_{D'}$.*

Unfortunately, some of the well known reductions between automata representations of regular $\omega$-languages do not work in polynomial time. For some cases one can even prove that this is not possible. In [Sa88] an overview on possibility or impossibility of polynomial time reductions between different types of $\omega$-automata is given.

**Theorem 3.** [Sa88] *The following figure represents some results on polynomial time reductions between different types of $\omega$-automata. A solid line means that here exists a polynomial time reduction from the notion below to the notion above. A dotted arc means that polynomial time reduction in this direction is not proved and not disproved. Moreover, there are no further polynomial time reductions between these types of $\omega$-automata which do not already follow from the solid lines and dotted lines.*

The following fact was established in [SW07]:

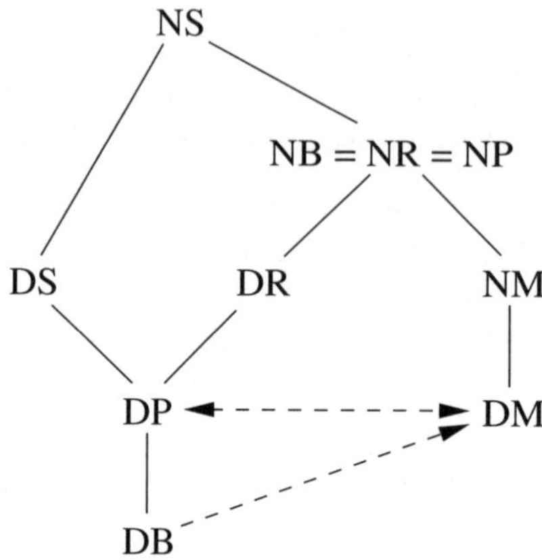

**Theorem 4.** *Let $\mathcal{C}$ be a Boolean combination of classes from $\{\mathcal{R}_\alpha \mid \alpha < \omega^\omega\} \cup \{\text{co-}\mathcal{R}_\alpha \mid \alpha < \omega^\omega\}$ where $\overline{\mathcal{C}} =_{\text{def}} \mathcal{R} \setminus \mathcal{C}$. Then $(\mathcal{C})_D \in \text{PSPACE}$ for $D \in \{\text{DB}, \text{NB}, \text{DM}, \text{NM}, \text{DR}, \text{NR}, \text{DP}, \text{NP}, \text{DS}\}$, i.e. the problem of deciding, given a deterministic or nondeterministic Büchi (Muller, Rabin, Mostowski) automaton or a deterministic Streett automaton $\mathcal{M}$, whether $L_\omega(\mathcal{M}) \in \mathcal{C}$ is in PSPACE.*

Apparently, the following result was shown only recently.

**Theorem 5. [DG08]** *The problem of deciding, given a nondeterministic Büchi automaton $\mathcal{M}$, whether the language $L_\omega(\mathcal{M})$ is aperiodic is PSPACE-complete, i.e., $(\mathcal{A})_{\text{NB}}$ is PSPACE-complete.*

To our knowledge, this is the first complexity result on aperiodicity of regular $\omega$-languages. The analogon for languages of finite words was already known:

**Theorem 6. [CH91]** *The problem of deciding, given a dfa $\mathcal{M}$, whether the language $L(\mathcal{M})$ is aperiodic is PSPACE-complete.*

As shown in [DG08], the same complexity estimation holds for the nfa's.

## 3   Complexity for Automata Representations

We first consider upper bounds to our problems.

**Lemma 1.** *For any $D \in \{\text{DB}, \text{NB}, \text{DM}, \text{NM}, \text{DR}, \text{NR}, \text{DP}, \text{NP}, \text{DS}\}$, the problem $(\mathcal{A})_D$ is in PSPACE, i.e., given a deterministic or nondeterministic Büchi (Muller, Rabin, Mostowski) automaton or a deterministic Streett automaton $\mathcal{M}$, it is decidable in polynomial space whether it accepts an aperiodic $\omega$-language.*

*Proof.* From Theorem 5, Proposition 3, and Theorem 3 we obtain the result for $D \in \{\mathrm{DB}, \mathrm{NB}, \mathrm{DM}, \mathrm{NM}, \mathrm{DR}, \mathrm{NR}, \mathrm{DP}, \mathrm{NP}\}$. The case $D = \mathrm{DS}$ then follows by Proposition 2. $\square$

From Lemma 1, Theorem 4, and Proposition 1 we also obtain

**Corollary 1.** $(\mathcal{A}_\alpha)_D, (\mathcal{A}|\mathcal{R}_\alpha)_D, (\text{co-}\mathcal{A}_\alpha)_D, (\mathcal{A}|\text{co-}\mathcal{R}_\alpha)_D, \in \mathrm{PSPACE}$ *for* $\alpha < \omega^\omega$ *and* $D \in \{\mathrm{DB}, \mathrm{NB}, \mathrm{DM}, \ \mathrm{NM}, \mathrm{DR}, \mathrm{NR}, \mathrm{DP}, \mathrm{NP}, \mathrm{DS}\}$.

Now we consider lower bounds for our problems. Define the block encoding (homomorphism) $c : \{0,1\}^+ \to \{0,1\}^*$ by $c(0) =_{\mathrm{def}} 11000$ and $c(1) =_{\mathrm{def}} 11010$.

**Lemma 2.** *For every regular language $L \subseteq \{0,1\}^+$, the language $L$ is aperiodic if and only if the $\omega$-language $c(L)111\{0,1\}^\omega$ is aperiodic. Moreover, given a dfa $\mathcal{M}$, one can compute in linear time a deterministic Muller (Büchi, Rabin, Mostowski, Streett) automaton $\mathcal{M}'$ over $A$ such that $L_\omega(\mathcal{M}')$ is an open set and $L(\mathcal{M})$ is aperiodic iff $L_\omega(\mathcal{M}')$ is aperiodic.*

*Proof.* Let $L$ be aperiodic, then $L = L_\phi$ for a first-order sentence $\phi$ of signature $\sigma_{\{0,1\}}$. It is easy to find (see [Se02] for details, were a slightly different encoding was used, but the argument applies also to our encoding here) a sentence $\phi'$ of signature $\sigma_{\{0,1\}}$ such that $\mathbf{w} \models \phi$ iff $\mathbf{c}(\mathbf{w}) \models \phi'$, for each $w \in \{0,1\}^+$. Moreover, it is easy to find a sentence $\psi$ of $\sigma_{\{0,1\}}$ such that $c(\{0,1\}^+) = L_\psi$. Obviously, $c(L) = c(\{0,1\}^+) \cap L_{\phi'}$ hence $c(L)$ is aperiodic. By remarks on the regular expressions in Section 2, the $\omega$-language $c(L)111\{0,1\}^\omega$ is also aperiodic.

Finally, let $L$ be non-aperiodic and let $\mathcal{M} = (Q, A, \delta, s_0, F)$ be the minimal dfa recognizing $L$. Then $\mathcal{M}$ is not aperiodic, hence there are states $s_1, \ldots, s_n$ (for some $n \geq 2$) and words $x, z, u, v$ such that: $\delta(s_0, x) = s_1$, $\delta(s_i, u) = s_i$ for all $i \in \{1, \ldots, n\}$, $\delta(s_i, v) = s_{i+1}$ for all $i \in \{1, \ldots, n-1\}$, $\delta(s_n, v) = s_1$, $\delta(s_1, y) \in F$, and $\delta(s_2, y) \notin F$. Then for any $k \geq 0$ we have $c(x)c(v)^{nk}c(z)1^\omega \in c(L)111\{0,1\}^\omega$ and $c(x)c(v)^{nk+1}c(z)1^\omega \notin c(L)111\{0,1\}^\omega$. It is well known (see e.g. Corollary 6.3 in [Th90]) that the last conditions imply that $c(L)111\{0,1\}^\omega$ is not aperiodic.

The second assertion is proved by a straightforward construction. $\square$

We are ready to give the

**Proof of Theorem 1.** By Lemma 1 and Theorem 4, the classes $\mathcal{A}_\alpha$ and co-$\mathcal{A}_\alpha$, for each $\alpha \in \omega^\omega$, are in PSPACE for any type of automata representations. Since PSPACE is closed under the Boolean operations, the class $\mathcal{C}$ is also in PSPACE. It remains to show that $\mathcal{C}$ is PSPACE-hard. By Theorem 6, it suffices to associate in polynomial time with any dfa $\mathcal{M}$ a deterministic $\omega$-automaton $\mathcal{M}_1$ of any considered type such that $L(\mathcal{M})$ is aperiodic iff $L_\omega(\mathcal{M}_1) \in \mathcal{C}$. W.l.o.g. we may assume that $\mathcal{A}_1 \leq_W \mathcal{C}$ for some $C \in \mathcal{C}$ (otherwise, $\mathcal{C} = \text{co-}\mathcal{A}_1$ by the structure of $(\mathcal{A}; \leq_W)$ and the assertion follows in the obvious way). Let $\mathcal{M}'$ be the automaton from the proof of Lemma 1. By remarks in Section 2 and Theorem 3, there is a deterministic aperiodic automaton (of any type) $\mathcal{M}_2$ with $L_\omega(\mathcal{M}_2) = C$. Let $\mathcal{M}_1$ be an automaton (of the corresponding type) satisfying

$L_\omega(\mathcal{M}_1) = 0L_\omega(\mathcal{M}')\cup 1L_\omega(\mathcal{M}_2)$. Then $L_\omega(\mathcal{M}_1) \in \mathcal{C}$ whenever $L(\mathcal{M})$ is aperiodic because $L_\omega(\mathcal{M}')$ is open and hence $L_\omega(\mathcal{M}') \leq_W A_1$ by [Wag79]. By Lemma 1, $L_\omega(\mathcal{M}_1)$ is not aperiodic whenever $L(\mathcal{M})$ is not aperiodic. Thus, the automaton $\mathcal{M}_1$ provides the desired reduction. Since it is constructed in a very easy and obvious way from the fixed automaton $\mathcal{M}_2$ and the easily computable automaton $\mathcal{M}'$, $\mathcal{M}_1$ is computable from $\mathcal{M}$ in polynomial time.    $\square$

The following assertion is an immediate corollary.

**Theorem 7.** *Let* $1 \leq \alpha < \omega^\omega$ *and* $D \in \{\mathrm{NB}, \mathrm{DM}, \mathrm{NM}, \mathrm{DR}, \mathrm{NR}, \mathrm{DP}, \mathrm{NP}, \mathrm{DS}\}$. *Then* $(\mathcal{A})_D$, $(\mathcal{A}_\alpha)_D$, $(\mathrm{co}\text{-}\mathcal{A}_\alpha)_D$, $(\mathcal{A}|\mathcal{R}_\alpha)_D$, $(\mathcal{A}|\mathrm{co}\text{-}\mathcal{R}_\alpha)_D$ *are* PSPACE-*hard for* $D = \mathrm{NS}$ *and* PSPACE-*complete for all the other cases.*

This completely solves the question of the complexity of aperiodicity for deterministic and nondeterministic Büchi, Muller, Rabin and Mostowski automata and for deterministic Streett automata. The nondeterministic Büchi case was known before, and the upper bound for nondeterministic Streett automata remains open.

# 4    Complexity for Logical Formula Representation

In this section we consider the complexity questions for the logical representation of regular $\omega$-languages, in particular we prove a generalization of Theorem 2.

For the upper bound of our result we need a result on the complexity of the weak second order theory of one successor. Let $\sigma_1$ be the signature which consists of only one binary predicate symbol $S$. A natural $\sigma_1$-structure is $(\mathbb{N}, S)$, i.e., the set of natural numbers with the successor relation $S$. Let WS1S be the set of monadic second-order $\sigma_1$-sentences $\phi$ true in $(\mathbb{N}, S)$ when the predicate variables are interpreted as finite subsets of $\mathbb{N}$ (in symbols, $(\mathbb{N}, S) \models_w \phi\})$.

For $k \geq 1$ define the tower-of-exponents function exp by $\exp(0, m) =_{\mathrm{def}} m$ and $\exp(k+1, m) =_{\mathrm{def}} 2^{\exp(k,m)}$. For sets $A$ and $B$, we write $A \leq_{\mathrm{m}}^{\text{p-lin}} B$ if there exists a polynomial time computable function $f$ such that $x \in A \leftrightarrow f(x) \in B$ and $|f(x)| \leq k \cdot |x|$ for all $x$. We will use the following result:

**Theorem 8 ([Me73]).** WS1S *is* $\leq_{\mathrm{m}}^{\text{p-lin}}$-*complete for* $\mathrm{DSPACE}(\exp(\mathcal{O}(n), 1))$.

We establish the lower complexity bound for our problems by reducing WS1S to them.

**Lemma 3.** *For any non-zero ordinal* $\alpha < \omega^\omega$, WS1S $\leq_{\mathrm{m}}^{\text{p-lin}} (\mathcal{A}|\mathcal{R}_\alpha)_{\mathrm{MSO}}$.

*Proof.* Let us start with explaining how to transform a $\sigma_1$-sentence into a "somehow equivalent" $\sigma_A$-sentence, for a given alphabet $A$. Let $\phi$ be a $\sigma_1$-sentence. We replace in $\phi$ equivalently any entry of the successor relation $S(x, y)$ with n $x < y \wedge \neg\exists z(x < z < y)$, any entry $\exists X \cdots$ of the existential predicate quantifier with $\exists X(Fin(X) \wedge \cdots)$ and any entry $\forall X \cdots$ of the universal predicate quantifier with $\forall X(Fin(X) \rightarrow \cdots)$ where $Fin(X)$ is $\exists x \forall y(X(y) \rightarrow y \leq x)$. In such a manner $\phi$ transforms in polynomial time into a $\sigma_A$-sentence $\phi'$ such that

$|\phi'| \leq k \cdot |\phi|$ for some universal constant $k > 0$. Observe that in $\phi'$ no relation symbol $\mathbf{Q_a}$ occurs, and hence either no structure $\overline{\xi}$ fulfills $\phi'$ or all structures $\overline{\xi}$ fulfill $\phi'$. More specific, $(\mathbb{N}, S) \models_w \phi$ implies $L_{\phi'} = A^\omega$ and $(\mathbb{N}, S) \not\models_w \phi$ implies $L_{\phi'} = \emptyset$.

Let $\psi$ be a $\sigma_A$-sentence such that $L_\psi \in \mathcal{R}_\alpha \setminus \mathcal{A}$. We conclude

$$\phi \in \text{WS1S} \implies (\mathbb{N}, S) \models_w \phi \implies L_{\phi'} = A^\omega \implies L_{\neg\phi'} = \emptyset \implies$$
$L_{\neg\phi' \wedge \psi} = \emptyset \in \mathcal{R}_\alpha \cap \mathcal{A}$ and

$$\phi \notin \text{WS1S} \implies (\mathbb{N}, S) \not\models_w \phi \implies L_{\phi'} = \emptyset \implies L_{\neg\phi'} = A^\omega \implies$$
$L_{\neg\phi' \wedge \psi} = L_\psi \in \mathcal{R}_\alpha - \mathcal{A}.$

Hence, the mapping $\phi \mapsto \neg\phi' \wedge \psi$ reduces WS1S to $(\mathcal{A}|\mathcal{R}_\alpha)_{\text{MSO}}$ in polynomial time and there holds $|\neg\phi' \wedge \psi| \leq k \cdot |\phi|$ for some universal constant $k > 0$.   $\square$

**Lemma 4.** *The sets $(\mathcal{R}_\alpha)_{\text{MSO}}$ and $(\mathcal{A})_{\text{MSO}}$ are in* $\text{DSPACE}(\exp(\mathcal{O}(n), 1))$.

*Proof.* As is well known (see e.g. [Bü60, Me73]), it is possible to translate a $\sigma_A$-sentence $\phi$ of length $n$ into an deterministic Muller automaton $\mathcal{M}$ of length $\exp(\mathcal{O}(n), 1)$ and check aperiodicity and membership in $\mathcal{R}_\alpha$ in polynomial space and polynomial time, resp., relative to length $\exp(\mathcal{O}(n), 1)$.   $\square$

From Lemma 3, Lemma 4, and Proposition 1 we get immediately

**Theorem 9.** *For any non-zero ordinal $\alpha < \omega^\omega$, the sets $(\mathcal{A}|\mathcal{R}_\alpha)_{\text{MSO}}$, $(\mathcal{A}_\alpha)_{\text{MSO}}$, and $(\mathcal{A})_{\text{MSO}}$ are $\leq_m^{\text{p-lin}}$-complete for* $\text{DSPACE}(\exp(\mathcal{O}(n), 1))$.

The proof of Theorem 2 is now an easy exercise.

The proof of Lemma 3 can be modified to show a dichotomic behavior of the complexity of problems on regular $\omega$-languages represented by logical formulas. We say that a problem $P$ of regular $\omega$-languages is *non-trivial* if there exist regular $\omega$-languages in $P$ and outside $P$. Theorem 9 is a particular case of the next result (proof is given in the appendix).

**Theorem 10. (Rice theorem for MSO representation)**
*If $P$ is a non-trivial problem of regular $\omega$-languages then $(P)_{\text{MSO}}$ is $\leq_m^{\text{p-lin}}$-hard for* $\text{DSPACE}(\exp(\mathcal{O}(n), 1))$. *If in addition $(P)_{\text{DM}}$ is in* $\text{DSPACE}(\exp(k, n))$ *for some $k > 0$ (i.e., if $(P)_{\text{DM}}$ is elementary recursive) then $(P)_{\text{MSO}}$ is even $\leq_m^{\text{p-lin}}$-complete for* $\text{DSPACE}(\exp(\mathcal{O}(n), 1))$.

## 5   Complexity for Regular Expression Representations

For the main result of this section we need the following lemmas where $c$ is the coding from Lemma 2.

**Lemma 5.** *Let $L$ be a regular language and $L'$ be a regular $\omega$-language.*

1. *If $L \neq \emptyset$ and $L'$ is not aperiodic then $c(L)111L'$ is not aperiodic.*
2. *For $\alpha < \omega^\omega$, if $L' \in \mathcal{R}_\alpha$ then $c(L)111L' \in \mathcal{R}_\alpha$, and if $L' \in \text{co-}\mathcal{R}_\alpha$ then $c(L)111L' \in \text{co-}\mathcal{R}_\alpha$*

*Proof.* 1) is checked just as its particular case in the proof of Lemma 2. 2) follows from the well known properties of the classes $\mathcal{R}_\alpha$ and co-$\mathcal{R}_\alpha$[Wag79].     $\square$

Set EMPTY $=_{\text{def}} \{\emptyset\}$.

**Lemma 6.** (EMPTY)$_\mathcal{O} \leq_{\text{m}}^{\text{p-lin}} (\mathcal{A}|\mathcal{R}_\alpha)_\mathcal{O}$ *and* (EMPTY)$_\mathcal{O} \leq_{\text{m}}^{\text{p-lin}} (\mathcal{A}|\text{co-}\mathcal{R}_\alpha)_\mathcal{O}$, *for* $1 \leq \alpha < \omega^\omega$ *and* $\{\cup,\cdot\} \subseteq \mathcal{O} \subseteq \{\cup,\cap,^-,\cdot,\,^2,^*\}$.

From the literature, lower and upper bounds for the emptyness problem for various sets $\mathcal{O} \subseteq \{\cup,\cap,^-,\cdot,\,^2,^*\}$ are known. We cite

**Theorem 11. [Sto74]** (EMPTY)$_{\{\cup,-,\cdot\}} \notin \text{DSPACE}(\exp(\log n, 1))$.

The results above imply the following:

**Theorem 12.** *Let* $\{\cup,^-,\cdot\} \subseteq \mathcal{O} \subseteq \{\cup,\cap,^-,\cdot,\,^2,^*\}$.
1. $(\mathcal{A})_\mathcal{O}, (\mathcal{A}_\alpha)_\mathcal{O}, (\text{co-}\mathcal{A}_\alpha)_\mathcal{O}, (\mathcal{A}|\mathcal{R}_\alpha)_\mathcal{O}, (\mathcal{A}|\text{co-}\mathcal{R}_\alpha)_\mathcal{O} \notin \text{DSPACE}(\exp(\log n, 1))$.
2. $(\mathcal{A})_\mathcal{O}, (\mathcal{A}_\alpha)_\mathcal{O}, (\text{co-}\mathcal{A}_\alpha)_\mathcal{O}, (\mathcal{A}|\mathcal{R}_\alpha)_\mathcal{O}, (\mathcal{A}|\text{co-}\mathcal{R}_\alpha)_\mathcal{O} \in \text{DSPACE}(\exp(n, 1))$.

*Proof.* The lower bound follows from the previous theorem, Lemma 6, and Proposition 1. For the upper bound converse the regular expressions of length $n$ into a deterministic Muller automaton of length $\exp(n, 1)$, and decide aperiodicity and membership in $\mathcal{R}_\alpha$ and co-$\mathcal{R}_\alpha$.     $\square$

**Acknowledgement.** The authors are grateful to Volker Diekert for the reference [DG08] and to Kai Salomaa for useful hints.

# References


[Bü60]     Büchi, J.R.: Weak second-order arithmetic and finite automata. Z. Math. Logic Grundl. Math. 6, 66–92 (1960)

[CH91]     Cho, S., Huynh, D.T.: Finite-automaton aperiodicity is PSPACE-complete. Theoretical Computer Science 88, 99–116 (1991)

[DG08]     Diekert, V., Gastin, P.: First-order definable languages. In: Flum, J., Grädel, E., Wilke, T. (eds.) Logic and Automata: History and Perspectives. Texts in Logic and Games, pp. 261–306. Amsterdam University Press (2008)

[KPB95]    Krishnan, S., Puri, A., Brayton, R.: Structural complexity of $\omega$-automata. LNCS, vol. 915, pp. 143–156. Springer, Berlin (1995)

[Me73]     Meyer, A.: Weak monadic second order theory of successor is not elementary-recursive. Project MAC TR 38 (1973)

[PP04]     Perrin, D., Pin, J.-E.: Infinite Words. Pure and Applied Mathematics, vol. 141. Elsevier, Amsterdam (2004)

[Sa88]     Safra, S.: On the complexity of $\omega$-automata. In: IEEE FOCS, pp. 319–327 (1988)

[Se02]     Selivanov, V.L.: Relating automata-theoretic hierarchies to complexity-theoretic hierarchies. Theoretical Informatics and Applications 36, 29–42 (2002)

[Se07]     Selivanov, V.L.: Fine hierarchy of regular aperiodic $\omega$-languages. In: Harju, T., Karhumäki, J., Lepistö, A. (eds.) DLT 2007. LNCS, vol. 4588, pp. 399–410. Springer, Heidelberg (2007)

[Sta97]   Staiger, L.: $\omega$-Languages. In: Handbook of Formal Languages, vol. 3, pp. 339–387. Springer, Berlin (1997)

[Sto74]   Stockmeyer, L.: The complexity of decision problems in automata theory and logic. Thesis, M.I.T. Project MAC TR-133 (1974)

[SW07]   Selivanov, V.L., Wagner, K.W.: Complexity of topological properties of regular $\omega$-languages. Fundamenta Informatica 20, 1–21 (2008)

[Th90]   Thomas, W.: Automata on infinite objects. In: Handbook of Theoretical Computer Science, vol. B, pp. 133–191 (1990)

[Th97]   Thomas, W.: Languages, automata and logic. In: Handbook of Formal Languages, vol. 3, pp. 389–455 (1997)

[Wag79]   Wagner, K.: On $\omega$-regular sets. Information and Control 43, 123–177 (1979)

[WY95]   Wilke, T., Yoo, H.: Computing the Wadge degree, the Lipschitz degree, and the Rabin index of a regular language of infinite words in polynomial time. LNCS, vol. 915, pp. 288–302. Springer, Berlin (1995)

# $\omega$-Degree Spectra

Alexandra A. Soskova[*]

Faculty of Mathematics and Computer Science,
Sofia University
asoskova@fmi.uni-sofia.bg

**Abstract.** We present a notion of a degree spectrum of a structure with respect to countably many sets, based on the notion of $\omega$-enumeration reducibility. We prove that some properties of the degree spectrum such as the minimal pair theorem and the existence of quasi-minimal degree are true for the $\omega$-degree spectrum.

## 1 Introduction

Let $\mathfrak{A} = (\mathbb{N}; R_1, \ldots, R_s)$ be a structure, where $\mathbb{N}$ is the set of all natural numbers, each $R_i$ is a subset of $\mathbb{N}^{r_i}$ and the equality $=$ and the inequality $\neq$ are among $R_1, \ldots, R_s$. *An enumeration $f$ of $\mathfrak{A}$* is a total mapping from $\mathbb{N}$ onto $\mathbb{N}$.

Given an enumeration $f$ of $\mathfrak{A}$ and a subset $A$ of $\mathbb{N}^a$ let $f^{-1}(A) = \{\langle x_1, \ldots, x_a\rangle \mid (f(x_1), \ldots, f(x_a)) \in A\}$. Denote by $f^{-1}(\mathfrak{A}) = f^{-1}(R_1) \oplus \ldots \oplus f^{-1}(R_s)$.

Given a set $X$ of natural numbers denote by $d_e(X)$ the enumeration degree of $X$ and by $d_T(X)$ the Turing degree of $X$.

The notion of *Turing degree spectrum* of $\mathfrak{A}$ is introduced by Richter [6]: $\mathrm{DS}_T(\mathfrak{A}) = \{d_T(f^{-1}(\mathfrak{A})) \mid f$ is an injective enumeration of $\mathfrak{A}\}$. Soskov [8] initiated the study of the properties of the degree spectra as sets of enumeration degrees. *The enumeration degree spectrum of $\mathfrak{A}$* (called shortly degree spectrum of $\mathfrak{A}$) is the set: $\mathrm{DS}(\mathfrak{A}) = \{d_e(f^{-1}(\mathfrak{A})) \mid f$ is an enumeration of $\mathfrak{A}\}$.

The benefit of considering all enumerations of the structure instead of only the injective ones is that every degree spectrum is upwards closed with respect to total enumeration degrees [8]. Soskov considered the notion of *co-spectrum* $\mathrm{CS}(\mathfrak{A})$ of $\mathfrak{A}$ as the set of all lower bounds of the elements of the degree spectrum of $\mathfrak{A}$ and proved several properties which show that the degree spectra behave with respect to their co-spectra very much like the cones of the enumeration degrees $\{\mathbf{x} \mid \mathbf{x} \geq \mathbf{a}\}$ behave with respect to the ideals $\{\mathbf{x} \mid \mathbf{x} \leq \mathbf{a}\}$. Further properties true of the degree spectra but not necessarily true of all upwards closed sets are: the minimal pair theorem for the degree spectrum and the existence of quasi-minimal degree for the degree spectrum.

In this paper we shall relativize Soskov's approach to degree spectra by considering multi-component spectra, i.e. a degree spectrum with respect to a given sequence of sets of natural numbers, considering the $\omega$-enumeration reducibility

---
[*] This work was partially supported by Sofia University Science Fund.

A. Beckmann, C. Dimitracopoulos, and B. Löwe (Eds.): CiE 2008, LNCS 5028, pp. 544–553, 2008.

introduced and studied in [10,11,12]. It is a uniform reducibility between sequences of sets of natural numbers. We shall prove that the so defined $\omega$-degree spectrum preserves almost all properties of the degree spectrum and generalizes the notion of relative spectrum, i.e multi-component spectrum of a structure with respect to finitely many structures, introduced in [13].

## 2  $\omega$-Enumeration Degrees

We assume the reader is familiar with enumeration reducibility and refer to [3] for further background. Denote by $\mathcal{D}_{\mathrm{e}}$ the set of all enumeration degrees. Recall that an enumeration degree $\mathbf{a}$ is total if $\mathbf{a}$ contains a total set, i.e. a set $A$ such that $A \equiv_{\mathrm{e}} A^+$, where $A^+ = A \oplus (\mathbb{N} \setminus A)$. If $X$ is a total set then $A \leq_{\mathrm{e}} X \iff A$ is c.e. in $X$. Cooper [2] introduced the jump operation $"'"$ for enumeration degrees. By $A'$ we denote the enumeration jump of the set $A$ and $d_{\mathrm{e}}(A)' = d_{\mathrm{e}}(A')$.

Denote by $\mathcal{S}$ the set of all sequences of sets of natural numbers. For each element $\mathcal{B} = \{B_n\}_{n<\omega}$ of $\mathcal{S}$ call *the jump class of $\mathcal{B}$* the set

$$J_{\mathcal{B}} = \{d_{\mathrm{T}}(X) \mid (\forall n)(B_n \text{ is c.e. in } X^{(n)} \text{ uniformly in } n)\} \ .$$

For every two sequences $\mathcal{A}$ and $\mathcal{B}$ let $\mathcal{A} \leq_{\omega} \mathcal{B}$ ($\mathcal{A}$ is $\omega$-*enumeration reducible to $\mathcal{B}$*) if $J_{\mathcal{B}} \subseteq J_{\mathcal{A}}$ and let $\mathcal{A} \equiv_{\omega} \mathcal{B}$ if $J_{\mathcal{A}} = J_{\mathcal{B}}$. The relation $\equiv_{\omega}$ is an equivalence relation on $\mathcal{S}$. Let the $\omega$-*enumeration degree of $\mathcal{B}$* be $d_{\omega}(\mathcal{B}) = \{\mathcal{A} \mid \mathcal{A} \equiv_{\omega} \mathcal{B}\}$ and $\mathcal{D}_{\omega} = \{d_{\omega}(\mathcal{B}) \mid \mathcal{B} \in \mathcal{S}\}$. If $\mathbf{a} = d_{\omega}(\mathcal{A})$ and $\mathbf{b} = d_{\omega}(\mathcal{B})$ then $\mathbf{a} \leq_{\omega} \mathbf{b}$ if $\mathcal{A} \leq_{\omega} \mathcal{B}$. Denote by $\mathbf{0}_{\omega} = d_{\omega}(\emptyset_{\omega})$, where $\emptyset_{\omega}$ is the sequence with all members equal to $\emptyset$. There is a natural embedding of the enumeration degrees into the $\omega$-enumeration degrees. Given a set $A$ denote by $A \uparrow \omega$ the sequence $\{A_n\}_{n<\omega}$, where $A_0 = A$ and for all $n > 0$ $A_n = \emptyset$. For every $A, B \subseteq \mathbb{N}$ we have that $A \leq_{\mathrm{e}} B \iff A \uparrow \omega \leq_{\omega} B \uparrow \omega$. So the mapping $\kappa(d_{\mathrm{e}}(A)) = d_{\omega}(A \uparrow \omega)$ gives an isomorphic embedding of $\mathcal{D}_{\mathrm{e}}$ to $\mathcal{D}_{\omega}$. We shall identify the enumeration degree $d_{\mathrm{e}}(A)$ with its representation $d_{\omega}(A \uparrow \omega)$ in $\mathcal{D}_{\omega}$. So when $\mathbf{a} = d_{\mathrm{e}}(A)$ and $\mathbf{b} \in \mathcal{D}_{\omega}$ then by writing $\mathbf{a} \leq_{\omega} \mathbf{b}$ ($\mathbf{b} \leq_{\omega} \mathbf{a}$) we mean $d_{\omega}(A \uparrow \omega) \leq_{\omega} \mathbf{b}$ ($\mathbf{b} \leq_{\omega} d_{\omega}(A \uparrow \omega)$).

Given a sequence of sets of natural numbers $\mathcal{B} = \{B_n\}_{n<\omega}$ we define the respective *jump sequence* $\mathcal{P}(\mathcal{B}) = \{\mathcal{P}_n(\mathcal{B})\}_{n<\omega}$ by induction on $n$:

(1) $\mathcal{P}_0(\mathcal{B}) = B_0$;
(2) $\mathcal{P}_{n+1}(\mathcal{B}) = (\mathcal{P}_n(\mathcal{B}))' \oplus B_{n+1}$ .

Note that if $X \subseteq \mathbb{N}$, then $\mathcal{P}_n(X \uparrow \omega) \equiv_{\mathrm{e}} X^{(n)}$ uniformly in $n$.

The following theorem of Soskov and Kovachev [10] gives an explicit characterization of the uniform reducibility.

**Theorem 1.** *Let $\mathcal{A} = \{A_n\}_{n<\omega}$ and $\mathcal{B} = \{B_n\}_{n<\omega}$ be elements of $\mathcal{S}$. The following conditions are equivalent:*

*(1) $\mathcal{A} \leq_{\omega} \mathcal{B}$, i.e. for every total set $X$, if $B_n \leq_{\mathrm{e}} X^{(n)}$ uniformly in $n$ then $A_n \leq_{\mathrm{e}} X^{(n)}$ uniformly in $n$.*

*(2) $A_n \leq_{\mathrm{e}} \mathcal{P}_n(\mathcal{B})$ uniformly in $n$, i.e. there is a computable function $g$ such that $A_n = \Gamma_{g(n)}(\mathcal{P}_n(\mathcal{B}))$ for every $n$.*

It follows that if $X \subseteq \mathbb{N}$ then for every sequence $\mathcal{A} = \{A_n\}_{n<\omega}$ we have: $A_n \leq_e X^{(n)}$ uniformly in $n$ if and only if $\mathcal{A} \leq_\omega \{X^{(n)}\}_{n<\omega}$ if and only if $\mathcal{A} \leq_\omega X \uparrow \omega$. It is clear also that $\mathcal{A} \equiv_\omega \mathcal{P}(\mathcal{A})$.

With a slight modification of the proof of Theorem 1 we have the following:

**Corollary 2.** *Let $\mathcal{A}_0, \ldots, \mathcal{A}_r, \ldots$ be sequences of sets such that for every $r$, $\mathcal{A}_r \not\leq_\omega \mathcal{B}$. There is a total set $X$ such that $\mathcal{B} \leq_\omega \{X^{(n)}\}_{n<\omega}$ and $\mathcal{A}_r \not\leq_\omega \{X^{(n)}\}_{n<\omega}$ for each $r$.*

The jump operator on the $\omega$-enumeration degrees is defined by Soskov [11]. For every $\mathcal{A} \in \mathcal{S}$ the $\omega$-*enumeration jump* of $\mathcal{A}$ is $\mathcal{A}' = \{\mathcal{P}_{n+1}(\mathcal{A})\}_{n<\omega}$ and $d_\omega(\mathcal{A})' = d_\omega(\mathcal{A}')$. Furthermore $\mathcal{A}^{(k+1)} = (\mathcal{A}^{(k)})'$ and $d_\omega(\mathcal{A})^{(k+1)} = d_\omega(\mathcal{A}^{(k+1)})$. Then $\mathcal{A}^{(k)} = \{\mathcal{P}_{n+k}(\mathcal{A})\}_{n<\omega}$ for each $k$.

## 3   The $\omega$-Degree Spectra of Structures

In this section we shall generalize the notion of degree spectrum of $\mathfrak{A}$ by considering a multi-component spectrum. The first step in this direction was the notion of relative spectrum of the structure $\mathfrak{A}$ with respect to finitely many given structures $\mathfrak{A}_1, \ldots, \mathfrak{A}_n$ studied in [13]. *The relative spectrum* $\mathrm{RS}(\mathfrak{A}, \mathfrak{A}_1, \ldots, \mathfrak{A}_n)$ of the structure $\mathfrak{A}$ with respect to $\mathfrak{A}_1, \ldots, \mathfrak{A}_n$ is the set

$$\{d_e(f^{-1}(\mathfrak{A})) \mid f \text{ is an enumeration of } \mathfrak{A} \text{ s. t. } (\forall k \leq n)(f^{-1}(\mathfrak{A}_k) \leq_e f^{-1}(\mathfrak{A})^{(k)})\}.$$

It turns out that all properties of the degree spectra obtained by Soskov [8] remain true for the relative spectra.

We shall define here the notion of a spectrum of the structure $\mathfrak{A}$ with respect to a given infinite sequence of sets using the $\omega$-enumeration reducibility.

Let $\mathcal{B} = \{B_n\}_{n<\omega}$ be a sequence of sets of natural numbers. An enumeration $f$ of $\mathfrak{A}$ is called *acceptable with respect to the sequence* $\mathcal{B}$ if for every $n$, $f^{-1}(B_n) \leq_e f^{-1}(\mathfrak{A})^{(n)}$ uniformly in $n$. Denote by $\mathcal{E}(\mathfrak{A}, \mathcal{B})$ the class of all acceptable enumerations of $\mathfrak{A}$ with respect to the sequence $\mathcal{B}$.

**Definition 3.** *The $\omega$-degree spectrum* of the structure $\mathfrak{A}$ with respect to the sequence $\mathcal{B}$ is the set $\mathrm{DS}(\mathfrak{A}, \mathcal{B}) = \{d_e(f^{-1}(\mathfrak{A})) \mid f \in \mathcal{E}(\mathfrak{A}, \mathcal{B})\}$ .

The notion of the $\omega$-degree spectrum is a generalization of the relative spectrum since $\mathrm{RS}(\mathfrak{A}, \mathfrak{A}_1, \ldots, \mathfrak{A}_n) = \mathrm{DS}(\mathfrak{A}, \mathcal{B})$, where $\mathcal{B} = \{B_k\}_{k<\omega}$, $B_0 = \emptyset$, $B_k$ is the positive diagram of the structure $\mathfrak{A}_k$ for $0 < k \leq n$ and $B_k = \emptyset$ for all $k > n$.

Given an enumeration $f$ of $\mathfrak{A}$ denote by $\mathcal{P}^f = \{\mathcal{P}_n^f\}_{n<\omega}$ the respective jump sequence of the sequence $\{f^{-1}(\mathfrak{A}) \oplus f^{-1}(B_0), f^{-1}(B_1), \ldots, f^{-1}(B_n), \ldots\}$ where $\mathcal{P}_n^f = \mathcal{P}_n(\{f^{-1}(\mathfrak{A}) \oplus f^{-1}(B_0), f^{-1}(B_1), \ldots, f^{-1}(B_n), \ldots\})$. Note that if $f$ is an acceptable enumeration of $\mathfrak{A}$ with respect to $\mathcal{B}$ then $\mathcal{P}^f \equiv_\omega \{f^{-1}(\mathfrak{A})^{(n)}\}_{n<\omega} \equiv_\omega f^{-1}(\mathfrak{A}) \uparrow \omega$. So $f \in \mathcal{E}(\mathfrak{A}, \mathcal{B})$ if and only if $\mathcal{P}^f \leq_\omega f^{-1}(\mathfrak{A}) \uparrow \omega$.

First we shall see that the $\omega$-degree spectrum of the structure $\mathfrak{A}$ with respect to $\mathcal{B}$ is upwards closed with respect to total enumeration degrees.

**Lemma 4.** *Let $f$ be an enumeration of $\mathfrak{A}$ and $F$ be a total set such that $f^{-1}(\mathfrak{A}) \leq_e F$ and $f^{-1}(B_n) \leq_e F^{(n)}$ uniformly in $n$. Then there exists an acceptable enumeration $g$ of $\mathfrak{A}$ with respect to $\mathcal{B}$ such that $g^{-1}(\mathfrak{A}) \equiv_e F$.*

*Proof.* The construction of $g$ is the following. Let $s \neq t \in \mathbb{N}$.

$$g(x) \simeq \begin{cases} f(x/2) & \text{if } x \text{ is even,} \\ s & \text{if } x = 2z+1 \text{ and } z \in F, \\ t & \text{if } x = 2z+1 \text{ and } z \notin F. \end{cases}$$

It is easy to see that $F \oplus f^{-1}(\mathfrak{A}) \equiv_e g^{-1}(\mathfrak{A})$ and hence $F \equiv_e g^{-1}(\mathfrak{A})$. Moreover for every set $B \subseteq \mathbb{N}$ we have that $g^{-1}(B) \leq_e F \oplus f^{-1}(B)$.

Then $g^{-1}(B_n) \leq_e F \oplus f^{-1}(B_n) \leq_e F \oplus F^{(n)} \equiv_e F^{(n)} \equiv_e g^{-1}(\mathfrak{A})^{(n)}$ uniformly in $n$. And thus $g$ is an acceptable enumeration of $\mathfrak{A}$ with respect to $\mathcal{B}$.    $\square$

Using Theorem 1 and the previous lemma one can find an acceptable enumeration $f \in \mathcal{E}(\mathfrak{A}, \mathcal{B})$ such that $f^{-1}(\mathfrak{A})$ is a total set.

Another corollary of Lemma 4 is the following:

**Proposition 5.** *The $\omega$-degree spectrum is upwards closed with respect to total e-degrees, i.e. if* $\mathbf{b}$ *is a total e-degree and for some* $\mathbf{a} \in \mathrm{DS}(\mathfrak{A}, \mathcal{B})$, $\mathbf{b} \geq_e \mathbf{a}$ *then* $\mathbf{b} \in \mathrm{DS}(\mathfrak{A}, \mathcal{B})$.

It is obvious that $\mathrm{DS}(\mathfrak{A}) \supseteq \mathrm{DS}(\mathfrak{A}, \mathcal{B})$ for every structure $\mathfrak{A}$ and every $\mathcal{B} \in \mathcal{S}$. It is easy to find a structure $\mathfrak{A}$ and a sequence of sets $\mathcal{B}$ so that $\mathrm{DS}(\mathfrak{A}) \neq \mathrm{DS}(\mathfrak{A}, \mathcal{B})$. For example consider the structure $\mathfrak{A} = \{\mathbb{N}, S, =, \neq\}$, where $S \subseteq \mathbb{N}^2$ is defined as $S = \{(n, n+1) \mid n \in \mathbb{N}\}$. It is clear that the structure $\mathfrak{A}$ admits an effective enumeration $f$, i.e. $f^{-1}(\mathfrak{A})$ is c.e. Thus $\mathbf{0}_e \in \mathrm{DS}(\mathfrak{A})$. By Proposition 5 all total enumeration degrees are elements of $\mathrm{DS}(\mathfrak{A})$. Consider now an arbitrary acceptable enumeration $f$ of $\mathfrak{A}$ with respect to $\mathcal{B} = \{B_n\}_{n<\omega}$. Fix a number $x_0$ such that $f(x_0) = 0$. Then $k \in B_n \iff (\exists x_1) \ldots (\exists x_k)(f^{-1}(S)(x_0, x_1) \& \ldots \& f^{-1}(S)(x_{k-1}, x_k) \& x_k \in f^{-1}(B_n))$. Then $B_n \leq_e f^{-1}(\mathfrak{A}) \oplus f^{-1}(B_n) \leq_e f^{-1}(\mathfrak{A})^{(n)}$. Let $B_0 = \emptyset'$ and let $B_n = \emptyset$ for each $n \geq 1$. Then $\emptyset' \leq_e B_0 \leq_e f^{-1}(\mathfrak{A})$. Thus $\mathbf{0}_e \notin \mathrm{DS}(\mathfrak{A}, \mathcal{B})$.

Let $k \in \mathbb{N}$. *The kth $\omega$-jump spectrum of $\mathfrak{A}$ with respect to $\mathcal{B}$* is the set $\mathrm{DS}_k(\mathfrak{A}, \mathcal{B}) = \{\mathbf{a}^{(k)} \mid \mathbf{a} \in \mathrm{DS}(\mathfrak{A}, \mathcal{B})\}$.

**Proposition 6.** *The kth $\omega$-jump spectrum of $\mathfrak{A}$ with respect to $\mathcal{B}$ is upwards closed with respect to total e-degrees, i.e. if* $\mathbf{b}$ *is a total e-degree,* $\mathbf{b} \geq_e \mathbf{a}^{(k)}$ *for some* $\mathbf{a} \in \mathrm{DS}(\mathfrak{A}, \mathcal{B})$ *then* $\mathbf{b} \in \mathrm{DS}_k(\mathfrak{A}, \mathcal{B})$.

*Proof.* Let $G$ be a total set, $d_e(G) = \mathbf{b}$ and let $f \in \mathcal{E}(\mathfrak{A}, \mathcal{B})$ such that $f^{-1}(\mathfrak{A}) \in \mathbf{a}$. Then $f^{-1}(\mathfrak{A})^{(k)} \leq_e G$ and $\mathcal{P}_k^f \leq_e G$ since $\mathcal{P}_k^f \leq_e f^{-1}(\mathfrak{A})^{(k)}$. By the jump inversion theorem from [7] there exists a total set $F$ such that $G \equiv_e F^{(k)}$, $f^{-1}(\mathfrak{A}) \leq_e F$ and $f^{-1}(B_i) \leq_e F^{(i)}$ for $i \leq k$. Moreover $f^{-1}(B_{n+k}) \leq_e f^{-1}(\mathfrak{A})^{(n+k)} \leq_e G^{(n)} \equiv_e F^{(n+k)}$ uniformly in $n$. By Lemma 4 there is an acceptable enumeration $g$ of $\mathfrak{A}$ with respect to $\mathcal{B}$ so that $g^{-1}(\mathfrak{A}) \equiv_e F$. Thus $d_e(g^{-1}(\mathfrak{A})) \in \mathrm{DS}(\mathfrak{A}, \mathcal{B})$ and $g^{-1}(\mathfrak{A})^{(k)} \equiv_e G$. Hence $d_e(G) \in \mathrm{DS}_k(\mathfrak{A}, \mathcal{B})$.    $\square$

For every $\mathcal{D} \subseteq \mathcal{D}_e$ denote by $co(\mathcal{D}) = \{\mathbf{b} \mid \mathbf{b} \in \mathcal{D}_\omega \ \& \ (\forall \mathbf{a} \in \mathcal{D})(\mathbf{b} \leq_\omega \mathbf{a})\}$.

**Definition 7.** *The $\omega$-co-spectrum of $\mathfrak{A}$ with respect to $\mathcal{B}$ is the set* $\mathrm{CS}(\mathfrak{A}, \mathcal{B}) = co(\mathrm{DS}(\mathfrak{A}, \mathcal{B}))$.

For each $\mathcal{A} \in \mathcal{S}$ it holds that $d_\omega(\mathcal{A}) \in \mathrm{CS}(\mathfrak{A}, \mathcal{B})$ if and only if $\mathcal{A} \leq_\omega \mathcal{P}^f$ for every acceptable enumeration $f$ of $\mathfrak{A}$ with respect to $\mathcal{B}$. Actually the $\omega$-co-spectrum of $\mathfrak{A}$ with respect to $\mathcal{B}$ is a countable ideal of $\omega$-enumeration degrees.

The $k$th $\omega$-co-spectrum of $\mathfrak{A}$ with respect to $\mathcal{B}$ is the set $\mathrm{CS}_k(\mathfrak{A}, \mathcal{B}) = co(\mathrm{DS}_k(\mathfrak{A}, \mathcal{B}))$. It is clear that $d_\omega(\mathcal{A}) \in \mathrm{CS}_k(\mathfrak{A}, \mathcal{B})$ if and only if $\mathcal{A} \leq_\omega \{\mathcal{P}^f_{n+k}\}_{n<\omega}$ for every acceptable enumeration $f$ of $\mathfrak{A}$ with respect to $\mathcal{B}$. As we shall see in section 4. Corollary 22, the $k$th $\omega$-co-spectrum of $\mathfrak{A}$ with respect to $\mathcal{B}$ is the least ideal containing all $k$th $\omega$-enumeration jumps of the elements of $\mathrm{CS}(\mathfrak{A}, \mathcal{B})$.

In order to obtain a forcing normal form of the sequences with $\omega$-enumeration degrees in $\mathrm{CS}(\mathfrak{A}, \mathcal{B})$ we shall define the notions of a forcing relation $\tau \Vdash_n F_e(x)$ and a relation $f \models_n F_e(x)$.

Let $f$ be an enumeration of $\mathfrak{A}$. For every $n$ and $e$, $x \in \mathbb{N}$, define the relations $f \models_n F_e(x)$ and $f \models_n \neg F_e(x)$ by induction on $n$:

$$
\begin{aligned}
&1.\ f \models_0 F_e(x) \iff && (\exists v)(\langle v, x \rangle \in W_e\ \&\ D_v \subseteq f^{-1}(\mathfrak{A}) \oplus f^{-1}(B_0)); \\
&2.\ f \models_{n+1} F_e(x) \iff && (\exists v)(\langle v, x \rangle \in W_e\ \&\ (\forall u \in D_v)( \\
& && (u = \langle 0, e_u, x_u \rangle\ \&\ f \models_n F_{e_u}(x_u))\ \vee \\
& && (u = \langle 1, e_u, x_u \rangle\ \&\ f \models_n \neg F_{e_u}(x_u))\ \vee \\
& && (u = \langle 2, x_u \rangle\ \&\ x_u \in f^{-1}(B_{n+1})))); \\
&3.\ f \models_n \neg F_e(x) \iff && f \not\models_n F_e(x)\ .
\end{aligned}
$$

**Lemma 8.** (a) *Let* $A \subseteq \mathbb{N}$, $n \in \mathbb{N}$. *Then* $A \leq_e \mathcal{P}^f_n$ *if and only if* $A = \{x \mid f \models_n F_e(x)\}$ *for some* $e \in \mathbb{N}$.

(b) *Let* $\mathcal{A} = \{A_n\}_{n<\omega}$. *Then* $\mathcal{A} \leq_\omega \mathcal{P}^f$ *if and only if there exists a computable function* $g$ *such that* $A_n = \{x \mid f \models_n F_{g(n)}(x)\}$ *for every* $n$.

The forcing conditions, called *finite parts*, are finite mappings $\tau$ of $\mathbb{N}$ to $\mathbb{N}$. We will denote the finite parts by letters $\delta, \tau, \rho$.

For each $n$ and $e$, $x \in \mathbb{N}$ and for every finite part $\tau$, define the forcing relations $\tau \Vdash_n F_e(x)$ and $\tau \Vdash_n \neg F_e(x)$ following the definition of the relation "$\models_n$".

$$
\begin{aligned}
&1.\ \tau \Vdash_0 F_e(x) \iff && (\exists v)(\langle v, x \rangle \in W_e\ \&\ D_v \subseteq \tau^{-1}(\mathfrak{A}) \oplus \tau^{-1}(B_0)); \\
&2.\ \tau \Vdash_{n+1} F_e(x) \iff && (\exists v)(\langle v, x \rangle \in W_e\ \&\ (\forall u \in D_v)( \\
& && (u = \langle 0, e_u, x_u \rangle\ \&\ \tau \Vdash_n F_{e_u}(x_u))\ \vee \\
& && (u = \langle 1, e_u, x_u \rangle\ \&\ \tau \Vdash_n \neg F_{e_u}(x_u))\ \vee \\
& && (u = \langle 2, x_u \rangle\ \&\ x_u \in \tau^{-1}(B_{n+1})))); \\
&3.\ \tau \Vdash_n \neg F_e(x) \iff && (\forall \rho \supseteq \tau)(\rho \not\Vdash_n F_e(x))\ .
\end{aligned}
$$

An enumeration $f$ of $\mathfrak{A}$ is *$k$-generic with respect to $\mathcal{B}$* if for every $j < k$ and $e, x \in \mathbb{N}$ it holds that $(\exists \tau \subseteq f)(\tau \Vdash_j F_e(x) \vee \tau \Vdash_j \neg F_e(x))$.

**Lemma 9.** (1) *If* $\tau \subseteq \rho$ *then* $\tau \Vdash_k (\neg)F_e(x) \Rightarrow \rho \Vdash_k (\neg)F_e(x)$.

(2) *For every* $(k+1)$-*generic enumeration* $f$ *of* $\mathfrak{A}$ *with respect to* $\mathcal{B}$
$$f \models_k (\neg)F_e(x) \iff (\exists \tau \subseteq f)(\tau \Vdash_k (\neg)F_e(x)).$$

**Definition 10.** Let $\mathcal{A} = \{A_n\}_{n<\omega}$. The sequence $\mathcal{A}$ is *forcing definable on* $\mathfrak{A}$ *with respect to* $\mathcal{B}$ if there exist a finite part $\delta$ and a computable function $g$ such that for every $n \in \mathbb{N}[\ x \in A_n \iff (\exists \tau \supseteq \delta)(\tau \Vdash_n F_{g(n)}(x))\ ]$.

**Proposition 11.** *Let $\mathcal{A} = \{A_n\}_{n<\omega}$ be a sequence not forcing definable on $\mathfrak{A}$ with respect to $\mathcal{B}$. Then there exists an enumeration $f$ of $\mathfrak{A}$ such that $\mathcal{A} \not\leq_\omega \mathcal{P}^f$.*

*Proof.* The enumeration $f$ is constructed on stages. On each stage $q$ we find a finite part $\delta_q$ so that $\delta_q \subseteq \delta_{q+1}$ and ultimately we define $f = \bigcup_q \delta_q$. We consider three kinds of stages. On stages $q = 3r$ we ensure that the mapping $f$ is total and surjective. On stages $q = 3r + 1$ we ensure that $f$ is $k$-generic for each $k > 0$ and on stages $q = 3r + 2$ we ensure that $f$ satisfies the omitting condition: $\mathcal{A} \not\leq_\omega \mathcal{P}^f$.

Let $g_0, g_1, \ldots$ be an enumeration of all computable functions. For each $n, e$, $x \in \mathbb{N}$ denote by $Y^n_{\langle e,x\rangle}$ the set of all finite parts $\rho$ such that $\rho \Vdash_n F_e(x)$.

Let $\delta_0 = \emptyset$. Suppose that we have already defined $\delta_q$.

(a) Case $q = 3r$. Let $x_0$ be the least natural number which does not belong to $\mathrm{dom}(\delta_q)$ and let $t_0$ be the least natural number which does not belong to the range of $\delta_q$. Set $\delta_{q+1}(x_0) \simeq t_0$ and $\delta_{q+1}(x) \simeq \delta_q(x)$ for $x \neq x_0$.

(b) Case $q = 3\langle e, n, x\rangle + 1$. Check whether there exists a finite part $\rho \in Y^n_{\langle e,x\rangle}$ that extends $\delta_q$. If there is such a finite part then let $\delta_{q+1}$ be the least extension of $\delta_q$ that belongs to $Y^n_{\langle e,x\rangle}$. Otherwise let $\delta_{q+1} = \delta_q$.

(c) Case $q = 3r + 2$. Consider the function $g_r$. For each $n$ denote by

$$C_n = \{x \mid (\exists \tau \supseteq \delta_q)(\tau \Vdash_n F_{g_r(n)}(x))\} \ .$$

Clearly $\mathcal{C} = \{C_n\}_{n<\omega}$ is forcing definable on $\mathfrak{A}$ with respect to $\mathcal{B}$ and hence $\mathcal{C} \neq \mathcal{A}$. Then $C_n \neq A_n$ for some $n$. Let $\langle x, n\rangle$ be the least pair such that

$$x \in C_n \ \& \ x \notin A_n \ \lor \ x \notin C_n \ \& \ x \in A_n \ .$$

(i) Suppose that $x \in C_n$. Then there exists a finite part $\tau$ such that

$$\delta_q \subseteq \tau \ \& \ \tau \Vdash_n F_{g_r(n)}(x) \ . \tag{1}$$

Let $\delta_{q+1}$ be the least $\tau$ satisfying (1).

(ii) If $x \notin C_n$ then set $\delta_{q+1}(x) \simeq \delta_q(x)$. Note that in this case we have that $\delta_{q+1} \Vdash_n \neg F_{g_r(n)}(x)$.

Let $f = \bigcup_q \delta_q$. The enumeration $f$ is total and surjective. Let $k \in \mathbb{N}$. In order to prove that $f$ is $(k+1)$-generic, suppose that $j \leq k$ and consider the stage $q = 3\langle e, j, x\rangle + 1$. If there is a finite part $\rho \supseteq \delta_q$ such that $\rho \Vdash_j F_e(x)$ then from the construction we have that $\delta_{q+1} \Vdash_j F_e(x)$. Otherwise $\delta_{q+1} \Vdash_j \neg F_e(x)$.

To prove that $f$ satisfies the omitting condition suppose for a contradiction that $\mathcal{A} \leq_\omega \mathcal{P}^f$. Then there exists a computable function $g_s$ such that for each $n$ we have that $A_n = \{x \mid f \models_n F_{g_s(n)}(x)\}$. Since the enumeration $f$ is $(n+1)$-generic, by Lemma 9 we have for each number $x$:

$$f \models_n (\neg) F_{g_s(n)}(x) \iff (\exists \tau \subseteq f)(\tau \Vdash_n (\neg) F_{g_s(n)}(x)). \tag{2}$$

Consider the stage $q = 3s + 2$. From the construction there are numbers $n$ and $x$ such that one of the following two cases holds:

(i) $x \notin A_n \ \& \ \delta_{q+1} \Vdash_n F_{g_s(n)}(x)$. By (2) $f \models_n F_{g_s(n)}(x)$ and hence $x \in A_n$. A contradiction.

(ii) $x \in A_n$ and $(\forall \rho \supseteq \delta_q)(\rho \not\Vdash_n F_{g_s(n)}(x))$. Then $\delta_q \Vdash_n \neg F_{g_s(n)}(x)$. So by (2), $f \not\Vdash_n F_{g_s(n)}(x)$ and hence $x \notin A_n$. A contradiction.    $\square$

**Corollary 12.** *Let* $\mathcal{A}_0, \mathcal{A}_1, \ldots, \mathcal{A}_i \ldots$ *be a sequence of elements of* $\mathcal{S}$ *such that each* $\mathcal{A}_i$ *is not forcing definable on* $\mathfrak{A}$ *with respect to* $\mathcal{B}$. *Then there exists an enumeration* $f$ *of* $\mathfrak{A}$ *such that* $\mathcal{A}_i \not\leq_\omega \mathcal{P}^f$ *for each* $i$.

The construction of the enumeration $f$ is very similar to that in Proposition 11. On stages of the form $q = 3\langle r, i \rangle + 2$ we consider the computable function $g_r$ and ensure that $\mathcal{A}_i \neq \mathcal{C}$, where the sequence $\mathcal{C}$ is defined by the same way.

**Proposition 13.** *Let* $\mathcal{A} = \{A_n\}_{n<\omega}$ *be a sequence not forcing definable on* $\mathfrak{A}$ *with respect to* $\mathcal{B}$. *Then there exists an enumeration* $g \in \mathcal{E}(\mathfrak{A}, \mathcal{B})$ *such that* $\mathcal{A} \not\leq_\omega \mathcal{P}^g$ *and the enumeration degree of* $g^{-1}(\mathfrak{A})$ *is total.*

*Proof.* By Proposition 11 there is an enumeration $f$ of $\mathfrak{A}$ such that $\mathcal{A} \not\leq_\omega \mathcal{P}^f$. Then by Theorem 1 there exists a total set $F$ such that $\mathcal{P}^f \leq_\omega \{F^{(n)}\}_{n<\omega}$ and $\mathcal{A} \not\leq_\omega \{F^{(n)}\}_{n<\omega}$. By Lemma 4 there exists an acceptable enumeration $g$ of $\mathcal{A}$ with respect to $\mathcal{B}$ such that $g^{-1}(\mathfrak{A}) \equiv_e F$ and hence $g^{-1}(\mathfrak{A})^{(n)} \equiv_e F^{(n)}$ uniformly in $n$. It is clear that $\mathcal{A} \not\leq_\omega \mathcal{P}^g$.    $\square$

**Corollary 14.** *For every sequence* $\mathcal{A}$ *if* $d_\omega(\mathcal{A}) \in \mathrm{CS}(\mathcal{A}, \mathcal{B})$ *then* $\mathcal{A}$ *is forcing definable on* $\mathfrak{A}$ *with respect to* $\mathcal{B}$.

*Proof.* If a sequence $\mathcal{A}$ is not forcing definable on $\mathfrak{A}$ with respect to $\mathcal{B}$ then by Proposition 13 there exists an acceptable enumeration $g$ of $\mathfrak{A}$ with respect to $\mathcal{B}$ such that $\mathcal{A} \not\leq_\omega \mathcal{P}^g$. Hence $d_\omega(\mathcal{A}) \notin \mathrm{CS}(\mathfrak{A}, \mathcal{B})$.    $\square$

**Definition 15.** *Let* $k \in \mathbb{N}$, $\mathcal{A} \in \mathcal{S}$ *and let* $\mathcal{A} = \{A_n\}_{n<\omega}$. *The sequence* $\mathcal{A}$ *is forcing $k$-definable on* $\mathfrak{A}$ *with respect to* $\mathcal{B}$ *if there exist a finite part* $\delta$ *and a computable function* $g$ *such that for every* $n \in \mathbb{N}$ $[x \in A_n \iff (\exists \tau \supseteq \delta)(\tau \Vdash_{n+k} F_{g(n)}(x))]$.

**Corollary 16.** *For every sequence* $\mathcal{A}$ *if* $d_\omega(\mathcal{A}) \in \mathrm{CS}_k(\mathcal{A}, \mathcal{B})$ *then* $\mathcal{A}$ *is forcing $k$-definable on* $\mathfrak{A}$ *with respect to* $\mathcal{B}$.

We shall give an explicit form of all sequences which are forcing $k$-definable on $\mathfrak{A}$ with respect to $\mathcal{B}$ by means of recursive $\Sigma_k^+$ formulae. These formulae can be considered as a modification of Ash's formulae [1] appropriate for their use on abstract structures presented by Soskov and Baleva [9].

Let $\mathcal{L} = \{T_1, \ldots, T_s\}$ be the first order language of the structure $\mathfrak{A}$. For each $n$ let $P_n$ be a new unary predicate representing the set $B_n$.

(1) An elementary $\Sigma_0^+$ formula with free variables among $W_1, \ldots, W_r$ is an existential formula of the form $\exists Y_1 \ldots \exists Y_m \Phi(W_1, \ldots, W_r, Y_1, \ldots, Y_m)$, where $\Phi$ is a finite conjunction of atomic formulae in $\mathcal{L} \cup \{P_0\}$;

(2) A $\Sigma_n^+$ formula is a c.e. disjunction of elementary $\Sigma_n^+$ formulae;

(3) An elementary $\Sigma_{n+1}^+$ formula is a formula of the form $\exists Y_1 \ldots \exists Y_m \Phi(W_1, \ldots, W_r, Y_1, \ldots, Y_m)$, where $\Phi$ is a finite conjunction of atoms of the form $P_{n+1}(Y_j)$ or $P_{n+1}(W_i)$ and $\Sigma_n^+$ formulae or negations of $\Sigma_n^+$ formulae in $\mathcal{L} \cup \{P_0\} \cup \ldots \cup \{P_n\}$.

**Definition 17.** Let $\mathcal{A} = \{A_n\}_{n<\omega}$ and $k \in \mathbb{N}$. The sequence $\mathcal{A}$ is *formally $k$-definable* on $\mathfrak{A}$ with respect to $\mathcal{B}$ if there exists a recursive sequence $\{\Phi^{\gamma(n,x)}\}_{n,x<\omega}$ of formulae such that for every $n$, $\Phi^{\gamma(n,x)}$ is a $\Sigma^+_{n+k}$ formula with free variables among $W_1, \ldots, W_r$ and elements $t_1, \ldots, t_r$ of $\mathbb{N}$ such that for every $x \in \mathbb{N}$, the following equivalence holds:

$$x \in A_n \iff (\mathfrak{A}, \mathcal{B}) \models \Phi^{\gamma(n,x)}(W_1/t_1, \ldots, W_r/t_r).$$

This means that $x \in A_n$ if and only if the formula $\Phi^{\gamma(n,x)}$ is true in $\mathfrak{A}$ with all sets $B_n$ added as new predicates, under the variable assignment $v$ such that $v(W_1) = t_1, \ldots, v(W_n) = t_n$. With a uniform variant of the proof given in [9] we obtain the following:

**Theorem 18.** *If a sequence $\mathcal{A}$ is forcing $k$-definable on $\mathfrak{A}$ with respect to $\mathcal{B}$ then $\mathcal{A}$ is formally $k$-definable on $\mathfrak{A}$ with respect to $\mathcal{B}$.*

**Corollary 19.** *Let $\mathcal{A} \in \mathcal{S}$ and let $k \in \mathbb{N}$. Then the following are equivalent:*

(1) $d_\omega(\mathcal{A}) \in \mathrm{CS}_k(\mathfrak{A}, \mathcal{B})$;
(2) $\mathcal{A}$ *is forcing $k$-definable on $\mathfrak{A}$ with respect to $\mathcal{B}$.*
(3) $\mathcal{A}$ *is formally $k$-definable on $\mathfrak{A}$ with respect to $\mathcal{B}$.*

## 4    Properties of the $\omega$-Degree Spectra

We prove that some properties of degree spectra from [8] remain true for $\omega$-degree spectra. The first property follows directly from Theorem 1 and Lemma 4.

**Proposition 20.** $\mathrm{CS}(\mathfrak{A}, \mathcal{B}) = co(\{\mathbf{a} \mid \mathbf{a} \in \mathrm{DS}(\mathfrak{A}, \mathcal{B}) \ \& \ \mathbf{a} \ \textit{is total e-degree}\})$.

The next property is an analogue of the minimal pair theorem for the degree spectrum of a structure $\mathfrak{A}$ by Soskov [8]: There exist $\mathbf{f}$ and $\mathbf{g}$ in $\mathrm{DS}(\mathfrak{A})$ such that
$\mathbf{a} \leq_e \mathbf{f}^{(k)} \ \& \ \mathbf{a} \leq_e \mathbf{g}^{(k)} \Rightarrow \mathbf{a} \in \mathrm{CS}_k(\mathfrak{A})$ for every $\mathbf{a} \in \mathcal{D}_e$ and each $k \in \mathbb{N}$.

**Theorem 21.** *For every structure $\mathfrak{A}$ and every sequence $\mathcal{B} \in \mathcal{S}$ there exist total enumeration degrees $\mathbf{f}$ and $\mathbf{g}$ in $\mathrm{DS}(\mathfrak{A}, \mathcal{B})$ such that for every $\omega$-enumeration degree $\mathbf{a}$ and $k \in \mathbb{N}$:*

$$\mathbf{a} \leq_\omega \mathbf{f}^{(k)} \ \& \ \mathbf{a} \leq_\omega \mathbf{g}^{(k)} \Rightarrow \mathbf{a} \in \mathrm{CS}_k(\mathfrak{A}, \mathcal{B}) \ . \tag{3}$$

*Proof.* First we shall construct total enumeration degrees $\mathbf{f}$ and $\mathbf{g}$ in $\mathrm{DS}(\mathfrak{A}, \mathcal{B})$ satisfying (3) for $k = 0$. Then we will show that $\mathbf{f}$ and $\mathbf{g}$ satisfy (3) for every $k$.

Let $f$ be an acceptable enumeration of $\mathfrak{A}$ with respect to $\mathcal{B}$, such that $f^{-1}(\mathfrak{A})$ is a total set. Then $\mathcal{P}^f \equiv_\omega \{f^{-1}(\mathfrak{A})^{(n)}\}_{n<\omega}$. Denote by $F = f^{-1}(\mathfrak{A})$. It is clear that $d_e(F) \in \mathrm{DS}(\mathfrak{A}, \mathcal{B})$.

Denote by $\mathcal{X}_0, \mathcal{X}_1, \ldots \mathcal{X}_r \ldots$ all sequences $\omega$-enumeration reducible to $\mathcal{P}^f$.

Consider the sequence $\mathcal{C}_0, \mathcal{C}_1, \ldots, \mathcal{C}_r \ldots$ of these elements of $\mathcal{X}_0, \mathcal{X}_1, \ldots \mathcal{X}_r \ldots$ which are not forcing definable on $\mathfrak{A}$ with respect to $\mathcal{B}$. By Proposition 12 there is an enumeration $h$ such that $\mathcal{C}_r \not\leq_\omega \mathcal{P}^h$ for all $r$. Then by Corollary 2 there is a total set $G$ such that $\mathcal{P}^h \leq_\omega \{G^{(n)}\}_{n<\omega}$ and $\mathcal{C}_r \not\leq_\omega \{G^{(n)}\}_{n<\omega}$ for all $r$. By

Lemma 4 there is an acceptable enumeration $g$ of $\mathfrak{A}$ with respect to $\mathcal{B}$ such that $g^{-1}(\mathfrak{A}) \equiv_e G$. Thus $d_e(G) \in \mathrm{DS}(\mathfrak{A}, \mathcal{B})$.

Suppose now that $\mathcal{A}$ is a sequence such that $\mathcal{A} \leq_\omega \{F^{(n)}\}_{n<\omega}$ and $\mathcal{A} \leq_\omega \{G^{(n)}\}_{n<\omega}$. Then $\mathcal{A} = \mathcal{X}_r$ for some $r$. If we assume that $\mathcal{A}$ is not forcing definable on $\mathfrak{A}$ with respect to $\mathcal{B}$ then $\mathcal{A} = \mathcal{C}_l$ for some $l$ and hence $\mathcal{A} \not\leq_\omega \{G^{(n)}\}_{n<\omega}$, which is a contradiction. Thus $\mathcal{A}$ is forcing definable on $\mathfrak{A}$ with respect to $\mathcal{B}$ and $d_\omega(\mathcal{A}) \in \mathrm{CS}(\mathfrak{A}, \mathcal{B})$ by Proposition 19. Then by setting $\mathbf{f} = d_e(F)$ and $\mathbf{g} = d_e(G)$ we obtain the desired minimal pair.

For each $\mathbf{a} \in \mathcal{D}_e$ denote by $I(\mathbf{a}) = \{\mathbf{b} \mid \mathbf{b} \in \mathcal{D}_\omega \ \& \ \mathbf{b} \leq_\omega \mathbf{a}\} = co(\{\mathbf{a}\})$ the principal ideal generated by $\mathbf{a}$. We have that $\mathrm{CS}(\mathfrak{A}, \mathcal{B}) = I(\mathbf{f}) \cap I(\mathbf{g})$, since $\mathbf{f}, \mathbf{g} \in \mathrm{DS}(\mathfrak{A}, \mathcal{B})$. We shall prove now that $I(\mathbf{f}^{(k)}) \cap I(\mathbf{g}^{(k)}) = \mathrm{CS}_k(\mathfrak{A}, \mathcal{B})$ for every $k$. Since $\mathbf{f}^{(k)}, \mathbf{g}^{(k)} \in \mathrm{DS}_k(\mathfrak{A}, \mathcal{B})$ it follows that $\mathrm{CS}_k(\mathfrak{A}, \mathcal{B}) \subseteq I(\mathbf{f}^{(k)}) \cap I(\mathbf{g}^{(k)})$. Suppose that $\mathcal{A} = \{A_n\}_{n<\omega}$, $\mathcal{A} \leq_\omega F^{(k)} \uparrow \omega$ and $\mathcal{A} \leq_\omega G^{(k)} \uparrow \omega$. Denote by $\mathcal{C} = \{C_n\}_{n<\omega}$ the sequence such that $C_n = \emptyset$ for $n < k$, and $C_{n+k} = A_n$ for each $n$. Clearly $\mathcal{A} \leq_\omega \mathcal{C}^{(k)}$ and $\mathcal{C} \leq_\omega \{F^{(n)}\}_{n<\omega}$, $\mathcal{C} \leq_\omega \{G^{(n)}\}_{n<\omega}$. So $d_\omega(\mathcal{C}) \in \mathrm{CS}(\mathfrak{A}, \mathcal{B})$. Consider an arbitrary acceptable enumeration $h$ of $\mathfrak{A}$ with respect to $\mathcal{B}$. Then $\mathcal{C} \leq_\omega \{h^{-1}(\mathfrak{A})^{(n)}\}_{n<\omega}$ and thus $\mathcal{C}^{(k)} \leq_\omega \{h^{-1}(\mathfrak{A})^{(n)}\}_{n<\omega}^{(k)}$. It follows that $\mathcal{A} \leq_\omega \{h^{-1}(\mathfrak{A})^{(n)}\}_{n<\omega}^{(k)}$ for every $h \in \mathcal{E}(\mathfrak{A}, \mathcal{B})$. Hence $d_\omega(\mathcal{A}) \in \mathrm{CS}_k(\mathfrak{A}, \mathcal{B})$. $\square$

**Corollary 22.** $\mathrm{CS}_k(\mathfrak{A}, \mathcal{B})$ *is the least ideal containing all kth $\omega$-jumps of the elements of* $\mathrm{CS}(\mathfrak{A}, \mathcal{B})$.

*Proof.* Ganchev [5] proved that if the enumeration degrees $\mathbf{f}$ and $\mathbf{g}$ form an exact pair for a countable ideal $I$ of $\omega$-enumeration degrees, i.e. $I = I(\mathbf{f}) \cap I(\mathbf{g})$ then for every $k$ the pair $\mathbf{f}^{(k)}, \mathbf{g}^{(k)}$ form an exact pair for the least ideal $I^{(k)}$ containing all $k$th $\omega$-jumps of the elements of $I$, i.e. $I^{(k)} = I(\mathbf{f}^{(k)}) \cap I(\mathbf{g}^{(k)})$. Let $\mathbf{f}$ and $\mathbf{g}$ be the minimal pair from Theorem 21. Since $I = \mathrm{CS}(\mathfrak{A}, \mathcal{B})$ is a countable ideal, $I \subseteq \mathcal{D}_\omega$ and $\mathrm{CS}(\mathfrak{A}, \mathcal{B}) = I(\mathbf{f}) \cap I(\mathbf{g})$ then $I^{(k)} = I(\mathbf{f}^{(k)}) \cap I(\mathbf{g}^{(k)})$ for each $k$. On the other hand $I(\mathbf{f}^{(k)}) \cap I(\mathbf{g}^{(k)}) = \mathrm{CS}_k(\mathfrak{A}, \mathcal{B})$ for each $k$. Thus $I^{(k)} = \mathrm{CS}_k(\mathfrak{A}, \mathcal{B})$. $\square$

Soskov [8] showed that for any structure $\mathfrak{A}$, there is a quasi-minimal e-degree $\mathbf{q}$ with respect to $\mathrm{DS}(\mathfrak{A})$, i.e. $\mathbf{q} \notin \mathrm{CS}(\mathfrak{A})$ and if $\mathbf{a}$ is a total e-degree and $\mathbf{a} \geq_e \mathbf{q}$ then $\mathbf{a} \in \mathrm{DS}(\mathfrak{A})$ and if $\mathbf{a}$ is a total e-degree and $\mathbf{a} \leq_e \mathbf{q}$ then $\mathbf{a} \in \mathrm{CS}(\mathfrak{A})$. We can give an analogue of this theorem.

**Theorem 23.** *For every structure $\mathfrak{A}$ and $\mathcal{B} \in \mathcal{S}$, there exists a set $F \subseteq \mathbb{N}$ such that for $\mathbf{q} = d_\omega(F \uparrow \omega)$ it holds:*
(1) $\mathbf{q} \notin \mathrm{CS}(\mathfrak{A}, \mathcal{B})$;
(2) *If $\mathbf{a}$ is a total e-degree and $\mathbf{a} \geq_\omega \mathbf{q}$ then $\mathbf{a} \in \mathrm{DS}(\mathfrak{A}, \mathcal{B})$;*
(3) *If $\mathbf{a}$ is a total e-degree and $\mathbf{a} \leq_\omega \mathbf{q}$ then $\mathbf{a} \in \mathrm{CS}(\mathfrak{A}, \mathcal{B})$.*

*Proof.* Let $\mathfrak{A} = (\mathbb{N}; R_1, \ldots, R_s)$ and $\mathcal{B} = \{B_n\}_{n<\omega}$. Consider the structure $\mathfrak{A}_0 = (\mathbb{N}; R_1, \ldots, R_s, B_0)$.

Soskov [8] proved that there is a partial generic enumeration $f$ of $\mathfrak{A}_0$ such that $d_e(f^{-1}(\mathfrak{A}_0))$ is quasi-minimal with respect to $\mathrm{DS}(\mathfrak{A}_0)$. Moreover if $i = \lambda x.x$ then $f^{-1}(\mathfrak{A}_0) \not\leq_e i^{-1}(\mathfrak{A}_0)$. Ganchev [4] showed that there is a set $F$ such that $f^{-1}(\mathfrak{A}_0) <_e F$, $f^{-1}(B_n) \leq_e F^{(n)}$ uniformly in $n$ and for any total set $X$, if

$X \leq_{\mathrm{e}} F$ then $X \leq_{\mathrm{e}} f^{-1}(\mathfrak{A}_0)$. We call the set $F$ quasi-minimal over $f^{-1}(\mathfrak{A}_0)$ with respect to $\{f^{-1}(B_n)\}_{n<\omega}$. The set $F$ is constructed as a partial regular enumeration of $\mathfrak{A}_0$. Set $\mathbf{q} = d_\omega(F \uparrow \omega)$. We will prove that $\mathbf{q}$ has the desired properties.     ·

Suppose for a contradiction that $\mathbf{q} \in \mathrm{CS}(\mathfrak{A}, \mathcal{B})$. Then $d_\omega(f^{-1}(\mathfrak{A}_0) \uparrow \omega) \in \mathrm{CS}(\mathfrak{A}, \mathcal{B})$ since $f^{-1}(\mathfrak{A}_0) <_{\mathrm{e}} F$. It follows that $f^{-1}(\mathfrak{A}_0) \uparrow \omega$ is forcing definable on $\mathfrak{A}$ with respect to $\mathcal{B}$. Then $f^{-1}(\mathfrak{A}_0) \leq_{\mathrm{e}} i^{-1}(\mathfrak{A}) \oplus B_0 \equiv_{\mathrm{e}} i^{-1}(\mathfrak{A}_0)$. A contradiction.

If $X$ is a total set and $X \leq_{\mathrm{e}} F$ then $X \leq_{\mathrm{e}} f^{-1}(\mathfrak{A}_0)$ as $F$ is quasi-minimal over $f^{-1}(\mathfrak{A}_0)$. Thus $d_{\mathrm{e}}(X) \in \mathrm{CS}(\mathfrak{A}_0)$ by the choice of $f^{-1}(\mathfrak{A}_0)$. But $\mathrm{DS}(\mathfrak{A}, \mathcal{B}) \subseteq \mathrm{DS}(\mathfrak{A}_0)$. So $d_\omega(X \uparrow \omega) \in \mathrm{CS}(\mathfrak{A}, \mathcal{B})$.

If $X$ is a total set and $X \geq_{\mathrm{e}} F$ then $X \geq_{\mathrm{e}} f^{-1}(\mathfrak{A}_0)$. Since "$=$" is among the predicates of $\mathfrak{A}$, $\mathrm{dom}(f) \leq_{\mathrm{e}} X$ and since $X$ is a total set, $\mathrm{dom}(f)$ is c.e. in $X$. Let $\rho$ be a recursive in $X$ enumeration of $\mathrm{dom}(f)$. Set $h = \lambda n.f(\rho(n))$. Thus $h^{-1}(\mathfrak{A}) \leq_{\mathrm{e}} X$ and $h^{-1}(B_n) \leq_{\mathrm{e}} X^{(n)}$ uniformly in $n$. By Lemma 4 there is an acceptable enumeration $g$ of $\mathfrak{A}$ such that $g^{-1}(\mathfrak{A}) \equiv_{\mathrm{e}} X$. And hence $d_{\mathrm{e}}(X) \in \mathrm{DS}(\mathfrak{A}, \mathcal{B})$. $\qquad\qquad\square$

**Acknowledgments.** The author would like to thank Hristo Ganchev and Ivan N. Soskov for the helpful comments and Mariya Soskova for editorial suggestions.

# References


1. Ash, C.J.: Generalizations of enumeration reducibility using recursive infinitary propositional senetences. Ann. Pure Appl. Logic 58, 173–184 (1992)
2. Cooper, S.B.: Partial degrees and the density problem. Part 2: The enumeration degrees of the $\Sigma_2$ sets are dense. J. Symb. Logic 49, 503–513 (1984)
3. Cooper, S.B.: Computability Theory. Chapman & Hall/CRC, Boca Raton (2004)
4. Ganchev, H.: A jump inversion theorem for the infinite enumeration jump. Ann. Sofia Univ ( to appear, 2005)
5. Ganchev, H.: Exact Pair Theorem for the $\omega$-Enumeration Degrees. In: Cooper, S.B., Löwe, B., Sorbi, A. (eds.) CiE 2007. LNCS, vol. 4497, pp. 316–324. Springer, Heidelberg (2007)
6. Richter, L.J.: Degrees of structures. J. Symb. Logic 46, 723–731 (1981)
7. Soskov, I.N.: A jump inversion theorem for the enumeration jump. Arch. Math. Logic 39, 417–437 (2000)
8. Soskov, I.N.: Degree spectra and co-spectra of structures. Ann. Sofia Univ. 96, 45–68 (2004)
9. Soskov, I.N., Baleva, V.: Ash's theorem for abstract structures. In: Chatzidakis, Z., Koepke, P., Pohlers, W. (eds.) Proc. of Logic Colloquium 2002, Muenster, Germany, 2002. Lect. Notes in Logic, vol. 27, pp. 327–341. ASL (2006)
10. Soskov, I.N., Kovachev, B.: Uniform regular enumerations. Mathematical Structures in Comp. Sci. 16(5), 901–924 (2006)
11. Soskov, I.N.: The $\omega$-enumeration degrees. J. Logic and Computation 17(6), 1193–1214 (2007)
12. Soskov, I.N., Ganchev, H.: The jump operator on the $\omega$-enumeration degrees. Ann. Pure Appl. Logic (to appear)
13. Soskova, A.A.: Relativized degree spectra. J. Logic and Computation 17(6), 1215–1233 (2007)

# Cupping Classes of $\Sigma_2^0$ Enumeration Degrees

Mariya Ivanova Soskova[*]

Department of Pure Mathematics
University of Leeds, Leeds LS2 9JT, U.K.
mariya@maths.leeds.ac.uk

**Abstract.** We prove that no subclass of the $\Sigma_2^0$ enumeration degrees containing the nonzero 3-c.e. enumeration degrees can be cupped to $\mathbf{0}_e'$ by a single incomplete $\Sigma_2^0$ enumeration degree.

## 1 Introduction

In an upper semi-lattice with greatest element $\langle \mathcal{A}, \leq, \vee, 1 \rangle$ we say that an element $\mathbf{a}$ is cuppable if there exists an element $\mathbf{b} \neq \mathbf{1}$ such that $\mathbf{a} \vee \mathbf{b} = \mathbf{1}$. Posner and Robinson showed that every nonzero degree in $\mathcal{D}_T(\leq 0')$ is cuppable. Cooper and Yates [5] showed the existence of a nonzero non-cuppable c.e. degree in the semi-lattice of the computably enumerable degrees. Meanwhile Cooper, Seetapun and (independently) Li proved that there exists a single incomplete $\Delta_2^0$ Turing degree that cups every nonzero c.e. degree.

In this paper we consider cupping properties of the local degree structure of the enumeration degrees below $\mathbf{0}_e'$. Intuitively we say that a set $A$ is *enumeration reducible* to a set $B$, denoted as $A \leq_e B$, if there is an effective procedure to enumerate $A$ given any enumeration of $B$. By identifying sets that are reducible to each other we obtain a degree structure, the structure of the enumeration degrees $\langle \mathcal{D}_e, \leq \rangle$. It is an upper semi-lattice with jump operator and least element $\mathbf{0}_e$, the collection of all computably enumerable sets. The semi-lattice of the enumeration degrees can be considered as an extension of the semi-lattice of the Turing degrees, as the second semi-lattice can be embedded in the first, via an order theoretic embedding $\iota$ preserving the least upper bound and the jump operator.

An important substructure of $\mathcal{D}_e$ is given by the $\Sigma_2^0$ enumeration degrees. Cooper [2] proved that the $\Sigma_2^0$ enumeration degrees are the enumeration degrees below $\mathbf{0}_e'$. There is a natural hierarchy of classes of enumeration degrees within this substructure. The $\Pi_1^0$ enumeration degrees, which are exactly the images of the c.e. Turing degrees under $\iota$, form the smallest class. Further classes can be obtained by considering the $n$-c.e. degrees for every $n \leq \omega$. Cooper [3] proved that the 2-c.e. enumeration degrees coincide with the $\Pi_1^0$ enumeration degrees. Thus the second class in our hierarchy consists of all 3-c.e. enumeration degrees. The

* Soskova is supported by the Marie Curie Early Training grant MATHLOGAPS (MEST-CT-2004-504029).

A. Beckmann, C. Dimitracopoulos, and B. Löwe (Eds.): CiE 2008, LNCS 5028, pp. 554–566, 2008.

last proper subclass of the $\Sigma_2^0$ enumeration degrees in this hierarchy comprises all $\Delta_2^0$ enumeration degrees.

In [6], Cooper, Sorbi and Yi proved that every nonzero $\Delta_2^0$ enumeration degree can be cupped by a total incomplete $\Delta_2^0$ enumeration degree, in contrast to the $\Sigma_2^0$ enumeration degrees where non-cuppable degrees exist. Soskova and Wu [11] examined the cupping properties of the $\Delta_2^0$ enumeration degrees further and showed that every nonzero $\Delta_2^0$ enumeration degree can be cupped by a 1-generic $\Delta_2^0$, hence partial and low, enumeration degree. Furthermore they showed that every nonzero $\omega$-c.e. enumeration degree can be cupped by a 3-c.e enumeration degree. These results exhibit the flexibility that we have when searching for cupping partners of enumeration degrees in each of the proper subclasses of the $\Sigma_2^0$ enumeration degrees and motivate the initial goal of this project: to find a single degree from a larger class that cups all nonzero enumeration degrees from a smaller class. One such example we obtain immediately by transferring Cooper, Seetapun and Li's result to the enumeration degrees via $\iota$, namely that there exists a single incomplete $\Delta_2^0$ enumeration degree which cups all nonzero $\Pi_1^0$ enumeration degrees. In this paper we prove that any other attempt at a result of this kind is doomed to failure as for every incomplete $\Sigma_2^0$ enumeration degree $\mathbf{a}$ there exists a nonzero member of the second class, a nonzero 3-c.e. enumeration degree $\mathbf{b}$, such that $\mathbf{b}$ is not cupped by $\mathbf{a}$ to $\mathbf{0'_e}$.

**Theorem 1.** *Let $\mathbf{a}$ be an incomplete $\Sigma_2^0$ enumeration degree. There exists a nonzero 3-c.e enumeration degree $\mathbf{b}$ such that $\mathbf{a} \vee \mathbf{b} \neq \mathbf{0'_e}$.*

Notation and terminology below are based on that of [4] and [10].

## 2    Requirements and Strategies

We shall start by giving a formal definition of enumeration reducibility and the $n$-c.e. degrees and then move on to establish the requirements and basic strategies for the proof of the main theorem.

**Definition 1.** *A set $A$ is enumeration reducible ($\leq_e$) to a set $B$ if there is a c.e. set $\Phi$ such that:*

$$n \in A \Leftrightarrow \exists u(\langle n, u \rangle \in \Phi \wedge D_u \subseteq B),$$

*where $D_u$ denotes the finite set with code $u$ under the standard coding of finite sets. We will refer to the c.e. set $\Phi$ as an enumeration operator and its elements will be called axioms.*

We say that an enumeration degree is $n$-c.e. ($n < \omega$) if it contains an $n$-c.e. set:

**Definition 2.** *$A$ is $n$-c.e. if there is a computable function $f$ such that for each $x$, $f(x,0) = 0$, $|\{s \mid f(x,s) \neq f(x,s+1)\}| \leq n$ and $A(x) = \lim_s f(x,s)$.*

Let $A$ be a representative of the given $\Sigma_2^0$ enumeration degree. Let $\{A_s\}_{s<\omega}$ be a good $\Sigma_2^0$ approximation to $A$. That every $\Sigma_2^0$ set has a good $\Sigma_2^0$ approximations is proved by Jockusch [7]. We use the definition given by Lachlan and Shore [8]. A good $\Sigma_2^0$ approximation to $A$ is a $\Sigma_2^0$ approximation with infinitely many good stages $s$ at which $A_s \subseteq A$.

We shall construct two 3-c.e sets $X$ and $Y$, so that ultimately the degree of one of them will have the requested properties. Consider the following requirements:

- Let $\{\Theta_i\}_{i<\omega}$ and $\{\Psi_i\}_{i<\omega}$ be effective enumerations of all enumeration operators. For every $i$ we will have a pair of requirements:

$$\mathcal{P}_i^0 : \Theta_i^{A,X} \neq \overline{K} \quad \text{and} \quad \mathcal{P}_i^1 : \Psi_i^{A,Y} \neq \overline{K}.$$

- Let $\{W_e\}_{e<\omega}$ be an enumeration of all c.e. sets. For every natural number $e$ we have a requirement:

$$\mathcal{N}_e : W_e \neq X \wedge W_e \neq Y.$$

We shall construct the sets $X$ and $Y$ so that for all $e$ the requirement $\mathcal{N}_e$ is satisfied, thus both $X$ and $Y$ have nonzero e-degree, and if $\mathcal{P}_i^j$ is not satisfied for some $i$ then for all $i'$ the requirement $\mathcal{P}_{i'}^{1-j}$ is satisfied, thus the degree of at least one of the sets $A \oplus X$ or $A \oplus Y$ is incomplete. The construction shall be carried out on a tree of strategies. Each node of the tree shall be assigned either an $\mathcal{N}$-requirement or a $\mathcal{P}^0$- and a $\mathcal{P}^1$-requirement. At each stage we shall construct a finite path in the tree of strategies visiting some of the nodes and allowing them to act towards satisfying one of the assigned requirements. The intention is that there will be an infinite leftmost path of nodes that is visited at infinitely many stages, providing a successful outcome to all strategies on it.

Each $\mathcal{P}$-node in the tree of strategies $\alpha$ is associated with a pair of requirements: $\mathcal{P}_\alpha^0$ and $\mathcal{P}_\alpha^1$. It will attempt at proving that at least one of them is satisfied. To do this the strategy constructs an e-operator $\Gamma_\alpha$, threatening to prove that $A \geq_e \overline{K}$. The outcomes of the strategy will be divided into two groups. There will be infinitely many infinitary outcomes - two for each number $n$ arranged from left to right by the order of the natural numbers: $\langle X, 0 \rangle <_L \langle Y, 0 \rangle < \langle X, 1 \rangle \ldots$ Then there will be two finitary rightmost outcomes $\langle X, w \rangle <_L \langle Y, w \rangle$. Thus all the outcomes of a $\mathcal{P}$-node are arranged as follows:

$$\langle X, 0 \rangle <_L \langle Y, 0 \rangle <_L \ldots <_L \langle X, n \rangle <_L \langle Y, n \rangle \ldots <_L \langle X, w \rangle <_L \langle Y, w \rangle$$

For each outcome the first element of the pair indicates which requirement has been satisfied. The next $\mathcal{P}$-strategy below outcomes $\langle X, - \rangle$ shall be associated with a new $\mathcal{P}^0$-requirement and the same $\mathcal{P}^1$-requirement. Similarly the next $\mathcal{P}$-strategy below outcomes $\langle Y, - \rangle$ will be associated with the same $\mathcal{P}^0$-requirement and a different $\mathcal{P}^1$-requirement. Thus if $\mathcal{P}_i^j$ never gets satisfied for some $i$ then all $\mathcal{P}_{i'}^{1-j}$ must be.

The strategy $\alpha$ performs cycles of increasing length. On the $k$-th cycle it examines all elements $n = 0, 1, \ldots, k$ in turn. The cycles do not necessarily correspond to a stage, in fact $\alpha$ can take any number of stages to complete a particular cycle. While it examines an element $n$ the strategy can choose to end its actions for the particular stage by choosing an outcome or move on to the next element in the cycle, possibly even starting a new cycle. Suppose $\alpha$ is examining the element $n$. If the element $n$ currently belongs to $\overline{K}$ then the only possible outcomes that it can choose for this element are the infinitary $\langle X, n \rangle$ or $\langle Y, n \rangle$. If the element $n$ is in both sets $\Theta_\alpha^{A,X}$ and $\Psi_\alpha^{A,Y}$ and has been there at all stages since $\alpha$ last looked at $n$ then it will enumerate an axiom for $n$ in $\Gamma_\alpha$ which comprises the $A$-parts of the two axioms for $n$ in $\Theta_\alpha$ and in $\Psi_\alpha$ that have been valid the longest and move on to the next element. Otherwise $\alpha$ will select the appropriate outcome corresponding to the set that has failed to provide a valid axiom and end its actions for this stage. When $\alpha$ is active again, it will start working with the next element of the cycle.

If the element $n$ has left the approximation of $\overline{K}$ then for each axiom in $\Gamma_\alpha$ for this element the strategy shall restore one of the axioms in either $\Theta_\alpha$ or $\Psi_\alpha$ by enumerating the corresponding $X$-part back in $X$ or $Y$-part back in $Y$ and have the corresponding finitary outcome $\langle X, w \rangle$ or $\langle Y, w \rangle$. The strategy shall then wait until it has observed a change in $A$ that rectifies the operator $\Gamma_\alpha$, i.e. it will not move on to the next element in the cycle until (if ever) this happens and it will keep having the same finitary outcome. Note that $\alpha$ will not only consider stages at which it is active, instead every time it is visited it will check if $\Gamma_\alpha^A(n)[s] = 0$ at any stage $s$ since the last stage at which $\alpha$ was active.

As $A$ is incomplete the strategy will eventually include in its cycles an element $n$ such that $\Gamma_\alpha^A(n) \neq \overline{K}(n)$. If there is an element $n$ such that $n \in \Gamma_\alpha^A \setminus \overline{K}$ then $n \in \Gamma_\alpha^A[s]$ at all but finitely many stages $s$. Thus eventually $\Gamma_\alpha^A(n)$ will not be rectified by any change in $A$ and $\alpha$ will have a finitary outcome proving the successful diagonalization. Otherwise $\alpha$ will have inifinitely many cycles and each element $n$ will be examined infinitely many times. Consider the least $n$ such that $n \in \overline{K} \setminus \Gamma_\alpha^A$. By the properties of a good approximation we have that at infinitely many stages $s$, in fact at all good stages, $n \notin \Gamma_\alpha^A[s]$. Thus infinitely often $\alpha$ will discover that at least one of the operators $\Theta_\alpha$ or $\Psi_\alpha$ has failed to provide it with an axiom that is permanently valid, i.e. infinitely often $\alpha$ will have proof that $\Theta_\alpha^{A,X}(n) = 0$ or $\Psi_\alpha^{A,Y}(n) = 0$ and have outcome $\langle X, n \rangle$ or $\langle Y, n \rangle$ respectively.

An $\mathcal{N}$-node $\beta$ working on $W_\beta$ would like to prove that $W_\beta \neq X$ and $W_\beta \neq Y$. The obvious strategy for $\beta$ would be to select a witness $x_\beta$ and wait until $x_\beta \in W_\beta$. We assume that the sets $X$ and $Y$ start off as $\omega$, then during the construction the strategies extract or enumerate back elements in the sets. Thus if $x_\beta$ never enters $W_\beta$ the strategy will be successful and will have outcome $w$. If the element does enter $W_\beta$ then the strategy will extract $x_\beta$ from both sets $X$ and $Y$, have outcome $d$, where $d <_L w$, and again will have proved a difference. This strategy is unfortunately incomplete as we shall see in the next section.

## 3   Elaborating the $\mathcal{N}$-Strategy to Avoid Conflicts

The naive $\mathcal{N}$-strategy described in the previous section is in conflict with the need of higher priority $\mathcal{P}$-strategies to restore axioms by enumerating elements back in one of the sets $X$ or $Y$. Therefore the strategy for an $\mathcal{N}$-node $\beta$ will have to be more elaborate. This conflict justifies the introduction of nonuniformity.

The elaborated strategy will start off as the original strategy: select a witness $x_\beta$ as a fresh number and wait until $x_\beta \in W_\beta$. If this never happens then the requirement will be satisfied with outcome $w$. Otherwise extract $x_\beta$ from both sets $X$ and $Y$. Suppose a higher priority strategy $\alpha$ requires that $x_\beta$ be enumerated back in $X$ or $Y$. In this case $\beta$ shall initialize all lower priority strategies, choose a new witness $y_\beta$ that has not been used in any axiom so far, restrain $X$ on $x_\beta$ and let $x_\beta$ be enumerated back in $Y$. From this point on any axiom that appears in the construction shall necessarily have $x_\beta \notin X$, thus $x_\beta$ and $y_\beta$ cannot appear in the same axiom. The strategy $\beta$ will wait again with outcome $w$ until $y_\beta$ enters $W_\beta$ and then extract it from $Y$ with outcome $d$ . Should a higher priority $\alpha$ require that $y_\beta$ be enumerated back in one of the sets then $\beta$ will only give permission to enumerate back in $X$.

This will resolve the central conflict between strategies. Note that as the only actions that the $\mathcal{P}$-strategies ever take is enumerating certain elements back in the sets $X$ and $Y$, the $\mathcal{P}$-strategies are not in conflict with each other.

Possible conflicts between $\mathcal{N}$-strategies are resolved via initialization. Whenever a higher priority $\mathcal{N}$-strategy $\beta$ decides to extract a number $n$ from $X$ or $Y$ all strategies below outcome $w$ are initialized and all strategies below outcome $d$ are in initial state. Thus lower priority strategies will operate at further stages under the assumption that $n$ is extracted, the axioms used by $\mathcal{P}$-strategies of lower priority will not include this element and the witnesses used by $\mathcal{N}$-strategies will be chosen as big numbers that do not appear in any axiom seen so far, thus cannot appear in an axiom that includes the element $n$.

## 4    Parameters and the Tree of Strategies

A $\mathcal{P}$-strategy $\alpha$ will have a parameter $\Gamma_\alpha$, the e-operator that it will construct when visited. At initialization $\Gamma_\alpha$ is set to the empty set. The strategy will also have parameters $k_\alpha$ denoting the current cycle of the strategy and $n_\alpha \leq k_\alpha$ denoting the current element of the cycle that $\alpha$ is working with. On initialization the values of the parameters are set to $k_\alpha = 0$ and $n_\alpha = 0$. Furthermore for each element $n < \omega$ the strategy $\alpha$ shall have one more parameter $D_\alpha(n)$, a list of all pairs of $X$- and $Y$-parts of axioms from $\Theta_\alpha$ and $\Psi_\alpha$ respectively, for which the $A$-parts are used in axioms for $n$ in $\Gamma_\alpha$. Initially the values of all such lists will be $\emptyset$. Finally it will have two parameters $Ax_\alpha^\theta(n)$ and $Ax_\alpha^\psi(n)$ denoting axioms in $\Theta_\alpha$ and $\Psi_\alpha$ respectively which will be candidates for the construction of a new axiom in $\Gamma_\alpha$, initially undefined.

An $\mathcal{N}$-strategy $\beta$ shall have parameters $x_\beta$, $y_\beta$, which will be undefined when $\beta$ is initialized. Furthermore on initialization $\beta$ will give up any restraint it has imposed so far.

Let $O_\mathcal{P}$ denote the set of all possible outcomes of a $\mathcal{P}$-strategy and $O_\mathcal{N} = \{d, w\}$. Let $O = O_\mathcal{P} \cup O_\mathcal{N}$ be the collection of all possible outcomes and $R$ the collection of all requirements. The tree of strategies is a computable function $T$

with domain a downwards closed subset of $O^{<\omega}$ and range a subset of $R^2 \cup R$ with the following inductive definition:

1. $T(\emptyset) = \langle \mathcal{P}_0^0, \mathcal{P}_0^1 \rangle$.
2. Let $\alpha$ be in the domain of $T$ and $\alpha$ be a $\langle \mathcal{P}_i^0, \mathcal{P}_j^1 \rangle$-node. Then $\alpha\hat{\ }o$, where $o \in O_\mathcal{P}$, is also in the domain of $T$ and $T(\alpha\hat{\ }o) = \mathcal{N}_{|\alpha|/2}$.
3. Let $\beta$ be an $\mathcal{N}$-node in the domain of $T$. Then $\beta = \alpha\hat{\ }o$, where $\alpha$ is a $\langle \mathcal{P}_i^0, \mathcal{P}_j^1 \rangle$-node for some $i$ and $j$. Then $\beta\hat{\ }o'$, where $o' \in O_\mathcal{N}$, is in the domain of $T$. If $o = \langle X, n \rangle$ for some $n \in \omega \cup \{w\}$ then $T(\beta\hat{\ }o') = \langle \mathcal{P}_{i+1}^0, \mathcal{P}_j^1 \rangle$. If $o = \langle Y, n \rangle$ for some $n \in \omega \cup \{w\}$ then $T(\beta\hat{\ }o') = \langle \mathcal{P}_i^0, \mathcal{P}_{j+1}^1 \rangle$.

## 5   Construction

We shall perform the construction in stages. At each stage $s$ we shall approximate the sets $X$ and $Y$ by constructing cofinite sets $X_s$ and $Y_s$. We shall also construct a string $\delta_s$ of length $s$ through the domain of $T$. We shall say that a node $\gamma \subset \delta_s$ is visited at stage $s$, also that $s$ is a $\gamma$-true stage. At true stages strategies will be allowed to modify their parameters and choose an outcome. At the end of stage $s$ we shall initialize all nodes to the right of $\delta_s$.

At stage 0 all nodes are initialized and $X_0 = Y_0 = \omega$, $\delta_0 = \emptyset$.

Suppose we have constructed $\delta_t$, $X_t$ and $Y_t$ for $t < s$. The sets $X_s$ and $Y_s$ shall be obtained by allowing the strategies visited at stage $s$ to modify the approximations $X_{s-1}, Y_{s-1}$ obtained at the previous stage. We construct $\delta_s(n)$ with an inductive definition. Define $\delta_s(0) = \emptyset$. Suppose that we have constructed $\delta_s \upharpoonright n$. If $n = s$, we end this stage and move on to $s + 1$. Otherwise we visit the strategy $\delta_s \upharpoonright n$ and let it determine its outcome $o$. Then $\delta_s(n) = o$. We have two cases depending on the type of the node $\delta_s \upharpoonright n$.

I. If $\delta_s \upharpoonright n = \alpha$ is a $\mathcal{P}$-node, we perform the following actions:
Let $s^-$ be the previous $\alpha$-true stage if $\alpha$ has not been initialized since and $s^- = s$ otherwise. The strategy $\alpha$ will inherit the values of its parameters from stage $s^-$ and during its actions it can change their values several times. Thus we will omit the subscript indicating the stage when we discuss $\alpha$'s parameters. If the current element $n_\alpha$ does not need further actions we shall move on to the next element. As this is a subroutine which is frequently performed in the construction, we define it here once and for all, and we refer to it with the phrase **reset the parameters**. Denote the current values of $n_\alpha$ by $n$ and of $k_\alpha$ by $k$. We *reset the parameters* by changing the values of the parameters as follows: $n_\alpha := n + 1$ if $n < k$, otherwise $n = k$ and we set $k_\alpha := k + 1$, $n_\alpha := 0$. In both cases we initialize the strategies extending $\alpha\hat{\ }\langle X, w \rangle$ and $\alpha\hat{\ }\langle Y, w \rangle$.

   1. Let $k = k_\alpha$ and $n = n_\alpha$. Let $s_n^-$ be the previous stage when $n$ was examined, if $\alpha$ has not been initialized since, $s_n^- = s$ otherwise.
   2. If $n \in \overline{K}[s]$ and $n \in \Gamma_\alpha^A[t]$ for all stages $t$ with $s_n^- < t \le s$ then *reset the parameters* and go to step 1.
   3. If $n \in \overline{K}[s]$, but $n \notin \Gamma_\alpha^A[t]$ at some stage $t$ with $s_n^- < t \le s$ then:

a.X If $Ax_\alpha^\theta(n)$ is not defined, then define it as the axiom that has been valid longest including at all stages $s_n^- < t \le s$ and move on to step $a.Y$. If there is no such axiom then let the outcome be $\langle X, n \rangle$ and *reset the parameters.*

b.X If $Ax_\alpha^\theta(n)$ is defined but was not valid at some stage $t$ with $s_n^- < t \le s$, then cancel its value (make it undefined) and let the outcome be $\langle X, n \rangle$, *reset the parameters.* Otherwise go to step $a.Y$.

a.Y If $Ax_\alpha^\psi(n)$ is not defined, then define it as the axiom that has been valid longest including at all stages $s_n^- < t \le s$ and move on to step $c$. If there is no such axiom then let the outcome be $\langle Y, n \rangle$ and *reset the parameters.*

b.Y If $Ax_\alpha^\theta(n)$ is defined but was not valid at some stage $t$ with $s_n^- < t \le s$, then cancel its value (make it undefined) and let the outcome be $\langle Y, n \rangle$, *reset the parameters.* Otherwise go to step $c$.

  c. If both $Ax_\alpha^\theta(n) = \langle n, A_\theta, X_\theta \rangle$ and $Ax_\alpha^\psi(n) = \langle n, A_\psi, Y_\psi \rangle$ are defined and have been valid at all stages $t$ with $s_n^- < t \le s$ then enumerate in $\Gamma_\alpha$ the axiom $\langle n, A_\theta \cup A_\psi \rangle$. Enumerate $\langle X_\theta, Y_\psi \rangle$ in $D_\alpha(n)$. *Reset the parameters* and go back to step 1.

4. If $n \notin \overline{K}[s]$ and $n \notin \Gamma_\alpha^A[t]$ at some stage $t$: $s_n^- < t \le s$ *reset the parameters* and go back to step 1.

5. Suppose $n \notin \overline{K}[s]$ but $n \in \Gamma_\alpha^A[t]$ at all $t$ such that $s_n^- < t \le s$. For every pair $\langle X_\theta, Y_\psi \rangle \in D_\alpha(n)$ find the highest priority $\mathcal{N}$-strategy $\beta \supset \alpha$ that has permanently restrained an element $x \in X_\theta$ out of $X$ or $y \in Y_\psi$ out of $Y$. If there is such a strategy $\beta$ and it has a permanent restraint on $X$, enumerate $Y_\psi$ in $Y[s]$; if it has a permanent restraint on $Y$, enumerate $X_\theta$ back in $X[s]$. Otherwise if there is no such strategy enumerate $Y_\psi$ back in $Y[s]$. Choose the axiom $\langle n, A_\theta \cup A_\psi \rangle$ in $\Gamma_\alpha^A$ that has been valid the longest. Let $X_\theta$ and $Y_\psi$ be the corresponding $X$ and $Y$ parts of the axioms $\langle n, A_\theta, X_\theta \rangle \in \Theta$ and $\langle n, A_\psi, Y_\psi \rangle \in \Psi$.

  a. If $X_\theta \subseteq X[s]$ then this will ensure that $n \in \Theta_\alpha^{A,X}[s]$. Let the outcome be $\langle X, w \rangle$. Note that we will not reset the parameters at this point, thus the construction will keep going through this step while there is no change in $A$.

  b. If $X_\theta \nsubseteq X[s]$ then $Y_\psi \subseteq Y[s]$ and this will ensure that $n \in \Psi_\alpha^{A,Y}[s]$. Let the outcome be $\langle Y, w \rangle$.

II. If $\delta_s \upharpoonright n = \beta$ is an $\mathcal{N}$-node, we perform the following actions: Let $s^-$ be the previous $\beta$-true stage if $\beta$ has not been initialized since, go to the step indicated at stage $s^-$. Otherwise $s^- = s$ and go to step 1.

1. Define $x_\beta$ as a fresh number, one that has not appeared in the construction so far and is bigger than $s$. Go to the next step.

2. If $x_\beta \notin W_\beta[s]$ then let the outcome be $o = w$, return to this step at the next stage. Otherwise go to the next step.

3. Extract $x_\beta$ from $X[s]$ and $Y[s]$. Restrain permanently $x_\beta$ out of $X$. Let the outcome be $o = d$, go to the next step at the next stage.

4. If $x_\beta \in Y[s]$ then define $y_\beta$ as a fresh number, initialize all strategies of lower priority than $\beta$ and go to the next step. Otherwise $o = d$, return to this step at the next stage.

5. If $y_\beta \notin W_\beta$ then let the outcome be $o = w$. Return to this step at the next stage. Otherwise go to the next step.
6. If $y_\beta$ is not yet restrained then restrain $y_\beta$ permanently out of $Y$ and extract $y_\beta$ from $Y[s]$. Let the outcome be $o = d$, return to this step at the next stage.

This completes the construction.

## 6  Proof

The tree is infinitely branching and therefore there is a risk that there might not be a path in the tree that is visited infinitely often. However we shall start the proof by establishing some basic facts about the relationship between strategies.

For clarity we shall define one more notation. Let $\alpha$ be a $\mathcal{P}$-strategy. To every axiom $Ax = \langle n, A_\theta \cup A_\psi \rangle \in \Gamma_\alpha$ we shall associate a corresponding entry $\langle n, A_\theta, X_\theta, A_\psi, Y_\psi \rangle$ so that $\langle n, A_\theta, X_\theta \rangle \in \Theta_\alpha$ and $\langle n, A_\psi, Y_\psi \rangle \in \Psi_\alpha$ are the corresponding axioms used to construct $Ax$.

**Lemma 1.** *Let $\beta$ be an $\mathcal{N}$-strategy, initialized for the last time at stage $s_i$. If $\beta$ has a witness $x_\beta$ that is extracted by $\beta$ at stage $s_x > s_i$ then $x_\beta \notin X[t]$ at all $t \geq s_x$. If $\beta$ has a witness $y_\beta$ that is extracted from $Y$ at stage $s_y > s_x$ then $y_\beta \notin Y[t]$ at all $t \geq s_y$.*

*Proof.* There are only finitely many $\mathcal{N}$-strategies of higher priority than $\beta$ that are ever visited in the construction as after stage $s_i$ no strategy to the left of $\beta$ is visited. Every higher priority strategy $\beta' < \beta$ that is ever visited is not initialized after stage $s_i$, as otherwise $\beta$ would be initialized after stage $s_i$ contrary to our assumption. We can inductively assume that the statement is valid for every higher priority strategy $\beta'$.

Suppose $\beta$ chooses the witness $x_\beta$ at stage $s_1 > s_i$. We can furthermore prove the following:

*Claim.* Any witness which is permanently extracted by a higher priority strategy $\beta'$ is extracted before or at stage $s_1$.

Indeed, suppose that $\beta'$ permanently extracts a new witness at stage $s_2 > s_1$. Then at stage $s_2$ the strategy $\beta'$ has outcome $d$. Thus if $\beta >_L \beta'$ or $\beta \supseteq \beta'^\frown w$ then $\beta$ would be initialized at stage $s_2$ contrary to assumption. This leaves us with the only possibility that $\beta \supseteq \beta'^\frown d$. Then at stage $s_1$, as $\beta$ was visited, $\beta'$ was visited and had outcome $d$. As $\beta'$ is not initialized after stage $s_1$ and permanently extracts a new witness at stage $s_2$ it must be the case that $\beta'$ permanently extracts a witness $y_{\beta'}$ from $Y$ and $x_{\beta'}$ was already extracted before or at stage $s_1$. It follows that between stages $s_1$ and $s_2$, $\beta'$ has selected this new witness $y_{\beta'}$ passing through $II.4$ of the construction and initializing all lower priority strategies including $\beta$. This leads again to a contradiction with the assumption that $\beta$ is not initialized after stage $s_1$ and hence the claim is correct.

Thus at stage $s_1$ all witnesses of higher priority strategies that are ever permanently restrained out of either set $X$ or $Y$ are already permanently restrained out of $X$ or $Y$. At stage $s_1$ the strategy $\beta$ selects $x_\beta$ as a fresh number, i.e. one that has not appeared in the construction so far. And at stage $s_x$ the witness $x_\beta$ is permanently restrained out of $X$.

Now we will prove again inductively but this time on the stage $t$, that $x_\beta \notin X[t]$ at all stages $t \geq s_x$.

So suppose this is true for $t < s_3$ and that at stage $s_3 > s_x$ a $\mathcal{P}$-strategy $\alpha$ is visited and reaches point $I.5$ of the construction. Suppose $\alpha$ wants to enumerate $X_\theta$ or $Y_\psi$ back in $X$ or $Y$ respectively for the axiom $\langle n, A_\theta \cup A_\psi \rangle$ in $\Gamma_\alpha$ with corresponding entry $\langle n, A_\theta, X_\theta, A_\psi, Y_\psi \rangle$. We have the following cases to consider:

1. Suppose $\alpha > \beta$. If $\alpha >_L \beta^\frown d$ then $\alpha$ is initialized at stage $s_x$. If $\alpha \subset \beta^\frown d$, then $\alpha$ was initialized at stage $s_i$ and was not accessible before stage $s_x$. Thus the axiom $\langle n, A_\theta \cup A_\psi \rangle$ was enumerated in $\Gamma_\alpha$ at stage $t$ with $s_x \leq t < s_3$, at which both $\langle n, A_\theta, X_\theta \rangle$ and $\langle n, A_\psi, Y_\psi \rangle$ were valid i.e. $X_\theta \subseteq X[t]$ and $Y_\psi \subseteq Y[t]$. By induction $x_\beta \notin X[t]$ hence $x_\beta \notin X_\theta$ and thus $\alpha$ does not enumerate $x_\beta$ back in $X$.

2. Suppose $\alpha < \beta$. If $\alpha <_L \beta$ then $\beta$ would be initialized at stage $s_3$, hence $\alpha \subset \beta$. Suppose the axiom $\langle n, A_\theta \cup A_\psi \rangle$ was enumerated in $\Gamma_\alpha$ at stage $t$. If $t \leq s_1$ then by the choice of $x_\beta$ as a fresh number at stage $s_1$ we have that $x_\beta \notin X_\theta$. If $t > s_1$ then both $\langle n, A_\theta, X_\theta \rangle$ and $\langle n, A_\psi, Y_\psi \rangle$ were valid at stage $t$ i.e. $X_\theta \subseteq X[t]$ and $Y_\psi \subseteq Y[t]$. By $I.5$ of the construction $\alpha$ will consider all $\mathcal{N}$-strategies that extend it and select the one with highest priority that has permanently restrained an element out of either set $X$ or $Y$.

   Consider any $\beta' < \beta$. By our *Claim* any witness $x_{\beta'}$ or $y_{\beta'}$ of $\beta'$ that is ever permanently restrained out of $X$ or $Y$ is already restrained out at stage $s_1$ and by induction at all stages $s \geq s_1$ including at stage $t$. Thus $X_\theta$ does not contain $x_{\beta'}$ and $Y_\psi$ does not contain $y_{\beta'}$. As this is true for an arbitrary strategy $\beta'$ of higher priority than $\beta$ that is ever visited, if $x_\beta \in X_\theta$ then $\beta$ will be the strategy selected by $\alpha$ and $\alpha$ will choose to enumerate $Y_\psi$ back in $Y$. Thus again $\alpha$ does not enumerate $x_\beta$ back in $X$.

To prove the second part of the lemma suppose $y_\beta$ is selected at stage $s_4$ and extracted at stage $s_y$. Because $s_1 < s_4$ and all strategies of lower priority than $\beta$ are initialized at stage $s_4$ the interactions between $\beta$ and other strategies are dealt with in the same way as in the case when we were considering $x_\beta$. The only thing left for us to establish is that $\beta$ does not come into conflict with itself. So suppose that at stage $s_5 > s_y$ a $\mathcal{P}$-strategy $\alpha$ is visited and reaches point $I.5$ of the construction. Suppose $\alpha$ wants to enumerate $X_\theta$ or $Y_\psi$ back in $X$ or $Y$ respectively for the axiom $\langle n, A_\theta \cup A_\psi \rangle$ in $\Gamma_\alpha$ with corresponding entry $\langle n, A_\theta, X_\theta, A_\psi, Y_\psi \rangle$. We will prove that if $x_\beta \in X_\theta$ then $y_\beta \notin Y_\psi$. Let $t$ be the stage at which the axiom $\langle n, A_\theta \cup A_\psi \rangle$ was enumerated in $\Gamma_\alpha$. If $t < s_4$ then $y_\beta \notin Y_\psi$ by the choice of $y_\beta$ at stage $s_4$ as a fresh number. If $t \geq s_4 > s_x$ then we have already proved that $x_\beta \notin X[t]$. The axiom $\langle n, A_\theta, X_\theta \rangle$ was valid at stage $t$, thus $X_\theta \subseteq X[t]$, and hence $x_\beta \notin X_\theta$.

This completes the induction step and the proof of the lemma.     $\square$

**Lemma 2.** *Let $\alpha$ be a $\mathcal{P}$-strategy, visited infinitely often and not initialized after stage $s_i$. If $\alpha$ performs finitely many cycles then:*

*(1)There is a stage $s_n \geq s_i$ after which the value of $n_\alpha$ does not change.*

*(2)At all $\alpha$-true stages $t > s_n$, $\alpha$ has either outcome $\langle X, w \rangle$ or outcome $\langle Y, w \rangle$.*

*(3)There is a stage $s_d \geq s_n$ such that at all $\alpha$-true stages $t > s_d$, $\alpha$ has the same outcome $o$ .*

*(4)If $o = \langle X, w \rangle$ then $\Theta_\alpha^{A,X} \neq \overline{K}$ and if $o = \langle Y, w \rangle$ then $\Psi_\alpha^{A,Y} \neq \overline{K}$.*

*Proof.* It follows from the construction and the definition of the action *reset the parameters* that if the value of $n_\alpha$ changes infinitely often, then there will be infinitely many cycles. Thus part (1) of the lemma is true. Let $s_n$ be the stage after which the value of $n_\alpha$ does not change. The only case when the value of the parameter $n_\alpha = n$ is not reset is when $n \notin \overline{K}$ and $n \in \Gamma_\alpha^A[t]$ at all stages $t$ since the last time $n$ was examined at stage $s_n^-$, thus $\alpha$ will have only outcomes $\langle X, w \rangle$ or $\langle Y, w \rangle$ at all stages after $s_n$ and part (2) is true. It follows from part $I.5$ of the construction and the fact that $n_\alpha$ does not change any longer that at all stage $t > s_n$, $n \in \Gamma_\alpha^A[t]$. By the properties of a good approximation and under these circumstances $n \in \Gamma_\alpha^A$. Then there will be an axiom $\langle n, A_\theta \cup A_\psi \rangle \in \Gamma_\alpha$ which is valid at all but finitely many stages. Select the axiom which is valid longest. This axiom has corresponding entry $\langle n, A_\theta, X_\theta, A_\psi, Y_\psi \rangle$. The strategy $\alpha$ will eventually be able to spot this precise axiom, after possibly finitely many wrong guesses. So after a stage $s_d \geq s_n$ the strategy $\alpha$ will consider this axiom to select its outcome.

At stage $s_n$ either $X_\theta \subset X[s_n]$ or $Y_\psi \subset Y[s_n]$. As we initialize all strategies below outcomes $\langle X, w \rangle$ and $\langle Y, w \rangle$ whenever we reset the parameters, we can be sure that $\mathcal{N}$-strategies visited at stages $t > s_n$ of lower priority than $\alpha$ will not extract any elements of $X_\theta \cup Y_\psi$ from $X$ or $Y$. Higher priority $\mathcal{N}$-strategies will not extract any elements at all, otherwise $\alpha$ would be initialized. Thus if $X_\theta \subseteq X[s_n]$ then for all stages $t \geq s_n$ we have $X_\theta \subseteq X[t]$ and similarly if $Y_\psi \subseteq Y[s_n]$ then for all stages $t \geq s_n$ we have $Y_\psi \subseteq Y[t]$.

Suppose $X_\theta \subseteq X[s_n]$. Then at stages $t \geq s_d$ the strategy $\alpha$ will always have outcome $\langle X, w \rangle$. The axiom $\langle n, A_\theta, X_\theta \rangle \in \Theta_\alpha$ will be valid at all stages $t \geq s_d$, thus $n \in \Theta^{A,X}$, and $n \notin \overline{K}$.

If $X_\theta \not\subseteq X[s_n]$ then there is a strategy $\beta \supset \alpha$ which is permanently restraining some element $x \in X_\theta$ out of $X$ at stage $s_n$. Then $\beta <_L \alpha^\smallfrown \langle X, w \rangle$ as strategies extending $\alpha^\smallfrown \langle X, w \rangle$ or to the right of it are in initial state at stage $s_n$ and do not have any restraints. This strategy $\beta$ will not be initialized at stages $t \geq s_n$ according to part (2) of this lemma and the choice of $s_n > s_i$. By Lemma 1 $x \notin X_t$ at all $t \geq s_n$.

Hence case $I.5.b$ of the construction is valid at all $t \geq s_d$. Thus $\alpha$ will have outcome $\langle Y, w \rangle$ at all stages $t \geq s_d$ and $n \in \Psi^{A,Y}$. This proves parts (3) and (4) of the lemma.                                                                 $\square$

**Lemma 3.** *Let $\alpha$ be a $\mathcal{P}$-strategy, visited infinitely often and not initialized after stage $s_i$. If $v$ is an element such that $\Gamma_\alpha^A(v) = \overline{K}(v)$ then there is a stage $s_v$ after which the outcomes $\langle X, v \rangle$ and $\langle Y, v \rangle$ are not accessible any longer.*

*Proof.* If $\alpha$ has finitely many cycles then by Lemma 2 there will be a stage $s_n$ after which $\langle X, v \rangle$ and $\langle Y, v \rangle$ are not accessible. Suppose there are infinitely many cycles.

If $v \notin \overline{K}$ then there is a stage $s_v$ at which $v$ exits $\overline{K}$. Then after stage $s_v$ the outcomes $\langle X, v \rangle$ and $\langle Y, v \rangle$ are not accessible.

If $v \in \Gamma_\alpha^A$ then there is an axiom in $\Gamma_\alpha$ that is valid at all but finitely many stages, say at all stages $t \geq s_v'$. If $\alpha$ is on its $k$-th cycle during stage $s_v'$ then let $s_v$ be the beginning of the $(k+2)$-nd cycle. Then after stage $s_v$, whenever $\alpha$ considers $v$, part $I.2$ of the construction holds and hence $\alpha$ will never have outcome $\langle X, v \rangle$ or $\langle Y, v \rangle$.     $\square$

**Lemma 4.** *Let $\alpha$ be a $\mathcal{P}$-strategy, visited infinitely often and not initialized after stage $s_i$. If $\alpha$ performs infinitely many cycles, then there is leftmost outcome $o$ that $\alpha$ has at infinitely many stages and*

*(1) If $o = \langle X, u \rangle$ then $\Theta_\alpha^{A,X}(u) \neq \overline{K}(u)$.*
*(2) If $o = \langle Y, u \rangle$ then $\Psi_\alpha^{A,Y}(u) \neq \overline{K}(u)$.*

*Proof.* The set $A$ is not complete by assumption, hence $\Gamma_\alpha^A \neq \overline{K}$. Let $u$ be the least difference between the sets. By Lemma 3 for every $v < u$ the outcomes $\langle X, v \rangle$ and $\langle Y, v \rangle$ are not visited at stages $t > s_v$. Let $s_0$ be a stage bigger than $\max\{s_v | v < u\}$. As $\alpha$ has infinitely many cycles there will be infinitely many stages $t > s_0$ at which $n_\alpha[t] = u$. If $u \notin \overline{K}$ and $u \in \Gamma_\alpha^A$ then there is a stage $s_1 > s$ such that at all stages $t > s_1$ we have $u \in \Gamma_\alpha^A[t]$ and $u \notin \overline{K}[t]$ and when $\alpha$ considers $u$ at the first stage after $s_1$, it will never move on to the next element, and $\alpha$ would have finitely many cycles. Hence $u \in \overline{K}$ and $u \notin \Gamma_\alpha^A$.

(1) If $u \notin \Theta_\alpha^{A,X}$ then all axioms for $u$ in $\Theta_\alpha$ are invalid at infinitely many stages. Let $t$ be any stage greater than or equal to $s_0$. We will prove that there is a stage $t' \geq t$ at which $\alpha$ has outcome $\langle X, u \rangle$. As $u \notin \Gamma_\alpha^A$ and $\{A_s\}_{s<\omega}$ is a good approximation to $A$ there are infinitely many stages $s$ at which $u \notin \Gamma_\alpha^A[s]$ and hence part $I.3$ of the construction holds at infinitely many stages at which we consider $u$. Let $t_1 \geq t$ at which $n_\alpha[t_1] = u$ and part $I.3$ of the construction is true. If $Ax_\alpha^\theta(u)$ is not defined and we are not able to define it as there is no appropriate axiom in $\Theta_\alpha$ valid for long enough then $\alpha$ will have outcome $\langle X, u \rangle$ at stage $t_1$, hence $t' = t_1$ proves the claim. Otherwise $Ax_\alpha^\theta(u)$ is defined at stage $t_1$ and by assumption there are infinitely many stages $t \geq t_1$ at which it is invalid. Let $t_2 > t_1$ be the next stage when $Ax_\alpha^\theta(u)$ is invalid and let $t' \geq t_2$ be the first stage after $t_2$ at which again $n_\alpha[t'] = u$ and part $I.3$ of the construction is true. By $I.3.b.X$ of the construction $\alpha$ will have outcome $\langle X, u \rangle$ at stage $t'$.

(2) Now assume that $u \in \Theta_\alpha^{A,X}$. Then there is an axiom $\langle u, A_\theta, X_\theta \rangle \in \Theta_\alpha$ valid at all but finitely many stages. Select the axiom, say $Ax$, that is valid the longest. Then $Ax_\alpha^\theta(u)$ will have a permanent value $Ax$ after a certain stage $s_1$. It follows that $u \notin \Psi_\alpha^{A,Y}$ as otherwise we would be able to find an axiom in $\Psi_\alpha^{A,Y}$ valid at all but finitely many stages, and construct an axiom in $\Gamma_\alpha$ valid at all but finitely many stages. Now a similar argument as the one used in part (1) of this lemma proves that $\alpha$ will have outcome $\langle Y, u \rangle$ at infinitely many stages.     $\square$

As an immediate corollary from Lemmas 2, 3 and 4 we obtain the existence of the true path:

**Corollary 1.** *There exists an infinite path through the tree of strategies with the following properties:*

*(1) $\forall n \exists^\infty s[f \upharpoonright n \subseteq \delta_s]$*
*(2) $\forall n \exists s_l(n) \forall t > s_l(n)[\delta_t \not\prec_L f \upharpoonright n]$*
*(3) $\forall n \exists s_i(n) \forall t > s_i(n)[f \upharpoonright n$ is not initialized at stage $t]$.*

**Corollary 2.** $X$ *and* $Y$ *are not c.e.*

*Proof.* For every requirement $\mathcal{N}_e$ there is an $\mathcal{N}_e$-strategy $\beta$ along the true path, visited infinitely often and not initialized at any stage $t > s_i$. Let $x_\beta$ and $y_\beta$ be the final values of $\beta$'s witnesses. If $\beta^\smallfrown w \subset f$ then there is an element $u \in \{x_\beta, y_\beta\}$ that never enters $W_e$. The way each $\mathcal{N}_e$-strategy chooses its witnesses ensures that only $\beta$ can extract $u$ form either of the sets $X$ or $Y$. The construction and the definition of the true path ensure that $\beta$ does not extract $u$ from $X$ and $Y$ at any stage. Hence $u \in X \cap Y$ and $u \notin W_e$.

If $\beta^\smallfrown d \subset f$ then $x_\beta \in W_e$ and there is a $\beta$-true stage $s_x$ at which $\beta$ extracts $x_\beta$ from $X$ and $Y$. By Lemma 1 $x_\beta \notin X[t]$ at all stages $t \geq s_x$. If at any stage $t \geq s_x$ we have that $x_\beta \in Y[t]$ then $\beta$ selects $y_\beta$ at its next true stage. As the true outcome is $d$, $y_\beta \in W_e[t']$ at some stage $t' \geq t$. Then at the next $\beta$-true stage $s_y \geq t'$ the strategy $\beta$ will permanently restrain $y_\beta$ out of $Y$ and by Lemma 1, we have that $y_\beta \notin Y$. $\qquad\square$

**Corollary 3.** $A \oplus X \not\equiv_e \overline{K}$ *or* $A \oplus Y \not\equiv_e \overline{K}$.

*Proof.* Consider the $\mathcal{P}$-nodes on the true path. From the definition of the tree it follows that either for every $\mathcal{P}_e^0$-requirement there is a node on the tree $\alpha$ which is associated with $\mathcal{P}_e^0$ or there is a fixed requirement $\mathcal{P}_e^0$ associated with all but finitely many nodes. In the latter case there is a node on the true path for every $\mathcal{P}_e^1$-requirement.

Suppose there is a node on the tree for each $\mathcal{P}_e^0$-requirement. We can show that $A \oplus X \not\equiv_e \overline{K}$. Assume for a contradiction $\Theta_e^{A,X} = \overline{K}$ and let $\alpha \subset f$ be the last node associated with $\mathcal{P}_e^0$. Then $\alpha$ has true outcome $\langle X, u \rangle$ for some $u \in \omega \cup \{w\}$. It follows from Lemma 2 and Lemma 4 that $\Theta_e^{A,X} \neq \overline{K}$.

The case when there is a node for every $\mathcal{P}_e^1$-requirement yields by a similar argument that $A \oplus Y \not\equiv_e \overline{K}$. $\qquad\square$

**Lemma 5.** *The sets $X$ and $Y$ are 3-c.e.*

*Proof.* We can easily obtain a 3-c.e. approximation of each of the sets $X$ and $Y$ from the one constructed. Define $\hat{X}_s = X_s \upharpoonright s$ and $\hat{Y}_s = Y_s \upharpoonright s$.

It follows from the construction that elements extracted from $X$ and $Y$ are necessarily witnesses of $\mathcal{N}$-strategies. Suppose therefore that $n$ is the witness $x_\beta$ for an $\mathcal{N}$-strategy $\beta$. Then $n$ appears in the defined approximations $\{\hat{X}_s\}_{s<\omega}$ and $\{\hat{Y}_s\}_{s<\omega}$ at stage $n+1$. If $\beta$ never extracts $x_\beta$ then we are done - as no other strategy can extract it. If $\beta$ extracts $x_\beta$ then it does so only once at stage

$s_x$ when it goes through $II.3$ and moves on to $II.4$ at the next stage. In order for $\beta$ to return to step $II.3$ of the construction it will have to be initialized and will select new witnesses. Thus after its extraction at stage $s_x$ from both $\hat{X}_{s_x}$ and $\hat{Y}_{s_x}$ , the number $x_\beta$ can only be enumerated back in either set and hence $|\{s \mid \hat{X}_{s-1}(x_\beta) \neq \hat{X}_s(x_\beta)\}| \leq 3$ and $|\{s \mid \hat{Y}_{s-1}(x_\beta) \neq \hat{Y}_s(x_\beta)\}| \leq 3$.

If $n$ is the witness $y_\beta$ then it will never be extracted from $X$. If it is ever extracted from $Y$ it is extracted only once by $\beta$ at the first stage it reaches step $II.6$. After that $y_\beta$ is already restrained by $\beta$ and whenever $\beta$ executes step $II.6$ it will ignore the first sentence of the instruction and just have outcome $o = d$. Thus again $|\{s \mid \hat{Y}_{s-1}(y_\beta) \neq \hat{Y}_s(y_\beta)\}| \leq 3$.     $\square$

**Acknowledgements.** Thanks are due to Prof. Guohua Wu and Prof. Yang Yue for suggesting this problem to me. I am grateful also to my anonymous reviewers for pointing out weaknesses of a previous version of this article and providing helpful suggestions.

# References


1. Cooper, S.B.: Partial degrees and the density problem. J. Symb. Log. 47, 854–859 (1982)
2. Cooper, S.B.: Partial Degrees and the density problem. part 2: the enumeration degrees of the $\Sigma_2$ sets are dense. J. Symb. Log. 49, 503–513 (1984)
3. Cooper, S.B.: Enumeration reducibility, nondeterminitsic computations and relative computability of partial functions. In: Recursion Theory Week, Oberwolfach 1989. Lecture Notes in Mathematics, vol. 1432, pp. 57–110 (1990)
4. Cooper, S.B.: Computability Theory. Chapman & Hall/CRC Mathematics, Boca Raton (2004)
5. Cooper, S.B.: On a theorem of C.E.M. Yates, handwritten notes (1973)
6. Cooper, S.B., Sorbi, A., Yi, X.: Cupping and noncupping in the enumeration degrees of $\Sigma_2{}^0$ sets. Ann. Pure Appl. Logic 82, 317–342 (1996)
7. Jockusch Jr., C.G.: Semirecursive sets and positive reducibility. Trans.Amer.Math.Soc. 131, 420–436 (1968)
8. Lachlan, A.H., Shore, R.A.: The n-rea enumeration degrees are dense. Arch. Math. Logic 31, 277–285 (1992)
9. Posner, D., Robinson, R.: Degrees joining to $0'$. J. Symbolic Logic 46, 714–722 (1981)
10. Soare, R.I.: Recursively enumerable sets and degrees. Springer, Heidelberg (1987)
11. Soskova, M., Wu, G.: Cupping $\Delta^0{}_2$ enumeration degrees to $0'$. In: Cooper, S.B., Löwe, B., Sorbi, A. (eds.) CiE 2007. LNCS, vol. 4497, pp. 727–738. Springer, Heidelberg (2007)

# Principal Typings for Explicit Substitutions Calculi


Daniel Lima Ventura[1,*], Mauricio Ayala-Rincón[1,**],
and Fairouz Kamareddine[2]

[1] Grupo de Teoria da Computação, Dep. de Matemática Universidade de Brasília,
Brasília D.F., Brasil
[2] School of Mathematical and Computer Sciences Heriot-Watt University, Edinburgh,
Scotland UK
{ventura,ayala}@mat.unb.br, fairouz@macs.hw.ac.uk



**Abstract.** Having principal typings (for short PT) is an important property of type systems. In simply typed systems, this property guarantees the possibility of a complete and terminating type inference mechanism. It is well-known that the simply typed $\lambda$-calculus has this property but recently J.B. Wells has introduced a system-independent definition of PT, which allows to prove that some type systems, e.g. the Hindley/Milner type system, do not satisfy PT. Explicit substitutions address a major computational drawback of the $\lambda$-calculus and allow the explicit treatment of the substitution operation to formally correspond to its implementation. Several extensions of the $\lambda$-calculus with explicit substitution have been given but some of which do not preserve basic properties such as the preservation of strong normalization. We consider two systems of explicit substitutions ($\lambda s_e$ and $\lambda\sigma$) and show that they can be accommodated with an adequate notion of PT. Specifically, our results are as follows:

- We introduce PT notions for the simply typed versions of the $\lambda s_e$- and the $\lambda\sigma$-calculi and prove that they agree with Wells' notion of PT.
- We show that these versions satisfy PT by revisiting previously introduced type inference algorithms.

**Keywords:** lambda-calculus, explicit substitution, principal typings.


## 1  Introduction

The development of well-behaved calculi of explicit substitutions is of great interest in order to bridge the formal study of the $\lambda$-calculus and its real implementations. Since $\beta$ contraction depends on the definition of the substitution operations, which is informally given in the theory of $\lambda$-calculus, they are in fact made explicit, but obscurely developed (that is, in an empirical manner), when most computational environments based on the $\lambda$-calculus are implemented. A

---


* Corresponding author, supported by a PhD scholarship of the CNPq.
** Partially supported by the CNPq.

A. Beckmann, C. Dimitracopoulos, and B. Löwe (Eds.): CiE 2008, LNCS 5028, pp. 567–578, 2008.
© Springer-Verlag Berlin Heidelberg 2008

remarkable exception is $\lambda$Prolog, for which its explicit substitutions calculus, the suspension calculus, has been extracted and formally studied [NaWi98].

In the study of making substitutions explicit, several alternatives rose out and all of them are directed to guarantee essential properties such as simulating beta-reduction, confluence, noetherianity (of the associated substitution calculus), subject reduction, having principal typings (for short PT), preservation of strong normalization etc. This is a non trivial task; for instance, the $\lambda\sigma$-calculus [ACCL91], which is one of the first proposed calculi of explicit substitutions, was reported to break the latter property some years after its introduction [Mel95]. This implies that infinite derivations starting from well-typed $\lambda$-terms are possible in this calculus, raising serious questions for any mechanism supposed to simulate the $\lambda$-calculus explicitly. In this paper, the focus is on the PT property, which means that for any typable term $M$, there exists a type judgment $A \vdash M : \tau$, representing all possible typings $\langle A', \tau' \rangle$ for $M$. For a discussion about the difference between principal type and principal typing see [Jim96]. In the simply typed $\lambda$-calculus this corresponds to the existence of *more representative* typings. PT guarantees compositional type inference and plays a crucial role in helping one to find a complete/terminating type inference algorithm.

In section 2 we present the type-free $\lambda$-calculus in de Bruijn notation, the $\lambda s_e$-calculus [KR97] and the $\lambda\sigma$-calculus [ACCL91]. In section 3 we present the relevant backgrounds for the type assignment systems we consider and then we present simply typed systems for each calculus we study. Then, we discuss the general notion of principal typings defined in [We2002] and present notions of principal typings for the $\lambda$-calculus in de Bruijn notation, the $\lambda\sigma$- and the $\lambda s_e$-calculi and prove that they are adequate. In section 4 we conclude and present future work. Detailed proofs and examples are included in an extended version of this work available at `www.mat.unb/~ayala/publications.html`.

## 2    The Type Free Calculi

### 2.1    The $\lambda$-Calculus in de Bruijn Notation

**Definition 1 (Set $\Lambda_{dB}$).** *The syntax of $\lambda$-calculus in de Bruijn notation, **the $\lambda dB$-calculus**, is defined inductively as*

> ***Terms** $M ::= \underline{n} \mid (M\ M) \mid \lambda.M$  where $n \in \mathbb{N}^* = \mathbb{N} \smallsetminus \{0\}$*

Terms like $(((((M_1\ M_2)\ M_3) \dots )\ M_n)$ are written as usual $(M_1\ M_2\ \dots\ M_n)$. Let $M$ be a $\lambda$-term. If, in the tree representation of $M$, there are exactly $n$ abstractors in the minimal path from the root position until the position of some subterm $M_1$, then $M_1$ is said to be **n-deep in** $M$. In other words, $M_1$ is in between $n$ abstractors.

**Definition 2.** *We say that $\underline{i}$  **occurs as free index** in a term $M$ if $\underline{i+n}$ is $n$-deep in $M$.*

The $\beta$-contraction definition for $\Lambda_{dB}$ needs a mechanism which detects and updates free indices. Below, we give an operator similar to the one in [ARKa2001a].

**Definition 3.** *Let $M \in \Lambda_{dB}$ and $i \in \mathbb{N}$. The **i-lift** of $M$, denoted as $M^{+i}$, is defined inductively as*

$$1.\ (M_1\,M_2)^{+i} = (M_1^{+i}\,M_2^{+i}) \qquad 3.\ \underline{n}^{+i} = \begin{cases} \underline{n+1}, & \text{if } n > i \\ \underline{n}, & \text{if } n \leq i. \end{cases}$$

$$2.\ (\lambda.M_1)^{+i} = \lambda.M_1^{+(i+1)}$$

The **lift** of a term $M$ is its 0-lift, denoted as $M^+$. Intuitively, the lift of $M$ corresponds to an increment by 1 of all free indices occurring in $M$. For instance, $(\lambda.(\underline{1}\ \underline{3}))^+ = \lambda.(\underline{1}\ \underline{4})$. Using the i-lift, we are able to present the definition of the substitution used by $\beta$-contractions, as done in [ARKa2001a].

**Definition 4.** *Let $m, n \in \mathbb{N}^*$. The $\beta$-**substitution** for free occurrences of $\underline{n}$ in $M \in \Lambda_{dB}$ by term $N$, denoted as $\{\underline{n}/N\}M$, is defined inductively by*

$$1.\ \{\underline{n}/N\}(M_1\,M_2) = (\{\underline{n}/N\}M_1\ \{\underline{n}/N\}M_2) \quad 3.\ \{\underline{n}/N\}\underline{m} = \begin{cases} \underline{m-1}, & \text{if } m > n \\ N, & \text{if } m = n \\ \underline{m}, & \text{if } m < n \end{cases}$$

$$2.\ \{\underline{n}/N\}\lambda.M_1 = \lambda.\{\underline{n+1}/N^+\}M_1$$

Observe that in item 2 of Def. 4, the lift operator is used to avoid captures of free indices in $N$. We present the $\beta$-contraction as defined in [ARKa2001a].

**Definition 5.** $\beta$-**contraction** *in $\lambda dB$ is defined by $(\lambda.M\ N) \to_\beta \{\underline{1}/N\}M$.*

Notice that item 3 in Definition 4, for $n = 1$, is the mechanism which does the substitution and updates the free indices in $M$ as consequence of the lead abstractor elimination.

## 2.2   The $\lambda s_e$-Calculus

The $\lambda s_e$-calculus is a proper extension of the $\lambda dB$-calculus. Two operators $\sigma$ and $\varphi$ are introduced for substitution and updating, respectively, to control the atomization of the substitution operation by arithmetic constraints.

**Definition 6 (Set $\Lambda_s$ of $\lambda s_e$-terms).**
  *The syntax of the $\lambda s_e$-calculus, where $n, i, j \in \mathbb{N}^*$ and $k \in \mathbb{N}$ is given by*

$$\textbf{\textit{Terms}}\ M ::= \underline{n}\,|\,(M\ M)\,|\,\lambda.M\,|\,M\sigma^i M\,|\,\varphi_k^j M$$

The term $M\sigma^i N$ represents the term $\{\underline{i}/N\}M$; i.e., the substitution of the free occurrences of $\underline{i}$ in $M$ by $N$, updating free variables in $M$ (and in $N$). The term $\varphi_k^j M$ represents $j-1$ applications of the $k$-lift to the term $M$; i.e., $M^{+k^{(j-1)}}$. Table 1 contains the rewriting rules of the $\lambda s_e$-calculus together with the rule (Eta), as given in [ARKa2001a]. The bottom seven rules on table 1 are those which extend the $\lambda s$-calculus to $\lambda s_e$ ([KR97]) with the rule (*Eta*) ([ARKa2001a]). They ensure confluence of the $\lambda s_e$-calculus on open terms and the application to the higher order unification problem. Hence, those rules are not the focus of this paper.

  $=_{s_e}$ denotes the equality for the associated substitution calculus, denoted as $s_e$, induced by all the rules except ($\sigma$-generation) and (Eta).

594    D.L. Ventura, M. Ayala-Rincón, and F. Kamareddine

**Table 1.** The rewriting system of the $\lambda s_e$-calculus with the Eta rule

$$
\begin{aligned}
(\lambda.M\ N) &\longrightarrow M\,\sigma^1 N & (\sigma\text{-generation})\\
(\lambda.M)\sigma^i N &\longrightarrow \lambda.(M\sigma^{i+1}N) & (\sigma\text{-}\lambda\text{-transition})\\
(M_1\ M_2)\sigma^i N &\longrightarrow ((M_1\sigma^i N)\ (M_2\sigma^i N)) & (\sigma\text{-app-trans.})\\
\underline{n}\,\sigma^i N &\longrightarrow
\begin{cases}
\underline{n-1} & \text{if } n>i\\
\varphi_0^i N & \text{if } n=i\\
\underline{n} & \text{if } n<i
\end{cases} & (\sigma\text{-destruction})\\
\varphi_k^i(\lambda.M) &\longrightarrow \lambda.(\varphi_{k+1}^i M) & (\varphi\text{-}\lambda\text{-trans.})\\
\varphi_k^i(M_1\ M_2) &\longrightarrow ((\varphi_k^i M_1)\ (\varphi_k^i M_2)) & (\varphi\text{-app-trans.})\\
\varphi_k^i\,\underline{n} &\longrightarrow
\begin{cases}
\underline{n+i-1} & \text{if } n>k\\
\underline{n} & \text{if } n\le k
\end{cases} & (\varphi\text{-destruction})\\
(M_1\sigma^i M_2)\sigma^j N &\longrightarrow (M_1\sigma^{j+1}N)\sigma^i(M_2\sigma^{j-i+1}N) \quad \text{if } i\le j & (\sigma\text{-}\sigma\text{-trans.})\\
(\varphi_k^i M)\sigma^j N &\longrightarrow \varphi_k^{i-1}M \quad \text{if } k<j<k+i & (\sigma\text{-}\varphi\text{-trans. 1})\\
(\varphi_k^i M)\sigma^j N &\longrightarrow \varphi_k^i(M\sigma^{j-i+1}N) \quad \text{if } k+i\le j & (\sigma\text{-}\varphi\text{-trans. 2})\\
\varphi_k^i(M\sigma^j N) &\longrightarrow (\varphi_{k+1}^i M)\sigma^j(\varphi_{k+1-j}^i N) \quad \text{if } j\le k+1 & (\varphi\text{-}\sigma\text{-trans.})\\
\varphi_k^i(\varphi_l^j M) &\longrightarrow \varphi_l^j(\varphi_{k+1-j}^i M) \quad \text{if } l+j\le k & (\varphi\text{-}\varphi\text{-trans. 1})\\
\varphi_k^i(\varphi_l^j M) &\longrightarrow \varphi_l^{j+i-1}M \quad \text{if } l\le k<l+j & (\varphi\text{-}\varphi\text{-trans. 2})\\
\lambda.(M\ \underline{1}) &\longrightarrow N \quad \text{if } M=_{s_e}\varphi_0^2 N & (Eta)
\end{aligned}
$$

## 2.3  The $\lambda\sigma$-Calculus

The $\lambda\sigma$-calculus is given by a first-order rewriting system, which makes substitutions explicit by extending the language with two sorts of objects: **terms** and **substitutions** which are called $\lambda\sigma$-expressions.

**Definition 7 (Set $\Lambda_\sigma$ of $\lambda\sigma$-expressions).** *The $\lambda\sigma$-expressions consist of:*

$$
\begin{aligned}
\textbf{\textit{Terms}} \qquad & M ::= \underline{1} \mid (M\ M) \mid \lambda.M \mid M[S]\\
\textbf{\textit{Substitutions}}\ & S ::= id \mid\ \uparrow\ \mid M.S \mid S\circ S
\end{aligned}
$$

Substitutions can intuitively be thought of as lists of the form $N/\underline{i}$ indicating that the index $\underline{i}$ should be changed to the term $N$. The expression $id$ represents a substitution of the form $\{\underline{1}/\underline{1}, \underline{2}/\underline{2}, \dots\}$ whereas $\uparrow$ is the substitution $\{\underline{i+1}/\underline{i}\,|\,i\in\mathbb{N}^*\}$. The expression $S\circ S$ represents the composition of substitutions. Moreover, $\underline{1}[\uparrow^n]$, where $n\in\mathbb{N}^*$, codifies the de Bruijn index $\underline{n+1}$ and $\underline{i}[S]$ represents the value of $\underline{i}$ through the substitution $S$, which can be seen as a function $S(i)$. The substitution $M.S$ has the form $\{M/\underline{1}, S(i)/\underline{i+1}\}$ and is called the **cons of** $M$ **in** $S$. $M[N.id]$ starts the simulation of the $\beta$-reduction of $(\lambda.M\ N)$ in $\lambda\sigma$. Thus, in addition to the substitution of the free occurrences of the index $\underline{1}$ by the corresponding term, free occurrences of indices should be decremented because of the elimination of the abstractor. Table 2 includes the rewriting system of the $\lambda\sigma$-calculus, as presented in [DoHaKi2000].

This system without (Eta) is equivalent to that of [ACCL91]. The associated substitution calculus, denoted by $\sigma$, is the one induced by all the rules except (Beta) and (Eta), and its equality is denoted as $=_\sigma$.

**Table 2.** The rewriting system for the $\lambda\sigma$-calculus with the Eta rule

| | | | | | |
|---|---|---|---|---|---|
| $(\lambda.M\ N)$ | $\longrightarrow M[N.id]$ | $(Beta)$ | $(\lambda.M)[S]$ | $\longrightarrow \lambda.(M[1.(S\circ\uparrow)])$ | $(Abs)$ |
| $(M\ N)[S]$ | $\longrightarrow (M[S]\ N[S])$ | $(App)$ | $\uparrow\circ(M.S)$ | $\longrightarrow S$ | $(ShiftCons)$ |
| $M[id]$ | $\longrightarrow M$ | $(Id)$ | $(S_1\circ S_2)\circ S_3$ | $\longrightarrow S_1\circ(S_2\circ S_3)$ | $(AssEnv)$ |
| $1[S].(\uparrow\circ S)$ | $\longrightarrow S$ | $(Scons)$ | $(M.S)\circ T$ | $\longrightarrow M[T].(S\circ T)$ | $(MapEnv)$ |
| $(M[S])[T]$ | $\longrightarrow M[S\circ T]$ | $(Clos)$ | $1.\uparrow$ | $\longrightarrow id$ | $(VarShift)$ |
| $id\circ S$ | $\longrightarrow S$ | $(IdL)$ | $1[M.S]$ | $\longrightarrow M$ | $(VarCons)$ |
| $S\circ id$ | $\longrightarrow S$ | $(IdR)$ | $\lambda.(M\ \underline{1})$ | $\longrightarrow N$ if $M=_\sigma N[\uparrow]$ | $(Eta)$ |

## 3  The Type Systems

**Definition 8.** *The syntax of the* **simple types** *and* **contexts** *is given by:*

$$\textbf{Types }\tau ::= \alpha\,|\,\tau\to\tau \qquad\qquad \textbf{Contexts } A ::= nil\,|\,\tau.A$$

*where $\alpha$ ranges over* **type variables**.

A **type assignment system** $\mathcal{S}$ is a set of rules, allowing some terms of a given system to be associated with a type. A **context** gives the necessary information used by $\mathcal{S}$ rules to associate a type to a term. In the simply typed $\lambda$-calculus [Hi97], the typable terms are strongly normalizing. The ordered pair $\langle A,\tau\rangle$, of a context and a type, is called a **typing in** $\mathcal{S}$. For a term $M$, $A\vdash M:\tau$ denotes that $M$ has type $\tau$ in context $A$, and $\langle A,\tau\rangle$ is called a **typing of** $M$. If $\Theta=\langle A,\tau\rangle$ is a typing in $\mathcal{S}$ then $\mathcal{S}\Vdash M:\Theta$ denotes that $\Theta$ is a typing of $M$ in $\mathcal{S}$.

The contexts for $\lambda$-terms in de Bruijn notation are sequences of types. If $A$ is some context and $n\in\mathbb{N}$ then $A_{<n}$ denotes the first $n-1$ types of $A$. Similarly we define $A_{>n}$, $A_{\leq n}$ and $A_{\geq n}$. Note that, for $A_{>n}$ and $A_{\geq n}$ the final $nil$ element is included. For $n=0$, $A_{\leq 0}.A=A_{<0}.A=A$. The length of $A$ is defined as $|nil|=0$ and, if $A$ is not $nil$, $|A|=1+|A_{>1}|$. The addition of some type $\tau$ at the end of a context $A$ is defined as $A.\tau=A_{\leq m}.\tau.nil$, where $|A|=m$.

Given a term $M$, an interesting question is whether it is typable in $\mathcal{S}$ or not. Note that, we are using a Curry-style/implicit typing, where in $\lambda.M$ we did not specify the type of the bound variable ($\underline{1}$). Such terms have many types, depending on the context. Another important question is whether given a term, its so-called most general typing can be found. An answer to this question, which represents any other answer, is called **principal typing**. Principal typing (which is context independent) is not to be confused with a principal type (which is context dependent). Let $\Theta$ be a typing in $\mathcal{S}$ and $\textbf{Terms}_{\mathcal{S}}(\Theta)=\{M|\mathcal{S}\Vdash M:\Theta\}$. J.B. Wells introduced in [We2002] a system-independent definition of PT and proved that it generalizes previous system-specific definitions.

**Definition 9 ([We2002]).** *A typing $\Theta$ in system $\mathcal{S}$ is principal for some term $M$ if $\mathcal{S}\Vdash M:\Theta$ and for any $\Theta'$ such that $\mathcal{S}\Vdash M:\Theta'$ we have that $\Theta\leq_{\mathcal{S}}\Theta'$, where $\Theta_1\leq_{\mathcal{S}}\Theta_2\iff Terms_{\mathcal{S}}(\Theta_1)\subseteq Terms_{\mathcal{S}}(\Theta_2)$.*

In simply typed systems the principal typing notion is tied to type substitution and weakening. **Weakening** allows one to add unnecessary information to contexts. **Type substitution** maps type variables to types. Given a type substitution $s$, the extension for functional types is straightforward as $s(\sigma{\to}\tau)= s(\sigma){\to}s(\tau)$ and the extension for sequential contexts as $s(nil)=nil$ and $s(\tau.A)= s(\tau).s(A)$. The extension for typings is given by $s(\Theta)=\langle s(A), s(\tau)\rangle$.

### 3.1  Principal Typings for the Simply Typed $\lambda$-Calculus in de Bruijn Notation $TA_{\lambda dB}$

**Definition 10.** *(The System $TA_{\lambda dB}$) The $TA_{\lambda dB}$ typing rules are given by:*

$$(Var) \qquad \tau.A \vdash \underline{1} : \tau \qquad\qquad (Varn) \quad \frac{A \vdash \underline{n} : \tau}{\sigma.A \vdash \underline{n+1} : \tau}$$

$$(Lambda) \;\; \frac{\sigma.A \vdash M : \tau}{A \vdash \lambda.M : \sigma \to \tau} \qquad (App) \;\; \frac{A \vdash M : \sigma \to \tau \quad A \vdash N : \sigma}{A \vdash (M\ N) : \tau}$$

This system is similar to $TA_\lambda$ ([Hi97]). The rule (Varn) allows the construction of contexts as sequences.

**Lemma 1.** *Let $M$ be a $\lambda dB$-term. If $A \vdash M : \tau$, then $A.\sigma \vdash M : \tau$. Hence, the rule $\dfrac{A \vdash M:\tau}{A.\sigma \vdash M:\tau}(\lambda dB\text{-weak})$ holds in the system $TA_{\lambda dB}$.*

Lemma 1 above is proved via the statement of a more general property of $TA_{\lambda dB}$, which justifies why the weakening for this type system has to be done only adding types at the end of contexts.

Using ($\lambda dB$-weak) and type substitution, we follow the definition of [We2002] for Hindley's Principal Typing to define principal typing for the $\lambda dB$-calculus.

**Definition 11.** *A **principal typing** in $TA_{\lambda dB}$ of a term $M$ is the typing $\Theta = \langle A, \tau \rangle$ such that*

1. *$TA_{\lambda dB} \Vdash M : \Theta$*
2. *If $TA_{\lambda dB} \Vdash M : \Theta'$ for any typing $\Theta' = \langle A', \tau' \rangle$, then there exists some substitution $s$ such that $s(A) = A'_{\leq |A|}.nil$ and $s(\tau) = \tau'$.*

Observe that, given a principal typing $\langle A, \tau \rangle$ of $M$, the context $A$ is the shortest context where $M$ can be typable. In contrast to the $\lambda$-calculus with names, where the context from a principal typing of $M$ is the smallest set because it declares types for exactly the free variables of $M$, the context from a principal typing in $\lambda dB$ may have some type declaration for variables not occurring in the term, to maintain the ordered structure of contexts. For example, a PT for $\underline{2}$ is $\langle \tau_1.\tau_2.nil, \tau_2 \rangle$.

As is the case for the simply typed $\lambda$-calculus with names, the best way to assure that Definition 11 is the correct translation of the PT concept, is to verify that Definition 11 corresponds to Definition 9.

**Theorem 1.** *A typing $\Theta$ is principal in $TA_{\lambda dB}$ according to Definition 11 iff $\Theta$ is principal in $TA_{\lambda dB}$ according to Definition 9.*

The proof is similar to the one in [We2002]. The 'sufficient' condition uses a substitution lemma as in [Hi97] 3A2.1(ii) and the weakening from Lemma 1. The 'necessary' condition is constructive by contraposition building a counter example: given a term $M$ with PT $\Theta$ one supposes a typing $\Theta'$ that is not PT of $M$ according to definition 11. From $M$ and the relation between $\Theta$ and $\Theta'$ given by definition 11, one builds a new term $N$ for which $\Theta'$ is a typing, but $\Theta$ is not. The main difference between the proof in [We2002] and this one is the recursive function used to give $N$ a structure exploring some specific $\Theta'$ feature, which has to be split, according to the order in which the term is bound during the recursive construction of the counter example.

We now present a type inference algorithm for $\lambda dB$-terms, similarly to the one in [AyMu2000] for $\lambda s_e$, to verify whether $TA_{\lambda dB}$ has PT according to Definition 11. Given any term $M$, decorate each subterm with a new type variable as subscript and a new context variable as superscript, obtaining a new term denoted by $M'$. For example, for term $\lambda.(\underline{2}\ \underline{1})$ we have the decorated term $(\lambda.(\underline{2}\,^{A_1}_{\tau_1}\ \underline{1}\,^{A_2}_{\tau_2})^{A_3}_{\tau_3})^{A_4}_{\tau_4}$. Then, rules from Table 3 are applied to pairs of the form $\langle\!\langle R, E\rangle\!\rangle$, where R is a set of decorated terms and E a set of equations on type and context variables.

Table 3. Rules for Type Inference in System $TA_{\lambda dB}$

| | | |
|---|---|---|
| (Var) | $\langle\!\langle R \cup \{\underline{1}^{A}_{\tau}\}, E\rangle\!\rangle$ | $\to \langle\!\langle R, E \cup \{A = \tau.A'\}\rangle\!\rangle$, where $A'$ is a fresh context variable; |
| (Varn) | $\langle\!\langle R \cup \{\underline{n}^{A}_{\tau}\}, E\rangle\!\rangle$ | $\to \langle\!\langle R, E \cup \{A = \tau'_1.\cdots.\tau'_{n-1}.\tau.A'\}\rangle\!\rangle$, where $A'$ and $\tau'_1,\ldots,\tau'_{n-1}$ are fresh context and type variables; |
| (Lambda) | $\langle\!\langle R \cup \{(\lambda.M^{A_1}_{\tau_1})^{A_2}_{\tau_2}\}, E\rangle\!\rangle$ | $\to \langle\!\langle R, E \cup \{\tau_2 = \tau^* \to \tau_1, A_1 = \tau^*.A_2\}\rangle\!\rangle$, where $\tau^*$ is a fresh type variable; |
| (App) | $\langle\!\langle R \cup \{(M^{A_1}_{\tau_1}\ N^{A_2}_{\tau_2})^{A_3}_{\tau_3}\}, E\rangle\!\rangle$ | $\to \langle\!\langle R, E \cup \{A_1 = A_2, A_2 = A_3, \tau_1 = \tau_2 \to \tau_3\}\rangle\!\rangle$ |

The inference rules in Table 3 are given according to the typing rules of $TA_{\lambda dB}$. Type inference for $M$ starts with $\langle\!\langle R_0, \varnothing\rangle\!\rangle$, where $R_0$ is the set of all $M'$ subterms. The rules from Table 3 are applied until one reaches $\langle\!\langle \varnothing, E_f\rangle\!\rangle$, where $E_f$ is a set of first-order equations over context and type variables.

*Example 1.* Let $M = \lambda.(\underline{2}\ \underline{1})$. Then $M' = (\lambda.(\underline{2}\,^{A_1}_{\tau_1}\ \underline{1}\,^{A_2}_{\tau_2})^{A_3}_{\tau_3})^{A_4}_{\tau_4}$ and $R_0 = \{\underline{2}\,^{A_1}_{\tau_1}, \underline{1}\,^{A_2}_{\tau_2}, (\underline{2}\,^{A_1}_{\tau_1}\ \underline{1}\,^{A_2}_{\tau_2})^{A_3}_{\tau_3}, (\lambda.(\underline{2}\,^{A_1}_{\tau_1}\ \underline{1}\,^{A_2}_{\tau_2})^{A_3}_{\tau_3})^{A_4}_{\tau_4}\}$. Using the rules in Table 3 we have the following reduction:

$\langle\!\langle R_0, \emptyset\rangle\!\rangle \to_{\text{Varn}}$

$\langle\!\langle R_1 = R_0 \smallsetminus \{\underline{2}\,^{A_1}_{\tau_1}\}, E_1 = \{A_1 = \tau'_1.\tau_1.A'_1\}\rangle\!\rangle \to_{\text{Var}}$

$\langle\!\langle R_2 = R_1 \smallsetminus \{\underline{1}\,^{A_2}_{\tau_2}\}, E_2 = E_1 \cup \{A_2 = \tau_2.A'_2\}\rangle\!\rangle \to_{\text{App}}$

$\langle\!\langle R_3 = R_2 \smallsetminus \{(\underline{2}\,^{A_1}_{\tau_1}\ \underline{1}\,^{A_2}_{\tau_2})^{A_3}_{\tau_3}\}, E_3 = E_2 \cup \{A_1 = A_2, A_2 = A_3, \tau_1 = \tau_2 \to \tau_3\}\rangle\!\rangle \to_{\text{Lambda}}$

$\langle\!\langle \emptyset = R_3 \smallsetminus \{(\lambda.(\underline{2}\,^{A_1}_{\tau_1}\ \underline{1}\,^{A_2}_{\tau_2})^{A_3}_{\tau_3})^{A_4}_{\tau_4}\}, E_4 = E_3 \cup \{\tau_4 = \tau^*_1 \to \tau_3, A_3 = \tau^*_1.A_4\}\rangle\!\rangle$

Thus, $E_4 = E_f$. Solving the trivial equation over context variables, i.e. $A_1 = A_2 = A_3$, and using variables of smaller subscripts, one gets $\{\tau_1 = \tau_2 \to \tau_3, \tau_4 =$

$\tau_1^* {\rightarrow} \tau_3, A_1 = \tau_1'.\tau_1.A_1', A_1 = \tau_2.A_2', A_1 = \tau_1^*.A_4\}$. Thus, simplifying one gets $\{\tau_1 = \tau_2 {\rightarrow} \tau_3, \tau_4 = \tau_1^* {\rightarrow} \tau_3, \tau_1'.\tau_1.A_1' = \tau_2.A_2' = \tau_1^*.A_4\}$. From these equations one gets the most general unifier (mgu for short) $\tau_4 = \tau_2 {\rightarrow} \tau_3$ and $A_4 = (\tau_2 {\rightarrow} \tau_3).A_1'$, for the variables of interest. Since the context must be the shortest one, $A_1' = nil$ and $\langle (\tau_2 {\rightarrow} \tau_3).nil, \tau_2 {\rightarrow} \tau_3 \rangle$ is a principal typing of $M$.

From Definition 11 and by the uniqueness of the solutions of the type inference algorithm, one deduces that $TA_{\lambda dB}$ satisfies PT. The next theorem says that every typable term has a principal typing.

**Theorem 2 (Principal Typings for $TA_{\lambda dB}$).** *$TA_{\lambda dB}$ satisfies the property of having principal typings.*

## 3.2   Principal Typings for $TA_{\lambda s_e}$, the Simply Typed $\lambda s_e$

The typed version of $\lambda s_e$ presented is in Curry style, which we have verified to have the same properties as the version in Church style presented in [ARKa2001a]. In particular, the properties in question being: weak normalisation (WN), confluence (CR) and subject reduction (SR). Thus, the syntax of $\lambda s_e$-terms and the rules are the same as the untyped version.

Since the syntax of $\lambda s_e$ remains close to the $\lambda dB$-calculus, to have a type assignment system for the $\lambda s_e$-calculus we only need to add typing rules to $TA_{\lambda dB}$ for the two new kinds of terms.

**Definition 12 (The System $TA_{\lambda s_e}$).** *$TA_{\lambda s_e}$ is given by (Var), (Varn), (App), (Lambda) from Definiton 10 and the following new rules.*

$$(Sigma) \quad \frac{A_{\geq i} \vdash N : \rho \quad A_{<i}.\rho.A_{\geq i} \vdash M : \tau}{A \vdash M \sigma^i N : \tau} \qquad (Phi) \quad \frac{A_{\leq k}.A_{\geq k+i} \vdash M : \tau}{A \vdash \varphi_k^i M : \tau}$$

*where in (Sigma) $|A| \geq i - 1$ and in (Phi) $|A| \geq k + i - 1$.*

Weakening for $\lambda s_e$ is done in the same way as for $\lambda dB$, adding types at the end of a context, giving the following lemma.

**Lemma 2 (Weakening for $\lambda s_e$).** *The rule ($\lambda s_e$-weak) holds in System $TA_{\lambda s_e}$, where $\dfrac{A \vdash M : \tau}{A.\sigma \vdash M : \tau}$($\lambda s_e$-weak).*

Consequently, the definition of principal typings in $\lambda s_e$ is the same as that for $TA_{\lambda dB}$. For the sake of completeness we repeat it here.

**Definition 13 (Principal Typings in $TA_{\lambda s_e}$).** *A **principal typing** of a term $M$ in $TA_{\lambda s_e}$ is a typing $\Theta = \langle A, \tau \rangle$ such that*

1. *$TA_{\lambda s_e} \Vdash M : \Theta$*
2. *If $TA_{\lambda s_e} \Vdash M : \Theta'$ for any typing $\Theta' = \langle A', \tau' \rangle$, then there exists a substitution $s$ such that $s(A) = A'_{\leq |A|}.nil$ and $s(\tau) = \tau'$.*

**Theorem 3.** *A typing $\Theta$ is principal in $TA_{\lambda s_e}$ according to Definition 13 iff $\Theta$ is principal in $TA_{\lambda s_e}$ according to Definition 9.*

**Table 4.** Type inference rules for the $\lambda s_e$-Calculus

---

(Sigma) $\langle\!\langle R \cup \{(M_{\tau_1}^{A_1} \sigma^i N_{\tau_2}^{A_2})_{\tau_3}^{A_3}\}, E\rangle\!\rangle \rightarrow$

$\qquad \langle\!\langle R, E \cup \{\tau_1 = \tau_3, A_1 = \tau_1'.\cdots.\tau_{i-1}'.\tau_2.A_2, A_3 = \tau_1'.\cdots.\tau_{i-1}'.A_2\}\rangle\!\rangle,$

$\qquad$ where $\tau_1', \ldots, \tau_{i-1}'$ are new type variables and the sequence is

$\qquad$ empty if $i = 1$;

(Phi) $\quad \langle\!\langle R \cup \{(\varphi_k^i M_{\tau_1}^{A_1})_{\tau_2}^{A_2}\}, E\rangle\!\rangle \rightarrow$

$\qquad \langle\!\langle R, E \cup \{\tau_1 = \tau_2, A_2 = \tau_1'.\cdots.\tau_{k+i-1}'.A', A_1 = \tau_1'.\cdots.\tau_k'.A'\}\rangle\!\rangle,$

$\qquad$ where $A'$ and $\tau_1', \ldots, \tau_{k+i-1}'$ are new variables of context and

$\qquad$ type. If $k + i - 1 = 0$ or $k = 0$, then the sequences $\tau_1', \ldots, \tau_{k+i-1}'$

$\qquad$ and $\tau_1', \ldots, \tau_k'$, respectively, are empty.

---

The proof of Theorem 3 is a straightforward extension of that of Theorem 1.

We now present a type inference algorithm for the $\lambda s_e$-calculus, similarly to that of [AyMu2000]. The algorithm is composed of the rules from Table 3 and the new rules in Table 4.

Similarly to the previous algorithm, the rules of Table 4 were developed according to the rules of Definition 12. The decorated term associated with $M$, denoted by $M'$, has a syntax close to that of decorated $\lambda dB$-terms: any subterm is decorated with its type and its context variables. The rules are applied to pairs $\langle\!\langle R, E\rangle\!\rangle$, starting from the pair $\langle\!\langle R_0, \varnothing\rangle\!\rangle$, as was done to $TA_{\lambda dB}$.

*Example 2.* For $M = \lambda.((\underline{1}\,\sigma^2\underline{2})\,(\varphi_0^2\,\underline{2}))$, one obtains the corresponding $R_0$ from $M' = (\lambda.((\underline{1}_{\tau_1}^{A_1}\sigma^2\underline{2}_{\tau_2}^{A_2})_{\tau_3}^{A_3}\,(\varphi_0^2\,\underline{2}_{\tau_4}^{A_4})_{\tau_5}^{A_5})_{\tau_6}^{A_6})_{\tau_7}^{A_7}$. Then, applying the rules in Table 3 and 4 to the pair $\langle\!\langle R_0, \emptyset\rangle\!\rangle$, obtaining the pair $\langle\!\langle \emptyset, E_f\rangle\!\rangle$, and simplifying $E_f$, in a similar fashion to example 1, one obtains the system of equations which lead to the mgu $\tau_7 = (\tau_2 \rightarrow \tau_6) \rightarrow \tau_6$ and $A_7 = \tau_1'.\tau_2.A_2'$ for the variables of interest.

**Theorem 4 (Principal Typings for $TA_{\lambda s_e}$).** $TA_{\lambda s_e}$ *satisfies the property of having principal typings.*


### 3.3  Principal Typings for $TA_{\lambda\sigma}$, the Simply Typed $\lambda\sigma$

The typing rules of the $\lambda\sigma$-calculus provide types for objects of sort term as well as for objects of sort substitution. An object of sort substitution, due to its semantics, can be viewed as a list of terms. Consequently, its type is a context. $S \rhd A$ denotes that the object of sort substitution $S$ has type $A$.

**Definition 14 (The System $TA_{\lambda\sigma}$).** $TA_{\lambda\sigma}$ *is given by the following typing rules.*

$(var) \qquad \tau.A \vdash \underline{1} : \tau \qquad\qquad (lambda) \quad \dfrac{\sigma.A \vdash M : \tau}{A \vdash \lambda.M : \sigma \rightarrow \tau}$

$(app) \quad \dfrac{A \vdash M : \sigma \rightarrow \tau \quad A \vdash N : \sigma}{A \vdash (M\,N) : \tau} \qquad (clos) \quad \dfrac{A \vdash S \rhd A' \quad A' \vdash M : \tau}{A \vdash M[S] : \tau}$

$(id) \qquad A \vdash id \rhd A \qquad\qquad\qquad (shift) \quad \tau.A \vdash\,\uparrow\,\rhd A$

$(cons) \quad \dfrac{A \vdash M : \tau \quad A \vdash S \rhd A'}{A \vdash M.S \rhd \tau.A'} \qquad (comp) \quad \dfrac{A \vdash S \rhd A'' \quad A'' \vdash S' \rhd A'}{A \vdash S' \circ S \rhd A'}$

Observe that the name of the typing rules begin with lower-case letters, while the rewriting rules with upper-case letters. As for $\lambda s_e$, the typed version of the $\lambda\sigma$-calculus is presented in Curry style. We have verified that the Curry style version has WN, CR and SR as the Church style version of [DoHaKi2000].

The notion of typing for $TA_{\lambda\sigma}$ has to be adapted because the $\lambda\sigma$-expression of sort substitution is decorated with contexts variables as types and as contexts. Thus, one may say that $\Theta = \langle A, \mathbb{T} \rangle$ is a typing of a $\lambda\sigma$-expression in $TA_{\lambda\sigma}$, where $\mathbb{T}$ can be either a type or a context. If the analysed expression belongs to the $\lambda$-calculus, the notion of typing corresponds to that of $TA_{\lambda dB}$.

**Lemma 3 (Weakening for $\lambda\sigma$).** *Let $M$ be a $\lambda\sigma$-term and $S$ a $\lambda\sigma$-substitution. If $A \vdash M : \tau$, then $A.\sigma \vdash M : \tau$, for any type $\sigma$. Similarly, if $A \vdash S \rhd A'$, then $A.\sigma \vdash S \rhd A'.\sigma$. Hence, the rules ($\lambda\sigma$-tweak) and ($\lambda\sigma$-sweak) hold in System $TA_{\lambda\sigma}$, where*

$$\frac{A \vdash M : \tau}{A.\sigma \vdash M : \tau}(\lambda\sigma\text{-}tweak) \qquad\qquad \frac{A \vdash S \rhd A'}{A.\sigma \vdash S \rhd A'.\sigma}(\lambda\sigma\text{-}sweak)$$

Lemma 3 and type substitutions allow us present a definition for PT in $TA_{\lambda\sigma}$.

**Definition 15 (Principal Typings in $TA_{\lambda\sigma}$).** *A **principal typing** of an expression $M$ in $TA_{\lambda\sigma}$ is a typing $\Theta = \langle A, \mathbb{T} \rangle$ such that*

1. *$TA_{\lambda\sigma} \Vdash M : \Theta$*
2. *If $TA_{\lambda\sigma} \Vdash M : \Theta'$ for any typing $\Theta' = \langle A', \mathbb{T}' \rangle$, then there exists a substitution $s$ such that $s(A) = A'_{\leq |A|}.nil$ and if $\mathbb{T}$ is a type, $s(\mathbb{T}) = \mathbb{T}'$, otherwise we have that $s(\mathbb{T}) = \mathbb{T}'_{\leq |\mathbb{T}|}.nil$.*

We might verify if this PT definition has a correspondence with Wells' system-independent definition [We2002].

**Theorem 5.** *A typing $\Theta$ is principal in $TA_{\lambda\sigma}$ according to Definition 15 iff $\Theta$ is principal in $TA_{\lambda\sigma}$ according to Definition 9.*

Despite the fact that the notion of typing is extended to include the sort substitution, the techniques used to prove Theorem 5 are the same applied to prove Theorems 1 and 3.

We now present an algorithm for type inference, to verify if $TA_{\lambda\sigma}$ has PT according to Definition 15. Thus, given an expression $M$, we will work with the decorated expression $M'$ but the type for substitutions is a context as well. We use the same syntax for decorated expressions as in [Bo95].

The inference rules presented in Table 5 are given according to the typing rules of the system $TA_{\lambda\sigma}$ presented in Definition 14. Similarly to the previous algorithm, the rules are applied to pairs $\langle\!\langle R, E \rangle\!\rangle$, where $R$ is a set of subexpressions of $M'$ and $E$ a set of equations over type and context variables.

*Example 3.* For $M = (\underline{2}.id) \circ \uparrow$ one has $M' = (((\underline{1}^{A_1}_{\tau_1}[\uparrow^{A_2}_{A_3}])^{A_4}_{\tau_2}.id^{A_5}_{A_6})^{A_7}_{A_8} \circ \uparrow^{A_9}_{A_{10}})^{A_{11}}_{A_{12}}$. Then $R_0 = \{(\underline{1}^{A_1}_{\tau_1}[\uparrow^{A_2}_{A_3}])^{A_4}_{\tau_2}, ((\underline{1}^{A_1}_{\tau_1}[\uparrow^{A_2}_{A_3}])^{A_4}_{\tau_2}.id^{A_5}_{A_6})^{A_7}_{A_8}, (((\underline{1}^{A_1}_{\tau_1}[\uparrow^{A_2}_{A_3}])^{A_4}_{\tau_2}.id^{A_5}_{A_6})^{A_7}_{A_8} \circ \uparrow^{A_9}_{A_{10}})^{A_{11}}_{A_{12}}, \underline{1}^{A_1}_{\tau_1}, \uparrow^{A_2}_{A_3}, id^{A_5}_{A_6}, \uparrow^{A_9}_{A_{10}}\}$. Applying the rules from Table 5 to the pair $\langle\!\langle R_0, \emptyset \rangle\!\rangle$ until the pair $\langle\!\langle \emptyset, E_f \rangle\!\rangle$ is reached, and simplifying $E_f$ as in example 1, one obtains the

**Table 5.** Type inference rules for the $\lambda\sigma$-calculus

| | | |
|---|---|---|
| (Var) | $\langle\!\langle R \cup \{\underline{1}_\tau^A\}, E\rangle\!\rangle$ | $\to \langle\!\langle R, E \cup \{A = \tau.A'\}\rangle\!\rangle$, where $A'$ is a fresh context variable; |
| (Lambda) | $\langle\!\langle R \cup \{(\lambda.M_{\tau_1}^{A_1})_{\tau_2}^{A_2}\}, E\rangle\!\rangle$ | $\to \langle\!\langle R, E \cup \{\tau_2 = \tau^* \to \tau_1, A_1 = \tau^*.A_2\}\rangle\!\rangle$, where $\tau^*$ is a fresh type variable; |
| (App) | $\langle\!\langle R \cup \{(M_{\tau_1}^{A_1}\ N_{\tau_2}^{A_2})_{\tau_3}^{A_3}\}, E\rangle\!\rangle$ | $\to \langle\!\langle R, E \cup \{A_1 = A_2, A_2 = A_3, \tau_1 = \tau_2 \to \tau_3\}\rangle\!\rangle$ |
| (Clos) | $\langle\!\langle R \cup \{(M_{\tau_1}^{A_1}[S_{A_3}^{A_2}])_{\tau_2}^{A_4}\}, E\rangle\!\rangle$ | $\to \langle\!\langle R, E \cup \{A_1 = A_3, A_2 = A_4, \tau_1 = \tau_2\}\rangle\!\rangle$ |
| (Id) | $\langle\!\langle R \cup \{id_{A_2}^{A_1}\}, E\rangle\!\rangle$ | $\to \langle\!\langle R, E \cup \{A_1 = A_2\}\rangle\!\rangle$ |
| (Shift) | $\langle\!\langle R \cup \{\uparrow_{A_2}^{A_1}\}, E\rangle\!\rangle$ | $\to \langle\!\langle R, E \cup \{A_1 = \tau'.A_2\}\rangle\!\rangle$, where $\tau'$ is a fresh type variable; |
| (Cons) | $\langle\!\langle R \cup \{(M_{\tau_1}^{A_1}.S_{A_3}^{A_2})_{A_5}^{A_4}\}, E\rangle\!\rangle$ | $\to \langle\!\langle R, E \cup \{A_1 = A_2, A_2 = A_4, A_5 = \tau_1.A_3\}\rangle\!\rangle$ |
| (Comp) | $\langle\!\langle R \cup \{(S_{A_2}^{A_1} \circ T_{A_4}^{A_3})_{A_6}^{A_5}\}, E\rangle\!\rangle$ | $\to \langle\!\langle R, E \cup \{A_1 = A_4, A_2 = A_6, A_3 = A_5\}\rangle\!\rangle$ |

set of equations $\{\tau_1 = \tau_2, A_{11} = A_{12} = \tau_2.A_2, A_2 = \tau_1'.A_1, A_1 = \tau_1.A_1'\}$. From this equational system one obtains the mgu $A_{11}=A_{12}=\tau_1.\tau_1'.\tau_1.A_1'$, for the variables of interest. Thus, $\langle \tau_1.\tau_1'.\tau_1.nil, \tau_1.\tau_1'.\tau_1.nil \rangle$ is a principal typing of $M$.

**Theorem 6 (Principal Typings for $TA_{\lambda\sigma}$).** $TA_{\lambda\sigma}$ *satisfies the property of having principal typings.*

## 4 Conclusions and Future Work

We considered for $\lambda s_e$ and $\lambda\sigma$ particular notions of principal typings and gave respective definitions which we proved to agree with the system-independent notion of Wells in [We2002]. The adaptation of this general notion of principal typings for the $\lambda\sigma$ requires special attention, since this calculus enlarges the language of the $\lambda$-calculus by introducing a new sort of *substitution* objects, whose types are contexts. Thus, the provided PT notion has to deal with the principality of substitution objects as well. Then, the property of having principal typings is straightforwardly proved by revisiting type inference algorithms for the $\lambda s_e$ and the $\lambda\sigma$, previously presented in [AyMu2000] and [Bo95], respectively. The result is based on the correctness, completeness and uniqueness of solutions given by adequate first-order unification algorithms (e.g. see the unification algorithm given in [Hi97]).

The investigation of this property for more elaborated typing systems of explicit substitutions is an interesting work to be done.

## References

[ACCL91]    Abadi, M., Cardelli, L., Curien, P.-L., Lévy, J.-J.: Explicit Substitutions. J. of Functional Programming 1(4), 375–416 (1991)

[ARMoKa2005]    Ayala-Rincón, M., de Moura, F., Kamareddine, F.: Comparing and Implementing Calculi of Explicit Substitutions with Eta-Reduction. Annals of Pure and Applied Logic 134, 5–41 (2005)

[ARKa2001a]     Ayala-Rincón, M., Kamareddine, F.: Unification via the $\lambda s_e$-Style of Explicit Substitution. The Logical Journal of the Interest Group in Pure and Applied Logics 9(4), 489–523 (2001)

[AyMu2000]      Ayala-Rincón, M., Muñoz, C.: Explicit Substitutions and All That. Revista Colombiana de Computación 1(1), 47–71 (2000)

[Bo95]          Borovanský, P.: Implementation of Higher-Order Unification Based on Calculus of Explicit Substitutions. In: Bartošek, M., Staudek, J., Wiedermann, J. (eds.) SOFSEM 1995. LNCS, vol. 1012, pp. 363–368. Springer, Heidelberg (1995)

[deBru72]       de Bruijn, N.G.: Lambda calculus notation with nameless dummies, a tool for automatic formula manipulation, with application to the Church-Rosser theorem. Indagationes Mathematicae 34, 381–392 (1972)

[DoHaKi2000]    Dowek, G., Hardin, T., Kirchner, C.: Higher-order Unification via Explicit Substitutions. Information and Computation 157(1/2), 183–235 (2000)

[Hi97]          Hindley, J.R.: Basic Simple Type Theory. Cambridge Tracts in Theoretical Computer Science, vol. 42. Cambridge University Press, Cambridge (1997)

[Jim96]         Jim, T.: What are principal typings and what are they good for? In: Proc. of POPL 1995: Symp. on Principles of Programming Languages, pp. 42–53. ACM, New York (1996)

[KR97]          Kamareddine, F., Ríos, A.: Extending a $\lambda$-calculus with Explicit Substitution which Preserves Strong Normalisation into a Confluent Calculus on Open Terms. J. of Func. Programming 7, 395–420 (1997)

[Mel95]         Melliès, P.-A.: Typed $\lambda$-calculi with explicit substitutions may not terminate. In: Dezani-Ciancaglini, M., Plotkin, G. (eds.) TLCA 1995. LNCS, vol. 902, pp. 328–334. Springer, Heidelberg (1995)

[NaWi98]        Nadathur, G., Wilson, D.S.: A Notation for Lambda Terms A Generalization of Environments. Theoretical Computer Science 198, 49–98 (1998)

[We2002]        Wells, J.: The essence of principal typings. In: Widmayer, P., Triguero, F., Morales, R., Hennessy, M., Eidenbenz, S., Conejo, R. (eds.) ICALP 2002. LNCS, vol. 2380, pp. 913–925. Springer, Heidelberg (2002)

# How We Think of Computing Today*

Jiří Wiedermann[1] and Jan van Leeuwen[2]

[1] Institute of Computer Science, Academy of Sciences of the Czech Republic,
Pod Vodárenskou věží 2, 182 07 Prague 8, Czech Republic
`jiri.wiedermann@cs.cas.cz`
[2] Department of Information and Computing Sciences, Utrecht University,
Padualaan 14, 3584 CH Utrecht, The Netherlands
`j.vanleeuwen@cs.uu.nl`

**Abstract.** Classical models of computation no longer fully correspond to the current notions of computing in modern systems. Even in the sciences, many natural systems are now viewed as systems that compute. Can one devise models of computation that capture the notion of computing as seen today and that could play the same role as Turing machines did for the classical case? We propose two models inspired from key mechanisms of current systems in both artificial and natural environments: evolving automata and interactive Turing machines with advice. The two models represent relevant adjustments in our apprehension of computing: the shift to potentially non-terminating interactive computations, the shift towards systems whose hardware and/or software can change over time, and the shift to computing systems that evolve in an unpredictable, non-uniform way. The two models are shown to be equivalent and both are provably computationally more powerful than the models covered by the old computing paradigm. The models also motivate the extension of classical complexity theory by non-uniform classes, using the computational resources that are natural to these models. Of course, the additional computational power of the models cannot in general be meaningfully exploited in concrete goal-oriented computations.

**Keywords:** Turing machines, evolving automata, interactive computation, non-uniform complexity.

## 1   Introduction

Can the Internet be simulated, at least in principle, by a Turing machine? Can the living cell, the brain, and any other natural information processing system be simulated likewise? The answer is not at all clear and depends very much on one's viewpoint. While the Turing machine paradigm is well suited for modeling stepwise computational processes, it may be less suited for modeling the behaviour of the computational systems as we know them today. What model

---

* This research was partially supported by project BRICKS in the Netherlands, and by Institutional Research Plan AV0Z10300504 and grants No. 1ET100300419 and 1ET100300517 within the Czech National Research Program 'Information Society'.

A. Beckmann, C. Dimitracopoulos, and B. Löwe (Eds.): CiE 2008, LNCS 5028, pp. 579–593, 2008.

of computation could replace the classical Turing machine and serve as the new paradigm?

The shift from Turing machines to a new computational model should correspond to the shift in thinking about computing in the systems of today. It is no longer the case that only isolated computers compute. We have no problem admitting that sensor nets, embedded control systems in all kinds of interacting devices and robots, 'always operating' information services, and in fact the Internet, as a whole, perform computations, albeit in some non-standard way. Moreover, it is no longer the case that only artificial gadgets compute. Biologists frequently speak of living cells or entire organisms as complex information processing systems, as do psychologists in the case of the human mind and sociologists in the case of animal or human societies. Some physicists even believe that the entire Universe can be viewed in this way [3]. Can there be a single model of computation covering all these cases, like the Turing machine did for early computing, or are we deemed to have many different models, tailored to each case at hand?

In this expository paper we describe a number of computational paradigms that have emerged in recent years and that lead to ingredients for new models of computation. We give the background for these paradigms and of some models that have been based on them. We show that some of these models are indeed, at least in principle, more powerful than classical computing in the sense that they are provably computationally more powerful models than those fitting the old paradigm. The models also enable extensions of classical complexity theory (cf. [18,19]), showing that our modern notions of computing can lead naturally into the domain of non-uniform complexity.

## 2     From Isolated to Interactive Computation

> *New technologies, from the telegraph to the World Wide Web, have expanded our abilities to communicate widely, flexibly, and efficiently. This urge to communicate will continue to drive the expanding technology with the advent of widespread two-way video, wireless connectivity, and high-bandwidth audio, video, 3-D imaging, and more yet to be imagined.*
> T. Winograd ([23], 1997).

The way we think of computing is closely related to what we consider to be a computation.

### 2.1   Classical Computing

Historically, when arithmetic was invented in the early days of mankind, computing seemed an ability by which only people are endowed. Later, when abaci and mechanical calculators appeared, it was taken for granted that 'computing devices' are artifacts designed by people. Consequently, computing was perceived as an activity intrinsic to people or to devices invented by people for that purpose. Computing was seen as an 'invented' process, in contrast to the 'natural' processes which were driven by natural laws and which worked 'by itself'. Computing was something artificial or non-natural: it had to be planned ahead, streamlined, powered and monitored. Computing devices had a rigorous, regular and highly organized structure and therefore, had to be engineered with great ingeniousness. Mainframe-, midi- and minicomputers and the many types of PCs confirm this view of computing. In theory, their functioning is suitably modeled by the Turing machine paradigm or by easily simulated models like the random access machine.

Ever since Turing's formulation of the model [12] in 1936, classical Turing machines have dominated the thinking of computing. However, the paradigm of Turing machines does not only suggest that any process which deserves to be called algorithmic can be modeled by a Turing machine. There is more to it: the paradigm also assumes a certain *computational scenario* which determines how the machine is used. In classical Turing machines, this scenario requires that a finite amount of input data is present prior to the start of a computation; during the computation no new data can be added. The result is to be extracted after a finite number of steps and only after, and when, the computation has terminated. This allows one to view a computation as a process that maps finite input data to finite output data and hence to view a computer as a device realizing standard mathematical functions, calculating their value given an input value. Computability theory has been based on this view.

## 2.2  'Always On' Computing

A different view of computing systems arose when the first automated control systems emerged. Here the computer was not used to compute function values. Instead it was used to monitor, to serve, or to process potentially infinite streams of data. Normally, as in the first computer operating systems and modern 'always on' systems, infinite input streams are presented as un-ending streams of finite chunks of data. Each chunk is processed according to the Turing machine paradigm. Aside from mathematical reasons, it inspired computability theory to study infinite computations, by using Turing machines (or restricted variants like finite automata) under the *generalized* scenario of infinite input strings and infinite output strings. The results were seen as a natural generalization of the finitary case, not as a revision of the Turing machine paradigm. After all, the machine model had remained the same, merely the computational scenario had changed. There is now a refined theory of $\omega$-automata [10], with many applications in e.g. process theory. In relativistic computing, infinite computations are seen from yet a different angle [4,21].

## 2.3  Interactive Computing

From computations over infinite streams of data it is only a small step to *interactive* computations, where a machine interacts continuously with its environment. The computational view of interaction was propagated by Wegner [20]. In interactive computing we have continuous on-line entry of input data and delivery of output data. Interactive machines do not have input and output tapes but input and output ports. Interactive computing principally differs from computing over infinite streams in two ways. First, in interactive computing we only consider *potentially infinite streams*, i.e., streams that are always finite but can be prolonged without limit, with unpredictable next inputs at any time. We include port symbols denoting 'no input' and 'no output', as valid inputs or outputs in streams, respectively. Second, a *finite delay condition* [17] may be required during computation, asserting that after any non-empty input symbol a non-empty output symbol must be produced sometime. Any infinite input string that can be fed to the interactive machine properly in this way, is called a 'valid' input.

The previous conditions mean e.g. that internal and external phases of computing alternate, depending on whether an interactive device needs to do some finite computation before outputting a non-empty response symbol (or taking in a new non-empty input) or not. Unlike the scenario of classical machines performing infinite computations over infinite streams [10], interactive machines cannot answer questions requiring the processing of infinite streams 'in the limit' (such as 'is there a finite number of 1's in the given infinite stream?'). The interactive use of standard stand-alone PCs corresponds well to this view of interactive machines. The corresponding change in the Turing machine model leads to so-called *interactive Turing machines* (ITMs) introduced in [16]. The processing of infinite streams of symbols by ITMs leads to so-called 'interactively realizable translations' on valid (infinite) input strings, see [18] for formal details.

From the viewpoint of computability theory, interactive computing e.g. with ITMs does not lead to super-Turing computing power. Interactive computing merely extends our view of classically computable functions over finite domains to computable functions (translations) defined over infinite domains. Interactive computers simply compute something different from non-interactive ones because they follow a different scenario. Remembering the respective inputs over time, a finite computation of an interactive machine can always be replayed *a posteriori* by a non-interactive machine giving the same outputs as the interactive machine [17].

## 3   From Interactive to Evolving Computation

*Every physical system registers information, and just by evolving in time, by doing its thing, it changes that information, transforms that information, or, if you like, processes that information.*
S. Lloyd ([3], 2002).

Data interaction with computers differs from classical computing, as reflected in the change of computational scenario. Historically it allowed the use of computers in many more applications than before and thus it extends our apprehension of computing even though, from a computability point, interactive Turing machines cannot compute more, i.e. other mappings than non-interactive machines. A change in computational power can only come from a change of view on the functioning of an isolated PC itself. In particular, does it make sense to change a computing device, or to let it change, *during* its computation?

Obviously, this feature could be especially useful in the case of interactive computing which potentially prolongs indefinitely: programs may be upgraded over time, and so can the hardware. For instance, the user could add more internal memory, upgrade the disk (while maintaining the original data), or exchange the processor for a newer one. One could couple it to several other computers, or connect it to a network like the *Internet*. Can changes like this be accommodated within the Turing machine paradigm? What happens, from a computational viewpoint?

This brings us back to a question posed in the beginning: can the Internet be simulated by a Turing machine? This is a difficult question, since it asks for comparing a piece of high-level computing and communication technology that exists in the real world with a highly simplified model of computation that exists in the abstract world. Therefore we will answer the question in two steps. In the first step, we propose an abstract model capturing the important features of the Internet (and as we will see later, those of many other computational gadgets in the sciences). In the second step, we compare this model with the classical Turing machine.

### 3.1   Modeling the Internet

What could an appropriate abstract model of the Internet be that has the same abstract simplicity as a Turing machine? The Internet has one important feature that we want to capture: *its structure evolves over time.* New sites are added to or deleted from the network all the time, possibly even connecting to it by means of wireless technologies. A 'site' can be anything: a workstation connected to the net via a cable, a notebook in an airplane, or a mobile phone. In order to capture all this variety in a simple model one must choose a fairly abstract viewpoint.

Consider the evolution of the Internet over time from its very beginning till now. Concentrate on the moments when it underwent some 'hardware changes' as mentioned above: a computer joined or left the network, or a computer on the net was upgraded. (We ignore cabling issues.) Between these moments of change, the structure of the Internet can be seen as stable. In these periods, one can view the entire Internet as a huge finite automaton, with finitely many input and output ports corresponding to all data entry and exit points (like keyboards, cameras, monitors, terminals, printers, and so on). In this automaton, the contents of the Internet is modeled by the (huge number of) states. Transitions between states correspond to operations taking place over the Internet, like changes in its contents by new inputs.

The automaton works in a 'parallel' interactive mode, receiving inputs through all its inputs ports and producing outputs over all its output ports. Of course, this abstraction neglects many other issues, like variable message transfer times. Allowing this simplification we go even farther: we merge all input streams into a single input stream while remembering the identity of the individual elements (i.e., we can always say which element belongs to which individual stream) and do the same with the outputs. What we get is an equivalent finite automaton with a single input and a single output port which, in principle, computes the same transformation of input to output as the Internet did *in a period in which it had a stable structure*. Note that in the same way we can model any single computer over its lifetime with consecutive upgrades. In-between two consecutive periods, the structure of the net, and hence of the modeling finite automaton, is said to *evolve*.

## 3.2   Evolving Automata

In order to model the evolution of a computational system like the Internet over time, we consider the (ordered) *sequence* of finite automata corresponding to the successive stable periods. The notion of sequence has been used in computational complexity theory before in different contexts e.g. to capture the computational power of non-uniform families of circuits (cf. [1]). In a sequence of automata, the $i$-th automaton corresponds to the Internet contents and computations during the $i$-th stable period of the Internet. In the course of this time, only the $i$-th automaton receives input and produces output.

We have arrived at the following computational model called the *evolving automaton*, introduced in [16], [19]. (In [18] the model is called a 'lineage' of automata but we give a simplified formulation for expository reasons.)

**Definition 1.** *For $i = 1, 2, \ldots$, let $A_i$ be a finite automaton with a single input and a single output port, let its alphabet be $\Sigma_i$, let $S_i$ be the set of states of $A_i$, and let $\emptyset \neq Q_i \subseteq S_i$, be a set of 'preserving' states in $A_i$. Let $T_\mathcal{A} = t_1, t_2, \ldots,$ with $t_i \in \mathbf{N}$, $t_1 = 1$ and $t_i < t_{i+1}$ be the sequence of switching times. The infinite sequence of finite automata $\mathcal{A} = A_1, A_2, \ldots$ is called an* evolving automaton *with schedule $T_\mathcal{A}$, or just an evolving automaton if $T_\mathcal{A}$ is understood, if $Q_i \subsetneq Q_{i+1}$, for $i = 1, 2, \ldots$ and the switching in processing from one automaton to the next takes place at the times given by $T_\mathcal{A}$.*

In the model, the condition $Q_i \subsetneq Q_{i+1}$ captures the persistence of the relevant data over time (cf. [6]). In the language of finite automata the condition ensures that some information available to $A_i$ and represented in the states in $Q_i$, is available also to $A_{i+1}$ after the 'change moment'. We require here that the transferred information can only grow, but this can easily be avoided by slightly modifying the definition, see [18]. The schedule $T_\mathcal{A}$ of an evolving automaton $\mathcal{A}$ determines the 'switching times' $t_i \in T_\mathcal{A}$ when the input stream to $A_i$ must be redirected to $A_{i+1}$, in a state in $Q_i$.

An evolving automaton $\mathcal{A}$ clearly is an infinite object, given by an explicit enumeration of its elements. However, at each time the computation is performed by only one element of $\mathcal{A}$, which is a finite object. In general, there need not exist an algorithm for computing $A_i$ given the previous elements in the sequence, the input and the schedule. A similar remark holds for the switching schedule; in general, its elements are non-computable from knowing $\mathcal{A}$ and the input sequence. Evolving automata are *non-uniform* systems just like families of circuits:

their development over time cannot be described by an algorithm. The Internet is a case in point: the decision to upgrade a computer or connect it to the Internet, depends entirely on the person owning the computer and has nothing to do with the computability.

### 3.3   Complexity of Evolving Automata

Given an evolving automaton $\mathcal{A} = A_1, A_2, \ldots$, it is natural to consider the number of states of the individual automata $A_i$ as a measure for the complexity of $\mathcal{A}$. Define the *size complexity* of an evolving automaton $\mathcal{A}$ as the function $g$ such that for every $i$, $g(i)$ is the number of states of $A_i$. Given this complexity measure, one can now try to categorize the translations realized by evolving automata $\mathcal{A}$ with any possible time schedule $T_\mathcal{A}$.

**Definition 2.** *A translation $\phi$ of infinite streams to infinite streams is said to be of complexity $g$ if there is an evolving automaton of complexity $g$ that realizes $\phi$. For any function $g : \mathbf{N} \to \mathbf{N}$, let $SIZE(g)$ be the class of all translations $\phi$ that can be realized by an evolving automaton of complexity $g$.*

By the non-uniformity of the model, the classes $SIZE(g)$ will in general contain an abundance of non-computable translations, even though all of them will be 'non-uniformly realizable' by evolving automata within 'growth bound' $g$. A precise characterization of the translations that are non-uniformly realizable by evolving automata was given in [18,19]. The complexity measure is a realistic one, as indicated by the following result from [18,19].

**Theorem 1.** *Let $g, h : \mathbf{N} \to \mathbf{N}$ be positive non-decreasing functions such that $g(i) \leq h(i)$ for all $i$ and $g(i) < h(i)$ for at least one $i$. Then $SIZE(g)$ is properly contained in $SIZE(h)$.*

In fact, in [18,19] it is shown that $SIZE(g)$ and $SIZE(h)$ differ whenever $g$ and $h$ do. Thus evolving automata have a fitting complexity theory, directly derived from the nature of the model.

### 3.4   Modeling the Internet 2

Finally, if one would want to build an evolving automaton simulating the existing Internet, then we could construct such an automaton only *a posteriori,* after watching the Internet's evolution and taking snapshots of it at the times of its changes, plus a recording of all input streams. The snapshots would then be used for constructing the sequence of automata which, on the recorded input streams, would produce the same translation as the Internet did. Of course, we are not seriously proposing to do it, it is only a *Gedankenexperiment,* serving as proof of principle.

Conversely, can some network simulate an evolving automaton $\mathcal{A} = A_1, A_2, \ldots$ with switching schedule $T_\mathcal{A} = t_1, t_2, \ldots$? Of course it can. To show it, we begin with a computer simulating $A_1$. At time $t_1$ we replace it by (or upgrade it to) a computer simulating $A_2$ and continue processing the inputs till time $t_2$, etc.

# 4    Two New Models of Computation

Evolving automata and Turing machines are both defined using the same formal language. This allows us to compare the computational power of both models.

**Proposition 1.** *Every classical Turing machine, or even an ITM, $\mathcal{T}$ can be simulated by an evolving automaton.*

*Proof.* (Sketch) Observe the computation of $\mathcal{T}$ on an input stream $\sigma$ and note the times $t_i$ when any of the Turing machine's heads moves past the 'next' rightmost symbol on its tape. These times define the switching schedule. Between times $t_i$ and $t_{i+1}$, the computation of $\mathcal{T}$ can be modeled by a finite automaton $A_i$. This leads to a sequence of automata $\mathcal{A}$ and a schedule $T_{\mathcal{A}}$ computing the same translation as $\mathcal{T}$.                                      □

The proposition establishes that computationally, evolving automata are at least as powerful as (interactive) Turing machines. Observe that, in order to simulate $\mathcal{T}$, the construction of $\mathcal{A}$ and that of the switching schedule depended, in a computable way, solely on $\mathcal{T}$ and on the input $\sigma$. Thus, $\mathcal{A}$ and $T_{\mathcal{A}}$ were computable from knowing $\mathcal{T}$ and $\sigma$. Note that in general, the definition of an evolving automaton does not require the latter to be the case. Thus, there seems to be some 'room' in the computational performance of evolving automata. Could they even simulate devices that are computationally more powerful than those modeled by ITMs?

As we shall see below, this is indeed the case: evolving automata are provably more powerful than Turing machines. Does it mean that the Turing machine is out of the game when looking for a new paradigm that captures the ideas of contemporary computing? Not entirely.

## 4.1    Computing with Advice

Rather than attempting a reverse simulation of evolving automata, let us try to simulate a yet more powerful model of a Turing machine by evolving automata: the so-called *interactive Turing machine with advice* (ITM/A). The model extends the well-known and well-studied model of (ordinary) Turing machines with advice in computational complexity theory (cf. [8]).

**Definition 3.** *An* interactive Turing machine with advice *(ITM/A) is an interactive Turing machine as described before, enhanced by an advice function $f : \mathbf{N} \to \Sigma^*$. Advice allows the insertion of external information $f(t)$ into the course of a computation at suitable times.*

A standard Turing machine with advice, with input of size $n$, is allowed to 'ask' for the value of its advice function only for that particular value of $n$. Similarly, an ITM/A can call its advice at time $t$ only for values $t_1 \leq t$. To realize such a call an ITM/A is equipped with a separate *advice tape* and among its states it has a distinguished *advice state*. By writing $t_1$ on the advice tape and by entering the advice state at time $t \geq t_1$ the value of $f(t_1)$ will appear on the advice tape (in a single step). By this action the original contents of the advice tape is completely rewritten. Note that the value of $f(t_1)$ does not depend on the input read before or after time $t$: the advice called at time $t$ with argument $t_1 \leq t$ is the same for all input streams. This makes advice different from oracles also considered in the computability theory: oracle values can depend on the current input (cf. [13]).

The mechanism of advise functions is very powerful and can provide an ITM/A with any non-computable 'assistance'. For theoretical and practical reasons it is useful to restrict the size of advice growth in ITM/As to polynomial functions. With advice functions that grow exponentially one could encode arbitrary oracles in advice.

**Proposition 2.** *Evolving automata can simulate interactive Turing machines with advice and vice versa.*

*Proof.* (Sketch) First we sketch how an evolving automaton, $\mathcal{A}$, can simulate an ITM/A $\mathcal{O}$. Follow the given simulation of an ITM without advice, but now also consider the actions of $\mathcal{O}$ with its advice: include the times of calling $\mathcal{O}$'s advice in the schedule of switching times as well. At each switching moment, the respective automaton will also encode the corresponding advice in its states. Note that now the members of $\mathcal{A}$ cannot be computed solely from knowing $\sigma$ and $\mathcal{O}$ as before. This time, we also have to know the advice at each calling time. Note that the automaton sizes in $\mathcal{A}$ grow proportionally with the space complexity and the advice size of the $\mathcal{O}$ in the simulated time segment.

The reverse simulation by means of an ITM/A is easy. An ITM/A $\mathcal{O}$ is supplied with the description of $\mathcal{A}$'s members and the times of $T_{\mathcal{A}}$ 'on demand', via its advice tape. The computation of $\mathcal{O}$ on input $\sigma$ starts by calling the advice. $\mathcal{O}$ gets the description of $A_1$ followed by the value of $t_1$. All $\mathcal{O}$ has to do is to simulate $A_1$ on the next $t_1$ input symbols. Then $\mathcal{O}$ calls its advice again, obtaining description of $A_2$ and the value of $t_2$, and $\mathcal{O}$ simulates $A_2$ for the next $t_2 - t_1$ steps. Then the process of calling advice repeats again, etc. Now the space complexity of $\mathcal{O}$ grows as fast as the automata size in $\mathcal{A}$. $\qquad\square$

As a corollary we obtain that *the Internet, modeled by an evolving automaton, can be simulated by an interactive Turing machine with advice.* By this, we have finished the second step of our plan: we have identified a model which is an extension of Turing machines and whose computational power matches exactly that of a highly simplified model of the (unrestricted) Internet.

## 4.2    Complexity of ITM/As

As for evolving automata we consider the question whether ITM/As admit an 'own' complexity theory.

**Proposition 3.** *ITM/As are more powerful than ITMs (without advice).*

*Proof.* (Sketch) We begin by exhibiting a translation $\kappa$ that can be realized by an ITM/A, but not by any ITM without an advice. The computation will ask for solving the halting problem (known to be undecidable) for all classical Turing machines.

As input stream, we consider a computable enumeration of all Turing machines. Given this enumeration, $\kappa$ should output with each valid machine description a 1 if and only if this machine accepts its own description, and 0 otherwise. We construct an ITM/A $\mathcal{I}$ that does this. In-between producing 0s or 1s, $\mathcal{I}$ will output only empty symbols.

$\mathcal{I}$ enumerates all TMs in the same order as they occur in the input stream. Then $\mathcal{I}$ can recognize whether a segment of the input stream is indeed a valid encoding of a TM. On segments that are not encodings of a TM, $\mathcal{I}$ produces 0. On a segment $w$ that is an encoding of length $n$ of some TM, $\mathcal{I}$ calls its advice with value $n$ (note that the advice is called for a value which does not depend on the particular input read thus far). The advice gives the encoding $\langle M \rangle$ of a TM whose running time is the longest from among the running time of all TMs of size $n$ that terminate on their own description. Running this machine on input $\langle M \rangle$ in parallel with the simulation of the $w$, $\mathcal{I}$ has an upper bound on the running time on $w$ within which the machine must halt on its own description when it does. In this way $\mathcal{I}$ can correctly answer the halting problem for $w$ in finite time, and proceed with the next segment of the input.

Now we sketch that no ITM without advice can solve the halting problem. Suppose there was an ITM $\mathcal{H}$ computing $\kappa$. Obviously, due to the properties of interactive machines, $\mathcal{H}$ should produce the answer to any particular halting problem in finite time. Thus, if we were interested in solving only a particular halting problem it would be enough to run a classical TM simulating $\mathcal{H}$ until it produces, in finite time, the solution of our decision problem. This contradicts the undecidability of the halting problem by classical TMs. $\qquad\square$

Given an ITM/A with advice function $f$, it is natural to consider the size of the advice $f(t)$ for each individual value of $t$ as a measure for the 'complexity' of the ITM/A (in addition to the usual measures of time and space for the Turing machine part). Define the *advice complexity* of an ITM/A with advice function $f$ as the function $\alpha : \mathbf{N} \to \mathbf{N}$ such that for every $t$, $\alpha(t) = |f(t)|$ (the length of the string $f(t)$). Given this measure one can try to distinguish between the computational power of different ITM/As. For example, Verbaan [18] proved the following interesting result, extending a similar result known for ordinary Turing machines with advice.

**Theorem 2.** *Consider ITM/As over input and advice alphabets with a fixed size bound $b$. Let $\alpha$ and $\beta$ be integer-valued functions such that $\alpha = o(\beta)$ and $\beta(t) \leq \frac{b^t}{\log b}$ for all $t$. Then there is a translation $\phi$ of infinite streams to infinite streams that can be realized by an ITM/A of advice complexity $\beta$, but not by any ITM/A of advice complexity $\alpha'$, for any function $\alpha'$ with $\alpha'(t) \leq \alpha(t)$ for all but finitely many $t$.*

In fact, as soon as $\alpha$ is strictly 'below' $\beta$ for all but finitely many values of $t$, ITM/As of advice complexity $\beta$ are more powerful than ITM/As of advice complexity $\alpha$ [18].

The computational equivalence between evolving automata and ITM/As also opens the question whether their complexity theories can be linked. An example of a result in this direction is the following.

**Theorem 3.** *Let $\phi$ be a translation of infinite streams to infinite streams. Let $\phi$ be realizable by an evolving automaton of size complexity $g$. Then $\phi$ can be realized by an ITM/A of advice complexity $O(g \log g)$ and space complexity $O(\log g)$.*

It follows e.g. that evolving automata of polynomially bounded size complexity can be simulated by an ITM/A of polynomially bounded advice complexity and logarithmic space. The converse result can be shown as well [18].

## 5 Extending the Turing Machine Paradigm

*[$\cdots$] a comprehensive theory of computation must reflect in a stylized way aspects of the underlying physical world.*
T. Toffoli ([11], 1982).

In our search for a new computational model, we have presented three important insights:

*(i)* we have devised a model of evolving automata capturing interactive and non-uniformly evolving computing,

*(ii)* we have shown the computational equivalence of evolving automata and interactive Turing machines with advice and, last but not least,

*(iii)* we have shown that these two models are computationally more powerful than interactive Turing machines (without advice) which only capture interaction.

This leads to the new computational paradigm that we have in mind:

**Extended Turing Machine paradigm:** *A computational process is any process whose evolution over time can be captured by evolving automata or, equivalently, by interactive Turing machines with advice.*

The new paradigm represents a new understanding of computing, motivated by developments like the Internet and even by the computational views of living systems. It innovates the classical view of computing in three ways: a shift from finite computations to potentially infinite interactive ones, a shift from rigid computing systems towards systems whose architecture and functionality evolve over time and, last but not least, an understanding that in general the latter process of evolution happens in an unpredictable, non-uniform, non-computable way.

In our view, both models mentioned in the extended paradigm, the ITM/A and evolving automata, have their use. Together they illustrate the dual view of a non-computable evolution. The metaphor of an ITM/A corresponds better to our intuition and experience in which computers are perceived as well engineered devices with a fixed architecture driven solely by input data, now with their 'evolution' driven by data as well (namely by those from the advice). In this way, an TM/A models non-uniform software evolution. On the other hand, the metaphor of an evolving automaton models a hardware evolution. This makes this model more suitable for modeling systems where a non-uniform hardware evolution is readily visible (as was the case of the Internet). Of course, both models are only different sides of the same coin.

## 5.1   Non-computability Issues in the Extended Paradigm

The new paradigm indirectly asserts that the 'new computing' is computationally more powerful than classical computing, since the models of computing serving in the new paradigm are provably computationally more powerful than those in the old paradigm. Does it mean that the new paradigm encompasses some form of super-Turing computing capability, and if so, can the extra power be used for 'solving' undecidable problems?

Of course the answer to both questions is negative, as seen from the proof of Proposition 3. In order to solve the halting problem, the constructed ITM/A had to be provided with non-computable information. In general, an ITM/A can solve classically undecidable problems if and only if its advice contains the respective non-computable information. In the proof we were not interested in how this information could be obtained, we just made use of the fact that such information in principle exists. Thus, there is nothing miraculous in our result: if a device has non-computable information at its disposal, it can solve non-computable tasks. This has been known since Turing's times (cf. [13], [2]).

In 'real' computational environments (such as in the Internet), the non-computability manifests itself, e.g., as the non-predictability of their evolution or in the unpredictable variance in message transfer times among the systems part. We do not know of any computational exploitation of these phenomena (except, perhaps, as a source of random numbers). In fact, in most of our computing activities we strive for being shielded from these phenomena. Hence, the hyper-Turing power implied by the extended paradigm is needed for the purposes of theoretical modeling, but, unfortunately, cannot be purposefully harnessed for any goal-oriented computational purposes.

## 5.2   The Scope of the Extended Paradigm

What remains is to see whether the other systems, man-made or natural, mentioned in the introduction as examples of systems 'that process information and compute' are covered by the extended paradigm. No doubt that, once we agree that the models in the paradigm capture the Internet, then they also capture all variations of this theme: wireless ad hoc networks, sensor nets, etcetera. Generalizing, one can say that to the extent to which finite automata mirror the data-processing capability of some entity (such as that of a biological cell or of a biological neuron), the extended paradigm also mirrors the data-processing capabilities and computations of ensembles (such as organisms or brains) and communities of such entities (such as swarms of ants or bees, or also communities of humans). Cf. [22] for a more in-depth, complexity-oriented study of such an approach.

The case of physical systems in general, and especially of the Universe itself, is interesting. Apparently, there are 'no external inputs' to the Universe. A current state of the universe completely determines its next state (albeit not in the deterministic way, as in the case of deterministic finite automata). Or, to quote Toffoli [11]: *In a sense, nature has been continually computing the 'next state' of the universe for billions of years; all we have to do - and actually, all we can do - is 'hitch a ride' on this huge ongoing computation.* What we 'observe' is the potentially infinite sequence of instances of the Universe. Could this be modeled by a kind of a gigantic, 'natural' evolving automaton (perhaps a quantum automaton?) whose evolution is governed by the laws of the Nature? According to Lloyd [3], the Universe computes its own evolution. This seems to be close to the spirit of our paradigm.

## 6    Conclusions

The contemporary perception of computing sees it as any act of information processing and transfer, occurring in both the local and global behavior of systems. In this view, computation encompasses communication, interaction, reaction, receiving, sending, storing, retrieving and transformation of information. The Extended Turing Machine paradigm captures it in an abstract manner.

The extended paradigm keeps the central position of Turing machines in our apprehension of computing, continuing in this way the tribute to A.M. Turing. In fact, the new paradigm also makes use of the language of classical Turing machines, upgraded this time, by the notions of interaction and advice. The new paradigm also encompasses non-uniform computing, which seems to be far more ubiquitous and less artificial than believed before. The complementary view of interactive Turing machines as that of evolving automata stresses the dual sides of both software and hardware evolution. In addition to the known cases of computing artifacts, the extended paradigm also covers the information processing occurring in the Nature. For informal use, there is no need to formally revise the good old Turing machine paradigm. What is needed is to be more liberal in understanding different variants of Turing machines and their scenarios. This was also concluded in a debate on Lance Fortnow's weblog: [5].

> *Call me a rationalist then as I continue to hold the belief that no matter how complicated the computational model, we can still use the simple Turing machine to capture its power.*
> L. Fortnow ([5], 2006).

There is some advantage in having the paradigms of science formulated in not very precise terms. Namely, in such a case, their rejection requires a real revolution to happen in the field. Otherwise, let our paradigms evolve along with the evolution of the notions they deal with. But it is good to know that when it comes to the details, we are able to make our paradigms more precise.

# References


1. Balcázar, J.L., Díaz, J., Gabarró, J.: Structural Complexity I. 2nd edn. Springer, Berlin (1995)
2. Davis, M.: The myth of hypercomputation. In: Teuscher, C. (ed.) Alan Turing: Life and Legacy of a Great Thinker, pp. 195–212. Springer, Heidelberg (2004)
3. Edge, The Computational Universe: Seth Lloyd [10.24.02] (October 24, 2002), http://www.edge.org/3rd_culture/lloyd2/lloyd2_index.html
4. Etesi, G., Németi, I.: Turing computability and Malament-Hogarth space-times. International Journal of Theoretical Physics 41(2), 342–370 (2002), http://arxiv.org/abs/gr-qc/0104023
5. Fortnow, L.: Principles of problem solving: A TCS Response, weblog Computational Complexity, Friday (July 14, 2006), http://weblog.fortnow.com/2006/07/principles-of-problem-solving-tcs.html
6. Goldin, D.Q., Smolka, S.A., Attie, P.C., Sonderegger, E.: Turing machines, transition systems, and interaction. Information and Computation 194(2), 101–128 (2004)
7. Goldin, D.Q., Smolka, S., Wegner, P. (eds.): Interactive Computing: The New Paradigm. Springer, Berlin (2006)
8. Karp, R.M., Lipton, R.: Turing machines that take advice, L'Enseignement Mathématique, $II^e$ Série, Tome XXVIII, pp. 191–209 (1982)
9. Lloyd, S.: The Computational Universe. Originally published on Edge, (October 24, 2002), http://www.edge.org/3rd_culture/lloyd2/lloyd2_p2.html
10. Thomas, W.: Automata on infinite objects. In: van Leeuwen, J. (ed.) Handbook of Theoretical Computer Science. Formal Models and Semantics, vol.B, ch. 4, pp. 133–192. Elsevier Science Publishers, Amsterdam (1990)
11. Toffoli, T.: Physics and computation. Int. Journal of Theor. Physics 21, 165–175 (1982)
12. Turing, A.M.: On computable numbers, with an application to the Entscheidungsproblem. Proc. London Math. Soc. Series 2 42, 230–265 (1936)
13. Turing, A.M.: Systems of logic based on ordinals. Proc. London Math. Soc. Series 2 45, 161–228 (1939)
14. van Leeuwen, J., Wiedermann, J.: The Turing machine paradigm in contemporary computing. In: Enquist, B., Schmidt, W. (eds.) Mathematics Unlimited - 2001 and Beyond, pp. 1139–1155. Springer, Berlin (2001)
15. van Leeuwen, J., Wiedermann, J.: A computational model of interaction in embedded systems, Technical Report UU-CS-2001-02, Dept.of Information and Computing Sciences, Utrecht University (2001)
16. van Leeuwen, J., Wiedermann, J.: Beyond the Turing limit: Evolving interactive systems. In: Pacholski, L., Ružička, P. (eds.) SOFSEM 2001: Theory and Practice of Informatics. LNCS, vol. 2234, pp. 90–109. Springer, Heidelberg (2001)
17. van Leeuwen, J., Wiedermann, J.: A Theory of Interactive Computation. In: Goldin, D., Smolka, S., Wegner, P. (eds.) Interactive Computing: The New Paradigm, ch. 6, pp. 119–142. Springer, Berlin (2006)

18. Verbaan, P.R.A.: The Computational Complexity of Evolving Systems, Ph.D.Thesis, Dept.of Information and Computing Sciences, Utrecht University (2006)
19. Verbaan, P.R.A., van Leeuwen, J., Wiedermann, J.: Complexity of evolving interactive systems. In: Karhumäki, J., et al. (eds.) Theory Is Forever. LNCS, vol. 3113, pp. 268–281. Springer, Berlin (2004)
20. Wegner, P.: Why interaction is more powerful than algorithms. C. ACM 40, 315–351 (1997)
21. Wiedermann, J., van Leeuwen, J.: Relativistic computers and non-uniform complexity theory. In: Calude, C., et al. (eds.) UMC 2002. LNCS, vol. 2509, pp. 287–299. Springer, Heidelberg (2002)
22. Wiedermann, J., van Leeuwen, J.: The emergent computational potential of evolving artificial living systems. AI Communications 15(4), 205–215 (2002)
23. Winograd, T.: From computing machinery to interaction design. In: Denning, P., Metcalfe, R. (eds.) Beyond Calculation: The Next Fifty Years of Computing, pp. 149–162. Springer, Berlin (1997), `http://hci.stanford.edu/~winograd/acm97.html`

# Author Index

Printing: Mercedes-Druck, Berlin
Binding: Stein+Lehmann, Berlin